# Chemical Reactor
# Analysis and Design

# Chemical Reactor Analysis and Design

Gilbert F. Froment
Rijksuniversiteit Gent, Belgium

Kenneth B. Bischoff
University of Delaware

John Wiley & Sons
New York    Chichester    Brisbane    Toronto

**Library of Congress Cataloging in Publication Data**

Froment, Gilbert F.
  Chemical reactor analysis and design.

  Includes index.
  1. Chemical reactors.  2. Chemical reactions.
3. Chemical engineering.  I. Bischoff, Kenneth B.,
joint author.  II. Title.

TP157.F76      660'.29'9      78-12465

ISBN 0-471-02447-3

Printed in the United States of America

10 9 8 7 6 5 4 3 2 1

To our wives:
Mia and Joyce

# Preface

This book provides a comprehensive study of chemical reaction engineering, beginning with the basic definitions and fundamental principles and continuing all the way to practical application. It emphasizes the real-world aspects of chemical reaction engineering encountered in industrial practice. A rational and rigorous approach, based on mathematical expressions for the physical and chemical phenomena occurring in reactors, is maintained as far as possible toward useful solutions. However, the notions of calculus, differential equations, and statistics required for understanding the material presented in this book do not extend beyond the usual abilities of present-day chemical engineers. In addition to the practical aspects, some of the more fundamental, often more abstract, topics are also discussed to permit the reader to understand the current literature.

The book is organized into two main parts: applied or engineering kinetics and reactor analysis and design. This allows the reader to study the detailed kinetics in a given "point," or local region first and then extend this to overall reactor behavior.

Several special features include discussions of chain reactions (e.g., hydrocarbon pyrolysis), modern methods of statistical parameter estimation and model discrimination techniques, pore diffusion in complex media, general models for fluid-solid reactions, catalyst deactivation mechanisms and kinetics, analysis methods for chemical processing aspects of fluid-fluid reactions, design calculations for plug flow reactors in realistic typical situations (e.g., thermal cracking), fixed bed reactors, fluidized bed reactor design, and multiphase reactor design. Several of these topics are not usually covered in chemical reaction engineering texts, but are of high current interest in applications.

Comprehensive and detailed examples are presented, most of which utilize real kinetic data from processes of industrial importance and are based on the authors' combined research and consulting experience.

We firmly believe, based on our experience, that this book can be taught to both undergraduate and graduate classes. If a distinction must be made between undergraduate and graduate material it should be in the extension and the depth of coverage of the chapters. But we emphasize that to prepare the student to solve the problems encountered in industry, as well as in advanced research, the approach must be the same for both levels: there is no point in ignoring the more complicated areas that do not fit into idealized schemes of analysis.

Several chapters of the book have been taught for more than 10 years at the

Rijksuniversiteit Gent, at the University of Maryland, Cornell University, and the University of Delaware. Some chapters were taught by G.F.F. at the University of Houston in 1973, at the Centre de Perfectionnement des Industries Chimiques at Nancy, France, from 1973 onwards and at the Dow Chemical Company, Terneuzen, The Netherlands in 1978. K.B.B. used the text in courses taught at Exxon and Union Carbide and also at the Katholieke Universiteit Leuven, Belgium, in 1976. Substantial parts were presented by both of us at a NATO-sponsored Advanced Study Institute on "Analysis of Fluid–Solid catalytic Systems" held at the Laboratorium voor Petrochemische Techniek, Rijksuniversiteit, Gent, in August 1974.

We thank the following persons for helpful discussions, ideas, and critiques: among these are dr. ir. L. Hosten, dr. ir. F. Dumez, dr. ir. J. Lerou, ir. J. De Geyter and ir. J. Beeckman, all from the Laboratorium voor Petrochemische Techniek of Rijksuniversiteit Gent; Prof. Dan Luss of the University of Houston and Professor W. D. Smith of the University of Rochester.

**Gilbert F. Froment**
**Kenneth B. Bischoff**

# Contents

## 2   Kinetics of Heterogeneous Catalytic Reactions    76

## 3   Transport Processes with Fluid-Solid Heterogeneous Reactions    141

# Notation

Two consistent sets of units are listed in the following pages: one that is currently the most common in engineering calculations (including, for example, m, hr, atm, kcal) and the S.I. units, which are only slowly penetrating into everyday use. In some formulas other units had to be used: the chemical engineering literature contains many correlations that are not based on dimensionless groups and they require the quantities to be expressed in certain given units only. This has been carefully indicated in the text, however.

All the numerical calculations in the text are in the above mentioned engineering units, but the intermediate and final results are also given in S.I. units. We feel that this reflects—and even simplifies—the practical reality that is going to last for many more years, and we have preferred this pragmatic approach to preserve the feeling for orders of magnitude gained from years of manipulation of the engineering units. Finally, great attention has been given to the detailed definition of the units of the different quantities: for example, when a dimension of length is used, it is always clarified as to whether this length concerns the catalyst or the reactor. We have found that this greatly promotes insight into the mathematical modeling of a phenomenon.

| | | Engineering units | S.I. units |
|---|---|---|---|
| $A$ | reaction component | | |
| $A_b$ | heat exchange surface, packed bed side | $m^2$ | $m^2$ |
| $A_j$ | reacting species in a reaction system | | |
| $A_k$ | heat exchange surface in a batch reactor, on the side of the reaction mixture | $m^2$ | $m^2$ |
| $A_m$ | logarithmic mean of $A_k$ and $A_r$ or of $A_b$ and $A_u$ | $m^2$ | $m^2$ |
| $A_r$ | heat exchange surface for a batch reactor on the side of the heat transfer medium | $m^2$ | $m^2$ |
| $A_t$ | total heat exchange surface | $m^2$ | $m^2$ |

| | | Engineering units | S.I. units |
|---|---|---|---|
| $A_u$ | heat exchange surface for a packed bed on the side of the heat transfer medium | $m^2$ | $m^2$ |
| $A_v$ | gas-liquid interfacial area per unit liquid volume | $m_i^2/m_L^3$ | $m_i^2/m_L^3$ |
| $A_{vt}$ | interfacial area per unit tray surface | $m_i^2/m^2$ | $m_i^2/m^2$ |
| $A_0$ | frequency factor | | |
| $A'$ | absorption factor, $L'/mF$ | | |
| $A_v'$ | gas-liquid interfacial area per unit gas + liquid volume | $m_i^2/m_{L+G}^3$ | $m_i^2/m_{L+G}^3$ |
| $a$ | stoichiometric coefficient | | |
| $a_0, a_1$ | parameters (Sec. 8.3.b) | | |
| $a_g$ | surface to volume ratio of a particle | $m_p^2/m_p^3$ | $m_p^2/m_p^3$ |
| $a_m$ | external particle surface area per unit catalyst mass | $m_p^2/kg\ cat.$ | $m_p^2/kg\ cat.$ |
| $a_v$ | external particle surface area per unit reactor volume | $m_p^2/m_r^3$ | $m_p^2/m_r^3$ |
| $a'$ | order of reaction with respect to $A$ | | |
| $a_j'$ | order of reaction with respect to $A_j$ | | |
| $a_v'$ | gas-liquid interfacial area per unit packed volume | $m_i^2/m_r^3$ | $m_i^2/m_r^3$ |
| $a_v''$ | liquid-solid interfacial area per unit packed volume | $m_i^2/m_r^3$ | $m_i^2/m_r^3$ |
| $B$ | reaction component | | |
| $B_m$ | fictitious component | | |
| $\mathbf{B}$ | vector of fictitious components | | |
| $b$ | stoichiometric coefficient | | |
| $b'$ | order of reaction with respect to $B$ | | |
| $C_A, C_B, C_j$ | molar concentration of species $A, B, j$ | $kmol/m^3$ | $kmol/m^3$ |
| $C_{Ab}, C_{Bb} \ldots$ | molar concentrations of species $A, B \ldots$ in the bulk fluid | $kmol/m_f^3$ | $kmol/m_f^3$ |
| $C_{Al}, C_{Bl} \ldots$ | molar concentrations of adsorbed $A, B \ldots$ | $kmol/kg\ cat.$ | $kmol/kg\ cat.$ |
| $C_D$ | drag coefficient for spheres | | |

| | | Engineering units | S.I. units |
|---|---|---|---|
| $C_S$ | molar concentration of reacting component $S$ of solid | $kmol/m_p^3$ | $kmol/m_p^3$ |
| $C_c$ | coke content of catalyst | kg coke/kg cat. | kg coke/kg cat. |
| $C_l$ | molar concentration of vacant active sites of catalyst | kmol/kg cat. | kmol/kg cat. |
| $C_t$ | total molar concentration of active sites | kmol/kg cat. | kmol/kg cat. |
| $C_0$ | inlet concentration | $kmol/m^3$ | $kmol/m^3$ |
| C | vector of concentrations | $kmol/m^3$ | $kmol/m^3$ |
| $C_{Aeq}$ | molar concentration of $A$ at equilibrium | $kmol/m^3$ | $kmol/m^3$ |
| $C_{A_i}$ | molar concentration of $A$ in front of the interface | $kmol/m_f^3$ | $kmol/m_f^3$ |
| $C_{A_s}, C_s$ | molar concentration of fluid reactant inside the solid | $kmol/m_f^3$ | $kmol/m_f^3$ |
| $C_{P_s}, C_{P_s}^c$ | molar concentration of sorbed poison inside catalyst, with respect to core boundary | $kmol/m_f^3$ | $kmol/m_f^3$ |
| $C_{P_{s\infty}}$ | equilibrium molar concentration of sorbed poison inside catalyst | $kmol/m_f^3$ | $kmol/m_f^3$ |
| $C_{s0}$ | reactant molar concentration at centerline of particle (Chapter 3) | $kmol/m_f^3$ | $kmol/m_f^3$ |
| $\overline{C_A}$ | Laplace transform of $C_A$ | | |
| $C_{A_s}^s, C_s^s$ | molar concentration of fluid reactant in front of the solid surface | $kmol/m_f^3$ | $kmol/m_f^3$ |
| $C'_{A_s}$ | molar concentration of $A$ inside completely reacted zone of solid | $kmol/m_f^3$ | $kmol/m_f^3$ |
| $c_p$ | specific heat of fluid | kcal/kg K | kJ/kg K |
| $c_{p_s}$ | specific heat of solid | kcal/kg K | kJ/kg K |
| $Da$ | Damköhler number for poisoning, $k_{sP} R/D_{eP}$ | | |
| $D_A, D_B$ | molecular diffusivities of $A$, $B$ in liquid film | $m^3/m_L$ hr | $m^3/m_L$ s |
| $D_{AB}$ | molecular diffusivity for $A$ in a binary mixture of $A$ and $B$ | $m_f^3/m$ hr | $m_f^3/m$ s |

| | | Engineering units | S.I. units |
|---|---|---|---|
| $D_K$ | Knudsen diffusivity | $m_f^3/m \cdot hr$ | $m_f^3/m \; s$ |
| $D_e, D_{eA}, D_{eB}$ | effective diffusivities for transport in a (pseudo-) continuum, or (Chapter 13) in emulsion phase | $m_f^3/m \; hr$ or $m_f^3/m_r \; hr$ | $m_f^3/m \; s$ or $m_f^3/m_r \; s$ |
| $D_{eG}$ | gas phase effective diffusivity in axial direction in a multiphase packed bed | $m_G^3/m_r \; hr$ | $m_G^3/m_r \; s$ |
| $D_{eL}$ | liquid phase effective diffusivity in axial direction in a multiphase packed bed | $m_L^3/m_r \; hr$ | $m_L^3/m_r \; s$ |
| $D_{eP}$ | effective pore diffusivity for poison | $m_f^3/m \; cat. \; hr$ | $m_f^3/m \; cat. \; s$ |
| $D_{ea}, D_{er}$ | effective diffusivities in axial, respectively radial directions in a packed bed | $m_f^3/m_r \; hr$ | $m_f^3/m_r \; s$ |
| $D_{eg}$ | effective diffusivity for transport of A through a grain (Sec. 4.4) | $m_f^3/m_p \; hr$ | $m_f^3/m_p \; s$ |
| $D_{ep}$ | effective diffusivity for transport of A in the pores between the grains (Sec. 4.4) | $m_f^3/m_p \; hr$ | $m_f^3/m_p \; s$ |
| $D_{i,n}$ | measure of divergence between rival models for the $n$th experiment in the $i$th grid point | | |
| $D_{j,l}$ | eddy diffusivity for species $j$ in the 1 direction | $m_f^3/m \; hr$ | $m_f^3/m \; s$ |
| $D_l$ | eddy diffusivity in the $l$ direction | $m_f^3/m \; hr$ | $m_f^3/m \; s$ |
| $D_e'$ | effective diffusivity for transport through completely reacted solid (Chapter 4) | $m_f^3/m_p \; hr$ | $m_f^3/m_p \; s$ |
| $D_{jm}$ | effective molecular diffusivity of $j$ in a multicomponent mixture | $m_f^3/m \; hr$ | $m_f^3/m \; s$ |
| $d$ | wall thickness | m | m |
| $d_b$ | bubble diameter | m | m |
| $d_c$ | coil diameter | m | m |
| $d_p$ | particle diameter | m | m |
| $d_r$ | reactor diameter | m | m |

|  |  | Engineering units | S.I. units |
|---|---|---|---|
| $d_s$ | stirrer diameter | m | m |
| $d_t$ | internal tube diameter also tower diameter (Chapter 14) | m | m |
| $E$ | activation energy | kcal/kmol | kJ/kmol |
| $E_C$ | Murphree tray efficiency corrected for entrainment |  |  |
| $E_i(x)$ | exponential integral |  |  |
| $E'$ | Murphree tray efficiency |  |  |
| $\bar{E}$ | overall tray efficiency |  |  |
| $\dot{E}$ | point tray efficiency along gas streamline |  |  |
| $\text{Eö}_b$ | Eötvös number, based on bubble diameter, $\dfrac{d_b \rho_L g}{\sigma}$ |  |  |
| $\text{erf}(\eta)$ | error function |  |  |
| $\text{erfc}(\eta)$ | complementary error function, $1-\text{erf}(\eta)$ |  |  |
| $F$ | total molar flow rate | kmol/hr | kmol/s |
| $F_A$ | enhancement factor |  |  |
| $F_{A_0}, F_{j_0}$ | molar feed rate of reactants $A$ and $j$ | kmol/hr | kmol/s |
| $F_k$ | force exerted per unit cross section | kgf/m$^2$ or atm | N/m$^2$ |
| $\mathscr{F}$ | objective function |  |  |
| $F'$ | volumetric gas flow rate | m$^3$/hr | m$^3$/s |
| $F'_0$ | volumetric gas feed rate | m$^3$/hr | m$^3$/s |
| $F''$ | volumetric gas flow rate (Chapter 14) | ft$^3$/ft$^2$ min | m$^3$/m$^2$ s |
| $f$ | friction factor in Fanning equation |  |  |
| $f_b$ | fraction of total fluidized bed volume occupied by bubble gas |  |  |
| $f_e$ | fraction of total fluidized bed volume occupied by emulsion gas |  |  |
| $G$ | superficial mass flow velocity | kg/m$_r^2$ hr | kg/m$_r^2$ s |
| $\mathbf{G}$ | matrix of partial derivatives of model with respect to the parameters |  |  |
| $\mathbf{G}^T$ | transpose of $\mathbf{G}$ |  |  |

| | | Engineering units | S.I. units |
|---|---|---|---|
| $g$ | acceleration of gravity | $m/hr^2$ | $m/s^2$ |
| $g_{j,l}$ | external force on species $j$ in the $l$ direction per unit mass of $j$ | kcal/kg m | N/kg |
| $g_{u,i}$ | partial derivative of reaction rate with respect to the parameter $K_i$ at the $u$th set of experimental conditions | | |
| $H$ | Henry's law coefficient | $m^3$atm/kmol | Nm/kmol |
| $H_{G,n}$ | enthalpy of gas on plate $n$ | kcal/kmol | kJ/kmol |
| $H_L$ | liquid height | m | m |
| $H_{L,n}$ | enthalpy of liquid on plate $n$ | kcal/kmol | kJ/kmol |
| $H_{fj}$ | heat of formation of species $j$ | kcal/kmol | kJ/kmol |
| $H_i$ | height of stirrer above bottom | m | m |
| $\bar{H}_j$ | molar enthalpy of species $j$ | kcal/kmol | kJ/kmol |
| $-\Delta H$ | heat of reaction | kcal/kmol | kJ/kmol |
| $h_f$ | heat transfer coefficient for film surrounding a particle | kcal/$m_p^2$ hr °C | kJ/$m_p^2$ s K |
| $I$ | initiator; also intermediate species; inert; | | |
| $\mathbf{I}$ | unit matrix | | |
| $J_{j,l}$ | molar flux of species $j$ in $l$ direction, with respect to mass average velocity | kmol/m² hr | kmol/m² s |
| $J_s$ | pressure drop in straight tubes | kgf/m² or atm | N/m² |
| $j_D$ | $j$-factor for mass transfer, $\dfrac{k_g M_m P_{fA}}{G} Sc^{2/3}$ | | |
| $j_H$ | $j$-factor for heat transfer, $\dfrac{h_f}{c_p G} (Pr)^{2/3}$ | | |
| $K, K_A, K_1 \ldots$ | equilibrium constants | $atm^{-1}$ or $m^3$/kmol | m²/N or $m^3$/kmol |
| $\mathbf{K}$ | matrix of rate coefficients | | |
| $\hat{K}$ | kinetic energy per unit mass | $m^2/hr^2$ | $m^2/s^2$ |
| $\hat{K}_1$ | flow averaged kinetic energy per unit mass | $m^2/hr^2$ | $m^2/s^2$ |
| $k$ | reaction rate coefficient | see $k_c, k_y, k_p$ | |

| | | Engineering units | S.I. units |
|---|---|---|---|
| $k$ | rate coefficient with respect to unit solid mass for a reaction with order $n$ with respect to fluid reactant $A$ and order $m$ with respect to solid component $S$ | $m_f^{3n}(\text{kmol } A)^{1-n}$ $(\text{kmol } S)^{-m}$ $m_p^{3(m-1)}\,\text{hr}^{-1}$ | $m_f^{3n}(\text{kmol } A)^{1-n}$ $(\text{kmol } s)^{-m}$ $m_p^{3(m-1)}\text{s}^{-1}$ |
| $k_C$ | coking rate coefficient | kg coke/kg cat. hr atm or hr$^{-1}$ | kg coke/kg cat. s(N/m$^2$) or s$^{-1}$ |
| $k_G$ | gas phase mass transfer coefficient referred to unit interfacial area | $m_G^3/m_i^2$ hr | $m_G^3/m_i^2$ s |
| $k_L$ | liquid phase mass transfer coefficient referred to unit interfacial area | $m_L^3/m_i^2$ hr | $m_L^3/m_i^2$ s |
| $k_T$ | mass transfer coefficient (including interfacial area) between flowing and stagnant liquid in a multiphase reactor | $m_L^3/m_r^3$ hr | $m_L^3/m_r^3$ s |
| $k_{T1}, k_{T2}$ | mass transfer coefficient (including interfacial area) beween regions 1 and 2 of flow model (Chapter 12) | $m_L^3/m_r^3$ hr | $m_L^3/m_r^3$ s |
| $k_c$ | rate coefficient based on concentrations | hr$^{-1}\{$kmol/ m$^3\}^{1-(a'+b'\ldots)}$ | s$^{-1}\{$kmol/ m$^3\}^{1-(a'+b'\ldots)}$ |
| $k_g$ | gas phase mass transfer coefficient; when based on concentrations; when based on mole fractions; when based on partial pressures; in a fluidized bed | $m_f^3/m_p^2$ hr; kmol/m$_p^2$ hr; kmol/m$_p^2$ hr atm $m_f^3/m_c^3$ hr | $m_f^3/m_p^2$ s; kmol/m$_p^2$ s; kmol/m$_p^2$ s (N/m$^2$); $m_f^3/m_c^3$ s |
| $k_{gP}$ | interfacial mass transfer coefficient for catalyst poison | $m_f^3/m_p^2$ hr | $m_f^3/m_p^2$ s |
| $k_l$ | mass transfer coefficient between liquid and catalyst surface, referred to unit interfacial area | $m_L^3/m_i^2$ hr | $m_L^3/m_i^2$ s |
| $k_p$ | reaction rate coefficient based on partial pressures | hr$^{-1}$kmol/m$^3$ atm$^{-(a'+b'\ldots)}$ | s$^{-1}$kmol/m$^3$ (N/m$^2$)$^{-(a'+b'\ldots)}$ |
| $k_{pr}$ | rate coefficient for propagation reaction in addition polymerization | m$^3$/kmol hr | m$^3$/kmol s |

| | | Engineering units | S.I. units |
|---|---|---|---|
| $k_r$ | reaction rate coefficient (Chapter 3) | $m_f{}^3/m^2$ cat. hr | $m_f{}^3/m^2$ cat. s |
| $k_{rA}, k_{rB}$ | rate coefficient for catalytic reaction subject to poisoning | $m_f{}^3/m^2$ cat. hr | $m_f{}^3/m^2$ cat. s |
| $k_{rP}$ | rate coefficient for first-order poisoning reaction at core boundary | $m_f{}^3/m^2$ cat. hr | $m_f{}^3/m^2$ cat. s |
| $k_s$ | surface-based rate coefficient for catalytic reaction (Chapter 5) | $m_f{}^3/m^2$ cat. hr | $m_f{}^3/m^2$ cat. s |
| $k_t, k_{tr}$ | rate coefficients for termination reactions | $m^3/kmol$ hr or $hr^{-1}$ | $m^3/kmol$ s or $s^{-1}$ |
| $k_v, k_v^\circ$ | volume-based rate coefficient for catalytic reaction during poisoning, resp. in absence of poison | $m_f{}^3/m^3$ cat. hr | $m_f{}^3/m^3$ cat. s |
| $k_y$ | rate coefficient based on mole fractions | $kmol/m^3$ hr | $kmol/m^3$ s |
| $k_1$ | elutriation rate coefficient (Chapter 13) | $kg/m^2$ hr | $kg/m^2$ s |
| $k_1, k_2 \ldots$ | reaction rate coefficients | see $k_c, k_y, k_p$ | |
| $k_A^\circ$ | rate coefficient of catalytic reaction in absence of coke | depending on rate dimensions | |
| $k_g^\circ$ | mass transfer coefficient in case of equimolar counterdiffusion, $k_g y_{f_A}$ | see $k_g$ | |
| $k_l'$ | mass transfer coefficient between stagnant liquid and catalyst surface in a multiphase reactor | $m_L{}^3/m_r{}^3$ hr | $m_L{}^3/m_r{}^3$ s |
| $k_s'$ | surface based reaction rate coefficient for gas-solid reaction | $\left(\dfrac{m_p}{hr}\right)\bigg/\left(\dfrac{kmol\ A}{m_f{}^3}\right)$ | $\left(\dfrac{m_p}{s}\right)\bigg/\left(\dfrac{kmol\ A}{m_f{}^3}\right)$ |
| $(k_{bc})_b$ | mass transfer coefficient from bubble to interchange zone, referred to unit bubble volume | $m_f{}^3/m_b{}^3$ hr | $m_f{}^3/m_b{}^3$ s |
| $(k_{be})_b$ | overall mass transfer coefficient from bubble to emulsion, referred to unit bubble volume | $m_G{}^3/m_b{}^3$ hr | $m_G{}^3/m_b{}^3$ s |

| | | Engineering units | S.I. units |
|---|---|---|---|
| $(k_{ce})_b$ | mass transfer coefficient from interchange zone to emulsion, referred to unit bubble volume | $m_f^3/m_b^3 \ hr$ | $m_f^3/m_b^2 \ s$ |
| $(k_{ce})_c$ | mass transfer coefficient from bubble + interchange zone to emulsion, referred to unit bubble + interchange zone volume | $m_G^3/m_c^3 \ hr$ | $m_G^3/m_c^3 \ s$ |
| $L$ | volumetric liquid flow rate | $m_L^3/hr$ | $m_L^3/s$ |
| | also distance from center to surface of catalyst pellet (Chapter 3) | m | m |
| | also distance between pores in a solid particle (Sec. 4.5) and thickness of a slab (Sec. 4.6) | m | m |
| $L_f$ | total height of fluidized bed | m | m |
| $L_{mf}$ | height of a fluidized bed at minimum fluidization | m | m |
| $L'$ | molar liquid flow rate | kmol/hr | kmol/s |
| $Lw'$ | modified Lewis number, $\lambda_e/\rho_s c_{p_s} D_e$ | | |
| $l$ | vacant active site | | |
| $M$ | ratio of initial concentrations $C_{B_0}/C_{A_0}$ | | |
| $M_j$ | molecular weight of species $j$ | kg/kmol | kg/kmol |
| $M_m$ | mean molecular weight | kg/kmol | kg/kmol |
| $M_1$ | monomer (Sec. 1.4-6) | | |
| $m$ | Henry's coefficient based on mole fractions, also order of reaction | | |
| $m_t$ | total mass | kg | kg |
| $\dot{m}$ | total mass flow rate | kg/hr | kg/s |
| $\dot{m}_j$ | mass flow rate of component $j$ | kg/hr | kg/s |
| $N$ | stirrer revolution speed; also runaway number, $2U/R_t \rho c_p k_v$ (Sec. 11.5.c) | $hr^{-1}$ | $s^{-1}$ |
| $\dot{N}_A$ | molar rate of absorption per unit gas-liquid interfacial area | $kmol/m_i^2 \ hr$ | $kmol/m_i^2 \ s$ |

| | | Engineering units | S.I. units |
|---|---|---|---|
| | also molar flux of $A$ with respect to fixed coordinates | kmol/m$^2$ hr | kmol/m$^2$ s |
| $\dot{N}_A(t)$ | instantaneous molar absorption rate in element of age $t$ per unit gas-liquid interfacial area | kmol/m$_i^2$ hr | kmol/m$_i^2$ s |
| $N_A, N_B, N_j \ldots$ | number of kmoles of reacting components $A$, $B$, $j \ldots$ in reactor | kmol | kmol |
| $N_s$ | dimensionless group, $$\frac{3D_{ep}t_{\mathrm{ref}}\,C_{P,\mathrm{ref}}}{R^2\rho_s C_{Pl\infty}} \; \text{(Sec. 5.2.c)}$$ | | |
| $N_t$ | total number of kmoles in reactor | kmol | kmol |
| $N_0$ | minimum stirrer speed for efficient dispersion | hr$^{-1}$ | s$^{-1}$ |
| $N_0^*$ | characteristic speed for bubble aspiration and dispersion | hr$^{-1}$ | s$^{-1}$ |
| $n$ | order of reaction | | |
| $P$ | reaction product | | |
| | also power input (Chapter 14) | kgf m/hr | Nm/s |
| Pr | Prandtl number, $c_p\mu/\lambda$ | | |
| $P_N$ | profit over $N$ adiabatic fixed beds | \$/hr | \$/s |
| $P_1, P_2 \ldots$ | active polymer | | |
| Pe$_a$ | Peclet number based on particle diameter, $u_i d_p/D_{ea}$ | | |
| Pe$_a'$ | Peclet number based on reactor length, $u_i L/D_{ea}$ | | |
| $\bar{P}_N$ | number averaged degree of polymerization | | |
| $\bar{P}_W$ | weight-averaged degree of polymerization | | |
| $p$ | probability of adding another monomer unit to a chain | | |
| $p_A, p_B, p_j \ldots$ | partial pressures of components $A$, $B$, $j \ldots$ | atm | N/m$^2$ |
| $p_{ac}$ | partial pressure of acetone (Chapter 9) | atm | N/m$^2$ |

| | | Engineering units | S.I. units |
|---|---|---|---|
| $p_c$ | critical pressure | atm | N/m² |
| $p_{fA}$ | film pressure factor | atm | N/m² |
| $p_t$ | total pressure | atm | N/m² |
| $Q$ | reaction component | | |
| $Q_{ox}, Q_a, Q_{abs}$ | heats of oxidation, adsorption, absorption | kcal/kmol | kJ/kmol |
| $q$ | stoichiometric coefficient; also heat flux | kcal/m² hr | kJ/m²s or kW/m² |
| $q'$ | order of reaction with respect to $Q$ | | |
| $q'_j$ | order of reaction with respect to $Aj$ | | |
| $R$ | gas constant | kcal/kmol K or atm m³/kmol K | kJ/kmol K |
| | also radius of a spherical particle (Chapters 4 and 5) also reaction component | m | |
| Re | Reynolds number, $d_p G/\mu$ or $d_t G/\mu$ | | |
| $R_j$ | total rate of change of the amount of component $j$ | kmol/m³ hr | kmol/m³ s |
| $R_p$ | pore radius in pore model of Szekely and Evans (Sec. 4.5) | m | m |
| $R_t$ | tube radius | m | m |
| $R_1^{\cdot}, R_2^{\cdot}$ | free radicals | | |
| $r$ | rate of reaction per unit volume | kmol/m³ hr | kmol/m³ s |
| | also pore radius (Chapter 3) | m | m |
| | also radial position in spherical particle (Chaper 4) also stoichiometric coefficient | m | m |
| $r_A$ | rate of reaction of component $A$ per unit volume | kmol/m³ hr | kmol/m³ s |
| | or per unit catalyst mass | kmol/kg cat. hr | kmol/kg. cat. s |
| $r_C$ | rate of coke deposition | kg coke/kg cat. hr | kg coke/kg cat. s |
| $r_P$ | rate of poison deposition | kmol/kg cat. hr | kmol/kg cat. s |
| $r_S$ | rate of reaction of $S$, reactive component of solid, in gas-solid reactions | kmol/kg solid hr | kmol/kg solid s |

| | | Engineering units | S.I. units |
|---|---|---|---|
| $r_{Ai}$ | rate of reaction of $A$ at interface | $kmol/m_i^2\ hr$ | $kmol/m_i^2\ s$ |
| $r_b$ | radius of bend of coil | m | m |
| $r_c$ | radial position of unpoisoned or unreacted core in a sphere | m | m |
| $r_v$ | reaction rate per unit pellet volume | $kmol/m_p^3\ hr$ | $kmol/m_p^3\ s$ |
| $\bar{r}$ | mean pore radius | m | m |
| $S$ | reaction component also dimensionless group, $\beta\gamma$ (Chapter 11) | | |
| Sc | Schmidt number, $\mu/\rho D$ | | |
| $S_g$ | internal surface area per unit mass of catalyst | $m^2cat./kg\ cat.$ | $m^2cat./kg\ cat.$ |
| $S_x$ | external surface area of a pellet | $m^2$ | $m^2$ |
| $Sh_m$ | modified Sherwood number for liquid film, $k_L/A_v D_A$ | | |
| $Sh'$ | modified Sherwood number, $k_g L/D_e$ (Chapter 3) | | |
| $Sh'_P$ | modified Sherwood number for poisoning, $k_{gP}R/D_{eP}$ | | |
| $s$ | stoichiometric coefficient also parameter in Danckwerts' age distribution function | $hr^{-1}$ | $s^{-1}$ |
| | also Laplace transform variable | $hr^{-1}$ | $s^{-1}$ |
| $s_i^2$ | experimental error variance of model $i$ | | |
| $s'$ | order of reaction with respect to $S$ | | |
| $\bar{s}^2$ | pooled estimate of variance | | |
| $T$ | temperature | K, °C | K |
| $T_R$ | bed temperature at radius $R_t$ | K, °C | K |
| $T_c$ | critical temperature | K, °C | K |
| $T_m$ | maximum temperature | K, °C | K |
| $T_r$ | temperature of surroundings | K, °C | K |
| $T_s, T_s^s$ | temperature inside solid, resp. at solid surface | K, °C | K |

|  |  | Engineering units | S.I. units |
|---|---|---|---|
| $t$ | clock time | hr | s |
|  | also age of surface element (Chapter 6) | hr | s |
| $t_{\text{ref}}$ | reference time | hr | s |
| $t'$ | reduced time |  |  |
| $t^*$ | time required for complete conversion (Chapter 4) | hr | s |
| $\bar{t}$ | contact time | hr | s |
| $\mathscr{T}(s, \boldsymbol{\alpha})$ | transfer function of flow model (Chapter 12) |  |  |
| $U$ | overall heat transfer coefficient | kcal/m$^2$ hr °C | kJ/m$^2$ s K |
| $u$ | linear velocity | m$_r$/hr | m$_r$/s |
| $u_b$ | bubble rising velocity, absolute | m$_r$/hr | m$_r$/s |
| $u_{br}$ | bubble rising velocity, with respect to emulsion phase | m$_f{}^3$/m$_r{}^2$ hr | m$_f{}^3$/m$_r{}^2$ s |
| $u_e$ | emulsion gas velocity, interstitial | m$_r$/hr | m$_r$/s |
| $u_i$ | interstitial velocity | m$_r$/hr | m$_r$/s |
| $u_{iG}, u_{iL}$ | interstitial velocity of gas, resp. liquid | m$_r$/hr | m$_r$/s |
| $u_1$ | fluid velocity in direction 1 | m/hr | m/s |
| $u_s$ | superficial velocity | m$_f{}^3$/m$_r{}^2$ hr | m$_f{}^3$/m$_r{}^2$ s |
| $u_{sG}$ | superficial gas velocity | m$_G{}^3$/m$_r{}^2$ hr | m$_G{}^3$/m$_r{}^2$ s |
| $u_t$ | terminal velocity of particle | m$_r$/hr | m$_r$/s |
| $V$ | reactor volume or volume of considered "point" | m$_r{}^3$ | m$_r{}^3$ |
| $V_p$ | volume of a particle | m$_p{}^3$ | m$_p{}^3$ |
| $V_R$ | equivalent reactor volume, that is, reactor volume reduced to isothermality | m$_r{}^3$ | m$_r{}^3$ |
| $V_b$ | bubble volume | m$_b{}^3$ | m$_b{}^3$ |
| $V_c$ | critical volume | m$_f{}^3$ | m$_f{}^3$ |
|  | also volume of bubble + interchange zone | m$_c{}^3$ | m$_c{}^3$ |
| $V_{iz}$ | volume of interchange zone | m$^3$ | m$^3$ |
| $V_s$ | product molar volume | m$^3$/kmol | m$^3$/kmol |
| $V_b'$ | bubble volume corrected for the wake | m$^3$ | m$^3$ |
| $V_c'$ | corrected volume of bubble + interchange zone | m$^3$ | m$^3$ |

| | | Engineering units | S.I. units |
|---|---|---|---|
| $V'_{iz}$ | volume of interchange zone, corrected for wake | $m^3$ | $m^3$ |
| $W$ | total catalyst mass | kg cat. | kg cat. |
| $W(d_p)$ | mass of amount of catalyst with diameter $d_p$ | kg | kg |
| $W(\theta)$ | increase in value of reacting mixture | \$ | \$ |
| We | Weber number, $\rho_L L^2 \, d_p / \Omega^2 \sigma_L$ | | |
| $W_j$ | amount of catalyst in bed $j$ of a multibed adiabatic reactor | kg | kg |
| $W_0, W_P,$ $W_Q, W_R$ | cost of reactor idle time, reactor charging time, reactor discharging time and of reaction time | \$/hr | \$/s |
| $w_{ij}$ | weighting factor in objective function (Sec. 1.6-2) | | |
| $w_j$ | price per kmole of chemical species $A_j$ | \$/kmol | \$/kmol |
| $x$ | fractional conversion | | |
| $x_A, x_B, x_j \ldots$ | fractional conversion of $A, B, j \ldots$ | | |
| $x_{Aeq}$ | fractional conversion of $A$ at equilibrium | | |
| $x_{aK}$ | conversion of acetone into ketene (Chapter 9) | | |
| $x_{at}$ | total conversion of acetone (Chapter 9) | | |
| $x_n$ | mole fraction in liquid phase on plate $n$ | | |
| $\mathbf{x}_m$ | eigenvector of rate coefficient matrix $\mathbf{K}$ (Ex. 1.4.1-1) | | |
| $x'_A, x'_B \ldots$ | conversion of $A, B \ldots$ | kmol | kmol |
| $x''_A, x''_B \ldots$ | conversion of $A, B \ldots$ for constant density | $kmol/m^3$ | $kmol/m^3$ |
| $Y$ | radius of grain in grain model of Sohn and Szekely (Chapter 4) | m | m |

| | | Engineering units | S.I. units |
|---|---|---|---|
| $\hat{Y}$ | calculated value of dependent variable (Sec. 1.6-2) also experimental value of dependent variable (Sec. 1.6-2) | | |
| $y$ | coordinate perpendicular to gas-liquid interface | m | m |
| | also radial position inside a grain in grain model of Sohn and Szekely (Chapter 4) | | |
| | also position of reaction front inside the solid in pore model of Szekely and Evans (Chapter 4) | m | m |
| $y_A, y_B, y_j \dots$ | mole reaction of species $A$, $B, j \dots$ | | |
| $y_G$ | gas film thickness | m | m |
| $y_L$ | liquid film thickness for mass transfer | m | m |
| $y_h$ | liquid film thickness for heat transfer | m | m |
| $y_n$ | mole fraction in gas phase leaving plate $n$ | | |
| $y_1, y_2$ | weight fractions of gasoil, gasoline (Sec. 5.3-c) | | |
| $\mathbf{y}$ | vector of mole fractions | | |
| $Z$ | compressibility factor also total reactor or column length | m | m |
| $Z_c$ | critical compressibility factor | | |
| $z$ | distance inside a slab of catalyst | m | $m_p$ |
| | also axial coordinate in reactor | $m_r{}^p$ | $m_r{}^p$ |
| $z_l$ | distance coordinate in $l$ direction | m | m |

# Greek Symbols

|  |  | Engineering units | S.I. units |
|---|---|---|---|
| $\alpha$ | convective heat transfer coefficient | kcal/m² hr °C | kJ/m² s K |
|  | also profit resulting from the conversion of 1 kmole of $A$ into desired product (Sec. 11.5.d) | \$/kmol | \$/kmol |
|  | also weighting factor in objective function (Sec. 2.3.c-2) |  |  |
| $\boldsymbol{\alpha}$ | vector of flow model parameters (Chapter 12) |  |  |
| $\alpha$, $\alpha_c$ | deactivation constants | kg cat./kg coke or hr$^{-1}$ | kg cat./kg coke or s$^{-1}$ |
| $\alpha_i$ | convective heat transfer coefficient, packed bed side | kcal/m² hr °C | kJ/m² s K |
| $\alpha_j$, $\alpha_{ij}$ | stoichiometric coefficient of component $j$ in a single, with respect to the $i$th, reaction |  |  |
| $\alpha_k$ | convective heat transfer coefficient on the side of the reaction mixture | kcal/m² hr °C | kJ/m² s K |
| $\alpha_r$ | convective heat transfer coefficient on the side of the heat transfer medium | kcal/m² hr °C | kJ/m² s K |
| $\alpha_u$ | convective heat transfer coefficient for a packed bed on the side of the heat transfer medium | kcal/m² hr °C | kJ/m² s K |
| $\alpha_w$ | convective heat transfer coefficient in the vicinity of the wall | kcal/m² hr °C | kJ/m² s K |
| $\alpha_w{}^s$ | wall heat transfer coefficient for solid phase | kcal/m² hr °C | kJ/m² s K |
| $\alpha_w{}^f$ | wall heat transfer coefficient for fluid | kcal/m² hr °C | kJ/m² s K |

|  |  | Engineering units | S.I. units |
|---|---|---|---|
| $\beta$ | radical involved in a bimolecular propagation step; also weighting factor in objective function (Sec. 2.3.c); stoichiometric coefficient (Chapter 5); cost of 1 kg of catalyst (Chapter 11); dimensionless adiabatic temperature rise, $T_{ad} - T_0/T_0$ (Sec. 11.5.c) also Prater number $= (-\Delta H)D_e C_s^s/\lambda_e T_s^s$ (Chapter 3) |  |  |
| $\Gamma_e$ | locus of equilibrium conditions in $x - T$ diagram |  |  |
| $\Gamma_m$ | locus of the points in $x - T$ diagram where the rate is maximum |  |  |
| $\Gamma_{\lambda m}$ | locus of maximum rate along adiabatic reaction paths in $x - T$ diagram |  |  |
| $\gamma$ | Hatta number, $$y_L\sqrt{\frac{kC_{Bb}}{D_A}} = \frac{\sqrt{kC_{Bb}/D_A}}{k_L}$$ also dimensionless activation energy, $E/RT$ (Section 11.5.c and Chapter 3) also weighting factor in objective function (Section 2.3.c) |  |  |
| $\delta$ | molar ratio steam/ hydrocarbon |  |  |
| $\delta_A$ | expansion per mole of reference component $A$, $(q + s - a - b)/a$ |  |  |
| $\varepsilon$ | void fraction of packing | $m_f^3/m_r^3$ | $m_f^3/m_r^3$ |
| $\varepsilon_A$ | expansion factor, $y_{A0}\delta_A$ |  |  |
| $\varepsilon_G$ | gas hold up | $m_G^3/m_r^3$ | $m_G^3/m_r^3$ |
| $\varepsilon_L$ | liquid holdup | $m_L^3/m_r^3$ | $m_L^3/m_r^3$ |

| | | Engineering units | S.I. units |
|---|---|---|---|
| $\varepsilon_{Lf}$ | liquid holdup in flowing fluid zone in packed bed | $m_L^3/m_r^3$ | $m_L^3/m_r^3$ |
| $\varepsilon_c$ | void fraction of cloud, that is, bubble + interchange zone | $m^3/m_c^3$ | $m^3/m_c^3$ |
| $\varepsilon_m$ | pore volume of macropores | $m_f^3/m_s^3$ | $m_f^3/m_s^3$ |
| $\varepsilon_{mf}$ | void fraction at minimum fluidization | $m_f^3/m_r^3$ | $m_f^3/m_r^3$ |
| $\varepsilon_s$ | internal void fraction or porosity | $m_f^3/m_s^3$ | $m_f^3/m_s^3$ |
| $\varepsilon_u$ | pore volume of micropores | $m_f^3/m_s^3$ | $m_f^3/m_s^3$ |
| $\varepsilon_L'$ | dynamic holdup | $m_f^3/m_r^3$ | $m_f^3/m_r^3$ |
| $\zeta$ | factor used in pressure drop equation for the bends; also correction factor in (Sec. 4.5-1) | | |
| $\zeta_m$ | quantity of fictitious component | | |
| $\eta$ | effectiveness factor for solid particle | | |
| $\eta_0$ | effectiveness factor for reaction in an unpoisoned catalyst | | |
| $\eta_L$ | utilization factor, liquid side | | |
| $\eta_G$ | global utilization factor | | |
| $\eta_b$ | effectiveness factor for particle + film | | |
| $\theta$ | fractional coverage of catalyst surface; also dimensionless time, $D_e t/L^2$ (Chapter 3), $ak'C_A t$ (Chapter 4); residence time | | |
| $\theta_P$ | reactor changing time | hr | s |
| $\theta_R$ | reaction time | hr | s |
| $\theta_Q$ | reactor discharging time | hr | s |
| $\theta_f$ | reaction time corresponding to final conversion | hr | s |
| $\theta_0$ | reactor idle time | hr | s |
| $\Lambda$ | angle described by bend of coil | ° | rad |
| $\Lambda$ | matrix of eigenvalues | | |

| | | Engineering units | S.I. units |
|---|---|---|---|
| $\lambda$ | thermal conductivity; also slope of the change of conversion versus temperature for reaction in an adiabatic reactor, $\dot{m}c_p/F_{A0}(-\Delta H)$ | kcal/m hr °C  $(°C)^{-1}$ | kW/m K  $(K)^{-1}$ |
| $\lambda_e$ | effective thermal conductivity in a solid particle | kcal/m hr °C | kJ/m s K |
| $\lambda_{ea}, \lambda_{er}$ | effective thermal conductivity in a packed bed in axial, with respect to radial direction | kcal/m$_r$ hr °C | kJ/m$_r$ s K |
| $\lambda_l$ | effective thermal conductivity in $l$ direction | kcal/m hr °C | kJ/m s K |
| $\lambda_m$ | negative of eigenvalue of rate coefficient matrix $\mathbf{K}$ | | |
| $\lambda_s$ | thermal conductivity of solid | kcal/m hr °C | kJ/m s K |
| $\lambda_{er}^{f}, \lambda_{er}^{s}$ | effective thermal conductivity for the fluid phase with respect to a solid phase in a packed bed | kcal/m$_r$ hr °C | kJ/m$_r$ s K |
| $\mu$ | dynamic viscosity; also radical in a unimolecular propagation step | kg/m hr | kg/m s |
| $\mu_s$ | viscosity at the temperature of the heating coil surface | kg/m hr | kg/m s |
| $\mu_w$ | viscosity at the temperature of the wall | kg/m hr | kg/m s |
| $\xi$ | extent of reaction; also reduced length, $z/L$ or reduced radial position inside a particle, $r/R$ | kmol | kmol |
| $\xi_c$ | reduced radial position of core boundary | | |
| $\xi_i$ | extent of $i$th reaction | kmol | kmol |
| $\xi'$ | radial coordinate inside particle | m$_p$ | m$_p$ |
| $\xi_i'$ | extent of $i$th reaction per unit mass of reaction mixture | kmol kg$^{-1}$ | kmol kg$^{-1}$ |
| $\pi_{i, n-1}$ | prior probability associated | | |

|  |  | Engineering units | S.I. units |
|---|---|---|---|
|  | with the $i$th model, used in the design of the $n$th experiment |  |  |
| $\rho_B$ | catalyst bulk density | kg cat./$m_r^3$ | kg cat./$m_r^3$ |
| $\rho_L$ | liquid density | kg/$m_L^3$ | kg/$m_L^3$ |
| $\rho_b$ | bulk density of bubble phase | kg/$m_b^3$ | kg/$m_b^3$ |
| $\rho_e$ | bulk density of emulsion phase | kg/$m_e^3$ | kg/$m_e^3$ |
| $\rho_f$ | fluid density | kg/$m_f^3$ | kg/$m_f^3$ |
| $\rho_g$ | gas density | kg/$m_G^3$ | kg/$m_G^3$ |
| $\rho_{mf}$ | bulk density of fluidized bed at minimum fluidization | kg/$m_r^3$ | kg/$m_r^3$ |
| $\rho_s$ | density of solid | kg sol/$m_p^3$ | kg sol/$m_p^3$ |
| $\sigma$ | standard deviation also active and alumina site (Sec. 2.2) |  |  |
| $\sigma^2$ | error variance |  |  |
| $\sigma_i^2$ | variance of response values predicted by the $i$th model |  |  |
| $\sigma_L$ | surface tension of liquid | kgf/m | N/m |
| $\sigma_{L.c}$ | critical surface tension of liquid | kgf/m | N/m |
| $\sigma_P$ | sorption distribution coefficient, Chap. 5 |  |  |
| $\tau$ | tortuosity factor (Chapter 3); also mean residence time (Chapter 10) | s | s |
| $\phi$ | Thiele modulus, $V_p/S_x\sqrt{k_v/D_e}$; (Chapters 3 and 5), $\sqrt{ak'C_{S0}/D_e}$; (Chapter 4), $V_p/S_x\sqrt{k(T_s^s)c_s/D_e}$; (Chapter 11); also partioning factor (Sec. 3.5.c) |  |  |
| $\Phi_c$ | deactivation function |  |  |
| $\chi_c$ | Bartlett's $\chi_c^2$ test |  |  |
| $\Psi$ | sphericity of a particle |  |  |
| $\Psi(t)$ | age distribution function |  |  |
| $\Omega$ | cross section of reactor or column | $m^2$ | $m^2$ |

# Subscripts

| | |
|---|---|
| $A, B \ldots$ | with respect to $A, B \ldots$ |
| $C$ | coke |
| $G$ | gas; also global (Chapters 6 and 14) or regenerator (Chapter 13) |
| $L$ | liquid |
| $P$ | poison |
| $R$ | reactor (Chapter 13) |
| $T$ | at actual temperature |
| $T_1$ | a reference temperature |
| $a$ | adsorption; also in axial direction |
| ad | adiabatic |
| $b$ | bulk; also bubble phase |
| $c$ | bubble + interchange zone; also critical value; based on concentration |
| $d$ | desorption |
| $e$ | emulsion phase; also effective or exit stream from reactor |
| eq | at chemical equilibrium |
| $f$ | fluid; also film; also at final conversion |
| $g$ | average; also grain or gas |
| $i$ | interface; also $i$th reaction |
| $j$ | with respect to $j$th component |
| $l$ | liquid; also in $l$ direction |
| $m$ | maximum; also measurement point (Chapter 12) |
| $n$ | tray number |
| $p$ | pellet, particle; also based on partial pressures |
| $r$ | reactor dimension; also surroundings also in radial direction |
| $s$ | inside solid; also surface based or superficial velocity |
| $sr$ | surface reaction |
| $t$ | total; also tube |
| $v$ | volume based |
| $w$ | at the wall |
| $y$ | based on mole fractions |
| $0$ | initial or inlet condition; also overall value |

# Superscripts

| | |
|---|---|
| $T$ | transpose |
| $d$ | stagnant fraction of fluid |
| $f$ | flowing fraction of fluid |
| $s$ | condition at external surface |
| $0$ | in absence of poison or coke |
| . | radical |
| ^ | calculated or estimated value |

# Part One

CHEMICAL
ENGINEERING
KINETICS

# 1

## ELEMENTS
## OF
## REACTION
## KINETICS

We begin the study of chemical reactor behavior by considering only "local" regions. By this we mean a "point" in the reactor in much the same way as is customary in physical transport phenomena, that is, a representative volume element. After we develop quantitative relations for the local rate of change of the amount of the various species involved in the reaction, they can be "added together" (mathematically integrated) to described an entire reactor.

In actual experiments, such local phenomena cannot always be unambiguously observed, but in principle they can be discussed. The real-life complications will then be added later in the book.

## 1.1 Reaction Rate

The rate of a homogeneous reaction is determined by the composition of the reaction mixture, the temperature, and the pressure. The pressure can be determined from an equation of state together with the temperature and composition; thus we focus on the influence of the latter factors.

Consider the reaction

$$aA + bB \ldots \longrightarrow qQ + sS \ldots \tag{1.1-1}$$

It can be stated that $A$ and $B$ react at rates

$$r'_A = -\frac{dN_A}{dt} \qquad r'_B = -\frac{dN_B}{dt}$$

and $Q$ and $S$ are formed at rates

$$r'_S = \frac{dN_S}{dt} \qquad r'_Q = \frac{dN_Q}{dt}$$

where $N_j$ represents the molar amount of one of the chemical species in the reaction, and is expressed in what follows in kmol, and $t$ represents time.

The following equalities exist between the different rates:

$$-\frac{1}{a}\frac{dN_A}{dt} = -\frac{1}{b}\frac{dN_B}{dt} = \frac{1}{q}\frac{dN_Q}{dt} = \frac{1}{s}\frac{dN_S}{dt} \tag{1.1-2}$$

Each term of these equalities may be considered as the rate of the reaction.

This can be generalized to the case of $N$ chemical species participating in $M$ independent[1] chemical reactions,

$$\alpha_{i1}A_1 + \alpha_{i2}A_2 + \cdots + \alpha_{iN}A_N = 0$$

or

$$0 = \sum_{j=1}^{N} \alpha_{ij}A_j \qquad i = 1, 2, \ldots M \tag{1.1-3}$$

with the convention that the stoichiometric coefficients, $\alpha_{ij}$, are taken positive for products and negative for reactants. A comparison with Eq. (1.1-1) would give $A_1 \equiv A$, $\alpha_1 \equiv -a$ (for only one reaction the subscript, $i$, is redundant, and $\alpha_{ij} \rightarrow \alpha_j$), $A_2 \equiv B$, $\alpha_2 \equiv -b$, $A_3 \equiv Q$, $\alpha_3 \equiv q$, $A_4 \equiv S$, $\alpha_4 \equiv s$.

The rate of reaction is generally expressed on an intensive basis, say reaction volume, so that when $V$ represents the volume occupied by the reaction mixture:

$$r_i \equiv \frac{1}{V}\frac{1}{\alpha_{ij}}\frac{dN_j}{dt} \tag{1.1-4}$$

For the simpler case:

$$r = \frac{-1}{aV}\frac{dN_A}{dt} = \frac{-1}{aV}\frac{d}{dt}(C_A V) = \frac{-1}{aV}\left(V\frac{dC_A}{dt} + C_A\frac{dV}{dt}\right) \tag{1.1-5}$$

where $C_A$ represents the molar concentration of $A$ (kmol/m$^3$). When the density remains constant, that is, when the reaction volume does not vary, Eq. (1.1-5) reduces to

$$r = \frac{-1}{a}\frac{dC_A}{dt} \tag{1.1-6}$$

In this case, it suffices to measure the change in concentration to obtain the rate of reaction.

---

[1] By independent is meant that no one of the stoichiometric equations can be derived from the others by a linear combination. Discussions of this are given by Denbigh [1], Prigogine and Defay [2], and Aris [3]. Actually, some of the definitions and manipulations are true for any set of reactions, but it is convenient to work with the minimum, independent set.

## 1.2 Conversion and Extent of Reaction _____

Conversions are often used in the rate expressions rather than concentrations, as follows:

$$x'_A = N_{A_0} - N_A \qquad x'_B = N_{B_0} - N_B \qquad (1.2\text{-}1)$$

For constant density,

$$x''_A = C_{A_0} - C_A \qquad x''_B = C_{B_0} - C_B \qquad (1.2\text{-}2)$$

Most frequently, fractional conversions are used:

$$x_A = \frac{N_{A_0} - N_A}{N_{A_0}} \qquad x_B = \frac{N_{B_0} - N_B}{N_{B_0}} \qquad (1.2\text{-}3)$$

which show immediately how far the reaction has progressed. One must be very careful when using the literature because it is not always clearly defined which kind of conversion is meant. The following relations may be derived easily from Eq. (1.2-1) to (1.2-3):

$$x'_j = N_{j_0} x_j \qquad (1.2\text{-}4)$$

$$\frac{x'_A}{a} = \frac{x'_B}{b} = \cdots = \frac{x'_Q}{q} \cdots \qquad (1.2\text{-}5)$$

$$x_B = \frac{b}{a} \frac{N_{A_0}}{N_{B_0}} x_A \qquad (1.2\text{-}6)$$

An alternate, but related, concept to the conversion is the extent or degree of advancement of the general reaction Eq. (1.1-3), which is defined as

$$\xi = \frac{N_j - N_{j0}}{\alpha_j} \qquad (1.2\text{-}7a)$$

a quantity that is the same for any species. Also

$$N_j = N_{j0} + \alpha_j \xi \qquad (1.2\text{-}7b)$$

where $N_{j0}$ is the initial amount of $A_j$ present in the reaction mixture. For multiple reactions,

$$N_j = N_{j0} + \sum_{i=1}^{M} \alpha_{ij} \xi_i \qquad (1.2\text{-}8)$$

Equations 1.2-3 and 1.2-7 can be combined to give

$$N_j = N_{j0} + \alpha_j \frac{N_{A_0}}{a} x_A$$

If species $A$ is the limiting reactant (present in least amount), the maximum extent of reaction is found from

$$0 = N_{A_0} + \alpha_A \xi_{max}$$

and the fractional conversion defined by Eq. 1.2-3 becomes

$$x_A = \frac{\xi}{\xi_{max}} \tag{1.2-9}$$

Thus, either conversion or extent of reaction can be used to characterize the amount of reaction that has occurred. For industrial applications, the conversion of a feed is usually of interest, while for other scientific applications, such as irreversible thermodynamics (Prigogine [4]), the extent is often more useful; both concepts should be known. Further details are given by Boudart [5] and Aris [6].

In terms of the extent of reaction, the reaction rate Eq. (1.1-4) can be written

$$r_i = \frac{1}{V} \frac{1}{\alpha_{ij}} \frac{dN_j}{dt} = \frac{1}{V} \frac{d\xi_i}{dt} \tag{1.2-10}$$

With this rate, the change in moles of any species is, for a single reaction,

$$\frac{dN_j}{dt} = \alpha_j V r \tag{1.2-11}$$

for multiple reactions,

$$\frac{dN_j}{dt} = \sum_{i=1}^{M} \alpha_{ij} V r_i \equiv V R_j \tag{1.2-12}$$

The last part of Eq. (1.2-12) is sometimes useful as a definition of the "total" rate of change of species $j$. The utility of these definitions will be illustrated later in the book.

## 1.3 Order of Reaction

From the law of mass action,[2] based on experimental observation and later explained by the collision theory, it is found that the rate of reaction (1.1-1) can often be expressed as

$$r = k_c C_A^{a'} C_B^{b'} \tag{1.3-1}$$

The proportionality factor $k_c$ is called the rate coefficient or rate constant. By definition, this rate coefficient is independent of the quantities of the reacting species, but dependent on the other variables that influence the rate. When the reaction mixture is thermodynamically nonideal, $k_c$ will often depend on the

---

[2] See reference [7] at the end of this chapter.

concentrations because the latter do not completely take into account the interactions between molecules. In such cases, thermodynamic activities need to be used in (1.3-1) as described in Sec. 1.7. When $r$ is expressed in $kmol/m^3hr$, then $k_c$, based on (1.3-1) has dimensions

$$hr^{-1}(kmol/m^3)^{[1-(a'+b'+\cdots)]}$$

It can also be verified that the dimensions of the rate coefficients used with conversions are the same as those given for use with concentrations. Partial pressures may also be used as a measure of the quantities of the reacting species,

$$r = k_p p_A{}^{a'} p_B{}^{b'} \tag{1.3-2}$$

In this case, the dimensions of the rate coefficient are

$$hr^{-1} \, kmol \, m^{-3} atm^{-(a'+b'+\cdots)}$$

With thermodynamically nonideal conditions (e.g., high pressures) partial pressures may have to be replaced by fugacities. When use is made of mole fractions, the corresponding rate coefficient has dimensions $hr^{-1} \, kmol \, m^{-3}$. According to the ideal gas law:

$$C_i = \frac{p_i}{RT} = \frac{p_t}{RT} y_i$$

so that

$$k_c = (RT)^{a'+b'+\cdots} k_p = \left(\frac{RT}{p_t}\right)^{a'+b'+\cdots} k_y \tag{1.3-3}$$

In the following, the subscript is often dropped, however. The powers $a', b', \ldots$ are called "partial orders" of the reaction with respect to $A, B, \ldots$ The sum $a' + b' \ldots$ may be called the "global order" or generally just "order" of the reaction.

The order of a reaction has to be determined experimentally since it only coincides with the molecularity for elementary processes that actually occur as described by the stoichiometric equation. Only for elementary reactions does the order have to be 1, 2, or 3. When the stoichiometric equation (1.1-1) is only an "overall" equation for a process consisting of several mechanistic steps, the order cannot be predicted on the basis of this stoichiometric equation. The order may be a fraction or even a negative number. In Sec. 1.4, examples will be given of reactions whose rate cannot be expressed as a simple product like Eq. (1.3-1).

Consider a volume element of the reaction mixture in which the concentrations have unique values. For an irreversible first-order constant density reaction, Eqs. (1.1-6) and (1.3-1) lead to

$$r_A = -\frac{dC_A}{dt} = kC_A \tag{1.3-4}$$

When the rate coefficient, $k(\mathrm{hr}^{-1})$, is known, Eq. 1.3-4 permits the calculation of the rate, $r_A$, for any concentration of the reacting component. Conversely, when the change in concentration is known as a function of time, Eq. (1.3-4) permits the calculation of the rate coefficient. This method for obtaining $k$ is known as the "differential" method; further discussion will be presented later.

Integration of Eq. (1.3-4) leads to

$$kt = \ln\left(\frac{C_{A_0}}{C_A}\right) \tag{1.3-5}$$

Thus, a semilog plot of $C_A/C_{A_0}$ versus $t$ permits one to find $k$. A more thorough treatment will be given in Sec. 1.6.

The integrated forms of several other simple-order kinetic expressions, obtained under the assumption of constant density, are listed in Table 1.3-1.

*Table 1.3-1  Integrated forms of simple kinetic expressions (constant density)*

Zero order

$$kt = C_{A_0} - C_A \qquad\qquad kt = C_{A_0} x_A$$

First order

$$A \longrightarrow Q$$

$$kt = \ln\frac{C_{A_0}}{C_A} \qquad\qquad kt = \ln\frac{1}{1 - x_A}$$

Second order

$$2A \longrightarrow Q + S$$

$$kt = \frac{1}{C_A} - \frac{1}{C_{A_0}} \qquad\qquad C_{A_0} kt = \frac{x_A}{1 - x_A}$$

$$A + B \longrightarrow Q + S$$

$$kt = \frac{1}{C_{A_0} - C_{B_0}} \ln\frac{C_{B_0}}{C_{A_0}}\frac{C_A}{C_B} \qquad C_{A_0} kt = \frac{1}{1 - M} \ln\frac{M(1 - x_A)}{M - x_A}$$

$$M = \frac{C_{B_0}}{C_{A_0}}$$

Third order

$$3A \longrightarrow Q$$

$$2kt = \frac{1}{C_A{}^2} - \frac{1}{C_{A_0}{}^2} \qquad 2kt = \frac{1}{C_{A_0}{}^2}\left[\frac{1}{(1 - x_A)^2} - 1\right]$$

Caddell and Hurt [8] presented Fig. 1.3-1, which graphically represents the various simple integrated kinetic equations of Table 1.3-1. Note that for a second-order reaction with a large ratio of feed components, the order degenerates to a pseudo first order.

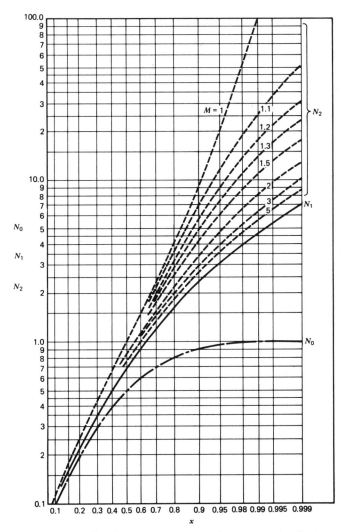

*Figure 1.3-1 Graphical representation of various simple integrated kinetic equations (from Caddell and Hurt [8]).*

$$N_0 = \frac{kt}{C_{A_0}} = x$$

$$N_1 = kt = \ln \frac{1}{1 - x}$$

$$N_2 = C_{B_0} kt = \frac{M}{1 - M} \ln \frac{M(1 - x)}{M - x} \qquad \left(M = \frac{C_{B_0}}{C_{A_0}}\right)$$

All reactions are, in principle, reversible, although the equilibrium can be sufficiently far toward the products to consider the reaction irreversible for simplicity. The above considerations can be used for the reverse reaction and lead to similar results.

For example, if we consider the simple reversible first-order reaction:

$$A \underset{2}{\overset{1}{\rightleftharpoons}} Q$$

$$r_A = -\frac{dC_A}{dt} = k_1 C_A - k_2 C_Q$$

From the stoichiometry,

$$C_A + C_Q = C_{A_0} + C_{Q_0}$$

and,

$$-\frac{dC_A}{dt} = k_1 C_A - k_2(C_{A_0} + C_{Q_0} - C_A)$$

$$= (k_1 + k_2)C_A - k_2(C_{A_0} + C_{Q_0})$$

The solution to this simple differential equation is

$$C_A = (C_{A_0} + C_{Q_0})\frac{k_2}{k_1 + k_2} + \frac{k_1 C_{A_0} - k_2 C_{Q_0}}{k_1 + k_2} e^{-(k_1 + k_2)t}$$

The equilibrium concentration of $A$ is given by,

$$C_{Aeq} = \frac{k_2}{k_1 + k_2}(C_{A_0} + C_{Q_0})$$

In terms of this, the equation can be written,

$$(C_A - C_{Aeq}) = (C_{A_0} - C_{Aeq})e^{-(k_1 + k_2)t}$$

or

$$\ln\frac{C_A - C_{Aeq}}{C_{A_0} - C_{Aeq}} = -(k_1 + k_2)t$$

Note that the last equation can be written in terms of conversions to give the result:

$$\ln\left(1 - \frac{x_A}{x_{Aeq}}\right) = -(k_1 + k_2)t$$

This result can also be found more simply by first introducing the conversion into the rate expression, and then integrating. Also, the rate expression can be alternately written as:

$$r_A = (k_1 + k_2)(C_A - C_{Aeq}) = (k_1 + k_2)C_{Ao}(x_{Aeq} - x_A)$$

Similarly, for a general second-order reversible reaction:

$$A + B \underset{k_2}{\overset{k_1}{\rightleftharpoons}} Q + S$$

The net rate, made up of forward and reverse rates, is given by

$$r_A = k_1 C_A^{a'} C_B^{b'} - k_2 C_Q^{q'} C_S^{s'}$$

or

$$r_A = k_1 \left( C_A^{a'} C_B^{b'} - \frac{1}{K} C_Q^{q'} C_S^{s'} \right) \tag{1.3.6}$$

where

$$K_C = \frac{k_1}{k_2} = \left( \frac{C_Q^{q'} C_S^{s'}}{C_A^{a'} C_B^{b'}} \right)_{eq}$$

represents the equilibrium constant.

Denbigh [1] showed that a more general relationship that satisfies both the kinetic and thermodynamic formulations is

$$\frac{k_1}{k_2} = \left( \frac{C_Q^{q'} C_S^{s'}}{C_A^{a'} C_B^{b'}} \right)_{eq} = \left( \frac{C_Q^q C_S^s}{C_A^a C_B^b} \right)^n = (K_c)^n$$

where

$$n = \frac{a'}{a} = \frac{b'}{b} = \frac{q'}{q} = \frac{s'}{s}$$

$$= \text{stoichiometric number}$$

However, since the stoichiometric equation is unchanged by multiplication with any positive constant, $\beta > 0$,

$$\sum_{j=1}^{N} \alpha_{ij} A_j = 0 = \sum_{j=1}^{N} \beta \alpha_{ij} A_j$$

one can choose $n = 1$ (Aris) [6]. Also see Boyd [9] for an extensive review. Laidler [10] also points out that if the overall reaction actually consists of several steps, the often-used technique of measuring the "initial" rate constants, starting with the reactants and then with the products, need not result in their ratio being equal to the equilibrium constant. For example, consider

$$A \underset{k_2}{\overset{k_1}{\rightleftharpoons}} Q \underset{k_4}{\overset{k_3}{\rightleftharpoons}} S$$

The two "initial" rate constants are $k_1$ and $k_4$, but the principle of microscopic reversibility shows that, at true equilibrium,

$$\left(\frac{C_S}{C_A}\right)_{eq} = \left(\frac{C_Q}{C_A} \cdot \frac{C_S}{C_Q}\right)_{eq} = \frac{k_1}{k_2}\frac{k_3}{k_4} \neq \frac{k_1}{k_4}$$

Therefore, caution must be used in the interpretation of combined kinetic and equilibrium results for complicated reaction systems.

Equation (1.3-6) can be written in terms of conversions in order to simply find the integrated form (for $a' = 1 = b' = q' = s'$):

$$C_{A_0}\frac{dx_A}{dt} = k_1\left[(C_{A_0} - C_{A_0}x_A)(C_{B_0} - C_{A_0}x_A) - \frac{1}{K}(C_{Q_0} + C_{A_0}x_A)(C_{S_0} + C_{A_0}x_A)\right]$$

or

$$\frac{dx_A}{dt} = \frac{k_1}{K}\left\{C_{A_0}(K-1)x_A^2 - [K(C_{A_0} + C_{B_0}) + C_{Q_0} + C_{S_0}]x_A\right.$$
$$\left. + KC_{B_0} - \left(\frac{C_{Q_0}C_{S_0}}{C_{A_0}}\right)\right\}$$
$$= \frac{k_1}{K}(\alpha x_A^2 + \beta x_A + \gamma)$$

Then

$$\frac{k_1 t}{K} = \frac{1}{q}\ln\frac{1 + 2\alpha x_A/(\beta - q)}{1 + 2\alpha x_A/(\beta + q)}$$

where

$$\alpha = C_{A_0}(K - 1)$$
$$\beta = -\{K(C_{A_0} + C_{B_0}) + C_{Q_0} + C_{S_0}\}$$
$$\gamma = KC_{B_0} - \frac{C_{Q_0}C_{S_0}}{C_{A_0}}$$
$$q^2 = \beta^2 - 4\alpha\gamma \geq 0$$

Other cases, such as $A \rightleftharpoons Q + S$, can be handled by similar techniques, and Hougen and Watson [11] present a table of several results.

## Example 1.3-1  The Rate of an Autocatalytic Reaction

An autocatalytic reaction has the form

$$A + Q \underset{2}{\overset{1}{\rightleftharpoons}} Q + Q$$

Here,

$$\frac{dC_A}{dt} = -k_1 C_A C_Q + k_2 C_Q^2$$

$$\frac{dC_Q}{dt} = -k_1 C_A C_Q + 2k_1 C_A C_Q - 2k_2 C_Q^2 + k_2 C_Q^2$$

$$= k_1 C_A C_Q - k_2 C_Q^2$$

Thus,

$$\frac{d}{dt}(C_A + C_Q) = 0$$

or

$$C_A + C_Q = \text{constant} = C_{A_0} + C_{Q_0} \equiv C_0$$

In this case it is most convenient to solve for $C_Q$:

$$\frac{dC_Q}{dt} = k_1(C_0 - C_Q)C_Q - k_2 C_Q^2$$

or

$$k_1 C_0 t = \ln \left| \frac{C_Q}{C_{Q_0}} \frac{C_0 - (1 - k_2/k_1)C_{Q_0}}{C_0 - (1 - k_2/k_1)C_Q} \right|$$

and $C_A$ would be found from

$$C_A(t) = C_0 - C_Q(t)$$

Note that initially some $Q$ *must* be present for any reaction to occur, but $A$ could be formed by the reverse reaction. For the irreversible case, $k_2 = 0$,

$$k_1 C_0 t = \ln \left| \frac{C_Q}{C_0 - C_Q} \frac{C_{A_0}}{C_{Q_0}} \right|$$

Here, *both* $A$ and $Q$ must be present initially for the reaction to proceed. These kinetic results can also be deduced from physical reasoning. A plot of $C_Q(t)$ gives an "S-shaped" curve, starting at $C_Q(0) = C_{Q_0}$ and ending at $C_Q(\infty) = C_0 = C_{A_0} + C_{Q_0}$; this is sometimes called a "growth curve" since it represents a buildup and then finally depletion of the reacting species. Figures 1 and 2 illustrate this.

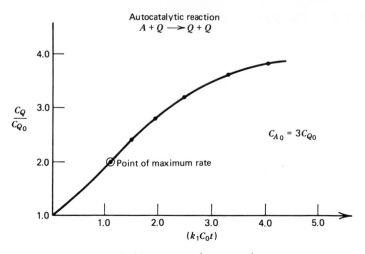

*Figure 1  $C_Q/C_{Q_0}$ versus dimensionless time.*

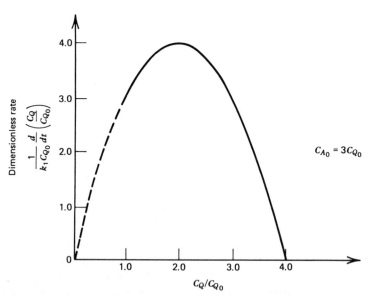

*Figure 2  Dimensionless rate versus $C_Q/C_{Q_0}$.*

Autocatalytic reactions can occur in homogeneously catalytic and enzyme systems, although usually with different specific kinetics.

---

For the general reaction (1.1-3), the following treatment is used (see Aris [3] for more details):

$$\sum_{j=1}^{N} \alpha_j A_j = 0$$

$$C_j = C_{j0} + \alpha_j \left(\frac{\xi}{V}\right)$$

$$r = k_1 \prod_{j=1}^{N} C_j^{a'_j} - k_2 \prod_{j=1}^{N} C_j^{q'_j}$$

$$= k_1 \left( \prod_{j=1}^{N} C_j^{a'_j} - \frac{1}{K} \prod_{j=1}^{N} C_j^{q'_j} \right) \tag{1.3-7}$$

In most cases the forward reaction depends only on the reactants and so the $a'_j$ corresponding to those $j$ with positive $\alpha_j$ are zero. Similarly, the reverse reaction usually depends only on the products. Aris [6] has given the relations for these for the case of simple reactions where the stoichiometric equation also represents the molecular steps:

$$a'_j = \tfrac{1}{2}(|\alpha_j| - \alpha_j) \qquad q'_j = \tfrac{1}{2}(|\alpha_j| + \alpha_j) \tag{1.3-8a}$$

There are cases, however, where this is not true, as in product inhibition or autocatalytic reactions. In the former, increasing product concentration decreases the rate, and so the $a'_j$ are negative when they correspond to positive $\alpha_j$; thus, for all these situations:

$$\alpha_j a'_j \leq 0 \qquad \text{and} \qquad \alpha_j q'_j \geq 0 \tag{1.3-8b}$$

which is useful in deducing certain mathematical features of the kinetics. The only exceptions are autocatalytic reactions where the $\alpha_j a'_j > 0$ for the species inducing the autocatalytic behavior.

Also note that the rate can be expressed in terms of only the extent (and other variables such as temperature, of course) and the initial composition. This is seen by substituting for the concentrations in Eq. 1.3-7,

$$r\left(\frac{\xi}{V}; C_{j0}; T\right) = k_1 \left[ \prod_{j=1}^{N} \left(C_{j0} + \alpha_j \frac{\xi}{V}\right)^{a'_j} - \frac{1}{K} \prod_{j=1}^{N} \left(C_{j0} + \alpha_j \frac{\xi}{V}\right)^{q'_j} \right] \tag{1.3-9}$$

Thus again we see that the progress of a reaction can be completely described by the single variable of extent/degree of advancement or conversion.

Among other derivations, Aris [6] has shown how Eq. 1.3-9 can be used to show that

$$\frac{\partial r}{\partial(\xi/V)} = k_1 \left[ \left( \prod_{j=1}^{N} C_j^{a_j'} \right) \left( \sum_{j=1}^{N} \frac{\alpha_j a_j'}{C_j} \right) - \left( \prod_{j=1}^{N} C_j^{q_j'} \right) \left( \sum_{j=1}^{N} \frac{\alpha_j q_j'}{C_j} \right) \right] \quad (1.3\text{-}10)$$

and since $k_1$, $C_j$, and the forward and reverse products ($\prod_j$) are all positive, the sign of the right-hand side of Eq. 1.3-10 depends on the signs of the forward and reverse sums ($\sum_j$). For the nonautocatalytic cases where Eq. (1.3-8b) are satisfied, it is clear that

$$\frac{\partial r}{\partial(\xi/V)} < 0 \quad (1.3\text{-}11)$$

which states that the rate *always* decreases with increasing extent as the reaction approaches equilibrium. For autocatalytic reactions this is not true and the rate may increase and then decrease. This general feature of any reaction with rate law (1.3-7) will be found useful later for some qualitative reasoning in reactor design.

## 1.4 Complex Reactions

The rate equations for complex reactions are constructed by combinations of terms of the type (1.3-1). For parallel reactions, all of the same order,

$$A \overset{1}{\underset{2}{\overset{\longrightarrow}{\rightrightarrows}}} \begin{array}{l} Q \\ S \\ R \end{array}$$

$$R_A = (k_1 + k_2 + k_3 + \cdots) C_A^{a'} \quad (1.4\text{-}1)$$

$$r_Q = k_1 C_A^{a'} \quad (1.4\text{-}2)$$

$$r_S = k_2 C_A^{a'} \quad (1.4\text{-}3)$$

The integrated forms of Equations 1.4-1 to 1.4-3 can easily be found from the following relations for first-order reactions:

$$-\frac{dC_A}{dt} = k_1 C_A + k_2 C_A$$

$$\frac{dC_Q}{dt} = k_1 C_A$$

$$\frac{dC_S}{dt} = k_2 C_A$$

CHEMICAL ENGINEERING KINETICS

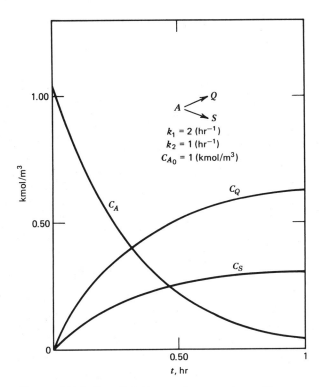

*Figure 1.4-1 Parallel first-order reactions. Concentration versus time.*

and then:

$$C_A = C_{A_0} e^{-(k_1 + k_2)t}$$

$$C_Q - C_{Q_0} = \frac{k_1}{k_1 + k_2} C_{A_0}[1 - e^{-(k_1 + k_2)t}]$$

Figure 1.4-1 illustrates the results.

The relative product concentrations can be simply found by formally dividing the rate equations (or the integrated results):

$$\frac{dC_Q}{dC_S} = \frac{k_1}{k_2}$$

This ratio is implicit in time, and yields, after integration,

$$\frac{C_Q - C_{Q_0}}{C_S - C_{S_0}} = \frac{k_1}{k_2}$$

For consecutive reactions:

$$A \xrightarrow{\phantom{x}1\phantom{x}} Q \xrightarrow{\phantom{x}2\phantom{x}} S$$

$$R_A = k_1 C_A^{a'} \tag{1.4-4}$$

$$R_Q = k_1 C_A^{a'} - k_2 C_Q^{q'} \tag{1.4-5}$$

$$R_S = k_2 C_Q^{q'} \tag{1.4-6}$$

Equations 1.4-4 to 1.4-6 can also be easily integrated for first-order reactions:

$$-\frac{dC_A}{dt} = k_1 C_A$$

$$\frac{dC_Q}{dt} = k_1 C_A - k_2 C_Q$$

and

$$C_A = C_{A_0} e^{-k_1 t}$$

$$C_Q = C_{Q_0} e^{-k_2 t} + \frac{k_1 C_{A_0}}{k_2 - k_1} (e^{-k_1 t} - e^{-k_2 t})$$

$$C_S = C_{A_0} + C_{Q_0} + C_{S_0} - C_A - C_Q$$

These results are illustrated in Fig. 1.4-2.

If experimental data of $C_A$, $C_Q$ are given as functions of time, the values of $k_1$ and $k_2$ can, in principle, be found by comparing the computed curves, as in Fig. 1.4-2, with the data. However, it is often more effective to use an analog computer to quickly generate many solutions as a function of $(k_1, k_2)$, and compare the outputs with the data.

The maximum in the $Q$ curve can be found by differentiating the equation for $C_Q$ and setting this equal to zero in the usual manner with the following result:

$$k_1 t_m = \frac{1}{(k_2/k_1) - 1} \ln\left\{\frac{k_2}{k_1}\left[1 - \frac{C_{Q_0}}{C_{A_0}}\left(1 + \frac{k_2}{k_1}\right)\right]\right\}$$

Again, it is often simpler to find the selectivity directly from the rate equations. Dividing gives

$$\frac{dC_Q}{dC_A} = -1 + \frac{k_2}{k_1}\frac{C_Q}{C_A}$$

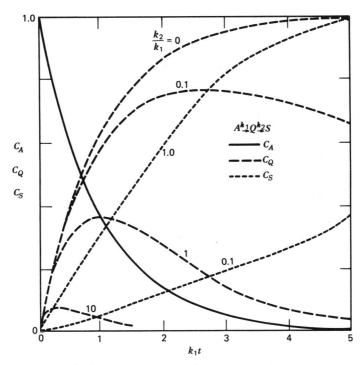

*Figure 1.4-2 Consecutive first-order reactions. Concentrations versus time for various ratios $k_2/k_1$.*

which has the solution

$$\frac{C_Q - C_{Q_0}}{C_{A_0} - C_A} = \frac{C_{Q_0}}{C_{A_0}} \frac{(1 - x_A)^\kappa - 1}{x_A} + \frac{1 - x_A}{\kappa - 1} \frac{1 - (1 - x_A)^{\kappa - 1}}{x_A}$$

where

$$x_A = 1 - \left(\frac{C_A}{C_{A_0}}\right) \qquad \kappa = \frac{k_2}{k_1}$$

## Example 1.4-1  Complex Reaction Networks

Many special cases are given in Rodigin and Rodigina [12]. The situation of general first-order reaction networks has been considered by Wei and Prater [13] in a particularly elegant and now classical treatment. Boudart [5] also has a more abbreviated discussion.

The set of rate equations for first-order reversible reactions between the $N$ components of a mixture can be written

$$\frac{dy_1}{dt} = \left(-\sum_{j=1}^{N}{}' k_{j1}\right) y_1 + k_{12} y_2 + \cdots + k_{1N} y_N$$

$$\frac{dy_2}{dt} = k_{21} y_1 + \left(-\sum_{j=1}^{N}{}' k_{j2}\right) y_2 + \cdots + k_{2N} y_N \qquad \text{(a)}$$

$$\vdots$$

$$\frac{dy_N}{dt} = k_{N1} y_1 + k_{N2} y_2 + \cdots + \left(-\sum_{j=1}^{N}{}' k_{jN}\right) y_N$$

where the $y_i$ are, say, mole fractions, $k_{ji}$ is the rate coefficient of the reaction $A_i \rightarrow A_j$ and

$$\sum_{j=1}^{N}{}' k_{ji} \equiv \sum_{\substack{j=1 \\ j \neq i}}^{N} k_{ji}$$

In matrix form:

$$\frac{d\mathbf{y}}{dt} = \mathbf{Ky} \qquad \text{(b)}$$

$$\mathbf{y} = \begin{bmatrix} y_1 \\ y_2 \\ \vdots \\ y_N \end{bmatrix} \qquad \mathbf{K} = \begin{bmatrix} -\sum{}' k_{ji} & k_{12} & \cdots & k_{1N} \\ k_{21} & -\sum{}' k_{j2} & \cdots & k_{2N} \\ \vdots & \vdots & & \vdots \\ k_{N1} & k_{N2} & \cdots & -\sum_{j=1}^{N}{}' k_{jN} \end{bmatrix} \qquad \text{(c)}$$

It is simplest to consider a three-component system, where the changes in composition with time—the reaction paths—can be followed on a triangular diagram. Figure 1 shows these for butene isomerization data from the work of Haag, Pines, and Lago (see Wei and Prater) [13].

We observe that the reaction paths all converge to the equilibrium value in a tangent fashion, and also that certain ones (in fact, two) are straight lines. This has important implications for the behavior of such reaction networks.

It is known from matrix algebra that a square matrix possesses $N$-eigenvalues, the negatives of which are found from

$$\det(\mathbf{K} + \lambda_m \mathbf{I}) = 0 \qquad \text{(d)}$$

where $\mathbf{I}$ is a unit matrix and $\lambda_m \geq 0$ for the rate coefficient matrix. Also, $N$-eigenvectors, $\mathbf{x}_m$, can then be found from

$$\mathbf{Kx}_m = -\lambda_m \mathbf{x}_m \qquad \text{(e)}$$

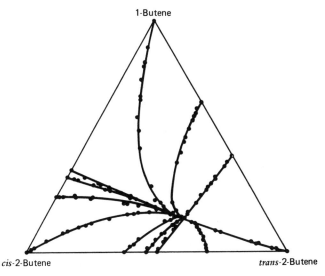

1-Butene

cis-2-Butene

trans-2-Butene

Figure 1 Comparison of calculated reaction paths with experimentally observed compositions for butene isomerization. The points are observed composition and the solid lines are calculated reaction paths. (Wei and Prater [13]).

and these combined into

$$\mathbf{KX} = \mathbf{X\Lambda}$$ (f)

where

$$\mathbf{X} = [\mathbf{x}_0 \mathbf{x}_1 \cdots], \Lambda = -\begin{bmatrix} \lambda_0 & 0 & \cdots & 0 \\ 0 & \lambda_1 & \cdots & 0 \\ \vdots & \vdots & & \vdots \\ 0 & 0 & \cdots & \lambda_{N-1} \end{bmatrix}$$

Wei and Prater found that a new set of fictitious components, **B**, can be defined that have the important property of being uncoupled from each other. The quantities of **B** are represented by $\zeta$. These components decay according to

$$\frac{d\zeta_0}{dt} = -\lambda_0 \zeta_0 \quad \text{so that} \quad \zeta_0 = \zeta_0(0)e^{-\lambda_0 t}$$

$$\frac{d\zeta_1}{dt} = -\lambda_1 \zeta_1 \quad \zeta_1 = \zeta_1(0)e^{-\lambda_1 t}$$ (g)

$$\vdots$$

where the $\zeta$ represent the quantities of **B**.

This can be readily shown from the above matrix equations; let

$$\mathbf{y} = \mathbf{X}\boldsymbol{\zeta} = \zeta_0 \mathbf{x}_0 + \zeta_1 \mathbf{x}_1 + \cdots \tag{h}$$

where

$$\boldsymbol{\zeta} = \begin{bmatrix} \zeta_0 \\ \zeta_1 \\ \vdots \end{bmatrix}$$

Then substituting Eq. h into Eq. b gives

$$\mathbf{X}\frac{d\boldsymbol{\zeta}}{dt} = \mathbf{K}\mathbf{X}\boldsymbol{\zeta}$$

and premultiplying each side by $\mathbf{X}^{-1}$ yields

$$\frac{d\boldsymbol{\zeta}}{dt} = \mathbf{X}^{-1}\mathbf{K}\mathbf{X}\boldsymbol{\zeta}$$

$$= \boldsymbol{\Lambda}\boldsymbol{\zeta} \tag{i}$$

where Eq. f was used for the last step. Equations g and i are the same, and so Eq. h shows that the fictitious components are special linear combinations of the real ones:

$$\boldsymbol{\zeta} = \mathbf{X}^{-1}\mathbf{y}$$

Now, at equilibrium,

$$\frac{d\mathbf{y}_{eq}}{dt} = \mathbf{0} = \mathbf{K}\mathbf{y}_{eq} = 0\mathbf{y}_{eq}$$

and comparing the last equality with Eq. e shows that one of the eigenvectors is the equilibrium composition and that the corresponding eigenvalue is zero:

$$\mathbf{x}_0 = \mathbf{y}_{eq} \qquad \lambda_0 = 0$$

Physically, this is obvious, since a reaction path starting at the equilibrium composition does not change with time:

$$\zeta_0 = \zeta_0(0) = \text{constant}$$

Eq. (h) can then be written as

$$\mathbf{y} - \mathbf{y}_{eq} = \zeta_1 \mathbf{x}_1 + \zeta_2 \mathbf{x}_2 + \cdots \zeta_{N-1}\mathbf{x}_{N-1}$$

$$= \zeta_1(0)e^{-\lambda_1 t}\mathbf{x}_1 + \cdots \zeta_{N-1}(0)e^{-\lambda_{N-1} t}\mathbf{x}_{N-1} \tag{j}$$

which gives the decay of deviations from equilibrium. Geometrically, each eigenvector, $\mathbf{x}_m$, represents a direction in space, and so the right-hand side of Eq. j represents all the contributions that make up the reaction paths. Special initial conditions of, say, $\zeta_1(0) \neq 0$, $\zeta_{m>1}(0) = 0$ leave only one term on the right-hand side, and this one direction thus is that of the special straight-line reaction paths. Thus, knowing the rate constants, $k_{ji}$, a series of matrix computations will permit one to determine the proper (real) starting compositions for straight-line reaction paths, which are $\zeta_m(t)$. The above figure shows an experimental determination of these paths. Wei and Prater also show that only $N - 1$ such paths need to be found, and the last can be computed from matrix manipulations.

In addition to illustrating many features of monomolecular reaction networks, Wei and Prater illustrated how these results, especially the straight line reaction paths, could be helpful in planning experiments for and the determination of rate constants, and this will be discussed later. Also, these same methods have been used in the "stochastic" theory of reaction rates, which consider the question of how simple macroscopic kinetic relations (e.g., the mass action law) can result from the millions of underlying molecular collisions—see Widom for comprehensive reviews [14].

---

Another common form of mixed consecutive-parallel reactions is the following:

$$A + B \xrightarrow{\;\;1\;\;} Q$$

$$Q + B \xrightarrow{\;\;2\;\;} S$$

Successive chlorinations of benzene, for example, fall into this category. The main feature is the common second reactant $B$, so that in a sense the reactions are also parallel. The rate expressions are

$$\frac{-dC_A}{dt} = k_1 C_A C_B$$

$$\frac{dC_Q}{dt} = k_1 C_A C_B - k_2 C_Q C_B$$

There is no simple solution of these differential equations as a function of time. However, the selectivities can again be found by dividing the equations:

$$\frac{dC_Q}{dC_A} = -1 + \frac{k_2}{k_1}\frac{C_Q}{C_A}$$

This is precisely the same as for the simpler first-order case considered above, and so would result in the same final results. Thus, the common reactant, $B$, has no effect on the selectivity, but will cause a different behavior with time. An important consequence of this is that a "selectivity diagram" or a plot of $C_Q, C_S, \ldots$ versus $C_A$, or conversion, $x_A$, is often rather insensitive to details of the reaction network other than the concentrations of the main chemical species. This concept is often used in complicated industrial process kinetics of catalytic cracking, for example, to develop good correlations of product distributions as a function of conversion.

### Example 1.4-2 Catalytic Cracking of Gasoil

An overall kinetic model for the cracking of gasoils to gasoline products was developed by Nace, Voltz, and Weekman [15]. The actual situation was a catalytic reaction and the data were from specific reactor types, but mass-action type rate expressions were used and illustrate the methods of this section.

The overall reaction is as follows:

where $A$ represents gasoil, $Q$ gasoline, and $S$ other products ($C_1 - C_4$, coke). For the conditions considered, the gasoil cracking reaction can be taken to be approximately second order and the gasoline cracking reaction to be first order (see Weekman for justification of this common approximation for the complicated cracking reaction) [16, 17]. Then, the kinetic equations are (where $y$ represents weight fractions):

$$\frac{dy_A}{dt} = -k_1 y_A^2 - k_3 y_A^2 = -(k_1 + k_3)y_A^2 \equiv -k_0 y_A^2 \tag{a}$$

$$\frac{dy_Q}{dt} = k_1 y_A^2 - k_2 y_Q \tag{b}$$

$$y_S = 1 - y_A - y_Q \tag{c}$$

This parallel-consecutive kinetic scheme can be integrated, but an expression for the important gasoline selectivity can also be found directly by formally dividing Eqs. a and b:

$$\frac{dy_Q}{dy_A} = -\frac{k_1}{k_0} + \frac{k_2}{k_0}\frac{y_Q}{y_A^2} \tag{d}$$

Integrating gives

$$y_Q = \frac{k_1 k_2}{k_0^2} e^{-k_2/k_0 y_A}\left(\frac{k_0}{k_2}e^{k_2/k_0} - \frac{k_0 y_A}{k_2}e^{k_2/k_0 y_A} + \int_{k_2/k_0}^{k_2/k_0} e^\eta \frac{d\eta}{\eta}\right)$$

$$= \frac{k_1 k_2}{k_0^2} e^{-k_2/k_0 y_A}\left[\frac{k_0}{k_2}e^{k_2/k_0} - \frac{k_0 y_A}{k_2}e^{k_2/k_0 y_A} + E_i(k_2/k_0 y_A) - E_i(k_2/k_0)\right]$$

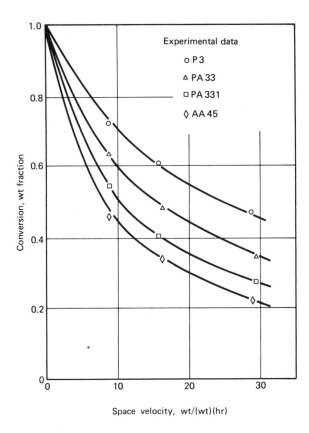

*Figure 1 Comparison of experimental conversions with model predictions for different charge stocks. Catalyst residence time: 1.25 min. (Nace, Voltz, and Weekman [15]).*

where

$$E_i(x) = \text{exponential integral (tabulated function)}$$

$$= \int_{-\infty}^{x} e^{\eta} \frac{d\eta}{\eta}$$

Figure 1 shows the conversion versus (reciprocal) time behavior for four different feedstocks, and a catalyst residence time of 1.25 min in the fluidized bed reactor

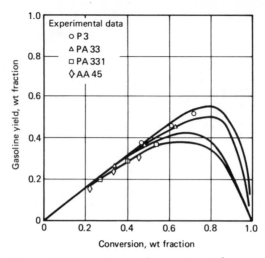

*Figure 2 Comparison of experimental gasoline selectivities with model predictions for different charge stocks. Catalyst residence time: 1.25 min. (Nace, Voltz, and Weekman [15]).*

and Fig. 2 shows the corresponding gasoline selectivities. The feedstock properties are as follows:

| | Weight percent | | | Rate Constants at 900°F hr$^{-1}$ (weight fraction)$^{-1}$ | | |
|---|---|---|---|---|---|---|
| Feedstock | Paraffins | Napthenes | Aromatics | $k_0$ | $k_1$ | $k_2$ |
| P3 | 46.2 | 35.1 | 18.6 | 34.0 | 28.0 | 1.86 |
| PA33 | 31.3 | 30.4 | 38.3 | 22.1 | 17.6 | 1.48 |
| PA331 | 17.7 | 26.2 | 56.1 | 15.5 | 12.6 | 2.66 |
| AA45 | 11.0 | 14.2 | 74.8 | 12.3 | 9.30 | 2.28 |

These results show the effects of different catalytic feedstock compositions on the rates of reaction—Nace, Voltz, and Weekman's paper contains additional valuable information of this type.

Further papers from the same group [18, 19] provide correlations of these overall rate constants with important feedstock properties. An example is given in Fig. 3, and illustrates how a large variety of practical data can often be correlated by using the properties of groups of similar chemical species as "pseudo-

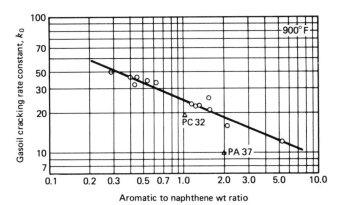

Figure 3 *Relationship between gasoil cracking rate constant and aromatic to naphthene ratio (Voltz, Nace, and Weekman [18]).*

species." Another interesting example of this was given by Anderson and Lamb [20]. Further aspects of the catalytic cracking data will be utilized in future chapters.

More comprehensive utilization of these techniques of "lumping" (the currently used terminology) groups of chemical species with similar kinetic behavior has been provided by Jacob et al. [21]. Based on heuristic reasoning from the rather well-known chemistry of catalytic cracking, plus availability of modern chemical analysis techniques, a 10-lump kinetic model was formulated. This model involved paraffins, naphthenes, aromatic rings, and aromatic substituent groups in light and heavy fuel oil fractions. Using the same data base as described above, the more detailed model was essentially able to predict correlations, such as in Fig. 3, as well as predict results for a much wider range of feedstocks.

The question of efficient techniques for performing these lumping analyses are a subject of current research. The successful applications to date, as above, have been based on heuristic reasoning, and more formal procedures are not available. Basic theoretical results for monomolecular reaction systems have been provided by Wei and Kuo [22] and Ozawa [23], and important other features are given by Luss et al. [24] (and other references provided there).

## Example 1.4-3 Rate Determining Step and Steady-State Approximation

Consider a simple reaction with one intermediate species:

$$A \xrightarrow{\;1\;} I \xrightarrow{\;2\;} P$$

The rate equations are:

$$\frac{dC_A}{dt} = -k_1 C_A \tag{a}$$

$$\frac{dC_I}{dt} = k_1 C_A - k_2 C_I \tag{b}$$

$$\frac{dC_P}{dt} = k_2 C_I \tag{c}$$

This is the same case solved earlier, and is illustrated in Fig. 1.4-2.

There are some interesting and useful features of this simple system that will illustrate the important concept of the rate determining step. Note from Fig. 1.4-2 that when $k_1 \gg k_2$, the two reactions are almost separate in time, and the overall rate of product formation is dominated by the slow reaction 2. Algebraically, from the integrated rate equations given above, after a certain time interval:

$$\frac{dC_P}{dt} = k_2 C_I$$

$$= \left( -\frac{dC_I}{dt} \right) \left( 1 - \frac{e^{-k_1 t}}{\dfrac{k_2}{k_2 - k_1} e^{-k_2 t} - \dfrac{k_1}{k_2 - k_1} e^{-k_1 t}} \right)$$

$$\sim \left( -\frac{dC_I}{dt} \right) \qquad \text{for} \quad \frac{k_1}{k_2} \gg 1 \tag{d}$$

For the opposite case of $k_2 \gg k_1$, the integrated rate equations in a different rearrangement give:

$$\frac{dC_P}{dt} = \left( -\frac{dC_A}{dt} \right) \frac{k_2}{k_2 - k_1} [1 - e^{+(k_1 - k_2)t}]$$

$$\sim \left( -\frac{dC_A}{dt} \right) \qquad \text{for} \quad \frac{k_2}{k_1} \gg 1 \tag{e}$$

again after a certain time interval. Thus, the overall rate of product formation is dominated by the slow reaction 1. This shows that the overall rate is always dominated by any slow steps in the reaction sequence;[3] this concept of a "rate limiting step" will be used many times in the ensuing discussions.

One of the most useful applications pertains to the notion of a stationary or steady state of the intermediate. If a stationary state between the main reactant and

---

[3] This material was adapted from Kondrat'ev [25].

product is to exist for this simple case, the rate of disappearance of $A$ must be approximately equal to the rate of production of $P$. This would make a plot of $C_P(t)$ the mirror image of $C_A(t)$. From Fig. 1.4-2, or from Eq. e, it is seen that this is almost true for large $k_2/k_1 > 10 \rightarrow \infty$. Physically, a large value of $k_2$, relative to $k_1$, means that as soon as any $I$ is formed from reaction of $A$, it is immediately transformed into $P$, and so the product formation closely follows the reactant loss. Thus, the intermediate is very short lived, and has a very low concentration; this can also be seen in Figure 1.4-2.

The sum of Eq. a, b, and c gives

$$\frac{dC_A}{dt} + \frac{dC_I}{dt} + \frac{dC_P}{dt} = 0 \tag{f}$$

If the stationary state exists, and the reactant loss and product formation are approximately equivalent,

$$\frac{dC_P}{dt} \simeq -\frac{dC_A}{dt} \tag{e}$$

and so

$$\frac{dC_I}{dt} \simeq 0 \tag{g}$$

which is the usual statement. Then, from Eq. b

$$C_I \simeq \left(\frac{k_1}{k_2}\right) C_A \tag{h}$$

which is indeed small for finite $C_A$ and $(k_2/k_1) \gg 1$. Also,

$$C_P \simeq C_{A_0} - C_A \tag{i}$$

and the exact details of the intermediate need not be known.

Rigorous justification of the steady-state approximation has naturally been of interest for many years, and Bowen, Acrivos, and Oppenheim [26] have resolved the conditions under which it can be properly used. The mathematical question concerns the correctness of ignoring the derivatives in some of a set of differential equations (i.e., changing some to algebraic equations), which is analogous to ignoring the highest derivatives in a single differential equation. These questions are answered by the rather complicated theory of singular perturbations, discussion of which is given in the cited article.

Predictions from the steady-state approximation have been found to agree with experimental results, where it is appropriate. This should be checked by using relations such as Eq. h to be sure that the intermediate species concentrations are, in fact, much smaller than those of the main reactants and products in the reaction. When valid, it permits kinetic analysis of systems that are too complicated to conveniently handle directly, and also permits very useful overall kinetic relationships to be obtained, as is seen in Ex. 1.4-4 to 1.4-6.

## *Example  1.4-4  Classical Unimolecular Rate Theory*

Another interesting example of complex reactions is in describing the chemical mechanism that may be the basis of a given overall observed kinetics. A question of importance in unimolecular decompositions (e.g., cyclohexane, nitrous oxide, azo methane—see Benson [27])—is how a single molecule becomes sufficiently energetic by itself to cause it to react. The theory of Lindemann [28] explains this by postulating that actually bimolecular collisions generate extraenergetic molecules, which then decompose:

$$A + A \underset{2}{\overset{1}{\rightleftharpoons}} A^* + A \tag{a}$$

$$A^* \overset{3}{\longrightarrow} Q + \cdots \text{(slow)} \tag{b}$$

Then, the rate of product formation observed is

$$\frac{dC_Q}{dt} = k_3 C_{A^*} \tag{c}$$

To find $A,^*$ its kinetics are given by:

$$\frac{dC_{A^*}}{dt} = k_1 C_A{}^2 - k_2 C_{A^*} C_A - k_3 C_{A^*} \tag{d}$$

To solve this differential equation in conjunction with a similar one for species $A$ would be very difficult, and recourse is usually made to the "steady-state approximation." This assumes that $dC_{A^*}/dt \simeq 0$, or that the right-hand side of Eq. d is in a pseudo-equilibrium or stationary state. Justification for this was provided in the last example.

With this approximation, Eq. d is easily solved:

$$C_{A^*} = \frac{k_1 C_A{}^2}{k_3 + k_2 C_A}$$

Then,

$$\frac{dC_Q}{dt} = \frac{k_1 k_3 C_A{}^2}{k_3 + k_2 C_A}$$

Now, at high concentrations (pressure), $k_2 C_A \gg k_3$ (recall reaction 3 is presumably slow), and so,

$$\frac{dC_Q}{dt} = \frac{k_1 k_3}{k_2} C_A$$

which is a first-order rate. Conversely, for low pressures, $k_2 C_A \ll k_3$, and,

$$\frac{dC_Q}{dt} = k_1 C_A{}^2$$

Thus, this theory indicates that simple decompositions that are first order at high pressures should change to second order at low pressures—many years of experimentation have shown this to be the case. Better quantitative agreement with the data is provided by more elaborate but similar theories—see Laidler [10] or Benson [27].

---

An important example of complex reactions are those involving *free radicals* in chain reactions. These reactions consist of three essential steps:

1. Initiation or formation of the free radicals.

2. Propagation, by reaction of the free radicals with reactants.

3. Termination by reaction of free radicals to form stable products.

Many types of reactions have mechanisms in this category: thermal cracking, some polymerizations, many liquid phase oxidations and combustion reactions, photochlorinations, and others.

In a review article, Benson [29] distinguishes two broad categories of chain reactions that have somewhat different kinetic features: pyrolytic chains, containing a unimolecular step, and metathetical chains involving two reactants and only bimolecular steps. We consider the interesting and practical case of thermal cracking, or pyrolysis, to illustrate the principles.

The Rice–Herzfeld [30] mechanism, or variations, can often be used to explain the kinetics. In addition to the concepts noted above, they postulated that the fastest mode of reaction of a free radical with a hydrogen-containing molecule is the abstraction of a hydrogen atom, followed by decomposition of the new radical into an olefin molecule and another radical. These steps are then the propagation part of the scheme.

Thus, the essential idea is that the overall reaction

$$A_1 \longrightarrow A_2 + A_3 \tag{1.4-7}$$

can be represented by a sequence of initiation, propagation, and termination steps:

1. Initiation by breaking weak chemical bond:

$$(\text{reactant}) \longrightarrow (\text{free radicals}) \tag{1.4-8}$$

**2.** Propagation, consisting of hydrogen abstraction:

$$\text{(free radical)} + \text{(reactant)} \longrightarrow \text{(free radical + abstracted hydrogen)} + \text{(large free radical)} \qquad (1.4\text{-}9)$$

and large free radical decomposition:

$$\text{(free radical)} \longrightarrow \text{(product)} + \text{(free radical)} \qquad (1.4\text{-}10)$$

**3.** Termination:

$$\text{(free radical)} + \text{(free radical)} \longrightarrow \text{(product)} \qquad (1.4\text{-}11)$$

There are certain general rules that are very helpful in constructing a mechanism, Laidler [10]. The initiation step can be considered from the viewpoint of classical unimolecular reaction rate theory and is first order if:

**1.** The degrees of freedom of the atoms in the reactant molecules are large; that is, the molecule is complicated.

**2.** The temperature is low.

**3.** The partial pressure is high.

For the opposite conditions, the initiation reaction can be second order, following unimolecular reaction rate theory (Ex. 1.4-4).

The termination step is determined by the following factors:

**1.** Relative rate constants of the propagation steps, which lead to relative radical concentrations.

**2.** Magnitude of rate constant of termination steps, which depend on the complexity of the radicals.

**3.** Degrees of freedom in the termination reaction; if these are large, no third body (external) is required and if small, a third body is involved.

Consider a simple example of a free radical reaction, which is represented by the following stoichiometric equation:

$$A_1 \longrightarrow A_2 + A_3 \qquad (1.4\text{-}7)$$

In reality, the reaction might proceed by the following steps:

$$A_1 \xrightarrow{k_1} 2R_1^{\cdot} \qquad \text{Initiation} \qquad (1.4\text{-}12)$$

$$R_1^{\cdot} + A_1 \xrightarrow{k_2} R_1 H + R_2^{\cdot} \qquad \begin{array}{l}\text{Hydrogen}\\ \text{abstraction}\end{array} \left.\vphantom{\begin{array}{l}a\\b\\c\\d\end{array}}\right\} \text{Propagation} \qquad (1.4\text{-}13)$$

$$R_2^{\cdot} \xrightarrow{k_3} A_2 + R_1^{\cdot} \qquad \begin{array}{l}\text{Radical}\\ \text{decomposition}\end{array} \qquad (1.4\text{-}14)$$

$$R_1^{\cdot} + R_2^{\cdot} \xrightarrow{k_4} A_3 \qquad \text{Termination} \qquad (1.4\text{-}15)$$

$R_1^{\cdot}$ and $R_2^{\cdot}$ are radicals (e.g., when hydrocarbons are cracked $CH_3^{\cdot}$, $C_2H_5^{\cdot}$, $H^{\cdot}$). The rate of consumption of $A_1$ may be written:

$$-\frac{dC_{A_1}}{dt} = k_1 C_{A_1} + k_2 C_{A_1} C_{R_1} \qquad (1.4\text{-}16)$$

The rate of initiation is generally much smaller than the rate of propagation so that in Eq. (1.4-16) the term $k_1 C_{A_1}$ may be neglected. The problem is now to express $C_{R_i}$, which are difficult to measure, as a function of the concentrations of species which are readily measurable. For this purpose, use is made of the hypothesis of the steady-state approximation in which rates of change of the concentrations of the intermediates are assumed to be approximately zero, so that

$$\frac{dC_{R_1}}{dt} \simeq 0 \qquad \frac{dC_{R_2}}{dt} \simeq 0$$

or, in detail,

$$\frac{dC_{R_1}}{dt} = 0 = 2k_1 C_{A_1} - k_2 C_{R_1} C_{A_1} + k_3 C_{R_2} - k_4 C_{R_1} C_{R_2}$$

$$\frac{dC_{R_2}}{dt} = 0 = k_2 C_{R_1} C_{A_1} - k_3 C_{R_2} - k_4 C_{R_1} C_{R_2}$$

These conditions must be fulfilled simultaneously. By elimination of $C_{R_2}$ one obtains a quadratic equation for $C_{R_1}$:

$$2k_1 C_{A_1} - k_2 C_{R_1} C_{A_1} + \frac{k_2 k_3 C_{R_1} C_{A_1}}{k_3 + k_4 C_{R_1}} - \frac{k_2 k_4 C_{R_1}{}^2 C_{A_1}}{k_3 + k_4 C_{R_1}} = 0$$

the solution of which is,

$$C_{R_1} = \frac{k_1 k_3}{4k_2 k_4} + \frac{1}{4k_2 k_4} \sqrt{(k_1 k_3)^2 + 16 k_1 k_2 k_3 k_4}$$

Since $k_1$ is very small, this reduces to

$$C_{R_1} \simeq \sqrt{\frac{k_1 k_3}{k_2 k_4}} \qquad (1.4\text{-}17)$$

so that Eq. (1.4-16) becomes:

$$-\frac{dC_{A_1}}{dt} = \sqrt{\frac{k_1 k_2 k_3}{k_4}} \, C_{A_1} \qquad (1.4\text{-}18)$$

which means that the reaction is essentially first order.

There are other possibilities for termination. Suppose that not (1.4-15) but the following is the fastest termination step:

$$R_1^{\cdot} + R_1^{\cdot} \xrightarrow{\ k_5\ } A_1$$

It can be shown by a procedure completely analogous to the one given above that the rate is given by

$$-\frac{dC_{A_1}}{dt} = k_2 \sqrt{\frac{2k_1}{k_5}} \, (C_{A_1})^{3/2} \qquad (1.4\text{-}19)$$

which means that the reaction is of order 3/2.

Goldfinger, Letort, and Niclause [31] (see Laidler [10]) have organized results of this type based on defining two types of radicals:

$\mu$—a radical involved as a reactant in a unimolecular propagation step.

$\beta$—a radical involved as a reactant in a bimolecular propagation step.

Usually the $\mu$ radical is larger than the $\beta$ radical, so that

$$\text{(termination rate constant magnitude)} (\mu\mu) < (\beta\mu) < (\beta\beta) \qquad (1.4\text{-}20)$$

This leads to the results shown in Table 1.4-1.

*Table 1.4-1 Overall Orders for Free Radical Mechanisms*

| First-Order Initiation | | Second-Order Initiation | | |
|---|---|---|---|---|
| **Simple Termination** | **Third Body** | **Simple Termination** | **Third Body** | **Overall Order** |
| | | $\beta\beta$ | | 2 |
| $\beta\beta$ | | $\beta\mu$ | $\beta\beta$M | $\frac{3}{2}$ |
| $\beta\mu$ | $\beta\beta$M | $\mu\mu$ | $\beta\mu$M | 1 |
| $\mu\mu$ | $\beta\mu$M | | $\mu\mu$M | $\frac{1}{2}$ |
| | $\mu\mu$M | | | 0 |

Note that in the above example a first-order initiation step was assumed, and with a termination step involving both $R_1(\beta)$ and $R_2(\mu)$, an overall first-order reaction was derived, in agreement with Table 1.4-1. The alternate $R_1 + R_1$ termination was of the $(\beta\beta)$ type, leading to a three-half-order reaction.

Franklin [32] and Benson [29] have summarized methods for predicting the rates of chemical reactions involving free radicals and Gavalas [33] has shown how the steady-state approximation and use of the chain propagation reactions alone (long-chain approximation) leads to reasonably simple calculation of the relative concentrations of the nonintermediate species. Also see Benson [34].

## Example 1.4-5  Thermal Cracking of Ethane

The overall reaction is

$$C_2H_6 = C_2H_4 + H_2$$

and can be considered to proceed by the following mechanism:
    Initiation:

$$\text{Eq. 1.4-12: } \underset{(A_1)}{C_2H_6} \xrightarrow{k_1} \underset{(R_1)}{2CH_3^{\cdot}} \tag{a}$$

Hydrogen abstraction:

$$\text{Eq. 1.4-13: } \underset{(R_1)}{CH_3^{\cdot}} + \underset{(A_1)}{C_2H_6} \xrightarrow{k_2} \underset{(R_1H)}{CH_4} + \underset{(R_2)}{C_2H_5^{\cdot}} \tag{b}$$

and:

$$\underset{(R_3)}{H^{\cdot}} + \underset{(A_1)}{C_2H_6} \xrightarrow{k_4} \underset{(R_3H)}{H_2} + \underset{(R_2)}{C_2H_5^{\cdot}} \tag{c}$$

Radical decomposition:

$$\text{Eq. 1.4-14: } \underset{(R_2)}{C_2H_5^{\cdot}} \xrightarrow{k_3} \underset{(A_2)}{C_2H_4} + \underset{(R_3)}{H^{\cdot}} \tag{d}$$

Termination:

$$\text{Eq. (1.14-15): } C_2H_5^{\cdot} + H^{\cdot} \xrightarrow{k_5} C_2H_6 \tag{e}$$

$$\text{or} \qquad\qquad 2C_2H_5^{\cdot} \xrightarrow{k_5} C_4H_{10} \tag{f}$$

$$\text{or} \qquad\qquad 2H^{\cdot} \xrightarrow{k_5} H_2 \tag{g}$$

$$\text{or} \qquad\qquad CH_3^{\cdot} + H^{\cdot} \longrightarrow CH_4 \tag{h}$$

By the above rules, since ethane is only a moderately complicated molecule (in terms of degrees of freedom), the initiation reaction (a) could be either first or second order. The classical Rice–Herzfeld [30] scheme would use the former, and with termination reaction (e), which is $(\beta\mu)$-$(H; C_2H_5)$, would lead to an over-all rate expression of first order. Using the above techniques gives

$$\frac{d[C_2H_4]}{dt} = k_3[C_2H_5^{\cdot}] \tag{i}$$

$$= \left(\frac{k_1 k_3 k_4}{k_5}\right)^{1/2} [C_2H_6]$$

This agrees with the overall rate data, which is first order. However, estimates of the concentrations of the ethyl and hydrogen radicals, as found from the steady-state approximation and the free radical rate expressions, indicate that the former is the larger, and thus that the alternate termination reaction (f) would be more appropriate.[4] Unfortunately, this is $(\mu\mu)$, and leads to an incorrect order of one-half. There are also other predictions of temperature coefficients of reaction and foreign gas effects that are not in agreement with the experiment. This is an il-lustration of how carefully one must check all the implications of an assumed mechanism.

By assuming that the unimolecular initiation step was in the second-order range, Küchler and Theile [35] developed an alternate free radical result using termination Eq. (f). From Table 1.4-1 this $(\mu\mu)$ termination for second-order initiation again leads to the proper overall first-order reaction rate:

$$\frac{d[C_2H_4]}{dt} = k_3\left(\frac{k_1}{k_5}\right)^{1/2} [C_2H_6] \tag{j}$$

The ratio of ethyl to hydrogen radicals can be found from the rate expression for hydrogen radicals:

$$\frac{d[H^{\cdot}]}{dt} = 0 = k_3[C_2H_5^{\cdot}] - k_4[H^{\cdot}][C_2H_6]$$

---

[4] Benson [27] presents the following estimates:

*Initial Free Radical Concentrations during Pyrolysis of $C_2H_6$*

| $T, K$ | $(CH_3)/(C_2H_5)$ | $(H)/(C_2H_5)$ | $(C_2H_5)$, mol/liter | $P_{C_2H_6}$, atm |
|--------|-------------------|----------------|-----------------------|-------------------|
| 850 | 0.03 | 0.0014 | $6 \times 10^{-9}$ | 1 |
|  | 0.3 | 0.14 | $6 \times 10^{-10}$ | 0.01 |
| 900 | 0.12 | 0.0041 | $2.5 \times 10^{-8}$ | 1 |
|  | 1.2 | 0.41 | $2.5 \times 10^{-9}$ | 0.01 |

or

$$\frac{[C_2H_5^\cdot]}{[H^\cdot]} = \frac{k_4}{k_3}[C_2H_6] \tag{k}$$

At moderate pressures, this expression gives larger ethyl than hydrogen radical concentrations, and is consistent with the use of termination (f). At lower pressures, the relative amount of hydrogen radicals is larger, and increases the importance of termination (e). This $(\beta\mu)$ step then leads to an overall order of $\frac{3}{2}$, which is what is experimentally observed at low pressures.

Other possible terminations are (g) and (h). The first would require a third body, because hydrogen is an uncomplicated radical, yielding a $(\beta\beta M)$ case with $\frac{3}{2}$-order reaction. This is usually not observed, however, because of the slowness of ternary reactions. Case h could be $(\beta\beta)$—second order—or $(\beta\beta M)$—$\frac{3}{2}$ order with second-order initiation—or it could be $(\beta\beta)$—$\frac{3}{2}$ order—or $(\beta\beta M)$—first order with first-order initiation. In any case, however, it would not predict the proper product distribution.

Quinn [36] has performed further experiments indicating that the first-order initiation is probably more correct. To obtain the proper overall first-order behavior, he had to assume that the radical decomposition step (d) has a rate intermediate between first- and second-order kinetics, approximately proportional to $[C_2H_5^\cdot][C_2H_6]^{1/2}$. This makes the ethyl radical have behavior between $\beta$ and $\mu$, say $(\beta\mu)$, and the table then indicates approximate first-order overall reaction, tending toward $(\beta\beta)$ termination—and $\frac{3}{2}$ order—for lower pressures. More recent data indicate that a wide range of observations is best represented by Quinn's mechanism.

The pyrolysis of larger hydrocarbons is somewhat simpler in choice of mechanism, since the hydrogen atoms play a less dominant role. Also, the molecules are sufficiently complicated so that the initiation step is usually first order. For example, Laidler [10] discusses the case of butane:

$$C_4H_{10} \longrightarrow 2C_2H_5^\cdot$$

$$C_2H_5^\cdot + C_4H_{10} \longrightarrow C_2H_6 + C_4H_9^\cdot$$

$$C_4H_9^\cdot \longrightarrow CH_3^\cdot + C_3H_6$$

$$C_4H_9^\cdot \longrightarrow C_2H_5^\cdot + C_2H_4$$

$$CH_3^\cdot + C_4H_{10} \longrightarrow CH_4 + C_4H_9^\cdot$$

$$C_2H_5^\cdot \longrightarrow C_2H_4 + H^\cdot$$

$$H^\cdot + C_4H_{10} \longrightarrow H_2 + C_4H_9^\cdot$$

$$2C_2H_5^\cdot \longrightarrow C_4H_{10} \text{ or } C_2H_4 + C_2H_6$$

Thus, the ethyl radical is both $\beta$ and $\mu$, although the slowness of its decomposition reaction tends to make the former more important. Thus, with first-order initiation and approximate ($\beta\mu$) behavior, the overall order is again approximately unity. Further details are given in Steacie [37] and Laidler [10] and Benson [27] among others.

This rather involved example illustrated the large amount of information that can be obtained from the general free radical reaction concepts.

## Example 1.4-6  Free Radical Addition Polymerization Kinetics

Many olefinic addition polymerization reactions, such as that of ethylene or styrene polymerization, occur by free radical mechanisms. The initiation step can be activated thermally or by bond breaking additives such as peroxides. The general reaction scheme is:

$$aM_1 + bI \xrightarrow{k_i} P_1 \qquad \text{Initiation} \qquad\qquad \text{(a)}$$

$$
\left.
\begin{aligned}
P_1 + M_1 &\xrightarrow{k_{pr}} P_2 \\
&\;\;\vdots \\
P_{n-1} + M_1 &\xrightarrow{k_{pr}} P_n
\end{aligned}
\right\}
\qquad \text{Propagation} \qquad \text{(b)}
$$

$$P_n + P_m \xrightarrow{k_t} M_{n+m} \qquad \text{Termination} \qquad \text{(c)}$$

where $M_1$ is the monomer, $I$ is any initiator, $P_n$ is active polymer, and $M_{n+m}$ is inactive. Note that all the propagation steps are assumed to have the same rate constant, $k_{pr}$, which seems to be reasonable in practice. Also, $a$ or $b$ can be zero, depending on the mode of initiation.

The rates of the reactions are

$$\frac{dM_1}{dt} = -ar_i - k_{pr} M_1 \sum P_n \qquad\qquad \text{(d)}$$

$$\frac{dP_1}{dt} = r_i - k_{pr} M_1 P_1 - k_t P_1 \sum P_n \qquad\qquad \text{(e)}$$

$$\vdots$$

$$\frac{dP_n}{dt} = k_{pr} M_1 P_{n-1} - k_{pr} M_1 P_n - k_t P_n \sum P_n, \quad n \geq 2 \qquad \text{(f)}$$

where $r_i$ is the initiation rate of formation of radicals. Aris [3] has shown how these equations may be analytically integrated to give the various species as a function of time for an initiation step first order in the monomer, $M_1$, and a simple termination step of an extension of Eq. (b), $P_n + M_1 \rightarrow M_{n+1}$. The more general case is most easily handled by use of the steady-state approximation, whereby $dP_n/dt = 0$, as discussed above. Then each of equations e to f are equal to zero and, when added together, give

$$0 = r_i - k_t(\sum P_n)^2 \tag{g}$$

which states that under the steady-state assumption, the initiation and termination rates are equal. Thus, Eq. d is changed to

$$\frac{dM_1}{dt} = -ar_i - k_{pr}M_1\left(\frac{r_i}{k_t}\right)^{1/2} \tag{h}$$

$$\simeq -k_{pr}\left(\frac{r_i}{k_t}\right)^{1/2}M_1 \tag{i}$$

for initiation independent of monomer, $a = 0$ in Eq. a, or for small magnitude of monomer used in the initiation step relative to the propagation or polymerization steps (usually the case).

There are several possibilities for initiation, as mentioned above: second order in monomer (thermal), first order in each monomer and initiator catalyst, $I$, or first order in $I$. For the latter, the initiation rate of formation of radicals is given by,

$$r_i = k_i I \tag{j}$$

so that

$$(\sum P_n) = \left(\frac{k_i}{k_t}\right)^{1/2} I^{1/2} \tag{k}$$

The rate of monomer disappearance is, then,

$$\frac{dM_1}{dt} = -k_{pr}\left(\frac{k_i}{k_t}\right)^{1/2}(I)^{1/2}M_1 \tag{l}$$

This expression for the overall polymerization rate is found to be generally true for such practical examples of free radical addition polymerization as polyethylene, and others.

Even further useful relations can be found by use of the above methods. Consider the case of reactions in the presence of "chain transfer" substances as treated by Alfrey in Rutgers [38] and Boudart [5]. This means a chemical species, $S$, that reacts with any active chain, $P_n$, to form an inactive chain but an active species, $S^{\cdot}$:

$$P_n + S \xrightarrow{k_{tr}} M_n + S^{\cdot} \tag{m}$$

This active species can then start a new chain by the reaction

$$S^{\cdot} + M_1 \longrightarrow P_1 \tag{n}$$

Thus, $S$ acts as a termination agent as far as the chain length of $P_n$, but does propagate a free radical $S\,\dot{}$ to continue the reaction. In other words, the average chain length is modified but not the overall rate of reaction.

These effects are most easily described by the number average degree of polymerization, $\bar{P}_N$, which is the average number of monomer units in the polymer chains. This can be found as follows.

For no chain transfer:

$$(\bar{P}_N)_0 = \frac{\text{rate of monomer molecules polymerized}}{\text{rate of new chains started}}$$

$$= \frac{k_{\mathrm{pr}} M_1 \sum P_n}{k_i I}$$

$$= \frac{k_{\mathrm{pr}}}{\sqrt{k_i k_t}} \frac{M_1}{I^{1/2}} \tag{o}$$

With a chain transfer agent present, this is changed to

$$\bar{P}_N = \frac{k_{\mathrm{pr}} M_1 \sum P_n}{k_i I + k_{\mathrm{tr}} S \sum P_n}$$

or

$$\frac{1}{\bar{P}_N} = \frac{1}{(\bar{P}_N)_0} + \frac{k_{\mathrm{tr}}}{k_{\mathrm{pr}}} \frac{S}{M_1} \tag{p}$$

and shows the decrease in average chain length with increasing $S$.

Further details about the molecular weight distribution of the polymer chains can be obtained by simple probability arguments. If the probability of adding another monomer unit to a chain is $p$, the probability of a chain length $P$ (number distribution) with random addition is

$$N(P) = (1 - p)p^{P-1} \text{ (starting with the monomer)} \tag{q}$$

which is termed the "most probable" or "Schultz–Flory" distribution. Note that $\sum_{P=1}^{\infty} N(P) = 1$, a normalized distribution. The number average chain length is, then,

$$\bar{P}_N = \sum_{P=1}^{\infty} P N(P)$$

$$= \frac{1}{1 - p} \tag{r}$$

The weight distribution is

$$W(P) = (1 - p)^2 P p^{P-1} \quad \text{(normalized)} \tag{s}$$

and the weight average chain length is

$$\bar{P}_W = \sum_{P=1}^{\infty} P W(P)$$

$$= \frac{1 + p}{1 - p} \tag{t}$$

$$\simeq \frac{2}{1 - p} \tag{t'}$$

The latter equation (t') is valid, since it will be shown below that $p \simeq 1.0^-$. Thus, for random addition, the ratio of weight to number average chain lengths is always essentially equal to 2:

$$\frac{\bar{P}_W}{\bar{P}_N} = 1 + p \simeq 2 \tag{u}$$

For the specific free radical mechanism, the probability of adding another monomer unit is:

$$p = \frac{k_{pr} M_1 P_n}{k_{pr} M_1 P_n + k_t P_n \sum P_n}$$

$$= \left[ 1 + \frac{(k_i k_t)^{1/2}}{k_{pr}} \frac{I^{1/2}}{M_1} \right]^{-1}$$

$$\simeq 1 - \frac{(k_i k_t)^{1/2}}{k_{pr}} \frac{I^{1/2}}{M_1} \tag{v}$$

since the ratio of initiation to propagation rates is small. Thus, the number average degree of polymerization is

$$\bar{P}_N = \frac{k_{pr}}{(k_i k_t)^{1/2}} \frac{M_1}{I^{1/2}} \left[ 1 + \frac{(k_i k_t)^{1/2}}{k_{pr}} \frac{I^{1/2}}{M_1} \right]$$

$$\simeq \frac{k_{pr}}{(k_i k_t)^{1/2}} \frac{M_1}{I^{1/2}} \tag{w}$$

The same type of result as above would also be found with chain transfer agents, but, in addition, the effects of the various kinetic constants on the molecular weight distributions can then be estimated. Finally, note that many of these results can also be obtained by directly solving Eqs. d to f rather than using the classical probability arguments; see Ray [39] for an extensive review.

# 1.5 Influence of Temperature

The rate of a reaction depends on the temperature, through variation of the rate coefficient. According to Arrhenius:

$$\ln k = -\frac{E}{R}\frac{1}{T} + \ln A_0 \tag{1.5-1}$$

where $T$: temperature (°K)

$R$: gas constant kcal/kmol K

$E$: activation energy kcal/kmol

$A_0$: a constant called the frequency factor

Consequently, when $\ln k$ is plotted versus $1/T$, a straight line with slope $-E/R$ is obtained.

Arrhenius came to this formula by thermodynamic considerations. Indeed for the reversible reaction, $A \underset{2}{\overset{1}{\rightleftharpoons}} Q$, the Van't Hoff relation is as follows:

$$\frac{d}{dT}\ln K = \frac{\Delta H}{RT^2} \tag{1.5-2}$$

As

$$K_c = \left(\frac{C_R}{C_A}\right)_{eq} = \frac{k_1}{k_2}$$

Eq. 1.5-2 may be written,

$$\frac{d}{dT}\ln k_1 - \frac{d}{dT}\ln k_2 = \frac{\Delta H}{RT^2}$$

This led Arrhenius to the conclusion that the temperature dependence of $k_1$ and $k_2$ must be analogous to Eq. 1.5-2:

$$\frac{d}{dT}\ln k_1 = \frac{E_1}{RT^2} \qquad \frac{d}{dT}\ln k_2 = \frac{E_2}{RT^2}$$

with

$$E_1 - E_2 = \Delta H \tag{1.5-3}$$

which is Eq. 1.5-1) Note that $E_2 > E_1$, for an exothermic and conversely for an endothermic reaction. Since then, this hypothesis has been confirmed many times experimentally, although, according to the collision theory, $k$ should be proportional to $T^{1/2}\exp[-E/RT]$ and, from the theory of the activated complex, to $T\exp[-E/RT]$. (Note that these forms also satisfy the Van't Hoff relation.) The influence of $T^{1/2}$ or even $T$ in the product with $e^{-E/RT}$ is very small, however, and to observe this requires extremely precise data.

The Arrhenius equation is only strictly valid for single reactions. If a reaction is accompanied by a parallel or consecutive side reaction, which is not accounted for in detail, deviations from the straight line may be experienced in the Arrhenius plot for the overall rate. If there is an influence of transport phenomena on the measured rate, deviations from the Arrhenius law may also be observed; this will be illustrated in Chapter 3.

From the practical standpoint, the Arrhenius equation is of great importance for interpolating and extrapolating the rate coefficient to temperatures that have not been investigated. With extrapolation, take care that the mechanism is the same as in the range investigated. Examples of this are given later.

## Example 1.5-1 Determination of the Activation Energy

For a first-order reaction, the following rate coefficients were found:

| Temperature (°C) | $k(hr^{-1})$ |
| --- | --- |
| 48.5 | 0.044 |
| 70.4 | 0.534 |
| 90.0 | 3.708 |

These values are plotted in Fig. 1, and it follows that:

$$-\frac{E}{2.3R} = \frac{1.651}{32 \times 10^{-5}} = -5.159 \text{ K}$$

$$E = 24500 \text{ kcal/kmol (98200 kJ/kmol)}$$

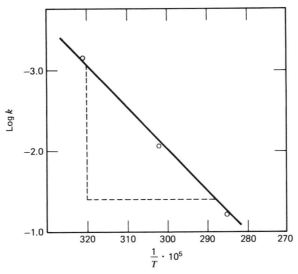

Figure 1  Determination of activation energy.

## Example 1.5-2 Activation Energy for Complex Reactions

The overall rate equation based on a complex mechanism often has an overall rate constant made up of the several individual constants for the set of reactions. The observed activation energy is then made up of those of the individual reactions and may be able to be predicted, or used as a consistency check of the mechanism.

For example, the Rice–Herzfeld mechanism for hydrocarbon pyrolysis has overall rate expressions such as Eq. 1.4-18:

$$-\frac{dC_{M_1}}{dt} = \sqrt{\frac{k_1 k_2 k_3}{k_4}}\, C_{M_1} \tag{a}$$

$$\equiv k_0 C_{M_1}$$

Thus,

$$k_0 = (k_1 k_2 k_3 / k_4)^{1/2} \tag{b}$$

and

$$\ln A_0 - \frac{E_0}{RT} = \frac{1}{2}\left( \ln \frac{A_1 A_2 A_3}{A_4} - \frac{E_1 + E_2 + E_3 - E_4}{RT}\right) \tag{c}$$

Equating the temperature coefficients:

$$\frac{d \ln k_0}{d(1/T)} = \frac{d}{d(1/T)}\,\text{(R.H.S.)}$$

gives the relationship:

$$E_0 = \tfrac{1}{2}(E_1 + E_2 + E_3 - E_4) \tag{d}$$

An order of magnitude estimate of the overall activation energy is given by using typical values for the initiation, hydrogen abstraction, radical decomposition, and termination steps:

$$E_0 = \tfrac{1}{2}(85 + 10 + 35 - 0)$$
$$= 65 \text{ kcal/mol (271.7 kJ/mol)}$$

This is the size of overall activation energy that is observed. Note that it is much lower than the very high value for the difficult initiation step, and is thus less than the nominal values for breaking carbon-carbon bonds.

For the specific Example 1.4-5 of ethane pyrolysis, Eq. i of that example shows that the overall rate constant is:

$$k_0 = \left(\frac{k_1 k_3 k_4}{k_5}\right)^{1/2}$$

and

$$E_0 = \tfrac{1}{2}(E_1 + E_3 + E_4 - E_5) \qquad (e)$$

Values from Benson [27], p. 354, give

$$E_0 = \tfrac{1}{2}(88.0 + 40.6 + 6.4 - 0)$$
$$= 67.5 \text{ kcal/mol } (282.15 \text{ kJ/mol})$$

Benson states that observed overall values range from 69.8 to 77 kcal/mol (291.8 to 321.9 kJ/mol) and so Eq. e provides a reasonable estimate. Laidler and Wojciechowski [40] present another table of values, which lead to $E_0 = 65.6$ kcal/mol (274.2 kJ/mol). Both estimates are somewhat low, as mentioned in Ex. 1.4-5.

For the second-order initiation mechanism, the rate constant is

$$k_0 = k_3(k_1'/k_5)^{1/2}$$

and

$$E_0 = E_3 + \tfrac{1}{2}(E_1' - E_5) \qquad (f)$$

Using Laidler and Wojciechowski's values

$$E_0 = 39.5 + \tfrac{1}{2}(70.2\text{-}0)$$
$$= 74.6 \text{ kcal/mol } (311.8 \text{ kJ/mol})$$

This seems to be a more reasonable value.

The exponential temperature dependency of the rate coefficient can cause enormous variations in its magnitude over reasonable temperature ranges.

Table 1.5-1 gives the magnitude of the rate coefficient for small values of $RT/E$. It follows then that the "rule" that a chemical reaction rate doubles for a 10 K

*Table 1.5-1 Variation of rate coefficient with temperature*

| $RT/E$ | $E/RT$ | $k/A_0$ |
|--------|--------|---------|
| 0.01 | 100 | $4 \times 10^{-44}$ |
| 0.02 | 50 | $2 \times 10^{-22}$ |
| 0.04 | 25 | $1.4 \times 10^{-11}$ |
| 0.06 | 16.7 | $5.7 \times 10^{-8}$ |
| 0.08 | 12.5 | $4 \times 10^{-6}$ |
| 0.10 | 10 | $4.5 \times 10^{-5}$ |
| 0.20 | 5 | $6.7 \times 10^{-3}$ |

rise in temperature often gives the correct order of magnitude, but is really only true for certain ranges of the parameter.

Theoretical estimates of the frequency factor, $A$, for various types of reactions can be found in Frost and Pearson [41].

## 1.6 Determination of Kinetic Parameters _____

### 1.6-1 Simple Reactions

For simple homogeneous reactions, there are two main characteristics to be determined: the reaction order and the rate coefficient. The latter can be found in several ways if the kinetics (order) is given, but the former is often quite difficult to unequivocally determine.

The case of a simple first-order, irreversible reaction was briefly discussed in Section 1.3. In principle, with Eq. 1.3-5, one value of $(C_A, t)$ suffices to calculate $k$ when $C_{A_0}$ is known. In practice, it is necessary to check the value of $k$ for a set of values of $(C_A, t)$. This method, called the "integral" method, is simpler than the differential method when the kinetic equation (1.3-4) can be integrated. When the order of the reaction is unknown, several values for it can be tried. The stoichiometric equation may be a guide for the selection of the values. The value for which $k$, obtained from Eq. 1.3-4 or Eq. 1.3-5, is found to be independent of the concentration is considered to be the correct order.

The trial-and-error or iterative procedure may be avoided by the use of the following method, which is, in fact, also a differential method. Taking the logarithm of Eq. 1.3-1 leads to

$$\log r = \log k + a' \log C_A + b' \log C_B \qquad (1.6\text{-}1a)$$

There are three unknowns in this equation: $k$, $a'$, and $b'$, so three sets of values of $r$, $C_A$, and $C_B$ are sufficient to determine them were it not for the random errors inherent in experimental data of this type. It is preferable to determine the best values of $a'$ and $b'$ by the method of least squares. Indeed, the above equation is of the type

$$y = a_0 + ax_1 + bx_2 \qquad (1.6\text{-}1b)$$

and eminently suited for application of the least squares technique.

Sometimes it may be worthwhile to check the partial orders obtained in this way by carrying out experiments in which all but one of the reacting species are present in large excess with respect to the component whose partial order is to be checked. This partial order is then obtained from

$$r = k'C_A^{a'} \qquad \text{where } k' \equiv kC_B C_C \cdots$$

By taking logarithms

$$\log r = \log k' + a' \log C_A$$

The slope of the straight line on a $\log r - \log C_A$ plot is the partial order $a'$.

For a given simple order, the rate expression can be integrated and special plots utilized to determine the rate coefficient. For example, the $k$ for a first-order irreversible reaction can be found from the slope of a plot of $\ln C_A/C_{A_0}$ versus $t$, as indicated in Section 1.3. A plot of $1/C_A$ versus $t$ or $x_A/(1 - x_A)$ versus $t$ is used similarly for a second-order irreversible reaction. For $1 - 1$ reversible reactions, a plot of $\ln(C_A - C_{Aeq})/(C_{A_0} - C_{Aeq})$ or $\ln(1 - x_A/x_{Aeq})$ versus $t$ yields $(k_1 + k_2)$ from the slope of the straight line, and with the thermodynamic equilibrium constant, $K = k_1/k_2$, both $k_1$ and $k_2$ can be found. Certain more complicated reaction rate forms can be rearranged into such linear forms, and Levenspiel [42] or chemical kinetics texts give several examples. These plots are useful for an estimate of the "quality" of the fit to the experimental data, and can also provide initial estimates to formal linear regression techniques, as mentioned above.

A more extensive discussion and comparison of various methods is presented in Chapter 2; they form the basis for many of the recent applications and can also be used for homogeneous reactions. Useful surveys are given by Bard and Lapidus [43], Kittrell [44], and by Froment [45]. However, methods primarily for mass action form rate laws are considered here.

## 1.6-2 Complex Reactions

Complex kinetic schemes cannot be handled easily, and, in general, a multidimensional search problem must be solved, which can be difficult in practice. This general problem has been considered for first-order reaction networks by Wei and Prater [13] in their now-classical treatment. As described in Ex. 1.4-1, their method defines fictitious components, $B_m$, that are special linear combinations of the real ones, $A_j$, such that the rate equations for their decay are uncoupled, and have solutions:

$$\zeta_m(t) = \zeta_m(0)e^{-\lambda_m t} \qquad (1.6.2\text{-}1)$$

Both the $\lambda_m$ and the coefficients in the linear combination relations are functions of the rate constants, $k_{ji}$, through the matrix transformations. Obviously, Eq. 1.6.2-1 is enormously easier to use in determination of the $\lambda_m$ than the full solutions for the $y_j$ which consist of $N$-exponential terms, and which would require nonlinear regression techniques. In fact, simple logarithmic plots, as just described, can be used. Once the straight-line reaction paths are used to determine the $\lambda_m$, numerical matrix manipulations can then be used to readily recover the $k_{ji}$.

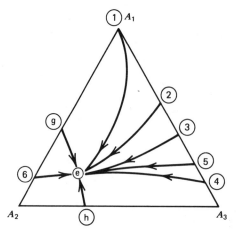

Figure 1.6.2-1 Reaction paths in three species reaction mixture.

Unfortunately, the method is not an automatic panacea to all problems of complex first-order kinetics. The only directly measured quantities are the $y_j$. The $\zeta_m$ are found by a matrix transformation using the $k_{ji}$. However, we don't yet know these since, in fact, this is what we are trying to find. Thus, a trial-and-error procedure is required, which makes the utilization of the method somewhat more complicated. Wei and Prater suggest an experimental trial-and-error scheme that is easily illustrated by a simple example and some sketches.

The three-species problem to be considered is (e.g., butene isomerization):

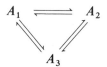

The compositions can be plotted on a triangular graph as shown in Fig. 1.6.2-1. The arrows indicate the course of the composition change in time and the point "$e$" is the equilibrium position. Thus, an experiment ① starts with pure $A_1$ and proceeds to equilibrium along the indicated curve. Now the above scheme for three components will give three $\lambda_m$, one of which is zero. It can be shown that the other two $\lambda_m$—each corresponding to a $\zeta_m$—will give a straight line reaction path on the above diagram, lines ⑤ ⑥ and ⑨ ⓗ. The first experiment didn't give a straight line and so one of the $\zeta_m$ is *not* pure $A_1$. Thus, a second experiment is done, ②, which again probably won't give a straight line. Finally, at experiment ⑤, a straight line is found and possibly confirmed by experiment ⑥ with the indicated initial composition (mixture of ~5 parts $A_2$ and 1 part $A_1$). The compositions for experiment ⑤ or ⑥ are plotted as $\ln[\zeta_m - \zeta_{me}]$ versus time and the slope will be $\lambda_1$. The other straight-line path ⑨ ⓗ can be found from matrix calculations, and then confirmed experimentally. For larger numbers of reacting

       CHEMICAL ENGINEERING KINETICS

species, more $(N - 1)$ of the straight line paths must be found experimentally by the iterative technique. Then the $k_{ji}$ are found. Obviously, this is a rather laborious procedure and is most realistically done with bench scale studies. However, as Wei and Prater strongly pointed out, extensive data must be taken if one really wants to find out about the kinetics of the process. Finally, the entire procedure is only good for first-order reactions, which is another restriction. However, many industrial reactions are assumed first order in any event, and so the method can have many applications. For example, see Chapter 10 in Boudart [5]. Gavalas [46] provides another technique for first-order systems that again estimates values for the eigenvalues of the rate coefficient matrix.

Another method that can be used is to take the $C_j$ measured as a function of time, and from them compute the various slopes, $dC_j/dt$. The general form of kinetic expressions can then be written, for $M$ reactions, as:

$$\frac{dC_1}{dt} = k_1 r_{11}(\mathbf{C}) + \cdots + k_p r_{1p}(\mathbf{C}) + \cdots + k_M r_{1M}(\mathbf{C})$$

$$\vdots$$

$$\frac{dC_j}{dt} = k_1 r_{j1}(\mathbf{C}) + \cdots + k_p r_{jp}(\mathbf{C}) + \cdots + k_M r_{jM}(\mathbf{C}) \tag{1.6.2-2}$$

$$\vdots$$

$$\frac{dC_N}{dt} = k_1 r_{N1}(\mathbf{C}) + \cdots + k_p r_{Np}(\mathbf{C}) + \cdots + k_M r_{NM}(\mathbf{C})$$

where $\mathbf{C}$ is the $N$-vector of concentrations. Then, since all the $k$'s appear in a *linear* fashion, at any one temperature, standard linear regression techniques can be used, even with the arbitrary rate forms $r_{jp}$, to determine the rate constants. Unfortunately, however, this differential method can only be used with very precise data in order to successfully compute accurate values for the slopes, $dC_j/dt$.

An alternate procedure was devised by Himmelblau, Jones, and Bischoff [47]. This was to take the basic equations (1.6.2-2) for the $C_j$ and directly integrate (*not* formally solve) them:

$$\frac{dC_j}{dt} = \sum_{p=1}^{M} k_p r_{jp}(\mathbf{C}) \qquad j = 1, 2, \ldots N \tag{1.6.2-3}$$

which leads to

$$\underbrace{C_j(t_i) - C_j(t_0)}_{\substack{\text{Directly} \\ \text{measured}}} = \sum_{p=1}^{M} k_p \underbrace{\int_{t_0}^{t_i} r_{jp}(\mathbf{C}(t)) dt}_{\substack{\text{Integrals of} \\ \text{measured data}}} \tag{1.6.2-4}$$

Notice again that the $k$'s occur linearly no matter what the functions $r_{jp}$ are, and so standard *linear* regression methods, including various weighting, and so on, can be used. Also, only *integration* of experimental data is necessary (*not* differentiation), which is a smoothing operation. Thus, it seems that the advantages of linear regression are retained without the problems arising with data differentiation.

Equation 1.6.2-4 can now be abbreviated as

$$C_j(t_i) - C_j(t_0) = \sum_{p=1}^{M} k_p X_{ijp} \qquad (1.6.2\text{-}5)$$

The standard least squares method would minimize the following relation:

$$\sum_{i=1}^{n} \sum_{j=1}^{N} [w_{ij}(Y_{ij} - \hat{Y}_{ij})]^2 \qquad (1.6.2\text{-}6)$$

where

$n$ = number of time data points

$Y_{ij} = C_j(t_i) - C_j(t_0)$, experimental value of dependent variable

$\hat{Y}_{ij} = \sum_{p=1}^{M} k_p X_{ijp}$, calculated value of dependent variable

$w_{ij}$ = any desired weighting function for the deviations

Standard routines can perform the computations for Eq. 1.6.2-6 and will not be further discussed here. The result would be least squares fit values for the kinetic parameters, $k_p$.

This latter technique of Himmelblau, Jones, and Bischoff (H-J-B) has proved to be efficient in various practical situations with few, scattered, data available for complex reaction kinetic schemes (see Ex. 1.6.2-1). Recent extensions of the basic ideas are given by Eakman, Tang, and Gay [48, 49, 50]. It should be pointed out, however, that the problem has been cast into one of linear regression at the expense of statistical rigor. The "independent variables", $X_{ijp}$, do not fulfill one of the basic requirements of linear regression: that the $X_{ijp}$ have to be free of experimental error. In fact, the $X_{ijp}$ are functions of the dependent variables $C_j(t_i)$ and this may lead to estimates for the parameters that are erroneous. This problem will be discussed further in Chapter 2, when the estimation of parameters in rate equations for catalytic reactions will be treated. Finally, all of the methods have been phrased in terms of batch reactor data, but it should be recognized that the same formulas apply to plug flow and constant volume systems, as will be shown later in this book.

### Example 1.6.2-1 Rate Constant Determination by the Himmelblau–Jones–Bischoff Method

To illustrate the operation of the H-J-B method described above, as well as gain some idea of its effectiveness, several reaction schemes were selected, rate constants

    

*Table 1 Application of the Himmelblau–Jones–Bischoff method to estimation of rate coefficients in a simple consecutive reaction system*

$$A_1 \underset{k_2}{\overset{k_1}{\rightleftharpoons}} A_2 \underset{k_4}{\overset{k_3}{\rightleftharpoons}} A_3$$

$$\frac{dc_1}{dt} = k_2 c_2 - k_1 c_1$$

$$\frac{dc_2}{dt} = k_1 c_1 + k_4 c_3 - (k_2 + k_3)c_2$$

$$\frac{dc_3}{dt} = k_3 c_2 - k_4 c_3$$

Data points: 31 at equal time intervals

| Coefficient | Original value | Calculated value | |
|---|---|---|---|
| | | *a* | *b* |
| **Run 1 (no error)** | | | |
| $k_1$ | 1.000 | 1.000 | 1.000 |
| $k_2$ | 1.000 | 1.000 | 1.000 |
| $k_3$ | 1.000 | 1.000 | 1.000 |
| $k_4$ | 1.000 | 1.000 | 1.000 |
| **Run 2 (no error)** | | | |
| $k_1$ | 1.00 | 1.013 | 1.012 |
| $k_2$ | 0.50 | 0.497 | 0.496 |
| $k_3$ | 10.0 | 10.125 | 10.112 |
| $k_4$ | 5.0 | 4.990 | 4.989 |
| **Run 3 (5% error randomized by sign)** | | | |
| $k_1$ | 1.00 | 0.968 | 0.962 |
| $k_2$ | 0.50 | 0.487 | 0.467 |
| $k_3$ | 10.0 | 9.730 | 9.687 |
| $k_4$ | 5.0 | 4.900 | 4.873 |
| **Run 4 (10% error randomized by sign)** | | | |
| $k_1$ | 1.00 | 1.025 | 1.000 |
| $k_2$ | 0.50 | 0.586 | 0.500 |
| $k_3$ | 10.0 | 10.226 | 10.042 |
| $k_4$ | 5.0 | 5.197 | 5.086 |
| **Run 5 (15% error randomized by sign)** | | | |
| $k_1$ | 1.00 | 1.009 | 0.977 |
| $k_2$ | 0.50 | 0.233 | 0.056 |
| $k_3$ | 10.0 | 9.766 | 9.534 |
| $k_4$ | 5.0 | 4.623 | 4.392 |

assumed, and hypothetical values of the dependent variables generated. The differential equations were solved for $C_j(t)$ at various times using analytical methods for simpler models and a Runge–Kutta numerical integration for the more complicated models. Error was added to the deterministic variables, and the resulting simulated data were processed with linear regression programs to yield estimates for the rate coefficients.

Tables 1 and 2 show a simple consecutive reaction scheme and a more complex one and compare the original rate coefficients with those calculated from the simulated data. Each of the simulated sets of data was run for two weights: $a$: equal weighting of deviations of concentrations; $b$: weights inversely proportional to the concentration.

For the relatively simple scheme of Table 1, the proposed method yielded constants in good agreement with the originally fixed constants, even as increasing

Table 2 Application of the Himmelblau–Jones–Bischoff method to estimation of rate coefficients in a more complex consecutive reaction system

$$A_1 \underset{k_2}{\overset{k_1}{\rightleftharpoons}} A_2 \underset{k_4}{\overset{k_3}{\rightleftharpoons}} A_3 \underset{k_6}{\overset{k_5}{\rightleftharpoons}} A_4$$

$$\frac{dc_1}{dt} = k_2 c_2 - k_1 c_1$$

$$\frac{dc_2}{dt} = k_1 c_1 + k_4 c_3 - (k_2 + k_3)c_2$$

$$\frac{dc_3}{dt} = k_3 c_2 + k_6 c_4 - (k_4 + k_5)c_3$$

$$\frac{dc_4}{dt} = k_5 c_3 - k_6 c_4$$

Data points: 34 data points, no error, equal time intervals. (Double precision arithmetic used)

| | | Calculated value | |
|---|---|---|---|
| Coefficient | Original value | a | b |
| $k_1$ | 2.0 | 1.819 | 1.857 |
| $k_2$ | 10.0 | 8.952 | 9.157 |
| $k_3$ | 15.0 | 13.338 | 12.195 |
| $k_4$ | 6.0 | 6.117 | 5.451 |
| $k_5$ | 4.0 | 4.034 | 4.025 |
| $k_6$ | 0.1 | 0.101 | 0.100 |

error was introduced, except for the value of $k_2$ in run 5. For the more complex model in Table 2, even without introducing random error, the values of $k_2$ and $k_3$ deviated as much as 10 percent from the original values.

After analyzing all of the computer results, including trials not shown, it was concluded that most of the error inherent in the method originates because of the sensitivity of the rate coefficients to the values obtained in the numerical integration step. If the concentration-time curves changed rapidly during the initial time increments, and if large concentration changes occurred, significant errors resulted in the calculated rate parameters. It has been found that data-smoothing techniques before the numerical integration step help to remedy this problem.

Another source of error is that errors in the beginning integrals tend to throw off all the predicted values of the dependent variables because the predicted values are obtained by summing the integrals up to the time of interest. Thus, it would seem that the use of unequal time intervals with more data at short times is important in obtaining good precision.

### *Example 1.6.2-2  Kinetics of Olefin Codimerization*

Paynter and Schuette [51] have utilized the above technique for the complex industrial process of the codimerization of propylene and butenes to hexene, heptene, octene, and some higher carbon number products of lesser interest. Not only are there a variety of products, but also many possible feed compositions. This is actually a catalytic process, but the mass-action kinetics used can serve to illustrate the principles of this section, as well as previous parts of this chapter.

The most straightforward reaction scheme to represent the main features of this system are:

$$2C_3 \xrightarrow{\ \ 1\ \ } C_6$$

$$C_3 + C_{4-1} \xrightarrow{\ \ 2\ \ } C_7$$

$$C_3 + C_{4-2} \xrightarrow{\ \ 3\ \ } C_7$$

$$2C_{4-1} \xrightarrow{\ \ 4\ \ } C_8$$

$$2C_{4-2} \xrightarrow{\ \ 5\ \ } C_8$$

where the concentrations are:

$C_3$-propylene; $C_{4-1}$-butene-1; $C_{4-2}$-butene-2 (both *cis* and *trans*); $C_6$-hexene; $C_7$-heptenes; $C_8$-octenes.

The $C_9{}^+$ compounds are not of primary interest, and so an approximate overall reaction was used to account for their formation:

$$(C_3 + C_{4-1} + C_{4-2}) + (C_6 + C_7 + C_8) \xrightarrow{\phantom{xx}6\phantom{xx}} C_9{}^+$$

To obtain the proper initial selectivity, a further overall reaction was introduced:

$$3C_3 \xrightarrow{\phantom{xx}8\phantom{xx}} C_9{}^+$$

Finally, the butene isomerization reaction was also accounted for:

$$C_{4-1} \underset{7'}{\overset{7}{\rightleftharpoons}} C_{4-2}, \text{ with equilibrium constant } K \simeq 12$$

The straightforward mass action rate equations then are

$$\frac{dC_3}{dt} = -2k_1 C_3{}^2 - k_2 C_3 C_{4-1} - k_3 C_3 C_{4-2} - k_6 C_3(C_6 + C_7 + C_8) - 3k_8 C_3{}^3$$

(a)

$$\frac{dC_{4-1}}{dt} = -k_2 C_3 C_{4-1} - 2k_4 C_{4-1}{}^2 - k_6 C_{4-1}(C_6 + C_7 + C_8)$$

$$- k_7(C_{4-1} - C_{4-2}/12) \qquad \text{(b)}$$

$$\frac{dC_{4-2}}{dt} = -k_3 C_3 C_{4-2} - 2k_5 C_{4-2}{}^2 - k_6 C_{4-2}(C_6 + C_7 + C_8)$$

$$+ k_7(C_{4-1} - C_{4-2}/12) \qquad \text{(c)}$$

$$\frac{dC_6}{dt} = k_1 C_3{}^2 - k_6 C_6(C_3 + C_{4-1} + C_{4-2}) \qquad \text{(d)}$$

$$\frac{dC_7}{dt} = k_2 C_3 C_{4-1} + k_3 C_3 C_{4-2} - k_6 C_7(C_3 + C_{4-1} + C_{4-2}) \qquad \text{(e)}$$

$$\frac{dC_8}{dt} = k_4 C_{4-1}{}^2 + k_5 C_{4-2}{}^2 - k_6 C_8(C_3 + C_{4-1} + C_{4-2}) \qquad \text{(f)}$$

$$\frac{dC_{9+}}{dt} = k_6(C_3 + C_{4-1} + C_{4-2})(C_6 + C_7 + C_8) + k_8 C_3{}^3 \qquad \text{(g)}$$

Certain aspects of these rate equations are obviously empirical, and illustrate the compromises often necessary in the analysis of complex practical industrial reacting systems.

Paynter and Schuette found that with a "practical" amount of data, the direct determination of the eight rate constants by the H-J-B method (or presumably by others) could adequately fit the data, but the constants were not consistent in all ways. Thus, several other types of data were also utilized to independently relate certain of the rate constants, and these concepts are considered here.

The initial selectivities of $C_6/C_7$ and $C_6/C_8$ are found by taking the ratios of Eqs. d, e, or f under initial conditions:

$$\frac{dC_6}{dC_7} \cong \left(\frac{k_1 C_3{}^2}{k_2 C_3 C_{4-1} + k_3 C_3 C_{4-2}}\right) \tag{h}$$

For pure butene-1 feed this reduces to

$$\frac{dC_6}{dC_7} = \frac{k_1}{k_2}\left(\frac{C_3}{C_{4-1}}\right) \tag{i}$$

$$\frac{dC_6}{dC_8} = \left(\frac{k_1 C_3{}^2}{k_4 C_{4-1}{}^2 + k_5 C_{4-2}{}^2}\right) \tag{j}$$

which again reduces, for pure butene-1 feed, to

$$\frac{dC_6}{dC_8} = \frac{k_1}{k_4}\left(\frac{C_3}{C_{4-1}}\right)^2 \tag{k}$$

Thus, with pure butene-1 feed, a plot of $C_6$ versus $C_7$ has an initial slope of $(k_1/k_2)(C_3/C_{4-1})$ Eq. (i), and knowing the feed composition yields $(k_1/k_2)$; see Fig. 1. Similarly, for a given ratio of $C_{4-2}$ and $C_{4-1}$, plus $(C_3/C_4)$, Eq. (h) yields

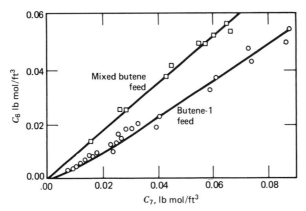

Figure 1  Hexenes  versus  heptenes,  $T = 240°F$. (Paynter and Schuette [51]).

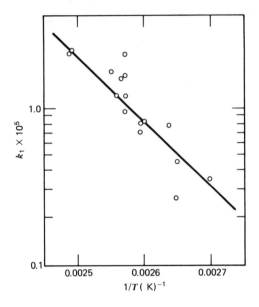

*Figure 2 Arrhenius plot of $k_1$.*

$$k_2 = 1.68 \, k_1$$
$$k_3 = 0.2 \, k_1$$
$$k_4 = 0.705 \, k_1$$
$$k_5 = 0.01 \, k_1$$

*(Paynter and Schuette [51]).*

$(k_3/k_1)$. After a similar treatment of Eq. (k) and (j) the following values were obtained at 240°F:

$$\frac{k_2}{k_1} = 1.68 \qquad \frac{k_3}{k_1} = 0.2 \qquad \frac{k_4}{k_1} = 0.7 \qquad \frac{k_5}{k_1} = 0.01$$

(all units so that rates are in pound moles/hr-ft$^3$ catalyst). Note that butene-2 is much less reactive than butene-1. Data at different temperatures give about the same ratio, indicating similar activation energies for reactions 1 to 5.

At this point, only four constants, $k_1, k_6, k_7, k_8$ need be determined by the H-J-B method. Figure 2 shows an Arrhenius plot for $k_1$. Figure 3 presents a final comparison of experimental data with model predictions using the determined rate constant values.

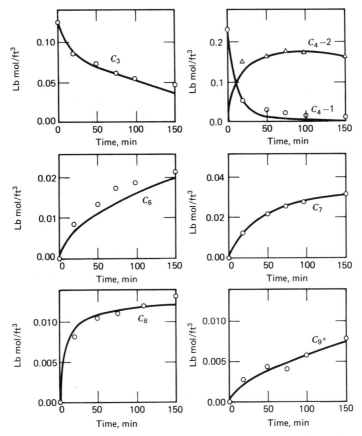

*Figure 3 Comparison of experimental data with model predictions (concentration versus time). (Paynter and Schuette [51]).*

## *Example 1.6.2-3  Thermal Cracking of Propane*

From a literature survey and from the experimental study of Van Damme et al. [52], Sundaram and Froment [53] developed the following so-called molecular reaction scheme for the thermal cracking of propane. Such a molecular scheme is an approximation for the true radical scheme. It is simpler and the corresponding set of rate equations is much easier to integrate, a great advantage when the integral method of kinetic analysis is adopted. The reaction scheme is given in Table 1.

All the reactions, except 4 and 5, are considered to be elementary, so that their order equals the molecularity. Reactions 4 and 5 are more complex and first order

*Table 1 Molecular scheme for the thermal cracking of propane*

| | Reaction | Rate Rate equation |
|---|---|---|
| 1 | $C_3H_8 \longrightarrow C_2H_4 + CH_4$ | $r_1 = k_1 C_{C_3H_8}$ |
| 2 | $C_3H_8 \rightleftharpoons C_3H_6 + H_2$ | $r_2 = k_2\left(C_{C_3H_8} - \dfrac{C_{C_3H_6}C_{H_2}}{K_{C_2}}\right)$ |
| 3 | $C_3H_8 + C_2H_4 \longrightarrow C_2H_6 + C_3H_6$ | $r_3 = k_3 C_{C_3H_8} C_{C_2H_4}$ |
| 4 | $2C_3H_6 \longrightarrow 3C_2H_4$ | $r_4 = k_4 C_{C_3H_6}$ |
| 5 | $2C_3H_6 \longrightarrow 0.5C_6 + 3CH_4$ | $r_5 = k_5 C_{C_3H_6}$ |
| 6 | $C_3H_6 \rightleftharpoons C_2H_2 + CH_4$ | $r_6 = k_6\left(C_{C_3H_6} - \dfrac{C_{C_2H_2}C_{CH_4}}{K_{C_6}}\right)$ |
| 7 | $C_3H_6 + C_2H_6 \longrightarrow C_4H_8 + CH_4$ | $r_7 = k_7 C_{C_3H_6} C_{C_2H_6}$ |
| 8 | $C_2H_6 \rightleftharpoons C_2H_4 + H_2$ | $r_8 = k_8\left(C_{C_2H_6} - \dfrac{C_{C_2H_4}\cdot C_{H_2}}{K_{C_8}}\right)$ |
| 9 | $C_2H_4 + C_2H_2 \longrightarrow C_4H_6$ | $r_9 = k_9 C_{C_2H_4} C_{C_2H_2}$ |

is assumed for these. The equilibrium constants $K_{C_2}$, $K_{C_6}$ and $K_{C_8}$ are obtained from thermodynamic data (F. Rossini et al.) [54]. It follows that the total rate of disappearance of propane $R_{C_3H_8}$ is given by

$$R_{C_3H_8} = -(r_1 + r_2 + r_3)$$

while the net rate of formation of propylene is given by

$$R_{C_3H_6} = r_2 + r_3 - 2r_4 - 2r_5 - r_6 - r_7$$

The experimental study of Froment et al. (loc. cit) was carried out in a tubular reactor with plug flow. The data were obtained as follows: total conversion of propane versus a measure of the residence time, $V_R/(F_{C_3H_8})_0$; conversion of propane into propylene versus $V_R/(F_{C_3H_8})_0$ and so on. $V_R$ is the reactor volume reduced to isothermal and isobaric conditions, as explained in Chapter 9 on tubular reactors and $(F_{C_3H_8})_0$ is the propane feed rate.

It will be shown in Chapter 9 that a mass balance on propane over an isothermal differential volume element of a tubular reactor with plug flow may be written

$$\frac{dF_j}{dV_R} = R_j = \sum \alpha_{ij} r_i \tag{a}$$

In Eq. (a) a more general notation is used. $\alpha_{ij}$ is the stoichiometric coefficient of the $j$th component in the $i$th reaction.

After integration over the total volume of an isothermal reactor, Eq. a yields the various flow rates $F_j$ at the exit of the reactor, for which $V_R/(F_{C_3H_8})_0$ has a certain value, depending on the propane feed rate of the experiment. If Eq. a is integrated with the correct set of values of the rate coefficients $k_1 \ldots k_9$ the experimental values of $F_j$ should be matched. Conversely, from a comparison of experimental and calculated $\hat{F}_j$ the best set of values of the rate coefficients may be obtained. The fit of the experimental $F_j$ by means of the calculated ones, $\hat{F}_j$, can be expressed quantitatively by computing the sum of squares of deviations between experimental and calculated exit flow rates, for example. These may eventually be weighted to account for differences in degrees in accuracies between the various $F_j$ so that the quantity to be minimized may be written, for $n$ experiments:

$$\sum_{i=1}^{n} \sum_{j=1}^{N} w_j (F_{ij} - \hat{F}_{ij})^2$$

Sundaram and Froment [loc. cit] systematized this estimation by applying nonlinear regression.

The results at 800°C are given in Table 2.

The estimation was repeated at other temperatures so that activation energies and frequency factors could be determined.

Figure 1 compares experimental and calculated yields for various components as a function of propane conversion at 800°C.

*Table 2 Values for the rate coefficients of the molecular scheme for propane cracking at 800°C*

| Rate coefficient | Value (s$^{-1}$ or $^+$ : m$^3$ kmol$^{-1}$ s$^{-1}$) |
|:---:|:---:|
| $k_1$ | 2.341 |
| $k_2$ | 2.12 |
| $k_3$ | 23.635$^+$ |
| $k_4$ | 0.721 |
| $k_5$ | 0.816 |
| $k_6$ | 0.305 |
| $k_7$ | $3.342.10^2$ $^+$ |
| $k_8$ | 2.416 |
| $k_9$ | $4.064.10^3$ $^+$ |

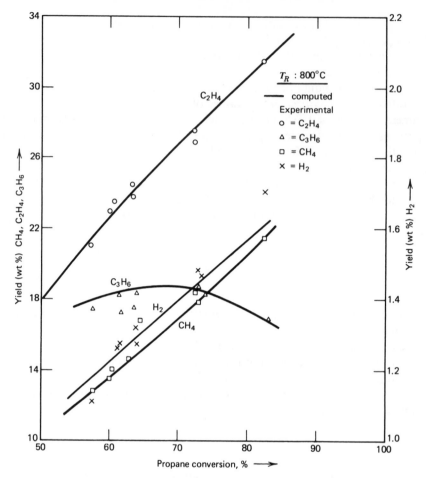

*Figure 1 Comparison of experimental and calculated yields for various components as a function of propane conversion at 800°C.*

## 1.7 Thermodynamically Nonideal Conditions _____

It was mentioned in Sec. 1.3 that the rate "constant" defined there is actually only constant for thermodynamically ideal systems, and that in general it may vary with composition. Also, the classical form of the mass action law gives for the reaction

$$A + B \xrightleftharpoons[2]{1} R + S \qquad (1.7\text{-}1)$$

the rate law

$$r_A = k_{1c} C_A C_B - k_{2c} C_R C_S \qquad (1.7\text{-}2)$$

At equilibrium, it was also shown there that

$$\frac{k_{1c}}{k_{2c}} = \left(\frac{C_R C_S}{C_A C_B}\right)_{eq} = K_c \qquad (1.7\text{-}3)$$

Now we know from thermodynamics that the concentration equilibrium constant is not the "proper" one in the sense that it can be a function of concentrations in addition to temperature, especially for liquids and for gases at high pressure. Thus, in thermodynamics, the "proper" variable of activity is introduced:

$$a_j \equiv \gamma_j C_j \qquad \gamma_j = \text{activity coefficient} \qquad (1.7\text{-}4)$$

This leads to an equilibrium constant that is a function only of temperature

$$K(T) = \left(\frac{a_R a_S}{a_A a_B}\right)_{eq} = \left(\frac{\gamma_R C_R \gamma_S C_S}{\gamma_A C_A \gamma_B C_B}\right)_{eq} \qquad (1.7\text{-}5)$$

How can this be extended into the kinetic equation so that it has a "proper" driving force?

A useful way to do this is to use the transition state theory of chemical reaction rates (e.g., see Glasstone, Laidler, and Eyring [55]; also, for a current review, see Laidler [56]). This is based on the hypothesis that all elementary reactions proceed through an *activated complex*:

$$A + B \; \rightleftharpoons \; X^{\ddagger} \; \longrightarrow \; \text{(products)} \qquad (1.7\text{-}6)$$

This activated complex is an unstable molecule, made up of the reactant molecules, and when it decomposes yields the products. For some simple reactions, the approximate structure of the activated complex can be estimated. It is also assumed that the activated complex is in thermodynamic equilibrium with the reactants even when the reaction as a whole is not in equilibrium. This assumption would be difficult to prove, but seems to be essentially correct in practice.

The rate of decomposition of the activated complex can be computed by the methods of statistical mechanics, and by utilizing the notion that one of the vibrational energy modes of the complex must be the one that allows dissociation to the products, leads to the following relation:

$$|\text{rate}| = k_0 C_{X^{\ddagger}} \qquad (1.7\text{-}7)[5]$$

where

$$k_0 = \frac{k_B T}{k_P} \sim 6 \times 10^{12} \text{sec}^{-1} \text{ at } T = 300 \text{ K} \qquad (1.7\text{-}8)$$

---

[5] That the concentration rather than the activity of the activated complex should be used here has been justified for certain cases by Emptage and Ross [57].

The factor, $k_0$, is a universal frequency that can be used for any reaction and relates the magnitude of the rate to the concentration of the activated complex.

Next using the assumption of equilibrium between the activated complex and the reactants,

$$K_{ABX}{}^{\ddagger}(T) = \frac{a_{X^{\ddagger}}}{a_A a_B} = \frac{\gamma^{\ddagger} C_{X^{\ddagger}}}{\gamma_A C_A \gamma_B C_B}$$

and Eq. (1.7-7) becomes:

$$|\text{rate}| = k_0 \frac{K_{ABX}{}^{\ddagger}}{\gamma^{\ddagger}} \gamma_A C_A \gamma_B C_B$$

$$= k_0 \frac{K_{ABX}{}^{\ddagger}}{\gamma^{\ddagger}} a_A a_B \tag{1.7-9}$$

$$\equiv \frac{k_1(T)}{\gamma^{\ddagger}} a_A a_B \tag{1.7-9a}$$

Similar considerations for the reverse reaction[6] give,

$$|\text{rate}|_r = k_0 \frac{K_{RSX}{}^{\ddagger}}{\gamma^{\ddagger}} a_R a_S \tag{1.7-10}$$

$$\equiv \frac{k_2(T)}{\gamma^{\ddagger}} a_R a_S \tag{1.7-10a}$$

Note that the $k_1$, $k_2$ defined by Eqs. 1.7-9a to 10a are dependent only on temperature,

$$k_i(T) = \frac{k_B T}{k_P} K^{\ddagger}(T)$$

The complete net rate can, therefore, be written

$$r_A = \frac{1}{\gamma^{\ddagger}} (k_1 a_A a_B - k_2 a_R a_S) \tag{1.7-11}$$

$$= \frac{1}{\gamma^{\ddagger}} (k_1 \gamma_A C_A \gamma_B C_B - k_2 \gamma_R C_R \gamma_S C_S) \tag{1.7-11a}$$

This equation properly reduces to the equilibrium Eq. 1.7-5 no matter what the value of $\gamma^{\ddagger}$, which could be a function of concentration just as any other activity coefficient. This equilibrium condition would also be true if $\gamma^{\ddagger}$ were ignored

---

[6] This is based on the principle of microscopic reversibility, which here means that the same activated complex is involved in both the forward and reverse reactions.

$(\gamma^{\ddagger} = 1)$, but the *kinetic* relation would not be the same. In other words, the simple expedient of merely replacing the concentrations in Eq. 1.7-2 with activities does *not* give the same result.

Comparing Eqs. 1.7-11 and 1.7-2 shows

$$k_{1c} = k_1 \frac{\gamma_A \gamma_B}{\gamma^{\ddagger}} \qquad k_{2c} = k_2 \frac{\gamma_R \gamma_S}{\gamma^{\ddagger}} \qquad (1.7\text{-}12)$$

These relations can now be used to relate the concentration rate constants under thermodynamically ideal conditions, $k_i$, to the values for any system. The utility of Eq. 1.7-11 to 12 will be illustrated by examples. Eckert [58] and Eckert et al. [59], [60], [61] have given reviews of several examples of the use of these results.

## Example 1.7-1 Reactions of Dilute Strong Electrolytes

A very interesting application of Eq. 1.7-12 is the Brønsted–Bjerrum equation for rate constants in solutions where the Debye–Hückel theory is applicable. The latter provides an equation for the activity coefficient, Rutgers [38]:

$$\log \gamma_j = -AZ_j^2 \sqrt{I} \qquad (a)$$

where

$Z_j$ = charge (valency) of ion $j$
$I$ = ionic strength of solution
$\quad = \frac{1}{2} \sum_j C_j Z_j^2$
$C_j$ = concentration of ion $j$
$A$ = constant $\simeq 0.51$ for water at 25°C

For the reaction of $A + B$, with charges $Z_A$ and $Z_B$, the activated complex must have charge $(Z_A + Z_B)$. Therefore, Eq. 1.7-12 gives

$$\log k_{1c} = \log k_1 + \log \gamma_A + \log \gamma_B - \log \gamma^{\ddagger}$$
$$= \log k_1 - A[Z_A^2 + Z_B^2 - (Z_A + Z_B)^2]\sqrt{I}$$
$$= \log k_1 + 2AZ_A Z_B \sqrt{I} \qquad (b)$$

Eq. b gives an excellent comparison with experimental data, and is very useful for correlating liquid phase reaction data. Boudart [5] points out that the naive result of taking $\gamma^{\ddagger} = 1$ would result in

$$\log k_{1c} = \log k_1 - A(Z_A^2 + Z_B^2)$$

which is neither qualitatively nor quantitatively correct.

## Example 1.7-2  Pressure Effects in Gas Phase Reactions

In the review of Eckert mentioned above, the study of Eckert and Boudart [61] on pressure effects in the decomposition of hydrogen iodide was summarized:

$$HI + HI \longrightarrow H_2 + I_2 \qquad (a)$$

This is one of the few gas phase reactions that seems to occur in the single bimolecular step as shown[7] and so can be handled directly with Eq. 1.7-12.

For gases, the activity can be expressed as the fugacity (with standard state of 1 atm), and so Eq. 1.7-4 shows

$$a_j = f_j = \phi_j P_j$$

$$= \gamma_j C_j = \frac{\gamma_j P_j}{ZRT} \qquad (b)$$

where

$$\phi_j = \text{fugacity coefficient}$$
$$Z = \text{compressibility factor}$$

Thus,

$$\gamma_j = ZRT\, \phi_j$$

and Eq. 1.7-12 becomes:

$$\frac{k_c}{k} = \frac{(ZRT\, \phi_{HI})^2}{(ZRT\, \phi^\ddagger)}$$

$$= RT \frac{\phi_{HI}^2}{\phi^\ddagger} Z \qquad (c)$$

At low pressures, $\phi_j \to 1$ and $z \to 1$, so Eq. c becomes

$$\left(\frac{k_c}{k}\right)_0 = RT$$

Thus, the ratio of the observed rate constant at high to that at low pressures is,

$$\frac{k_c}{k_{c_0}} = \frac{\phi_{HI}^2}{\phi^\ddagger} Z \qquad (d)$$

If the activated complex were not considered, a similar derivation would lead to

$$\frac{k_c}{k_{c_0}} = \phi_{HI}^2 Z^2 \qquad (e)$$

The variation of the thermodynamic properties with pressure was calculated using the virial equation of state, with the constants for HI taken from data and

---

[7] See Amdur and Hammes [62]. Above about 600 K, the reaction is dominated by the usual halogen-hydrogen chain reaction mechanism.

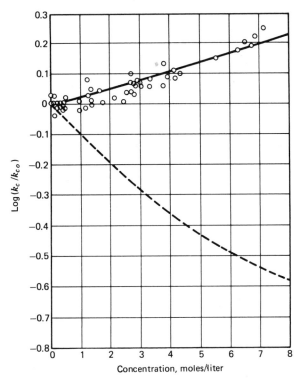

*Figure 1 Variation in rate of HI decomposition at 321.4°C. Points taken from Kistiakowsky's data; line represents Eq. d. (Adapted from Eckert and Boudart [61]).*

estimated from a model of the activated complex. Figure 1 shows excellent agreement with the data of Kistiakowsky [63] at pressures up to 250 atm, leading to a density variation of 300, for Eq. d but *not* for Eq. e. Thus, proper use of the thermodynamic corrections allows prediction of the nonideal effects.

## Problems

1.1  For the thermal cracking of ethane in a tubular reactor, the following data were obtained for the rate coefficient at different reference temperatures:

| $T(°C)$ | 702 | 725 | 734 | 754 | 773 | 789 | 803 | 810 | 827 | 837 |
|---------|-----|-----|-----|-----|-----|-----|-----|-----|-----|-----|
| $k(s^{-1})$ | 0.15 | 0.273 | 0.333 | 0.595 | 0.923 | 1.492 | 2.138 | 2.718 | 4.137 | 4.665 |

Determine the corresponding activation energy and frequency factor.

1.2  Derive the result given in Table 1.3-1 for the reaction $A + B \rightarrow Q + S$.

1.3 Derive the solutions to the rate equations for the first order reversible reaction given in Sec. 1.3.

1.4 A convenient laboratory technique for measuring the kinetics of ideal gas phase single reactions is to follow the change in total pressure in a constant volume and temperature container. The concentration of the various species can be calculated from the total pressure change.
Consider the reaction

$$aA + bB + \cdots \longrightarrow qQ + sS + \cdots$$

(a) Show that the extent can be found from:

$$\xi = \frac{V}{RT} \frac{p_t - p_{t0}}{\Delta \alpha}$$

where $\Delta \alpha = q + s + \cdots - a - b - \cdots$

(Note that the method can only be used for $\Delta \alpha \neq 0$.)

(b) Next show that the partial pressure for the $j$th species can be found from

$$p_j = p_{j0} + \frac{\alpha_j}{\Delta \alpha}(p_t - p_{t0})$$

(c) Use the method to determine the rate coefficient for the first-order decomposition of di-$t$-butyl peroxide

$$(CH_3)_3COOC(CH_3)_3 \longrightarrow 2(CH_3)_2CO + C_2H_6$$

The data given below are provided by J. H. Raley, F. E. Rust, and W. E. Vaughn [*J.A.Ch.S.*, **70**, 98 (1948)]. They were obtained at 154.6°C under a 4.2-mmHg partial pressure of nitrogen, which was used to feed the peroxide to the reactor. Determine the rate coefficient by means of the differential and integral method of kinetic analysis.

| $t$(min) | $p_t$(mm Hg) |
| --- | --- |
| 0 | 173.5 |
| 2 | 187.3 |
| 3 | 193.4 |
| 5 | 205.3 |
| 6 | 211.3 |
| 8 | 222.9 |
| 9 | 228.6 |
| 11 | 239.8 |
| 12 | 244.4 |
| 14 | 254.5 |
| 15 | 259.2 |
| 17 | 268.7 |
| 18 | 273.9 |
| 20 | 282.0 |
| 21 | 286.8 |
| ∞ | 491.8 |

1.5 The results of Problem 1.4 can be generalized for the measurement of any property of the reaction mixture that is linear in the concentration of each species:

$$\lambda_j = K_j C_j$$

The $\lambda_j$ could be partial pressures (as in Problem 1.4), various spectral properties, ionic conductivity in dilute solutions, and so on. Then the total observed measurement for the mixture would be:

$$\lambda = \sum_j \lambda_j = \sum_j K_j C_j$$

(a) For the general single reaction,

$$\sum_j \alpha_j A_j = 0$$

show that the relation between the extent of reaction and $\lambda$ is

$$\lambda = \lambda_0 + \left(\sum_j \alpha_j K_j\right)\frac{\xi}{V}$$

where

$$\lambda_0 = \sum_j K_j C_{j0}$$

(b) After a long ("infinite") time, the extent $\xi_\infty$ can be evaluated for irreversible reactions from the limiting reagent, and for reversible reactions from thermodynamics. Use this to formulate the desired relation containing only measured or determined variables (see Frost and Pearson [41]):

$$\frac{\lambda - \lambda_0}{\lambda_\infty - \lambda_0} = \frac{\xi}{\xi_\infty}$$

1.6 Show that the general expression for the concentration at which the autocatalytic reaction of Ex. 1.3-1 has a maximum rate is

$$\left(\frac{C_Q}{C_{Q0}}\right)_{max} = \frac{1}{2}\left(1 + \frac{C_{A0}}{C_{Q0}}\right)$$

Note that this agrees with the specific results in the example.

1.7 Derive Eq. 1.3-10.

1.8 Derive the concentration as a function of time for the general three species first order reactions:

These should reduce to all the various results for first order reactions given in Secs 1.3 and 1.4. Also determine the equilibrium concentrations $C_{Aeq}$, $C_{Qeq}$, $C_{Seq}$ in terms of the equilibrium constants for the three reactions.

1.9 Show that if a solution $\zeta = Ay_A + By_B + Cy_C$ is assumed for the network of Problem 1.8, such that

$$\frac{d\zeta}{dt} = -\lambda\zeta$$

the values of $\lambda$ are found from

$$\lambda(\lambda^2 - \alpha\lambda + \beta) = 0$$

where $\alpha > 0, \beta > 0$ are to be expressed in terms of the individual rate constants. Demonstrate how this is consistent with the Wei–Prater treatment. Show that the root $\lambda = 0$ gives the equilibrium concentrations as found from the three coupled equilibria, and that the other roots are real and positive.

1.10 For the complex reactions

$$aA + bB \quad \xrightarrow{\;\;1\;\;} \quad qQ$$

$$q'Q + b'B \quad \xrightarrow{\;\;2\;\;} \quad sS$$

(a) Use Eqs. 1.2-10 and 12 to express the time rates of change of $N_A$, $N_B$, $N_Q$, and $N_S$ in terms of the two extents of reaction and the stoichiometric coefficients $a, b, b', q, q'$, and $s$; for example,

$$\frac{dN_A}{dt} = -a\frac{d\xi_1}{dt} + (0)\frac{d\xi_2}{dt}$$

(b) In practical situations, it is often useful to express the changes in all the mole numbers in terms of the proper number of independent product mole number changes—in this case, two. Show that the extents in part (a) can be eliminated in terms of $dN_Q/dt$ and $dN_S/dt$ to give

$$\frac{dN_A}{dt} = -\frac{a}{q}\left(\frac{dN_Q}{dt}\right) - \frac{a}{s}\frac{q'}{q}\left(\frac{dN_S}{dt}\right)$$

$$\frac{dN_B}{dt} = -\frac{b}{q}\left(\frac{dN_Q}{dt}\right) - \left(\frac{b}{s}\frac{q'}{q} + \frac{b'}{s}\right)\left(\frac{dN_S}{dt}\right)$$

This alternate formulation will be often used in the practical problems to be considered later in the book.

(c) For the general reaction

$$\sum_{j=1}^{N} \alpha_{ij}A_j = 0 \qquad i = 1, 2, \dots, M$$

The mole number changes in terms of the extents are:

$$\frac{dN_j}{dt} = \sum_{i=1}^{M} \alpha_{ij} \frac{d\xi_i}{dt}$$

or

$$\frac{d\mathbf{N}}{dt} = \boldsymbol{\alpha}^T \frac{d\boldsymbol{\xi}}{dt}$$

where $\mathbf{N}$ is the $N$-vector of numbers of moles, $\boldsymbol{\xi}$ is the $M$-vector of extents, and $\boldsymbol{\alpha}^T$ is the transpose of the $M \times N$ stoichiometric coefficient matrix $\boldsymbol{\alpha}$. Show that if an alternate basis of mole number changes is defined as an $M$-vector

$$\frac{d\mathbf{N}^b}{dt}$$

that the equivalent expressions for all the mole number changes are

$$\frac{d\mathbf{N}}{dt} = \boldsymbol{\alpha}^T \{[\boldsymbol{\alpha}^b]^T\}^{-1} \frac{d\mathbf{N}^b}{dt}$$

where $\boldsymbol{\alpha}^b$ is the $M \times M$ matrix of the basis species stoichiometric coefficients.

Finally, show that these matrix manipulations lead to the same result as in part (b) if the basis species are chosen to be $Q$ and $S$.

1.11 Show that the overall orders for a free radical reaction mechanism with a first-order initiation step are $\frac{3}{2}$ and $\frac{1}{2}$ for a $\beta\beta$, respectively $\mu\mu$ termination.

1.12 The thermal decomposition of dimethyl ether

$$CH_3OCH_3 \longrightarrow CH_4 + CO + H_2$$

or

$$CH_3OCH_3 \longrightarrow CH_4 + HCHO$$

is postulated to occur by the following free radical chain mechanism:

$$CH_3OCH_3 \xrightarrow{k_1} CH_3^{\cdot} + OCH_3^{\cdot}$$

$$CH_3^{\cdot} + CH_3OCH_3 \xrightarrow{k_2} CH_4 + CH_2OCH_3^{\cdot}$$

$$CH_2OCH_3^{\cdot} \xrightarrow{k_3} CH_3^{\cdot} + HCHO$$

$$CH_3^{\cdot} + CH_2OCH_3^{\cdot} \xrightarrow{k_4} C_2H_5OCH_3$$

(a) For a first-order initiation step, use the Goldfinger–Letort–Niclause table to predict the overall order of reaction.

(b) With the help of the steady-state assumption and the usual approximations of small initiation and termination coefficients, derive the detailed kinetic expression for the overall rate:

$$\frac{-d[CH_3OCH_3]}{dt} = k_0[CH_3OCH_3]^n$$

and verify that the overall order, $n$, is as predicted in part (a). Also find $k_0$ in terms of $k_1, k_2, k_3$, and $k_4$.

(c) If the activation energies of the individual steps are $E_1 = 80$, $E_2 = 15$, $E_3 = 38$, $E_4 = 8$ kcal/mol, show that the overall activation energy is $E_0 = 62.5$ kcal/mol.

1.13 Laidler and Wojciechowski [40] provide the following table of individual rate constants for ethane pyrolysis:

| Reaction | $A_0^*$ | $E$ (kcal/mol) | |
|---|---|---|---|
| 1 | $1.0 \times 10^{17}$ | 85.0 | 1st-order initiation |
| 1a | $2(6.5) \times 10^{17}$ | 70.2 | 2nd-order initiation |
| 2 | $2.0 \times 10^{11}$ | 10.4 | hydrogen abstraction |
| 3 | $3.0 \times 10^{14}$ | 39.5 | radical decomposition |
| 4 | $3.4 \times 10^{12}$ | 6.8 | $H^{\cdot} + C_2H_6 \rightarrow$ |
| 5 | $1.6 \times 10^{13}$ | 0 | $H^{\cdot} + C_2H_5^{\cdot} \rightarrow$ termination |
| 6 | $1.6 \times 10^{13}$ | 0 | $C_2H_5^{\cdot} + C_2H_5^{\cdot} \rightarrow$ termination |

* In $s^{-1}$ or $cm^3\ mol^{-1}s^{-1}$.

(a) Derive the overall kinetic expressions for the four combinations of the two possible initiation steps (1 or 1a) and the termination steps (5 or 6).

(b) Compare the overall rate constants at $T = 873$ K with the experimental value of $8.4\ 10^{-4}\ s^{-1}$.

(c) Show that the ratio of the rates of reaction 5 and 6 is given by

$$\frac{r_5}{r_6} = \frac{k_3 k_5}{k_4 k_6} \frac{1}{[C_2H_6]}$$

(d) Calculate the "transition pressure level" where terminations (5) and (6) are equivalent ($r_5 = r_6$) at $T = 640°C$, and compare with the measured value of 60 mmHg. At this point, the overall reaction is changing from 1 to $\frac{3}{2}$ order.

*1.14 The overall reaction for the decomposition of nitrogen pentoxide can be written as:

$$2N_2O_5 \longrightarrow 4NO_2 + O_2$$

* These problems were contributed by Prof. W. J. Hatcher, Jr., University of Alabama.

The following reaction mechanism is proposed:

$$N_2O_5 \longrightarrow NO_3 + NO_2$$

$$NO_2 + NO_3 \longrightarrow N_2O_5$$

$$NO_2 + NO_3 \longrightarrow NO_2 + O_2 + NO$$

$$NO + NO_3 \longrightarrow 2NO_2$$

If the steady-state approximation for the intermediates is assumed, prove that the decomposition of $N_2O_5$ is first order. [See R. A. Ogg, *J. Ch. Phys.*, **15**, 337 (1947)].

*1.15 The previous reaction was carried out in a constant volume and constant temperature vessel to allow the application of the "total pressure method" outlined in Problem 1.4. There is one complication however: the dimerization reaction $2NO_2 \rightleftharpoons N_2O_4$ also occurs. It may be assumed that this additional reaction immediately reaches equilibrium, the dimerization constant being given by

$$\log K_p = \frac{2866}{T} - \log T - 9.132 \ (T \text{ in K}; K_p \text{ in mm}^{-1})$$

The following data were obtained by F. Daniels and E. H. Johnson [*J. Am. Chem. Soc.*, **43**, 53 (1921)] at 35°C, with an initial pressure of 308.2 mmHg:

| $t$(min) | $p_t$(mmHg) |
|---|---|
| 40 | 400.2 |
| 50 | 414.0 |
| 60 | 426.5 |
| 70 | 438.0 |
| 80 | 448.1 |
| 90 | 457.2 |
| 100 | 465.2 |
| 120 | 480.0 |
| 140 | 492.3 |
| 160 | 503.2 |
| 180 | 512.0 |
| 200 | 519.4 |
| 240 | 531.4 |
| 280 | 539.5 |
| 320 | 545.2 |
| 360 | 549.9 |
| $\infty$ | 565.3 |

Determine the first-order rate coefficient as a function of time. What is the conclusion?

*1.16 Reconsider the data of Problem 1.15. Determine the order of reaction together with the rate coefficient that best fits the data. Now recalculate the value of the rate coefficient as a function of time.

1.17 The catalytic oxidation of a hydrocarbon A by means of air into the desired product G is assumed to occur according to the mechanism

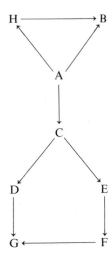

The following conversion data of the different species were collected for an inlet partial pressure of $A$ equal to 0.00252 atm.

| t (kg cat. hr/kmol) | A | B | C | D | E | F | G | H |
|---|---|---|---|---|---|---|---|---|
| 409 | 0.147 | 0.062 | 0.024 | 0.0091 | 0.0035 | 0.0067 | 0.0089 | 0.0033 |
| 748 | 0.202 | 0.06 | 0.025 | 0.0112 | 0.0056 | 0.0091 | 0.0205 | 0.0092 |
| 1619 | 0.388 | 0.148 | 0.0476 | 0.0304 | 0.0144 | 0.0364 | 0.0724 | 0.0209 |
| 1667 | 0.355 | 0.128 | 0.0452 | 0.0257 | 0.0128 | 0.0287 | 0.053 | 0.017 |
| 1751 | 0.375 | 0.133 | 0.0437 | 0.0232 | 0.0118 | 0.0219 | 0.0665 | 0.021 |
| 3807 | 0.674 | 0.271 | 0.0576 | 0.0388 | 0.0224 | 0.0588 | 0.125 | 0.031 |
| 4906 | 0.8 | 0.34 | 0.0525 | 0.0404 | 0.0252 | 0.08 | 0.16 | 0.0227 |

Each of these reactions are considered to be pseudo first order.
Determine the rate coefficients by means of the method of Himmelblau, Jones, and Bischoff.

# References

[1] Denbigh, K. G., *The principles of Chemical Equilibrium*, Cambridge University Press, Cambridge (1955).

[2] Prigogine, I., Defay, R.; Everett, D. H., Transl., *Chemical Thermodynamics*, Longmans, London (1954).

[3] Aris, R., *Introduction to the Analysis of Chemical Reactors*, Prentice-Hall, Englewood Cliffs, N.J. (1965).

[4] I. Prigogine, ed., *Advances in Chemical Physics*, Vol. 11, Interscience, New York (1967).

[5] Boudart, M., *Kinetics of Chemical Processes*, Prentice-Hall, Englewood Cliffs, N.J. (1968).

[6] Aris, R., *Elementary Chemical Reactor Analysis*, Prentice-Hall, Englewood Cliffs, N.J. (1969).

[7] *The Law of Mass Action—A Centenary Volume 1864–1964*, Det Norse Videnskaps—Akademi I. Oslo Universitetsforlaget, Oslo (1964).

[8] Caddell, J. R. and Hurt, D. M., *Chem. Eng. Prog.*, **47**, 333 (1951).

[9] Boyd, R. K., *Chem. Rev.*, **77**, 93 (1977).

[10] Laider, K. J., *Chemical Kinetics*, McGraw-Hill, New York (1965).

[11] Hougen, O. A. and Watson, K. M., *Chemical Process Principles*, Vol. III, Wiley, New York (1947).

[12] Rodigin, N. M. and Rodigina, E. N., *Consecutive Chemical Reactions*, Van Nostrand, New York (1964).

[13] Wei, J. and Prater, C. D., "The Structure and Analysis of Complex Reaction Systems," *Advances in Catalysis*, **13**, Academic Press, New York (1962).

[14] Widom, B., *Science*, **148**, 1555 (1965); *J. Chem. Phys.* **61**, 672 (1974).

[15] Nace, D. M., Voltz, S. E., and Weekman, V. W., *I.E.C. Proc. Des. Devt.*, **10**, 530 (1971).

[16] Weekman, V. W., *Ind. Eng. Chem. Proc. Des. Devpt.*, **7**, 90 (1968).

[17] Weekman, V. W., *Ind. Eng. Chem. Proc. Des. Devpt.*, **8**, 385 (1969).

[18] Voltz, S. E., Nace, D. M., and Weekman, V. W., *Ind. Eng. Chem. Proc. Des. Devpt.*, **10**, 538 (1971).

[19] Voltz, S. E., Nace, D. M., Jacob, S. M., and Weekman, V. W., *Ind. Eng. Chem. Proc. Des. Devpt.*, **11**, 261 (1972).

[20] Anderson, J. D. and Lamb, D. E., *Ind. Eng. Chem. Proc. Des. Devpt.*, **3**, 177 (1964).

[21] Jacob, S. M., Gross, B., Voltz, S. E., and Weekman, V. W., *A.I.Ch.E.J.*, **22**, 701 (1976). Also see U.S. Patent 3,960,707 (June 1, 1976).

[22] Wei, J. and Kuo, J. C. W., *Ind. Eng. Chem. Fundam.*, **8**, 114, 124 (1969).

[23] Ozawa, Y., *Ind. Eng. Chem. Fundam.*, **12**, 191 (1973).

[24] Luss, D. and Golikeri, S. V., *A.I.Ch.E.J.*, **21**, 865 (1975).

[25] Kondrat'ev, V. N., *Chemical Kinetics of Gas Reactions*, Pergamon Press, Oxford (1964).

[26] Bowen, J. R., Acrivos, A., and Oppenheim, A. K., *Chem. Eng. Sci.*, **18**, 177 (1963).

[27] Benson, S. W., *Foundations of Chemical Kinetics*, McGraw-Hill, New York (1960).

[28] Lindemann, F. A., *Trans. Faraday Soc.*, **17**, 598 (1922).

[29] Benson, S. W., *Ind. Eng. Chem. Proc. Des. Devpt.*, **56**, No. 1, 19 (1964).

[30] Rice, F. O. and Herzfeld, K. F., *J. Am. Chem. Soc.*, **56**, 284 (1944).

[31] Goldfinger, P., Letort, M., and Niclause, M., *Contribution à l'étude de la structure moléculaire*, Victor Henri Commemorative Volume, Desoer, Liège (1948).

[32] Franklin, J. L., *Brit. Chem. Eng.*, **7**, 340 (1962).

[33] Gavalas, G. R., *Chem. Eng. Sci.*, **21**, 133 (1966).

[34] Benson, S. W., *Thermochemical Kinetics*, Wiley, New York (1968).

[35] Küchler, L. and Theile, H., *Z. Physik. Chem.*, **B42**, 359 (1939).

[36] Quinn, C. P., *Proc. Roy. Soc. London*, Ser. **A275**, 190 (1963a); *Trans. Faraday Soc.*, **59**, 2543 (1963b).

[37] Steacie, E. W. R., *Free Radical Mechanisms*, Reinhold, New York (1946).

[38] Rutgers, A. J., *Physical Chemistry*, Interscience, New York (1953).

[39] Ray, W. H., *J. Macromolec. Sci. Rev. Macromol. Chem.*, **c8**, 1 (1972).

[40] Laidler, K. J. and Wojciechowski, B. W., *Proc. Roy. Soc. London*, **A260**, 91 (1961).

[41] Frost, A. A. and Pearson, R. G., *Kinetics and Mechanisms*, 2nd ed., Wiley, New York (1961).

[42] Levenspiel, O., *J. Catal.*, **25**, 265 (1972).

[43] Bard, Y. and Lapidus, L., "Kinetic Analysis by Digital Parameter Estimation," *Catal. Rev.*, **2**, 67 (1968).

[44] Kittrell, J. R., *Advan. Chem. Eng.*, **8**, 97 (1970).

[45] Froment, G. F., *A.I.Ch.E.J.*, **21**, 1041 (1975).

[46] Gavalas, G. R., *A.I.Ch.E.J.*, **19**, 214 (1973).

[47] Himmelblau, D. M., Jones, C. R., and Bischoff, K. B., *Ind. Eng. Chem. Proc. Des. Devpt.*, **6**, 536 (1967).

[48] Eakman, J. M., *Ind. Eng. Chem. Fundam.*, **8**, 53 (1969).

[49] Tang, Y. P., *Ind. Eng. Chem. Fundam.*, **10**, 321 (1971).

[50] Gay, I. D., *J. Phys. Chem.*, **75**, 1610 (1971).

[51] Paynter, J. D. and Schuette, W. L., *Ind. Eng. Chem. Proc. Des. Devpt.*, **10**, 250 (1971).

[52] Van Damme, P., Narayanan, S., and Froment, G. F., *A.I.Ch.E.J.*, **21**, 1065 (1975).

[53] Sundaram, K. M. and Froment, G. F., *Chem. Eng. Sci.*, **32**, 601 (1977).

[54] Rossini, F., *Selected Values of Thermodynamic Properties of Hydrocarbons and Related Compounds*, Carnegie Press, Pittsburgh, Pa. (1953).

[55] Glasstone, S., Laidler, K. J., and Eyring, H., *The Theory of Rate Processes*, McGraw-Hill, New York, 1941.

[56] Laidler, K. J., *Theories of Chemical Reaction Rates*, McGraw-Hill, New York (1969).

[57] Emptage, M. R. and Ross, J., *J. Chem. Phys.*, **51**, 252 (1969)

[58] Eckert, C. A., *Ind. Eng. Chem.*, **59**, No. 9, 20 (1967).

[59] Eckert, C. A., *Ann. Rev. Phys. Chem.*, **23**, 239 (1972).

[60] Eckert, C. A., Hsieh, C. K., and McCabe, J. R., *A.I.Ch.E.J.*, **20** (1974).

[61] Eckert, C. A. and Boudart, M., *Chem. Eng. Sci.*, **18**, 144 (1963).

[62] Amdur, I. and Hammes, G. G., *Chemical Kinetics*, McGraw-Hill, New York (1966).

[63] Kistiakowsky, G., *J. Amer. Chem. Soc.*, **50**, 2315 (1928).

# 2

## KINETICS
## OF
## HETEROGENEOUS
## CATALYTIC
## REACTIONS

## 2.1 Introduction

The principles of homogeneous reaction kinetics and the equations derived there remain valid for the kinetics of heterogeneous catalytic reactions, provided that the concentrations and temperatures substituted in the equations are really those prevailing at the point of reaction. The formation of a surface complex is an essential feature of reactions catalyzed by solids and the kinetic equation must account for this. In addition, transport processes may influence the overall rate: heat and mass transfer between the fluid and the solid or inside the porous solid, so that the conditions over the local reation site do not correspond to those in the bulk fluid around the catalyst particle. Figure 2.1-1 shows the seven steps involved when a molecule moves into the catalyst, reacts, and the product moves back to the bulk fluid stream. To simplify the notation the index $s$, referring to concentrations inside the solid, will be dropped in this chapter.

The seven steps are:

1. Transport of reactants $A, B \ldots$ from the main stream to the catalyst pellet surface.

2. Transport of reactants in the catalyst pores.

3. Adsorption of reactants on the catalytic site.

4. Surface chemical reaction between adsorbed atoms or molecules.

5. Desorption of products $R, S \ldots.$

6. Transport of the products in the catalyst pores back to the particle surface.

7. Transport of products from the particle surface back to the main fluid stream.

76

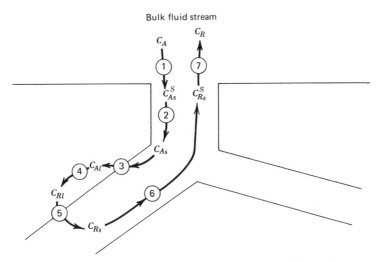

Figure 2.1-1 *Steps involved in reactions on a solid catalyst.*

Steps 1, 3, 4, 5, and 7 are strictly consecutive processes and can be studied separately and then combined into an overall rate, somewhat analogous to a series of resistances in heat transfer through a wall. However, steps 2 and 6 cannot be entirely separated: active centers are spread all over the pore walls so that the distance the molecules have to travel and therefore the resistance they encounter, is not the same for all of them. This chapter concentrates on steps 3, 4, and 5 and ignores the complications induced by the transport phenomena, which is treated in detail in Chapter 3.

The main goal in this chapter is to obtain suitable expressions to represent the kinetics of catalytic processes. Many details of the chemical phenomena are still obscure, and so, just as in Chapter 1, we will only briefly discuss the mechanistic aspects of catalysis. Further details are presented in several books in this area— an entree to this area is provided in books on chemical kinetics and catalysis; some texts specifically intended for chemical engineers are by Thomas and Thomas [1], Boudart [2], and a useful brief introduction by Thomson and Webb [3] and a discussion of several important industrial catalytic processes is given in Gates, Katzer and Schuit [62]. For further comprehensive surveys, see Emmett [4] and, for current progress, the series *Advances in Catalysis* [5].

Even though we won't consider catalytic mechanisms in detail, there are certain principles that are useful in developing rate expressions. The most obvious is that the catalytic reaction is often much more rapid than the corresponding homogeneous reaction. From the principle of microscopic reversibility, the reverse reaction will be similarly accelerated, and so the overall equilibrium will *not* be affected. As an example of this acceleration, Boudart [6] compared the homogeneous versus catalytic rates of ethylene hydrogenation. The first route involves a chain mechanism, with the initiation step (Chapter 1) involving hydrogen and ethyl radicals—a usual difficult first step. The catalytic reaction, on the other hand, has as a first step the formation of a solid surface—ethylene complex, that is apparently energetically a more favorable reaction. Using the available data for both types of reactions, and knowing the surface area per volume of the (CuO-MgO) catalyst, Boudart showed that the two rates were

Homogeneous:

$$r = 10^{27} \exp\left(-\frac{43000}{RT}\right) p_{H_2}$$

Catalytic:

$$r = 2.10^{27} \exp\left(-\frac{13000}{RT}\right) p_{H_2}$$

For example, at 600 K the ratio of catalytic to homogeneous rate is $1.44.10^{11}$.

The above equations show that the principal reason for the much higher catalytic rate is the decrease in activation energy. This feature is the commonly accepted special feature of catalytic versus homogeneous reactions.

The exact nature of the reasons for and the ease of formation of the surface complex are still not entirely known. One can visualize certain structural requirements of the underlying solid surface atoms in order to accomodate the reactants, and this has led to one important set of theories. Also, as will be seen, various electron transfer steps are involved in the formation of the complex bonds, and so the electronic nature of the catalyst is also undoubtedly important. This has led to other important considerations concerning the nature of catalysts. The classification of catalysts of Table 2.1-1 gives some specific examples (Innes; see Moss [7]). Recent compilations also give very useful overviews of catalytic activity: Thomas [8] and Wolfe [9]. Burwell [10] has discussed the analogy between catalytic and chain reactions:

| Reaction | Overall Reaction | |
| --- | --- | --- |
| | Chain Terminology | Catalysis Terminology |
| $A + B \rightarrow R$ Catalyst (cat.) | Chain initiation | Preparation and introduction of catalyst; sorption |
| $A + \text{cat.} \rightarrow A \text{ cat.}$ $\left.\begin{array}{c} \\ B + A \text{ cat.} \rightarrow R + \text{cat.} \end{array}\right\}$ | Chain propagation | Catalytic reaction |
| Cat. $+ P \rightarrow P$ cat. | Chain termination | Desorption; poisoning by $P$ |

*Table 2.1-1 Classification of heterogeneous catalysts*

| Primary Class | Examples of Reactions | Some Catalysts |
|---|---|---|
| Hydrogenation–dehydrogenation | Of multiple carbon–carbon bonds (e.g., butadiene synthesis) | Chromia, iron oxide, calcium–nickel phosphate |
| | Hydrogenation of aromatics and aromatization | Platinum–acid alumina and chromium or molybdenum oxides |
| | Of oxy-organic compounds (e.g., ethanol → acetaldehyde) | Copper (generally transition metals and oxides Group 1B metals for first three reactions) |
| | Hydrogenation of oxides of carbon and the reverse reaction (e.g., methane reforming with steam) | Nickel |
| | Methanol synthesis from $CO + H_2$ | Zinc oxide with chromia; copper |
| | Hydrocarbon synthesis (Fischer–Tropsch) | Promoted iron oxide; cobalt |
| | $CO + H_2 +$ olefin (oxo-process) | Cobalt–thoria |
| | Amonia synthesis | Iron promoted with potash and alumina |
| | Hydrodesulphurization | Cobalt–molybdenum oxide; sulphides of nickel, tungsten |
| Oxidation | $SO_2 \rightarrow SO_3$; naphthalene to phthalic anhydride | Vanadium pentoxide |
| | Ammonia to oxides of nitrogen | Platinum |
| | Ethylene to ethylene oxide | Silver |
| | Water gas shift | Iron oxide |
| Acid catalyzed | Cracking; alkylation; isomerization; polymerization | Synthetic silica–aluminas, acid-treated montmorillonite and other clays; aluminium chloride, phosphoric acid |
| Hydration–dehydration | Ethanol ⇄ ethylene, also dehydration of higher alcohols | Alumina; phosphoric acid on a carrier |
| Halogenation–dehalogenation | Methane chlorination (to methyl chloride) | Cupric chloride (generally chlorides, fluorides of copper, zinc, mercury, silver) |

From Moss [7]

79

*Table 2.1-2  Products of thermal and catalytic cracking*

| Hydrocarbon | Thermal Cracking | Catalytic Cracking |
|---|---|---|
| n-Hexadecane (cetane) | Major product is $C_2$ with much $C_1$ and $C_3$; much $C_4$ to $C_{15}$ n-α-olefins; few branched aliphatics | Major product is $C_2$ to $C_6$, few n-α-olefins above $C_4$; aliphatics mostly branched |
| Alkyl aromatics | Cracked within side chain | Cracked next to ring |
| Normal olefins | Double bond shifts slowly; little skeletal isomerization | Double bond shifts rapidly; extensive skeletal isomerization |
| Olefins | Hydrogen transfer is a minor reaction and is nonselective for tertiary olefins | Hydrogen transfer is an important reaction and is selective for tertiary olefins |
|  | Crack at about same rate as corresponding paraffins | Crack at much higher rate than corresponding paraffins |
| Naphthenes | Crack at lower rate than paraffins | Crack at about same rate as paraffins with equivalent structural groups |
| Alkyl aromatics (with propyl or larger substituents) | Crack at lower rate than paraffins | Crack at higher rate than paraffins |
| Aliphatics | Small amounts of aromatics formed at 500°C | Large amounts of aromatics formed at 500°C |

From Oblad, Milliken, and Mills [11].

One or two examples of the use of these concepts will illustrate the ideas and help to formulate appropriate rate equations. The acidic catalysts, such as silica-alumina, can apparently act as Lewis (electron acceptor) or Brønsted (proton donor) acids, and thus form some sort of carbonium ion from hydrocarbons, for example. Note the analogy between this hydrogen deficient entity and a free radical. However, the somewhat different rules for the reactions of carbonium ions apply from organic chemistry and permit semiquantitative predictions of the products expected; see Table 2.1-2 from Oblad, et al. [11].

Greensfelder, Voge, and Good [12] in a classic work, used the following concepts for n-hexadecane cracking: (1) the initial carbonium ions formed are dominated by secondary ions because of the ratio of 28 to 6 possible hydrogen atoms, (2) the carbonium ion splits at a beta-position to the original ionic carbon atom, forming an alpha-olefin and another primary carbonium ion, (3) this new ion rearranges and again reacts as in (2) until a difficult-to-form fragment of 3 or more carbon atoms might be formed (e.g., from n-sec-$C_5^+$), and (4) this final carbonium ion reacts with a new hexadecane molecule, thereby propagating the chain plus yielding a small paraffin. A final assumption, based on separate cracking

CHEMICAL ENGINEERING KINETICS

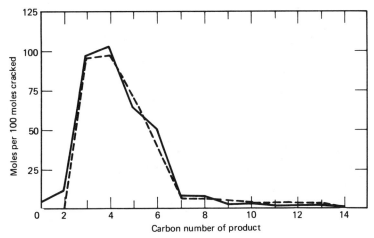

*Figure 2.1-2 Catalytic cracking of n-hexadecane. Solid line: experimental products, 24 per cent conversion over alumina-zirconia-silica at 500°C. Dotted line: Calculated products, carbonium ion mechanism (from Greensfelder, Voge, and Good* [12]).

studies, was that the olefins were highly reactive, and that, half formed as in step (2), they crack according to the same scheme. Figure 2.1-2 illustrates the predictions resulting from this method. In a later comprehensive review, Voge [13] indicated that different catalysts, in fact, gave somewhat different product distributions; these could be approximately accounted for by altering the last assumption about the fraction of olefins that crack.

More extensive discussions for several reaction types is provided by Germain [14]. Most catalytic cracking today utilizes zeolite catalysts. These are crystalline aluminosilicates that contain "cages," often of molecular dimensions, that can physically "block," branched chain molecules, for example (often called molecular sieves). Some of the above ideas undoubtedly apply, but the prediction of the selectivity is now much more complicated. (They are also much more active as catalysts.) Some aspects of their properties are reviewed by Venuto [15], where more than 50 different reactions catalyzed by zeolites are listed.

Metal catalysts are primarily concerned with hydrogenations and dehydrogenations. (Note that, except for noble metals, they would not actually survive in a severe oxidizing environment.) The classical example of the difference in behavior of acid and metal catalysts is the ethanol decomposition:

$$C_2H_5OH \xrightarrow[\text{catalyst}]{\text{acid}} C_2H_4 + H_2O \qquad \text{(dehydration)}$$

or

$$C_2H_5OH \xrightarrow[\text{catalyst}]{\text{metal}} C_2H_4O + H_2 \qquad \text{(dehydrogenation)}$$

With hydrocarbons, the two types of catalysts cause cracking or isomerization versus hydrogenation or dehydrogenations.

An interesting and very practical example of these phenomena concerns catalysts composed of both types of materials—called "dual function," or bifunctional (in general, polyfunctional) catalysts. A lucid discussion is provided by Weisz [16], and a few examples indicate the importance of these concepts, not only to catalysis, but also to the kinetic behavior. Much of the reasoning is based on the concept of reaction sequences involving the surface intermediates. Consider the scheme where the species within the dashed box are the surface intermediates.

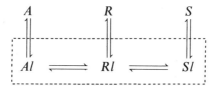

The amount of $R$ in the fluid phase now depends not only on the relative rates between $Al$, $Rl$, $Sl$, as in homogeneous kinetics, but also on the relative rates of desorption to reaction. For irreversible surface reactions, and very slow desorption rates, no fluid phase $R$ will even be observed! A detailed experimental verification of this general type of behavior was provided by Dwyer, Eagleton, Wei, and Zahner [17] for the successive deuterium exchanges of neopentane. They obtained drastic changes in product distributions as the ratio (surface reaction rate)/(desorption rate) increase.

If the above successive reactions were each catalysed by a different type of site (e.g., a metal and an acid), a bifunctional catalytic system results:

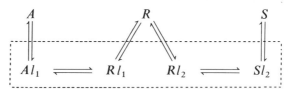

The essential difference here is that the true intermediate, $R$, *must* desorb, move through the fluid phase, and adsorb on the new site if any product $S$ is to be formed. As will be seen, this can allow an extra degree of freedom in the catalyst behavior.

Weisz defines a "nontrivial" polystep sequence as one where a unique conversion or selectivity can be achieved relative to the usual type of sequence. Thus,

$$A \xrightarrow[\phantom{xx}]{\overset{\text{site}}{①}} R \xrightarrow[\phantom{xx}]{\overset{\text{site}}{②}} S$$

would be considered "trivial," since the results obtained from a bifunctional catalyst would be essentially similar to those from the two reactions successively carried out one after the other. Now for the sequence

$$A \xrightleftharpoons{\overset{\text{site}}{①}} R \xrightarrow{\overset{\text{site}}{②}} S$$

the maximum conversion to $S$ would be limited by the equilibrium amount of $R$ formed when the steps were successively performed. However, if the second site were intimately adjacent to the first, the $Rl_1$ intermediate would be continuously "bled off," thus shifting the equilibrium toward higher overall conversion. This is extremely important for cases with very adverse equilibrium.

This appears to be the situation for the industrially important isomerization of saturated hydrocarbons (reforming), which are generally believed to proceed by the following sequence:

$$(\text{saturate}) \underset{\substack{\text{metal} \\ \text{cat.}}}{\overset{-H_2}{\rightleftharpoons}} (\text{unsaturate}) \xrightarrow{\substack{\text{acid} \\ \text{cat.}}} (\text{isounsaturate})$$

$$+H_2 \updownarrow \text{ metal cat.}$$

$$(\text{isosaturate})$$

[See also Sinfelt [18] and Haensel [19]. The isomerization step is usually highly reactive (recall the cracking discussion), and so the first part of the reaction has exactly the above sequence. Weisz and co-workers performed imaginative experiments to prove this conjecture. They made small particles of acid catalyst and small particles containing platinum. These particles were then formed into an overall pellet for reaction. Weisz et al. found that a certain intimacy of the two catalysts was required for appreciable conversion of $n$-heptane to isoheptane, as seen in Fig. 2.1-3. Particles larger than about 90 $\mu$m forced the two steps to proceed successively, since the intermediate unsaturates resulting from the metal site dehydrogenation step could not readily move to the acid sites for isomerization. This involves diffusion steps, which would carry us too far afield for now, but the qualitative picture is clear. Further evidence that olefinic intermediates are involved was from experiments showing that essentially similar product distributions occur with dodecane or dodecene feeds.

Another example presented was for cumene cracking, which is straightforward with acidic (silica-alumina) catalyst:

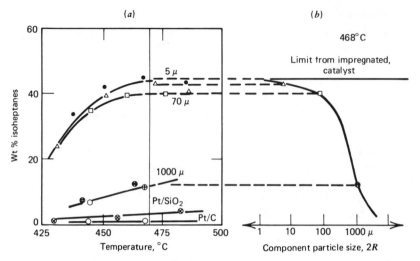

*Figure 2.1-3 Isomerization of n-heptane over mixed component catalyst, for varying size of the component particles: (a) conversion versus temperature; (b) conversion at 468°C versus component particle diameter (from Weisz [17]).*

However, a drastic change in product distribution occurred with a $Pt/Al_2O_3$ catalyst, which mainly favors the reaction:

$$CH_3 - CH - CH_3 \longrightarrow CH_3 - C = CH_2 + H_2$$
(Methylstyrene)

The presumed sequence was:

Cumene $\xrightarrow{(SiAl)}$ | cumene/catalyst | $\xrightarrow{(SiAl)}$ benzene + $C_3H_6$
$\updownarrow$ (SiAl)
| (intermediate) | $\xrightarrow{(Pt)}$ styrene + $H_2$

With only acid sites, the intermediate actually plays no role, but the metal sites permit the alternate, and then apparently dominant, reaction. Many further aspects of polyfunctional catalyst conversion and selectivity behavior were also discussed by Weisz [16], but our main goal is to develop kinetic rate expressions.

84 _____ CHEMICAL ENGINEERING KINETICS

The above discussion should provide some basis for construction of rate equations. We usually assume that we have a given catalyst from which experimental data will be obtained. However, the above considerations should always be kept in mind if changes are made in the catalyst formulation, or if changes occur during the process—obviously the kinetic expressions could be qualitatively, and certainly quantitatively, different in certain cases.

In all of the above, we have been rather nonquantitative about the surface intermediates. In fact, their nature is a subject of current research, and so only a fairly general quantitative treatment is possible. It is generally conceded that an adsorption step forms the surface intermediate, and so a brief discussion of this subject is useful before proceeding to the actual rate equations.

Some useful references are Brunauer [20], de Boer [21], Flood [22], Gregg and Sing [23], Clark [24], and Hayward and Trapnell [25].

There are two broad categories of adsorption, and the important features for our purposes are:

| Physisorption | Chemisorption |
|---|---|
| van der Waals forces | covalent chemical bonds |
| more than single layer coverage possible | only single layer coverage |

For a surface-catalyzed reaction to occur, chemical bonds must be involved, and so our interest is primarily with chemisorption. Again, some general classifications of various metals for chemisorption of gases are possible, as shown in Table 2.1-3 from Coughlin [26], and similar properties are involved. Note that the transition elements of the periodic table are frequently involved, and this appears to be based on the electronic nature of their d-orbitals.

The classical theory of Langmuir is based on the following hypotheses:

1. Uniformly energetic adsorption sites.

2. Monolayer coverage.

3. No interaction between adsorbed molecules.

Thus, it is most suitable for describing chemisorption (except possibly for assumption 1) and low-coverage physisorption where a single layer is probable. For higher-coverage physisorption, a theory that accounts for multiple layers is the Brunauer–Emmett–Teller (B-E-T) isotherm (see [20–24], [63]).

Langmuir also assumed that the usual mass-action laws could describe the individual steps. Thus, calling "$l$" an adsorption site, the reaction is:

$$A + l \; \rightleftharpoons \; Al \tag{2.1-1}$$

*Table 2.1-3  Classification of metals as to chemisorption*

| Group | Metals | $O_2$ | $C_2H_2$ | $C_2H_4$ | CO | $H_2$ | $CO_2$ | $N_2$ |
|---|---|---|---|---|---|---|---|---|
| A | Ca, Sr, Ba, Ti, Zr, Hf, V, Nb, Ta, Cr, Mo, W, Fe, Re[a] | + | + | + | + | + | + | + |
| $B_1$ | Ni, Co[a] | + | + | + | + | + | + | − |
| $B_2$ | Rh, Pd, Pt, Ir[a] | + | + | + | + | + | − | − |
| C | Al, Mn, Cu, Au[b] | + | + | + | + | − | − | − |
| D | K | + | + | − | − | − | − | − |
| E | Mg, Ag, Zn, Cd, In, Si, Ge, Sn, Pb, As, Sb, Bi | + | − | − | − | − | − | − |
| F | Se, Te | − | − | − | − | − | − | − |

From Coughlin [26].
[a] Behavior is not certain as to group.
[b] Au does not adsorb $O_2$.

where "$Al$" represents adsorbed $A$. The rates are:

$$r_a = k_a C_A C_l \qquad k_a = A_a e^{-E_a/RT} \qquad (2.1\text{-}2)$$

$$r_d = k_d C_{Al} \qquad k_d = A_d e^{-E_d/RT} \qquad (2.1\text{-}3)$$

where $C_l$ and $C_{Al}$ are surface concentrations, kmols/kg catalyst. Also, the total sites are either vacant or contain adsorbed $A$:

$$C_t = C_l + C_{Al} \qquad (2.1\text{-}4)$$

At equilibrium, the "adsorption isotherm" is found by equating the rates:

$$k_d C_{Al} = k_a C_A C_l$$
$$= k_a C_A (C_t - C_{Al})$$

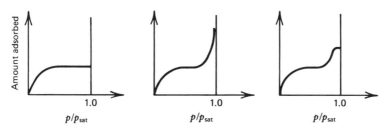

*Figure 2.1-4 Types of adsorption isotherm (after Brunauer, Deming, Deming and Teller [28]).*

Thus, the amount adsorbed is given by:

$$C_{Al} = \frac{C_t K_A C_A}{1 + K_A C_A} \qquad (2.1-5)$$

where

$$K_A = k_a/k_d$$
$$= \text{adsorption equilibrium constant}$$

An alternate way to write Eq. 2.1-5, (often used by chemists) is in terms of the fractional coverage:

$$\theta = \frac{C_{Al}}{C_t} = \frac{K_A C_A}{1 + K_A C_A} \qquad (2.1\text{-}5a)$$

The shape of Eq. 2.1-5 is a hyperbola.

There are three forms of isotherm commonly observed, although others occasionally occur, and they are shown in Figure 2.1-4. Here, $p_{sat}$ refers to the saturation pressure of the gas at the given temperature. Type I is the Langmuir isotherm, and Type II results from multilayer physisorption at higher coverages. Type IV is the same as Type II, but in a solid of finite porosity, giving the final level portion as $p \rightarrow p_{sat}$. The "heat of adsorption" is

$$Q_a = E_d - E_a \qquad (2.1\text{-}6)$$

and for chemisorption, can have a magnitude similar to that for other chemical reactions—more than 10 of kcal/mol.

The Langmuir treatment can be extended to other situations, and we consider two that will be of use for constructing kinetic expressions. For two species adsorbing on the same sites:

$$A + l \;\rightleftharpoons\; Al$$
$$B + l \;\rightleftharpoons\; Bl$$

$$\frac{dC_{Al}}{dt} = k_{aA}C_A C_l - k_{dA}C_{Al} \qquad (2.1\text{-}7a)$$

$$\frac{dC_{Bl}}{dt} = k_{aB}C_B C_l - k_{dB}C_{Bl} \qquad (2.1\text{-}7b)$$

$$C_t = C_l + C_{Al} + C_{Bl} \qquad (2.1\text{-}7c)$$

At equilibrium:

$$C_{Al} = K_A C_A C_l \qquad (2.1\text{-}8a)$$

$$C_{Bl} = K_B C_B C_l \qquad (2.1\text{-}8b)$$

$$C_t = C_l + K_A C_A C_l + K_B C_B C_l \qquad (2.1\text{-}8c)$$

and

$$C_l = \frac{C_t}{1 + K_A C_A + K_B C_B} \cdot \qquad (2.1\text{-}9)$$

Thus, the adsorbed amounts are given by

$$C_{il} = \frac{C_t K_i C_i}{1 + K_A C_A + K_B C_B} \qquad (2.1\text{-}10)$$

If the molecule dissociates on adsorption:

$$A_2 + 2l \; \rightleftharpoons \; 2Al$$

and at equilibrium:

$$C_{Al}{}^2 = K_A C_{A_2} C_l{}^2 \qquad (2.1\text{-}11)$$

Then,

$$C_t = C_l + \sqrt{K_A C_{A_2}} \, C_l$$

and finally:

$$C_{Al} = \frac{C_t \sqrt{K_A C_{A_2}}}{1 + \sqrt{K_A C_{A_2}}} \qquad (2.1\text{-}12)$$

Another way to state the assumptions of the classical Langmuir theory is that the heat of adsorption, $Q_a$, is independent of surface coverage, $\theta$. This is not always the case, and more general isotherms for nonuniform surfaces can be developed by summing (integrating) over the individual sites, $\theta_i$, (e.g., see Clark [24] and Rudnitsky and Alexeyev [64]).

$$\theta = \int_0^1 \frac{(A_a/A_d)\exp[Q_a(\theta_i)/RT]C_A}{1 + (A_a/A_d)\exp[Q_a(\theta_i)/RT]C_A} \, d\theta_i$$

If $Q_a$ depends logarithmically on $\theta$ over a range of surface coverages greater than zero,

$$Q_a = -Q_{am} \ln \theta$$

$$\theta = \exp(-Q_a/Q_{am})$$

$$d\theta = \frac{d\theta}{dQ_a} dQ_a$$

$$= -\frac{1}{Q_{am}} \exp(-Q_a/Q_{am})dQ_a$$

Then,

$$\theta = \frac{1}{Q_{am}} \int_0^\infty \frac{\exp(-Q_a/Q_{am})dQ_a}{1 + (A_d/A_a)C_A^{-1} \exp(-Q_a/RT)}$$

$$\simeq \left(\frac{A_a}{A_d} C_A\right)^{RT/Q_{am}}, \qquad Q_{am} \gg RT$$

$$\equiv aC_A{}^m$$

This has the form of the Freundlich isotherm, which often empirically provides a good fit to adsorption data, especially in liquids, that cannot be adequately fit by a Langmuir isotherm. Using a linear dependence of $Q_a$ on $\theta$,

$$Q_a = Q_{a0}(1 - \alpha\theta)$$

approximately gives the Temkin isotherm:

$$\theta \simeq \left(\frac{RT}{\alpha Q_{a0}}\right) \ln\left[\frac{A_a}{A_d} C_A\right]$$

This has been extensively used for ammonia synthesis kinetics.

Even though these isotherms presumably account for nonuniform surfaces, they have primarily been developed for single adsorbing components. Thus, the rational extensions to interactions in multicomponent systems is not yet possible, as with the Langmuir isotherm. This latter point is important for our further applications, and so we essentially use only the Langmuir isotherms for developing kinetic rate expressions. However, not all adsorption data can be represented by a Langmuir isotherm, and this is still an unresolved problem in catalytic kinetics.

## 2.2 Rate Equations

Any attempt to formulate a rate equation for solid-catalyzed reactions starts from the basic laws of chemical kinetics encountered in the treatment of homogeneous reactions. However, care has to be taken to substitute in these laws the concentrations and temperatures at the locus of reaction itself. These do not necessarily correspond to those just above the surface or the active site, due to the adsorption characteristics of the system. In order to develop the kinetics, an expression is required that relates the rate and amount of adsorption to the concentration of the component of the fluid in contact with the surface.

The application of Langmuir isotherms for the various reactants and products was begun by Taylor, in terms of fractional coverage, and the more convenient use of surface concentrations for complex reactions by Hougen and Watson [27]. Thus, the developments below are often termed Langmuir–Hinshelwood–Hougen–Watson (L-H-H-W) rate equations.

Consider the simple overall reaction:

$$A \rightleftharpoons R$$

The chemisorption step will be written as,

$$A + l \rightleftharpoons Al$$

where "$l$" represents a vacant site.

Assuming a simple mass action law:

$$r_a = k_A \left( C_A C_l - \frac{C_{Al}}{K_A} \right) \tag{2.2-1}$$

where

$$
\begin{aligned}
k_A &= \text{chemisorption rate coefficient} \\
C_l &= \text{concentration of vacant site} \\
C_{Al} &= \text{concentration of chemisorbed } A \\
K_A &= \text{adsorption equilibrium constant}
\end{aligned}
$$

The surface chemical reaction step is

$$Al \rightleftharpoons Rl$$

If both reactions are assumed to be of first order, the net rate of reaction of $Al$ is:

$$r_{sr} = k_{sr} \left( C_{Al} - \frac{C_{Rl}}{K_{sr}} \right) \tag{2.2-2}$$

where

$$
\begin{aligned}
k_{sr} &= \text{surface reaction rate coefficient} \\
K_{sr} &= \text{surface reaction equilibrium constant}
\end{aligned}
$$

Finally, the desorption step is

$$Rl \rightleftharpoons R + l$$

with rate

$$r_d = k_R' \left( C_{Rl} - \frac{C_R \cdot C_l}{K_d} \right)$$

or

$$r_d = k_R \left( \frac{C_{Rl}}{K_R} - C_R C_l \right) \tag{2.2-3}$$

where

$$k_R = \text{rate constant for desorption step}$$
$$K_R = \text{adsorption equilibrium constant} = 1/K_d$$

Note that *adsorption* equilibrium constants are customarily used, rather than both adsorption and desorption constants.

Since the overall reaction is the sum of the individual steps, the ordinary thermodynamic equilibrium constant for the overall reaction is

$$K = \frac{K_A K_{sr}}{K_R} \tag{2.2-4}$$

This relation can be used to eliminate one of the other equilibrium constants, often the unknown $K_{sr}$.

If the total number of sites, $C_t$, is assumed constant, it must again consist of the vacant plus occupied sites, so that

$$C_t = C_l + C_{Al} + C_{Rl} \tag{2.2-5}$$

The total sites may not always remain constant during use, and this will be discussed further in Chapter 5 on catalyst deactivation.

The rigorous combination of these three consecutive rate steps leads to a very complicated expression, but this needs to be done only in principle for transient conditions, although even then a sort of steady-state approximation is often used for the surface intermediates in that it is assumed that conditions on the surface are stationary. The rates of change of the various species are

$$\frac{dC_A}{dt} = -r_a$$

$$\frac{dC_{Al}}{dt} = r_a - r_{sr}$$

$$\frac{dC_{Rl}}{dt} = r_{sr} - r_d$$

$$\frac{dC_R}{dt} = r_d$$

Thus, a steady-state approximation on the middle two equations, as in Chapter 1, indicates that the three surface rates will be equal:

$$r_a = r_{sr} = r_d = r_A \qquad (2.2\text{-}6)$$

Combining Eq. 2.2-1, 2, 3, 5, and 6 permits us to eliminate the unobservable variables $C_1$, $C_{Al}$, $C_{Rl}$ in terms of the fluid phase compositions $C_A$ and $C_R$, as shown by Aris [28]:

$$r_A = \cfrac{C_t(C_A - C_R/K)}{\left[\left(\cfrac{1}{K_A k_{sr}} + \cfrac{1}{k_A} + \cfrac{1}{Kk_R}\right) + \left(\cfrac{1}{K_A k_{sr}} + \cfrac{1 + K_{sr}}{Kk_R}\right)K_A C_A \right.}$$
$$\left. + \left(\cfrac{1}{K_A k_{sr}} + \cfrac{1 + K_{sr}}{K_{sr} k_A}\right)K_R C_R\right] \qquad (2.2\text{-}7)$$

Equation 2.2-7 thus gives the reaction rate in terms of fluid phase compositions and the parameters of the various steps. Even for this very simple reaction, the result is rather complicated for the general case. Quite often it is found that one of the steps is much slower than the others and it is then termed the "rate controlling step." For example, suppose the surface reaction was very slow compared to the adsorption or desorption steps:

$$k_A, k_R \gg k_{sr}$$

Then Eq. 2.2-7 approximately reduces to

$$r_A = \frac{K_A k_{sr} C_t(C_A - C_R/K)}{1 + K_A C_A + K_R C_R} \qquad (2.2\text{-}8)$$

which is much simpler than the general case. Another example would be adsorption of $A$ controlling:

$$k_R, k_{sr} \gg k_A$$

which leads to:

$$r_A = \frac{k_A C_t(C_A - C_R/K)}{1 + \left(1 + \cfrac{1}{K_{sr}}\right)K_R C_R} \qquad (2.2\text{-}9a)$$

$$= \frac{k_A C_t(C_A - C_R/K)}{1 + \cfrac{K_A}{K} C_R + K_R C_R} \qquad (2.2\text{-}9b)$$

For other than simple first-order reactions, the general expression similar to Eq. 2.2-7 is exceedingly tedious, or even impossible, to derive, and so a rate-controlling step is usually assumed right from the beginning. This can be dangerous, however, in the absence of knowing the correct mechanism, and more than one rate-controlling step is certainly feasible. For example, if one step is controlling in one region of the variables and another for different conditions, there must obviously be a region between the two extremes where both steps have roughly equal importance. The resulting kinetic equations are not as complicated as the general result, but still quite a bit more involved than for one rate-controlling step and will not be discussed further here; see Bischoff and Froment [29] and Shah and Davidson [30].

As an example of this procedure, let us derive the rate equation for $A \rightleftharpoons R$ when surface reaction is rate controlling. This means that in Eq. 2.2-1, $k_A \to \infty$, and since from Eq. 2.2-6 the rate must remain finite, this shows that

$$(C_A C_l - C_{Al}/K_A) \longrightarrow 0$$

or

$$C_{Al} \cong K_A C_A C_l \qquad (2.2\text{-}10)$$

Eq. 2.2-10 does *not* mean that the adsorption step is in true equilibrium, for then the rate would be identically zero, in violation of Eq. 2.2-6. The proper interpretation is that for very large $k_A$, the surface concentration of $A$ is very close to that of Eq. 2.2-10. Similarly, from the desorption Eq. 2.2-3,

$$C_{Rl} \cong K_R C_R C_l \qquad (2.2\text{-}11)$$

If Eqs. 2.2-10 and 11 are substituted into Eq. 2.2-5, we obtain

$$C_t = C_l(1 + K_A C_A + K_A C_R)$$

or

$$C_l = \frac{C_t}{1 + K_A C_A + K_R C_R} \qquad (2.2\text{-}12)$$

Thus, finally substituting Eqs. 2.2-10, 11, and 12 into Eq. 2.2-2 gives

$$r_A = k_{sr}\left(K_A C_A - \frac{K_R C_R}{K_{sr}}\right)C_l$$

$$= \frac{K_A k_{sr} C_t(C_A - C_R/K)}{1 + K_A C_A + K_R C_R}$$

where Eq. 2.2-4 was also used. This final result is exactly the same as Eq. 2.2-8, which was found by reducing the general Eq. 2.2-7. This direct route, however, avoided having to derive the general result at all.

The total active sites concentration, $C_t$, is not measurable. Note from Eqs. 2.2-7, 2.2-8, and 2.2-9 and the other expressions that $C_t$ always occurs in combination with the rate constants $k_A$, $k_{sr}$, and $k_R$. Therefore, it is customary to absorb $C_t$ into these rate coefficients so that new coefficients $k$ are used, where $k = k_i C_t$.

Even the simpler, one rate-controlling step equations still contain a large number of parameters that must be experimentally determined. This important subject is discussed in detail in Section 2.3. It has been suggested several times that, for design and correlation purposes, the whole adsorption scheme is unnecessary and should be eliminated in favor of a strictly empirical approach, using, say, simple orders. For some purposes this is indeed a reasonable alternative, but should be justified as a permissible simplification of the adsorption mechanisms. These are still the only reasonably simple, comprehensive results we have for describing catalytic kinetics and sometimes provide valuable clues to qualitative behavior in addition to their use in quantitative design. The following example illustrates this.

### Example 2.2-1 Competitive Hydrogenation Reactions

This application of the foregoing concepts was discussed by Boudart [31]. The following data on the liquid phase catalytic cohydrogenation of $p$-xylene ($A$) and tetraline ($B$) were given by Wauquier and Jungers [32]. As a simulation of a practical situation, a mixture of $A$ and $B$ was hydrogenated, giving the following experimental data:

| Composition of Mixture | | | Total Hydrogenation Rate | |
|---|---|---|---|---|
| $C_A$ | $C_B$ | $C_A + C_B$ | Exp. | Calc. |
| 610 | 280 | 890 | 8.5 | 8.3 |
| 462 | 139 | 601 | 9.4 | 9.0 |
| 334 | 57 | 391 | 10.4 | 9.8 |
| 159 | 10 | 169 | 11.3 | 11.3 |

Note that the common simple procedure of correlating total rate with total reactant concentration would lead to the rate *increasing* with *decreasing* concentration (i.e., a negative order). This effect would be rather suspect as a basis for design. In order to investigate this closer, data on the hydrogenation rates of $A$ and $B$ alone were measured, and they appeared to be zero order reactions with rate constants:

Hydrogenation rate of $A$ alone:

$$(r_A) = 12.9 \tag{a}$$

Hydrogenation rate of $B$ alone:

$$(r_B) = 6.7 \tag{b}$$

Also, $B$ is more strongly adsorbed than $A$, and the ratio of equilibrium constants is

$$K_A/K_B = 0.18 \tag{c}$$

Our problem is to explain all of these features with a consistent rate equation.

Consider a simple chemisorption scheme with the surface reaction controlling. For $A$ reacting alone,

$$A + l \rightleftharpoons Al \qquad C_{Al} = K_A C_A C_l \tag{d}$$

where concentrations have been used for the bulk liquid composition measure.

If the reaction product is weakly adsorbed, the total sites equation becomes

$$C_t \cong C_l + C_{Al} = C_l(1 + K_A C_A) \tag{e}$$

For a simple first-order, irreversible surface reaction:

$$Al \longrightarrow \text{product}, (r_A)_1 = k_1' C_{Al} \tag{f}$$

The use of Eq. (d) and (e) gives:

$$(r_A)_1 = \frac{k_1' C_t K_A C_A}{1 + K_A C_A} \tag{g}$$

In liquids, an approximately full coverage of adsorption sites is common (i.e., very large adsorbed concentrations), which means that $K_A C_A \gg 1$, and Eq. (g) becomes

$$(r_A)_1 = k_1' C_t = k_1$$
$$= 12.9 \, [\text{Eq. (a)}] \tag{h}$$

Thus, the zero order behavior of $A$ alone is rationalized. Similarly, for $B$ alone,

$$(r_B)_1 = \frac{k_2' C_t K_B C_B}{1 + K_B C_B} \tag{i}$$

$$k_2' C_t = k_2$$
$$= 6.7 \, [\text{Eq. (b)}] \tag{j}$$

Now for both reactions occuring simultaneously,

$$C_t = C_l + C_{Al} + C_{Bl}$$
$$= C_l(1 + K_A C_A + K_B C_B)$$
$$\cong C_l(K_A C_A + K_B C_B) \tag{k}$$

and

$$r_A = k_1' C_{Al}$$

$$= \frac{k_1 K_A C_A}{K_A C_A + K_B C_B} \tag{l}$$

$$r_B = k_2' C_{Bl}$$

$$= \frac{k_2 K_B C_B}{K_A C_A + K_B C_B} \tag{m}$$

The total rate is given by

$$r = r_A + r_B = \frac{k_1 K_A C_A + k_2 K_B C_B}{K_A C_A + K_B C_B} \tag{n}$$

$$= \frac{(k_1) \dfrac{K_A}{K_B} \dfrac{C_A}{C_B} + (k_2)}{\dfrac{K_A}{K_B} \dfrac{C_A}{C_B} + 1}$$

$$= \frac{12.9(0.18) \dfrac{C_A}{C_B} + 6.7}{0.18 \dfrac{C_A}{C_B} + 1} \tag{o}$$

If the values of $C_A$ and $C_B$ given in the cohydrogenation data table are substituted into Eq. o, it is found that the total rate values given in that table are predicted. In addition to illustrating an adsorption scheme for a real reaction, this example also shows that for some cases the observed phenomena can only be rationally explained by these ideas. Some parts of the data could be empirically correlated [zero and negative (?) orders] without any theory, but the adsorption scheme can explain all the data.

---

Let us now consider a more complicated reaction and devise the chemisorption reaction rate form. Dehydrogenation reactions are of the form

$$A \rightleftharpoons R + S$$

and a specific example will be discussed later. The fluid phase composition will here be expressed in partial pressures rather than concentrations, as is the custom in adsorption work for gases. Assume that the adsorption of $A$ is rate controlling, so that for the chemisorption step,

$$A + l \; \rightleftharpoons \; Al \qquad C_{Al} \neq C_l K_A p_A \qquad (2.2\text{-}13)$$

for the reaction step,

$$Al + l \; \rightleftharpoons \; Rl + Sl \qquad K_{sr} = \frac{C_{Rl} C_{Sl}}{C_{Al} C_l} \qquad (2.2\text{-}14)$$

and for the desorption steps,

$$Rl \; \rightleftharpoons \; R + l \qquad C_{Rl} = C_l K_R p_R \qquad (2.2\text{-}15)$$

$$Sl \; \rightleftharpoons \; S + l \qquad C_{Sl} = C_l K_S p_S \qquad (2.2\text{-}16)$$

The total concentration of active sites is

$$C_t = C_l + C_{Al} + C_{Rl} + C_{Sl}$$

$$= C_l + \frac{C_l K_R p_R C_l K_S p_S}{K_{sr} C_l} + C_l K_R p_R + C_l K_S p_S$$

$$= C_l \left( 1 + \frac{K_A}{K} p_R p_S + K_R p_R + K_S p_S \right) \qquad (2.2\text{-}17)$$

where the overall equilibrium relation $K = K_A K_{sr}/K_R K_S$ was used in the last step.

Equations 2.2-14 to 17 are now substituted into the rate equation for adsorption,

$$r_A = k'_A \left( p_A C_l - \frac{C_{Al}}{K_A} \right)$$

to give

$$r_A = \frac{k_A(p_A - p_R p_S / K)}{1 + \dfrac{K_A}{K} p_R p_S + K_R p_R + K_S p_S} \qquad (2.2\text{-}18)$$

Equation 2.2-18 is the kinetic equation of the reaction $A \rightleftharpoons R + S$ under the assumption that the adsorption is of the type $A + l \rightleftharpoons Al$ (i.e., without dissociation of $A$), and is of second order to the right, first order to the left, and is the rate determining step of the process. The form of the kinetic equation would be different if it had been assumed that step 2—the reaction itself—or step 3—the desorption— is the rate-determining step. The form would also have been different had the mechanism of adsorption been assumed different.

When the reaction on two adjacent sites is rate determining, the kinetic equation is as follows:

$$r_A = \frac{k_{sr} K_A(p_A - p_R p_S/K)}{(1 + K_A p_A + K_R p_R + K_S p_S)^2} \qquad (2.2\text{-}19)$$

where $k_{sr} = k'_{sr} s C_t$ and where $s$ = number of nearest neighbor sites.[1]

When the desorption of $R$ is the rate-determining step:

$$r_A = \frac{k_R K\left(\dfrac{p_A}{p_S} - \dfrac{p_R}{K}\right)}{1 + K_A p_A + K K_R \dfrac{p_A}{p_S} + K_S p_S} \qquad (2.2\text{-}20)$$

Kinetic equations for reactions catalyzed by solids based on the chemisorption mechanism may always be written as a combination of three groups:

a kinetic group: [e.g., in Eq. 2.2-18], $k'_A C_t = k_A$

a driving-force group: $(p_A - p_R p_S/K)$

an adsorption group: $1 + \dfrac{K_A}{K} p_R p_S + K_R p_R + K_S p_S$

such that the overall rate is:

$$= \frac{(\text{kinetic factor})(\text{driving-force group})}{(\text{adsorption group})} \qquad (2.2\text{-}21)$$

Summaries of these groups for various kinetic schemes are given in Table 2.2-1. (See Yang and Hougen [33].) The various kinetic terms $k$ and $kK$ all contain the

---

[1] For a reaction $A + B \rightarrow$, the proper driving force is based on the adsorbed concentration of $B$ that is *adjacent* to the adsorbed $A$:

$$C_{Bl}|_{adj} = (\text{no. nearest neighbors}) (\text{probability of } B \text{ adsorbed})$$

$$= s\left(\frac{C_{Bl}}{C_t}\right)$$

Then,

$$r_A = k'_{sr} C_{Al} C_{Bl}|_{adj}$$

$$= \frac{k'_{sr}(C_t K_A p_A)\left(\dfrac{s}{C_t} C_t K_B p_B\right)}{(1 + K_A p_A + K_B p_B + \cdots)^2}$$

$$= \frac{k'_{sr} s C_t K_A K_B p_A p_B}{(1 + K_A p_A + K_B p_B + \cdots)^2}$$

See Hougen and Watson [27] for further details. Similar reasoning leads to Eq. 2.2-19.

*Table 2.2-1  Groups in kinetic equations for reactions on solid catalysts*

### Driving-Force Groups

| Reaction | $A \rightleftharpoons R$ | $A \rightleftharpoons R + S$ | $A + B \rightleftharpoons R$ | $A + B \rightleftharpoons R + S$ |
|---|---|---|---|---|
| Adsorption of $A$ controlling | $p_A - \dfrac{p_R}{K}$ | $p_A - \dfrac{p_R p_S}{K}$ | $p_A - \dfrac{p_R}{K p_B}$ | $p_A - \dfrac{p_R p_S}{K p_B}$ |
| Adsorption of $B$ controlling | $0$ | $0$ | $p_B - \dfrac{p_R}{K p_A}$ | $p_B - \dfrac{p_R p_S}{K p_A}$ |
| Desorption of $R$ controlling | $p_A - \dfrac{p_R}{K}$ | $\dfrac{p_A}{p_S} - \dfrac{p_R}{K}$ | $p_A p_B - \dfrac{p_R}{K}$ | $\dfrac{p_A p_B}{p_S} - \dfrac{p_R}{K}$ |
| Surface reaction controlling | $p_A - \dfrac{p_R}{K}$ | $p_A - \dfrac{p_R p_S}{K}$ | $p_A p_B - \dfrac{p_R}{K}$ | $p_A p_B - \dfrac{p_R p_S}{K}$ |
| Impact of $A$ controlling ($A$ not adsorbed) | $0$ | $0$ | $p_A p_B - \dfrac{p_R}{K}$ | $p_A p_B - \dfrac{p_R p_S}{K}$ |
| Homogeneous reaction controlling | $p_A - \dfrac{p_R}{K}$ | $p_A - \dfrac{p_R p_S}{K}$ | $p_A p_B - \dfrac{p_R}{K}$ | $p_A p_B - \dfrac{p_R p_S}{K}$ |

### Replacements in the General Adsorption Groups
$$(1 + K_A p_A + K_B p_B + K_R p_R + K_S p_S + K_I p_I)^n$$

| Reaction | $A \rightleftharpoons R$ | $A \rightleftharpoons R + S$ | $A + B \rightleftharpoons R$ | $A + B \rightleftharpoons R + S$ |
|---|---|---|---|---|
| Where adsorption of $A$ is rate controlling, replace $K_A p_A$ by | $\dfrac{K_A p_R}{K}$ | $\dfrac{K_A p_R p_S}{K}$ | $\dfrac{K_A p_R}{K p_B}$ | $\dfrac{K_A p_R p_S}{K p_B}$ |
| Where adsorption of $B$ is rate controlling, replace $K_B p_B$ by | $0$ | $0$ | $\dfrac{K_B p_R}{K p_A}$ | $\dfrac{K_B p_R p_S}{K p_A}$ |
| Where desorption of $R$ is rate controlling, replace $K_R p_R$ by | $K K_R p_A$ | $K K_R \dfrac{p_A}{p_S}$ | $K K_R p_S p_B$ | $K K_R \dfrac{p_A p_B}{p_S}$ |
| Where adsorption of $A$ is rate controlling with dissociation of $A$, replace $K_A p_A$ by | $\sqrt{\dfrac{K_A p_R}{K}}$ | $\sqrt{\dfrac{K_A p_R p_S}{K}}$ | $\sqrt{\dfrac{K_A p_R}{K p_B}}$ | $\sqrt{\dfrac{K_A p_R p_S}{K p_B}}$ |
| Where equilibrium adsorption of $A$ takes place with dissociation of $A$, replace $K_A p_A$ by and similarly for other components adsorbed with dissociation | $\sqrt{K_A p_A}$ | $\sqrt{K_A p_A}$ | $\sqrt{K_A p_A}$ | $\sqrt{K_A p_A}$ |

Table 2.2-1 (Continued)

| Where $A$ is not adsorbed, replace $K_A p_A$ by | 0 | 0 | 0 | 0 |
|---|---|---|---|---|
| and similarly for other components that are not adsorbed | | | | |

## Kinetic Groups

| Adsorption of $A$ controlling | $k_A$ |
|---|---|
| Adsorption of $B$ controlling | $k_B$ |
| Desorption of $R$ controlling | $k_R K$ |
| Adsorption of $A$ controlling with dissociation | $k_A$ |
| Impact of $A$ controlling | $k_A K_B$ |
| Homogeneous reaction controlling | $k$ |

### Surface Reaction Controlling

| | $A \rightleftharpoons R$ | $A \rightleftharpoons R + S$ | $A + B \rightleftharpoons R$ | $A + B \rightleftharpoons R + S$ |
|---|---|---|---|---|
| Without dissociation | $k_{sr} K_A$ | $k_{sr} K_A$ | $k_{sr} K_A K_B$ | $k_{sr} K_A K_B$ |
| With dissociation of $A$ | $k_{sr} K_A$ | $k_{sr} K_A$ | $k_{sr} K_A K_B$ | $k_{sr} K_A K_B$ |
| $B$ not adsorbed | $k_{sr} K_A$ | $k_{sr} K_A$ | $k_{sr} K_A$ | $k_{sr} K_A$ |
| $B$ not adsorbed, $A$ dissociated | $k_{sr} K_A$ | $k_{sr} K_A$ | $k_{sr} K_A$ | $k_{sr} K_A$ |

## Exponents of Adsorption Groups

| Adsorption of $A$ controlling without dissociation | $n = 1$ |
|---|---|
| Desorption of $R$ controlling | $n = 1$ |
| Adsorption of $A$ controlling with dissociation | $n = 2$ |
| Impact of $A$ without dissociation $A + B \rightleftharpoons R$ | $n = 1$ |
| Impact of $A$ without dissociation $A + B \rightleftharpoons R + S$ | $n = 2$ |
| Homogeneous reaction | $n = 0$ |

### Surface Reaction Controlling

| | $A \rightleftharpoons R$ | $A \rightleftharpoons R + S$ | $A + B \rightleftharpoons R$ | $A + B \rightleftharpoons R + S$ |
|---|---|---|---|---|
| No dissociation of $A$ | 1 | 2 | 2 | 2 |
| Dissociation of $A$ | 2 | 2 | 3 | 3 |
| Dissociation of $A$ ($B$ not adsorbed) | 2 | 2 | 2 | 2 |
| No dissociation of $A$ ($B$ not adsorbed) | 1 | 2 | 1 | 2 |

From Yang and Hougen [33].

total number of active sites, $C_t$. Some of them also contain the number of adjacent active sites, $s$ or $s/2$ or $s(s - 1)$. Both $C_t$ and $s$ are usually not known and therefore they are not explicitly written in these groups. They are characteristic for a given catalytic system, however. An example of the use of the Yang–Hougen tables would be for the bimolecular reaction

$$A + B \rightleftharpoons R + S$$

For surface-reaction controlling:

$$r_A = \frac{C_t k_{sr} K_A K_B (p_A p_B - p_R p_S / K)}{(1 + K_A p_A + K_B p_B + K_R p_R + K_S p_S + K_I p_I)^2}$$

where $I$ = any adsorbable inert.

Finally, schemes alternate to the L-H-H-W mechanisms are the Rideal–Eley mechanisms, where one adsorbed species reacts with another species in the gas phase:

$$Al + B \longrightarrow Rl$$

These yield similar kinetic expressions, but they are somewhat different in detail.

## Example 2.2-2 Kinetics of Ethylene Oxidation on a Supported Silver Catalyst

Klugherz and Harriott [34] provide an interesting example of an extension of the standard L-H-H-W kinetic schemes. Based on several types of evidence, including lack of qualitative or quantitative fit of the experimental data with the usual kinetic equation forms, they postulated that the bare metal was not, in fact, the location of the active sites. For example, ethylene does not particularly adsorb on metallic silver. They further postulated that a certain portion of the silver metal contained one type of chemisorbed oxygen, which then provided the active sites for the main reaction. Further evidence for this type of behavior was provided by Marcinkowsky and Berty [35], and more detailed mechanism studies by Kenson [36].

The kinetic scheme was:

oxygen chemisorption:

$$2\,Ag + O_2 \rightleftharpoons 2l \text{ (equilibrium)} \qquad (a)$$

ethylene oxidation:

$$C_2H_4 + l \rightleftharpoons C_2H_4 \cdot l$$
$$O_2 + l \rightleftharpoons O_2 \cdot l \qquad (b)$$

or

$$O_2 + 2l \; \rightleftharpoons \; 2O \cdot l \tag{$b_1$}$$

$$C_2H_4 \cdot l + O_n \cdot l \; \longrightarrow \; C_2H_4O + 2l \tag{c}$$
$$\longrightarrow \; (CO_2 + H_2O)$$

Define:

$$C_t^* = \text{silver surface with atomic oxygen}$$

$$C_t - C_t^* = \text{silver surface that is bare}$$

$$p_O, p_P = \text{partial pressure of oxygen, respectively reaction products}$$

$$K_S, K_E, K_O, K_P = \text{adsorption equilibrium constants}$$

Then, if Eq. (a) is assumed to be in (dissociative) equilibrium, the results of Sec. 2.1 give

$$C_t^* = \frac{C_t\sqrt{K_S p_O}}{1 + \sqrt{K_S p_O}} \tag{d}$$

Based on various evidence about adsorption and desorption rates, a surface reaction controlling relation was chosen. (Sec. 2.3 presents more formal methods for such decisions.) Then, the other steps yield:

$$C_{El} = K_E p_E C_l^* \tag{e}$$

$$C_{Ol} = K_O p_O C_l^* \qquad \text{for Eq. (b)} \tag{f}$$

$$C_{Pl} = K_P p_P C_l^* \tag{g}$$

Finally, the total *active* site concentration, $C_t^*$, is

$$C_t^* = C_l^* + C_{El} + C_{Ol} + C_{Pl}$$
$$= C_l^*(1 + K_E P_E + K_O p_O + K_P p_P) \tag{h}$$

The rate equation is then found from

$$r = k'_{sr} C_{El} C_{Ol}|_{\text{adj}} \tag{i}$$

$$= \frac{s k'_{sr} C_t^* K_E P_E K_O p_O}{(1 + K_E P_E + K_O p_O + K_P p_P)^2} \tag{j}$$

$$= \frac{k_{Sr} K_E K_O K_S p_E p_O^2}{(1 + K_E P_E + K_O p_O + K_P p_P)^2 (1 + \sqrt{K_S p_O})^2} \tag{k}$$

Note that Eq. (k) has some different features from the usual L-H-H-W forms. At high ethylene pressures and low oxygen pressures, reaction orders for oxygen greater than unity are possible—this seems to be often observed in hydrocarbon

_____ CHEMICAL ENGINEERING KINETICS

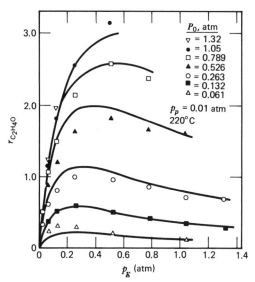

*Figure 1 Comparison of predicted relative rate of ethylene oxide formation based on Equation (k) with experimental data (lines are predicted rates) (from Klugherz and Harriott [37]).*

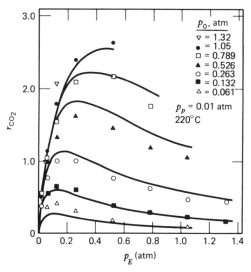

*Figure 2 Comparison of predicted relative rate of carbon dioxide formation with experimental data (lines are predicted rates) (from Klugherz and Harriott [37]).*

103

oxidation systems. Also, maxima in rates are predicted. Figures 1 and 2 illustrate the use of equations of the form of Eq. (k) for both ethylene and by-product $CO_2$ production.

For transformations consisting of sequences of reversible reactions, it is frequently possible to take advantage of the concept of the rate-determining step to simplify the kinetic equations. This is similar to the approach used above for single reactions consisting of a sequence of adsorption-, reaction- and desorption steps. Boudart [37] has discussed this approach and shown that catalytic sequences comprised of a large number of steps can frequently be treated as if they took place in at most two steps.

An example of this is provided by Hosten and Froment's study of the kinetics of $n$-pentane isomerization on a dual function Pt-$Al_2O_3$ reforming catalyst, carried out in the presence of hydrogen [38]. As discussed earlier in Sec. 2.1 of this chapter. this reaction involves a three-step sequence consisting of dehydrogenation, isomerization, and hydrogenation. The dehydrogenation and hydrogenation steps occur on platinum sites, represented by $l$; the isomerization step occurs on the acidic alumina sites, represented by $\sigma$. Each of these steps involves adsorption, surface reaction, and desorption so that the following mechanistic scheme can be written for the overall reaction:

Dehydrogenation

$$A + l \rightleftharpoons Al \qquad\qquad K_1 = c_{Al}/p_A \cdot c_l$$
$$Al + l \rightleftharpoons Ml + H_2l \qquad K_2 = c_{Ml} \cdot c_{H_2l}/c_{Al} \cdot c_l$$
$$H_2l \rightleftharpoons H_2 + l \qquad\qquad K_3 = p_{H_2} \cdot c_l/c_{H_2l}$$
$$Ml \rightleftharpoons M + l \qquad\qquad K_4 = p_M \cdot c_l/c_{Ml}$$

Isomerization

$$M + \sigma \rightleftharpoons M\sigma \qquad\qquad K_5 = c_{M\sigma}/p_M \cdot c_\sigma$$
$$M\sigma \rightleftharpoons N\sigma \qquad\qquad K_6 = c_{N\sigma}/c_{M\sigma}$$
$$N\sigma \rightleftharpoons N + \sigma \qquad\qquad K_7 = p_N \cdot c_\sigma/c_{N\sigma}$$

Hydrogenation

$$N + l \rightleftharpoons Nl \qquad\qquad K_8 = c_{Nl}/p_N \cdot c_l$$
$$H_2 + l \rightleftharpoons H_2l \qquad\qquad K_9 = c_{H_2l}/p_{H_2} \cdot c_l$$
$$Nl + H_2l \rightleftharpoons Bl + l \qquad K_{10} = c_{Bl} \cdot c_l/c_{Nl} \cdot c_{H_2l}$$
$$Bl \rightleftharpoons B + l \qquad\qquad K_{11} = p_B \cdot c_l/c_{Bl}$$

It was observed experimentally that the overall rate was independent of total pressure, and this provides a clue as to which step might be rate determining, When one of the steps of the dehydrogenation or hydrogenation reactions is considered to be rate determining, the corresponding overall rate equation is always pressure dependent. This results from the changing of the number of moles and was illustrated already by means of the treatment of dehydrogenation reactions given above. Since these pressure dependent rate equations are incompatible with the experimental results, it may be concluded that the isomerization step proper determines the rate of the overall reaction. Additional evidence for this conclusion was based on the enhancement of the overall rate by addition of chlorine, which only affects the acid site activity.

When the surface reaction step in the isomerization is rate determining, the overall reaction rate is given by

$$r = \frac{kK_5\left(p_M - \dfrac{p_N}{K_5 K_6 K_7}\right)}{1 + K_5 p_M + \dfrac{1}{K_7} p_N}$$

The total pressure dependence of the rate is only apparent. Provided the isomerization is rate controlling, $n$-pentene is in equilibrium with $n$-pentane/hydrogen and $i$-pentene with $i$-pentane/hydrogen. When the equilibrium relations are used, the partial pressures of the pentenes can be expressed in terms of the partial pressures of the pentanes and hydrogen, leading to

$$r = \frac{kK_5 K_D\left(p_A - \dfrac{p_B}{K}\right)}{p_{H_2} + K_5 K_D p_A + \dfrac{1}{K_7 K_H} p_B}$$

It is clear that written in this form the rate is independent of total pressure.[2]

For the case of adsorption of $n$-pentene on the acid sites rate determining, a similar derivation leads to

$$r = \frac{kK_D\left(p_A - \dfrac{p_B}{K}\right)}{p_{H_2} + \dfrac{1}{K_7 K_H}\left(\dfrac{1}{K_6} + 1\right)p_B}$$

and for desorption of $i$-pentene rate controlling:

$$r = \frac{kK_5 K_D K_6\left(p_A - \dfrac{p_B}{K}\right)}{p_{H_2} + K_5 K_D(1 + K_6)p_A}$$

[2] Where $K_D = K_1 K_2 K_3 K_4$ is the equilibrium constant for dehydrogenation, and $K_H = K_8 K_9 K_{10} K_{11}$ is the equilibrium constant for hydrogenation.

These two equations are also independent of total pressure. The discrimination between these three rate equations is illustrated in the next section.

## 2.3 Model Discrimination and Parameter Estimation _____

In a kinetic investigation it is not known *a priori* which is the rate-controlling step and therefore the form of the rate equation or the model. Also unknown, of course, are the values of the rate coefficient $k$ and of the adsorption coefficients $K_A, K_R, K_S, \ldots$, or, in other words, of the parameters of the model. A kinetic investigation, therefore, consists mainly of two parts: model discrimination and parameter estimation. This can ultimately only be based on experimental results.

### 2.3.a Experimental Reactors

Kinetic experiments on heterogeneous catalytic reactions are generally carried out in flow reactors.

This flow reactor may be of the tubular type illustrated schematically in Fig. 2.3.a-1 and generally operated in single pass. To keep the interpretation as simple as possible the flow is considered to be perfectly ordered with uniform velocity (of the "plug flow" type, as discussed in Chapter 9). This requires a sufficiently high velocity and a tube to particle diameter ratio of at least 10, to avoid too much short circuiting along the wall, where the void fraction is higher than in the core of the bed. The tube diameter should not be too large either, however, to avoid radial gradients of temperature and concentration, which again lead to complications in the interpretation, as will be shown in Chapter 11. For this reason, temperature gradients in the longitudinal (i.e., in the flow direction) should also be avoided. Although computers have enabled to handle nonisothermal situations up to a certain extent, determining the functional form of the rate equation is possible only on the basis of isothermal data. Isothermal conditions are not easily achieved with reactions having important heat effects. Care should be taken to minimize heat transfer resistance at the outside wall (for very exothermic reactions, for example, through the use of molten salts). Ultimately, however, no further gain can be realized since the most important resistance then becomes that at the inside wall, and this cannot be decreased at will, tied as it is to the process conditions. If isothermicity is still not achieved the only remaining possibility is to dilute the catalyst bed.

Excessive dilution has to be avoided as well: all the fluid streamlines should hit the same number of catalyst particles. Plug flow tubular reactors are generally operated in an integral way, that is, with relatively large conversion. This is achieved by choosing an amount of catalyst, $W(\text{kg})$, which is rather large with respect to the flow rate of the reference component $A$ at the inlet, $F_{A0}(\text{kmol/hr})$. By varying the ratio $W/F_{A0}$ a wide range of conversions $(x)$ may be obtained. To determine the

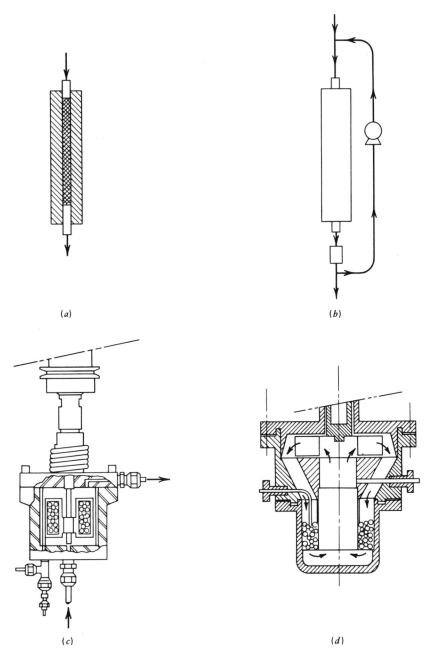

*Figure 2.3.a-1 Various types of experimental reactors. (a) Tubular reactor, (b) tubular reactor with recycle, (c) spinning basket reactor, and (d) reactor with internal recycle.*

reaction rate, the conversion versus $W/F_{A0}$ data pertaining to the same temperature have to be differentiated, as can be seen from the continuity equation for the reference component $A$ in this type of reactor (see Chapter 9)

$$F_{A0}\, dx_A = r_A\, dW$$

and over the whole reactor:

$$\frac{W}{F_{A0}} = \int_{x_{A1}}^{x_{A2}} \frac{dx_A}{r_A}$$

Plug flow reactors can also be operated in a differential way. In that case, the amount of catalyst is relatively small so that the conversion is limited and may be considered to occur at a nearly constant concentration of $A$. The continuity equation for $A$ then becomes

$$F_{A0}\Delta x_A = r_A W \tag{2.3-1}$$

and $r_A$ follows directly from the measured conversion.

Very accurate analytical methods are required in this case, of course. Furthermore, it is always a matter of debate how small the conversion has to be to fulfill the requirements. Figure 2.3.a-1 also shows a reactor with recycle. In kinetic investigations such a reactor is applied to come to a differential way of operation without excessive consumption of reactants. The recirculation may be internal too, also shown in Fig. 2.3.a-1. It is clear that in both cases it is possible to come to a constant concentration of the reactant over the catalyst bed. These conditions correspond to those of complete mixing, a concept that will be discussed in Chapter 10 and whereby the rate is also derived from Eq. 2.3.a-1. Another way of achieving complete mixing of the fluid is also shown in Fig. 2.3.a-1. In this reactor the catalyst is inserted into a basket which spins inside a vessel. Recycle reactors or spinning basket reactors present serious challenges of mechanical nature when they have to operate at high temperatures and pressures, as is often required with petrochemical and petroleum refining processes.

Transport phenomena can seriously interfere with the reaction itself and great care should be taken to eliminate these as much as possible in kinetic investigations.

Transfer resistances between the fluid and the solid, which will be discussed more quantitatively in Chapter 3, may be minimized by sufficient turbulence. With the tubular reactor this requires a sufficiently high flow velocity. This is not so simple to realize in laboratory equipment since the catalyst weight is often restricted to avoid a too-high consumption of reactant or to permit isothermal operation. With the spinning basket reactor the speed of rotation has to be high.

Transport resistances inside the particle, also discussed in detail in Chapter 3, can also obscure the true rate of reaction. It is very difficult to determine the true reaction kinetic equation in the presence of this effect. Suffice it to say here that

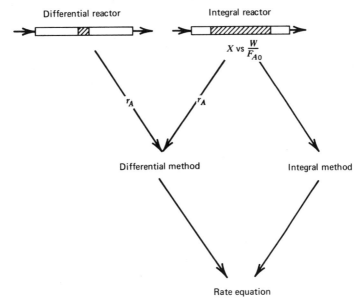

*Figure 2.3.a-2  Relation between differential and integral methods of kinetic analysis and differential and integral reactors.*

internal resistance can be decreased, for a given catalyst, by crushing the catalyst to reduce its dimensions. If the industrial reactor is to operate with a catalyst with which internal resistances are of importance the laboratory investigation will involve experiments at several particle diameters. The experimental results may be analyzed in two ways, as mentioned already in Chapter 1—by the differential method of kinetic analysis or by the integral method, which uses the $x$ versus $W/F_{A0}$ data. The results obtained in an integral reactor may be analyzed by the differential method provided the $x$ versus $W/F_{A0}$ curves are differentiated to get the rate, as illustrated by Fig. 2.3.a-2. An excellent review of laboratory reactors and their limitations is by Weekman [65]. Both methods will be discussed in the following section.

## 2.3.b  The Differential Method of Kinetic Analysis

A classical example of this method is the study of the hydrogenation of isooctenes of Hougen and Watson [27]. By considering all possible mechanisms and rate determining steps they set up 18 possible rate equations. Each equation was confronted with the experimental data and the criterion for acceptance of the model was that the parameters $k_1$, $K_A$, $K_R$, ..., had to be positive. In this way 16 of the 18 possible models could be rejected. The choice between the seventeenth and eighteenth was based on the goodness of fit. The way Hougen and Watson determined the parameters deserves further discussion. Let us take the reaction $A \rightleftarrows R + S$, with the surface reaction on dual sites the rate-controlling step, as an example. The Eq. 2.2-19 may be transformed into

$$y = a + bp_A + cp_R + dp_S \tag{2.3.b-1}$$

where

$$y = \sqrt{\dfrac{p_A - \dfrac{p_R p_S}{K}}{r_A}} \qquad a = \dfrac{1}{\sqrt{kK_A}} \qquad c = \dfrac{K_R}{\sqrt{kK_A}}$$

$$b = \dfrac{K_A}{\sqrt{kK_A}} \qquad d = \dfrac{K_S}{\sqrt{kK_A}}$$

Eq. 2.3.b-1 lends itself particularly well for determining $a$, $b$, $c$, and $d$, which are combinations of the parameters of Eq. 2.2-19, by linear regression. This method has been criticized: it is not sufficient to estimate the parameters but it also has to be shown that they are statistically significant. Furthermore, before rejecting a model because one or more parameters are negative it has to be shown that they are significantly negative. This leads to statistical calculations (e.g., of the confidence intervals).

Later, Yang and Hougen [33] proposed to discriminate on the basis of the total pressure dependence of the initial rate. Initial rates are measured, for example, with a feed consisting of only $A$ when no products have yet been formed (i.e., when $p_R = p_S = 0$). Nowadays this method is only one of the so-called "intrinsic parameter methods." (See Kittrell and Mezaki [39].) Equations 2.2-19, 2.2-18, and 2.2-20 are then simplified:

$$r_{A0} = \dfrac{k_{sr} K_A p_t}{(1 + K_A p_t)^2} \tag{2.3.b-2}$$

$$r_{A0} = k_A p_t \tag{2.3.b-3}$$

$$r_{A0} = \dfrac{k_R K p_t}{K K_R p_t} = \dfrac{k_R}{K_R} \tag{2.3.b-4}$$

Clearly these relations reveal by mere inspection which one is the rate-determining step (see Fig. 2.3.b-1). A more complete set of curves encountered when $r_{A0}$ is plotted versus the total pressure or versus the feed composition can be found in Yang and Hougen [39].

These methods are illustrated in what follows on the basis of the data of Franckaerts and Froment [40]. They studied the dehydrogenation of ethanol into acetaldehyde in an integral type flow reactor over a Cu-Co on asbestos catalyst.

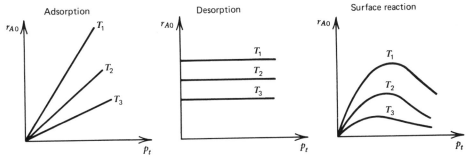

*Figure 2.3.b-1  Initial rate versus total pressure for various rate controlling steps.*

In most of the experiments, the binary azeotropic mixture ethanol-water, containing 13.5 mole percent of water was used. This was called "pure feed." A certain number of experiments were also carried out with so-called "mixed feed" containing ethanol, water and one of the reaction products, acetaldehyde, for reasons which will become obvious from what follows. Figure 2.3.b-2 shows an example of a conversion-$W/F_{A0}$ diagram at 1 atm with pure feed. Analogous diagrams were established at 3, 4, 7, and 10 atm, with both pure and mixed feed. From these results the initial rates were obtained by numerically differentiating the data at $x = 0$ and $W/F_{A0} = 0$. The temperature and total pressure dependence of this is shown in Fig. 2.3.b-3. This clearly shows that the surface reaction on dual sites is the rate-determining step. An even more critical test results from rearranging Eq. 2.3.b-2.

$$\sqrt{\frac{p_t}{r_{A0}}} = \frac{1}{\sqrt{kK_A}} + \frac{K_A}{\sqrt{kK_A}} p_t \qquad (2.3.b-5)$$

which leads to the plot shown in Fig. 2.3.b-4. $k$ and $K_A$ may be calculated from the intercept and the slope. Of course, it is even better to use linear regression methods. It is evident that the other parameters $K_R$ and $K_S$ can only be determined from the complete data, making use of the full equation 2.2-19:

$$r_A = \frac{kK_A\left(p_A - \dfrac{p_R p_S}{K}\right)}{(1 + K_A p_A + K_R p_R + K_S p_S + K_W p_W)^2} \qquad (2.3.b-6)$$

where the additional term $K_W p_W$ takes into account the presence of water in the feed and its possible adsorption. In order to determine all the constants from Eq. 2.3.b-6, it is transformed into

$$y = a + b p_A + c p_R + d p_S + e p_W \qquad (2.3.b-7)$$

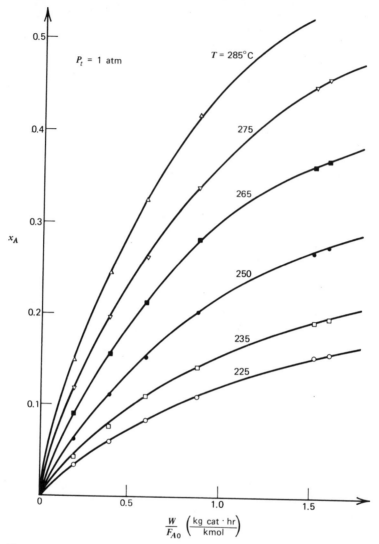

*Figure 2.3.b-2 Ethanol dehydrogenation. Conversion versus space time at various temperatures. $(W/F_{A_0})(kg\ cat.\ hr/kmol)$.*

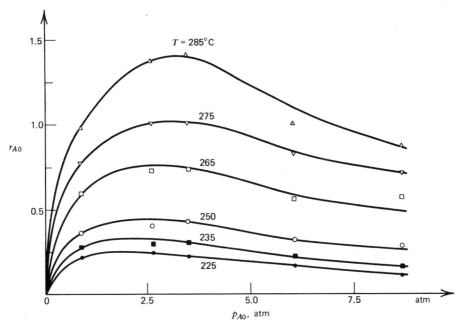

*Figure 2.3.b-3 Ethanol dehydrogenation. Initial rate versus total pressure at various temperatures.*

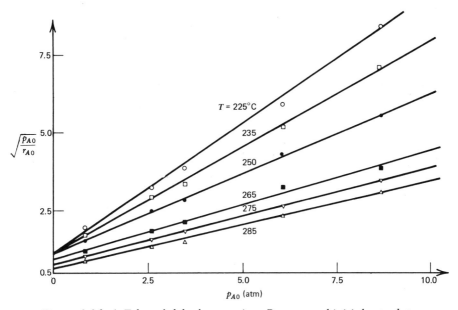

*Figure 2.3.b-4 Ethanol dehydrogenation. Rearranged initial rate data.*

where $y$, $a$, $b$, $c$, and $d$ have the form given in Eq. 2.3.b-1 and where

$$c = \frac{K_W}{\sqrt{kK_A}}$$

Note that for pure feed of $A$, the reaction stoichiometry dictates that $p_R = p_S$, and so from this type of data only the sum of $c + d = (K_R + K_S)/\sqrt{kK_A}$ can be determined. $K_R$ and $K_S$ can only be obtained individually when experimental results are available for which $p_R \neq p_S$. This requires mixed feeds containing $A$ and either $R$ or $S$ or both in unequal amounts. The equilibrium constant $K$ was obtained from thermodynamic data, and the partial pressure and rates were derived directly from the data. The groups $a$, $b$, $c$, $d$, and $e$ may then be estimated by linear regression. Further calculations lead to the 95 percent confidence limits, the $t$-test, which tests for the significance of a regression coefficient and an $F$-test, which determines if the regression is adequate. Franckaerts and Froment [40] performed these estimations and the statistical calculations for different sets of experimental data as shown in Fig. 2.3.b-5 in order to illustrate which kind of experiments should be performed to determine all parameters significantly.

Franckaerts and Froment also found $K_W$ to be nonsignificant so that they deleted it from the equations without affecting the values of the other parameters. The final results are shown in the Arrhenius plot of Fig. 2.3.b-6.

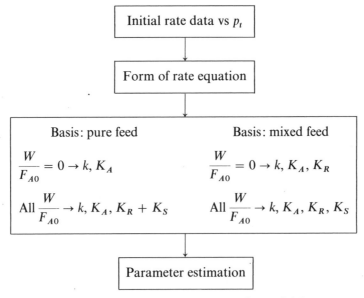

*Figure 2.3.b-5 Strategy of experimentation for model discrimination and parameter estimation.*

CHEMICAL ENGINEERING KINETICS

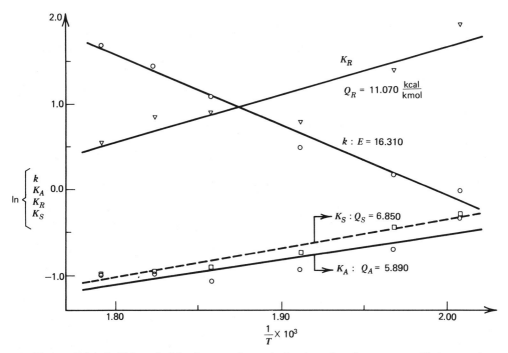

*Figure 2.3.b-6 Ethanol dehydrogenation. Arrhenius plot for rate coefficient and adsorption constants.*

From the standpoint of statistics, the transformation Eq. 2.2-19 into Eq. 2.3.b-1 and the determination of the parameters from this equation may be criticized. What is minimized by linear regression are the $\sum$(residuals)$^2$ between experimental and calculated $y$-values. The theory requires the error to be normally distributed. This may be true for $r_A$, but not necessarily for the group $\sqrt{(p_A - p_R p_S/K)/r_A}$ and this may, in principle, affect the values of $k$, $K_A$, $K_R$, $K_S$, .... However, when the rate equation is not rearranged, the regression is no longer linear, in general, and the minimization of the sum of squares of residuals becomes iterative. Search procedures are recommended for this (see Marquardt [41]). It is even possible to consider the data at all temperatures simultaneously. The Arrhenius law for the temperature dependence then enters into the equations and increases their nonlinear character.

## 2.3.c The Integral Method of Kinetic Analysis
The integration of the rate equation leads to

$$\frac{W}{F_{A0}} = f(x, k, K_A \cdots)$$

*Table 2.3.c-1  Comparison of the differential and integral methods at 285°C*

|  | $k$(kmol/kg cat.hr) | $K_A$(atm$^{-1}$) | $K_R$(atm$^{-1}$) | $K_S$(atm$^{-1}$) |
|---|---|---|---|---|
| Differential method with linear regression | 1.66 | 0.40 | 2.23 | 0.49 |
| Integral method with nonlinear regression | 2.00 | 0.39 | 3.17 | 0.47 |

What can be minimized in this case is either $\sum[(W/F_{A0}) - (\widehat{W/F_{A0}})]^2$ or $\sum(x - \hat{x})^2$.

The regression is generally nonlinear and in the second case the computations are even more complicated because the equation is implicit in $x$. Peterson and Lapidus [42] used the integral method with nonlinear regression on Franckaerts and Froment's data and found excellent agreement, as shown by Table 2.3.c-1. A further illustration of such agreement is based on Hosten and Froment's data on the isomerization of $n$-pentane [38] as analyzed by Froment and Mezaki [43].

The data indicated that the overall rate was independent of total pressure, supporting the conclusion that the isomerization step was rate controlling. Within this step, three partial steps may be distinguished: surface reaction, adsorption, or desorption, which could be rate controlling. The first was rejected because of (significant) negative parameter values. The adsorption and desorption rate expressions each contained two parameters—with values given in Table 2.3-c-2. Note here that discrimination based on the Yang–Hougen total pressure criterion is impossible in this case, since both rate equations are independent of total pressure.

In this case the expression $W/F_{A0}$ versus $f(x)$ was linear in two groups containing the parameters, so that linear regression was possible when the sum of squares on $W/F_{A0}$ was minimized. When the objective function was based on the conversion itself, an implicit equation had to be solved and the regression was nonlinear. Only approximate confidence intervals can then be calculated from a linearization of the model equation in the vicinity of the minimum of the objective function.

Again the agreement between the linear and nonlinear regression is excellent, which is probably due to the precision of the data. Poor data may give differences, but they probably do not deserve such a refined treatment, in any event.

The problem of estimation in algebraic equations that are nonlinear in the parameters was recently reviewed by Seinfeld [44] and by Froment [45] and [46], who give extensive lists of references. Standard textbooks dealing with this topic are by Wilde and Beightler [47], Beveridge and Schechter [48], Hoffmann and Hofmann [49], Himmelblau [50], and Rosenbrock and Storey [51].

The kinetic analysis of complex reaction systems requires more than one rate

*Table 2.3.c-2 Isomerization of* n-*pentane: comparison of methods for parameter estimation*

$$\text{n-pentane} \underset{\text{Pt}}{\rightleftharpoons} \text{n-pentene} \underset{\text{Al}_2\text{O}_3 + \text{Cl}_2}{\rightleftharpoons} \text{i-pentene} \underset{\text{Pt}}{\rightleftharpoons} \text{i-pentane}$$

**Integral Method**

Desorption rate controlling: $r = \dfrac{k\left(p_A - \dfrac{p_B}{K}\right)}{p_{H_2} + K_A p_A} \rightarrow \dfrac{W}{F_{A0}} = \dfrac{1}{k}(\alpha_1 + \alpha_2 K_A)$

| **Regression** | **Linear** | **Nonlinear** |
|---|---|---|
| $k$(kmol/kg cat. atm hr) | $0.93 \pm 0.21$ | $0.92 \pm 0.09$* |
| $K_A$(atm$^{-1}$) | $2.20 \pm 1.94$ | $2.28 \pm 0.95$* |

Sum of squares of residuals: $1.05 \left(\text{on } \dfrac{W}{F_{A0}}\right)$  $2.82 \times 10^{-3}$ (on $x$)

Adsorption rate controlling: $r = \dfrac{k\left(p_A - \dfrac{p_B}{K}\right)}{p_{H_2} + K_B p_B} \rightarrow \dfrac{W}{F_{A0}} = \dfrac{1}{k}(\alpha_1 + \alpha_3 K_B)$

| **Regression** | **Linear** | **Nonlinear** |
|---|---|---|
| $k$(kmol/kg cat. atm hr) | $0.89 \pm 0.10$ | $0.89 \pm 0.07$* |
| $K_B$(atm$^{-1}$) | $6.57 \pm 3.47$ | $8.50 \pm 2.78$* |

Sum of squares of residuals: $0.70 \left(\text{on } \dfrac{W}{F_{A0}}\right)$  $1.25 \times 10^{-3}$ (on $x$)

($A$ represents $n$-pentane; $B$ is: $i$-pentane. $\alpha_1$, $\alpha_2$, and $\alpha_3$ are functions of the feed composition, of $K$, $x$ and $\eta$, given in the original paper of Hosten and Froment [38]. $K$ is the equilibrium constant, $x$ the conversion, and $\eta$ the selectivity for the isomerization, accounting for a small fraction of the pentane converted by hydrocracking.)
* approximate 95 percent confidence interval.

or more than one exit concentration or conversion to be measured. It is then advisable to determine the parameters of the different rate equations by minimizing an objective function that is a generalization of the sum of squares of residuals used in the "single response" examples discussed so far, that is, the weighted least squares.

Several degrees of sophistication can be considered. Let it suffice to mention here the relatively simple case of the following objective function:

$$\mathscr{F} = \alpha\sum(x - \hat{x})^2 + \beta\sum(y - \hat{y})^2 + \gamma\sum(z - \hat{z})^2 + \ldots$$

$x$, $y$, $z$, ..., are the measured conversions (or "responses") and $\alpha$, $\beta$, $\gamma$, ..., are weighting factors that are inversely proportional with the variance of the corresponding response. To determine these variances requires replicate experiments, however. In the absence of these experiments, the weighting factors have to be chosen on the basis of sound judgment, providing that it is checked if the parameter estimates are independent of this choice. An example of multiresponse analysis of kinetic data of the complex $o$-xylene oxidation on a $V_2O_5$ catalyst using the integral method is given by Froment [52]. There are cases in which the continuity equation cannot be integrated analytically, but only numerically, in particular when several reactions are occurring simultaneously. Parameter estimation still remains possible, although it is complicated by the numerical integration of the differential equations in each iteration of the parameter matrix. Various techniques used to estimate parameters in algebraic equations that are nonlinear in these parameters may be used to optimize the iterations. One positive aspect of the numerical integration is that it yields the conversions directly, but this does not compensate for the increase in computing effort with respect to that required for the solution of an implicit algebraic equation.

Another approach, called "indirect," is often applied for estimation in the process control area but is equally applicable here. It proceeds with the necessary conditions for minimizing what is now an objective functional instead of an objective function and then attempts to determine parameter estimates that satisfy these conditions. The above-mentioned references also deal with these methods.

An example in which the kinetic equations had to be integrated numerically is given by De Pauw and Froment [53]. It concerns the isomerization of $n$-pentane accompanied by coke deposition. Another example is given by Emig, Hofmann, and Friedrich [54] and concerns the oxidation of methanol.

### 2.3.d Sequential Methods for Optimal Design of Experiments

Mechanistic model studies of the type discussed here have not always been convincing. Often the data were too scanty or not sufficiently precise, but, even more often, the design was poor so that the variables were not varied over a sufficient range. There is no fitting technique that can compensate for a poor experimental design.

In the design of experiments, much is just common sense. However, when the cases are complex, a rigorous, systematic approach may be required to achieve maximum efficiency. Until recently, most designs were of the factorial (i.e., of the a priori) type. During the last few years, however, sequential methods have been proposed that design an experiment taking advantage of the information and insight obtained from the previous experiments. Two types of sequential methods for optimal design have been proposed: optimal discrimination and optimal estimation.

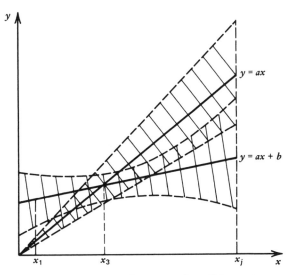

*Figure 2.3.d.1-1  Overlapping of confidence intervals.*

### 2.3.d.1  Optimal Sequential Discrimination

Suppose one has to discriminate between two models $y^{(1)} = ax + b$ and $y^{(2)} = ax$, where $y$ is a dependent variable that can be a conversion or a rate. At first sight it is logical to plan an experiment where a maximum difference or "divergence" can be expected. It can be seen from Fig. 2.3.d.1-1 that for the given example this would be for values of the independent variable $x$ close to zero and $x_j$, but surely not in the vicinity of $x_3$. Suppose $n - 1$ experiments have been performed at $n - 1$ settings of $x$, so that estimates for $a$ and $b$ can be obtained. To plan the $n$th experiment the region of interest ("operability region") on the $x$-axis is divided into a certain number of intervals. The grid points are numbered $i$. Then the estimates $\hat{y}^{(1)}$ and $\hat{y}^{(2)}$ are computed for each grid point. Then the divergence between the estimates of the function $y$ for each of the two models

$$D_{i,n} = (\hat{y}_i^{(1)} - \hat{y}_i^{(2)})^2 \qquad (2.3.d.1\text{-}1)$$

is calculated and the $n$th experiment is performed with settings corresponding to the grid point on the $x$-axis where $D_{i,n}$ is maximum.

The criterion is easily extended to more than two models, as follows:

$$D_{i,n} = \sum_{k=1}^{m} \sum_{l=k+1}^{m} (\hat{y}_i^{(k)} - \hat{y}_i^{(l)})^2 \qquad (2.3.d.1\text{-}2)$$

where $k$ and $l$ stand for the models and the index $i$ for the grid point. The double summation ensures that each model is taken consecutively as a reference.

Box and Hill [55] argued that the criterion would have to account for the uncertainties associated with the model predictions, that is, the variances $\sigma_j{}^2$, since the divergence might be obscured by eventual large uncertainties in the model predictions in a given range of the settings (Fig. 2.3.d.1-1). Starting from information theory Box and Hill derived the following expression for the divergence between two rival models:

$$
D_{i,n} = \pi_{1,n-1}\pi_{2,n-1}\left[\frac{(\sigma_2{}^2 - \sigma_1{}^2)^2}{(\sigma^2 + \sigma_1{}^2)(\sigma^2 + \sigma_2{}^2)}\right.
$$

$$
\left. + (\hat{y}_i{}^{(1)} - \hat{y}_i{}^{(2)})^2\left(\frac{1}{\sigma^2 + \sigma_1{}^2} + \frac{1}{\sigma^2 + \sigma_2{}^2}\right)\right] \qquad (2.3.d.1-3)
$$

$\sigma^2$ is the variance of the observations $y$ and $\sigma_1{}^2$, respectively. $\sigma_2{}^2$, is the variance of the estimated value of the dependent variable for the $i$th grid point under model 1, respectively model 2. $\pi_{1,n-1}$ is the prior probability of the model 1 after $n-1$ experiments. The product $\pi_{i,n-1}\pi_{2,n-1}$ is a factor that gives a greater weight to the model with the greatest probability after $n-1$ experiments. After the $n$th experiment has been performed at the settings of the independent variables where $D_{i,n}$ is a maximum the adequacy of each of the models remains to be tested. Box and Hill [55] and Box and Henson [66] expressed the adequacy in terms of the posterior probabilities. These will serve as prior probabilities in the design of the $n+1$th experiment. We will not go any further into this concept, which requires an insight into Bayesian probability theory. The approach proposed by Hosten and Froment [56] uses elementary statistical principles. The underlying idea is that the minimum sum of squares of residuals divided by the appropriate number of degrees of freedom is an unbiased estimate of the experimental error variance for the *correct* mathematical model only. For all other models this quantity is biased due to a lack of fit of the model, The criterion for adequacy, therefore, consists in testing the homogeneity of the estimates of the experimental error variance obtained from each of the rival models. This is done by means of Bartlett's $\chi^2$-test.

$$
\chi_c{}^2 = \frac{(\ln \bar{s}^2)\sum\limits_{i=1}^{m}(\text{D.F.})_i - \sum\limits_{i=1}^{m}(\text{D.F.})_i \ln s_i{}^2}{1 + \dfrac{1}{3(m-1)}\left[\sum\limits_{i=1}^{m}\dfrac{1}{(\text{D.F.})_i} - \dfrac{1}{\sum\limits_{i=1}^{m}(\text{D.F.})_i}\right]} \qquad (2.3.d.1-4)
$$

In Eq. 2.3.d.1-4, $\bar{s}^2$ is the pooled estimate of variance plus lack of fit; $(\text{D.F.})_i$ is the degrees of freedom associated with the $i$th estimate of error variance plus lack of fit, $s_i{}^2$; and $m$ is the number of rival models.

Whenever $\chi_c{}^2$ exceeds the tabulated value the model corresponding to the largest estimate of error variance is discarded and $\chi_c{}^2$ is recalculated. Another

_____ CHEMICAL ENGINEERING KINETICS

model may be discarded when $\chi_c^2$ exceeds the tabulated value and so on. Applying statistics to nonlinear models requires the model to be locally linear. For the particular application considered here this means that the residual mean square distribution is approximated to a reasonable extent by the $\chi^2$ distribution. Furthermore, care has to be taken with outliers, since $\chi^2$ appears to be rather sensitive to departures of the data from normality. In the example given below this was taken care of by starting the elimination from scratch again after each experiment. Finally, the theory requires the variance estimates that are tested on homogeneity to be statistically independent. It is hard to say how far this restriction is fulfilled. From the examples given, which have a widely different character, it would seem that the procedure is efficient and reliable.

### Example 2.3.d.1-1 Model discrimination in the dehydrogenation of 1-butene into butadiene

Dumez and Froment studied the dehydrogenation of 1-butene into butadiene on a chromium-aluminium oxide catalyst in a differential reactor [57]. This work is probably the first in which the experimental program was actually and uniquely based on a sequential discrimination procedure. The reader is also referred to a more detailed treatment, Dumez, Hosten, and Froment [58]. The following mechanisms were considered to be plausible:

(a) *Atomic Dehydrogenation; Surface Recombination of Hydrogen*

1. $B + l \rightleftharpoons Bl$      $a_1$
2. $Bl + l \rightleftharpoons Ml + Hl$      $a_2$
3. $Ml + l \rightleftharpoons Dl + Hl$      $a_3$
4. $Dl \rightleftharpoons D + l$      $a_4$
5. $2Hl \rightleftharpoons H_2l + l$
6. $H_2l \rightleftharpoons H_2 + l$

where $B = n$-butene; $D = $ butadiene; $H_2 = $ hydrogen, $M = $ an intermediate complex

(b) *Atomic Dehydrogenation; Gas Phase Hydrogen Recombination*

1. $B + l \rightleftharpoons Bl$      $b_1$
2. $Bl + l \rightleftharpoons Ml + Hl$      $b_2$
3. $Ml + l \rightleftharpoons Dl + Hl$      $b_3$
4. $Dl \rightleftharpoons D + l$      $b_4$
5. $2Hl \rightleftharpoons H_2 + 2l$

(c) *Molecular Dehydrogenation*

1. $B + l \rightleftharpoons Bl$ $\qquad c_1$
2. $Bl + l \rightleftharpoons Dl + H_2l$ $\qquad c_2$
3. $Dl \rightleftharpoons D + l$ $\qquad c_3$
4. $H_2l \rightleftharpoons H_2 + l$

(d) *Atomic Dehydrogenation; Intermediate Complex with Short Lifetime; Surface Recombination of Hydrogen*

1. $B + l \rightleftharpoons Bl$ $\qquad d_1$
2. $Bl + 2l \rightleftharpoons Dl + 2Hl$ $\qquad d_2$
3. $Dl \rightleftharpoons D + l$
4. $2Hl \rightleftharpoons H_2l + l$
5. $H_2l \rightleftharpoons H_2 + l$

(e) *As in (d) but with Gas Phase Hydrogen Recombination*

1. $B + l \rightleftharpoons Bl$ $\qquad e_1$
2. $Bl + 2l \rightleftharpoons Dl + 2Hl$ $\qquad e_2$
3. $Dl \rightleftharpoons D + l$
4. $2Hl \rightleftharpoons H_2 + 2l$

For each of these mechanisms several rate equations may be deduced, depending on the rate-determining step that is postulated. Fifteen possible rate equations were retained, corresponding to the rate-determining steps $a_1 \ldots a_4, b_1 \ldots b_4,$ $c_1 \ldots c_3, d_1, d_2, e_1$ and $e_2$ respectively. These equations will not be given here, except the finally retained one, by way of example.

$$r = \frac{k_1 K_1 s C_t \left( p_B - \dfrac{p_{H_2} \cdot p_D}{K} \right)}{\left( 1 + K_1 p_B + \dfrac{p_D}{K_3} + \dfrac{p_{H_2}}{K_4} \right)^2}$$

The discrimination was based on the divergence criterion of Eq. 2.3.d.1-2 in which $y$ is replaced by $r$ and model adequacy criterion Eq. 2.3.d.1-4 utilized. Since the experiments were performed in a differential reactor the independent variables were the partial pressures of butene, $p_B$, butadiene $p_D$ and hydrogen, $p_{H_2}$. The operability region for the experiments at 525°C is shown in Fig. 1. The equilibrium surface is also represented in this figure by means of hyperbola parallel to the $p_D p_{H_2}$-plane and straight lines parallel to the $p_B p_{H_2}$- and $p_B p_D$-plane respectively. Possible experiments are marked with a white dot. Experimental settings too close to the equilibrium were avoided, for obvious reasons. The maximum number of parameters in the possible models is six, so that at least seven preliminary

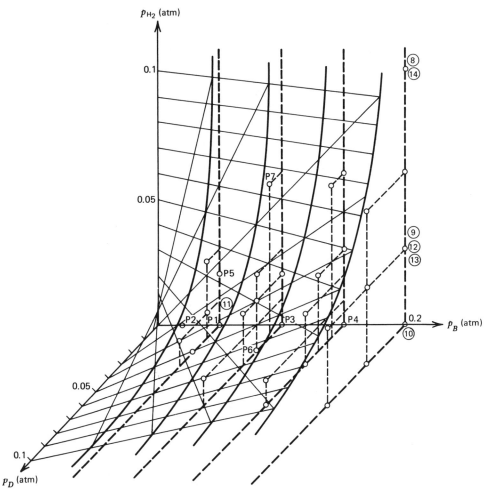

*Figure 1 Model discrimination in butene dehydrogenation. Operability region, equilibrium surface, location of preliminary and designed experiments at 525°C.*

experiments are required to estimate the parameters and start the discrimination procedure with Eq. 2.3.d.1-2. As can be seen from Table 1 after these seven preliminar experiments already the models $a_3$, $b_3$, $a_4$, $b_4$, and $c_3$ may be eliminated. The eighth experiment, which is the first of the designed ones, is carried out at the conditions represented by 8 in Fig. 1. The model adequacies are then recalculated. Note that after each experiment the elimination was started from scratch again to avoid discarding a model on the basis of one or more experiments with a biased error, especially in the early stages of discrimination.

Table 1 *Dehydrogenation of 1-butene. Evolution of sequential model discrimination*

| Number of Designed Experiments | 0 | 1 | 2 | 3 | 4 | 5 | 6 | 7 |
|---|---|---|---|---|---|---|---|---|
| **Total Number of Experiments** | 7 | 8 | 9 | 10 | 11 | 12 | 13 | 14 |
| $(\chi^2)_t$ | | | | $\chi_c^2$ | | | | |
| 23.68 | 50.59 | 68.80 | 85.31 | 84.80 | 98.38 | 114.14 | 129.44 | 139.13 |
| Eliminated model (EM) | $a_3$ | $a_3$ | $a_3$ | $a_3$ | $a_3$ | $a_3$ | $a_3$ | $a_3$ |
| 22.36 | 43.66 | 58.78 | 72.61 | 69.55 | 80.84 | 93.82 | 106.40 | 113.60 |
| EM | $b_3$ | $b_3$ | $b_3$ | $b_3$ | $b_3$ | $b_3$ | $b_3$ | $b_3$ |
| 21.03 | 32.50 | 42.19 | 50.55 | 43.37 | 50.86 | 58.81 | 66.39 | 69.17 |
| EM | $a_4$ | $a_4$ | $a_4$ | $a_4$ | $a_4$ | $a_4$ | $a_4$ | $a_4$ |
| 19.68 | 27.36 | 35.44 | 42.47 | 36.67 | 43.81 | 50.87 | 57.58 | 59.70 |
| EM | $b_4$ | $b_4$ | $b_4$ | $b_4$ | $b_4$ | $b_4$ | $b_4$ | $b_4$ |
| 18.31 | 20.18 | 25.82 | 30.72 | 28.03 | 34.75 | 40.40 | 45.75 | 46.92 |
| EM | $c_3$ | $c_3$ | $c_3$ | $c_3$ | $c_3$ | $c_3$ | $c_3$ | $c_3$ |
| 16.92 | 1.91 | 3.37 | 4.56 | 14.04 | 20.59 | 23.74 | 26.66 | 26.39 |
| EM | | | | | $a_1$ | $a_1$ | $a_1$ | $a_1$ |
| 15.51 | | | | | 18.83 | 21.79 | 24.53 | 24.39 |
| EM | | | | | $b_1$ | $b_1$ | $b_1$ | $b_1$ |
| 14.07 | | | | | 16.59 | 19.20 | 21.78 | 21.45 |
| EM | | | | | $d_1$ | $d_1$ | $d_1$ | $d_1$ |
| 12.59 | | | | | 13.95 | 16.23 | 18.32 | 18.18 |
| EM | | | | | $e_1$ | $e_1$ | $e_1$ | $e_1$ |
| i1.07 | | | | | 9.38 | 10.97 | 12.42 | 12.06 |
| EM | | | | | | | $c_1$ | $c_1$ |
| 9.49 | | | | | | | 0.44 | 1.28 |

After seven designed experiments or after a total of 14 experiments no further discrimination was possible between the dual-site rate-determining models $a_2$, $b_2$, $c_2$, $d_2$, and $e_2$, since the differences between these models were smaller than the experimental error. The models $a_2$, $b_2$, and $d_2$ were then eliminated because they contained at least one parameter that was not significantly different from zero at the 95 percent confidence level. It is interesting to note that none of the designed feed compositions contains butadiene. From the preliminary experiments it follows already that butadiene is strongly adsorbed. Consequently, it strongly reduces the rate of reaction and therefore the divergence. The design is based upon maximum divergence. Finally, it should be stressed how efficient sequential design procedures are for model discrimination. A classical experimental program, less conscious of the ultimate goal, would no doubt have involved a much more extensive experimental program. It is true that, at first sight, the limited number of experiments provides less feeling for the influence of the process variables on the rate or conversion, which is of course of great importance for practical application. Such information is easily generated a posteriori, however; the detailed response surface can be obtained by means of the computer, starting from the retained model.

---

### Example 2.3.d.1-2 Ethanol Dehydrogenation. Sequential Discrimination Using the Integral Method of Kinetic Analysis

The above example dealt with the design of an experimental program carried out in a differential reactor. When the data are obtained in an integral reactor it is more convenient to deal with the integrated form of the rate equation. This is illustrated in the present example, that also deals with real data, although the design is only applied a posteriori.

In the work of Franckaerts and Froment on ethanol dehydrogenation [40] three rate equations were retained. They were already referred to in Eqs. 2.2-18, 2.2-19, and 2.2-20. The authors discriminated between these models on the basis of a classical experimental program. This allowed the calculation of the initial rates and these were then plotted versus the total pressure.

Assuming the tubular reactor to be ideal and isothermal the continuity equation for ethanol may be written:

$$F_{A0}\, dx = r_A dW \tag{a}$$

where $r_A$ may be given by either Eqs. 2.2-18, 2.2-19, or 2.2-20, in which the partial pressures are expressed in terms of the conversion of ethanol. Equation 2.2-19 then takes the form

$$r_A = \frac{kK_A\left[\dfrac{1-x}{a+x}p_t - \dfrac{x^2}{(a+x)^2 K}p_t^{\,2}\right]}{\left[1 + K_A\dfrac{1-x}{a+x}p_t + (K_R + K_S)\dfrac{x}{a+x}p_t\right]^2}$$

where $a = 1 + 0.155$ and $0.155$ is the molar ratio of water to ethanol in the feed. What is measured in an integral reactor is the exit conversion, so that Eq. (a) has to be integrated for the three rival rate equations to give an expression of the form

$$\frac{W}{F_{A0}} = f(k, K_A, K_R, K_S, p_t, x) \tag{b}$$

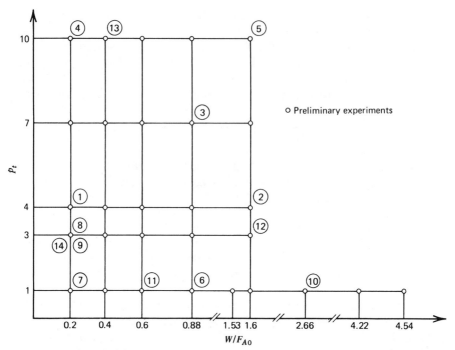

*Figure 1 Ethanol dehydrogenation. Operability region, location of preliminary and of designed experiments for optimal discrimination. Preliminary experiments 1, 2, 3, 4, 5.*

which is implicit in the dependent variable, the conversion, $x$. The independent variables are $W/F_{A0}$ and $p_t$. Equations like (b) are generally rather complex. By way of example, for the rate equation Eq. 2.2-19, the integrated continuity equation becomes

$$\frac{W}{F} = \frac{1}{kK_A}(D_1 + D_2 + D_3)$$

where

$$D_1 = \frac{2A_1}{\sqrt{4A_2 C_2 - B_2{}^2}}\left[\text{arc } tg\left(\frac{2C_2 x + B_2}{\sqrt{4A_2 C_2 - B_2{}^2}}\right) - \text{arc } tg\left(\frac{B_2}{\sqrt{4A_2 C_2 - B_2{}^2}}\right)\right]$$

$$D_2 = \frac{B_1}{2C_2}\left[\log\left(\frac{A_2 + B_2 x + C_2 x^2}{A_2}\right) - \frac{B_2 D_1}{A_1}\right]$$

$$D_3 = C_1\left[\frac{x}{C_2} - \frac{B_2}{2C_2{}^2}\log\left(\frac{A_2 + B_2 x + C_2 x^2}{A_2}\right) + \left(\frac{B_2{}^2 - 2A_2 C_2}{2C_2{}^2}\right)\cdot\frac{D_1}{A_1}\right] \qquad \text{(c)}$$

when $4A_2 C_2 - B_2{}^2$ is positive.

CHEMICAL ENGINEERING KINETICS

$$A_1 = (1.155 + K_A p_t)^2$$
$$B_1 = 2.31 - 2K_A^2 p_t^2 - 0.31 K_A p_t + 2K_A(K_R + K_S)p_t^2$$
$$C_1 = [1 - K_A p_t + (K_R + K_S)p_t]^2$$
$$A_2 = 1.155 \, p_t$$
$$B_2 = -0.155 \, p_t$$

$$C_2 = -p_t\left(1 + \frac{p_t}{K}\right)$$

Figure 1 shows the operability region in the $p_t - (W/F_{A0})$ plane at 275°C.

Since Eq. (b) contains four parameters, at least five preliminary runs have to be performed. Then the parameters are calculated by means of nonlinear regression, minimizing the sum of squares of residuals of the true dependent variable, $x$— preferably not of $W/F_{A0}$, as mentioned already. This requires a routine for solving the implicit equation for $x$, of course.

Next, the first experiment is designed using the criterion Eq. 2.3.d.1-2 in which $y$ now stands for the conversion, $x$. Then the adequacy criterion Eq. 2.3.d.1-4 is applied. The design is given in Table 1. Here too the adsorption and desorption

Table 1 Sequential design for optimum discrimination in the de-hydrogenation of ethanol into acetaldehyde, using integral reactor data as such

| Experiment Number | $W/F_{A0}$ | $p_t$ | $x$ | $\chi_c^2$ | $\chi_{tab}^2$ | Delete Model |
|---|---|---|---|---|---|---|
| 1 | 0.2 | 4 | 0.14 | | | |
| 2 | 1.6 | 4 | 0.32 | | | |
| 3 | 0.88 | 7 | 0.214 | | | |
| 4 | 0.2 | 10 | 0.1 | | | |
| 5 | 1.6 | 10 | 0.229 | 2.81 | 5.99 | |
| 6 | 0.88 | 1 | 0.339 | 5.83 | | |
| 7 | 0.2 | 1 | 0.118 | 2.0 | | |
| 8 | 0.2 | 3 | 0.14 | 3.75 | | |
| 9 | 0.2 | 3 | 0.14 | 5.40 | | |
| 10 | 2.66 | 1 | 0.524 | 7.59 | 5.99 | Adsorption |
| | | | | 2.01 | 3.84 | |
| 11 | 0.6 | 1 | 0.262 | 3.07 | | |
| 12 | 1.6 | 3 | 0.352 | 3.42 | | |
| 13 | 0.4 | 10 | 0.148 | 3.64 | | |
| 14 | 0.2 | 3 | 0.14 | 4.60 | 3.84 | Desorption |

models are rejected and the model with surface reaction as rate determining step is retained. Again the designed experiments, encircled on the figure, are located on the borderline of the operability region. Note that the design procedure for sequential discrimination is applicable even when the continuity equation (a) cannot be integrated analytically, but only numerically. This problem is encountered quite often when dealing with complex reactions.

*2.3.d.2 Sequential Design Procedure for Optimal Parameter Estimation*

Even if model discrimination has been accomplished and one test model has been selected as being adequate, it is frequently necessary to obtain more precise estimates of the parameters than those determined from the discrimination procedure. Or the model may be given, from previous experience, so that only estimation is required. Box and co-workers developed a sequential design procedure for decreasing the amount of uncertainty associated with estimates of parameters. It aims at reducing the joint confidence volume associated with the estimates. An example of such a joint confidence region is shown in Fig. 2.3.d.2-1 for a rate equation with three parameters (from Kittrell [67]).

If the model is linear in the parameters each point on the surface of this volume, that is, each set of parameter values corresponding to a point on the surface, will lead to the same sum of squares of residuals. The example given in the figure is typical for rate equations of the Hougen and Watson type. The long, narrow

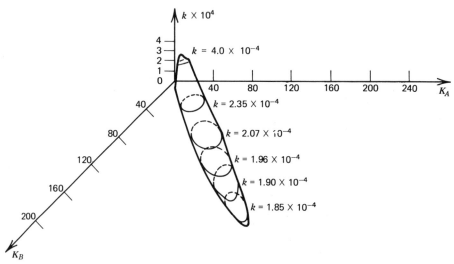

*Figure 2.3.d.2-1 Confidence region: heterogeneous rate equation with three parameters (after Kittrell [67]).*

CHEMICAL ENGINEERING KINETICS

shape results primarily from important covariance terms, that is, a high degree of correlation among the various parameter estimates. Widely varying values of the estimates will lead to the same overall fit of the equation to the data. The problem now is to choose the experimental settings in such way that the volume of the confidence region is minimized by a minimum number of experiments.

Let the rate be given by:

$$r = g(p_A, p_B, p_S \ldots ; k, K_A, K_B, \ldots)$$

or, more compactly,

$$r = g(\mathbf{p}, \mathbf{K})$$

Let the partial derivatives of $r$ with respect to any parameter, $K_i$, evaluated at the $u$th set of experimental conditions and taken at some set of parameter values $\mathbf{K}_0$ be given by $g_{u,i}$. Then,

$$g_{u,i} = \left. \frac{\partial g(\mathbf{p}_u, \mathbf{K})}{\partial K_i} \right|_{\mathbf{K} = \mathbf{K}_0}$$

After $n - 1$ experiments the matrix of these derivatives, $\mathbf{G}$, contains $n - 1$ rows and $\nabla$ columns ($\nabla$ parameters). When $\mathbf{G}^T$ is the transpose matrix of $\mathbf{G}$ the product $\mathbf{G}^T \cdot \mathbf{G}$ is a ($\nabla \times \nabla$) matrix. Box and Lucas [59] have shown that, under certain plausible assumptions, a choice of experimental settings for the $n$th experiment, which maximize the determinant of $\mathbf{G}^T\mathbf{G}$, will minimize the volume of the joint confidence region of the parameter estimates. The matrix $\mathbf{G}$, used in the planning of the $n$th experiment contains $n$ rows. The $n$th row is different for each of the grid points of the operability region. The $n$th experiment has to be carried out in that experimental setting where the determinant $\mathbf{G}^T\mathbf{G}$ is maximum. Then the parameters are reestimated. If the experimenter is not satisfied with the confidence volume another experiment is designed.

## Example 2.3.d.2-1 Sequential Design of Experiments for Optimal Parameter Estimation in n-pentane Isomerization. Integral Method of Kinetic Analysis

The method is illustrated for the adsorption rate controlling model for $n$-pentane isomerization. This rate equation contains two independent variables $p_A$ and $p_{H_2}$ or the $n$-pentane conversion and the ratio hydrogen/$n$-pentane. In reality these experiments were not planned according to this criterion. Thirteen experiments were carried out, shown in Fig. 1. This figure shows the limits on the experimental settings, that is, it shows the so-called operability region.

A grid is chosen through, or close to, the experimental settings to use the experimental results. Three preliminary, unplanned experiments are "performed"

Table 1 *n-pentane isomerization adsorption model. Sequential experimental design for optimal parameter determination*

| Case | Preliminary Runs | Planned Runs | $k$ | $2s(k)$ | $K_B$ | $2s(K_B)$ | $G^T \cdot G$ |
|------|------------------|--------------|-----|---------|-------|-----------|----------------|
| 1 | 108 | | | | | | |
|   | 121 | | | | | | |
|   | 111 | | 0.79 | 0.39 | 3.35 | 27.57 | $3.87 \times 10^{-3}$ |
|   | | 105 | 0.82 | 0.08 | 6.15 | 2.99 | $4.09 \times 10^{-1}$ |
|   | | 114 | 0.89 | 0.08 | 8.20 | 3.55 | $8.15 \times 10^{-1}$ |
|   | | 105 | 0.89 | 0.07 | 8.21 | 2.54 | 1.62 |
|   | | 114 | | | | | |
| 2 | 106 | | | | | | |
|   | 120 | | | | | | |
|   | 116 | | 0.79 | 0.18 | 2.39 | 7.62 | $2.54 \times 10^{-1}$ |
|   | | 105 | 0.87 | 0.19 | 6.32 | 6.92 | $5.98 \times 10^{-1}$ |
|   | | 114 | 0.89 | 0.12 | 7.31 | 4.78 | 1.27 |
|   | | 105 | | | | | |
| 3 | 106 | | | | | | |
|   | 120 | | | | | | |
|   | 116 | | | | | | |
|   | 109 | | | | | | |
|   | 105 | | 0.82 | 0.14 | 5.33 | 5.10 | 2.80 |
|   | | 114 | 0.87 | 0.12 | 6.96 | 5.01 | 1.85 |
|   | | 105 | 0.87 | 0.10 | 7.36 | 5.90 | 3.02 |
|   | | 114 | 0.87 | 0.08 | 7.86 | 3.42 | 4.11 |

13 unplanned experiments:
$$k = 0.89; \; 2s(k) = 0.10$$
$$K_B = 6.57; \; 2s(K_B) = 3.47$$

to calculate first estimates for the parameters. Then the fourth experiment is planned. The value of $G^T G$ is calculated in each point of the grid. The fourth experiment is performed at these values of the independent variables where the determinant is maximum.

The results are shown in Table 1 for three cases. The preliminary experiments for each case were chosen in a somewhat arbitrary manner in an attempt to investigate the sensitivity of the experimental design to the settings of the preliminary runs (i.e., the parameter estimates obtained from these runs). It can be seen that the designed experiments always fall on either of the two settings 105 and 114, both on the limits of the operability region. The design seems to be insensitive to

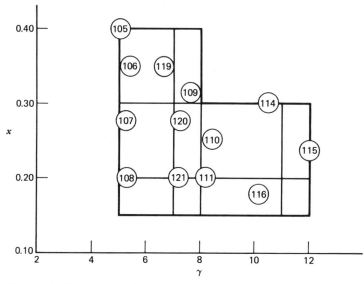

*Figure 1 Experimental settings for n-pentane isomerization at 425°C and with 0.0121 mol % chlorine.*

the choice of the preliminary runs and consequently to the preliminary estimates of the parameters. Also, it is shown that only three designed experiments suffice to reduce the standard deviation of the parameter estimates to that based on all 13 experiments of Fig. 1. The drop in the standard deviations experienced in Case 1 after only one designed experiment is really spectacular. This is due to the poor choice of the preliminary runs, of course.

Juusola et al. applied this procedure to the design of experiments on *o*-xylene oxidation in a differential reactor [60]. Hosten [61] recently proposed a different criterion than that discussed here. Instead of minimizing the volume of the joint confidence volume associated with the estimates, he used a criterion aimed at a more spherical shape for this confidence volume. The results are close to those described above.

To summarize, the approach followed in Section 2.3.d on optimal sequential design is illustrated in Fig. 2.3.d.2-2 by means of a kind of flow diagram (from Froment [45, 46]).

Finally, the sequential methods for the design of an experimental program permit a substantial saving in experimental effort for equal significance or a greater significance for comparable experimental effort, with respect to classical procedures. Automatic application of these methods, no matter how powerful they are, should

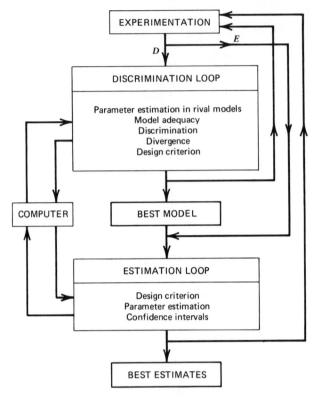

*Figure 2.3.d.2-2 Sequential procedure for optimal design of experiments (from Froment [46]).*

not be substituted for sound judgment. Mere visual inspection of the rate equations may already reveal regions of maximum divergence, although it has to be added that this may become more difficult, or perhaps impossible, with complex multivariable models.

## Problems

2.1 Derive the basic Eq. 2.2-7 for a single reversible catalytic reaction.

2.2 Consider the catalytic reaction

$$A + B \; \rightleftharpoons \; R + S$$

(a) Derive the Langmuir–Hinshelwood–Hougen–Watson kinetic rate expression, assuming that adsorption is rate controlling.

(b) Compare the result of part (a) with that found from Yang and Hougen, Table 2.2-1.

2.3 In a study of the dehydrogenation over a brass catalyst of *sec*-butyl alcohol to methyl ethyl ketone,

$$CH_3CHOHC_2H_5 \longrightarrow CH_3COC_2H_5 + H_2$$

L. H. Thaller and G. Thodos [*A.I.Ch.E.J.*, **6**, 369 (1960)] obtained data that appeared to show two different steps controlling, depending on the temperature level. At low temperatures, surface reaction was controlling, while at high temperatures desorption of (perhaps) hydrogen seemed controlling. A selection of their initial rate data is given below.

(a) Using the data at $T = 371°C$, determine the parameters occurring in the appropriate initial rate expression.

(b) Using the data at $T = 288°C$ and $302°C$, again determine the appropriate parameters.

Note that the intermediate temperature level results should presumably depend upon both surface reaction and desorption steps, since at some point both steps will have equal rates (see Problem 2.5).

Data:

| $T(°C)$ | $p_t(atm)$ | $r_{A0}(kmol/kg\ cat.hr)$ |
|---------|-----------|---------------------------|
| 371     | 1.0       | 0.195                     |
| 371     | 2.0       | 0.189                     |
| 371     | 4.0       | 0.188                     |
| 371     | 9.0       | 0.198                     |
| 371     | 12.0      | 0.190                     |
| 315.5   | 1.0       | 0.0392                    |
| 315.5   | 7.0       | 0.0416                    |
| 315.5   | 4.0       | 0.0416                    |
| 315.5   | 10.0      | 0.0326                    |
| 315.5   | 14.6      | 0.0247                    |
| 315.5   | 5.5       | 0.0415                    |
| 315.5   | 8.5       | 0.0376                    |
| 315.5   | 3.0       | 0.0420                    |
| 315.5   | 0.22      | 0.0295                    |
| 315.5   | 1.0       | 0.0410                    |
| 302     | 1.0       | 0.0227                    |
| 302     | 3.0       | 0.0277                    |
| 302     | 5.0       | 0.0255                    |
| 302     | 7.0       | 0.0217                    |
| 302     | 9.6       | 0.0183                    |
| 288     | 1.0       | 0.0115                    |
| 288     | 3.0       | 0.0161                    |
| 288     | 2.0       | 0.0146                    |

2.4 The Michaelis–Menten (Briggs–Haldane) mechanism in enzyme kinetics is based upon the following reaction scheme between the reactant (substrate $S$), and the catalyst (enzyme $E$) to give the product, $P$:

$$S + E \underset{2}{\overset{1}{\rightleftharpoons}} ES \underset{4}{\overset{3}{\rightleftharpoons}} P + E$$

(a) Use the steady-state hypothesis for the enzyme-substrate complex, $ES$, to derive the Michaelis–Menten kinetic expression:

$$-\frac{d[S]}{dt} = \frac{(k_3[E_0])([S] - [P]/K)}{K_m + [S] + (k_4/k_1)[P]} = \frac{d[P]}{dt}$$

where $[E_0] = [E] + [ES]$ represents the measurable total enzyme concentration

$$K_m = (k_2 + k_3)/k_1 \text{ is the "Michaelis constant"}$$
$$K = k_1 k_3/k_2 k_4$$

(b) Show that the maximum initial rate is given by

$$\left| -\frac{d[S]}{dt} \right|_{max} = k_3[E_0]$$

2.5 (a) For the reaction in Problem 2.3, show that the initial rate expression, assuming that *both* surface reaction and desorption of $R$ are rate controlling, is

$$r_{A0} = \left[ k_R + \frac{k_R{}^2}{2k_{sr}} \frac{(1 + K_A p_A)^2}{K_A p_A} \right]$$
$$- \left\{ \left[ k_R + \frac{k_R{}^2}{2k_{sr}} \frac{(1 + K_A p_A)^2}{K_A p_A} \right]^2 - k_R{}^2 \right\}^{1/2}$$

(See Bischoff and Froment [29].)

(b) Show that the result reduces to the proper Yang and Hougen Table 2.2-1 results for each of the special cases $(k_R/k_{sr}) \to \infty$ and $(k_{sr}/k_R) \to \infty$

(c) Using the combined results of Problem 2.3 and the above results, compare the model with the data at the intermediate temperature level, $T = 315.5°C$ (also see Shah and Davidson [30] and R. W. Bradshaw and B. Davidson, *Chem. Eng. Sci.* **24**, 1519 (1969).

2.6 Consider the reaction $A \rightleftharpoons R + S$, occurring on dual sites. Determine the rate equation in the case that all four elementary steps are simultaneously rate determining.

2.7 The following data were obtained by Sinfelt et al. [Sinfelt, J. H., Hurwitz, H., and Shulman, R. A. *J. Phys. Chem.*, **64**, 1559 (1960)] for the dehydrogenation of methylcyclohexane to toluene. In addition, they found that the product toluene had essentially no effect on the rate.

| $T(°C)$ | $p_M(atm)$ | $p_{H_2}(atm)$ | $r_D(kmol/kg\ cat.hr)$ |
|---------|------------|----------------|------------------------|
| 315 | 0.36 | 1.1 | 0.012 |
| 315 | 0.36 | 3.0 | 0.012 |
| 315 | 0.07 | 1.4 | 0.0086 |
| 315 | 0.24 | 1.4 | 0.011 |
| 315 | 0.72 | 1.4 | 0.013 |
| 344 | 0.36 | 1.1 | 0.030 |
| 344 | 0.36 | 3.1 | 0.032 |
| 344 | 0.08 | 1.4 | 0.020 |
| 344 | 0.24 | 1.4 | 0.034 |
| 344 | 0.68 | 1.4 | 0.034 |
| 372 | 0.36 | 1.1 | 0.076 |
| 372 | 0.36 | 4.1 | 0.080 |
| 372 | 1.1 | 4.1 | 0.124 |
| 372 | 2.2 | 4.1 | 0.131 |

(a) Discuss which of the steps—adsorption, surface reaction, and desorption—might be rate controlling in view of the above data.

(b) Show that a rate expression based on the mechanism

$$A + l \rightleftharpoons Al$$

$$Al \rightleftharpoons Rl$$

$$Rl \longrightarrow Q + S + l$$

fits the data; also estimate the activation energies.

(c) Discuss the results of b in view of a.

2.8 The isomerization of $n$-pentane was considered in the text, where several rate expressions were stated. Derive the final result for desorption of $i$-pentene controlling:

$$r = \frac{kK_5K_DK_6(p_A - p_B/K)}{p_{H_2} + K_5K_D(1 + K_6)p_A}$$

2.9 For the isomerization of $n$-pentane, derive the rate expression if the surface reaction step of the dehydrogenation reaction were rate controlling. Contrast this with the correct rate of Problem 2.8, especially regarding variations with total pressure.

2.10 For the isomerization of $n$-pentane, the following experimental data were collected by Hosten and Froment [38]:

| $x$ | $\gamma$ | $W/F_{A0}$(kg cat.hr/kmol) |
|------|--------|----------------------------|
| 0.4025 | 4.853 | 5.92 |
| 0.35 | 5.253 | 3.84 |
| 0.2784 | 5.29 | 2.84 |
| 0.2001 | 5.199 | 1.75 |
| 0.3529 | 6.833 | 5.74 |
| 0.2728 | 7.33 | 3.84 |
| 0.2038 | 7.344 | 2.66 |
| 0.3248 | 7.638 | 5.28 |
| 0.2571 | 8.514 | 3.9 |
| 0.2011 | 8.135 | 2.65 |
| 0.3017 | 10.598 | 5.73 |
| 0.2413 | 11.957 | 4.37 |
| 0.1734 | 10.227 | 2.65 |

$\gamma$ is the molar ratio $H_2$/hydrocarbon. The pentane feed consisted of 92.65 mole $\% \, n - C_5$ and 6.37 mole $\% \, i - C_5$. The overall equilibrium constant is 2.07, while the selectivity for isomerization is nearly constant and equal to 0.91. Estimate the parameters in the adsorption model by means of the integral method of kinetic analysis. Both $W/F_{A0}$ and $x$ can be used as dependent variables. Comment on this choice. Compare the results and the computational effort for both cases.

2.11 A catalytic reaction $A \rightleftharpoons B$ is carried out in a fixed bed reactor. Comment on the concentration profiles of adsorbed species as a function of bed depth for various rate determining steps.

2.12 The dehydrogenation of ethanol was carried out in an integral reactor at 275°C with the following results:

| $x$ | $p_t$(atm) | $W/F_{A0}$(kg cat.hr/kmol) | $x$ | $p_t$ | $W/F_{A0}$ |
|------|-----------|----------------------------|------|-------|-----------|
| 0.118 | 1 | 0.2 | 0.14 | 3 | 0.2 |
| 0.196 | 1 | 0.4 | 0.2 | 3 | 0.4 |
| 0.262 | 1 | 0.6 | 0.25 | 3 | 0.6 |
| 0.339 | 1 | 0.88 | 0.286 | 3 | 0.88 |
| 0.446 | 1 | 1.53 | 0.352 | 3 | 1.6 |
| 0.454 | 1 | 1.6 | 0.14 | 4 | 0.2 |
| 0.524 | 1 | 2.66 | 0.196 | 4 | 0.4 |
| 0.59 | 1 | 4.22 | 0.235 | 4 | 0.6 |
| 0.60 | 1 | 4.54 | 0.271 | 4 | 0.88 |
| | | | 0.32 | 4 | 1.6 |
| | | | 0.112 | 7 | 0.2 |
| | | | 0.163 | 7 | 0.4 |

| $x$ | $p_t$(atm) | $W/F_{A0}$(kg cat.hr/kmol) | $x$ | $p_t$ | $W/F_{A0}$ |
|---|---|---|---|---|---|
| | | | 0.194 | 7 | 0.6 |
| | | | 0.214 | 7 | 0.88 |
| | | | 0.254 | 7 | 1.6 |
| | | | 0.1 | 10 | 0.2 |
| | | | 0.148 | 10 | 0.4 |
| | | | 0.175 | 10 | 0.6 |
| | | | 0.188 | 10 | 0.88 |
| | | | 0.229 | 10 | 1.6 |

The overall equilibrium constant is 0.589. The feed consisted of the azeotropic mixture ethanol-water, containing 13.5 mole % water. Water is not adsorbed on the catalyst. Estimate the parameters of the adsorption, surface reaction, and desorption models, using conversion as the regression variable. Comment on the feasibility for the estimation of the parameters. Which model is the best? On what basis?

# References

[1] Thomas, J. M. and Thomas, W. J. *Introduction to the Principles of Heterogeneous Catalysis*, Academic Press, New York (1967).

[2] Boudart, M. *Kinetics of Chemical Processes*, Prentice-Hall, Englewood Cliffs, N. J. (1968).

[3] Thomson, S. J. and Webb, G. *Heterogeneous Catalysis*, Wiley, New York (1968).

[4] Emmett, P. H., ed. *Catalysis*, Vol. 1–7, Reinhold, New York (1954–1960).

[5] *Advances in Catalysis*, Academic Press, New York (1949–19  ).

[6] Boudart, M. *Ind. Chem. Belg.*, **23**, 383 (1958).

[7] Moss, R. L. *The Chemical Engineer (IChE)*, No. 6, CE 114 (1966).

[8] Thomas, C. L. *Catalytic Processes and Proven Catalysists*, Academic Press, New York (1970).

[9] *Catalyst Handbook*, Wolfe Scientific Books (1970).

[10] Burwell, R. *Chem. Eng. News*, Aug. 22, p. 58 (1966).

[11] Oblad, A. G., Milliken, T. H., and Mills, G. A. *The Chemistry of Petroleum Hydrocarbons*, Reinhold, New York (1955).

[12] Greensfelder, B. S., Voge, G. M., and Good, H. H. *Ind. Eng. Chem.*, **41**, 2573 (1949).

[13] Voge, G. M. *Catalysis*, Emmett, P. H. ed., Vol. VI, Reinhold, New York (1958).

[14] Germain, J. E. *Catalytic Conversion of Hydrocarbons*, Academic Press, New York (1969).

[15] Venuto, P. B. *Chem. Tech.*, April (1971).

[16] Weisz, P. B. *Adv. Catal.*, **13**, 137 (1962).

[17] Dwyer, F. G., Eagleton, L. C., Wei, J., and Zahner, J. C. *Proc. Roy. Soc.* London, **A302**, 253 (1968).

[18] Sinfelt, J. H. *Adv. Chem. Eng.*, **5**, 37 (1964).

[19] Haensel, V. *Ind. Eng. Chem.* **57**, No. 6, 18 (1965).

[20] Brunauer, S. *The Adsorption of Gases and Vapors*, Princeton University Press, Princeton, N. J. (1945).

[21] De Boer, J. H. *The Dynamical Character of Adsorption*, Oxford University Press, 2nd ed., Oxford (1968).

[22] Flood, E. A., ed. *The Solid-Gas Interface*, 2 vols., Marcel Dekker, New York (1967).

[23] Gregg, S. J. and Sing, K. S. W. *Adsorption, Surface Area, and Porosity*, Academic Press, New York (1967).

[24] Clark, A. *The Theory of Adsorption and Catalysis*, Academic Press, New York (1970).

[25] Hayward, D. O. and Trapnell, B. M. W. *Chemisorption*, Butterworths, London (1964).

[26] Coughlin, R. W. *Ind. Eng. Chem.*, **59**, No. 9, 45 (1967).

[27] Hougen, O. A. and Watson, K. M. *Chemical Process Principles*, Vol. III, Wiley, New York (1947).

[28] Aris, R. *Introduction to the Analysis of Chemical Reactors*, Prentice-Hall, Englewood Cliffs, N. J. (1965).

[29] Bischoff, K. B. and Froment, G. F. *Ind. Eng. Chem. Fund.*, **1**, 195 (1965).

[30] Shah, M. J., Davidson, B. *Ind. Eng. Chem.*, **57**, No. 10, 18 (1965).

[31] Boudart, M. *Chem. Eng. Prog.*, **58**, No. 73 (1962).

[32] Wauquier, J. P. and Jungers, J. C. *Bull. Soc. Chim.*, France, 1280 (1957).

[33] Yang, K. H. and Hougen, O. A. *Chem. Eng. Prog.*, **46**, 146 (1950).

[34] Klugherz, P. D. and Harriott, P. *A. I. Ch. E. J.*, **17**, 856 (1971).

[35] Marcinkowsky, A. E. and Berty, J. M. *J. Catal.*, **29**, 494 (1973).

[36] Kenson, R. E. *J. Phys. Chem.*, **74**, 1493 (1970).

[37] Boudart, M. *A. I. Ch. E. J.*, **18**, 465 (1972).

[38] Hosten, L. H. and Froment, G. F. *Ind. Eng. Chem. Proc. Des. Devpt.*, **10**, 280 (1971).

[39] Kittrell, J. R. and Mezaki, R. *A. I. Ch. E. J.*, **13**, 389 (1967).

[40] Franckaerts, J. and Froment, G. F. *Chem. Eng. Sci.*, **19**, 807 (1964).

[41] Marquardt, D. W. *J. Soc. Ind. Appl. Math.*, **2**, 431 (1963).

[42] Peterson, T. I. and Lapidus, L. *Chem. Eng. Sci.*, **21**, 655 (1965).

[43] Froment, G. F. and Mezaki, R. *Chem. Eng. Sci.*, **25**, 293 (1970).

[44] Seinfeld, J. H. *Ind. Eng. Chem.*, **62**, 32 (1970).

[45] Froment, G. F. Proc. 7th Eur. Symp. "Computer Application in Process Development," Erlangen, Dechema (April 1974).

[46] Froment, G. F. *A. I. Ch. E. J.*, **21**, 1041 (1975).

[47] Wilde, D. G. and Beightler, C. S. *Foundations of Optimization*, Prentice-Hall, Englewood Cliffs, N. J. (1967).

[48] Beveridge, G. S. G. and Schechter, R. S. *Optimization Theory and Practice*, McGraw-Hill, New York (1970).

[49] Hoffmann, U. and Hofmann, H. *Einführung in die Optimierung*, Verlag Chemie, Weinheim BRD (1971).

[50] Himmelblau, D. M. *Applied Non linear Programming*, McGraw-Hill, New York (1972).

[51] Rosenbrock, H. H. and Storey, C. *Computational Techniques for Chemical Engineers*, Pergamon Press, New York (1966).

[52] Froment, G. F. Proc. 4th Int. Symp. Chem. React. Engng, Heidelberg 1976, Dechema 1976.

[53] De Pauw, R. P. and Froment, G. F. *Chem. Eng. Sci.*, **30**, 789 (1975).

[54] Emig, G. Hofmann, H. and Friedrich, F. Proc. 2nd Int. Symp. Chem. React. Engng., Amsterdam 1972, Elsevier, B5-23 (1972).

[55] Box, G. E. P. and Hill, W. J. *Technometrics*, **9**, 57 (1967).

[56] Hosten, L. H. and Froment, G. F. Proc. 4th Int. Symp. Chem. React. Engng., Heidelberg 1976, Dechema 1976.

[57] Dumez, F. J. and Froment, G. F. *Ind. Eng. Chem. Proc. Des. Devt.*, **15**, 291 (1976).

[58] Dumez, F. J., Hosten, L. H., and Froment, G. F. *Ind. Eng. Chem. Fundam.*, **16**, 298 (1977).

[59] Box, G. E. P. and Lucas, H. L. *Biometrika*, **46**, 77 (1959).

[60] Juusola, J. A., Bacon, D. W., and Downie, J. *Can. J. Chem. Eng.*, **50**, 796 (1972).

[61] Hosten, L. H. *Chem. Eng. Sci.*, **29**, 2247 (1974).

[62] Gates, B. C., Katzer, J. R., and Schuit, G. C. A. *Chemistry of Catalytic Processes*, McGraw-Hill, New York (1978).

[63] Brunauer, S., Deming, L. S., Deming, W. E., and Teller, E. J. *J. Am. Soc.*, **62**, 1723 (1940).

[64] Rudnitsky, L. A. Alexeyev, A. M. *J. Catal.*, **37**, 232 (1975).

[65] Weekman, V. W. *A. I. Ch. E. J.*, **20**, 833 (1974).

[66] Box, G. E. P. and Henson, T. L., M.B.R. Tech. Rept. No. 51, University of Wisconsin, Madison, Wisconsin, January 1969.

[67] Kittrell, J. R., *Advan. Chem. Eng.* **8**, 97 (1970).

# 3

## TRANSPORT PROCESSES WITH FLUID-SOLID HETEROGENEOUS REACTIONS

The fact that various transport steps of the reactants and products must be considered was briefly described at the beginning of Chapter 2. This chapter provides a quantitative treatment of these aspects of the overall problem, called steps 1,7 and 2,6 in Chapter 2.

### Part One
### Interfacial Gradient Effects

### 3.1 Surface Reaction Between a Solid and a Fluid

Consider a reactive species $A$ in a fluid solution, in contact with a reactive solid. It is convenient for the present to define a rate based on the interfacial surface area, and if it is first order:

$$r_{Ai} = k_r C_{Ai} \qquad (3.1\text{-}1)$$

where

$$r_{Ai} = \text{rate of reaction of } A \text{ at surface, } \mathrm{kmol/m}_p{}^2 \text{ hr}$$
$$k_r = \text{rate coefficient for the reaction, } \mathrm{m}_f{}^3/\mathrm{m}_p{}^2 \text{ hr}$$
$$C_{Ai} = \text{concentration of } A \text{ at the interface, } \mathrm{kmol/m}_f{}^3$$

The consumption of $A$ at the interface has to be compensated for by transport from the bulk fluid. This is described by the usual mass transfer coefficient in terms of an appropriate driving force:

$$N_A = k_g(C_A - C_{Ai}) \qquad (3.1\text{-}2)$$

where

$N_A$ = mass flux with respect to the fixed solid surface, kmol/$m_p{}^2$ hr
$k_g$ = mass transfer coefficient, $m_f{}^3$/$m_p{}^2$ hr
$C_A$ = concentration of $A$ in bulk stream, kmol/$m_f{}^3$

For steady state, the two rates must be equal, and this is used to eliminate the unmeasured surface concentration, $C_{Ai}$

$$r_{Ai} = N_A = r_A$$

Thus,

$$C_{Ai} = \frac{k_g}{k_r + k_g} C_A$$

and

$$r_A = \left(\frac{1}{k_g} + \frac{1}{k_r}\right)^{-1} C_A \tag{3.1-3}$$

$$= k_0 C_A$$

where an "overall" rate coefficient can be defined as

$$\frac{1}{k_0} = \frac{1}{k_g} + \frac{1}{k_r} \tag{3.1-4}$$

There are two limiting cases: when the mass transfer step is much more rapid than the surface reaction step, $k_g \gg k_r$, and Eq. 3.1-4 gives: $k_0 \simeq k_r$. Also, $C_{Ai} \simeq C_A$, and so the reactant concentration at the surface is the same as that measured in the bulk. The observed rate corresponds to the actual reaction—this is termed "reaction controlling." The other limit is that of almost instantaneous reaction, $k_r \gg k_g$, and Eq. 3.1-4 gives $k_o \simeq k_g$. Also, $C_{Ai} \simeq 0$, and the observed rate corresponds to the fluid phase mass transfer step, not the reaction—this is termed "diffusion controlling."

The same procedure may be followed for a second-order reaction:

$$r_{Ai} = k_r C_{Ai}{}^2 \tag{3.1-5}$$

which, with Eq. 3.1-2, leads to

$$r_A = k_g\left\{\left[1 + (\tfrac{1}{2})\frac{k_g}{k_r C_A}\right] - \sqrt{\left[1 + (\tfrac{1}{2})\frac{k_g}{k_r C_A}\right]^2 - 1}\right\}C_A \tag{3.1-6}$$

A totally different form of concentration dependence is found, that is neither first order or second order (or one could state that the "overall" coefficient is not constant). Equation 3.1-6 reduces to the proper form in the two limiting situations:

$$r_A \simeq k_r C_A^2 \qquad k_g \gg k_r$$
$$r_A \simeq k_g C_A \qquad k_r \gg k_g$$

For an $n$th-order reaction, one finds

$$r_A = k_r \left( C_A - \frac{r_A}{k_g} \right)^n \qquad (3.1\text{-}7)$$

$r_A$ cannot be solved explicitly from Eq. 3.1-7 for arbitrary $n$, so that no equation equivalent to Eqs. 3.1-3 or 3.1-6 is obtained; $r_A$ may be obtained by iterative methods (see Frank–Kamenetskii [1]). Thus, in general, consecutive rate processes of different order cannot easily be combined into an overall expression, but can be handled by numerical techniques.

It follows that the occurrence of consecutive steps does not lead to serious complications when the rate is to be predicted, provided, of course, the rate coefficients and the order of the reaction are given. The reverse problem, that is, the determination of the order and the rate coefficients is much more complicated, however. Sometimes transport coefficients may be found in the literature for the case at hand so that it becomes possible to calculate the mass transfer effect. Generally, however, it will be necessary to derive the mass transfer coefficients from specific experiments. Therefore, the experiments have to be performed under conditions for which the global rate is preferably entirely determined by the mass transfer rate. This is generally achieved by operating at higher temperatures since the reaction rate coefficient is enhanced much more by a temperature increase than the mass transfer coefficient; that is, the activation energy of the reaction is much higher than that of the transport phenomenon. The other extreme situation whereby the global rate is entirely determined by the rate of reaction may be reached by increasing the turbulence or by operating at a lower temperature. Finally, it is evident that such experiments should be performed under isothermal conditions to avoid further complications such as the need to include a heat transfer rate equation in the treatment.

## 3.2 Mass and Heat Transfer Resistances

### 3.2.a Mass Transfer Coefficients

Section 3.1 described how the mass transfer coefficient can be combined with the rate coefficient for simple reactions. This section gives more detailed discussion of how to obtain values for the mass transfer coefficients.

As mentioned previously, the mass transfer coefficient is defined as in transport processes (e.g., Bird, Stewart, and Lightfoot [2]) and several driving force units are in common use:

$$N_A = k_g(y_A - y_{As}^s)$$
$$= k_g(C_A - C_{As}^s) \qquad (3.2.a\text{-}1)$$
$$= k_g(p_A - p_{As}^s)$$

The units and numerical values of $k_g$ are different, of course, for each of these equations, but to avoid complicating the notation only one symbol is used here, as was already done for the rate coefficient in Chapter 1.

It will be recalled from transport phenomena, we know that it is most useful to define a mass transfer coefficient to describe only the diffusive transport and not the total diffusive plus convective. The coefficients are identical only for the special case of equimolar counter-diffusion and this is the value of the coefficient $k_g^0$, which is actually correlated in handbooks.

For example, a very common situation in unit operations is diffusion of species $A$ through a stagnant film of $B$, for which the film theory, together with the proper solution of the diffusion equations, give

$$k_g = \frac{k_g^0}{(y_B)_{\text{log mean}}}$$

the driving force being expressed in mole fractions.

An analogous treatment using the relative flux ratios from the stoichiometry of the general reaction,

$$aA + bB + \cdots \quad \rightleftharpoons \quad rR + sS + \cdots$$

yields the result for transport of species $A$:

$$k_g = k_g^0/y_{fA} \qquad (3.2.a\text{-}2)$$

where

$$y_{fA} = \frac{(1 + \delta_A y_A) - (1 + \delta_A y_{As}^s)}{\ln \dfrac{1 + \delta_A y_A}{1 + \delta_A y_{As}^s}} \qquad (3.2.a\text{-}3)$$

with

$$\delta_A = \frac{(r + s + \cdots) - (a + b + \cdots)}{a} \qquad (3.2.a\text{-}4)$$

This expression is often written in terms of partial and total pressures and is then called the "film pressure factor, $p_{fA}$." The basis for Eqs. 3.2.a-2 to 4 is considered in Example 3.2.c-1. Then, correlations of the mass transfer coefficients can be presented in terms of the $j_D$-factor, for example,

$$j_D = \frac{k_g{}^0 M_m}{G} \text{Sc}^{2/3}$$

$$= \frac{k_g M_m y_{fA}}{G} \text{Sc}^{2/3} \qquad\qquad (3.2.\text{a-}5)$$

$$= \frac{k_g M_m p_{fA}}{G} \text{Sc}^{2/3} \text{ (gases)}$$

$$j_D = f(\text{Re})$$

where Sc = Schmidt number = $(\mu/\rho_f D)$ and the $k_g$ differ in numerical value, depending on the driving force.

Of particular interest for the following chapters is the mass transfer coefficient between a fluid and the particles of a packed bed. Figure 3.2.a-1 shows some of

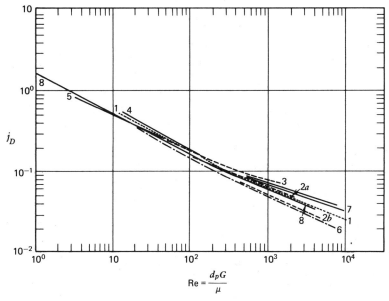

Figure 3.2.a-1 *Mass transfer between a fluid and a bed of particles.*
*Curve 1: Gamson et al.* [3], *Wilke and Hougen* [4]. *Curve 2: Taecker and Hougen* [5]. *Curve 3: McCune and Wilhelm* [6]. *Curve 4: Ishino and Otake* [7]. *Curve 5: Bar Ilan and Resnick* [8]. *Curve 6: De Acetis and Thodos* [9]. *Curve 7: Bradshaw and Bennett* [10]. *Curve 8: Hougen* [11]; *Yoshida, Ramaswami, and Hougen* [12] (*spheres*; ε = 0.37).

the most significant experimental results for this situation. For use in calculations it is convenient to have a numerical expression for the relation $j_D$ versus Re. The following relations are fairly representative for the results shown in Fig. 3.2.a-1. The intersection of the lines at Re = 190 has no physical meaning, merely representing the correlation of Hougen, et al. [11, 12]. For packed beds of spheres with $\varepsilon = 0.37$, for Re $= d_p G/\mu < 190$

$$j_D = 1.66 \, (\text{Re})^{-0.51} \qquad (3.2.a\text{-}6a)$$

and for Re $> 190$

$$j_D = 0.983 \, (\text{Re})^{-0.41} \qquad (3.2.a\text{-}6b)$$

The use of these correlations for calculating values for $k_g$ is illustrated below.

### 3.2.b Heat Transfer Coefficients

Fluid-to-particle interfacial heat transfer resistances also need to be considered. These are described by

$$(-\Delta H)r_A = h_f a_m (T_s^s - T) \qquad (3.2.b\text{-}1)$$

The heat transfer coefficient, $h_f$, is also correlated with respect to the Reynolds number by means of a $j$-factor expression:

$$j_H = \frac{h_f}{c_p G} \, \text{Pr}^{2/3} \qquad (3.2.b\text{-}2)$$

The most representative experimental results for the case of interfacial heat transfer between a fluid and the particles of a packed bed are shown in Fig. 3.2.b-1.

### 3.2.c Multicomponent Diffusion in a Fluid

For a binary mixture, the single diffusivity, $D_{AB}$, is used in the Schmidt number. However, most practical problems involve multicomponent mixtures, whose rigorous treatment is much more complicated.

In general, the flux of a given chemical species can be driven not only by its own concentration gradient, but also by those of all the other species; see Toor [17], for example:

$$\mathbf{N}_j = -\sum_{k=1}^{N-1} C_t D_{jk} \nabla y_k + y_j \sum_{k=1}^{N} \mathbf{N}_k \qquad j = 1, 2, \ldots, N-1 \qquad (3.2.c\text{-}1)$$

The last term accounts for bulk flow of the mixture. The exact form of the $D_{jk}$ depends on the system under study. For ideal gases, the kinetic theory leads to the Stefan–Maxwell equations, which can be rearranged into the form of Eq. 3.2.c-1–a treatment using matrix methods is given by Stewart and Prober [18].

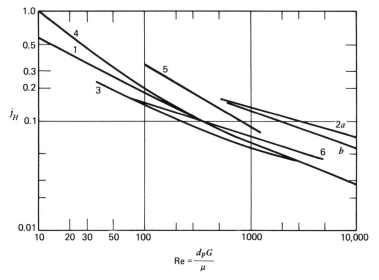

*Figure 3.2.b-1 Heat transfer between a fluid and a bed of particles.*
*Curve 1: Gamson et al., Wilke and Hougen [3, 4]. Curve 2: Bau-*
*meister and Bennett (a) for $d_t/d_p > 20$, (b) mean correlation [13].*
*Curve 3: Glaser and Thodos [14]. Curve 4: de Acetis and Thodos*
*[9]. Curve 5: Sen Gupta and Thodos [15]. Curve 6: Handley and*
*Heggs [16] ($\varepsilon = 0.37$).*

For liquids, there is no complete theory yet available—for a discussion of corrections for thermodynamic nonidealities, and other matters, see Bird, Stewart, and Lightfoot [2]. A comprehensive review of available information on gas diffusion is by Mason and Marrero [19], and for liquids see Dullien, Ghai, and Ertl [20, 21].

The form of Eq. 3.2.c-1 is too complex for many engineering calculations, and a common approach is to define a mean effective binary diffusivity for species $j$ diffusing through the mixture:

$$\mathbf{N}_j = -C_t D_{jm} \nabla y_j + y_j \sum_{k=1}^{N} \mathbf{N}_k \qquad (3.2.c\text{-}2)$$

Using Eq. 3.2.c-1, Toor [17] and Stewart and Prober [18] showed that the matrix of the $D_{jk}$ could be diagonalized, which then gives the form of Eq. 3.2.c-2, and the many solutions available for binary systems can be adapted for multicomponent mixtures.

Considering the case of ideal gases, the Stefan–Maxwell equations are given in Bird, Stewart, and Lightfoot [2]:

$$-C_t \nabla y_j = \sum_{\substack{k=1 \\ k \neq j}}^{N} \frac{1}{D_{jk}} (y_k \mathbf{N}_j - y_j \mathbf{N}_k) \qquad (3.2.c\text{-}3)$$

where the $D_{jk}$ are the usual binary diffusivities. For a binary system

$$-C_t \nabla y_1 = \frac{1}{D_{12}} [\mathbf{N}_1 - y_1(\mathbf{N}_1 + \mathbf{N}_2)] \qquad (3.2.c\text{-}4)$$

where $y_1 + y_2 = 1$ was utilized. Solving for the flux:

$$\mathbf{N}_1 = -C_t D_{12} \nabla y_1 + y_1(\mathbf{N}_1 + \mathbf{N}_2) \qquad (3.2.c\text{-}5)$$

For equimolar counter diffusion, $\mathbf{N}_2 = -\mathbf{N}_1$ and:

$$\mathbf{N}_1 = -C_t D_{12} \nabla y_1 \qquad (3.2.c\text{-}6)$$

Thus, for the multicomponent gas mixture, an effective binary diffusivity for species $j$ diffusing through the mixture is found by equating the driving force $\nabla y_j$ in Eqs. 3.2.c-2 and 3.2.c-3, with this result:

$$\frac{1}{D_{jm}} = \frac{\displaystyle\sum_{k=1}^{N} \frac{1}{D_{jk}} \left( y_k - y_j \cdot \frac{N_k}{N_j} \right)}{1 - y_j \displaystyle\sum_{k=1}^{N} N_k / N_j} \qquad (3.2.c\text{-}7)$$

The classical use of Eq. 3.2.c-7 in unit operations is the so-called "Wilke (1950) equation" for diffusion of species 1 through stagnant 2, 3, .... Here, all the flux ratios are zero for $k = 2, 3, \ldots$, and Eq. 3.2.c-7 reduces to

$$\frac{1}{D_{1m}} = \frac{1}{1 - y_1} \sum_{k=2,3\ldots}^{N} \frac{y_k}{D_{1k}} \qquad (3.2.c\text{-}8)$$

Even though Eq. (3.2.c-8) is often recommended for computing an effective diffusivity in reacting systems, it is not really the appropriate equation, except for very dilute solutions. In other cases, the other species are not necessarily stagnant, but rather the steady state flux ratios are determined by the reaction stoichiometry. Thus, for a general chemical reaction,

$$\frac{N_j}{\alpha_j} = \text{constant}$$

CHEMICAL ENGINEERING KINETICS

and so Eq. 3.2.c-7 becomes

$$\frac{1}{D_{jm}} = \frac{\sum\limits_{k \neq j}^{N} \frac{1}{D_{jk}} \left( y_k - y_j \frac{\alpha_k}{\alpha_j} \right)}{1 - y_j \sum\limits_{k=1}^{N} \alpha_k / \alpha_j} \tag{3.2.c-9a}$$

$$= \frac{1}{1 + \delta_j y_j} \sum\limits_{k \neq j}^{N} \frac{1}{D_{jk}} \left( y_k + y_j \frac{\alpha_k}{|\alpha_j|} \right) \tag{3.2.c-9b}$$

The last equation is for species $j$ a reactant. In a theoretical study, Hsu and Bird [22] have compared various uses of Eq. 3.2.c-9 in a ternary system with surface reaction versus the exact solution of the Stefan–Maxwell equations; the most straightforward is to use merely some mean composition, $\bar{y}_j$, to compute an average value of $D_{jm}$.

It is also useful, for certain applications, to define an alternate effective binary diffusivity with the flux relative to the fixed solid—any bulk flow is then included in the values for $D'_{jm}$:

$$\mathbf{N}_j = -C_t D'_{jm} \nabla y_j \tag{3.2.c-10}$$

then the same procedure results in

$$\frac{1}{D'_{jm}} = \sum\limits_{k \neq j}^{N} \frac{1}{D_{jk}} \left( y_k - \frac{N_k}{N_j} y_j \right) \tag{3.2.c-11}$$

which is essentially just the numerator of Eq. 3.2.c-7. Kubota, Yamanaka, and Dalla Lana [23] solved the same problem as Hsu and Bird, and stated that the results indicated that, using constant mean compositions, Eq. 3.2.c-11 provided somewhat more accurate representation of the exact Stefan–Maxwell results than did Eq. 3.2.c-7. However, there is not really enough experience at the present time to choose between them.

### Example 3.2.c-1 Use of Mean Effective Binary Diffusivity

For a chemical reaction

$$aA + bB + \cdots \rightleftharpoons rR + sS + \cdots \tag{a}$$

Eq. (3.2.c-9) gives for the mean binary diffusivity:

$$\frac{1}{D_{Am}} = \frac{1}{1 + \delta_A y_A} \left[ \frac{1}{D_{AB}} \left( y_B - \frac{b}{a} y_A \right) + \frac{1}{D_{AR}} \left( y_R + \frac{r}{a} y_A \right) + \frac{1}{D_{AS}} \left( y_S + \frac{s}{a} y_A \right) \cdots \right] \tag{b}$$

with

$$\delta_A = \frac{r + s + \cdots - a - b - \cdots}{a}$$

The flux expression Eq. 3.2.c-2 can be written for one-dimensional diffusion as

$$N_A = -C_t D_{Am} \frac{dy_A}{dz} + y_A (N_A + N_B + N_R + N_S \cdots) \tag{c}$$

$$= -C_t D_{Am} \frac{dy_A}{dz} + y_A N_A \left(1 + \frac{b}{a} - \frac{r}{a} - \frac{s}{a} \cdots\right) \tag{d}$$

or

$$N_A = \frac{-C_t D_{Am}}{1 + \delta_A y_A} \frac{dy_A}{dz} \tag{e}$$

When integrated for steady state diffusion, with $N_A = $ constant, and with an average constant value for $D_{Am}$, Eq. e gives:

$$N_A = \frac{C_t D_{Am}}{L \delta_A} \ln \frac{1 + \delta_A y_{A0}}{1 + \delta_A y_A(L)} = \left(\frac{C_t D_{Am}}{L}\right) \frac{y_{A0} - y_A(L)}{y_{fA}} \tag{f}$$

where $y_{fA}$ is the "film factor" of Eq. 3.2.a-3, which is defined relative to the equimolar counter diffusion case with $\delta_A = 0$, $y_{fA} = 1$.

## 3.3 Concentration or Partial Pressure and Temperature Differences Between Bulk Fluid and Surface of a Catalyst Particle

One of the most important uses of the above mass and heat transfer relationships is in determining external mass and heat transfer resistances for catalyst particles. Here, the rate is usually expressed in terms of catalyst mass (kmol/kg cat. hr), and using $a_m = $ external surface per weight of catalyst ($m_p^2$/kg cat.) gives

$$r_A = a_m k_g (C_A - C_{As}^s) \quad (k_g : m_f^3/m_p^2 \cdot hr)$$
$$= a_m k_g (p_A - p_{As}^s) \quad (k_g : kmol/m_p^2.hr.atm) \tag{3.3-1}$$

In experimental kinetic studies in particular, the question often arises if the partial pressure drop $\Delta p_A$ over the so-called external film may be neglected. One has to check whether or not it is allowed to substitute $p_A$, the partial pressure of $A$ in the bulk fluid stream, into the rate equation for the reaction. The value of $k_g$ is determined from a correlation, such as Eq. 3.2.a-5 with Eq. 3.2.a-2, 3.

The calculation of $\Delta p_A$ is not straightforward, since the calculation of the film pressure factor $p_{fA}$ requires the knowledge of $p_{As}^s$.

The iteration cycle then is as follows:

**1.** Start with the assumption that $p_{As}^s = p_A$ or $\Delta p_A = 0$. It can be shown by L'Hopital's rule that in this case $p_{fA} = p_t + \delta_A p_A$. With this value of $p_{fA}$, $k_g$

is calculated by means of Eq. 3.2.a-5 and with this $k_g$ the partial pressure drop $\Delta p_A$ is obtained from relation Eq. 3.3-1.

**2.** Substitution of $\Delta p_A$ in Eq. 3.2.a-3 gives a better estimate for $p_{fA}$ with which a new value for $k_g$ and $\Delta p_A$ are computed. The cycle is continued until convergence of the $\Delta p_A$ values is obtained.

It is usually found that $\Delta p_A$ is rather small, although exceptions occur. It is more common to find fairly large $\Delta T$. Significant $\Delta T$, or $\Delta p_A$, is especially likely in laboratory reactors, which are likely to have rather low flow rates through the reactor, whereas commercial reactors commonly have very high flow rates and thereby small external film resistance. The only positive check, of course, is to compute the actual values.

A simple estimate of the temperature difference in terms of the concentration drop is provided by dividing Eq. 3.2.b-1 by Eq. 3.3-1, as shown by Smith [24]:

$$(T_s^s - T) = \frac{k_g}{h_f}(-\Delta H)(C_A - C_{As}^s) \tag{3.3-2a}$$

$$= \left[\frac{j_D}{j_H}\left(\frac{\mathrm{Pr}}{\mathrm{Sc}}\right)^{2/3}\right]\left[\frac{(-\Delta H)}{\rho_f c_p}\right]\frac{\Delta C_A}{y_{fA}} \tag{3.3-2b}$$

For gases flowing in packed beds the values of the groups are such that

$$(T_s^s - T) \simeq 0.7\left[\frac{(-\Delta H)}{M_m c_p}\right]\frac{\Delta p_A}{p_{fA}} \tag{3.3-2c}$$

The maximum possible actual temperature difference would occur for complete, very rapid reaction and heat release, $p_{As}^s \simeq 0$:

$$(\Delta T)_{\max} \simeq 0.7\left[\frac{(-\Delta H)}{M_m c_p}\right]\left[\frac{\ln(1 + \delta_A p_A/p_t)}{\delta_A}\right] \tag{3.3-3}$$

Thus, use of the physical properties, the reaction stoichiometry, and the bulk fluid phase composition permits a quick estimate of $(\Delta T)_{\max}$.

### *Example 3.3-1 Interfacial Gradients in Ethanol Dehydrogenation Experiments*

The dehydrogenation of ethanol into acetaldehyde

$$C_2H_5OH \rightleftharpoons CH_3CHO + H_2$$
$$A \rightleftharpoons R + S$$

is studied in a tubular reactor with fixed catalytic bed at 275°C and atmospheric pressure.

The molar feed rate of ethanol, $F_{A0}$ is 0.01 kmol/hr, the weight of catalyst, $W : 0.01$ kg. At this value of $W/F_{A0}$ the measured conversion is 0.362 and the reaction rate, $r_A : 0.193$ kmol/kg cat.hr. The inside diameter of the reactor is 0.035 m. The catalyst particles are of cylindrical shape with diameter = height = $d = 0.002$ m. The bulk density of the bed, $\rho_B$ amounts to 1500 kg/m³ and the void fraction, $\varepsilon$, to 0.37. From these, $a_m = 1.26$ m²/kg. Estimate the partial pressure and temperature difference between the bulk gas stream and catalyst surface.

In a calculation of this type it is frequently encountered that physicochemical data concerning the reacting components are lacking. Excellent estimates may then be obtained through the use of general correlations for the transport properties, however. In this example only correlations that can be found in Reid and Sherwood [25] are used. They also explain the background of these correlations.

*Estimation of the Partial Pressure Drop over the Film*

Estimation of Viscosities

$H_2$ : Use the Lennard–Jones potential, with

$$\sigma = 2.827 \text{ Å} \qquad \frac{\varepsilon_0}{k} = 59.7 \text{ K}, \qquad \Omega_v = 0.8379$$

$$\mu_{H_2} = \frac{0.002669\sqrt{2 \times 548}}{(2.827)^2 \times 0.8379} = 0.013195 \text{ cp} \quad \text{or} \quad 0.0475 \frac{\text{kg}}{\text{m} \cdot \text{hr}}$$

$C_2H_5OH$: Use the Stockmayer potential, with

$$\sigma = 4.31 \text{ Å}, \qquad \frac{\varepsilon_0}{k} = 431 \text{ K}, \qquad \delta = 0.3, \qquad \Omega_v = 1.422$$

$$\mu_{Eth} = \frac{0.002669\sqrt{46 \times 548}}{(4.31)^2 \times 1.422} = 0.01604 \text{ cp} \quad \text{or} \quad 0.05775 \frac{\text{kg}}{\text{m} \cdot \text{hr}}$$

$CH_3CHO$: Use the method of corresponding states, since the potential parameters are not available.

$$T_c = 461 \text{ K}, \qquad p_c = 54.7 \text{ atm (55.4 bars)}, \qquad Z_c = 0.257$$

$$\mu_{Ac}\xi = (1.9 \, T_r - 0.29) \times 10^{-4} \, Z_c^{-2/3}$$

with

$$\xi = \frac{T_c^{1/6}}{M^{1/2}P_c^{2/3}} = \frac{461^{1/6}}{\sqrt{44}(54.7)^{2/3}} = 0.029078$$

$$\mu_{Ac} = \frac{1}{0.029078}\left(1.9 \times \frac{548}{461} - 0.29\right) \times 10^{-4} \frac{1}{(0.257)^{2/3}} = 0.016748 \text{ cp}$$

$$\text{or} \quad 0.060293 \frac{\text{kg}}{\text{m} \cdot \text{hr}}$$

## Viscosity of the Gas Mixture

Composition of the reaction mixture:

$$p_A = \frac{1 - x_A}{1 + x_A} p_t = 0.4684 \text{ atm} = 0.4745 \text{ bar}$$

$$p_R = p_S = \frac{x_A}{1 + x_A} p_t = 0.2658 \text{ atm} = 0.2693 \text{ bar}$$

Since the hydrogen content cannot be neglected, Wilke's method may yield too high a value for the viscosity of the mixture. Therefore, the viscosity is computed as

$$\mu_m = \sum y_j \mu_j$$

or $\mu_m = 0.4684 \times 0.05775 + 0.2658(0.0475 + 0.060293) = 0.0557 \text{ kg/m} \cdot \text{hr}$. From Wilke's method, a value of $0.06133 \text{ kg/m} \cdot \text{hr}$ is obtained.

$$M_m = \sum y_j M_j = 0.4684 \times 46 + 0.2658 \times (44 + 2) = 33.77 \frac{\text{kg}}{\text{kmol}}$$

$$\rho_m = \frac{M_{m0}}{V_0} \frac{T_0}{T} = \frac{33.77}{22.4} \times \frac{273}{548} = 0.7510 \frac{\text{kg}}{\text{m}^3}$$

### Diffusion Coefficients

Since some of the required potential parameters are not known, the semiempirical relation of Fuller–Schettler–Giddings will be applied.

$$(\sum v)_{H_2} = 7.07$$

$$(\sum v)_{C_2H_5OH} = 2v_C + 6v_H + v_0 = 2 \times 16.5 + 6 \times 1.98 + 5.48 = 50.4$$

$$(\sum v)_{CH_3CHO} = 2v_C + 4v_H + v_0 = 2 \times 16.5 + 4 \times 1.98 + 5.48 = 46.4$$

$$D_{Eth-H_2} = D_{AS} = \frac{0.001 \times (548)^{1.75}\sqrt{\frac{1}{46} + \frac{1}{2}}}{(1)[(50.36)^{1/3} + (7.07)^{1/3}]^2} = 1.4235 \frac{\text{cm}^2}{s}$$

### Note

$D_{AS}$ has been experimentally measured at 340 K as $0.578 \text{ cm}^2/\text{s}$. The Fuller–Schettler–Giddings formula yields for $D_{AS}$ at 340 K:

$$D_{AS} = 1.4235 \times \left(\frac{340}{548}\right)^{1.75} = 0.6174 \frac{\text{cm}^2}{s}$$

*Error*

$$\frac{0.61742 - 0.578}{0.578} \times 100 = 6.82\%$$

$$D_{\text{Eth}-\text{Ac}} = D_{AR} = \frac{0.001 \times (548)^{1.75} \times \sqrt{\frac{1}{44} + \frac{1}{46}}}{(1)[(46.4)^{1/3} + (50.36)^{1/3}]^2} = 0.2466 \frac{\text{cm}^2}{s}$$

From Eq. 3.2.c-9:

$$D_{Am} = \left( \frac{\dfrac{y_R + y_A}{D_{AR}} + \dfrac{y_S + y_A}{D_{AS}}}{1 + y_A} \right)^{-1}$$

$$= \left( \frac{\dfrac{0.2658 + 0.4684}{0.2466} + \dfrac{0.2658 + 0.4684}{1.4235}}{1 + 0.4684} \right)^{-1}$$

$$= 0.4203 \frac{\text{cm}^2}{s} = 0.1512 \frac{\text{m}^2}{\text{hr.}}$$

Now the Schmidt and Reynolds numbers may be calculated.

$$\text{Sc} = \frac{\mu_m}{\rho_m D_{Am}} = \frac{0.0557}{0.7510 \times 0.1512} = 0.490 \text{ from which } (\text{Sc})^{2/3} = 0.622$$

$$G = \frac{0.01 \times 46}{\frac{\pi}{4}(0.035)^2} = 478.1 \frac{\text{kg}}{\text{m}^2 \cdot \text{hr}}$$

$$\text{Re} = \frac{d_p G}{\mu_m} = \frac{2.289 \cdot 10^{-3} \times 478.1}{0.0557} = 19.65$$

Since Re < 190 the following $j_D$ correlation should be used:

$$j_D = 1.66 \,(\text{Re})^{-0.51} \quad \text{and} \quad j_D = 0.3635$$

Now the partial pressure drop can be calculated.

Assuming that $\Delta p_A = 0$ and with $\delta_A = 1$ the film pressure factor for a reaction $A \rightleftharpoons R + S$ becomes:

$$p_{fA} = p_t + \delta_A p_A = 1 + 1 \times 0.4684 = 1.4684 \text{ atm.}$$

$$\Delta p_A = \frac{r_A M_m p_{fA}}{a_m G j_D} (\text{Sc})^{2/3} = \frac{0.193 \times 33.77 \times 1.4684}{1.26 \times 478.1 \times 0.3635} \times 0.622$$

$$\Delta p_A = 0.02718 \text{ atm} = 0.02753 \text{ bar}$$

CHEMICAL ENGINEERING KINETICS

Substitution of this estimate for $\Delta p_A$ in Eq. (3.2.a-3), written in terms of partial pressures leads to a better estimate for $p_{fA}$

$$p_{fA} = \frac{0.02718}{\ln \dfrac{1.4684}{1.4412}} = 1.4537$$

This new estimate for the film pressure factor may be considered sufficiently close to the starting value 1.4684, so that no further iterations on $\Delta p_A$ need to be performed.

*Estimation of the Temperature Drop over the Film*

The calculation of $\Delta T$ requires two further properties of the reaction mixture to be calculated: the specific heat $c_p$, and the thermal conductivity, $\lambda$.

$c_p$-values for the pure components can be found in the literature or can be estimated accurately from the correlation of Rihani and Doraiswamy [26]. The $c_p$ values are given in the following table. The heat capacity of the mixture may be computed accurately by means of

$$c_{pm} = \sum y_j c_{pj}$$

$$c_{pm} = 0.4684 \times 25.43 + 0.2658(19.39 + 6.995) = 18.92 \frac{\text{kcal}}{\text{kmol K}}$$

$$= 79.085 \text{ kJ/kmol K}$$

The thermal conductivities of the pure components are estimated by Bromley's method.

|  | Ethanol | Acetaldehyde | Hydrogen |
|---|---|---|---|
| $T_b(\text{K})$ | 351.7° | 294° | 20.4° |
| $T_c(\text{K})$ | 516.3° | 461° | 33.3° |
| $\Delta H_v\left(\dfrac{\text{kcal}}{\text{kmol}}\right)$ | 9220 | 8919 | — |
| $\Delta S_v\left(\dfrac{\text{kcal}}{\text{kmol K}}\right)$ | 26.22 | 30.34 | — |
| $\rho_b\left(\dfrac{\text{kmol}}{\text{m}^3}\right)$ | $\dfrac{1000}{63}$ | $\dfrac{783}{44}$ | — |
| $c_p\left(\dfrac{\text{kcal}}{\text{kmol K}}\right)$ | 25.43 | 19.39 | 6.995 |
| $\lambda$ kcal/m.s.K | $1.102 \times 10^{-5}$ | $0.8989 \times 10^{-5}$ | $6.499 \times 10^{-5}$ |

The following details provide the basis for the numbers in the table:

$C_2H_5OH$ (*polar nonlinear molecule*)

From Perry [27]:

$$\Delta H_{vb} = 9220 \frac{kcal}{kmol} = 38600 \text{ kJ/kmol}$$

Consequently,

$$\Delta S_{vb} = \frac{\Delta H_{vb}}{T_b} = \frac{9220}{351.7} = 26.22 \frac{kcal}{kmol\ K} = 109.78 \frac{kJ}{kmol\ K}$$

$\rho_b$, the density of liquid ethanol at the normal boiling point is estimated using Schroeders' rule:

$$V_b = 9 \times 7 = 63 \frac{cm^3}{mol}$$

$$\rho_b = \frac{1000}{63} \frac{kmol}{m^3}$$

$$\alpha = 3\rho_b(\Delta S_{vb} - 8.75 - R \ln T_b)$$

$$\alpha = 3 \times \frac{1}{63}(26.22 - 8.75 - 1.986 \times \ln 351.7) = 0.277$$

$$c_{int\ rot} = \underset{\underset{-CH_2-OH}{\downarrow}}{1.19} + \underset{\underset{CH_3CH_2-}{\downarrow}}{2.03} = 3.22 \frac{kcal}{kmol\ K}$$

$$= 13.48 \text{ kJ/kmol K}$$

$$c_v = c_p - 2 = 25.43 - 2 = 23.43 \frac{kcal}{kmol\ K} = 91.8 \text{ kJ/kmol K}$$

$$\frac{M\lambda}{\mu} = 1.3c_v + 3.6 - 0.3\ c_{int\ rot} - 0.69 \frac{T_c}{T} - 3\alpha$$

$$\frac{46 \times \lambda}{1.604 \times 10^{-5}} = 1.3 \times 23.43 + 3.6 - 0.3 \times 3.22 - 0.69 \times \frac{516.3}{548} - 3 \times 0.277$$

$$= 31.1611 \text{ kcal/mol K}$$

$$= 130.46 \text{ kJ/kmol K}$$

$$\lambda = 1.102 \times 10^{-5} \frac{kcal}{m\ s\ K} = 4.61 \times 10^{-5} \frac{kJ}{m\ s\ K}$$

CH$_3$CHO (*polar nonlinear molecule*)

$\Delta H_{vb}$ has to be estimated

Giacalone's simple method is used.

$$\Delta H_{vb} = \frac{2.303 \, R T_b \, T_c \log p_c}{T_c - T_b} = \frac{2.303 \times 1.986 \times 294 \times 461 \times \log 54.7}{461 - 294}$$

$$\Delta H_{vb} = 8919 \, \frac{\text{kcal}}{\text{kmol}} = 37342 \text{ kJ/kmol}$$

$$\Delta S_{vb} = \frac{\Delta H_{vb}}{T_b} = \frac{8919}{294} = 30.34 \, \frac{\text{kcal}}{\text{kmol K}} = 127.03 \text{ kJ/kmol K}$$

$\rho_b$ is found in the literature:

$$\frac{783}{44} \, \frac{\text{kmol}}{\text{m}^3}$$

$$\alpha = 3 \times \frac{0.783}{44} \, (30.34 - 8.75 - 1.986 \times \ln 294) = 0.55$$

$$c_{\text{int rot}} = 1.21 \text{ kcal/kmol K} = 5.07 \text{ kJ/kmol K}$$

$$c_v = c_p - 2 = 19.39 - 2 = 17.39 \, \frac{\text{kcal}}{\text{kmol K}} = 72.81 \text{ kJ/kmol K}$$

$$\frac{44 \times \lambda}{1.675 \times 10^{-5}} = 1.3 \times 17.39 + 3.6 - 0.3 \times 1.21 - 0.69 \times \frac{461}{548} - 3 \times 0.55$$

$$= 23.614 \, \frac{\text{kcal}}{\text{kmol K}} = 98.867 \text{ kJ/kmol K}$$

$$\lambda = 0.8989 \times 10^{-5} \, \frac{\text{kcal}}{\text{m s K}} = 3.763 \times 10^{-5} \text{kJ/m s K}$$

H$_2$ (*nonpolar linear molecule*)

$$\frac{M\lambda}{\mu} = 1.3 \, c_v + 3.4 - 0.7 \frac{T_c}{T}$$

$$\frac{2 \times \lambda}{1.32 \times 10^{-5}} = 1.3 \times 4.995 + 3.4 - 0.7 \times \frac{33.3}{548} = 9.851 \text{ kcal/kmol K}$$

$$= 41.244 \text{ kJ/kmol K}$$

$$\lambda = 6.499 \times 10^{-5} \, \frac{\text{kcal}}{\text{m s K}} = 27.21 \times 10^{-5} \text{kJ/m s K}$$

*Thermal Conductivity of the Gas Mixture*

To estimate the factors $A_{ij}$, the Lindsay–Bromley equation is appropriate and will be applied here.

The required Sutherland constants are

$$S_{Eth} = 1.5 \times 351.7 = 527.55 \text{ K} \qquad S_{Eth-Ac} = 482.34 \text{ K}$$
$$S_{Ac} = 1.5 \times 294 = 441 \text{ K} \qquad S_{Eth-H_2} = 204.15 \text{ K}$$
$$S_{H_2} = 79 \text{ K} \qquad S_{Ac-H_2} = 186.65 \text{ K}$$

The Lindsay–Bromley formula yields

$$A_{12} = 0.9615 \qquad A_{21} = 1.038$$
$$A_{13} = 0.3653 \qquad A_{31} = 3.1565$$
$$A_{23} = 0.3872 \qquad A_{32} = 3.0988$$

$$\lambda_m = \frac{1.102 \times 10^{-5}}{1 + 0.9615 \times \dfrac{0.2658}{0.4684} + 0.3654 \times \dfrac{0.2658}{0.4684}}$$

$$+ \frac{0.8989 \times 10^{-5}}{1 + 1.038 \times \dfrac{0.4684}{0.2658} + 0.387 \times 1}$$

$$+ \frac{6.499 \times 10^{-5}}{1 + 3.156 \times \dfrac{0.4684}{0.2658} + 3.0988 \times 1}$$

$$\lambda_m = 0.6286 \times 10^{-5} + 0.2795 \times 10^{-5} + 0.6727 \times 10^{-5} = 1.5808 \times 10^{-5} \frac{\text{kcal}}{\text{m s K}}$$

$$= 6.618 \times 10^{-5} \text{ kJ/m s K}$$

***Note***

If the thermal conductivity of the mixture would have been considered as linear in the composition, $\lambda_m$ would be given by

$$\lambda_m = \sum y_j \lambda_j$$

or

$$\lambda_m = [0.4684 \times 1.102 + 0.2658 \times (0.8989 + 0.6499)] \times 10^{-5}$$

$$\lambda_m = 2.4825 \times 10^{-5} \frac{\text{kcal}}{\text{m s K}} = 10.394 \times 10^{-5} \text{ kJ/m s K}$$

This value is 50 percent higher than the more correct estimate.
Then, the Prandtl number is:

$$Pr = \frac{\dfrac{c_p}{M_m}\mu_m}{\lambda_m} = \frac{\dfrac{18.9}{33.8} \times \dfrac{0.0557}{3600}}{1.5808 \times 10^{-5}} = 0.5473$$

$$(Pr)^{2/3} = 0.670$$

From Fig. 3.2.b-1, a value of 0.60 may be chosen for $j_H$ at Re = 19.65. The heat of reaction is calculated as follows:

$$(-\Delta H) = (\Delta H)_{\text{Eth}} - (\Delta H)_{\text{Ac}} - (\Delta H)_{\text{H}_2} = 16800\,\frac{\text{kcal}}{\text{kmol}} = 70338 \text{ kJ/kmol}$$

so that

$$\Delta T = \frac{r_A(-\Delta H)(Pr)^{2/3}}{a_m j_H c_p G} = \frac{0.193 \times 16800 \times 0.670}{1.26 \times 0.6 \times \dfrac{18.92}{33.77} \times 478.1} = 10.7°$$

This is a difference between bulk and surface temperatures that may be considered as significant.

---

## Part Two
## Intraparticle Gradient Effects

Now that we have discussed various aspects of external mass transfer and surface reactions, the remaining problem of transport and reaction when the catalytic surface is not directly accessible to the bulk fluid needs to be described.

## 3.4 Catalyst Internal Structure

From the discussion of surface rates, it is seen that the total rate of reaction is proportional to the amount of catalytic surface present. The usual way to obtain a very large amount of catalytic surface area is to use a porous material with many small pores. The reason that this provides an enormous increase in area can be simply seen by considering a given volume of space filled with successively smaller tubes. For a cylinder

$$\frac{\text{surface area}}{\text{volume}} = \frac{2\pi r L}{\pi r^2 L} = \frac{2}{r} \tag{3.4-1}$$

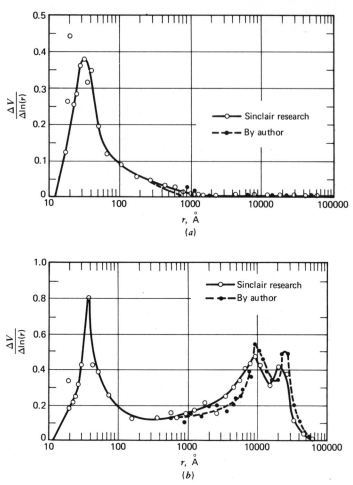

Figure 3.4-1  Pore-size distribution in catalyst pellets. (a) Pellet 2. (b) Pellet 1. (From Cunningham and Geankoplis [28].)

If a volume were filled with cylinders (idealized pores) of radius 2 cm, $2\mu m$, 20 Å, Eq. 3.4-1 gives

| Size of Cylinder, cm | Total Surface Area/Unit Volume, $cm^2/cm^3$ |
| --- | --- |
| 2.0 | 1.0 |
| $2 \times 10^{-4}$ | $10^4$ |
| $2 \times 10^{-7}$ | $10^7$ |

Thus, the amount of area in the unit volume is very much larger when it contains small pores and so most practical catalysts are manufactured in this form. Typical values of the amount of internal surface area available range from 10 m²/g cat. to 200 m²/g cat. with most toward the larger value.

A typical catalyst pellet will have a pore size distribution as shown in Fig. 3.4-1, given by Cunningham and Geankoplis [28].

The major pore sizes in pellet "2" (Fig. 3.4-1a) are between ~20 to 200 Å, although depending on the specific manufacturing details, many other distribution curves are possible. One important special case is where the pellet is made by compressing smaller particles together, for which the second peak in Fig. 3.4-1b (pellet "1") represents the so-called "macropores" between the particles while the usual peak represents the "micropores." Based on the above arguments, most of the catalytic surface is contained in the micropores, but all of the pores can contribute to diffusion resistances. Both pellets were made from 90 μm grains of alumina, but pellet 1 was not as highly compressed in manufacture; thus pellet 1 would be expected to have significant macropore structure, but not pellet 2. The physical properties were

| Pellet | Length [cm] | Dia [cm] | $S_g$ [m²/g cat] | $V_g$ [cm³/g cat] | $\rho_s$ [g cat/cm³ cat] |
|---|---|---|---|---|---|
| 1 | 1.705 | 2.623 | 314 | 1.921 | 0.441 |
| 2 | 1.717 | 2.629 | 266 | 0.528 | 1.115 |

The internal void fraction, or porosity, of each pellet is given by

$$\varepsilon_s = \rho_s V_g \qquad \varepsilon_{s1} = 0.85 \qquad \varepsilon_{s2} = 0.59$$

An excellent reference that discusses the methods for determination of pore area, volume, and size distributions is Gregg and Sing [29], where they show how to utilize nitrogen adsorption data for these purposes.

The "pore size distribution" can be defined:

$$f(r)dr = \text{fraction open volume of } (r, r + dr) \text{ pores} \tag{3.4-2}$$

Thus,

$$\varepsilon_s = \int f(r)dr \tag{3.4-3}$$

The above data were plotted with a logarithmic abcissa because of the large range of sizes covered, and the ordinate is such that

$$\int \left(\frac{\Delta V}{\Delta \ln r}\right) d(\ln r) = V_g = \varepsilon_s/\rho_s$$

directly from the graph. Thus, it is easily seen that the pore size distribution is found from

$$f(r) = \rho_s \left( \frac{\Delta V}{\Delta r} \right) \tag{3.4-4}$$

It is often very convenient not to have to utilize the entire curve, $f(r)$, by defining a mean pore size,

$$\bar{r} \equiv \frac{1}{\varepsilon_s} \int r f(r) dr \tag{3.4-5}$$

which can be computed from the $f(r)$ data. If the pores can be considered to be cylinders, the total internal surface area (which can also be measured directly) per pellet volume would be

$$\rho_s S_g = \int \frac{2}{r} f(r) dr \tag{3.4-6}$$

Now if there were really a single pore size,

$$f(r) = \varepsilon_s \delta(r - r_m)$$

where $\delta(\ )$ is the Dirac delta function, and then:

$$\bar{r} = r_m$$

$$\rho_s S_g = 2\varepsilon_s/r_m$$

Therefore, $\bar{r}$ could be found from:

$$\bar{r} = 2\varepsilon_s/\rho_s S_g \tag{3.4-7}$$

Since $\rho_s S_g$ and $\varepsilon_s$ can be measured more simply than the complete $f(r)$, this is a commonly-used approach.

It can be seen from the above derivation that the use of the average pore radius, $\bar{r}$, would be best for a fairly narrow pore size distribution, and possibly not very accurate for a wide one. Also, for a bimodal distribution, $\bar{r}$ occurs in the "valley," or the pore size present in least amount. These results can be seen from the above data where $\bar{r}_1 = 123$ Å and $\bar{r}_2 = 40$ Å. Therefore, except for a narrow pore size distribution, the more complete characterization by $f(r)$ should be used. Actually, automated equipment is now available to measure $f(r)$, and this should be done if there is any question of a complicated pore structure.

The assumption of (infinitely long) cylindrical pores is obviously not always going to be true for all porous solid structures, but Brunauer, Mikhail, and Bodor [30] argue that this shape is intermediate between flat-plate and spherical shapes, and, on the average for the (usually) unknown pore structure, would probably give the best results.

162 _____ CHEMICAL ENGINEERING KINETICS

## 3.5 Pore Diffusion _____

### 3.5.a Definitions and Experimental Observations

Let us first consider diffusion in an idealized single cylindrical pore. Fick's law for a binary system with equimolar counter diffusing occurring is:

$$N_A = -D_{AB} \frac{dC_A}{dz} \tag{3.5.a-1}$$

where $N_A$ is expressed in moles of $A$ diffusion per unit pore cross-section and unit time and where $z$ is the diffusion path length along the pore. The diffusivity, $D_{AB}$, is the ordinary fluid molecular diffusivity as used in other transport phenomena studies, and values for it can be found in handbooks. $D_{AB}$ varies as $T^{1.50}$ and $p^{-1}$ for gases. It is the result of fluid-fluid intermolecular collisions as considered in the kinetic theory of gases. When the pore size gets so small that its dimensions are less than the mean path of the fluid, however, fluid-fluid collisions are no longer the dominant ones. Instead, fluid-wall collisions are important, and the mode of diffusive transport is altered. This can occur for gases at less than atmospheric pressure, although not usually for liquids, in typical pellets. From the kinetic theory of gases, the so-called Knudsen diffusivity can be formulated to take the place of $D_{AB}$ in Eq. 3.5.a-1:

$$D_{KA} = \frac{4}{3} r \left( \frac{2}{\pi} \frac{RT}{M_A} \right)^{1/2} \tag{3.5.a-2}$$

where $M_A$ = molecular weight of the diffusing species. Note that $D_{KA}$ is a function of the pore radius, $r$, and varies with $T^{0.5}$, but is independent of $p_t$ (Strider and Aris [31], have generalized these results to more complicated shapes, for example, overlapping spheres structures).

Equation 3.5.a-2 was derived assuming totally random, or diffusive, collisions of the gas molecules with the wall, which is reasonable when the pore size is still large with respect to molecular *dimensions* (but much smaller than the mean free path). A further extension of this reasoning is to the case where the pore size is, in fact, of the same order of magnitude as the molecules themselves. Weisz [32] terms this region "configurational" diffusion, and Fig. 3.5.a-1 presents his estimate of the order of magnitude of the observed diffusivities.

The events here would be expected to be very complicated since specific details of the force-fields and so on of the molecules making up the walls and their interactions with the diffusing molecules would have to be accounted for. These situations can arise from considering very large molecules in the usual catalysts, such as in petroleum desulfurization processes, from solids with very small pores, such as zeolite catalysts, and in many biological situations such as diffusion across cell walls. Fairly large molecules in small capillaries can also undergo surface migration

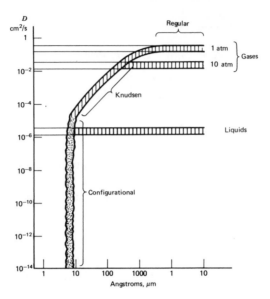

*Figure 3.5.a-1 Diffusivity and size of aperture (pore); the classical regions of regular and Knudsen and the new regime of configurational diffusion (adapted from Weisz, [32]).*

and other complications. There is no comprehensive theory yet available for these problems, but because of the recent importance and interest in zeolites, they are being intensively investigated (see Brown, Sherry, and Krambeck [33]; reviews are by Riekert [34] and Barrer [35].

## Example 3.5.a-1 Effect of Pore Diffusion in the Cracking of Alkanes on Zeolites

An interesting semiquantitative illustration of the possible strong effects of pore diffusion on a chemical reaction was provided by Gorring [36]. Hydrocarbon cracking was briefly discussed in Chapter 2, where a typical product distribution from silica-zirconia or silica-alumina catalyst was described. The cracking of $n$-tricosane over the zeolite H-erionite (Chen, Lucki, and Mower [37]) yielded a strikingly different result, shown in Fig. 1.

There are almost no $C_7$–$C_9$ products and maxima at $C_3$ and $C_{11}$. In the absence of any reason for the catalytic reaction to have this behavior inherently, it was postulated that diffusion in the rather restricted pores or "cages" of erionite might provide the answer (Figs. 2 and 3).

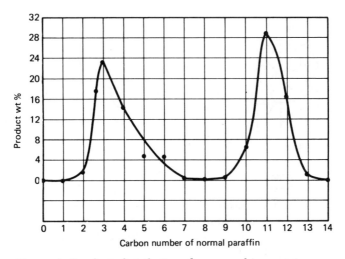

*Figure 1 Product distribution from cracking n-tricosane over H-erionite at 340°C (from Chen, Lucki, and Mower [37], after Gorring [36]).*

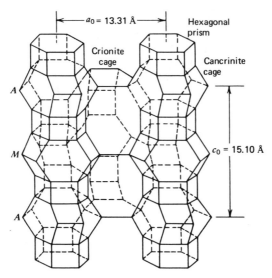

*Figure 2 View of erionite framework (from Gorring [36]).*

165

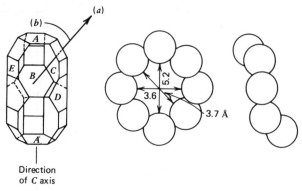

Direction
of *C* axis

*Figure 3 (a) Erionite cage viewed approximately 20°
from direction of δ-axis. (b) Erionite 8-membered ring
front and side views. View of offretite framework (from
Gorring [36]).*

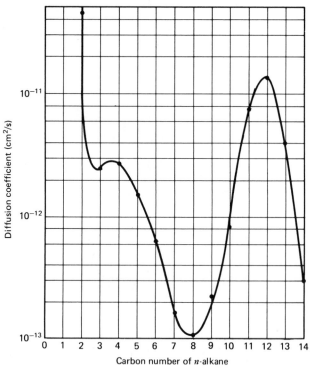

*Figure 4 Diffusion coefficients of n-alkanes in potassium
T zeolite at 300°C (from Gorring [36]).*

**166**

Gorring therefore measured effective diffusivities (the exact physical meaning of these "diffusivities" computed from the experimental data is not completely clear) for the *n*-alkanes, with the following results (shown in Fig. 4).

Note that the diffusivities change by order of magnitude, thus having a great effect on the relative concentrations and reaction rates. The underlying quantitative reasons for this so-called "window effect" are not completely clear, beyond estimations of the close dimensional fits of the molecules in the cages provided by Gorring. As mentioned above, this area of configurational diffusion needs additional work. Further applications of "shape-selective" catalysis have been reviewed by Chen and Weisz [38].

## 3.5.b General Quantitative Description of Pore Diffusion

In an actual solid, with its complicated pore structure, the concept of an effective diffusivity is defined by the equation:

$$N_A = -D_e \frac{dC_A}{dz} \tag{3.5.b-1}$$

where $N_A$ is expressed in moles of $A$ diffusing per unit pellet surface area and unit time. This measurable diffusion flux is per unit area of pellet, consisting of both pores and solid. It is therefore related to that of Eq. 3.5.a-1 by the ratio of surface holes/total area, which, for random pores using Dupuit's law, is equivalent to the internal void fraction, $\varepsilon_s$ (usually with values between 0.3 and 0.8). Also, the diffusion path length along the pores is greater than the measurable pellet thickness due to their "zigzag" nature and to constrictions, and so on. The concentration gradient must also thus be corrected by a "tortuosity factor," $\tau$, leading to:

$$N_A = -\varepsilon_s D_A \frac{dC_A}{\tau dz} \tag{3.5.b-2}$$

The definition of tortuosity factor in Eq. 3.5.b-2 includes both the effect of altered diffusion path length as well as changing cross-sectional areas in constrictions; for some applications, especially with two-phase fluids in porous media, it may be better to keep the two separate (e.g., Van Brakel and Heertjes [39]). This tortuosity factor should have a value of approximately $\sqrt{3}$ for loose random pore structures, but measured values of 1.5 up to 10 or more have been reported. Satterfield [40] states that many common catalyst materials have a $\tau \simeq 3$ to 4; he also gives further data.

Thus, the effective diffusivity would have the form

$$D_{eA} = \frac{\varepsilon_s}{\tau} D_A \tag{3.5.b-3}$$

The units of $D_e$ are

$$\frac{m_f{}^2}{m_p{}^2} \cdot \frac{m_p}{m_f} \cdot \frac{m_f{}^2}{s} = \frac{m_f{}^3}{m_p \cdot s}$$

Turning to a general description of pore diffusion, the "dusty gas" theory of Mason et al. [41, 42] utilizes the results from the formal kinetic theory of gases, with one "species," the "dust," having a very large "molecular weight." Their final results can be clearly visualized in the form utilized by Feng and Stewart [43].

$$\mathbf{N}_j = \text{(diffusive flux)} + \text{(viscous flow flux)}$$
$$+ \text{(fluxes caused by other driving forces)}$$
$$= \mathbf{N}_j{}^{(D)} + \mathbf{N}_j{}^{(v)} + \cdots \qquad (3.5.b\text{-}4)$$

where the viscous flow flux is found from

$$\mathbf{N}_j{}^{(v)} = -y_j \left( \frac{B_0 p_t}{RT\mu} \right) \nabla p_t \qquad (3.5.b\text{-}5)$$

with $B_0$ = D'Arcy constant, a function of porous media geometry

$$= r^2/8 \text{ for a long cylinder of radius } r$$

and the diffusive flux is found from the extended Stefan–Maxwell form:

$$\frac{-1}{RT} \nabla p_j = \sum_{k=1}^{N} \frac{1}{D_{e,jk}} (y_k \mathbf{N}_j{}^{(D)} - y_j \mathbf{N}_k{}^{(D)}) + \frac{\mathbf{N}_j{}^{(D)}}{D_{e,Kj}} \qquad (3.5.b\text{-}6)$$

Equations 3.5.b-4 to 6 can also be combined to give a single equation containing only the total flux resulting from both diffusive and viscous flow mechanisms:

$$\frac{-1}{RT} \nabla p_j = \sum_{k=1}^{N} \frac{1}{D_{e,jk}} (y_k \mathbf{N}_j - y_j \mathbf{N}_k) + \frac{\mathbf{N}_j}{D_{e,Kj}} + \frac{y_j}{D_{e,Kj}} \left( \frac{p_t B_0}{RT\mu} \right) \nabla p_t \qquad (3.5.b\text{-}7)$$

The use of these full equations involves the same complexity as described earlier in Section 3.2 for the ordinary Stefan–Maxwell equations. In a binary system, the above Eq. 3.5.b-7 gives, using $y_B = 1 - y_A$

$$\mathbf{N}_A = -D_{eA} \frac{\nabla p_A}{RT} - \frac{D_{eA}}{D_{e,KA}} y_A \left( \frac{p_t B_0}{RT\mu} \right) \nabla p_t \qquad (3.5.b\text{-}8)$$

where

$$\frac{1}{D_{eA}} = \frac{1 - y_A(1 + N_B/N_A)}{D_{e,AB}} + \frac{1}{D_{e,KA}} \qquad (3.5.b\text{-}9)$$

CHEMICAL ENGINEERING KINETICS

For equimolar counterdiffusion, $N_B = -N_A$, and then,

$$\frac{1}{D_{eA}} = \frac{1}{D_{e,AB}} + \frac{1}{D_{e,KA}} \qquad (3.5.b\text{-}10)$$

This additive resistance relation is often called the "Bosanquet formula."

For large pore materials (i.e., micron size pores) such as some carbons and glass, or for very high pressure drops, the forced flow term can be important (e.g., Gunn and King [44]). A detailed study of the effects of pressure gradients was presented by Di Napoli, Williams, and Cunningham [45], including criteria for when the isobaric equations are adequate; for less than 10 percent deviations, the following must be true:

$$\frac{B_0 p_t}{\mu D_{e,KA}} > 10 - 20 \qquad (3.5.b\text{-}11)$$

For isobaric and isothermal conditions, Eq. 3.5.b-8 gives

$$\mathbf{N}_A = -D_{eA} \nabla C_A \qquad (3.5.b\text{-}12)$$

which was also derived by Scott and Dullien and Rothfeld [46, 47] by a somewhat different method.

Similarly, for the second component,

$$\mathbf{N}_B = -D_{eB} \nabla C_B$$
$$= +D_{eB} \nabla C_A$$

for pure diffusion and steady-state conditions. Thus, the ratio of fluxes then always is given by

$$-\frac{N_B}{N_A} = \frac{D_{eB}}{D_{eA}} = \frac{D_{e,KB}}{D_{e,KA}} = \left(\frac{M_A}{M_B}\right)^{1/2} \qquad (3.5.b\text{-}13)$$

where the penultimate expression utilized Eq. 3.5.b-9, and the last, Eq. 3.5.a-2. Equation 3.5.b-13 is true for all pressure levels (if Eq. 3.5.b-11 is satisfied), not just in the Knudsen region, and is known as Graham's law.

To use Eqs. 3.5.b-7, 8, or 12 to predict the pore diffusivity requires knowledge of two parameters: the porosity/tortuosity ratio, $\varepsilon_s/\tau$, and the average pore radius, $\bar{r}$. The major difficulty resides in obtaining values for the tortuosity, $\tau$. The porosity, $\varepsilon_s$, is usually readily measured, as is a mean pore radius, $\bar{r}$, and values for the molecular and Knudsen diffusivities, $D$, can be estimated or found in data tabulations. Since real solids are normally quite complex in their internal structure, the tortuosity must usually be obtained from data on the actual solid of interest— Satterfield [40] gives typical values. Often this means performing a pore diffusion measurement at one pressure level, to define $\tau$, and then the above equations can be utilized to predict values for other conditions. For examples of this see Satterfield and Cadle [48, 49], Brown, Haynes, and Manogue [50], Henry, Cunningham, and Geankoplis [51] and Cunningham and Geankoplis [28].

For steady-state one-dimensional diffusion experiments (see Satterfield [40]), $\nabla \cdot \mathbf{N}_A = 0$ or $N_A = $ constant, and so Eq. 3.5.b-12 can be directly integrated between $z = 0$ and $z = L$:

$$N_A = \frac{C_t D_{e,AB}}{L(1 + N_B/N_A)} \ln\left[\frac{1 - (1 + N_B/N_A)y_A(L) + D_{e,AB}/D_{e,KA}}{1 - (1 + N_B/N_A)y_{A0} + D_{e,AB}/D_{e,KA}}\right] \quad (3.5.b\text{-}14)$$

with Eq. 3.5.b-13

$$\frac{N_B}{N_A} = -\left(\frac{M_A}{M_B}\right)^{1/2}$$

Since $(D_{e,AB}/D_{e,KA}) = \alpha P^{-1}$, measurements at various pressure levels permits determination of both $D_{e,AB}$ and $D_{e,KA}$. This experimental approach is usually termed the Wicke–Kallenbach method (Wicke and Kallenbach [52] or Weisz [53]; also see Satterfield [40]) and has been widely used to measure effective diffusivities. Transient methods are also available (e.g., Dogu and Smith [54]).

Again, for multicomponent systems, a practical method is to define an effective binary diffusivity as was done in Sec. 3.2. Using fluxes with respect to the pellet.

$$N_j = -D_{e,jm} \frac{dC_j}{dz} \quad (3.5.b\text{-}15)$$

Then, as in Section 3.2, the concentration gradient from Eq. 3.5.b-15 is equated to that of Eq. 3.5.b-7 (with $dp_t/dz = 0$), to give

$$\frac{1}{D_{e,jm}} = \sum_{k=1}^{N} \frac{1}{D_{e,jk}}\left(y_k - \frac{N_k}{N_j} y_j\right) + \frac{1}{D_{e,Kj}} \quad (3.5.b\text{-}16)$$

(Also see Butt [55].)

For chemical reactions, the steady-state flux ratios in Eq. 3.5.b-16 are determined by the stoichiometry $N_k/N_j = \alpha_k/\alpha_j$. As pointed out by Feng, Kostrov, and Stewart [56], however, this only leads to simple results for single reactions, since there is no simple relation between the species fluxes for complex networks.

### 3.5.c  The Random Pore Model

For the actual pore-size distribution to be taken into account, the above relations for single pore sizes are usually assumed to remain true, and they are combined with the pore size distribution information. The "random pore" model, or micro–macro pore model, of Wakao and Smith [57, 58]) is useful for compressed particle type pellets. The pellet pore-size distribution is, somewhat arbitrarily, broken up into macro ($M$) and micro ($\mu$) values for the pore volume and average pore radius: $\varepsilon_M$, $r_M$ and $\varepsilon_\mu$, $r_\mu$ (often a pore radius of $\sim 100$ Å is used as the dividing point). Based on random placement of the microparticles within the macropellet pores,

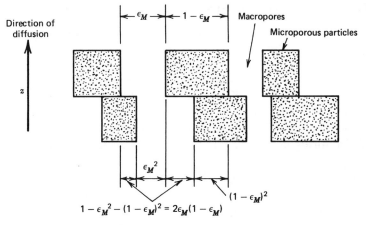

*Figure 3.5.c-1 Diffusion areas in random pore model. (Adapted from Smith [24].)*

a probabilistic argument for diffusion through the macroregions, the microregions, and series interconnections gives the indicated areas (see Fig. 3.5.c-1): The various parallel contribution are added up as follows:

$$
D_e = \varepsilon_M{}^2 D_M + (1 - \varepsilon_M)^2 \frac{\varepsilon_\mu{}^2}{(1 - \varepsilon_M)^2} D_\mu + 2[2\varepsilon_M(1 - \varepsilon_M)] \frac{\varepsilon_\mu{}^2}{(1 - \varepsilon_M)^2} D_\mu
$$

$$
= \varepsilon_M{}^2 D_M + \frac{\varepsilon_\mu{}^2(1 + 3\varepsilon_M)}{1 - \varepsilon_M} D_\mu \tag{3.5.c-1}
$$

where in the second and third terms the $D_\mu$ is based on the microvoid area, and so the ratio (microvoid/particle) area is required, and in the last term it is also assumed that in the macro–micro series part, the microdiffusion is the dominant resistance. In Eq. 3.5.c-1 $D_M$ and $D_\mu$ are found from Eq. 3.5.b-10, but not correcting for porosity and tortuosity which are already accounted for in Eq. 3.5.c-1:

$$
\frac{1}{D_{M \text{ or } \mu}} = \frac{1}{D_{AB}} + \frac{1}{D_{KM} \text{ or } D_{K\mu}} \tag{3.5.c-2}
$$

Note again that no tortuosity factor appears in Eq. 3.5.c-1; for either $\varepsilon_\mu = 0$ or $\varepsilon_M = 0$, it reduces to

$$
D_e = (\varepsilon_s{}^2 D)_{M \text{ or } \mu} \tag{3.5.c-3}
$$

which implies that $\tau = 1/\varepsilon_s$. This is often a reasonable approximation—see Weisz and Schwartz [59] and Satterfield [40].

For catalysts without unambiguous micro and macro pores, a different approach is required.

### 3.5.d The Parallel Cross-Linked Pore Model

More general models for the porous structure have also been developed by Johnson and Stewart [60] and by Feng and Stewart [43], called the parallel cross-linked pore model. Here, Eqs. 3.5.b-4 to 6 or Eq. 3.5.b-7 are considered to apply to a single pore of radius $r$ in the solid, and the diffusivities interpreted as the actual values rather than effective diffusivities corrected for porosity and tortuosity. A pore size and orientation distribution function $f(r, \Omega)$, similar to Eq. 3.4-2, is defined. Then $f(r, \Omega)drd\Omega$ is the fraction open area of pores with radius $r$ and a direction that forms an angle $\Omega$ with the pellet axis. The total porosity is then

$$\varepsilon_s = \iint f(r, \Omega)drd\Omega \qquad (3.5.d-1)$$

and the total internal surface area

$$(\rho_s S_g) = \iint \frac{2}{r} f(r, \Omega)drd\Omega \qquad (3.5.d-2)$$

The pellet flux is found by integrating the flux in a single pore with orientation $l$ and, by accounting for the distribution function;

$$\mathbf{N}_j = \iint \boldsymbol{\delta}_l N_{j,l} f(r, \Omega)drd\Omega \qquad (3.5.d-3)$$

where $\boldsymbol{\delta}_l$ represents a unit vector or direction cosine between the $l$-direction and the coordinate axes. Feng, Kostrov, and Stewart [56] utilize the complete Stefan–Maxwell formulation, Eq. 3.5.b-7, but we will only give results for the simpler mean binary diffusivity. Applied to a single pore and therefore excluding the porosity and tortuosity corrections, Eq. (3.5.b-15) may be written

$$N_{j,l} = -D_{jm} \frac{dC_j}{dl}$$

$$= -D_{jm} \boldsymbol{\delta}_l \cdot \nabla C_j \qquad (3.5.d-4)$$

where

$$\frac{1}{D_{jm}} = \sum_{k=1}^{N} \frac{1}{D_{jk}} \left( y_k - \frac{N_k}{N_j} y_j \right) + \frac{1}{D_{Kj}} \qquad (3.5.d-5)$$

Then, Eq. 3.5.d-3 becomes

$$\mathbf{N}_j = -\iint D_{jm} \boldsymbol{\delta}_l \boldsymbol{\delta}_l \cdot \nabla C_j f(r, \Omega)drd\Omega \qquad (3.5.d-6)$$

The term $\boldsymbol{\delta}_l \boldsymbol{\delta}_l$ is the tortuosity tensor.

      

Two limiting cases can be considered:

1. Perfectly communicating pores, where the concentrations are identical at a given position $z$—that is, $C_j(z; r, \Omega) = C_j(z)$.

2. Noncommunicating pores, where the complete profile $C_j(z; r, \Omega)$ is first found for a given pore, and then averaged.

   For pure diffusion at steady state, $dN_j/dz = 0$ or $N_j = $ constant, as used earlier for Eq. 3.5.b-14, and thus Eq. 3.5.d-6 can be directly integrated

$$N_{jz} \int_0^L dz = \iint dr d\Omega \left[ - \int_{C_{jo}}^{C_{jL}} D_{jm} dC_j \right] \delta_l \delta_l f(r, \Omega) \qquad (3.5.d\text{-}7)$$

where the square bracket in Eq. 3.5.d-7 would integrate to the same form as Eq. 3.5.b-14. Therefore, for steady-state pure diffusion, no assumption need be made about the communication of the pores, and Eq. 3.5.d-7 will always result. For other situations, however, the two extremes give different results, as will be discussed later for chemical reactions.

It would seem that for the usual types of catalyst pellets with random pore structure, the situation would be closest to the communicating pore case; then, since $C_j$ is now independent of $r$ and $\Omega$, Eq. 3.5.d-6 can be written as,

$$\mathbf{N}_j = - \left[ \int D_{jm}(r) \kappa(r) d\varepsilon_s(r) \right] \cdot \nabla C_j \qquad (3.5.d\text{-}8)$$

where $\kappa(r)$ is a reciprocal tortuosity that results from the $\Omega$-integration, and also the differential form of Eq. 3.5.d-1 was used. Thus, Eq. 3.5.d-8 provides the result that the proper diffusivity to use is one weighted with respect to the measured pore-size distribution.

Finally, if the pore size and orientation effects are uncorrelated,

$$f(r, \Omega) = f(r) f_\Omega(\Omega)$$

where $f(r)$ is exactly the distribution function of Eq. 3.4-2, and

$$\int f_\Omega(\Omega) d\Omega = 1$$

Then, Eq. 3.5.d-8 becomes

$$\mathbf{N}_j = - \left[ \int D_{jm}(r) d\varepsilon_s(r) \right] \left[ \int f_\Omega \kappa d\Omega \right] \cdot \nabla C_j \qquad (3.5.d\text{-}9)$$

For completely random pore orientations, the tortuosity depends only on the vector component $\cos \Omega$, and

$$\int f_\Omega \kappa d\Omega = \int f_\Omega \cos^2 \Omega d\Omega = \frac{1}{3}$$

so that in the notation of Eq. 3.5.b-3, $\tau = 3$. Recall that this value is commonly, but not always, found (Satterfield [40]).

Satterfield and Cadle [48, 49] and Brown, Haynes, and Manogue [50] have tested the various models against experimental data from several types of solids, pressures, and the like. Both the macro-micro and the parallel path models are often superior to the simple mean pore-size model, as might be expected; the former two are more or less equivalent, where applicable, but the parallel path model seems to be slightly more general in its predictive abilities. These theoretical models do not completely describe all aspects of pore diffusion, and some complex interactions have recently been described by Brown et al. [61, 62] and by Abed and Rinker [63].

Feng et al. [56] and Patel and Butt [64] have compared the fit of several of the above models to extensive experimental, multicomponent pore diffusion data, with resulting standard deviations in the range of 0.1.

In summary, a fairly narrow unimodal pore-size distribution can be adequately described by the simple mean pore-size model. A broad pore-size distribution, $f(r)$, requires a more extensive treatment, such as the parallel path model. A bimodal pore-size distribution can also be described by the micro-macro random pore model.

### 3.5.e Pore Diffusion with Adsorption; Surface Diffusion; Configurational Diffusion

When sorption of the diffusing species occurs, two additional complications may arise. One, the sorbed phase can have a sufficiently large accumulation of solute that it must be included in the mass balance equations. Second, the sorbed phase could be mobile, which would add to the diffusion flux. The former case has been extensively considered in a series of papers by Weisz, Zollinger, and Rys et al. [65]. The mass balance becomes

$$\varepsilon_s \frac{\partial C_A}{\partial t} + \rho_s \frac{\partial C_{Al}}{\partial t} = \frac{\partial}{\partial z} D_e \frac{\partial C_A}{\partial z} \tag{3.5.e-1}$$

where $C_{Al}(\text{kmol/kgsol}) = C_{Al}(C_A)$ through the adsorption process. If instantaneous adsorption equilibrium is assumed, the functional form is found from the isotherm, and (for constant $D_e$)

$$\frac{\partial C_A}{\partial t} = \left( \frac{D_e}{\varepsilon_s + \rho_s dC_{Al}/dC_A} \right) \frac{\partial^2 C_A}{\partial z^2} \tag{3.5.e-2}$$

The usual diffusion results are then used, but with a modified effective diffusivity, that does not have the same value as the steady-state value, $D_e$.

The second situation of "surface diffusion" is less well understood. It is usually represented by a Fickian-type flux expression, using the adsorbed concentration as the driving force:

$$N_s = -D_{s,e}\rho_s \frac{dC_{Al}}{dz} \tag{3.5.e-3}$$

If instantaneous adsorption equilibrium is again assumed, the total flux is

$$N_A = -\left(D_e + D_{s,e}\rho_s \frac{dC_{Al}}{dC_A}\right)\frac{dC_A}{dz} \tag{3.5.e-4}$$

Thus, except for a simple linear isotherm, $C_{Al} = (C_t K_A)C_A$, the diffusivity is concentration dependent. The mass balance now becomes

$$\frac{\partial C_A}{dt} = \frac{1}{\varepsilon_s + \rho_s dC_{Al}/dC_A}\frac{\partial}{\partial z}\left(D_e + D_{s,e}\rho_s \frac{dC_{Al}}{dC_A}\right)\frac{\partial C_A}{\partial z} \tag{3.5.e-5}$$

which, for a linear isotherm, reduces to

$$= \left(\frac{D_e + D_{s,e}\rho_s C_t K_A}{\varepsilon_s + \rho_s C_t K_A}\right)\frac{\partial^2 C_A}{\partial z^2} \tag{3.5.e-5a}$$

Some recent discussions of the theoretical bases are by Yang, Fenn, and Haller [66] and Sladek, Gilliland, and Baddour [67] for gases, and by Dedrick and Beckmann [149] and Komiyama and Smith [68] for liquids. Values of $D_{s,e}$ have been collected by Schneider and Smith [69] and by Sladek et al. [67], and for hydrocarbon gases in the usual catalyst substrate materials have values in the range $10^{-5}$–$10^{-3}$ cm$_p^2$/s. The contribution to the mass flux is most important for microporous solids, and can be appreciable under some conditions, especially for liquids.

### *Example 3.5.e-1 Surface Diffusion in Liquid-Filled Pores*

Komiyama and Smith [70] have studied intraparticle mass transport of benzaldehyde in polymeric porous amberlite particles. With methanol as the solvent, there was very little adsorption of the benzaldehyde, and the uptake data could be accurately represented by the usual constant diffusivity equation, Eq. 3.5.e-5a with a linear adsorption isotherm, as seen in Fig. 1. The porosity was about $\varepsilon_s = 0.5$, and the tortuosity about $\tau = 2.7$, which is a reasonable value based on earlier discussion.

However, with water as the solvent, there is much more adsorption, leading to both nonlinear isotherms and to significant surface diffusion. The uptake data now could only be adequately represented by the complete Eq. 3.5.e-5, as seen in Fig. 2. Line 1 is the result of assuming no surface diffusion and using the above

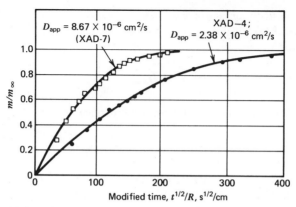

*Figure 1 Desorption curves of benzaldehyde from amberlite particles into methanol.* ● *Experimental (XAD-4).* □ *Experimental (XAD-7). Transient uptake solution of Eq. 3.5.e-5a for the indicated values of $D_{app}$, the apparent diffusivity. $m_\infty$ is the total amount of benzaldehyde desorbed at infinite time (from Komiyama and Smith [70]).*

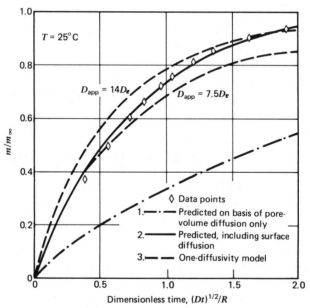

*Figure 2 Adsorption curve of benzaldehyde (in water) for amberlite (XAD-4); $C_\infty/C_0 = 0.0804$ (from Komiyama and Smith [70]).*

176

porosity and tortuosity values. Lines 3 were an attempt to fit the curve with a single diffusivity value, and it is seen that the entire set of data has a definitely different shape. Line 2 is the result of utilizing Eq. 3.5.e-5 along with the measured adsorption isotherm data, and it can be seen that excellent agreement is obtained. Note that for this system, the surface diffusion flux was about 5 to 14 times the pore volume diffusion flux.

---

As described in Sec. 3.5a, there are still many puzzling aspects of configurational diffusion that remain to be explained. About the only theoretical information available concerns the motion of spherical particles in liquids through cylindrical pores. Anderson and Quinn [71] have shown that the effective diffusivity in straight, round pores (tortuosity $\tau = 1.0$) is given by:

$$\frac{D_e}{\varepsilon_s D} = \Phi \overline{K^{-1}} \tag{3.5.e-6}$$

where

$$\Phi = \text{partitioning factor}$$

$$= \left(1 - \frac{a}{r}\right)^2$$

$$a = \tfrac{1}{2} \text{ (molecular size)}$$

and

$$\overline{K^{-1}} = \text{wall-particle interaction}$$

$$\simeq \left(1 - \frac{a}{r}\right)^n, \qquad n = \begin{cases} 2, \text{ sphere on center line} \\ 4^+, \text{ sphere off center} \end{cases}$$

Thus, as an approximation, including a tortuosity factor,

$$\frac{D_e}{(\varepsilon_s D/\tau)} \simeq \left(1 - \frac{a}{r}\right)^{2+n} \tag{3.5.e-7}$$

Satterfield and Colton et al. [72, 73] have studied diffusion of several sugars and other types of molecules in microporous catalyst support solids and correlated their data with the relation

$$\log_{10}\left[\frac{D_e}{(\varepsilon_s D/\tau)}\right] = -2\left(\frac{a}{r}\right) \tag{3.5.e-8}$$

The tortuosity factor, $\tau$, was estimated by extrapolating to $(a/r) \to 0$, together with known $D$ and $\varepsilon_s$, and reasonable values were obtained. Even though Eqs. 3.5.e-7 and 3.5.e-8 appear to be quite different, they result in similar numerical values of the hindrance effects.

## 3.6 Reaction with Pore Diffusion

### 3.6.a Concept of Effectiveness Factor

When reaction occurs on the pore walls simultaneously with diffusion, the process is not a strictly consecutive one, and both aspects must be considered together. Comprehensive discussions are available in Satterfield [40] and in Aris [74]. We first consider the simplest case of a first-order reaction, equimolar counter-diffusion, and isothermal conditions—generalizations will be discussed later. Also, the simplest geometry of a slab of catalyst will be used. When the $z$-coordinate is oriented from the center line to the surface, the steady-state diffusion equation is

$$D_e \frac{d^2 C_s}{dz^2} - k_v C_s = 0 \qquad (3.6.a-1)$$

where

$$
\begin{aligned}
k_v &= \text{reaction-rate coefficient based on pellet volume, } m_f{}^3/m_p{}^3 \text{ hr} \\
&= \rho_s S_g k_s \qquad (3.6.a-2) \\
k_s &= \text{surface rate coefficient } m_f{}^3/m_p{}^2 \text{ hr}
\end{aligned}
$$

The boundary conditions are

$$C_s(L) = C_s{}^s \text{ (surface concentration)}$$

$$\frac{dC_s(0)}{dz} = 0 \text{ (symmetry at center line)}$$

and the solution is

$$\frac{C_s(z)}{C_s{}^s} = \frac{\cosh \sqrt{k_v/D_e}\, z}{\cosh \sqrt{k_v/D_e}\, L} \qquad (3.6.a-3)$$

This then leads to the concentration profiles as shown in Fig. 3.6.a-1.

The physical meaning of the results is that the diffusion resistance causes a concentration profile to exist in the pellet since reactants cannot diffuse in from the bulk sufficiently rapidly. A small diffusion resistance (say, large $D_e$) gives a rather flat curve and conversely for a large diffusion resistance. Since the rate of reaction at any point in the pore is $k_v C_s(z)$, this profile causes a decreased averaged rate relative to that if the concentration were everywhere $C_s{}^s$. In a practical situation however, this slight penalty of loss of average reaction rate in a porous catalyst pellet is more than offset by the enormous increase in surface area of the pores, and the net result is still favorable for these catalyst formulations.

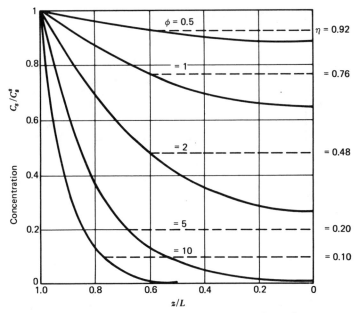

*Figure 3.6.a-1 Distribution and average value of reactant concentration within a catalyst pore as a function of the parameter φ. (Adapted from Levenspiel [75].)*

The above curves could be used to directly characterize the diffusion limitations, but it is more convenient to have a "rating factor" for the effect. This was provided by Thiele [76] and Zeldowich [77], who defined the *effectiveness factor*:

$$\eta \equiv \frac{\text{rate of reaction with pore diffusion resistance}}{\text{rate of reaction with surface conditions}}$$

$$= \frac{\dfrac{1}{V_p} \int r_A(C_s) dV_p}{r_A(C_s^s)} \qquad (3.6.a\text{-}4)$$

Thus, the actual reaction rate that would be observed is:

$$(r_A)_{\text{obs}} = \eta r_A(C_s^s) \qquad (3.6.a\text{-}5)$$

When the concentration profile found from the diffusion equation is substituted into the numerator of Eq. 3.6.a-4, this becomes:

$$\eta = \frac{\tanh \phi}{\phi} \qquad (3.6.a\text{-}6)$$

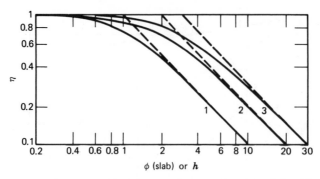

*Figure 3.6.a-2 Effectiveness factors for (1) slab, (2) cylinder, and (3) sphere (from Aris [78].)*

where

$$\phi = \text{modulus} = L\sqrt{k_v/D_e}$$

A plot of Eq. 3.6.a-6 is shown in Fig. 3.6.a-2. It shows results similar to our physical reasoning above concerning diffusion resistance. For $\phi \to 0$, $\eta \to 1$, which means no appreciable resistance, and conversely for $\phi \to \infty$. Note that the latter can occur for small diffusivity, large pellet size, $L$, or very rapid reaction rate. From the mathematical properties of tanh $\phi$, the asymptotic relation as $\phi \to \infty$ (i.e., exceeds 3) is

$$\eta_\infty \sim \frac{1}{\phi} \qquad \qquad (3.6.a-8)$$

These results can be extended to more practical pellet geometries, such as cylinders or spheres, by solving the diffusion equation in these geometries. For the sphere,

$$D_e \frac{1}{r^2} \frac{d}{dr} r^2 \frac{dC_s}{dr} - k_v C_s = 0 \qquad \qquad (3.6.a-9)$$

and for the same boundary conditions one finds

$$\eta = \frac{3}{h} \frac{h \coth h - 1}{h} \qquad \qquad (3.6.a-10)$$

where

$$h = R\sqrt{k_v/D_e} \qquad \qquad (3.6.a-11)$$
$$R = \text{sphere radius}$$

The $(h, \eta)$ plot has roughly the same shape as the result from Eq. 3.6.a-6, but the asymptote for $h \to \infty$ is:

$$\eta_\infty \sim \frac{3}{h} \qquad (3.6.a\text{-}12)$$

In other words, the curve has a similar shape, but is shifted on a log–log plot by a factor of three (see Fig. 3.6.a-2). Therefore, if a new spherical modulus were defined with a characteristic length of $R/3$, the following would result:

$$\phi = \frac{R}{3} \sqrt{k_v/D_e} = \frac{h}{3} \qquad (3.6.a\text{-}13)$$

$$\eta = \frac{1}{\phi} \frac{(3\phi)\coth(3\phi) - 1}{(3\phi)} \qquad (3.6.a\text{-}14)$$

$$\eta_\infty \sim \frac{1}{\phi} \qquad \phi \to \infty \qquad (3.6.a\text{-}15)$$

Now the curve for spheres exactly coincides with that for slabs when $\phi \to \infty$ (and $\phi \to 0$, of course), and almost coincides ($\sim 10$ to $15$ percent) for the whole range of $\phi$.

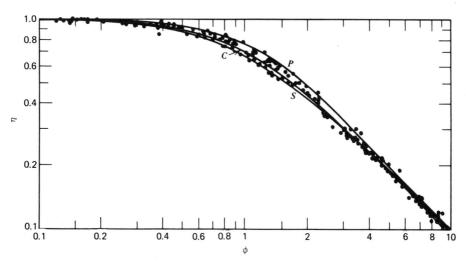

Figure 3.6.a-3 *Effectiveness factors for slab, cylinder and sphere as functions of the Thiele modulus $\phi$. The dots represent calculations by Amundson and Luss (1967) and Gunn (1967). (From Aris [74].)*

Aris [79] noted this, and from similar results for cylinders and other geometries found that a general modulus for all shapes could be defined with $\phi - \eta$ curves practically superimposed:

$$\phi = \frac{V_p}{S_x} \sqrt{k_v/D_e} \tag{3.6.a-16}$$

where $S_x$ is the external surface area of the pellet. Note that for spheres,

$$L \equiv \frac{V_p}{S_x} = \frac{\frac{4}{3}\pi R^3}{4\pi R^2} = \frac{R}{3} \tag{3.6.a-17}$$

as found above. Thus, Fig. 3.6.a-1 is approximately true for any shape of catalyst pellet, even irregular ones, if the proper modulus is used, Eq. 3.6.a-16—see Fig. 3.6.a-3.

It should be mentioned that the differences between values for various shapes in the intermediate range of $\phi \simeq 1$ can be larger for non-first-order reactions, particularly when tending toward zero order and/or with Langmuir–Hinshelwood rate forms; see Knudsen et al. [80] and Rester and Aris [81].

### 3.6.b Generalized Effectiveness Factor

All of the above was for first-order reactions. Since many catalytic reactions are not in this category, other cases must also be considered. It would seem that the diffusion equations must be solved for each new case, but a brief discussion of a method of Stewart et al. [82], Aris [83], Bischoff [84], and Petersen [85] will be given and will show that a generalized or normalized modulus can be defined that approximately accounts for all such cases.

Since the geometry can be handled by Eq. 3.6.a-16 consider the simple slab problem with a coordinate system such that $z = 0$ at the center. The general problem is then:

$$\frac{d}{dz}\left[D_e(C_s)\frac{dC_s}{dz}\right] = r_v(C_s) \tag{3.6.b-1}$$

where $r_v \equiv \rho_s r_A$ = reaction rate per pellet *volume*

$$C_s(L) = C_s^s \tag{3.6.b-2a}$$

$$\frac{dC_s(0)}{dz} = 0 \tag{3.6.b-2b}$$

Equation 3.6.b-1 with 3.6.b-2b can be integrated once to obtain

$$D_e(C_s)\frac{dC_s}{dz} = +\left[2\int_{C_{s0}}^{C_s} D_e(c')r_v(c')dc'\right]^{1/2} \tag{3.6.b-3}$$

where

$C_{s0}$ = reactant concentration at centerline (unknown as yet)

A second integration of Eq. (3.6.b-3) from center to surface gives

$$L = \int_{C_{s0}}^{C_s^s} \frac{D_e(c'')dc''}{\left[2 \int_{C_{s0}}^{C''} D_e(c')r_v(c')dc'\right]^{1/2}}$$ (3.6.b-4)

This equation gives $C_{s0}$ (implicitly) for given values of $L$, $C_s^s$, $D_e$, $r_v$.

The effectiveness factor is found from Eq. 3.6.a-4 by noting that for the steady-state situation Eq. 3.6.b-1 shows that

$$\int_0^L r_v dz = \int_0^L \frac{d}{dz} D_e \frac{dC_s}{dz} dz$$

$$= + D_e(C_s^s) \frac{dC_s(L)}{dz}$$

so that with Eq. 3.6.a-4:

$$\eta = \frac{+D_e(C_s^s) \dfrac{dC_s(L)}{dz}}{Lr_v(C_s^s)}$$ (3.6.b-5)

Combining Eqs. 3.6.b-3 and 5 gives:

$$\eta = \frac{\sqrt{2}}{Lr_v(C_s^s)} \left[\int_{C_{s0}}^{C_s^s} D_e(c')r_v(c')dc'\right]^{1/2}$$ (3.6.b-6)

The $C_{s0}$ in Eq. 3.6.b-6 is found from Eq. 3.6.b-4. Equation 3.6.b-6 thus gives the effectiveness factor for any reaction rate form and any effective diffusivity [such as in Eq. 3.5.b-9 or 16]. For a simple first-order reaction, Eq. 3.6.b-6 reduces to Eq. 3.6.a-6.

In order to match the asymptotic portions of the curves that could be generated from Eq. 3.6.b-6, let us consider briefly the physical meaning of $\phi \to \infty$, or strong diffusion limitations. Under these conditions, very little reactant would be able to diffuse to the center of the pellet, and any that did would be in equilibrium for a reversible reaction or zero for an irreversible one. Thus, the asymptotic effectiveness factor, which will be defined as $1/\phi$ for *all* cases, becomes from Eq. 3.6.b-6,

$$\eta_\infty \sim \frac{\sqrt{2}}{Lr_v(C_s^s)} \left[\int_{C_{s,eq}}^{C_s^s} D_e(c')r_v(c')dc'\right]^{1/2}$$

$$\equiv 1/\phi$$ (3.6.b-7)

Thus, a generalized or normalized modulus can be defined that will lead to approximately the same curve, Fig. 3.6.a-3, for *any* geometrical shape, *any* reaction rate form, and *any* diffusivity relationship:

$$\phi \equiv \frac{V_p}{S_x} \frac{r_v(C_s^{\,s})}{\sqrt{2}} \left[ \int_{C_{s,\,eq}}^{C_s^{\,s}} D_e(c) r_v(c) dc \right]^{-1/2} \tag{3.6.b-8}$$

Equation 3.6.b-8 and Fig. 3.6.a-3 then can be used for all of the above cases. See the books of Petersen [86] and Aris [74] for more extensive examples.

Specific applications have recently been provided by Dumez and Froment [150] and by Frouws, Vellenga, and De Wilt [151].

Certain reaction rate forms can lead to unusual behavior, that is not well represented by the general modulus approach, over the entire range of modulus values. For isothermal systems, these are associated with rate equations that can exhibit empirical or approximate negative order behavior. For example, Satterfield, Roberts, and Hartman [87, 88] have shown that rate equations of the form

$$r_v = \frac{k_v C_s (C_s + E)}{(1 + K C_s)^2}$$

can lead to more than one solution to the steady-state mass balance differential equations, for certain ranges of the parameters—primarily large values of $K$. These multiple steady states involve the transient stability of the catalyst particle, but are more commonly found in conjunction with thermal effects, and will be more thoroughly discussed in Section 3.7.

Luss [89] has derived a necessary and sufficient condition for *uniqueness*:

$$(C_s - C_s^{\,s}) \frac{d \ln r_v(C_s)}{dC_s} \le 1 \tag{3.6.b-9}$$

Thus, for an $n$th order reaction,

$$r_v = k_v C_s^{\,n}$$

one finds from the criterion Eq. 3.6.b-9 that there will be a unique steady state if

$$n \ge (n - 1)(C_s/C_s^{\,s})$$

From this result, it is readily seen that only orders $n < 0$ can possibly violate the criterion for certain values of $(C_s/C_s^{\,s})$. Luss also showed that for rate forms

$$r_v = \frac{k_v C_s}{[1 + (K C_s^{\,s})(C_s/C_s^{\,s})]^2}$$

uniqueness is guaranteed by

$$K C_s^{\,s} \le 8$$

Since this pathological behavior can only occur for very special ranges of the reaction rate parameters, (e.g., see Lee and Luss [90], we will not discuss it further here. Aris [91] has presented a comprehensive review of these questions of uniqueness and stability.

CHEMICAL ENGINEERING KINETICS

### Example 3.6.b-1 Generalized Modulus for First-Order Reversible Reaction

For this case, the reaction rate is

$$r_v(C_s) = \frac{k_v}{K}[(1 + K)C_s - C_{ts}^s] \tag{a}$$

where

$K$ = equilibrium constant

$C_{ts}^s$ = sum of surface concentration of reactant and product

Also,

$$C_{s,eq} = \frac{C_{ts}^s}{1 + K} \tag{b}$$

and with constant $D_e$,

$$\int_{C_{s,eq}}^{C_s^s} D_e r_v(c') dc' = \frac{k_v D_e}{K} \int_{C_{s,eq}}^{C_s^s} [(1 + K)c' - C_{ts}^s] dc'$$

$$= \frac{k_v D_e}{K}\left[\frac{1+K}{2}(C_s^{s2} - C_{s,eq}^2) - C_{ts}^s(C_s^s - C_{s,eq})\right] \tag{c}$$

Then,

$$\phi = \frac{V_p}{S_x}\sqrt{\frac{k_v}{D_e}} \frac{(1+K)C_s^s - C_{ts}^s}{(2K)^{1/2}\left[\frac{1+K}{2}(C_s^{s2} - C_{s,eq}^2) - C_{ts}^s(C_s^s - C_{s,eq})\right]^{1/2}} \tag{d}$$

$$= \frac{V_p}{S_x}\sqrt{\frac{k_v}{D_e}}\frac{1+K}{K} \tag{e}$$

For an irreversible reaction with $K \to \infty$, Eq. e reduces to Eq. (3.6.a-16). (See Carberry [92].)

---

For an $n$th order irreversible reaction, the generalized modulus becomes

$$\phi = \frac{V_p}{S_x}\sqrt{\frac{n+1}{2}\frac{k_v(C_s^s)^{n-1}}{D_e}} \qquad n > -1 \tag{3.6.b-10}$$

which with Fig. 3.6.a-3 gives a good estimate of $\eta$ for this case—also see Fig. 3.6.b-1. Equation 3.6.b-10 can also be used to show how the observed kinetic

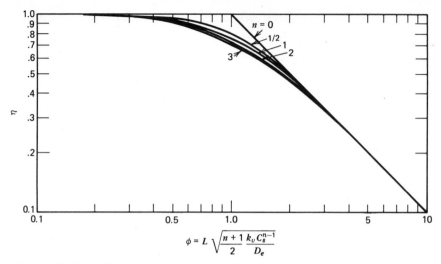

Figure 3.6.b-1 Generalized plot of effectiveness factor for simple order reactions.

parameters are related to the true ones in the region of strong pore diffusion resistance—so-called "diffusional falsification." From Eq. 3.6.a-5:

$$(r_v)_{obs} = \eta k_v (C_s^s)^n$$

$$\sim \frac{1}{\phi} k_v (C_s^s)^n \qquad (3.6.b\text{-}11)$$

$$= \frac{S_x}{V_p} \sqrt{\frac{2}{n+1} D_e k_v (C_s^s)^{(n+1)/2}}$$

Thus, the observed order is $(n + 1)/2$, and is only equal to the true order, $n$, for first-order reactions, $n = 1$. Also,

$$(k_v)_{obs} = \eta k_v$$

$$= \frac{S_x}{V_p} \sqrt{\frac{2}{n+1}} [A_D e^{-E_D/RT} \cdot A_0 e^{-(E/RT)}]^{1/2}$$

and so

$$(E)_{obs} = -\frac{d \ln (k_v)_{obs}}{d(1/T)} = \frac{E + E_D}{2} \simeq \frac{E}{2} \qquad (3.6.b\text{-}12)$$

186  CHEMICAL ENGINEERING KINETICS

Thus, with strong pore diffusion limitations, the observed activation energy is one-half the true one. This provides one possible experimental test for the presence of pore diffusion problems, since if the observed $E$ is $\simeq 5$–$10$ kcal/mol ($21$–$42$ kJ/mol), it is probably one-half of a true chemical activation energy value. However, if the observed $E$ is $\simeq 20$ kcal/mol ($84$ kJ/mol) it could be the true one or one-half of $\sim 40$ kcal/mol ($168$ kJ/mol), and so the test is inconclusive in this case.

Weisz and Prater [93] showed that over the entire range of $\phi$, the falsified activation energy is given by

$$\frac{E_{\text{obs}}}{E} = 1 + (1/2)\frac{d \ln \eta}{d \ln \phi} \tag{3.6.b-13}$$

or considering the diffusivity:

$$E_{\text{obs}} = E + (1/2)(E - E_D)\frac{d \ln \eta}{d \ln \phi} \tag{3.6.b-14}$$

Languasco, Cunningham, and Calvelo [94] derived a similar relation for the falsified order:

$$(n)_{\text{obs}} = n + \frac{n - 1}{2}\frac{d \ln \eta}{d \ln \phi} \tag{3.6.b-15}$$

(They actually extended these to nonisothermal situations; the full relations are given below.) From the shape of the effectiveness factor curves, it is seen that the logarithmic slopes vary from $0$ to $-1$, which give the simple limits given above.

Experimental verification of the above concepts will be given in the following examples. Of course, practical difficulties often prevent perfect agreement, and it is sometimes necessary to perform direct experiments with a catalyst pellet; see Balder and Petersen [95] for a useful single-pellet-reactor technique and a review by Hegedus and Petersen [152].

## Example 3.6.b-2 Effectiveness Factors for Sucrose Inversion in Ion Exchange Resins

This first-order reaction

$$\underset{\text{(sucrose)}}{C_{12}H_{22}O_{11}} + H_2O \xrightarrow{H^+} \underset{\text{(glucose)}}{C_6H_{12}O_6} + \underset{\text{(fructose)}}{C_6H_{12}O_6}$$

was studied in several different size particles by Gilliland, Bixler, and O'Connell [96]. Pellets of Dowex resin with diameters of $d_p = 0.77$, $0.55$, and $0.27$ mm were used, and in addition crushed Dowex with $d_p \simeq 0.04$ mm, all at $50°C$. By techniques to be discussed later in this chapter, the crushed resin was shown not to essentially have any diffusion limitations, the rate coefficient thus obtained was the chemical rate coefficient. The results were:

| $d_p$, mm | $k_v{}^a(s^{-1})$ | $k/k_{0.04}$ | $h = R\sqrt{k_v/D_e}$ | $\eta$ |
|---|---|---|---|---|
| 0.04 | 0.0193 | 1.0 | 0.53 | 1.0 |
| 0.27 | 0.0110 | 0.570 | 3.60 | 0.600 |
| 0.55 | 0.00664 | 0.344 | 7.35 | 0.352 |
| 0.77 | 0.00487 | 0.252 | 10.3 | 0.263 |

$^a$ Calculated on the basis of approximate normality of acid resin = 3 N.

The pellet diffusivity was also separately measured, in the $Na^+$ resin form: $D_e = 2.69 \times 10^{-7}$ cm$^2$/s. More exact values for the $H^+$ resin form were also computed from the reaction data, but as an example of the estimation of an effectiveness factor, the $Na^+$ value will be used here.

The Thiele modulus for spheres can now be computed from

$$3\phi = h = R\sqrt{k_v/D_e}$$

and the values are shown in the above table. It is seen that the effectiveness factor values thus determined give a good estimation to the decreases in the observed rate constants.

Gilliland et al. also determined rate constant values at 60° and 70°C, and the observed activation energies were

| $d_p$, mm | $E$, kcal/mol | (kJ/mol) |
|---|---|---|
| 0.04 | 25 | (105) |
| 0.27 | 20 | (84) |
| 0.55 | 18 | (75) |
| 0.77 | 18 | (75) |
| Homogeneous acid solution | 25 | (105) |

The activation energy for diffusion was about 6 to 10 kcal/mol (25–42 kJ/mol) and so in the strong pore diffusion limitation region with $d_p = 0.77$ mm., the predicted falsified activation energy is

$$E_{obs} = \frac{25 + (6-10)}{2} = 15\text{–}17 \text{ kcal/mol } (63\text{–}71 \text{ kJ/mol})$$

which is close to the experimental value of 18 kcal/mol (75 kJ/mol).

188 _____ CHEMICAL ENGINEERING KINETICS

Thus, the major features of the effectiveness factor concept have been illustrated. Several other aspects, including such practical complications as resin bead swelling with sucrose sorption and the like are discussed in Gilliland et al.'s original article.

### *Example 3.6.b-3 Methanol Synthesis*

Brown and Bennett [97] have studied this reaction using commercial 0.635 cm ($\frac{1}{4}$ in.) catalyst pellets, with $\varepsilon_s = 0.5$, $S_g = 130$ m$^2$/g, dominant pore size $\simeq 100$ to 200 Å. The reaction is

$$CO + 2H_2 \; \rightleftharpoons \; CH_3OH \tag{a}$$

The mean binary diffusivity for the reaction mixture was computed from Eq. 3.2.c-11:

$$\frac{1}{D_{CO, m}} = \frac{1}{D_{CO-H_2}} (y_{H_2} - 2y_{CO}) + \frac{1}{D_{CO-CH_3OH}} (y_{CH_3OH} + y_{CO}) \tag{b}$$

At the experimental conditions of $p_t = 207$ bars $= 204$ atm and $T = 300$ to 400°C, the Knudsen diffusion can be neglected, and ordinary bulk diffusivities can be used in Eq. b. It was stated that "the variation (of $D_{Am}$, with composition dependence from Eq. b) was *not* negligible."

The intrinsic rate was determined by crushing the catalyst pellets into small particles; the rate data could be represented by the Natta rate equation

$$r_A = \frac{f_{CO} f_{H_2}{}^2 - f_{CH_3OH}/K}{(\alpha + \beta f_{CO} + \gamma f_{H_2} + \delta f_{CH_3OH})^3} \tag{c}$$

where the $f_j$ are fugacities. However, it was also found that the pellet rate data was well represented by the empirical equation

$$r_v = k_v (C_{s, CO} - C_{s, CO, eq})^n \tag{d}$$

for any one temperature, and for the high ratio of H$_2$/CO $= 9^+$ used.

The results of Eqs. b and d were then introduced into the general modulus, Eq. 3.6.b-8,

$$\phi = \frac{d_p}{6} \frac{k_v (C_{s, CO}{}^s - C_{s, CO, eq})^n}{\sqrt{2}} \left[ \int_{C_{s, CO, eq}}^{C_{s, CO}{}^s} D_{e, CO, m}(c) r_v(c) dc \right]^{-1/2} \tag{e}$$

The value of the tortuosity factor was $\tau = 7.2$ for the effective diffusivity, which corresponds to results in similar pellets of Satterfield and Cadle [48, 49].

The results are given in Fig. 1, which shows good results—complete agreement for the higher temperature data, and within about 25 percent for lower temperatures. If the latter data were decreased to agree with the theoretical effectiveness factor line, a smoother Arrhenius plot is obtained, and so these points may contain some

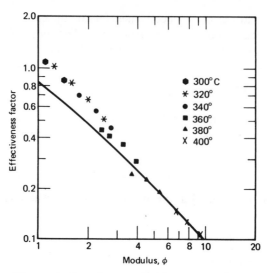

*Figure 1  Correlation of effectiveness factors for the 6.3-mm pellets at 207 bars. The solid line is calculated from the data for the 0.4-mm particles by the method of Bischoff. A value of 7.2 was used for τ (from Brown and Bennett [97]).*

systematic error. However, considering the precision of all the various data that must be utilized, the overall agreement is satisfactory.

A recent review on intraparticle diffusion in multicomponent catalytic reactions is by Schneider [153].

### 3.6.c  Criteria for Importance of Diffusional Limitations

We have now seen how possible pore diffusion problems can be evaluated: compute the generalized modulus and use Figs. 3.6.a-3 or 3.6.b-1 to see if the value of $\eta$ is less than unity. In the design situation this procedure can be used since $k_v$ is presumably known, but when determining kinetic constants from laboratory or pilot plant data this can't be done since $k_v$ is what is being sought. Thus, criteria for the importance of pore diffusion, independent of $k_v$ are also useful. There are two main types that are generally used.

The first is a classical test and involves performing experiments with two sizes of catalyst. Then from Eq. 3.6.b-8, if one assumes that $k_v$ and $D_e$ are the same (this may not be true if the smaller pellet is made by cutting a larger one; see Cadle and Satterfield [98]:

$$\frac{\phi_1}{\phi_2} = \frac{L_1}{L_2} \qquad (3.6.c\text{-}1)$$

($L$ is understood to be $V_p/S_x$, of course.)

Also,

$$\frac{(r_{\mathrm{obs}})_1}{(r_{\mathrm{obs}})_2} = \frac{\eta_1}{\eta_2} \qquad (3.6.c\text{-}2)$$

Thus, the two sizes will give two different values to the moduli (only the ratio can be determined from Eq. 3.6.c-1) and so if the two observed rates are the same, $\eta_1 = \eta_2$ and the operation must be on the horizontal part of the $\phi - \eta$ curve (i.e., no pore diffusion limitations). At the other extreme, $\eta = 1/\phi$, and so

$$\frac{(r_{\mathrm{obs}})_1}{(r_{\mathrm{obs}})_2} = \frac{\phi_2}{\phi_1} = \frac{L_2}{L_1} \qquad (3.6.c\text{-}3)$$

Therefore, in this case, the observed rates are inversely proportional to the two pellet sizes. For intermediate degrees of pore diffusion limitation, the ratio of rates will be less than proportional to $L_2/L_1$. A graphical procedure was given by Hougen and Watson [99] in which a line of coordinates $[L_1, (r_1)_{\mathrm{obs}}]$ to $[L_2, (r_2)_{\mathrm{obs}}]$, which is equivalent on a log–log graph to $(\phi_1, \eta_1)$ to $(\phi_2, \eta_2)$, except for additive scale factors, is plotted on the effectiveness factor graph. Then the position where the slope of the line segment is equal to the slope of the $(\phi, \eta)$ curve gives the region of operation. Then, the points $(L_1, r_1)$ and $(L_2, r_2)$ correspond to the points $(\phi_1, \eta_1)$ and $(\phi_2, \eta_2)$. This scheme also includes the two limiting cases just discussed, of course. If more than two particle sizes are used in the experiments, a method proposed by Stewart et al. [82] is somewhat more accurate. This involves plotting several of the observed rates versus particle size on a separate sheet of graph paper, shifting this on the $(\phi - \eta)$ plot, and then comparing the relative position of the two sets of coordinate axes. The advantage of the latter method is that more of the curves are matched rather than just the slopes.

The other method, which can be used for a single particle size, is called the Weisz–Prater criterion [93] and is found for a first-order reaction by solving Eq. 3.6.a-16 for $k_v$:

$$k_v = \frac{\phi^2}{L^2} D_e$$

and substituting into

$$(r_v)_{\mathrm{obs}} = \eta k_v C_s^{\,s}$$

$$= \eta \frac{\phi^2}{L^2} D_e C_s^{\,s}$$

Now if all the directly *observable* quantities are put on one side of the equation,

$$\frac{(r_v)_{\text{obs}} L^2}{D_e C_s^s} = \eta\phi^2 \equiv \Phi \tag{3.6.c-4}$$

At this point, we have just performed some algebraic manipulations, since we still can't find the right-hand side (RHS) of Eq. 3.6.c-4. However, from the two limiting cases of Fig. 3.6.a-3, the following is true:

**1.** For $\phi \ll 1$, $\eta = 1$ (no pore diffusion limitation) and so RHS $= (1)\,\phi^2 \ll 1$.

**2.** For $\phi \gg 1$, $\eta = 1/\phi$ (strong pore diffusion limitation) and so

$$\text{RHS} = \frac{1}{\phi} \cdot \phi^2 \gg 1.$$

Thus the criterion becomes, if

$$\Phi = \frac{(r_v)_{\text{obs}} L^2}{D_e C_s^s} \ll 1 \tag{3.6.c-5}$$

then there are no pore diffusion limitations.[1]

### Example 3.6.c-1 Minimum Distance Between Bifunctional Catalyst Sites for Absence of Diffusional Limitations

Weisz [100] utilized Eq. 3.6.c-5 to determine the minimum distance between the two types of active sites in a bifunctional catalyst so that the reactive intermediates could be sufficiently mobile to result in an appreciable overall reaction rate (this was qualitatively discussed in Sec. 2.1). For the important case of hydrocarbon isomerizations, the first step on catalytic site 1 has an adverse equilibrium, but the second step on site 2 is essentially irreversible:

$$A \underset{}{\overset{1}{\rightleftharpoons}} R \overset{2}{\longrightarrow} S$$

The maximum concentration of $R$, from the first-order kinetics, is

$$(C_R)_{\text{max}} = C_A \frac{1}{(1/K_1) + (k_2/k_1)} \tag{a}$$

---

[1] A more general way of stating this (P. B. Weisz, personal communication, 1973) is that if $\Phi \gg 1$, then there *are* pore diffusion limitations; if this is not the case, there usually will not be limitations, but for special cases, such as strong product inhibition, a more detailed analysis is required—see Ex. 3.6.c-2.

where $K_1 \ll 1$ is the equilibrium constant for the first reaction. Thus, if $(k_2/k_1) \gg 1$ or $1 \ll (1/K_1)$, the value of $(C_R)_{max}$ can be arbitrarily small. The overall rate, however, can still be large

$$r_s = k_2 C_R \sim \frac{k_2}{(1/K_1) + (k_2/k_1)} C_A \qquad \text{(b)}$$

which is caused by the constant removal of $R$ by reaction 2 (site 2), thereby shifting the equilibrium point of reaction 1 (site 1).

The above purely kinetic rates would be observed only in the absence of diffusional limitations. It is reasoned that if the first step is rate controlling, then the critical rate is

$$-r_R = k_1 \left( \frac{1 + K_1}{K_1} \right) (C_R - C_{R, eq}) \qquad \text{(c)}$$

where the same form for reversible reaction rate is used as in Ex. 3.6.b-1. Also, as shown in that example, the modulus is:

$$\Phi = \eta \phi^2 = \frac{(-r_R)_{obs}}{k_1 \left( \dfrac{1 + K_1}{K_1} \right) (C_R - C_{R, eq})} \frac{L^2 k_1}{D_e} \frac{1 + K_1}{K_1}$$

$$= \frac{(-r_R)_{obs} L^2}{D_e (C_R - C_{R, eq})} \qquad \text{(d)}$$

When rate determining, $C_R \ll C_{R, eq}$, and so Eq. d becomes, together with criterion Eq. 3.6.c-5:

$$\Phi \simeq \frac{(r_R)_{obs} L^2}{D_e C_{R, eq}} \ll 1 \qquad \text{(e)}$$

If on the other hand, the second step is rate controlling, the critical rate is (from Eq. b):

$$r_s = k_2 C_R \simeq k_2 C_{R, eq} \qquad \text{(f)}$$

which when combined with the criterion Eq. (3.6.c-5) leads to:

$$\Phi = \frac{(r_s)_{obs}}{k_2 C_{R, eq}} \cdot \frac{L^2 k_2}{D_e}$$

$$= \frac{(r_s)_{obs} L^2}{D_e C_{R, eq}} \qquad \text{(g)}$$

Equations (e) and (g) are the same, in terms of the observed limiting rate,

$$\frac{(r_v)_{obs} L^2}{D_e C_{R, eq}} \ll 1 \qquad \text{(h)}$$

for either situation.

Weisz [100] utilized typical values of parameters and observed reaction rates for hydrocarbon isomerizations to show that grain sizes of less than 1 $\mu$m for catalyst one or two can have extremely low intermediate ($R$) partial pressures of $10^{-7}$ atm, which would be unobservable even though the overall reaction had a finite rate. Also, for $n$-heptane isomerization, the thermodynamic equilibrium concentration under typical conditions, $C_{R,eq}$ can be used in Eq. h to compute the minimum catalyst grain size, or intimacy, required for appreciable reaction to occur, not influenced by diffusional limitations. At a 40 percent conversion to isoheptanes at 470°C, the result was

$$L < 20 - 40 \ \mu\text{m}$$

which is in good agreement with the data shown on Fig. 2.1-4.

The above treatment assumes the only diffusional limitations are in the catalyst (one or two) grains, and not in the pellet as a whole—more general grain models are described in Chapter 4 in another application to fluid–solid heterogeneous reactions. Also, other extensions have been provided by Gunn and Thomas [101].

---

Using the generalized modulus, the criterion Eq. 3.6.c-5 was extended by Petersen [85] and by Bischoff [102] to the case where the reaction rate may be written as

$$r_v(C) = k_v g(C) \tag{3.6.c-6}$$

where $k_v$ is the unknown rate constant and $g(C)$ contains the concentration dependency. Following the same procedure, the extended criterion is:

$$\Phi = \frac{(r_v)_{\text{obs}} L^2 g(C_s^{\ s})}{2 \displaystyle\int_{C_{s,\ eq}}^{C_s^{\ s}} D_e(C) g(C) dC} \ll 1 \tag{3.6.c-7}$$

The same idea can also be accomplished by replotting the effectiveness factor curve as $\eta$ versus $\Phi \equiv \eta\phi^2$, so that the abcissa contains only directly observed quantities—for example, see Fig. 3.6.c-1.

If Weisz and Prater's original observable group is retained, the extended criterion, Eq. 3.6.c-7 can also be written as

$$\Phi = \frac{(r_v)_{\text{obs}} L^2}{\bar{D}_e C_s^{\ s}} \ll \frac{2 \displaystyle\int_{C_{s,\ eq}}^{C_s^{\ s}} (D_e/\bar{D}_e) g(C) dC}{C_s^{\ s} g(C_s^{\ s})} \tag{3.6.c-8}$$

where $\bar{D}_e$ is an average value of the pore diffusivity. Alternate, but similar, criteria have been derived using perturbation techniques about the surface concentration

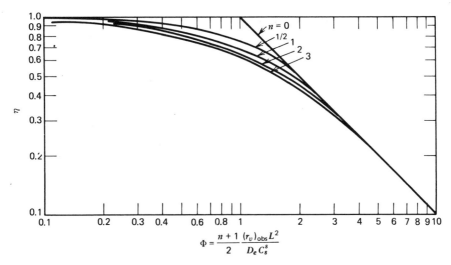

*Figure 3.6.c-1 Effectiveness factor plot in terms of observable modulus.*

by Hudgins [103] and using collocation methods by Stewart and Villadsen [104]; both results are essentially given by the following, using our notation:

$$\Phi = \frac{(r_v)_{obs} L^2}{D_e C_s^s} \ll \frac{g(C_s^s)}{C_s^s g'(C_s^s)} \tag{3.6.c-9}$$

where $g' \equiv dg/dC$.

Mears [105] discusses several criteria (and also heat effects—to be presented below). For example, a simple order reaction, $g(C) = C^n$, gives

$$\Phi \ll \frac{2}{n+1} \tag{3.6.c-8a}$$

(Again note that only $n > -1$ is meaningful.)

$$\ll \frac{1}{|n|} \tag{3.6.c-9a}$$

Calculation shows that these are roughly equivalent, and if the inequality is not taken too literally, it is not really much different from the original Weisz–Prater criterion. However, for certain situations such as strong product inhibition, this is not the case—see Ex. 3.6.c-2. Finally, Brown [106] has considered macro–micro pore systems. For typical types of catalyst structure and diffusivities, the conclusion was that normally there will be no diffusional limitations in the micropores if there is none in the macropores. Thus, use of the standard criteria for the macropores should be sufficient to detect any pore diffusion problems; however, the assumptions and calculations were probably not valid for zeolite molecular sieves, and so this case still needs special consideration.

## Example 3.6.c-2 Use of Extended Weisz–Prater Criterion

The use of the general criterion, Eq. 3.6.c-7 is illustrated by applying it to the case of the carbon–carbon dioxide reaction

$$C + CO_2 \longrightarrow 2CO \qquad (a)$$

Petersen [85] used some of the data of Austin and Walker [107] to show how the first-order criterion would not be correct; here the data are recalculated on the basis of Eq. 3.6.c-7. This reaction appears to be very strongly inhibited by adsorption of the product, carbon monoxide, which leads to large deviations from first-order behavior. The rate equation, in concentration units, was of the standard adsorption type.

$$r_v = \frac{k_v C_{CO_2}}{1 + K_2 C_{CO} + K_3 C_{CO_2}} \qquad (b)$$

where

$$C_{CO_2}, C_{CO} = \text{concentrations of } CO_2 \text{ and } CO \text{ inside the solid}$$
$$k_v = \text{rate constant } (s^{-1})$$
$$K_2, K_3 = \text{adsorption constants}$$

From the reaction stoichiometry, and assuming equal diffusivities and zero carbon monoxide concentration at the particle surface, Eq. (b) becomes

$$r_v = \frac{k_v C_{CO_2}}{(1 + 2K_2 C_{CO_2}{}^s) + (K_3 - 2K_2)C_{CO_2}} \qquad (c)$$

where

$$C_{CO_2}{}^s = C_{obs} = \text{concentration of } CO_2 \text{ at particle surface}$$

If Eq. c is substituted into the general criterion, Eq. 3.6.c-7, one obtains

$$\Phi = \frac{r_{v,\,obs} L^2}{2D_e} \left\{ \frac{1 + K_3 C_{CO_2}{}^s}{K_3 - 2K_2} \left[ 1 - \frac{1 + 2K_2 C_{CO_2}{}^s}{C_{CO_2}{}^s(K_3 - 2K_2)} \ln \frac{1 + K_3 C_{CO_2}{}^s}{1 + 2K_2 C_{CO_2}{}^s} \right] \right\}^{-1} \ll 1 \qquad (d)$$

At 1000 K, the following data were used (recalculated on concentration basis):

$$C_{CO_2}{}^s(1 \text{ atm}) = 1.22 \times 10^{-5} \text{ mol/cm}^3$$
$$K_2 = 4.15 \times 10^9 \text{ cm}^3/\text{mol}$$
$$K_3 = 3.38 \times 10^5 \text{ cm}^3/\text{mol}$$
$$D_e = 0.1 \text{ cm}^2/\text{s}$$
$$L = 0.7 \text{ cm}$$
$$r_{v,\,obs} = 4.67 \times 10^{-9} \text{ mol/cm}^3 \text{ s}$$

Substituting these values into the Weisz–Prater first-order criterion, Eq. 3.6.c-5 gives

$$\Phi = \frac{r_{v,\,obs} L^2}{C_{CO_2}^{\,s} D_e} = 2.28 \times 10^{-3} \ll 1.0$$

Thus the criterion is apparently satisfied ($\ll 1.0$) and would indicate that pore diffusion limitations did not exist. However, by cutting apart the particles and observing the profiles of reacted carbon, and from other tests, Austin and Walker found that there were indeed large diffusion effects present, and so the first-order criterion did not predict the behavior correctly.

If the data are now substituted into the general criterion, Eq. (d), one obtains

$$\Phi = \text{LHS (Eq. d)} = 2.5 > 1.0$$

Since the value of Eq. d is greater than 1, the criterion is not satisfied and so pore diffusion effects should be present, as Austin and Walker found. Therefore, the general criterion, Eq. 3.6.c-7, indicated the proper situation; it proves useful for similar tests for any reaction type, although for first-order reactions the original Weisz–Prater criterion is identical.

---

### 3.6.d Combination of External and Internal Diffusion Resistance

The addition of fluid phase resistance is relatively easy for a first-order reaction, just as for the simple consecutive surface reaction. The only change is that $C_s^{\,s}$ is now not known and must be found with the mass transfer coefficient. This means that the boundary condition at the surface becomes

$$k_g(C - C_s^{\,s}) = D_e\left(\frac{dC_s}{dz}\right)_s \tag{3.6.d-1}$$

which leads to the solution in terms of $C$; the bulk concentration:

$$C_s = C\,\frac{\cosh \phi z/L}{\cosh \phi + \dfrac{D_e \phi}{L k_g} \sinh \phi} \tag{3.6.d-2}$$

Equation 3.6.d-2, when used in defining the effectiveness factor based on the *bulk fluid concentration*, then gives

$$\eta_G = \frac{D_e C\,\dfrac{(\phi/L)\sinh \phi}{\cosh \phi + (D_e \phi/L k_g)\sinh \phi}}{L k_v C}$$

$$= \frac{\tanh \phi/\phi}{1 + (D_e \phi/L k_g)\tanh \phi} \tag{3.6.d-3}$$

The subscript, $G$, refers to a "global"—particle + film—effectiveness factor, which includes *both* resistances, and which reduces to Eq. 3.6.a-6 for $k_g \to \infty$. Equation 3.6.d-3 is more conveniently written as (see Aris [78]):

$$\frac{1}{\eta_G} = \frac{1}{\eta} + \frac{\phi^2}{Sh'} \tag{3.6.d-4}$$

$$= \frac{\phi}{\tanh \phi} + \frac{k_v}{k_g}\left(\frac{V_p}{S_x}\right) \tag{3.6.d-4a}$$

where $Sh' = k_g L/D_e = $ *modified* Sherwood number (note that the particle *half-width* and the *effective* diffusivity are used rather than the usual parameters), which is also called the Biot number for mass transfer ($Bi_m$).

Again, Eq. 3.6.d-4 clearly shows the additivity of resistances for first-order reactions. Note that in the asymptotic region, where $\phi$ is large,

$$\frac{1}{\eta_G} \sim \phi + \frac{\phi^2}{Sh'} \tag{3.6.d-5}$$

and for sufficiently large $\phi$ and finite $Sh'$:

$$\simeq \frac{\phi^2}{Sh'}$$

Thus, the ultimate log slope in this situation could be $-2$ rather than $-1$ for only internal diffusion.

Petersen [86] has demonstrated that with realistic values of the mass transfer and diffusion parameters, external transport limitations will never exist unless internal diffusion limitations are also present. This is most easily seen by comparing the reduction in reaction rate caused by internal limitations alone, $\eta$, with that caused by the additional external transport limitations, ($\eta_G/\eta$). Using Eq. 3.6.d-4

$$\frac{\eta_G}{\eta} = \left(1 + \frac{\phi^2 \eta}{Sh'}\right)^{-1} = \left(1 + \frac{\Phi^2}{Sh'}\right)^{-1} \tag{3.6.d-6}$$

Now smaller values of $Sh'$ tend to decrease this ratio, and Petersen used the minimum value of external mass transfer from a sphere through a stagnant fluid:

$$Sh = \frac{k_g R}{D} = 1 + 0.3 Re^{1/2} Sc$$

and so

$$Sh' = \frac{k_g R}{D} \cdot \frac{D}{D_e} \cdot \frac{L}{R}$$

$$= (1)\left(\frac{D}{D_e}\right)\left(\frac{1}{3}\right), \text{ minimum value}$$

Thus,

$$\left(\frac{\eta_G}{\eta}\right)_{\min} = \left(1 + 3\frac{D_e}{D}\phi^2\eta\right)^{-1} \qquad \text{(3.6.d-7)}$$

From Sec. 3.5, it was seen that realistic values of $D_e$ are approximately

$$\frac{D_e}{D} \leq \frac{\varepsilon_s}{\tau} \simeq \frac{0.5}{3} = \frac{1}{6}$$

and with this, it can be shown from Eq. 3.6.d-7 that the following is the case in the range when diffusional limitations could be of concern:

$$\left(\frac{\eta_G}{\eta}\right)_{\min} > \eta \qquad \text{(3.6.d-8)}$$

(In the asymptotic region where $\phi$ is large, the requirement is only $D_e/D < \frac{1}{3}$). Thus, the original assertion is true for first-order reactions.

### Example 3.6.d-1 *Experimental Differentiation Between External and Internal Diffusion Control*

Koros and Nowak [108] have proposed the following scheme using Eq. 3.6.d-4. The observed rate is

$$(r_v)_{obs} = \eta_G k_v C \qquad \text{(a)}$$
$$= \eta_G \rho_s S_g k_s C \qquad \text{(b)}$$

For possible strong pore diffusion limitation, Eq. 3.6.d-4 becomes

$$\frac{1}{\eta_G} = \phi + \frac{\rho_s S_g k_s}{k_g}\frac{V_p}{S_x}$$

$$= \left[\sqrt{\frac{\rho_s S_g k_s}{D_e}} + \frac{\rho_s S_g k_s}{k_g}\right]\frac{V_p}{S_x} \qquad \text{(c)}$$

Thus,

$$(r_v)_{obs} = \left\{[(\rho_s S_g k_s D_e)^{-1/2} + k_g^{-1}]\frac{V_p}{S_x}\right\}^{-1} C \qquad \text{(d)}$$

If the $(r_v)_{obs}$ does not vary with $(\rho_s S_g)$, external mass transfer is dominant and if $(r_v)_{obs}$ varies with $(\rho_s S_g)^{1/2}$, pore diffusion is limiting. If there are no mass transfer limitations, $\eta_G = 1$, and $(r_v)_{obs}$ would, of course, vary directly with $\rho_s S_g$. The actual implementation assumes that $(\rho_s S_g)$ can be changed for a given catalyst (e.g., change the amount of active catalyst in the pellet), but this is much more difficult in practice than changing pellet size. Also, it may be difficult to distinguish between variations of $(\rho_s S_g)^0$, $(\rho_s S_g)^{1/2}$, or $(\rho_s S_g)^{1.0}$ or intermediate values. However, in situations where it is difficult to estimate the external mass transfer coefficient, $k_g$, this method could be the only feasible one.

---

As would be expected from earlier discussions, the combination of resistances for non-first-order reactions is more complicated. Aris [109] has presented the rather remarkable result that in the large $\phi$ asymptotic region, Eq. 3.6.d-5 is true for arbitrary reaction rate forms, if there is not too large a difference between surface and bulk concentrations, $C_s^s \simeq C$, or $\phi_G < \text{Sh}'$:

$$\frac{1}{\eta_G} \sim \phi_G + \frac{(\phi_G)^2}{\text{Sh}'} \qquad (3.6.\text{d-}9)$$

where $\phi_G$ is the generalized modulus of Eq. 3.6.b-8, but using the observable bulk concentration in place of the surface concentration:

$$\phi_G \equiv \frac{V_p}{S_x} \frac{r_v(C)}{\sqrt{2D_e}} \left[ \int_0^C r_v(c')dc' \right]^{-1/2} \qquad (3.6.\text{d-}10)$$

A comprehensive study by Mehta and Aris [110] provides graphs for $n$th order reactions. A brief summary of their results follows:

$$\phi_G = \phi \left[ 1 + \frac{2}{n+1} \frac{\eta \phi^2}{\text{Sh}'} \right]^{(n-1)/2}$$

$$\frac{1}{\eta_G} = \frac{1}{\eta} \left[ 1 + \frac{2}{n+1} \frac{\eta \phi^2}{\text{Sh}'} \right]^n$$

where

$$\phi = L \sqrt{\frac{n+1}{2} k_v(C_s^s)^{n-1}/D_e} \qquad \eta = \eta(\phi) - \text{as usual}$$

$$\phi_G = L \sqrt{\frac{n+1}{2} k_v C^{n-1}/D_e} \qquad \eta_G = \eta_G(\phi_G)$$

For a given situation, $\phi_G$ and Sh′ can be computed, and an iterative solution is required to find $\phi$ and $\eta$, and thus $\eta_G$ (charts given in Mehta and Aris simplify this).

## 3.7 Thermal Effects

### 3.7.a Thermal Gradients Inside Catalyst Pellets

The final complication that must be introduced into the discussion is the fact that thermal conductivity limitations may cause temperature gradients in addition to concentration gradients within the pellet. To analyze these, the combined heat and mass balances must be solved; the balances for slab geometry are

$$\frac{d}{dz} D_e \frac{dC_s}{dz} = r_v(C_s, T_s) \tag{3.7.a-1}$$

$$-\frac{d}{dz} \lambda_e \frac{dT_s}{dz} = (-\Delta H)r_v(C_s, T_s) \tag{3.7.a-2}$$

where $\lambda_e$ is the effective thermal conductivity of the pellet (see Satterfield [40] or Smith [24] for further details); an order of magnitude value is $\lambda_e \simeq 10^{-3}$ cal/s cm °C]. Because of the coupling caused by the rate term, these equations must be solved simultaneously for the complete solution. However, some information can be obtained without the full solution.

If Eq. 3.7.a-2 is divided by $(-\Delta H)$ and subtracted from Eq. 3.7.a-1, the following results:

$$\frac{d}{dz} \left[ D_e \frac{dC_s}{dz} + \frac{\lambda_e}{(-\Delta H)} \frac{dT_s}{dz} \right] = 0$$

which when integrated from the center to a point $z$ gives:

$$D_e \frac{dC_s}{dz} + \frac{\lambda_e}{(-\Delta H)} \frac{dT_s}{dz} = \text{constant} = 0 \tag{3.7.a-3}$$

Another, integration gives, for constant $D_e$ and $\lambda_e$:

$$D_e C_s + \frac{\lambda_e}{(-\Delta H)} T_s = \text{constant} = D_e C_s{}^s + \frac{\lambda_e}{(-\Delta H)} T_s{}^s$$

or

$$T_s - T_s{}^s = \frac{D_e(-\Delta H)}{\lambda_e} (C_s{}^s - C_s) \tag{3.7.a-4}$$

Thus, Eq. 3.7.a-4 can be used to eliminate either $C_s$ or $T_s$ from one of the differential equations with the result that in general, only one (nonlinear) with one dependent variable must be solved.

The maximum temperature difference in a particle (without further complications of external mass and heat transfer resistances) is for complete reaction, $C_s = 0$, as pointed out by Prater [111]:

$$\frac{(\Delta T_s)_{\text{max}}}{T_s{}^s} = \frac{(-\Delta H)D_e C_s{}^s}{\lambda_e T_s{}^s} \tag{3.7.a-5}$$

$$\equiv \beta \tag{3.7.a-5a}$$

This result is actually true for any particle geometry, under steady-state conditions.

If the complete transient equations are considered, the important dimensionless groups of parameters can be formulated. Consider a first-order reaction with Arrhenius form of rate constant:

$$\frac{\partial C_s}{\partial t} = D_e \frac{\partial^2 C_s}{\partial z^2} - A_0 e^{-E/RT_s} C_s \tag{3.7.a-6}$$

$$\rho_s c_{ps} \frac{\partial T_s}{\partial t} = \lambda_e \frac{\partial^2 T_s}{\partial z^2} + (-\Delta H) A_0 e^{-E/RT_s} C_s \tag{3.7.a-7}$$

Define the following dimensionless variables:

$$u = C_s/C_s^s \qquad v = T_s/T_s^s \qquad \xi = z/L \qquad \theta = D_e t/L^2$$

Then Eq. 3.7.a-6, 7 become

$$\frac{\partial u}{\partial \theta} = \frac{\partial^2 u}{\partial \xi^2} - \phi_s^2 u \exp[\gamma(1 - 1/v)] \tag{3.7.a-8}$$

$$\frac{1}{Lw'} \frac{\partial v}{\partial \theta} = \frac{\partial^2 v}{\partial \xi^2} + \beta\phi_s^2 u \exp[\gamma(1 - 1/v)] \tag{3.7.a-9}$$

where

$$\phi_s^2 = \frac{L^2 A_0}{D_e} e^{-E/RT_s^s} \tag{3.7.a-10}$$

$$\gamma = E/RT_s^s \tag{3.7.a-11}$$

$$Lw' = \lambda_e/\rho_s c_{ps} D_e = Sc'/Pr' \tag{3.7.a-12}$$

The latter group is the *modified* Lewis number.

The steady-state solution will then only be a function of the modulus evaluated at the surface conditions, $\phi_s$, and also $\beta$ and $\gamma$. A full set of computations was performed by Weisz and Hicks [112], and Fig. 3.7.a-1 shows some results for a spherical pellet with $\gamma = 20$; they also presented graphs for other values of $\gamma$.

One of the most interesting features is that for $\beta > 0$ (exothermic), there are regions where $\eta > 1$. This behavior is based on the physical reasoning that with sufficient temperature rise caused by heat transfer limitations, the increase in the rate constant, $k_v$, more than offsets the decrease in reactant concentration, $C_s$, so that the internal rate is actually larger than that at surface conditions of $C_s^s$ and $T_s^s$, leading to an effectiveness factor greater than unity. The converse is, of course, true for endothermic reactions.

The other rather odd feature of Fig. 3.7.a-1 is that for large $\beta$, and a narrow range of $\phi_s$ values, three possible values of $\eta$ could be obtained. This behavior is caused by the fact that the heat generation term on the right-hand side of Eq. 3.7.a-2 is a strongly nonlinear function of $T_s$, which can lead to multiple solutions of the equations. This is an example of physicochemical instability, and will be

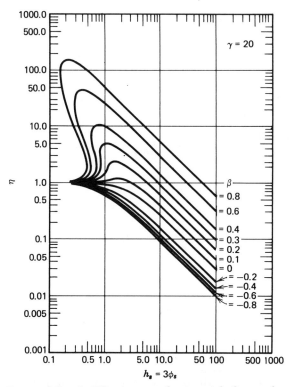

*Figure 3.7.a-1 Effectiveness factor with first-order reaction in a spherical nonisothermal catalyst pellet (from Weisz and Hicks [112]).*

discussed in detail later. Detailed study shows that only the highest or lowest values of $\eta$ are actually attained, depending on the direction of approach, and the center value represents an unstable state. Also notice that in certain regions close to the vertical parts of the curve, a small change in $\phi_s$ could cause a very large jump in $\eta$.

It is useful at this point to discuss some typical values of the various parameters, in order to determine the extent of further analysis that is important, and also any appropriate simplifications that might be made. A collection of parameter values for several industrial reactions was prepared by Hlavacek, Kubicek, and Marek [113], as shown in Table 3.7.a-1.

We see that $\beta$ is typically a small number, usually less than 0.1. Therefore, Eq. 3.7.a-5a indicates that the temperature change from the surface to inside the particle is, for steady-state conditions and standard catalysts, usually rather small. Also, the multiple steady-state behavior of Fig. 3.7.a-1 is not likely to be observed in common catalytic reactions.

Table 3.7.a-1 *Parameters of Some Exothermic Catalytic Reactions (after Hlavacek, Kubicek, and Marek* [113]*).*

| Reaction | $\beta$ | $\gamma$ | $\gamma\beta$ | Lw' | $\phi_s$ |
|---|---|---|---|---|---|
| $NH_3$ synthesis | 0.000061 | 29.4 | 0.0018 | 0.00026 | 1.2 |
| Synthesis of higher alcohols from CO and $H_2$ | 0.00085 | 28.4 | 0.024 | 0.00020 | — |
| Oxidation of $CH_3OH$ to $CH_2O$ | 0.0109 | 16.0 | 0.175 | 0.0015 | 1.1 |
| Synthesis of vinylchloride from acetylene and HCl | 0.25 | 6.5 | 1.65 | 0.1 | 0.27 |
| Hydrogenation of ethylene | 0.066 | 23–27 | 2.7–1 | 0.11 | 0.2–2.8 |
| Oxidation of $H_2$ | 0.10 | 6.75–7.52 | 0.21–2.3 | 0.036 | 0.8–2.0 |
| Oxidation of ethylene to ethylenoxide | 0.13 | 13.4 | 1.76 | 0.065 | 0.08 |
| Dissociation of $N_2O$ | 0.64 | 22.0 | 1.0–2.0 | — | 1–5 |
| Hydrogenation of benzene | 0.12 | 14–16 | 1.7–2.0 | 0.006 | 0.05–1.9 |
| Oxidation of $SO_2$ | 0.012 | 14.8 | 0.175 | 0.0415 | 0.9 |

Further insight into the magnitude of possible temperature gradients inside a catalyst pellet is provided by the experimental study of Kehoe and Butt [114] on the exothermic ($-\Delta H \simeq 50$ kcal/mol = 209 kJ/mol) benzene hydrogenation. The conditions are given in the tables:

| Property | Pellet 1 58% Ni on Kieselguhr (Harshaw Ni-0104P) | Pellet 2 25% Ni-0104P 25% graphite 50% $\gamma$-Al$_2$O$_3$ (Harshaw Al-0104T) |
|---|---|---|
| Pellet radius (cm) | 0.66 | 0.69 |
| Length $L'$ (cm) | 5.75 | 6.10 |
| Density (g/cm$^3$) | 1.88 | 1.57 |
| Heat capacity (cal/g°C) | 0.152 | 0.187 |
| Effective thermal conductivity (cal/cm s°C) | $3.6 \times 10^{-4}$ | $3.5 \times 10^{-3}$ |
| Effective diffusivity (cm$^2$/s) | 0.052 | 0.035 |
| Characteristic length* (cm) | 0.296 | 0.310 |

$$* L = V_p'/S_x = \frac{R/2}{1 + (R/L')}$$

204 ————————————————— CHEMICAL ENGINEERING KINETICS

| Run | Observed Rate mol/g cat. s | Modified Sherwood Number, $k_g L/D_e$ | Modified Nusselt Number $h_f L/\lambda_e$ |
|---|---|---|---|
| Pellet 1 | | | |
| 21 | $0.820 \times 10^{-6}$ | 215 | 10.8 |
| 24 | 1.506 | 215 | 10.8 |
| 27 | 2.258 | 215 | 10.8 |
| Pellet 2 | | | |
| 209 | $11.15 \times 10^{-6}$ | 401 | 1.35 |
| 212 | 22.4 | 401 | 1.35 |

Figure 3.7.a-2 shows that for the standard type pellet, No. 2, there is essentially no internal temperature gradient. Figure 3.7.a-3, for a pellet with a 10-times smaller effective thermal conductivity indicates that there were about 35°C maximum internal temperature differences. These are certainly important for kinetic studies and for reactor design predictions, but were still too small to cause any catalyst pellet instabilities.

Note that Eq. 3.7.a-5 would give an estimate for Run 27:

$$\beta = \frac{(-\Delta H)D_e C_s^s}{\lambda_e T_s^s} = \frac{(50,000)(0.052)[\sim(3.5 \times 10^{-5})(0.1348)]}{(3.6 \times 10^{-4})(340)}$$

$$= 0.100$$

or

$$\Delta T_{s,\,max} = T_s^s \beta = (340)(0.100)$$

$$= 34 \text{ K}$$

Thus, for most reactions, which are not highly exothermic, there would be only a very small temperature difference inside the catalyst pellet, although certain systems can have appreciable values (e.g., hydrogenations).

A final simplification is possible for small values of $\beta$, where it can be readily shown that the two parameters $\beta$ and $\gamma$ essentially only appear as the single product $(\beta\gamma)$ (see Tinkler and Metzner [115] and Carberry [116]). The dimensionless form of Eq. 3.7.a-4 is:

$$v - 1 = \beta(1 - u) \qquad (3.7.a-13)$$

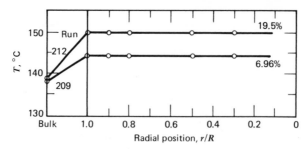

*Figure 3.7.a-2 Measured internal and external profiles for pellet 2, as a function of feed composition (from Kehoe and Butt [114]).*

If Eq. 3.7.a-13 is substituted into Eq. 3.7.a-8 to eliminate $v$, the result is:

$$\frac{\partial u}{\partial \theta} = \frac{\partial^2 u}{\partial \xi^2} - \phi_s^2 u \exp\left[\frac{\beta\gamma(1 - u)}{1 + \beta(1 - u)}\right] \qquad (3.7.\text{a-}14)$$

$$\simeq \frac{\partial^2 u}{\partial \xi^2} - \phi_s^2 u \exp[\beta\gamma(1 - u)] \qquad (3.7.\text{a-}14\text{a})$$

the latter equation 3.7.a-14a being true for $\beta < 0.1$. This is equivalent to approximating the Arrhenius temperature dependency with an exponential form.

Thus, the Weisz and Hicks curves can be collapsed into one set with the single parameter of $(\beta\gamma)$, as shown in Fig. 3.7.a-4, which gives a complete summary

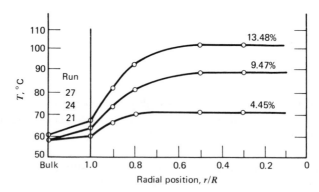

*Figure 3.7.a-3 Measured internal and external profiles for pellet 1 with feed temperature of 52°C (from Kehoe and Butt [114]).*

CHEMICAL ENGINEERING KINETICS

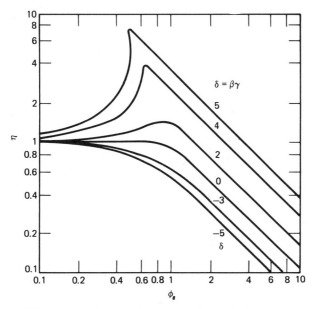

*Figure 3.7.a-4 Effectiveness factors for nonisothermal first-order reaction in the slab. (Adapted from Aris [109] and Petersen [86].)*

of nonisothermal effectiveness factors for this situation. Liu [117] gave the useful formula valid for the most important range of $(\beta\gamma)$:

$$\eta \simeq \frac{1}{\phi_s} \exp[\beta\gamma/5] \qquad (3.7.a\text{-}15)$$

Bischoff [102] showed that the generalized modulus concept of Eq. 3.6.b-8 can be extended by substituting the right hand side of Eq. 3.7.a-14 or Eq. 3.7.a-14a as the rate form. This then asymptotically unified all the isothermal and endothermic curves, and the suitable portions of the exothermic curves, but still would not permit prediction of the maxima or stability aspects.

A thorough computational study was made by Drott and Aris [118], and it was found that the uniqueness criterion of Luss [119]

$$\beta\gamma < 4(1 + \beta) \qquad (3.7.a\text{-}16)$$

provided a good estimate of conditions for *stability*. Also, the ranges of the Thiele modulus, over which multiple steady states could conceivably occur were quite narrow—for rather drastic parameter values, only between $\phi = 0.47$ and 0.49. Therefore, considering this and the information in Table 3.7.a-1 in practical situations, internal gradients are unlikely to cause particle instability.

If the full transient equations are considered, Wei [120] pointed out that temperatures exceeding the steady-state maximum temperature of Eq. 3.7.a-5 can exist, particularly at isolated points; Georgakis and Aris [121] have extended this discussion.

### 3.7.b External and Internal Temperature Gradients

If the experimental results of Kehoe and Butt [114] in Figs. 3.7.a-2 and 3.7.a-3 are studied, note that the external heat transfer resistance can be appreciable, and, especially for the isothermal pellet, must be considered. The same mass and heat balance equations (3.7.a-1, 2) [or (3.7.a-8, 9)] are used, but surface boundary conditions expressed in terms of the finite external heat and mass transfer resistances are used.

The determination of the maximum temperature differences between bulk fluid, catalyst pellet surface, and catalyst pellet interior in terms of directly observable quantities is a very useful tool in the study of catalytic reactions. Only if these temperature differences are significant need one be concerned with further extensive analysis of the transport phenomena.

Lee and Luss [122] provided such results in terms of the observable (Weisz) modulus and the external effective Sherwood and Nusselt numbers. The steady-state mass and heat balances for an arbitrary reaction, using slab geometry, are

$$D_e \frac{d^2 C_s}{dz^2} = r(C_s, T_s) \tag{3.7.b-1}$$

$$-\lambda_e \frac{d^2 T_s}{dz^2} = (-\Delta H) r(C_s, T_s) \tag{3.7.b-2}$$

The particle surface boundary conditions are:

$$D_e \frac{dC_s}{dz} = k_g (C - C_s^s) \tag{3.7.b-3}$$

$$\text{at } z = L$$

$$\lambda_e \frac{dT_s}{dz} = h_f (T - T_s^s) \tag{3.7.b-4}$$

Following Prater's [111] procedure, Eqs. 3.7.b-1 and 3.7.b-2 can be combined:

$$\frac{d^2}{dz^2} \left[ D_e C_s + \frac{\lambda_e}{(-\Delta H)} T_s \right] = 0 \tag{3.7.b-5}$$

which when integrated once from the pellet center to surface gives, utilizing Eqs. 3.7.b-3 and 4:

$$D_e \frac{dC_s}{dz} \bigg|_L + \frac{\lambda_e}{(-\Delta H)} \frac{dT_s}{dz} \bigg|_L = 0 = k_g (C - C_s^s) - \frac{h_f}{(-\Delta H)} (T_s^s - T) \tag{3.7.b-6}$$

A second integration and rearrangement gives the overall temperature difference:

$$T_s - T = (T_s^s - T) - \frac{D_e(-\Delta H)}{\lambda_e}(C_s - C_s^s)$$

$$= (-\Delta H)\frac{k_g}{h_f}(C - C_s^s) + (-\Delta H)\frac{D_e}{\lambda_e}(C_s^s - C_s) \qquad (3.7.b\text{-}7)$$

The right-hand side of Eq. 3.7.b-7 is the sum of the external and internal temperature differences, as pointed out by Hlavacek and Marek [123]. The maximum temperature difference is for complete reaction, when $C_s = 0$:

$$\frac{T_{s,\,\max} - T}{T} = \beta_G \frac{\text{Sh}'}{\text{Nu}'}\left(1 - \frac{C_s^{\,s}}{C}\right) + \beta_G \frac{C_s^{\,s}}{C} \qquad (3.7.b\text{-}8)$$

where $\beta_G \equiv (-\Delta H)D_e C/\lambda_e T$, at *bulk fluid* conditions.

The final step is to obtain $C_s^s/C$ in terms of an observable rate, which is the volume-averaged rate in the pellet:

$$r_{v,\,\text{obs}} = \frac{1}{L}\int_0^L r_v\,dz = +\frac{D_e}{L}\frac{dC_s}{dz}\bigg|_L$$

$$= \frac{k_g}{L}(C - C_s^s) \qquad (3.7.b\text{-}9)$$

Using the observable (Weisz) modulus:

$$\Phi_G \equiv \frac{L^2 r_{v,\,\text{obs}}}{D_e C} = \text{Sh}'\left(1 - \frac{C_s^{\,s}}{C}\right) \qquad (3.7.b\text{-}10)$$

Substituting Eq. 3.7.b-10 into Eq. 3.7.b-8 gives the result of Lee and Luss [122] [their Eq. (11), which was in spherical geometry]:

$$\frac{T_{s,\,\max} - T}{T} = \beta_G\left[1 + \Phi_G\left(\frac{1}{\text{Nu}'} - \frac{1}{\text{Sh}'}\right)\right] \qquad (3.7.b\text{-}11)$$

Lee and Luss also presented results for the maximum surface-to-interior temperature difference. Recall from Eq. 3.6.c-4 that the observable modulus can also be written in terms of the usual modulus and effectiveness factor:

$$\Phi_G = \eta_G \phi^2$$

where $\phi = L\sqrt{k_v/D_e}$ for a first-order reaction and, from Eq. 3.6.d-4,

$$\frac{1}{\eta_G} = \frac{1}{\eta} + \frac{\phi^2}{\text{Sh}'}$$

Thus, either type of modulus can be used in the analysis.

Carberry [124] presented an analysis showing that the fraction of the total temperature difference external to the pellet can be found in terms of a new observable quantity and the *ratio* of the effective Sherwood to Nusselt numbers, thus obviating the need to have precise values of both of them. He also defined a new observable group:

$$Ca \equiv \frac{Lr_{v,\text{obs}}}{k_g C} = \frac{\Phi_G}{\text{Sh}'} \tag{3.7.b-12}$$

Then, Eq. 3.7.b-11 can be written in terms of $Ca$ and only the *ratio* $\text{Sh}'/\text{Nu}'$:

$$\frac{T_{s,\text{max}} - T}{T} = \beta_G\left[1 + Ca\left(\frac{\text{Sh}'}{\text{Nu}'} - 1\right)\right] \tag{3.7.b-13}$$

Similarly, the interior temperature difference is:

$$\frac{T_{s,\text{max}} - T_s^s}{T} = \beta_G\left(1 - \Phi_G\frac{1}{\text{Sh}'}\right) \tag{3.7.b-14}$$

$$= \beta_G(1 - Ca) \tag{3.7.b-15}$$

and the external temperature difference is:

$$\frac{T_s^s - T}{T} = \beta_G\Phi_G\frac{1}{\text{Nu}'} \tag{3.7.b-16}$$

$$= \beta_G Ca\left(\frac{\text{Sh}'}{\text{Nu}'}\right) \tag{3.7.b-17}$$

Finally, the fractional external temperature difference is the ratio of Eq. 3.7.b-17 to Eq. 3.7.b-13:

$$\frac{T_s^s - T}{T_{s,\text{max}} - T} = \frac{Ca(\text{Sh}'/\text{Nu}')}{1 + Ca[(\text{Sh}'/\text{Nu}') - 1]} \tag{3.7.b-18}$$

Equations 3.7.b-10 to 18 are then a summary of the various temperature differences in terms of two possible observable groups.

### Example 3.7.a-1 Temperature Gradients with Catalytic Reactions

Kehoe and Butt's data, [114] given in Figure 3.7.a-2, is an example of the use of Eqs. 3.7.b-10 to 18. Considering Run 212 with pellet No. 2, the measured external temperature difference can be seen from Fig. 3.7.a-2 to be about 11 + °C, based on the fluid bulk temperature of 139°C = 412 K. The observed rate for this run was $22.4 \times 10^{-6}$ mol/g cat. s.

Then, the dimensionless parameters are:

$$\beta_G = (-\Delta H)D_e C/\lambda_e T$$

$$= \frac{(50{,}000)(0.035)[(2.9 \times 10^{-5})(0.195)]}{(3.5 \times 10^{-3})(412)}$$

$$= 0.0070$$

$$\Phi_G = \frac{L^2 r_{v,\,obs}}{D_e C} = \frac{(0.31)^2[(22.4 \times 10^{-6})(1.57 \text{ g/cm}^3)]}{(0.035)(2.96 \times 10^{-5} \times 0.195)}$$

$$= 16.7$$

The maximum external temperature difference is then estimated, from Eq. 3.7.b-16,

$$T_s^s - T = T\beta_G \Phi_G \frac{1}{Nu'}$$

$$= 412(0.0070)(16.7)\left(\frac{1}{1.35}\right)$$

$$= 412(0.0866) = 36 \text{ K}$$

The actual value of 11°C indicates that at the high reactant concentration of $y_B = 0.195$, the internal pellet concentration was not quite zero, as for maximum heat release conditions. The maximum overall temperature difference is estimated, from Eq. 3.7.b-11

$$T_{s,\,max} - T = T\beta_G\left[1 + \Phi_G\left(\frac{1}{Nu'} - \frac{1}{Sh'}\right)\right]$$

$$= 412(0.0070)\left[1 + 16.7\left(\frac{1}{1.35} - \frac{1}{401}\right)\right]$$

$$= 412(0.0933) = 38 \text{ K}$$

Finally, the internal temperature difference could be $38 - 36 = 2$ K; the value can also be estimated from Eq. 3.7.b-14 or Eq. 3.7.a-5, the latter using the measured surface temperature. Thus,

$$T_{s,\,max} - T_s^s = T\beta_G\left(1 - \Phi_G\frac{1}{Sh'}\right)$$

$$= (412)(0.0070)\left(1 - \frac{16.7}{401}\right)$$

$$= (412)(0.00671) = 2.8 \text{ K}$$

Again, this maximum value bounds the actual results of very little interior temperature differences for pellet 2.

The same type of results can be computed for pellet 1, run 27:

$$\beta_G = 0.1052 \qquad\qquad \Phi_G = 1.474$$
$$T_{s,\,max} - T = 40 \text{ K} \qquad (\sim 42°C \text{ experimental})$$
$$T_{s,\,max} - T_s^s = 35 \text{ K} \qquad (\sim 35°C \text{ experimental})$$
$$T_s^s - T = 5 \text{ K} \qquad (\sim 6 - 7°C \text{ experimental})$$

These results are also good estimates of the experimentally measured values.

---

Mears [125] showed that Eq. 3.7.b-11 (in spherical coordinates) could be combined with a perturbation expansion of the rate about $T_s = T$, to yield an experimental criterion for a 5 percent deviation from the rate at bulk temperature:

$$\frac{(-\Delta H)r_{v,\,obs}(R/3)}{h_f T} < 0.05 \frac{RT}{E} \qquad (3.7.b\text{-}19)$$

Mears [105] compares these and other criteria for diffusional effects.

Combining external and internal gradients also has an effect on the possible unstable behavior of the catalyst pellet. This could be studied by solving the complete transient Eq. 3.7.a-8, 9 together with the boundary conditions Eq. 3.7.b-3, 4. However, because of the mathematical complexity, most information concerns the steady-state situation.

McGreavy and Cresswell [126] and Kehoe and Butt [127] have presented computations for the effectiveness factor that illustrate the complicated behavior that can occur. There is more chance for multiplicity at reasonable values of the parameters. Criteria for these events to occur, similar to Eq. 3.7.a-16, have been derived by several investigators. Luss [128], for example, concludes that for first-order reactions, the proper sufficient criterion for uniqueness of the steady state, for all values of Sh', is:

$$\beta_G \gamma_G < 4\left(\frac{\text{Nu}'}{\text{Sh}'} + \beta_G\right) \qquad (3.7.b\text{-}20)$$

where

$$\beta_G = \frac{D_e(-\Delta H)C}{\lambda_e T}$$

$$\gamma_G = E/RT$$

are evaluated at bulk fluid conditions. The sufficient condition for existence of multiple steady states, for certain value of Sh', is:

$$\beta_G \gamma_G > 8\left(\frac{\text{Nu}'}{\text{Sh}'} + \beta_G\right) \qquad (3.7.b\text{-}21)$$

The intermediate region is complicated by various internal concentration gradient effects. Comparing Eq. 3.7.b-20 with Eq. 3.7.a-16 for typical values of $(Nu'/Sh') \simeq 0.1 - 0.2$, it is seen that multiple steady states are more likely to be caused by external transport resistances.

The situation is more complicated for other orders of reaction. Luss [128] shows that for order $n > 1$, there is less likelihood of multiplicity, and the converse is true for $n < 1$. As might be expected, the situation for reaction orders approaching zero or negative order behavior could combine the complications of possible concentration and thermal instability; for example, see Smith, Zahradnik, and Carberry [129].

Typical values for the modified Sherwood and Nusselt numbers have been estimated by Carberry [130], and a ratio of $Sh'/Nu' > 10$ seems to be true of many practical situations (with gases). Mercer and Aris [131] have considered possible (generous) maximum ranges that might be attained by the various parameters in physical systems:

| Parameter | Lower Bound | Upper Bound |
|:---:|:---:|:---:|
| $\beta$ | 0 (exothermic) | 1.0 |
| $\gamma$ | 0 | 60.0 |
| $Lw'$ | 0.001 | 100.0 |
| $Sh'$ | 0.1 | 5000.0 |
| $Nu'$ | 0.01 | 50.0 |
| $Sh'/Nu'$ | 1.0 | 2000.0 |

It is seen that some of these extreme values could cause pathologic phenomena like multiple steady states, and so on, but recall that most actual catalysts are rather far from these extremes (e.g., Table 3.7.a-1).

Solutions of the complete transient equations (with the additional parameter, Lewis number $Lw'$) have not been studied very much because of the mathematical complexity. Lee and Luss [90] have shown for some cases that $Lw' > 1$ (an author's definition of the Lewis number must be carefully checked—some use the reciprocal of our $Lw'$) can lead to limit-cycle and other complex behavior. However, Ray [132] estimates that for this to occur, considering reasonable values of $\beta$ and $\gamma$, the critical values are $Lw' > 5\text{-}10$ or larger. Thus, the conclusion from Table 3.7.a-1 is that this is not at all likely, except perhaps for very high-pressure reactors.

To conclude, an overall summary of calculations based on the above results indicates that the usual order of events as transport limitations occur is to begin with no limitations—chemical reaction controls throughout the pellet. Next, internal pore diffusion begins to have an effect, followed by external heat transfer resistance. Finally, for extremely rapid reactions, there is the possibility of external mass transfer resistance and some particle temperature profiles. Only for unrealistic situations is it likely that particle instabilities might occur, and even then only for narrow ranges of the parameters.

## 3.8 Complex Reactions with Pore Diffusion

As is true for many industrial situations, the question of diffusional effects on multiple reaction selectivity is equally as important as the effectiveness of conversion considerations. The basic concepts were provided by Wheeler [133], through consideration of three categories of situations.

The simplest is that of parallel, independent reactions (Wheeler Type I):

$$A \xrightarrow{\;1\;} R, \text{ with order } a_1$$

$$B \xrightarrow{\;2\;} S, \text{ with order } a_2$$

In the absence of pore diffusion, Chapter 1 gives the selectivity ratio as

$$\left(\frac{r_R}{r_S}\right) = \frac{k_1 (C_{As})^{a_1}}{k_2 (C_{Bs})^{a_2}} \tag{3.8-1}$$

Now with pore diffusion, the two independent rates are each merely multiplied by their own effectiveness factor to give

$$\left(\frac{r_R}{r_S}\right)_{obs} = \frac{\eta_1 \, k_1 \, (C_{As})^{a_1}}{\eta_2 \, k_2 \, (C_{Bs})^{a_2}} \tag{3.8-2}$$

The difference between Eqs. 3.8.-2 and 3.8-1 is not readily seen, although the former is clearly the same as the latter when $\eta_i \to 1.0$. For strong pore diffusion limitations, where $\eta_i \sim 1/\phi_i$, the following is the situation:

$$r_i \sim \left(\frac{2}{a_i + 1} \frac{k_i D_{ei}}{L^2}\right)^{1/2} (C_{is}{}^s)^{(a_i + 1)/2}$$

Thus, Eq. 3.8-2 becomes

$$\left(\frac{r_R}{r_S}\right)_{obs} \sim \left[\frac{a_2 + 1}{a_1 + 1} \frac{k_1}{k_2} \frac{D_{eA}}{D_{eB}} \frac{(C_{As}{}^s)^{a_1 + 1}}{(C_{Bs}{}^s)^{a_2 + 1}}\right]^{1/2} \tag{3.8-3}$$

and for both first order and $D_{eA} = D_{eB}$

$$= \sqrt{\frac{k_1}{k_2} \frac{C_{As}{}^s}{C_{Bs}{}^s}} \tag{3.8-3a}$$

Comparing Eq. 3.8-3a with Eq. 3.8-1 shows that the effect of strong pore diffusion limitations is to change the ratio of rate constants, $k_1/k_2$, to the square root of the ratio. Thus, when $k_1$ exceeds $k_2$, other conditions being equal, a given selectivity ratio will be reduced by the diffusional resistance.

The next case to be considered is that of consecutive first-order reactions (Wheeler Type III):

$$A \xrightarrow{\quad 1 \quad} R \xrightarrow{\quad 2 \quad} S$$

Here, the selectivity in the absence of pore diffusion is

$$\frac{r_R}{r_A} = 1 - \frac{k_2}{k_1}\frac{C_{Rs}}{C_{As}} \tag{3.8-4}$$

The diffusion-reaction equations are:

$$D_{eA}\frac{d^2C_{As}}{dz^2} = k_1 C_{As}$$

$$D_{eR}\frac{d^2C_{Rs}}{dz^2} = -k_1 C_{As} + k_2 C_{Rs}$$

The first obviously leads to the standard solution, which is then used to solve the second. The results are:

$$\left(\frac{r_R}{r_A}\right)_{obs} = \frac{\int_0^L r_R\, dz}{\int_0^L r_A\, dz} = \frac{\int_0^L (k_1 C_{As} - k_2 C_{Rs})dz}{\int_0^L k_1 C_{As}\, dz}$$

$$= \frac{1 - \sigma\eta_2/\eta_1}{1 - \sigma} - \left(\frac{\eta_2}{\eta_1}\right)\frac{k_2}{k_1}\frac{C_{Rs}{}^s}{C_{As}{}^s} \tag{3.8-5}$$

where

$$\eta_i = \frac{\tanh\phi_i}{\phi_i} \qquad \phi_i = L\sqrt{k_i/D_{ei}}$$

$$\sigma = (\phi_2/\phi_1)^2 = k_2 D_{eA}/k_1 D_{eR}$$

Again, it is most instructive to look at the strong pore diffusion asymptotic region:

$$\left(\frac{r_R}{r_A}\right)_{obs} \sim \frac{1}{1 + \sqrt{\sigma}} - \frac{1}{\sqrt{\sigma}}\frac{k_2}{k_1}\frac{C_{Rs}{}^s}{C_{As}{}^s} \tag{3.8-6}$$

$$= \frac{1}{1 + \sqrt{k_2/k_1}} - \sqrt{\frac{k_2}{k_1}}\frac{C_{Rs}{}^s}{C_{As}{}^s} \qquad D_{eA} = D_{eR} \tag{3.8-6a}$$

Again notice that the main difference between Eqs. 3.8-6a and 3.8-4 is that the ratio of rate constants, $k_2/k_1$, is effectively reduced by a square root factor, although there are now also several other complications. The effect is to reduce the selectivity that would be observed—recall the integrated curves in Chapter 1 as functions of $(k_2/k_1)$. Finally, in the region between strong and no pore diffusion effects, there would naturally be intermediate effects.

The third case of parallel reactions with a common reactant (Wheeler Type II) is more complicated mathematically, since the only situation of interest is when the reaction orders are different; otherwise the selectivity ratio is only a function of the ratio of rate constants.

$$
A \quad
\begin{array}{c}
\xrightarrow{\;1\;} R \quad \text{with order } a_1 \\
\xrightarrow{\;2\;} S \quad \text{with order } a_2
\end{array}
$$

With no diffusional limitations,

$$
\frac{r_R}{r_S} = \frac{k_1}{k_2} (C_{As}{}^s)^{a_1 - a_2} \tag{3.8-7}
$$

The selectivity ratio with pore diffusion limitations is found by solving the diffusion-reaction equation:

$$
D_{eA} \frac{d^2 C_{As}}{dz^2} - (k_1 C_{As}{}^{a_1} + k_2 C_{As}{}^{a_2}) = 0
$$

Then the selectivity ratio is found from

$$
\left(\frac{r_R}{r_S}\right)_{\text{obs}} = \frac{\displaystyle\int_0^L r_R \, dz}{\displaystyle\int_0^L r_S \, dz} \tag{3.8-8}
$$

$$
= \frac{k_1 \displaystyle\int_0^L C_{As}{}^{a_1} \, dz}{k_2 \displaystyle\int_0^L C_{As}{}^{a_2} \, dz} \tag{3.8-8a}
$$

The mathematical solutions of interest are quite involved, but Roberts [134] has presented several useful cases. The main simple result was in the strong diffusional limitation asymptotic region, where an approximate solution gave:

$$
\frac{(r_R/r_A)_{\text{obs}}}{(r_R/r_A)} \sim \frac{a_2 + 1}{2a_1 - a_2 - 1} \tag{3.8-9}
$$

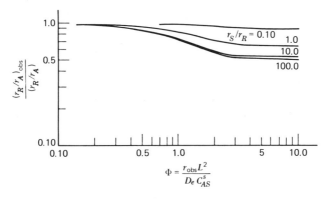

Fig. 3.8-1 Relative yield ratio versus the modulus $\phi$ for various values of $r_S/r_R$—second- and first-order reactions (from Roberts [134].)

for

$$(r_R/r_S) = k_1(C_{As}{}^s)^{a_1 - a_2}/k_2 \ll 1$$

and with

$$(r_R/r_A) = r_R/(r_R + r_S)$$
$$= \{[k_2/k_1(C_{As}{}^s)^{a_1 - a_2}] + 1\}^{-1} \qquad (3.8\text{-}10)$$

The selectivity then is:

$$\left(\frac{r_R}{r_S}\right)_{\text{obs}} = \left[\left(\frac{r_R}{r_S}\right)_{\text{obs}}^{-1} - 1\right]^{-1} \qquad (3.8\text{-}11)$$

We can see that this case apparently does not result in a simple square root of rate constant alteration as for the other two. Thus, for $a_1 = 2$, $a_2 = 1$, the largest deviation from the ratio with no diffusional effects in Eq. 3.8-9 is $= \frac{1}{2}$. For $a_1 = 2$, $a_2 = 0$, this becomes $\frac{1}{5}$. However, for less severe restriction on the ratio of rate constants and/or less severe diffusional limitations, the deviation from ideal selectivities is not so great. Figure 3.8-1 shows the results for the (2, 1) case:

## Example 3.8-1 Effect of Catalyst Particle Size on Selectivity in Butene Dehydrogenation

An experimental investigation of this industrially significant process was reported by Voge and Morgan [135]. Equation 3.8-5 for the local selectivity was used for a given conversion:

$$\frac{r_R}{r_A} = -\frac{dC_R}{dC_A}$$

$$= \left(\frac{1 - \sigma \eta_2/\eta_1}{1 - \sigma}\right) - \left(\frac{\eta_2}{\eta_1}\frac{k_2}{k_1}\right)\frac{C_{Rs}{}^s}{C_{As}{}^s} \tag{a}$$

where $R$ represents butadiene, $A$ butene, and

$$\sigma = \frac{k_2}{k_1}\left(\frac{D_{eA}}{D_{eR}}\right)$$

Equation a can then be simply integrated for $C_{Rs}$ as a function of $C_{As}$, and the butadiene yield thus predicted as a function or conversion. The effectiveness factors for spheres were actually used:

$$\frac{\eta_2}{\eta_1} = \frac{1}{\sigma}\frac{h_2 \coth h_2 - 1}{h_1 \coth h_1 - 1}$$

with

$$h_i = R\sqrt{k_i/D_{ei}}$$

Separate diffusion experiments gave value of

$$D_{eA} \simeq D_{eR} = D_e = 0.12D$$

and $D_{\text{butene-steam}} \simeq 0.720$ cm$^2$/s. The Thiele modulus was estimated from reaction data at 620°C in $\frac{3}{16}$ in. pellets to be

$$h_1 = 1.5$$

and then values for other pellet sizes could be obtained by ratio. Also, $k_2/k_1 = 0.9$. Figure 1 and Table 1 indicate the good agreement between the data and predictions. The exception for $\frac{3}{8}$ in. pellets was apparently caused by their looser structure—doubling $D_e$ to 0.144 cm$^2$/s would produce agrement.

---

Wei [136] considered the case of complex networks of first-order reactions when he used the Wei–Prater matrix decomposition method discussed in Chapter 1 to generalize the effectiveness factor concept. For a matrix diffusion-reaction

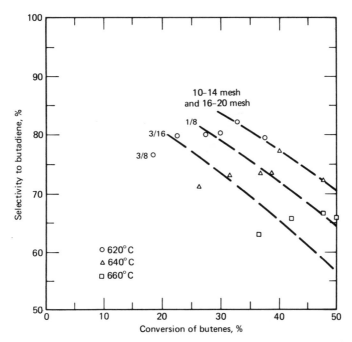

*Figure 1  Butadiene selectivity for different particle sizes and temperatures (from Voge and Morgan [135]).*

*Table 1 summary comparing experimental and calculated selectivities (from Voge and Morgan [135]).*

| Particle size | Temp, °C | $\phi$, calcd | Conversion, % | Selectivity, % | |
|---|---|---|---|---|---|
| | | | | Exptl | Calcd |
| 10–14 mesh | 620 | 0.38 | 33.1 | 82.0 | 81.8 |
| | 640 | 0.46 | 40.4 | 77.2 | 77.2 |
| $\frac{1}{8}$ in. | 620 | 1.00 | 27.7 | 79.9 | 80.5 |
| | 640 | 1.20 | 36.9 | 73.4 | 74.0 |
| | 660 | 1.43 | 48.0 | 66.4 | 66.1 |
| $\frac{3}{16}$ in. | 620 | 1.50 | 22.9 | 79.7 | 78.6 |
| | 640 | 1.80 | 31.7 | 73.0 | 71.5 |
| | 660 | 2.14 | 42.4 | 65.4 | 63.0 |
| $\frac{3}{8}$ in. | 620 | 2.80 | 18.7 | 76.4 | 68.5 |
| | 640 | 3.36 | 26.5 | 71.3 | 62.5 |
| | 660 | 4.00 | 36.8 | 63.0 | 54.0 |

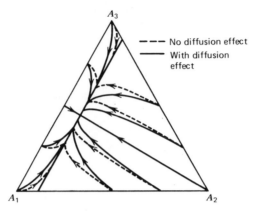

*Figure 3.8-2 The effect of diffusion on the reaction paths in an integral reactor; equal diffusivities (from Wei [136]).*

equation representation,

$$D\nabla^2 C = KC$$

where the diffusivities have been assumed to be concentration independent. Then the solution can be written in a familiar form for spherical pellets:

$$K_{obs} = K\eta$$

where

$$\eta = 3h^{-2}(h \coth h - I)$$

with

$$h^2 = R^2 D^{-1} K$$

All of the above operations and functions are understood to be in matrix form.

Figures 3.8-2 and 3.8-3 show the results of Wei's calculations and illustrate how the reaction paths are altered by diffusional limitations. We see that the differences between paths are decreased, meaning that the selectivity differences are decreased. In other words, selectivity is usually harmed by diffusional effects in the sense that it is more difficult to have products differing in composition. This is generally true of any diffusional step, either external or internal. Wei also points out that sufficient modification of reaction paths is easily possible such that a consecutive mechanism appears as a consecutive-parallel mechanism, and other similar drastic problems.

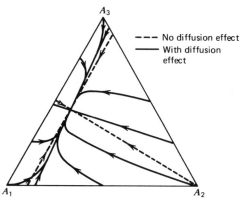

Figure 3.8-3 The effect of diffusion in a system where the diffusivities of the molecular species are not equal. Note altered straight line reaction paths (from Wei [136]).

## 3.9 Reaction with Diffusion in Complicated Pore Structures ___

### 3.9.a Particles with Micro- and Macropores

Most of the previous discussion was based on taking the solid catalyst pellet to have a simple pore structure—one average pore radius. For the case of a micro-macro pore size distribution, Mingle and Smith [137] and Carberry [92] have derived expressions for the overall effectiveness factor for both micro and macro diffusion. As a simple example, consider a first-order reaction, with intrinsic rate constant $k_s$ per surface area of catalyst. Then the mass balance for the micropores is:

$$D_\mu \frac{d^2 C_\mu}{dr^2} = \frac{2}{r_\mu} k_s C_\mu \qquad (3.9.a\text{-}1)$$

which can be solved in terms of its surrounding concentration in the macropores, by the usual methods:

$$\eta_\mu = \frac{\tanh \phi_\mu}{\phi_\mu} \qquad (3.9.a\text{-}2)$$

where

$$\phi_\mu = L_\mu \sqrt{2k_s/r_\mu D_\mu} \qquad (3.9.a\text{-}3)$$

It is understood, of course, that the size of the micropores, $L_\mu$, is the volume/ external surface of the micro particle.

The mass balance for the macropores becomes:

$$D_M \frac{d^2 C_M}{dz^2} = \rho_s s_g k_s \eta_\mu C_M \tag{3.9.a-4}$$

with the usual solution:

$$C_M = C_s{}^s \frac{\cosh \phi z/L}{\cosh \phi}$$

where

$$\phi = L\sqrt{\rho_s s_g k_s \eta_\mu / D_M} \tag{3.9.a-5}$$

$$= L\sqrt{k_v \eta_\mu / D_M} \tag{3.9.a-5a}$$

Then, the overall effectiveness factor is

$$\eta = \frac{1/L \int_0^L k_v \eta_\mu C_M \, dz}{k_v C_s{}^s}$$

$$= \eta_\mu \frac{\tanh \phi}{\phi} \tag{3.9.a-6}$$

$$= \eta_\mu \eta_M \tag{3.9.a-7}$$

Thus, the effect of the microparticles is to possibly have an effectiveness factor, $\eta_\mu$, less than unity, based on microparticle properties, Eq. 3.9.a-3. The overall effectiveness factor then consists of the product of $\eta_\mu$ and a macroeffectiveness factor, $\eta_M$, and the latter is based on macropellet properties plus the micro-effectiveness factor—Eq. 3.9.a-5. Often the microparticles are sufficiently small so that $\eta_\mu \simeq 1.0$; recall from above that there is usually not any micropore diffusional limitations unless these exist for the macropores.

This was extended to complex consecutive reactions by Carberry [138]. The result, neglecting external transport, was similar to Eq. 3.8-5, but with the same type of changes noted above:

$$\left(\frac{r_R}{r_A}\right)_{obs} = \frac{1 - \sigma \left(\frac{\eta_2}{\eta_1}\right)_M \left(\frac{\eta_2}{\eta_1}\right)_\mu}{1 - \sigma} - \left(\frac{\eta_2}{\eta_1}\right)_M \left(\frac{\eta_2}{\eta_1}\right)_\mu \frac{k_2}{k_1} \frac{C_{Rs}{}^s}{C_{As}{}^s} \tag{3.9.a-8}$$

where

$$\sigma = \frac{k_2}{k_1} \left(\frac{D_{eR}}{D_{eA}}\right)_M = \frac{k_2}{k_1} \left(\frac{D_{eR}}{D_{eA}}\right)_\mu$$

$$\eta_{i,\,\mu \, or \, M} = \frac{\tanh \phi_{i,\,\mu \, or \, M}}{\phi_{i,\,\mu \, or \, M}}$$

$$\phi_{i,\,\mu} = L_\mu \sqrt{k_i / D_{ei,\,\mu}}$$

$$\phi_{i,\,M} = L\sqrt{k_i \eta_{i,\,\mu} / D_{ei,\,M}}$$

Again, it is easiest to visualize the results in the large $\phi_i$ asymptotic region and equal $D_{eA} = D_{eR}$:

$$\left(\frac{r_R}{r_A}\right)_{\text{obs}} = \frac{1 - (k_2/k_1)^{1/4}}{1 - (k_2/k_1)} - \left(\frac{k_2}{k_1}\right)^{1/4} \frac{C_{Rs}^s}{C_{As}^s} \tag{3.9.a-9}$$

Comparing Eq. 3.9.a-9 with Eq. 3.8.6-a and with Eq. 3.8-4 shows that the micropores add another square root factor to the rate constant ratio, thereby further decreasing the selectivity. If $\eta_{i,\mu} \simeq 1.0$, the results reduce to those for macropore diffusion.

### 3.9.b Parallel Cross-Linked Pores

It was discussed in Sec. 3.5.d that the most realistic version of this model for catalyst pellets is the communicating pores limiting case. With uncorrelated tortuosity, Eq. 3.5.d-9 gives the diffusion flux:

$$N_j = -\frac{1}{\tau}\left[\int D_{jm}(r)f(r)dr\right]\frac{\partial C_j}{\partial z} \tag{3.9.b-1}$$

It will be postulated that the pore-size distribution does not vary with time, and is position independent (i.e., a macroscopically uniform pellet—not always true, see Satterfield [40]). The communicating pore limit, with concentrations only a function of position, $z$, as discussed in Sec. 3.5.d, then leads to the mass balance

$$\varepsilon_s \frac{\partial C_j}{\partial t} + \frac{\partial N_j}{\partial z} = \rho_s r_j(\mathbf{C}) \tag{3.9.b-2}$$

In Eq. 3.9.b-2, the flux $N_j$ is from Eq. 3.9.b-1, and the rate term is also averaged over the pore-size distribution:

$$\rho_s r_j \equiv \int \rho_s r_j(\mathbf{C}, r)f(r)dr \tag{3.9.b-3}$$

For example, if the rate equation in any pore can be written as

$$\rho_s r_j(\mathbf{C}, r) = \frac{2k_s}{r}g(\mathbf{C}) \tag{3.9.b-4}$$

and if the intrinsic surface rate constant, $k_s$, is independent of $r$ (e.g., no configurational effects on the molecular reaction) then Eqs. 3.9.b-3 and 4 give

$$\rho_s r_j = k_s g(\mathbf{C})\int \frac{2}{r}f(r)dr \tag{3.9.b-5}$$

$$= (\rho_s S_g)k_s g(\mathbf{C}) \tag{3.9.b-5a}$$

$$= k_v g(\mathbf{C})$$

where Eq. 3.5.d-3 was used. Again, this is for the communicating pores limit.

Finally, for steady state, and if the concentration dependency of $D_{jm}(r)$ is ignored, Eqs. 3.9.b-1,2,5 can be combined to give

$$\left[\frac{1}{\tau} \int D_{jm}(r) f(r) dr\right] \frac{d^2 C_j}{dz^2} = [(\rho_s S_g) k_s] g(C) \qquad (3.9.b\text{-}6)$$

For a simple first-order reaction, the usual solution would then be found,

$$\eta = \frac{\tanh \phi}{\phi} \qquad (3.9.b\text{-}7)$$

but with the pore-size distribution averaged diffusivity in the modulus:

$$\phi = L \left[ \frac{(\rho_s S_g) k_s}{(1/\tau) \int D_A(r) f(r) dr} \right]^{1/2} \qquad (3.9.b\text{-}8)$$

There is not really any experience with the use of Eq. 3.9.b-8 as yet (although see Steisel and Butt [139]) and whether it, or less restrictive, versions of the parallel cross-linked pore model are adequate representations of reactions in complex pore systems is not known.

### 3.9.c  Reaction with Configurational Diffusion

Applying the configurational diffusion results of Sec. 3.5.e has not been done to a significant extent, but the principles can be stated. The effective diffusivity would be given by Eq. 3.5.e-7

$$D_e = \frac{\varepsilon_s D}{\tau} \left(1 - \frac{a}{r}\right)^4 \qquad a < r \qquad (3.9.c\text{-}1)$$

and the equation solved in the usual way:

$$\frac{d}{dz} D_e \frac{dC_s}{dz} = r_v(C_s) \qquad (3.9.c\text{-}2)$$

For example, with a first-order reaction,

$$r_v = k_v C_s \qquad (3.9.c\text{-}3)$$

the effectiveness factor would be

$$\eta = \frac{\tanh \phi}{\phi} \qquad (3.9.c\text{-}4)$$

with

$$\phi = L \left\{ k_v \left[ \frac{\varepsilon_s D}{\tau} \left(1 - \frac{a}{r}\right)^4 \right]^{-1} \right\}^{1/2} \qquad (3.9.c\text{-}5)$$

An example of the use of these relations will be given below in Ex. 3.9.c-1.

It is more crucial here to consider specifically the pore-size distribution, since the large molecules will presumably not fit into the smaller pores. The parallel cross-linked pore model can be combined with the above to yield the following steady-state mass balance:

$$\left[\frac{D_{Am}}{\tau} \int \left(1 - \frac{a}{r}\right)^4 f(r)dr\right] \frac{d^2 C_{As}}{dz^2} = \left[\int \frac{2}{r} k_s(r) f(r)dr\right] g(C_{As}), \qquad a < r \quad (3.9.c\text{-}6)$$

Again, the formal solution with a first-order reaction would be Eq. (3.9.c-4), but with the properly pore-size distribution averaged parameters used in the modulus:

$$\phi = L \left[\frac{\int \frac{2}{r} k_s(r) f(r)dr}{\frac{D_{Am}}{\tau} \int \left(1 - \frac{a}{r}\right)^4 f(r)dr}\right]^{1/2} \qquad a < r \qquad (3.9.c\text{-}7)$$

Little is known about configurational effects on the surface rate coefficient, $k_s(r)$, and if it were taken to be constant, only the configurational diffusion effect would be used, with a final formula similar to Eq. 3.9.b-8.

## Example 3.9.c-1 Catalytic Demetallization (and Desulfurization) of Heavy Residuum Petroleum Feedstocks

Spry and Sawyer [140] discussed the peculiar problems associated with this process. Figure 1 shows the nature of the reacting species. We see that the molecular sizes range from $\simeq 25 - 150$ Å, which is precisely the size range of the pores in typical catalysts (Sec. 3.4). Thus, we would expect strong configurational diffusion effects on the observed rate.

The rate will be approximated by a first-order expression for Co-Mo catalysts:

$$r_v \simeq k_v C_s \qquad (a)$$

where the rate constant can be related to the intrinsic rate and the internal surface area:

$$k_v = \rho_s \cdot S_g \cdot k_s \qquad (b)$$

The surface area of a given size pore was given by the usual formula for cylinders Eq. 3.4-7:

$$S_g = \frac{2\varepsilon_s}{r\rho_s} \qquad (c)$$

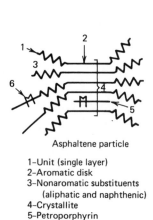

Asphaltene particle

1–Unit (single layer)
2–Aromatic disk
3–Nonaromatic substituents
 (aliphatic and naphthenic)
4–Crystallite
5–Petroporphyrin
6–Metal atom

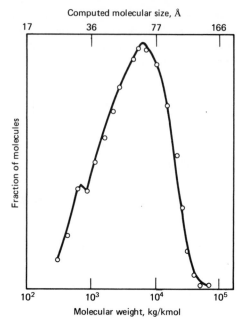

*Figure 1  Model of asphaltene particle and measured molecular weight distribution of asphaltenes in heavy Venezuelan crude. Asphaltene molecular size distribution computed by configurational diffusion model (from Spry and Sawyer [140]).*

The effective diffusivity was based on the equation discussed above for configurational diffusion, Eq. 3.5.e-7:

$$D_e = \frac{\varepsilon_s D}{\tau}\left(1 - \frac{a}{r}\right)^4 \qquad a < r \tag{d}$$

where $a = \frac{1}{2}$ (molecular size) from Figure 1.

In the strong diffusion control range, the effectiveness factor is given by the asymptotic value,

$$\eta \sim 1/\phi$$

$$= \left(\frac{R}{3}\sqrt{\frac{k_s \rho_s S_g}{D_e}}\right)^{-1} \tag{e}$$

As a simplified approach, only mean values of the molecular size and pore diameter distribution were used. The above results were then substituted into the

     CHEMICAL ENGINEERING KINETICS

relation for the observed pellet rate constant:

$$k_{v,\,\mathrm{obs}} = \eta k_v$$

$$= \frac{3\varepsilon_s}{R} \left(2k_s \frac{D}{\tau}\right)^{1/2} \frac{(\bar{r} - \bar{a})^2}{(\bar{r})^{2.5}} \tag{f}$$

Equation f exhibits a maximum value with respect to pore size, and so there is an optimum catalyst pore size that should be used, as shown in Figure 2. The reason is that very small pores hinder, or even block the diffusion of reactant into the catalyst pellet, but they do contain a large surface area. Very large pores, on the other hand, do not cause any hindered diffusion, but they do not have much surface area. Figure 3 shows the agreement of the model with data from a particular system.

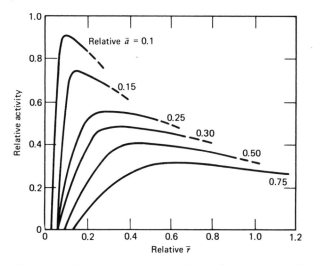

*Figure 2 Functional relationship of k on r̄ with simplified model assuming pore size distribution characterized by r̄ and molecular size distribution by ā. The activity values are the rate coefficients relative to a presumed intrinsic catalytic activity in the absence of diffusional effects, and is proportional to $(\bar{r} - \bar{a})^2/(\bar{r})^{2.5}$; the relative r̄ is based on a large average pore size where configurational diffusional hinderance becomes negligible; the relative ā is based on the largest molecular size in Fig. 1 ( from Spry and Sawyer [140]).*

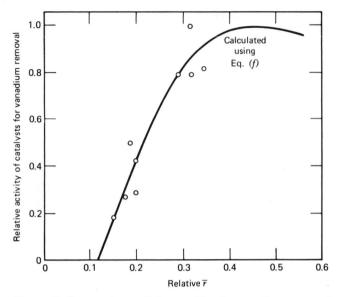

*Figure 3 Comparison of demetallization performance of nine hydrotreating catalysts with model predictions using simplified model (from Spry and Sawyer [140]).*

Finally, the parallel pore model was used to account for the actual distribution of molecular size and pore size. Eq. 3.5.d-9 for communicating pores, with an average value for tortuosity, was utilized:

$$D_e = \sum_i \sum_j \left(\frac{\varepsilon_s D}{\tau}\right) f(a_j) f(r_i) \left[1 - \frac{a_j}{r_i}\right]^4 \qquad a_j < r_i \qquad \text{(g)}$$

where

$$f(a_j) = \text{molecular-size distribution}$$
$$f(r_i) = \text{pore-size distribution}$$

Also, the internal surface area was determined from

$$S_g = \frac{\varepsilon_s}{\rho_s} \sum_i \sum_j f(a_j) f(r_i) \frac{2}{r_i} \qquad a_j < r_i \qquad \text{(h)}$$

The sums over the molecular sizes are semiempirical, but the complete formulation

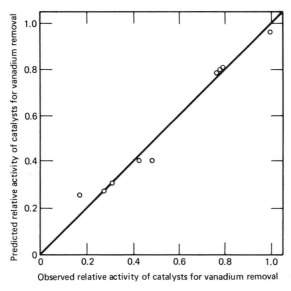

*Figure 4 Comparison of demetallization performance of nine hydrotreating catalysts with model predictions using more rigorous model (from Spry and Sawyer [140]).*

seems to do a good job of predicting the catalyst activity for different catalysts. Figure 4 shows this over a range of a factor of 5.

There are several other complicating effects on pore diffusion with reactions that are just beginning to be studied:

- Position dependent diffusivity and/or catalytic activity. (Kasaoka and Sakata; Corbett and Luss; Becker and Wei [141, 142, 143].)

- Supported liquid-phase catalysts. (Rony; Livbjerg, Sorensen, and Villadsen [144, 145, 146].)

- Pore-blockage effects in zeolites, for example, by catalyst-fouling compounds. (Butt, Delgado-Diaz, and Muno; Butt [147, 148].)

Since these are only in beginning stages of development, however, we do not consider them further here.

# Problems

3.1 The cracking of cumene into benzene and propylene was carried out in a fixed bed of zeolite particles at 362°C and atmospheric pressure, in the presence of a large excess of nitrogen. At a point in the reactor where the cumene partial pressure was 0.0689 atm, a reaction rate of 0.153 kmol/kgcat.hr was observed.

Further data: $M_m = 34.37$ kg/kmol; $\rho_g = 0.66$ kg/m$^3$

$\mu = 0.094$ kg/m.hr; $\lambda_g = 0.037$ kcal/m.hr.°C

$c_p = 0.33$ kcal/kg°C; Re $= 0.052$

Pr $= 0.846$; $D_{Am} = 0.096$ m$^2$/hr

$a_m = 45$ m$^2$cat/kgcat; $G = 56.47$ kg/m$^2$.hr

$(-\Delta H) = -41816$ kcal/kmol

Under these conditions, show that the partial pressure and temperature drops over the external film surrounding the particles are negligible.

3.2 The solid density of an alumina particle is 3.8 g/cm$^3$, the pellet density is 1.5 g/cm$^3$, and the internal surface is 200 m$^2$/g. Compute the pore volume per gram, the porosity, and the mean pore radius.

3.3 Carefully watching how the various fluxes combine, derive Eq. (3.5.b-7) for the molar flux in a porous medium.

3.4 A catalyst considered by Satterfield [40] has a void fraction of 0.40, an internal surface area of 180 m$^2$/g, and a pellet density of 1.40 g/cm$^3$. Estimate the effective diffusivity of thiophene with hydrogen at $T = 660$ K.

3.5 Calculate the diffusion flux for ethylene diffusing in hydrogen at 298 K in a porous medium with the following properties: thickness $= 1$ cm, $\varepsilon_s = 0.40$, $\rho_s = 1.4$ g/cm$^3$, $S_g = 105$ m$^2$/g. The conditions are steady pressure $p$ of ethylene on one side and hydrogen on the other, for $0.1 < p < 40$ atm.

3.6 The data given below, on diffusion of nitrogen ($A$) and helium ($B$) in porous catalyst pellets, have been provided by Henry, Cunningham, and Geankoplis [51], who utilized

| $p_t$ (mm Hg) | $y_{AL}$ | $y_{AO}$ | $N_A \times 10^8$ (mol/cm$^2$.s) | $-N_B/N_A$ |
|---|---|---|---|---|
| 0.500 | 0.0407 | 0.911 | 0.449 | 2.52 |
| 1.506 | 0.0358 | 0.900 | 0.881 | 3.06 |
| 3.25 | 0.0888 | 0.863 | 2.62 | 2.22 |
| 8.00 | 0.176 | 0.735 | 4.78 | 2.65 |
| 30.72 | 0.214 | 0.658 | 11.3 | 2.78 |
| 100.6 | 0.164 | 0.708 | 22.2 | 2.62 |
| 299.7 | 0.134 | 0.769 | 42.1 | 2.37 |
| 600.1 | 0.146 | 0.723 | 43.9 | 2.86 |

the steady-state Wicke–Kallenbach–Weisz technique. An alumina pellet with the following properties was used:

Length = 1.244 cm; pore volume = 0.5950 cm³/g
Porosity = 0.233 (macro); 0.492 (micro)
Pore radius = 20,000 Å (macro); 37 Å (micro)
Internal surface = 202 m²/g

(a) Compare the flux ratios with the theoretical prediction.
(b) Compute the experimental diffusivities, and plot $D_e \cdot p_t$ versus $p_t$. At what pressure is there a transition between Knudsen and bulk diffusion?
(c) Use the dusty-gas model, assuming one dominant pore size, to predict the changes of $D_e$ with pressure up to 2 atm. What value of tortuosity is required?
(d) Repeat the calculations of part (c) with the random pore model.
(e) Repeat the calculations of part (c) with the parallel cross-linked pores model.

3.7 Derive Eq. (3.6.a-10) for the effectiveness factor for a first-order reaction in a spherical catalyst pellet.

3.8 A series of experiments were performed using various sizes of crushed catalyst in order to determine the importance of pore diffusion. The reaction may be assumed to be first order and irreversible. The surface concentration of reactant was $C_s^s = 2 \times 10^{-4}$ mol/cm³.

Data

| Diameter of sphere (cm) | 0.25 | 0.075 | 0.025 | 0.0075 |
|---|---|---|---|---|
| $r_{obs}$ (mol/hr.cm³) | 0.22 | 0.70 | 1.60 | 2.40 |

(a) Determine the "true" rate constant, $k_v$, and the effective diffusivity, $D_e$, from the data.
(b) Predict the effectiveness factor and the expected rate of reaction ($r_{obs}$) for a commercial *cylindrical* catalyst pellet of dimensions 0.5 cm × 0.5 cm.

3.9 The following rates were observed for a first-order irreversible reaction, carried out on a spherical catalyst:

for $d_p = 0.6$ cm; $r_{obs} = 0.09$ mol/gcat.hr
for $d_p = 0.3$ cm; $r_{obs} = 0.162$ mol/gcat.hr

Strong diffusional limitations were observed in both cases. Determine the true rate of reaction. Is diffusional resistance still important with $d_p = 0.1$ cm?

3.10 A second-order gas phase reaction, $A \rightarrow R$, occurs in a catalyst pellet, and has a rate coefficient

$$k_v = 3.86 \text{ m}^3/\text{kmol.s}$$

The reactant pressure is one atmosphere, the temperature is 600 K, the molecular diffusivity is $D_{AR} = 0.10$ cm²/s, and the reactant molecular weight is $M_A = 60$. The catalyst pellets have the following properties:

Radius of sphere, $R = 9$ mm
Pellet density, $\rho_s = 1.2$ g/cm³
Internal surface area, $S_g = 100$ m²/g
Internal void fraction, $\varepsilon_s = 0.60$

(a) Estimate the effective diffusivity.

(b) Determine if there may be pore diffusion limitations.

(c) If part (b) results in pore diffusion limitations, what might be done to eliminate them? Justify your answer(s) with quantitative calculations.

3.11 A gas oil is cracked at 630°C and 1 atm by passing vaporized feed through a packed bed of spheres of silica-alumina catalyst with radius = 0.088 cm. For a feed rate of 60 cm³ liquid/cm$_r^3$.hr, a 50 percent conversion is found. The following data are also known:

Liquid density = 0.869 g/cm³

Feed molecular weight = 255 g/mol

Bulk density of packed bed = 0.7 g cat./cm$_r^3$

Solid density of catalyst = 0.95 g cat./cm³ cat.

Effective diffusivity in catalyst = $8 \times 10^{-4}$ cm²/s.

*Average* reactant concentration = $0.6 \times 10^{-5}$ mol/cm³.

Assume a first-order reaction and treat data as being average data of a differential reactor.

(a) Show that the *average* reaction rate is

$$3.9 \times 10^{-5} \ \text{mol/cm}^3\text{cat.s}$$

(b) Determine from the data whether or not pore diffusion was important.

(c) Find the value of the effectiveness factor.

(d) Determine the value of the rate coefficient.

3.12 Derive Eq. (3.8-5) for the selectivity with first-order consecutive reactions.

3.13 Verify the numerical results in Ex. 3.8-1.

# References

[1] Frank–Kamenetskii, D. *Diffusion and Heat Exchange in Chemical Kinetics*, 2nd ed., Plenum Press (1969).

[2] Bird, R. B., Stewart, W. E., and Lightfoot, E. N. *Transport Phenomena*, Wiley, New York (1960).

[3] Gamson, B. W., Thodos, G., and Hougen, O. A. *Trans. A. I. Ch. E.*, **39**, 1 (1943).

[4] Wilke, C. R. and Hougen, O. A. *Trans. A. I. Ch. E.*, **41**, 445 (1945).

[5] Taecker, R. G. and Hougen, O. A. *Chem. Eng. Prog.*, **45**, 188 (1949).

[6] McCune, L. K. and Wilhelm, R. H. *Ind. Eng. Chem.*, **41**, 1124 (1949).

[7] Ishino, T. and Otake, T. *Chem. Eng.* (Tokyo), **15**, 258 (1951).

[8] Bar Ilan, M. and Resnick, W. *Ind. Eng. Chem.*, **49**, 313 (1957).

[9] de Acetis, J. and Thodos, G. *Ind. Eng. Chem.*, **52**, 1003 (1960).

[10] Bradshaw, R. D. and Bennett, C. O. *A. I. Ch. E. J.*, **7**, 50 (1961).

[11] Hougen, O. A. *Ind. Eng. Chem.*, **53**, 509 (1961).

[12] Yoshida, F., Ramaswami, D., and Hougen, O. A. *A. I. Ch. E. J.*, **8**, 5 (1962).

[13] Baumeister, E. B. and Bennett, C. O. *A. I. Ch. E. J.*, **4**, 70 (1958).

[14] Glaser, M. B. and Thodos, G. *A. I. Ch. E. J.* ,**4**, 63 (1958).

[15] Sen Gupta, A. and Thodos, G. *A. I. Ch. E. J.*, **9**, 751 (1963).

[16] Handley, D. and Heggs, P. J. *Trans. Inst. Chem. Eng.*, **46**, 251 (1968).

[17] Toor, H. L. *A. I. Ch. E. J.*, **10**, 448, 460 (1964).

[18] Stewart, W. E. and Prober, R. *Ind. Eng. Chem. Fund.*, **3**, 224 (1964).

[19] Mason, E. A. and Marrero, T. R. *Adv. Atomic Mole Physics*, Vol. 6, Academic Press, New York (1970).

[20] Dullien, F. A. L., Ghai, R. K., and Ertl, H. *A. I. Ch. E. J.*, **19**, 881 (1973).

[21] Dullien, F. A. L., Ghai, R. K., and Ertl, H. *A. I. Ch. E. J.*, **20**, 1 (1974).

[22] Hsu, H. W. and Bird, R. B. *A. I. Ch. E. J.*, **6**, 516 (1960).

[23] Kubota, H., Yamanaka, Y., and Dalla Lana, I. G. *J. Chem. Eng. Japan*, **2**, 71 (1969).

[24] Smith, J. M. *Chemical Engineering Kinetics*, 2nd. ed., McGraw-Hill, New York (1970).

[25] Reid, R. C. and Sherwood, T. K. *The Properties of Gases and Liquids*, 2nd. ed., McGraw-Hill, New York (1966).

[26] Rihani, D. N. and Doraiswamy, L. K. *Ind. Eng. Chem. Fund.*, **4**, 17 (1965).

[27] Perry, R. H. and Chilton, C. H. *Chemical Engineers Handbook*, 5th ed., McGraw-Hill, New York (1973).

[28] Cunningham, R. S. and Geankoplis, C. J. *Ind. Eng. Chem. Fund.*, **7**, 535 (1968).

[29] Gregg, S. J. and Sing, K. S. W. *Adsorption, Surface Area, and Porosity*, Academic Press, New York (1967).

[30] Brunauer, S., Mikhail, R. Sh. and Bodor, E. E. *J. Coll. Interface Sci.*, **25**, 353 (1967).

[31] Strider, W. and Aris, R. *Variational Methods Applied to Problems of Diffusion and Reaction*, Springer Verlag, New York (1973).

[32] Weisz, P. B. *Chem. Techn.*, 504 (1973).

[33] Brown, L. M., Sherry, H. S., and Krambeck, F. J. *J. Phys. Chem.*, **75**, 3846, 3855 (1971).

[34] Riekert, L. *Adv. Catal.*, **21**, 281 (1970).

[35] Barrer, R. M. *Adv. Chem.*, **102**, 1 (1971).

[36] Gorring, R. L. *J. Catal.*, **31**, 13 (1973).

[37] Chen, N. Y., Lucki, S. J., and Mower, E. B. *J. Catal.*, **13**, 329 (1969).

[38] Chen, N. Y. and Weisz, P. B. *Chem. Eng. Progr. Symp. Ser.*, **67**, No. 73, 86 (1967).

[39] Van Brakel, J. and Heertjes, P. M. *Int. J. Heat Mass Trans.*, **17**, 1093 (1974).

[40] Satterfield, C. N. *Mass Transfer in Heterogeous Catalysis*, MIT Press, Cambridge, Mass. (1970).

[41] Storvick, T. S. and Mason, E. A. *J. Chem. Phys.*, **46**, 3199 (1967).

[42] Mason, E. A. and Evans, R. B. *J Chem. Ed.*, **46**, 358 (1969).

[43] Feng, C. F. and Stewart, W. E. *Ind. Eng. Chem. Fund.*, **12**, 143 (1973).

[44] Gunn, R. D. and King, C. J. *A. I. Ch. E. J.*, **15**, 507 (1969).

[45] Di Napoli, N. M., Williams, R. J. J., and Cunningham, R, E. *Lat. Am. J. Chem. Eng. Appl. Chem.*, **5**, 101 (1975).

[46] Scott, D. S. and Dullien, F. A. L. *A. I. Ch. E. J.*, **8**, 29 (1962).

[47] Rothfeld, L. B. *A. I. Ch. E. J.*, **9**, 19 (1963).

[48] Satterfield, C. N. and Cadle, P. J. *Ind. Eng. Chem. Fund.*, **7**, 202 (1968).

[49] Satterfield, C. N. and Cadle, P. J. *Ind. Eng. Chem. Proc. Des. Devpt.*, **7**, 256 (1968).

[50] Brown, L. F., Haynes, H. W., and Manogue, W. H. *J. Catal.*, **14**, 220 (1969).

[51] Henry, J. P., Cunningham, R. S., and Geankoplis, C. J. *Chem. Eng. Sci.*, **22**, 11 (1967).

[52] Wicke, E. and Kallenbach, R. *Kolloid Z.*, **97**, 135 (1941).

[53] Weisz, P. B. *Z. Physik Chem.* (Frankfurt), **11**, 1 (1957).

[54] Dogu, G. and Smith, J. M. *A. I. Ch. E. J.*, **21**, 58 (1975).

[55] Butt, J. B. *A. I. Ch. E. J.*, **9**, 707 (1963).

[56] Feng, C. F., Kostrov, V. V., and Stewart, W. E. *Ind. Eng. Chem. Fund.*, **13**, 5 (1974).

[57] Wakao, N. and Smith, J. M. *Chem. Eng. Sci.*, **17**, 825 (1962).

[58] Wakao, N. and Smith, J. M. *Ind. Eng. Chem. Fund.*, **3**, 123 (1964).

[59] Weisz, P. B. and Schwartz, A. B. *J. Catal.*, **1**, 399 (1962).

[60] Johnson, M. F. L. and Stewart, W. E. *J. Catal.*, **4**, 248 (1965).

[61] Omata, H. and Brown, L. F. *A. I. Ch. E. J.*, **18**, 1063 (1972).

[62] Stoll, D. R. and Brown, L. F. *J. Catal.*, **32**, 37 (1974).

[63] Abed, R. and Rinker, R. G. *J. Catal.*, **34**, 246 (1974).

[64] Patel, P. V. and Butt, J. B. *Ind. Eng. Chem. Proc. Des. Devpt.*, **14**, 298 (1974).

[65] Weisz, P. B., Zollinger, H. et al. *Trans. Farad. Soc.* **63**, 1801, 1807, 1815 (1967); **64**, 1693, (1968); **69**, 1696, 1705 (1973).

[66] Yang, R. T., Fenn, J. B., and Haller, G. L. *A. I. Ch. E. J.*, **19**, 1052 (1973).

[67] Sladek, K. J., Gilliland, E. R., and Baddour, R. F. *Ind. Eng. Chem. Fund.*, **13**, 100 (1974).

[68] Komiyama, H. and Smith, J. M. *A. I. Ch. E. J.*, **20**, 1110 (1974).

[69] Schneider, P. and Smith, J. M. *A. I. Ch. E. J.*, **14**, 886 (1968).

[70] Komiyama, H. and Smith, J. M. *A. I. Ch. E. J.*, **20**, 728 (1974).

[71] Anderson, J. L. and Quinn, J. *Biophys. J.*, **14**, 130 (1974).

[72] Satterfield, C. N., Colton, C. K., and Pitcher, W. H. Jr. *A. I. Ch. E. J.*, **19**, 628 (1973).

[73] Colton, C. K., Satterfield, C. N., and Lai. C. J. *A. I. Ch. E. J.*, **21**, 289 (1975).

[74] Aris, R. *The Mathematical Theory of Diffusion and Reaction in Permeable Catalysts*, Vols. I and II, Oxford University Press, London (1975).

[75] Levenspiel. O. *Chemical Reaction Engineering*, Wiley, New York (1962).

[76] Thiele, E. W. *Ind. Eng. Chem.*, **31**, 916 (1939).

[77] Zeldowich, Ia. B., *Zhur. Fiz. Khim.*, **13**, 163 (1939).

[78] Aris, R. *Elementary Chemical Reactor Analysis*, Prentice-Hall, Englewood Cliffs, N.J. (1969).

[79] Aris, R. *Chem. Eng. Sci.*, **6**, 262 (1957).

[80] Knudsen, C. W., Roberts, G. W., and Satterfield, C. N. *Ind. Eng. Chem. Fund.*, **5**, 325 (1966).

[81] Rester, S. and Aris, R. *Chem. Eng. Sci.*, **24**, 793 (1969).

[82] Bird, R. B., Stewart, W. E., and Lightfoot, E. N. *Notes on Transport Phenomena*, Wiley, New York (1958).

[83] Aris, R. *Ind. Eng. Chem. Fund.*, **4**, 227 (1965).

[84] Bischoff, K. B. *A. I. Ch. E. J.* **11**, 351 (1965).

[85] Petersen, E. E. *Chem. Eng. Sci.*, **20**, 587 (1965).

[86] Petersen, E. E. *Chemical Reaction Analysis*, Prentice-Hall, Englewood Cliffs, N.J. (1965).

[87] Satterfield, C. N., Roberts, G. W., and Hartman, J. *Ind. Eng. Chem. Fund.*, **5**, 317 (1966).

[88] Satterfield, C. N., Roberts, C. W., and Hartman, J. *Ind. Eng. Chem. Fund.*, **6**, 80 (1967).

[89] Luss, D. *Chem. Eng. Sci.*, **26**, 1713 (1971).

[90] Lee, J. and Luss, D. *Chem. Eng. Sci.*, **26**, 1433 (1971).

[91] Aris, R. *Chem. Eng. Sci.*, **24**, 149 (1969).

[92] Carberry, J. J. *A. I. Ch. E. J.*, **8**, 557 (1962).

[93] Weisz, P. B. and Prater, C. D. *Adv. Catal.*, **6**, 143 (1954).

[94] Languasco, J. M., Cunningham, R. E., and Calvelo, A. *Chem. Eng. Sci.*, **27**, 1459 (1972).

[95] Balder, J. R. and Petersen, E. E. *J. Catal.*, **11**, 195, 202 (1968).

[96] Gilliland, E. R., Bixler, H. J., and O'Connell, J. E. *Ind. Eng. Chem. Fund.*, **10**, 185 (1971).

[97] Brown, C. E. and Bennett, C. O. *A. I. Ch. E. J.*, **16**, 817 (1970).

[98] Cadle, P. J. and Satterfield, C. N. *Ind. Eng. Chem. Proc. Des. Devpt.*, **7**, 189, 192 (1968).

[99] Hougen, O. A. and Watson, K. M. *Chemical Process Principles*, Wiley, New York (1947).

[100] Weisz, P. B. *Adv. Catal.*, **13**, 148 (1962).

[101] Gunn, D. J. and Thomas, W. J. *Chem. Eng. Sci.*, **20**, 89 (1965).

[102] Bischoff, K. B. *Chem. Eng. Sci.*, **22**, 525 (1967).

[103] Hudgins, R. R. *Chem. Eng. Sci.*, **23**, 93 (1968).

[104] Stewart, W. E. and Villadsen, J. *A. I. Ch. E. J.*, **15**, 28 (1969).

[105] Mears, D. E. *Ind. Eng. Chem. Proc. Des. Devpt.*, **10**, 541 (1971).

[106] Brown, L. F. *Chem. Eng. Sci.*, **27**, 213 (1972).

[107] Austin, L. G. and Walker, P. L. *A. I. Ch. E. J.*, **9**, 303 (1963).

[108] Koros, R. M. and Nowak, E. J. *Chem. Eng. Sci.*, **22**, 470 (1967).

[109] Aris, R. *Introduction to the Analysis of Chemical Reactors*, Prentice-Hall, Englewood Cliffs, N.J. (1965).

[110] Mehta, B. N. and Aris, R. *Chem. Eng. Sci.*, **26**, 1699 (1971).

[111] Prater, C. D. *Chem. Eng. Sci.*, **8**, 284 (1958).

[112] Weisz, P. B. and Hicks, J. S. *Chem. Eng. Sci.*, **17**, 265 (1962).

[113] Hlavacek, V., Kubicek, M., and Marek, M. *J. Catal.*, **15**, 17, 31 (1969).

[114] Kehoe, J. P. G. and Butt, J. B. *A. I. Ch. E. J.*, **18**, 347 (1972).

[115] Tinkler, J. D. and Metzner, A. B. *Ind. Eng. Chem.*, **53**, 663 (1961).

[116] Carberry, J. J. *A. I. Ch. E. J.*, **7**, 350 (1961).

[117] Liu, S. L. *A. I. Ch. E. J.*, **15**, 337 (1969).

[118] Drott, D. W. and Aris, R. *Chem. Eng. Sci.*, **24**, 541 (1969).

[119] Luss, D. *Chem. Eng. Sci.*, **23**, 1249 (1968).

[120] Wei, J. *Chem. Eng. Sci.*, **21**, 1171 (1966).

[121] Georgakis, C. and Aris, R. *Chem. Eng. Sci.*, **29**, 291 (1974).

[122] Lee, J. C. M. and Luss, D. *Ind. Eng. Chem. Fund.*, **8**, 597 (1969).

[123] Hlavacek, V. and Marek, M. *Chem. Eng. Sci.*, **25**, 1537 (1970).

[124] Carberry, J. J. *Ind. Eng. Chem. Fund.*, **14**, 129 (1975).

[125] Mears, D. E. *J. Catal.*, **20**, 127 (1971).

[126] McGreavy, C. and Cresswell, D. *Chem. Eng. Sci.*, **24**, 608 (1969).

[127] Kehoe, J. P. G. and Butt, J. B. *Chem. Eng. Sci.*, **25**, 345 (1970).

[128] Luss, D. *Proc. 4th Intern. Symp. Chem. React. Engng.*, Heidelberg, (April 1976).

[129] Smith, T. G., Zahradnik, J., and Carberry, J. J. *Chem. Eng. Sci.*, **30**, 763 (1975).

[130] Carberry, J. J. *Ind. Eng. Chem.*, **58**, No. 10, 40 (1966).

[131] Mercer, M. C. and Aris, R. *Rev. Lat. Am. Ing. Quim, y Quim Apl.*, **2**, 149 (1971).

[132] Ray, W. H. *Proc. 2nd Int. Symp. Chem. React. Engng.*, Amsterdam (1972).

[133] Wheeler, A. *Adv. Catal.*, **3**, 313 (1951).

[134] Roberts, G. *Chem. Eng. Sci.*, **27**, 1409 (1972).

[135] Voge, H. H. and Morgan, C. Z. *Ind. Eng. Chem. Proc. Des. Devpt.*, **11**, 454 (1972).

[136] Wei, J. *J. Catal.*, **1**, 526, 538 (1962).

[137] Mingle, J. O. and Smith, J. M. *A. I. Ch. E. J.*, **7**, 243 (1961).

[138] Carberry, J. J. *Chem. Eng. Sci.*, **17**, 675 (1962).

[139] Steisel, N. and Butt, J. B. *Chem. Eng. Sci.*, **22**, 469 (1967).

[140] Spry, J. C. and Sawyer, W. H. Paper presented at 68th Annual A. I. Ch. E. Meeting, Los Angeles (November 1975).

[142] Kasaoko, S. and Sakata, Y. *J. Chem. Eng. Japan*, **1**, 138 (1968).

[142] Corbett, W. E. and Luss, D. *Chem. Eng. Sci.*, **29**, 1473 (1974).

[143] Becker, E. R. and Wei, J. *Proc. 4th Int. Symp. Chem. React. Engng.*, Heidelberg (April 1976).

[144] Rony, P. R. *Chem. Eng. Sci.*, **23**, 1021 (1968).

[145] Rony, P. R. *J. Catal.*, **14**, 142 (1969).

[146] Livbjerg, H., Sorensen, B., and Villadsen, J. *Adv. Chem. Ser.*, No. 133, 242 (1974).

[147] Butt, J. B., Delgado-Diaz, S. and Muno, W. E. *J. Catal.*, **37**, 158 (1975).

[148] Butt, J. B. *J. Catal.*, **41**, 190 (1976).

[149] Dedrick, R. L. and Beckmann, R. B. *Chem. Eng. Progr. Symp.* Ser. No. 74, **63**, 68 (1967).

[150] Dumez, F. J. and Froment, G. F. *Ind. Eng. Chem. Des. Devpt.*, **15**, 297 (1976).

[151] Frouws, M. J.,Vellenga, K., and De Wilt, H. G. J. *Biotech. Bioeng.*, **18**, 53 (1976).

[152] Hegedus, L. L. and Petersen, E. E. *Catal. Rev.*, **9**, 245 (1974).

[153] Schneider, P. *Catal. Rev.*, **12**, 201 (1975).

[154] Lee, J. C. M. and Luss, D., *A. I. Ch. E. J.* **16**, 620 (1970).

# 4

## NONCATALYTIC
## GAS-SOLID
## REACTIONS

### 4.1  A Qualitative Discussion of Gas- Solid Reactions _____

In this chapter the reaction between a fluid and a solid or a component of a solid is discussed in quantitative terms. Such a reaction is frequently encountered in the process industry (e.g., in coal gasification, in ore processing, iron production in the blast-furnace, and roasting of pyrites). There are no true gas-solid production processes in the petrochemical industry but gas-solid reactions are encountered (e.g., in the regeneration of coked catalysts by means of oxygen containing gases or in the reduction or reoxidation of nickel-reforming or iron-ammonia-synthesis catalysts prior to or after their use in the production proper). For all these examples the knowledge of the rate of reaction is a prerequisite to the analysis of an existing process, to the design of a new reactor, or to the safe conduct of the regeneration or reoxidation.

Gas-solid reactions have several aspects in common with reactions catalyzed by a porous solid, discussed already in Chapter 3. In the present case too, transport effects and reaction have to be considered simultaneously. Again it depends on the relative magnitudes of the rate of transport and the rate of reaction whether or not important gradients inside and around the particle are built up or not. There is one essential difference, however: with gas-solid reactions the conditions inside the particle change with time, since the solid itself is involved in the reaction. The process is therefore essentially of a non-steady-state nature.

In this chapter rate equations are set up for fluid-solid reactions. In Chapter 11 an example of such a reaction carried out in a fixed bed of particles is worked out and illustrates the difference in behavior of the reactor as compared with a fixed bed catalytic reactor.

Concentrating now on the phenomena inside a particle, an easily visualized situation is that of a gas reacting with a solid of low porosity to yield a porous reacted layer, often called "ash" layer. The reaction then takes place in a narrow zone that moves progressively from the outer surface to the center of the particle. Such a situation is described by the so-called heterogeneous shrinking-core model: heterogeneous because there are two distinct layers inside the particle, with clearly

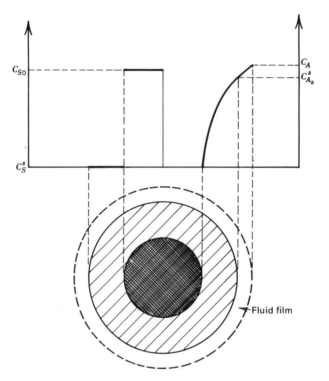

*Figure 4.1-1 Heterogeneous shrinking core model with sharp interface. Concentration profiles of gas and solid reactants (from Wen [2]).*

distinct properties. Figure 4.1-1 illustrates such a situation. When the transport rates through the two layers are not too different and the true rate of reaction is not infinitely fast, the situation is no longer as clear cut, and the sharp boundary between reacted and unreacted zone no longer exists. Such a case is illustrated in Fig. 4.1-2. In the extreme, with very porous material, when the transport through both reacted and unreacted structures is usually fast compared with the true reaction rate, the latter is governing the rate of the overall phenomenon. Then there are no gradients whatever inside the particle and the situation could be called truly homogeneous. Such a situation is represented in Fig. 4.1-3. Thus it is clear that the transport inside the particle plays an important role. Again, as in the case of transport inside a catalyst particle (dealt with in Chapter 3), the transport is evidently not a true diffusion, but rather an effective diffusion. In Chapter 3 also it was pointed out that a considerable effort has been made to relate the effective diffusivity to a detailed picture of the catalyst structure. The goal is to avoid having to

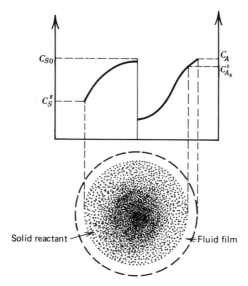

*Figure 4.1-2 General model (from Wen [2]).*

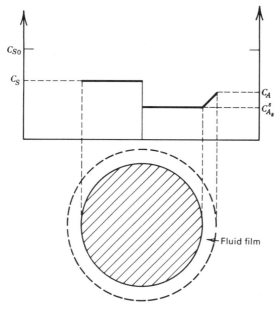

*Figure 4.1-3 Truly homogeneous model. Concentration profiles (from Wen [2]).*

241

determine the effective diffusivity for each reacting system and each catalyst and to rely on easily determined catalyst properties. With an effective diffusion (and an effective conduction) concept the particle is then considered as a continuum. This alters the problem to one of accounting for the detailed solid structure only through the effective diffusivity and not through the model equations proper. The same approach is used for gas-solid reactions too, although probably with considerably less accuracy, since the solid structure is modified by the reaction. Recently, models have been proposed that do account explicitly for the solid structure through the mathematical model, albeit in a rather simplified way.

In some cases, particularly at relatively low temperatures, the curves of conversion versus time have a sigmoidal shape. When the rate is plotted as a function of conversion the curve shows a maximum. This behavior has been explained in terms of nucleation. In a first stage, nuclei are being formed and the rate is low. In a second stage, a reaction front develops starting from the nuclei and growing into the surrounding solid. Such an approach has been developed in great detail by Delmon [1].

## 4.2 A General Model with Interfacial and Intraparticle Gradients

When the particle is assumed to be isothermal only differential material balances have to be written. The differential balance on the reacting gaseous component $A$— the continuity equation for $A$—contains an accumulation term accounting for the transient nature of the process, a term arising from the transport by effective diffusion and a reaction term:

$$\frac{\partial}{\partial t}(\varepsilon_s C_{As}) = \frac{1}{r^2}\frac{\partial}{\partial r}\left(D_e r^2 \frac{\partial C_{As}}{\partial r}\right) - r_A \rho_s \tag{4.2-1}$$

while the continuity equation for the reacting component of the solid is:

$$\frac{\partial C_S}{\partial t} = -r_S \rho_s \tag{4.2-2}$$

$C_A$, $r_A$, and $r_S$ are defined as in the preceding chapters on catalytic reactions and $C_S$ has units kmol/m$_p^3$. The initial and boundary conditions are:

$$\text{at} \quad t = 0 \quad C_{As} = C_{As_0} \quad \text{and} \quad C_S = C_{S0} \tag{4.2-3}$$

in the center of the sphere, $r = 0$:

$$\frac{\partial C_{As}}{\partial r} = 0 \tag{4.2-4}$$

for reasons of symmetry; at the surface, $r = R$

$$D_e\left(\frac{\partial C_{As}}{\partial r}\right)_{r=R} = k_g(C_A - C_{As}^s) \tag{4.2-5}$$

$C_{As}{}^s$ is the concentration of $A$ at the particle surface. The effective diffusivity of $A$, represented by $D_e$, is considered to vary with position if there is a change in porosity resulting from the reaction.

If it is assumed that the porosity depends linearly on the solid reactant conversion:

$$\varepsilon_s = \varepsilon_{s0} + C_{S0}(v_{s0} - v_s)(1 - \varepsilon_{s0})\left(1 - \frac{C_S}{C_{S0}}\right) \qquad (4.2\text{-}6)$$

where $\varepsilon_{s0}$ is the initial porosity, $C_{S0}$ is the initial concentration of the reacting component of the solid, $v_{s0}$ and $v_s$ are the reactant and product molar volumes (see Problem 4.1). It can be seen from Eq. 4.2-6 that the porosity increases if the reaction product has a lower molar volume than the reactant—that is, it is more dense. Wen [2] related the effective diffusivity to $\varepsilon_s$ by means of the relation:

$$\frac{D_e}{D_{e0}} = \left(\frac{\varepsilon_s}{\varepsilon_{s0}}\right)^\beta \qquad (4.2\text{-}7)$$

where $\beta$ lies between 2 and 3, the value of 2 corresponding to the random pore model. Alterations in the porous structure itself, (e.g., through sintering) can also affect the diffusivity—an example is given by Kim and Smith [3].

Equation 4.2-1 may often be simplified. Indeed, it is justified to neglect

$$\varepsilon_s \frac{\partial C_{As}}{\partial t} \qquad \text{when} \qquad \frac{C_{As}}{C_S} \leq 10^{-3}$$

as was shown by Bischoff [4, 5], Luss [22], Theofanous and Lim [23], and extensions have been given by Yoshida, Kunii, and Shimizu [24]. This condition is always satisfied for gas-solid reactions, but not necessarily for liquid-solid reactions. Physically, neglecting the transient term in Eq. 4.2-1 means that the rate at which the reaction layer moves is small with respect to the rate of transport of $A$. This assumption has frequently been referred to as the pseudo-steady-state approximation.

In Eq. 4.2-1 the rates of reaction $r_A$ and $r_S$ may be of the type encountered in Chapter 1 or Chapter 2. In general, the system of Eqs. 4.2-1 and 4.2-2 cannot be integrated analytically.

A transformation of the dependent variables $C_{As}$ and $C_S$ allowed DelBorghi, Dunn, and Bischoff [9] and Duduković [25] to reduce the coupled set of partial differential equations for reactions first-order in the fluid concentration and with constant porosity and diffusivity, into a single partial differential equation. With the pseudo-steady-state approximation, this latter equation is further reduced to an ordinary differential equation of the form considered in Chapter 3 on diffusion and reaction (see Problem 4.2). An extensive collection of solutions of such equations has been presented by Aris [7].

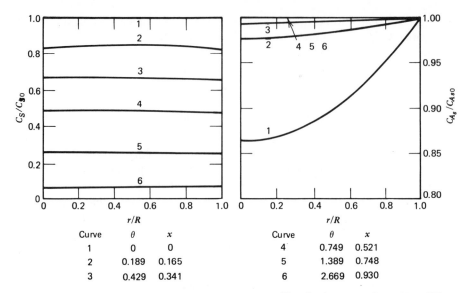

| Curve | $\theta$ | $x$ |
|-------|----------|------|
| 1 | 0 | 0 |
| 2 | 0.189 | 0.165 |
| 3 | 0.429 | 0.341 |

| Curve | $\theta$ | $x$ |
|-------|----------|------|
| 4 | 0.749 | 0.521 |
| 5 | 1.389 | 0.748 |
| 6 | 2.669 | 0.930 |

*Figure 4.2-1 General model. Concentration profiles for $h_s = 1$ (from Wen [2]).*

Wen [2] integrated Eq. (4.2-1,2) numerically for the following rate equations:

$$r_A \rho_s = ak C_{As}{}^n C_S{}^m \qquad (4.2\text{-}8a)$$

where $a$ is the number of moles of $A$ reacting with one mole of $S$

$$r_S \rho_s = k C_{As}{}^n C_S{}^m \qquad (4.2\text{-}8b)$$

Note that in Eqs. 4.2-7 and 4.2-8 $k$ has dimensions

$$[m_f{}^{3n}(\text{kmol } A)^{1-n}(\text{kmol } S)^{-m}.(m_p{}^3)^{m-1}.\text{hr}^{-1}].$$

The results are represented in Figs. 4.2-1 and 4.2-2 for $n = 2$ and $m = 1$. Figure 4.2-1 shows $C_A$ and $C_S$ profiles at various reduced times $\theta = k C_{As0}{}^2 t$ in the absence of interfacial gradients (Sh' $= \infty$) and for

$$\frac{C_{S0}(v_{s0} - v_s)(1 - \varepsilon_{s0})}{\varepsilon_{s0}} = 9, \qquad \beta = 2 \qquad \text{and} \qquad h_s = R\sqrt{\frac{ak C_{As}{}^s C_{S0}}{D_{e0}}} = 1$$

$x$ is the fractional conversion of $S$. The latter group is the Thiele modulus already encountered in Chapter 3. One is a low value for $h_s$ so that the chemical reaction is rate controlling and there are practically no gradients inside the particle. This is a situation that could be described satisfactorily by the homogeneous model and that is encountered at low temperatures. Figure 4.2-2 corresponds to a case for which the modulus $\phi$ is high and for which the diffusion of $A$ through the solid is rate controlling. This is a situation that could be described by the heterogeneous model with shrinking core.

CHEMICAL ENGINEERING KINETICS

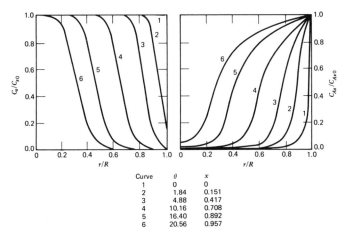

| Curve | $\theta$ | $x$ |
|-------|----------|-------|
| 1 | 0 | 0 |
| 2 | 1.84 | 0.151 |
| 3 | 4.88 | 0.417 |
| 4 | 10.16 | 0.708 |
| 5 | 16.40 | 0.892 |
| 6 | 20.56 | 0.957 |

*Figure 4.2-2 General model. Concentration profiles for $h_s = 70$ (from Wen [2]).*

When the form of the kinetic equation is such that the concentration of the reacting solid component, $C_S$, drops to zero in a finite time, two stages have to be considered in solving Eq. 4.2-1,2. In the first stage, which extends until the time at which $C_S$ becomes zero at the surface of the particle, the complete equations Eqs. 4.2-1,2 are solved directly. The second stage involves only diffusion through the region with completely exhausted solid reactant, up to the front where reaction is occurring, from which location onward the complete equations are used again. This is illustrated below for the useful special case of a zero-order reaction with respect to the reacting solid concentration, a good approximation for the situation that all of the solid is reactive.

Wen also has simplified Eq. 4.2-1 somewhat by allowing only two values for $D_e$, instead of letting it vary according to Eq. 4.2-7: a value $D_e$ for diffusion through unreacted or partially reacted solid and a value $D'_e$ for diffusion of $A$ through completely reacted solid, Wen [2]. This scheme is most useful, of course, only for those kinetic forms leading to complete conversion in a finite time. As mentioned already, this then means that two stages must be considered. In the first stage, Eq. 4.2-1 reduces to

$$\varepsilon_s \frac{\partial C_{As}}{\partial t} = D_e \left( \frac{\partial^2 C_{As}}{\partial r^2} + \frac{2}{r} \frac{\partial C_{As}}{\partial r} \right) - r_A \rho_s \qquad (4.2-9)$$

with Eq. 4.2-2 and the boundary conditions Eqs. 4.2-3, 4.2-4, and 4.2-5 unchanged and with $\partial C_{As}/\partial t = 0$ when the pseudo-steady-state hypothesis is valid. For an isothermal particle and a single reaction with simple kinetics, at least, the location where $C_S$ drops to zero is obviously the surface, where $C_A$ is highest. The second stage sets in when $C_S$ has reached zero at the surface. In the outer zone, which is originally very thin and gradually moves to extend to the center of the particle, there is no reaction any longer, only transport, and Eqs. 4.2-1 and 4.2-2 reduce to

$$D_e'\left(\frac{\partial^2 C_{As}'}{\partial r^2} + \frac{2}{r}\frac{\partial C_{As}'}{\partial r}\right) = 0 \qquad (4.2\text{-}10)$$

where the prime denotes conditions and properties related to the completely reacted zone. The boundary condition at the surface is unchanged. The boundary condition on the side of the inner zone, at some distance $r_m$ from the center, expresses the continuity in the $C_A$ profile and the equality of fluxes on both sides of that boundary at $r = r_m$

$$C_{As}' = C_{As}$$

$$D_e'\frac{\partial C_{As}'}{\partial r} = D_e\frac{\partial C_{As}}{\partial r} \qquad (4.2\text{-}11)$$

For the inner zone, in which both transport and reaction occurs, the differential equations are those of the first stage, but the boundary conditions are $\partial C_{As}/\partial r = 0$ at $r = 0$ and Eq. 4.2-11 at the boundary with the outer zone. This model corresponds to that set up by Ausman and Watson, to describe the rate of burning of carbon deposited inside a catalyst particle [8]. Analytical integration of this fairly general two-stage model is only possible for a zero-order, first-order or pseudo-first-order rate law, whereby Eq. 4.2-8 reduces to

$$r_A \rho_s = akC_{As} = ak'C_{As}C_{S0} \qquad (4.2\text{-}12)$$

The equations are developed in the paper by Ishida and Wen [9]. The gas concentration profile during the first stage is found by solving Eq. 4.2-9 accounting for Eq. 4.2-12, and the boundary condition Eq. 4.2-5. This leads to

$$\frac{C_{As}}{C_A} = \frac{1}{\theta_e}\frac{\sinh(\phi\xi)}{\xi\sinh\phi} \qquad (4.2\text{-}13)$$

with

$$\xi = r/R \quad\text{and}\quad \phi = R\sqrt{\frac{ak'C_{S0}}{D_e}}$$

$$\theta_e = 1 + \left(\frac{D_e}{k_g R}\right)(\phi\coth\phi - 1)$$

The solid concentration profile is found by integrating Eq. 4.2-2 with Eq. 4.2-12:

$$\frac{C_S}{C_{S0}} = 1 - \frac{\sinh(\phi\xi)}{\xi\sinh\phi}\frac{\theta}{\theta_e} \qquad (4.2\text{-}14)$$

with $\theta = ak'C_A t$

Finally, the solid conversion is found as follows:

$$x = 1 - 3 \int_0^1 \frac{C_S}{C_{SO}} \xi^2 \, d\xi$$

$$= \frac{3}{\phi^2} (\phi \coth \phi - 1) \frac{\theta}{\theta_c} \tag{4.2-15}$$

The second stage begins when $C_S(R, t) = 0$, which from Eq. 4.2-14 is, at times,

$$t_e = \frac{\theta_e}{ak'C_A} = \frac{1}{ak'C_A} \left[ 1 + \left( \frac{D_e}{k_g R} \right) (\phi \coth \phi - 1) \right]$$

Then, the concentrations during the second stage are found from Eq. 4.2-10,11, and the solid conversion is given by:

$$x = 1 - \xi_m{}^3 + \frac{3\xi_m}{\phi^2} [(\phi\xi_m)\coth(\phi\xi_m) - 1] \tag{4.2-16}$$

where the position of the moving boundary of completely exhausted solid, $\xi_m = \xi_m(\theta)$, is found from the implicit equation:

$$\theta = 1 + \left( 1 - \frac{D_e}{D'_e} \right) \ln \left[ \frac{\xi_m \sinh \phi}{\sinh(\phi\xi_m)} \right] + \frac{\phi'^2}{6} (1 - \xi_m)^2 (1 + 2\xi_m)$$

$$+ \left[ \left( \frac{D_e}{D'_e} \right) (1 - \xi_m) + \left( \frac{D_e}{k_g R} \right) \xi_m \right] [(\phi\xi_m)\coth(\phi\xi_m) - 1]$$

$$+ \frac{D'_e}{k_g R} \frac{\phi'^2}{3} (1 - \xi_m^3)$$

Figure 4.2-3 illustrates how the conversion progresses with time, and also indicates the boundary between the first and second stages (see Ishida and Wen [9]). Notice that with diffusional limitations, say $\phi > 5$, the first stage ends at less than 50 percent conversion of solid, and so the rather complicated second-stage description is used over a considerable range of final reaction. The homogeneous and heterogeneous models mentioned above may be considered as special cases of the two-stage model.

A homogeneous model *latu sensu* (i.e., with intraparticle concentration gradients but without distinct zones) requires $D_e = D'_e$. A truly homogeneous model, strictly speaking, requires $D_e = D'_e$ and the reaction to be rate controlling. The truly homogeneous model utilizes Eq. 4.2-2, with $C_{As} = C_{As}{}^s$:

$$\frac{\partial C_S}{\partial t} = -r_S \rho_s = f(C_S)$$

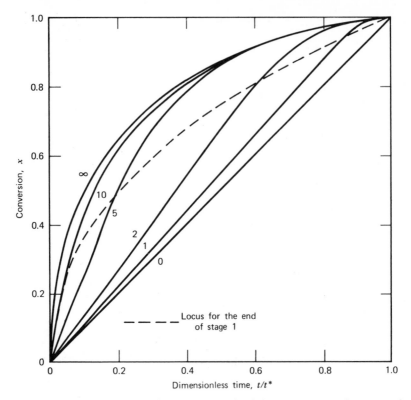

*Figure 4.2-3 Fractional conversion of solid reactant as a function of dimensionless time for homogeneous model with zero-order solid kinetics (sphere).*

$$\frac{k_g R}{D_e} = \infty; \, D_e = D'_e; \, h_s = R\sqrt{\frac{ak'C_{S0}}{D_e}} = 0, 1, 2, 5, 10, \infty;$$

$t^* = $ *time for complete conversion (from Ishida and Wen [9]).*

This relation can then be directly integrated:

$$\int_0^{C_S} \frac{dC_S}{(-r_s \rho_s)} = t \tag{4.2-17}$$

yielding expressions identical to those given in Table 1.3-1. For example, the result, when the order with respect to the fluid concentration of $A$ is 1 and to the solid concentration is zero, gives

$$C_S - C_{S0} = -kC_{As}^{\,s}t = -k'C_{As}^{\,s}C_{S0}t$$

or

$$x = k'C_{As}^{\,s}t \tag{4.2-18}$$

The heterogeneous model is obtained when $D_e \ll D'_e$ (i.e., when the effective diffusivity of $A$ in the unreacted solid is much smaller than in the reacted layer, so that the reaction is confined to a very narrow zone). When $D_e = D'_e$ a narrow reaction zone will of course also be obtained when $D_e = D'_e \ll k$ (i.e., when $\phi \gg 1$). The equations are easily directly derived, as is shown in Sec. 4.3.

## 4.3 Heterogeneous Model with Shrinking Unreacted Core ___

The model equation is again Eq. 4.2-1 in which the time derivative is set zero, as implied by the pseudo-steady-state approximation:

$$\frac{1}{r^2}\frac{\partial}{\partial r}\left(D'_e r^2 \frac{\partial C'_{As}}{\partial r}\right) = 0 \qquad (4.3\text{-}1)$$

while the continuity equation for the reacting component of the solid is exactly Eq. 4.2-2. A prime is used in Eq. 4.3-1, in accordance with the notation in Sec. 4.2, to denote conditions in a completely reacted zone. Also, since the reaction is confined to a front—which supposes that the true reaction rate is relatively large—the reaction rate term does not appear in the right-hand side, but only in the boundary condition at the reaction front:

$$\text{at} \quad r = r_e \quad D'_e\left(\frac{\partial C'_{As}}{\partial r}\right) = ak'_s C'_{As} C_{SO} \qquad (4.3\text{-}2)$$

The boundary condition at the surface is unchanged:

$$\text{at} \quad r = R \quad D'_e\left(\frac{\partial C'_{As}}{\partial r}\right) = k_g(C_A - C_{As}{}^s) \qquad (4.3\text{-}3)$$

The rate coefficient, $k'_s$, is based on the reacting surface. It can be related to the volume-based coefficient used in Sec. 4.2 through the boundary condition Eq. 4.3-2, which is valid also at $t = 0$, when the reaction plane is at the surface (i.e., when $r = r_c = R$). The concentration gradient $\partial C'_{As}/\partial r$ at $t = 0$ can be obtained from the general model with two stages. For the first stage the concentration profile of $A$ is given by Eq. 4.2-13, which may be rewritten as

$$C_{As} = C_{As}{}^s \frac{\sinh(\phi\xi)}{\xi \sinh \phi} \qquad (4.3\text{-}4)$$

since for $r = R$ (i.e., $\xi = 1$):

$$\frac{C_{As}{}^s}{C_A} = \frac{1}{\theta_e}$$

Differentiating Eq. 4.3-4 with respect to $r$ at $r = R$ and multiplying by $D_e$ leads to

$$D_e \left.\frac{\partial C_{As}}{\partial r}\right|_R = \frac{D_e C_{As}{}^s}{R}(\phi \coth \phi - 1) \qquad (4.3\text{-}5)$$

Substituting Eq. 4.3-5 into Eq. 4.3-2 taken at $r = R$ and considering that for large $\phi$ the expression $\phi \coth \phi - 1$ reduces to $\phi$ yields:

$$k'_s = \sqrt{\frac{D_e k'}{a C_{so}}} \qquad (4.3\text{-}6)$$

This rate coefficient has the dimension $\left(\dfrac{m_p}{hr} \bigg/ \dfrac{kmol\ A}{m_f^3}\right)$.

The continuity equation for $A$ Eq. 4.3-1 can be integrated twice to yield the following expression for the concentration profile of $A$:

$$C'_{As} = (C'_{As})_{r=r_c} + B\left(\frac{1}{r_c} - \frac{1}{r}\right)$$

where $B$ is an integration constant and the index $c$ refers to conditions at the reaction front. Accounting for the boundary conditions easily leads to

$$\frac{C'_{As}}{C_A} = \frac{\left(1 + \dfrac{D'_e}{a k'_s C_{so} r_c}\right) \dfrac{1}{r_c} - \dfrac{1}{r}}{\left(1 + \dfrac{D'_e}{a k'_s C_{so} r_c}\right) \dfrac{1}{r_c} - \left(1 - \dfrac{D'_e}{k_g R}\right) \dfrac{1}{R}} \qquad (4.3\text{-}7)$$

The concentration of $A$ at the reaction front is obtained by setting $r = r_c$ in Eq. 4.3-7.

The time required for the reaction front to move from the surface to a distance $r_c$ from the center of a spherical particle is obtained from Eq. 4.2-2, combined with Eq. 4-2-11:

$$a r'_s = D'_e \left(\frac{\partial C'_{As}}{\partial r}\right)_{r=r_c} \qquad (4.3\text{-}8)$$

The transition from the surface-based rate $r'_s$ to the change with time of the volume-based solid concentration requires a slight adaptation of Eq. 4.3-8 and yields:

$$-a \frac{d\left(C_{so} \dfrac{4\pi r_c^3}{3}\right)}{dt} = D'_e \left(\frac{\partial C'_{As}}{\partial r}\right)\bigg|_{r=r_c} 4\pi r_c^2$$

and with $\partial C'_{As}/\partial r$ derived from Eq. 4.3-7

$$a C_{so} \frac{dr_c}{dt} = - \frac{D'_e C_A \dfrac{1}{r_c^2}}{\left(1 + \dfrac{D'_e}{a k'_s C_{so} r_c}\right) \dfrac{1}{r_c} - \left(1 - \dfrac{D'_e}{k_g R}\right) \dfrac{1}{R}}$$

and finally

$$t = \frac{aRC_{S0}}{C_A}\left[\frac{1}{3}\left(\frac{1}{k_g} - \frac{R}{D'_e}\right)\left(1 - \frac{r_c^3}{R^3}\right) + \frac{R}{2D'_e}\left(1 - \frac{r_c^2}{R^2}\right) + \frac{1}{ak'_s C_{S0}}\left(1 - \frac{r_c}{R}\right)\right]$$

(4.3-9)

The time $t^*$ required for complete conversion is found by setting $r_c = 0$ in this formula, so that

$$t^* = \frac{aRC_{S0}}{C_A}\left(\frac{1}{3k_g} + \frac{R}{6D'_e} + \frac{1}{ak'_s C_{S0}}\right)$$

(4.3-10)

The three terms inside the parentheses of Eq. 4.3-10 represent the three resistances involved in the process. They are purely in series in this case. When the mass transfer through the external film is rate controlling, $3k_g \ll ak'_s C_{S0}$ and $k_g \ll 2D'_e/R$ so that Eq. 4.3-9 becomes

$$t = \frac{aRC_{S0}}{3C_A k_g} x$$

(4.3-11)

where $x$ is the conversion, defined by

$$x = 1 - \frac{(\frac{4}{3})\pi r_c^3}{(\frac{4}{3})\pi R^3} = 1 - \left(\frac{r_c}{R}\right)^3$$

When the effective diffusion through the reacted core is rate controlling, $2D_e/R \ll k_g$ and $6D_e/R \ll ak'_s C_{S0}$ so that Eq. 4.3-9 becomes in that case:

$$t = \frac{aR^2 C_{S0}}{6D'_e C_A}[1 - 3(1 - x)^{2/3} + 2(1 - x)]$$

(4.3-12)

The third limiting case of chemical reaction rate controlling is not consistent with the concept of a shrinking core model with a *single* diffusivity throughout the particle: the existence of a sharp boundary implies transport by effective diffusion that is potentially slow with respect to the reaction.

From plots of $x$ versus time it is possible to find out which is the rate-determining step. Also, from experiments with particles having different radii a comparison of the time required to reach the same conversion will reveal a dependence on the ratio of the particle sizes that is different for each rate-controlling step, as is clear from a scrutiny of Eq. 4.3-11 and Eq. 4.3-12. Evidently, both the formulas Eq. 4.3-11 or Eq. 4.3-12 could have been obtained directly from specific models considering only one step rate controlling, in contrast with the more general approach outlined in this section. White and Carberry [26] have considered situations where the particle size changes with reaction.

Park and Levenspiel [10] have proposed an extension of the basic shrinking core model, called the crackling core model. This arose from the observation that the initial state of many reacting solids is essentially nonporous and that a first step, either physical or chemical, is required to form a porous and reactive intermediate. The model essentially makes use of various combinations of the models discussed above.

### Example 4.3-1 Combustion of Coke within Porous Catalyst Particles

An examination of this problem was provided by Weisz and Goodwin [11, 12]. The pellets were silica-alumina cracking catalyst, and the coke resulted from the cracking of light gas oil and naphtha. Measurements of the burning rate were followed by oxygen consumption rates, as shown in Fig. 1.

It is evident that the pellets must have had significant diffusional resistance at the higher ($>450°$C) temperatures. Using the Weisz–Prater criterion discussed in Chapter 3, with values of $C_A = 3 \times 10^{-6}$ mol/cm$^3$ and $D_e \simeq 5 \times 10^{-3}$ cm$^2$/s

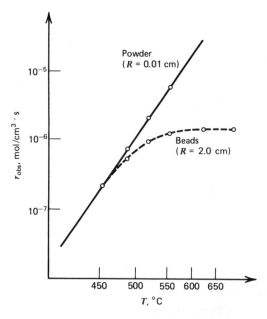

Figure 1 *Average observed burning rates of conventional silica-alumina cracking catalyst. Initial carbon content, 3.4 wt%. Beads (dashed line), and ground-up catalyst (full curve) (from Weisz and Goodwin [11]).*

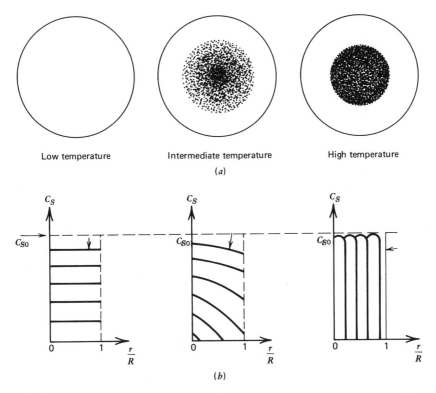

*Figure 2 Appearance after partial burnoff (a), and coke concentration versus radius in beads for successive stages of burnoff (b), for three temperature levels (from Weisz and Goodwin [11]).*

(for oxygen under combustion conditions), one can determine the rate below which diffusional limitations should be absent:

$$(r)_{obs} < \frac{C_A D_e}{R^2} = 4 \times 10^{-7} \text{ mol/cm}^3 \text{ s}$$

We see that this agrees very well with the results on the figure.

By submerging the silica-alumina pellets in a high refractive index liquid (carbon tetrachloride), they are rendered transparent, and the coke profiles for various temperature levels can be observed, as shown in Fig. 2. We see that these range from almost a homogeneous situation (as defined above) to the sharp-boundary shell-progressive situation. For the latter, Eq. 4.3-12 can be used:

$$\tfrac{1}{6} - \tfrac{1}{2}(1-x)^{2/3} + \tfrac{1}{3}(1-x) = \left(\frac{D_e C_A}{aR^2 C_{S0}}\right)t \tag{a}$$

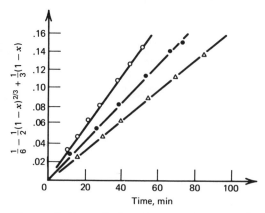

Figure 3 Burnoff function versus time for three different diameter beads (from Weisz and Goodwin [11]).

Figure 3 illustrates the agreement of the data with the form of Eq. (a) at 700°C. The slopes of the lines provide values of $D_e C_A / aR^2 C_{S0}$. Alternatively, the time for complete combustion can be obtained for $x = 1$:

$$t^* = \frac{1}{6} \frac{aR^2 C_{S0}}{D_e C_A} \tag{b}$$

Actually, this complete combustion time is often hard to determine unambiguously from (scattered) experimental data, and so the 85 percent completion time was more convenient:

$$t_{85} = 0.0755 aR^2 C_{S0} / D_e C_A \tag{c}$$

If all the bases of the model are correct, this 85 percent time should vary (1) linearly with initial coke level, (2) with the square of the particle size, (3) inversely with the effective diffusivity, and (4) inversely with oxygen partial pressure. Figures 4, 5, 6

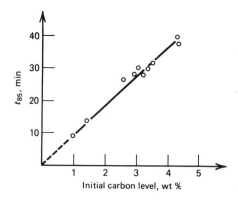

Figure 4 Dependence of burnoff time on initial carbon level, for diffusion controlled combustion (silica-alumina cracking catalyst, 700°C) (from Weisz and Goodwin [11]).

CHEMICAL ENGINEERING KINETICS

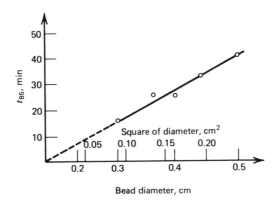

*Figure 5 Dependence of burnoff time on bead size for diffusion-controlled combustion (from Weisz and Goodwin [11]).*

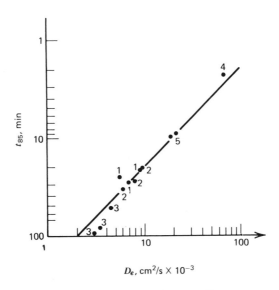

*Figure 6 Dependence of burnoff time on structural diffusivity, of various types of spherical particles, for diffusion-controlled combustion region (from Weisz and Goodwin [11]).*

255

Table 1  Comparison of burnoff times in air and oxygen

| Catalyst | Air | | | Oxygen | | | |
|---|---|---|---|---|---|---|---|
| | $C_{So}$ (%) | $t_{85}$ (min) | $t_{85}$ corr. (min) | $C_{So}$ (%) | $t_{85}$ (min) | $t_{85}$ corr. (min) | Ratio $t_x/t_{air}$ |
| Silica-alumina (lab. prep.) | | | | | | | |
| $R = 0.24$ cm.; temp. 630°C | 4.0 | 45 | 45 | 4.6 | 10.5 | 9.1 | — |
| $t_{85}$ corrected to | | | | | | | |
| $C_{s0} = 4.8\%$ wt | 3.7 | 48 | 52 | 4.1 | 10.6 | 10.3 | — |
| Average | | | 48.5 | | | 9.7 | 0.20 |
| Silica-alumina (0.15% $Cr_2O_3$, commercial) | | | | | | | |
| $R = 0.19$ cm.; temp. 690°C | 3.0 | 30 | 29.7 | 3.35 | 6.3 | 5.55 | — |
| $t_{85}$ corrected to | 3.4 | 29 | 25.6 | — | — | — | — |
| $C_{S0} = 3\%$ wt | 3.2 | 27 | 25.2 | — | — | — | — |
| Average | | | 26.6 | | | 5.55 | 0.21 |

From Weisz and Goodwin [11].

and Table 1 indicate that all of these are verified by the data. Thus, for high temperatures, the shrinking-core model provides a good description. At lower temperatures, the more general models would be required, however.

## 4.4 Grain Model Accounting Explicitly for the Structure of the Solid

Sohn and Szekely [13] developed a model in which the particle is considered to consist of a matrix of very small grains between which the fluid has easy access through the pores. Figure 4.4-1 illustrates how the reactive component of the grains is converted throughout the particle, which has a fluid reactant concentration gradient caused by the resistance to diffusion in the particle. This situation can be described on the basis of the models developed in Secs. 4.2 and 4.3.

The fluid reactant concentration in the particle of any geometry is obtained from

$$\varepsilon_s \frac{\partial C_{As}}{\partial t} \doteq D_{ep} \frac{\partial^2 C_{As}}{\partial r^2} - D_{eg} \frac{\partial C_{Ag}}{\partial y}\bigg|_Y (1 - \varepsilon_s)a_g \qquad (4.4\text{-}1)$$

with boundary conditions analogous to Eqs. 4.2-3,4,5, where $r$ refers to the particle, $y$ to the grain coordinate. $Y$ is the radius of the grain, oriented from the center

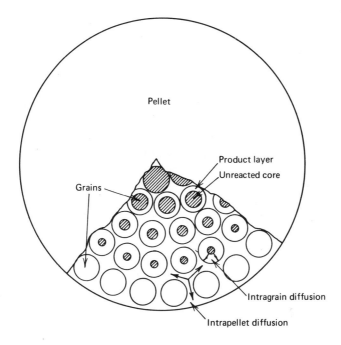

Figure 4.4-1 *Schematic representation of the grain model (from Sohn and Szekely* [13]).

to the surface for transport of $A$ inside the grain. $D_{eg}$ is the grain effective diffusivity, $D_{ep}$ is an effective diffusivity for transport through the pores between the grains and is, therefore, different from the $D_e$ and $D'_e$ used in the models described above, which do not distinguish between pores and solid (i.e., consider the particle as a pseudo-homogeneous solid), $C_{Ag}$ the fluid reactant concentration in the grain, and $a_g$ the surface to volume ratio of the grain. The factor $(1 - \varepsilon_s)a_g$ arises from the fact that Eq. 4.4-1 is written per unit particle volume, whereas the flux $D_{eg}(\partial C_{Ag}/\partial y)|_Y$ is per unit grain surface area. Equation 4.4-1 is a particular form of Eq. 4.2-1, expressing the fact that $A$ reacts only in the grains.

To obtain the concentration profile in the grain the general model of Sec. 4.2 could be used:

$$\varepsilon_g \frac{\partial C_{Ag}}{\partial t} = D_{eg} \frac{\partial^2 C_{Ag}}{\partial y^2} - r_A \rho_{sg} \tag{4.4-2}$$

$$\frac{\partial C_{Sg}}{\partial t} = -r_S \rho_{sg} \tag{4.4-3}$$

NONCATALYTIC GAS-SOLID REACTIONS ———————— 257

with boundary conditions: at

$$t = 0 \qquad C_{Ag} = C_{Ago} \qquad \text{and} \qquad C_{Sg} = C_{Sgo}$$

$$y = 0 \qquad \frac{\partial C_{Ag}}{\partial y} = 0$$

$$y = Y \qquad C_{Ag} = C_{As}(r)$$

As before, various rate laws could be substituted into these equations. Numerical integration would normally be required to solve the system Eqs. 4.4-1,2,3.

For the special case that the phenomena in the grain can be represented by the shrinking-core model, which is plausible since the grains are often very dense, Eqs. 4.4-2,3 lead to the same types of solutions as given in Sec. 4.3. Note that the shrinking-core models permit the concentration of the reactive solid component in the grains to become zero in a finite time, so that the solution of the particle equation Eq. 4.4-1 may involve two stages. For example, for grains with slab geometry and pseudo steady state, Eq. 4.4-2 without the rate term can be integrated twice, using the boundary conditions across the completely reacted shell, to give

$$\frac{C_{Ag}}{C_{As}} = 1 - \frac{\dfrac{Y - y}{Y - y_c}}{1 + \left(\dfrac{D_{eg}}{ak'_s C_{Sgo}}\right)\dfrac{1}{Y - y_c}} \tag{4.4-4}$$

from which

$$\left.\frac{\partial C_{Ag}}{\partial y}\right|_Y = \frac{C_{As}}{\left(\dfrac{D_{eg}}{ak'_s C_{Sgo}}\right) + (Y - y_c)} \tag{4.4-5}$$

Substituting Eq. 4.4-5 into the pseudo-steady-state form of Eq. 4.4-1 for particle slab geometry leads to

$$D_{ep}\frac{\partial^2 C_{As}}{\partial r^2} - \frac{ak'_s C_{Sgo}(1 - \varepsilon_s)a_g C_{As}}{1 + \dfrac{ak'_s C_{Sgo}}{D_{eg}}(Y - y_c)} = 0 \tag{4.4-6}$$

with

$$\frac{\partial y_c}{\partial t} = \frac{-k'_s C_{As}}{1 + \dfrac{ak'_s C_{Sgo}}{D_{eg}}(Y - y_c)} \tag{4.4-7}$$

Even these equations are not amenable to a complete analytical solution.

Sohn and Szekely [14] developed a very useful approximate solution, valid for various geometries of grain and particle, based on the additivity of times to reach a given conversion for different limiting processes—a concept analogous to that discussed after Eq. 4.3-10. The concept states that the time required to attain a given conversion is the sum of the times required to attain the same conversion (1) in the absence of any diffusion resistance, (2) with intraparticle diffusion controlling, and (3) with intragrain diffusion controlling. In mathematical terms:

$$\frac{k'_s C_A a_g l}{F_g} \simeq g_{F_g}(x) + \left[\frac{(1 - \varepsilon_s)ak'_s C_{Sg0} F_p}{2D_{ep}a_p^2}\left(\frac{a_g}{F_g}\right)\right]p_{F_p}(x) + \frac{ak'_s C_{Sg0}}{2D_{eg}a_g}p_{F_g}(x) \quad (4.4\text{-}8)$$

$F_g$ and $F_p$ are geometric factors for grains and particle, respectively, and have the values: 1 for slabs, 2 for cylinders, and 3 for spheres. The conversion $x$ in the grain attained by a shrinking-core mechanism is written

$$x = 1 - \left(\frac{a_g y_c}{F_g}\right)^{F_g}$$

$g_{F_g}$, $p_{F_p}$ and $p_{F_g}$ are functions corresponding to the limiting situations mentioned above and are defined as follows:

$$g_{F_g} = 1 - (1 - x)^{1/F_g}$$

$$
\begin{aligned}
p_{F_p} = p_{F_g} &= x^2 && \text{for } F_p \text{ or } F_g = 1 \\
&= x + (1 - x)\ln(1 - x) && \text{for } F_p \text{ or } F_g = 2 \\
&= 1 - 3(1 - x)^{2/3} + 2(1 - x) && \text{for } F_p \text{ or } F_g = 3
\end{aligned}
$$

Comparing the last expression with Eq. 4.3-7, obtained for the shrinking-core model with diffusion rate controlling, shows that $p_{F_p} = p_{F_g}$ is nothing but the ratio of the time required to reach a given conversion to the time required to reach complete conversion. Sohn and Szekely showed that Eq. 4.4-8 leads to a remarkably accurate approximation to the results obtained by numerical integration.

An analysis of experimental results on the reaction of $SO_2$ with limestone using a grain model similar to the one discussed in this section was published by Pigford and Sliger [15].

## 4.5 Pore Model Accounting Explicitly for the Structure of the Solid

Szekely and Evans [16] have developed equations for a model of a porous solid that considers the solid to have parallel pores as represented schematically in Fig. 4.5-1.

To simplify the mathematical treatment the particle is considered to be infinitely thick and isothermal. The pores are parallel, all have the same radius and are spaced

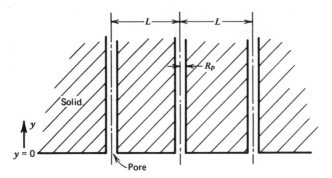

*Figure 4.5-1 Schematic representation of the pore model (from Szekely and Evans [16]).*

at equal distance, $L$. It is assumed that the initial structure is not modified by the reaction. To allow analytical solutions the reaction is considered to be of first order with respect to the fluid component $A$ and of zero order with respect to the solid component $S$. Also, to focus completely on the effect of the structure, external transport is not included in the model. Furthermore, the concentration of $A$ in the gas phase, $C_A$, is kept constant, as was also done in the models discussed in the previous section. The reactant $A$ diffuses inside the pore and then inside the solid, where it reacts. The progression through the solid is also shown in Fig. 4.5-2. It is clear from this figure that three zones have to be considered, depending on the depth.

First zone:     for depths extending from zero to a value $y_1$ the solid component has completely reacted.

Second zone:   for depths between $y_1$ and $y_2$ there is interaction between neighboring reaction fronts.

Third zone:     for depths between $y_2$ and infinity there is no interaction yet.

Continuity equations for $A$ in the pore and in the solid itself have to be set up.

The steady-state continuity equation for $A$ in the pore, accounting for diffusion in the pore axial direction and effective diffusion inside the solid at the pore wall, is

$$D_{ep}\frac{\partial^2 C_{Ap}}{\partial y^2} + 2\frac{\zeta}{R_p}D_{eg}\left(\frac{\partial C_{As}}{\partial r}\right)_{r=R_p} = 0 \qquad (4.5\text{-}1)$$

with boundary conditions: $C_{Ap} = C_A$ at $y = 0$

$$C_{Ap} = 0 \quad \text{at} \quad y = \infty$$

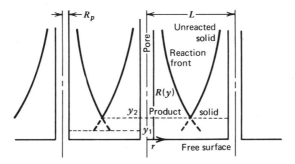

*Figure 4.5-2 The reaction front in the pore model
( from Szekely and Evans [16]).*

$C_{Ap}$ is the fluid reactant concentration in the pore, $R_p$ the pore radius. $D_{ep}$ in this model may be a harmonic mean of the bulk and Knudsen diffusion coefficient; with real geometries it would be a true effective diffusivity including the tortuosity factor and an internal void fraction. $D_{eg}$ is an effective diffusivity for the mass transfer inside the solid and $\zeta$ is a correction factor accounting for the restricted availability of reactant surface in the region where the partially reacted zones interfere. For $R(y) < L/2$ (shown in Fig. 4.5-2) or $y_2 < y$ the factor $\zeta = 1$; for $L/\sqrt{2} > R(y) > L/2$ or $y_1 < y < y_2$ the factor $\zeta = 1 - (4\theta/\pi)$ where $tg\theta = (2/L)\sqrt{R^2(y) - (L/2)^2}$; for $y < y_1$ the factor $\zeta = 0$, where $R(y)$ is the radial position of the reaction front. It is clear from Eq. 4.5-1 that no radial concentration gradient of $A$ is considered within the pore.

The continuity equation for $A$ in the completely reacted solid is written as in the previous section

$$\frac{\partial}{\partial r}\left( r^2 \frac{\partial C_{As}}{\partial r} \right) = 0 \qquad \text{for } R_p \leq r < R(y) \qquad (4.5\text{-}2)$$

for pseudo steady state and only radial diffusion inside the solid. The boundary conditions are: at $r = R_p$, $C_{Ap}(y)$; at the reaction front $R(y)$:

$$-D_{eg} \frac{\partial C_{As}}{\partial r} = ak'_s C_{As} C_{S0}$$

The reaction is considered to be of first order with respect to $A$, zero order with respect to $S$. Analytical integration of E. 4.5-2 leads to

$$C_{As} = \frac{C_{Ap}(y)}{1 + \dfrac{R(y)}{D_{eg}} ak'_s C_{S0} \ln \dfrac{R(y)}{R_p}} \qquad \text{at } r = R(y) \qquad (4.5\text{-}3)$$

From Eq. 4.5-3 it follows that, at $r = R_p$,

$$-\frac{\partial C_{As}}{\partial r} = \frac{C_{Ap}(y)}{R_p \left[\dfrac{D_{eg}}{R(y)ak'_s C_{S0}} + \ln \dfrac{R(y)}{R_p}\right]} \tag{4.5-4}$$

Equation 4.5-4 is now inserted into the equation for the concentration of $A$ inside the pore, Eq. 4.5-1, leading to a second-order differential equation linear in $C_{Ap}$, but containing $R(y)$ in the group multiplying $C_{Ap}$. With $R(y) = R_p$ at $t = 0$, the equation can be solved for the initial concentration profile of $A$ in the pore.

$$C_{Ap} = C_A \exp\left[-\sqrt{\frac{2ak'_s C_{S0}}{D_{ep} R_p}}\, y\right] \tag{4.5-5}$$

The evolution of $R(y)$ with time follows from a balance of $S$ per unit solid surface.

$$C_{S0} \frac{\partial R(y)}{\partial t} = k'_s C_{S0} C_{As}(y)\Big|_{r=R(y)} \tag{4.5-6}$$

Substituting Eq. 4.5-3 into Eq. 4.5-6 leads to

$$\frac{\partial R(y)}{\partial t} = \frac{k'_s C_{Ap}(y)}{1 + \left[\dfrac{ak'_s C_{S0} R(y)}{D_{eg}} \ln \dfrac{R(y)}{R_p}\right]} \tag{4.5-7}$$

with $R(y) = R_p$ at $t = 0$.

Equation 4.5-1 [with $\partial C_{As}/\partial r$ given by Eq. 4.5-4 and Eq. 4.5-7] with the corresponding initial and boundary conditions represent a complete statement of the system. This system was integrated numerically by Szekely and Evans to yield the position of the reaction front $R(y)$ as a function of time.

To allow comparison with other models and experimental data, Szekely and Evans recast the results in an alternative form. They defined an equivalent penetration, which is a direct measure of the conversion:

$$\text{E.P.} = Y_{max} + \frac{\pi \int_{Y_{max}}^{\infty} [\gamma R^2(y) - R_p^2]\, dy}{L^2 - \pi R_p^2} \tag{4.5-8}$$

$Y_{max}$ is the ordinate value corresponding to the height where the solid is converted over the complete $L$ distance. After a sufficient time, $Y_{max}$ is equal to $y_1$ as defined above. However, for short times the diffusion directly from the particle surface cannot really be ignored, as was done in Eq. 4.5-2. Szekely and Evans assumed that this effect could be analyzed independently from the radial diffusion of $A$ from the pore. The shrinking-core model applied to rectangular coordinates can be solved as in Sec. 4.3 with the result that the position of the moving boundary is located at a distance $y_c$ from the particle surface:

$$y_c = \frac{-D_{eg} + \sqrt{D_{eg}^2 + 2a(k_s')^2 C_{S0} C_A D_{eg} t}}{ak_s' C_{S0}}$$

Consequently, for short times $Y_{max}$ is chosen to be the largest of the values $y_1$, $y_c$. The quantity under the integral of Eq. 4.5-8 is the volume reacted within a zone minus the pore volume. $\gamma$ accounts for the overlapping of the reacted zones.

$$\gamma = 1 \quad \text{for } R(y) < \frac{L}{2} \text{ (i.e., } y_2 < y)$$

$$\gamma = \zeta + \frac{2L}{\pi R} \sin \theta \quad \text{for } \frac{L}{2} < R(y) < \frac{L}{\sqrt{2}} \text{ (i.e., } y_1 < y < y_2)$$

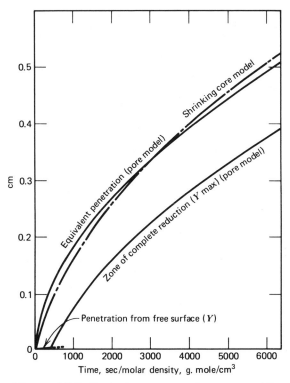

*Figure 4.5-3 Equivalent penetration versus $t/C_{S0}$ for the pore model with the following parameters: $D_A = 2.8 \ m^2/hr$; $D_e = 7.2 \times 10^{-5} \ m^2/hr$; $k_s' C_{S0} = 720 \ m/hr$; $R_p = 5.10^{-6} \ m$; $\varepsilon_s = \pi R_p^2/L^2 = 0.126$; also shown is shrinking core model with $k_s' C_{S0} = 720 \ m/hr$ (from Szekely and Evans [16]).*

Figure 4.5-3 is a plot of E.P. versus $t/\rho_s'$. It also contains the results obtained with the shrinking-core model. For equal parameter values both curves are almost coinciding. Consequently it would be hard to distinguish between these models, although the location of the $Y_{max}$ curve with respect to the E.P. curve would indicate that the reaction is certainly not restricted to a sharp boundary.

## 4.6 Reaction Inside Nonisothermal Particles

In the preceding sections we assumed that the particles were isothermal, although situations might occur where this condition is not fulfilled. Certainly, when the reaction is more or less homogeneously distributed throughout the particle, the temperature will no doubt be very nearly uniform, as was shown in Chapter 3 for catalytic reactions. However, when the reaction is very fast and takes place in a narrow zone, as described by the shrinking-core model, localizing the heat source may lead to temperature gradients, especially when the reactive solid component is present in high concentrations.

The mathematical description of such a situation would comprise the continuity equations for the fluid and solid reactants encountered in Sec. 4.3 for the unreacted shrinking-core model and a heat balance that assumes pseudo steady state in the shell and an integral averaged temperature in the core up to the front.

For the shell:

$$\frac{\partial^2 T_s}{\partial r^2} + \frac{2}{r}\frac{\partial T_s}{\partial r} = 0 \tag{4.6-1}$$

with boundary conditions:

$$\text{at } r = R \qquad -\lambda_e \frac{\partial T_s}{\partial r}\bigg|_{r=R} = h_f(T_s^s - T) \tag{4.6-2}$$

$$\text{at } r = r_c \qquad T_s = T_{sc} \tag{4.6-3}$$

For the core:

$$\tfrac{4}{3}\pi r_c^3 \rho_{sc} C_{psc} \frac{\partial T_{sc}}{\partial t} = 4\pi r_c^2 a k_s' C_{Asc} C_{S0c}(-\Delta H) + 4\pi r_c^2 \lambda_e \frac{\partial T_s}{\partial r}\bigg|_{r=r_c} \tag{4.6-4}$$

with initial condition:

$$\text{at } t = 0 \qquad T_s = T_{sc} = (T_{sc})_0 \tag{4.6-5}$$

Wang and Wen [17] used this model to simulate the burning of coke from fire clay particles of 1.2 cm radius with up to 41 percent by weight of carbon. Figures 4.6-1 and 4.6-2 show the agreement between calculated and experimental conversions and temperatures. Similar balances were used by Costa and Smith [18] to analyze experimental results concerning hydrofluorination of uranium dioxide.

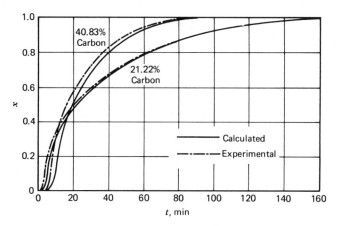

*Figure 4.6-1 Typical experimental and calculated time-conversion curves (from Wang and Wen [17]).*

Luss and Amundson [19] discussed alternate theoretical models for moving boundary reactions in nonisothermal particles, and concluded that they all gave similar predictions.

Sohn [20] has developed several analytical solutions to the combined heat and mass balances for the shrinking-core model, using the following assumptions: (1) slab geometry, also applicable to other curved geometries for the practical case

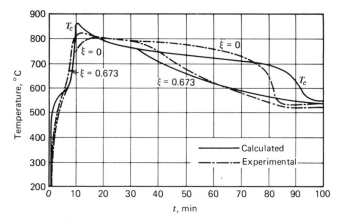

*Figure 4.6-2 Experimental and calculated time-temperature curves for high-carbon run for two points within the particle ($\xi = 0$ and $\xi = 0.673$). Bulk phase temperature: 518°C. Initial particle temperature: 30°C (from Wang and Wen [17]).*

of maximum temperature close to the surface, (2) uniform temperature in the pellet, $Nu' \to 0$ (also valid for $Nu' \simeq 0$-3). The maximum temperature $T_m$, is then:

For $Sh' \to \infty$

$$\frac{a[T_m - T]}{[(-\Delta H)D_e C_A / h_f L]} \sim 0.765 \sqrt{\frac{h_f L}{\rho_s c_{ps}} \cdot \frac{aC_{so}}{D_e C_A}}$$

where $(-\Delta H)$ is the heat of reaction per mole $S$ transformed.

For $\alpha = \dfrac{1}{Sh'} \cdot \sqrt{\dfrac{1}{2} \cdot \dfrac{h_f L}{\rho_s c_{ps}} \cdot \dfrac{aC_{so}}{D_e C_A}} \to \infty$ (e.g., $Sh' \to 0$)

$$\frac{a[T_m - T]}{[(-\Delta H)D_e C_A / h_f L]} \sim Sh'\left(1 + \frac{1}{\alpha^2}\right)^{-1}$$

An interpolation formula was also provided, but an estimate is able to be sketched from the above two results, especially for the most important range of

$$\frac{h_f L}{\rho_s c_{ps}} \cdot \frac{aC_{so}}{D_e C_A} > 10$$

Further aspects of this problem have also been discussed by Sampath and Hughes [21].

## 4.7 A Concluding Remark

This chapter has briefly described a series of models for gas-solid reactions. The literature contains several more and many more could be developed. It would be hard, if not impossible, to assess these models as to their respective merits since careful and detailed experimentation is seriously lagging behind. In the few cases in which it was possible to check the theoretical results with experimental data the lack of fit has mainly been ascribed to inaccuracies in the models. Insufficient attention has been devoted to the kinetic equations proper: there is no reason for limiting the kinetics of the reaction between a fluid and the component of a solid to zero- or first-order expressions.

## Problems

4.1 Derive Eq. 4.2-3 by using simple geometric arguments. (Also see Kim and Smith [3] and Wen [2].)

4.2 Consider the general model with the reaction first-order in fluid phase concentration:

$$\varepsilon_s \frac{\partial C_{As}}{\partial t} = D_e \nabla^2 C_{As} - kC_{As} f(C_S)$$

$$\frac{\partial C_S}{\partial t} = -kC_{As} f(C_S)$$

where $f(C_S)$ is the rate dependency on the solid reactant concentration; (e.g., a grain model or mass action form). The simplest boundary conditions would be:

$$C_{As} = 0$$
$$C_S = C_{S0} \qquad \text{at } t = 0$$

and

$$C_{As} = C_{As}^{\ s} \qquad \text{on the pellet surface.}$$

(a) Show that the new variable

$$\psi(\mathbf{x}, t) \equiv k \int_0^t C_{As}(\mathbf{x}, t') dt' \qquad \text{(cumulative concentration)}$$

is also defined by the formal integral

$$\psi = -\int_{C_{S0}}^{C_S} \frac{dC_S'}{f(C_S')} dC_S'$$

This result can be solved, in principle, for

$$C_S = H_S[\psi(\mathbf{x}, t); C_{S0}]$$

(b) Then, the new variable can be differentiated in space ($\nabla$), and these results combined with the original mass balance differential equation to yield:

$$\varepsilon_s \frac{\partial \psi}{\partial t} = D_e \nabla^2 \psi - k[C_{S0} - H_S(\psi; C_{S0})]$$

Prove this result.

(c) The boundary conditions can similarly be transformed to:

$$\psi = 0 \qquad \text{at} \qquad t = 0$$
$$\psi = C_{As}^{\ s} k t \qquad \text{on the surface}$$

Prove these additional results.

(d) The results of parts (b) and (c) show that the original two coupled partial differential equations can be reduced to solving one diffusion type equation, with a time-dependent boundary condition—a much simpler problem. For the special case of rectangular (slab) geometry, and where the pseudo steady state approximation is valid (gas-solid reaction), show that the mathematical problem is reduced to:

$$D_e \frac{\partial^2 \psi}{\partial z^2} = k[C_{S0} - H_s(\psi; C_{S0})]$$

with

$$\psi = kC_{As}^{\ s} \cdot t \qquad \text{on the surface } (z = L)$$

and

$$\frac{\partial \psi}{\partial z} = 0 \qquad \text{at the center, } z = 0 \text{ (symmetry)}$$

Thus notice that the results of Chapter 3 can be utilized to solve the transformed problem.

For a zero order solid concentration dependency ($f(C_S) = C_{S0}$), show that the following results are obtained:

$$C_{As} = C_{As}{}^s \frac{\cosh(\phi z/L)}{\cosh(\phi)} \qquad \phi = L\sqrt{kC_{S0}/D_e}$$

$$C_S = C_{S0}\left[1 - \frac{\cosh(\phi z/L)}{\cosh(\phi)}(kC_{As}{}^s t)\right]$$

which are the type of results obtained by Ishida and Wen [9] in Eqs. 4.2-13 and 14.

(e) Finally, for the slab geometry of part (e), show that the conversion is given by

$$x_A(t) = \frac{1}{\phi}\left\{2\int_{\psi_c}^{kC_{As}{}^s t}[C_{S0} - H(\psi)]d\psi\right\}^{1/2}$$

which is based on the generalized modulus concept of Chapter 3. Thus, it is seen that the complicated gas-solid reaction problem can be reduced to an analogy with the simpler effectiveness factor problem of Chapter 3. For more general results, see Del-Borghi, Dunn, and Bischoff [6] and for extensive results for first-order solid reactions, $f(C_S) = C_S$, see Dudoković [25].

4.3 (a) Derive the results of Eqs. 4.2-13 to 15 by directly solving the appropriate differential mass balances.

(b) Compute the conversion-time results of Fig. 4.2-3 for $\phi = 2.0$ (first stage only).

4.4 Equation 4.3-6 related the surface rate coefficient of the shrinking core model with the volumetric rate coefficient of the general model for the special case of first order in fluid concentration and zero order in solid concentration (two-stage model).

(a) For the more general situation, when

$$D_e \nabla^2 C_{As} = akC_{As}f(C_S) \qquad \text{per unit volume)}$$
$$C_{As}(L) = C_{As}{}^s$$

the corresponding shrinking core approximation would be:

$$D'_e \nabla^2 C'_{As} = 0 \qquad R \leq R \leq r_c$$

$$C'_{As}(L) = C_{As}{}^s$$

$$D'_e \nabla C'_{As}|_{r_c} = ak_s\, C'_{As}f(C'_S) \qquad \text{(per unit surface)}$$

With large $\phi$, only a small "penetration" zone exists when the reaction occurs, and so for the approximate "slab geometry," show that

$$C_{As} \simeq C_{As}{}^s \frac{\cosh(\phi z/L)}{\cosh(\phi)} \sim C_{As}{}^s \exp[-\phi(1 - z/L)]$$

with

$$\phi^2 = L^2 akf(C_{S0})/D_e$$

(b) To compare the two types of models, equate the total amount reacted [as for Eq. (4.3-5)]:

$$D_e \nabla C_{As}|_L = \int_0^L ak C_{As} f(C_S) dz \sim L \frac{ak C_{As}{}^s f(C_{S0})}{\phi} \quad \text{(general model)}$$

(derive this) with

$$D'_e \nabla C'_{As}|_L = D'_e \nabla C'_{As}|_{r_c} = ak_s C_{As} f(C_S)|_{r_c} \quad \text{(shrinking core model)}$$

to give the result

$$k_s = [kD_e/af(C_{S0})]^{1/2}$$

Note that for the zero-order reaction, Eq. 4.3-6 is recovered.

(c) Show that for the general rate $r_{As}(C_{As}, C_S)$ a similar "penetration distance" derivation gives $(r_{As})^s = (D_e C_{As}{}^s)^{1/2}(r_A{}^s)^{1/2}$.

4.5 Trace through the details leading to Eq. 4.3-9 and thus find the result.

4.6 Determine the various rate parameters in Ex. 4.3-1.

# References

[1] Delmon, B. *Introduction à la cinétique hétérogène*, Technip, Paris (1969).

[2] Wen, C. Y. *Ind. Eng. Chem.*, **60**, No. 9, 34 (1968).

[3] Kim, K. K. and Smith, J. M. *A. I. Ch. E. J.*, **20**, 670 (1974).

[4] Bischoff, K. B. *Chem. Eng. Sci.*, **18**, 711 (1963).

[5] Bischoff, K. B. *Chem. Eng. Sci.*, **20**, 783 (1965).

[6] DelBorghi, M., Dunn, J. C., and Bischoff, K. B. *Chem. Eng. Sci.*, **31**, 1065 (1976).

[7] Aris, R. *The Mathematical Theory of the Diffusion Reaction Equation*, Oxford University Press, London (1974).

[8] Ausman, J. M. and Watson, C. C. *Chem. Eng. Sci.*, **17**, 323 (1962).

[9] Ishida, M. and Wen, C. Y. *A. I. Ch. E. J.*, **14**, 311 (1968).

[10] Park, J. Y. and Levenspiel, O. *Chem. Eng. Sci.*, **30**, 1207 (1975).

[11] Weisz, P. B. and Goodwin, R. D. *J. Catal.*, **2**, 397 (1963).

[12] Weisz, P. B. and Goodwin, R. D. *J. Catal.*, **6**, 227, 425 (1966).

[13] Sohn, H. Y. and Szekely, J. *Chem. Eng. Sci.*, **27**, 763 (1972).

[14] Sohn, H. Y. and Szekely, J. *Chem. Eng. Sci.*, **29**, 630 (1974).

[15] Pigford, R. L. and Sliger, G. *Ind. Eng. Chem. Proc. Des. Devpt.*, **12**, 85 (1973).

[16] Szekely, J. and Evans, J. W. *Chem. Eng. Sci.*, **25**, 1091 (1970).

[17] Wang, S. C. and Wen, C. Y. *A. I. Ch. E. J.*, **18**, 1231 (1972).

[18] Costa, E. C. and Smith, J. M. *A. I. Ch. E. J.*, **17**, 947 (1971).

[19] Luss, D. and Amundson, N. R. *A. I. Ch. E. J.*, **15**, 194 (1969).

[20] Sohn, H. Y. *A. I. Ch. E. J.*, **19**, 191 (1973).

[21] Sampath, B. S. and Hughes, R. *The Chemical Engineer*, No. 278, 485 (1973).

[22] Luss, D. *Can. J. Chem. Eng.* **46**, 154 (1968).

[23] Theofanous, T. G. and Lim, H. C. *Chem. Eng. Sci.*, **26**, 1297 (1971).

[24] Yoshida, K., Kunii, D., and Shimizu, F. *J. Chem. Eng. Japan*, **8**, 417 (1975).

[25] Dudoković, M. P. *A. I. Ch. E. J.*, **22**, 945 (1976).

[26] White, D. E. and Carberry, J. J., *Can J. Chem. Eng.*, **43**, 334 (1965).

# 5

## CATALYST
## DEACTIVATION

## 5.1 Types of Catalyst Deactivation _____

Catalysts frequently lose an important fraction of their activity while in operation. There are primarily three causes for deactivation:

a. Structural changes in the catalyst itself. These changes may result from a migration of components under the influence of prolonged operation at high temperatures, for example, so that originally finely dispersed crystallites tend to grow in size. Or, important temperature fluctuations may cause stresses in the catalyst particle, which may then disintegrate into powder with a possible destruction of its fine structure. Refer to the comprehensive review of Butt [1] for further discussion of this topic.

b. Essentially irreversible chemisorption of some impurity in the feed stream, which is termed poisoning.

c. Deposition of carbonaceous residues from a reactant, product or some intermediate, which is termed coking.

This chapter discusses the local (i.e., up to the particle size) effects of deactivation by poisoning and by coking. The effect on the reactor scale is dealt with in Chapter 11.

## 5.2 Kinetics of Catalyst Poisoning _____

### 5.2.a Introduction

Metal catalysts are poisoned by a wide variety of compounds, as is evidenced by Fig. 5.2.a-1. The sensitivity of Pt-reforming catalysts and of Ni-steam reforming catalysts is well known. To protect the catalyst, "guard" reactors are installed in industrial operation. They contain Co-Mo-catalysts that transform the sulfur

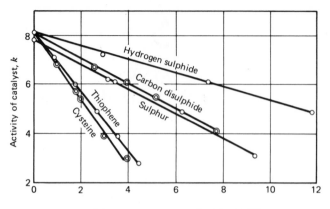

*Figure 5.2.a-1 Hydrogenation of crotonic acid on a Pt-catalyst. Catalyst activity (measured by rate coefficient k) as a function of poison content. (After Maxted and Evans [2].)*

compounds into easily removable components. Acid catalysts can be readily poisoned by basic compounds, as shown in Fig. 5.2.a-2. Poisoning by metals in the feed is also encountered. For example, in hydrofining petroleum residuum fractions, parts per million of iron-, nickel- and vanadium compounds in the feed suffice to completely deactivate the catalyst after a few months of operation. A review paper by Maxted [2] is still useful for a basic introduction to this area.

When an impurity in the feed is irreversibly chemisorbed on the catalyst, the latter acts very similarly to an adsorbent or an ion-exchange resin in an adsorption or ion-exchange process. The impurity naturally does not necessarily act like the reactants (or products) and could be deposited into the solid completely independently of the main chemical reaction and have no effect on it. The latter situation would have no bearing on the kinetics. More often, however, the active sites for the main reaction are also active for the poison chemisorption, and the interactions need to be considered. Since the poison species is separate from the reactants or products, its chemisorption can be treated by the mathematical methods used in adsorption, ion-exchange, or chromatography. Several results based on various assumptions concerning the chemisorption, diffusion, and deactivation or poisoning effects on the main catalytic reaction will be described. Within the context of the assumptions, these results give a rational form for the function expressing the deactivation in the case of poisoning, and also valuable clues for possible functions to use for coking, about which less is known quantitatively.

CHEMICAL ENGINEERING KINETICS

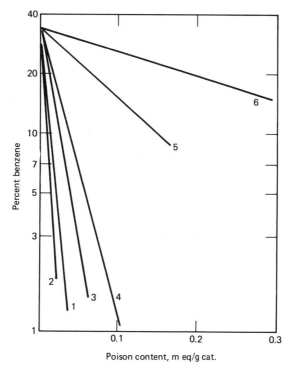

*Figure 5.2.a-2  Cumene dealkylation (1) Quinoline;*
*(2) quinaldine; (3) pyrrole; (4) piperidine; (5)*
*decyclamine; (6) aniline (after Mills et al. [3]).*

### 5.2.b  Kinetics of Uniform Poisoning

An early analysis by Wheeler [4] treated poisoning in an idealized pore, and
can be generalized to a catalyst particle as shown in Chapter 3. Fundamental to his
development, and the others of this section, is the assumption that the catalytic
site that has adsorbed poison on it is completely inactive. If $C_{Pl}$ is the concentration
of sites covered with poison the fraction of sites remaining active, called the deactivation or activity function, is represented by

$$\Phi = \frac{C_t - C_{Pl}}{C_t} \qquad (5.2.b-1)$$

This deactivation function is based on the presumed chemical events occurring on
the active sites, and can be related to various chemisorption theories. The overall
observed activity changes of a catalyst pellet can also be influenced by diffusional

effects, etc., but the deactivation function utilized here will refer only to the deactivation chemistry, to which these other effects can then be added.

Since $C_{Pl}$ is not normally measured, it must be expressed in terms of the poison concentration, $C_{Ps}$ in the gas phase inside the catalyst. Wheeler used a linear relation

$$C_{Pl} = \sigma_P C_{Ps} \tag{5.2.b-2}$$

that can be a reasonable approximation over an appreciable fraction of the total saturation level. Since the rate coefficient of the reaction, $k_{rA}$, is proportional to the number of available active sites, its value at the poison level $C_{Ps}$ is given by

$$k_{rA} = k_{rA}{}^0\left(1 - \sigma_P \frac{C_{Ps}}{C_t}\right) = k_{rA}{}^0(1 - \sigma C_{Ps}) = \Phi k_{rA}{}^0 \tag{5.2.b-3}$$

and the activity decreases linearly with the poison concentration. Consider now the case whereby diffusion limitations are felt in the pore and let the reaction be of first order. At the poison level $C_{Ps}$:

$$r_A \rho_s = \eta k_{rA} C_A \tag{5.2.b-4}$$

$$= \frac{1}{\phi}\left[\coth(3\phi) - \frac{1}{3\phi}\right]k_{rA} C_A \tag{5.2.b-5}$$

where, as usual

$$\phi = \frac{R}{3}\sqrt{\frac{k_{rA}}{D_{eA}}} \tag{5.2.b-6}$$

Substituting $k_{rA}$ in Eqs. 5.2.b-5 and 5.2.b-6 by its value given by Eq. 5.2.b-3, so as to account for the effect of the poison, yields the rate in the form:

$$r_A \rho_s = \frac{\coth\left(3\phi^0\sqrt{1 - \sigma C_{Ps}}\right) - \dfrac{1}{3\phi^0\sqrt{1 - \rho C_{Ps}}}}{\phi^0\sqrt{1 - \sigma C_{Ps}}}(1 - \sigma C_{Ps})k_r{}^0 C_A \tag{5.2.b-7}$$

The ratio of this rate to that at zero poison level, taken at identical $C_A$ values, can be written:

$$\frac{r_A}{r_A{}^0} = \frac{3\phi^0\sqrt{1 - \sigma C_{Ps}}\coth(3\phi^0\sqrt{1 - \sigma C_{Ps}}) - 1}{3\phi^0\coth(3\phi^0) - 1} \tag{5.2.b-8}$$

Two limiting cases are of interest. For virtually no diffusion limitations to the main reaction, $\phi^0 \to 0$ and

$$\frac{r_A}{r_A{}^0} \to (1 - \sigma C_{Ps}) \tag{5.2.b-9}$$

so that this ratio is just the deactivation function as defined by Eq. 5.2.b-3. The opposite extreme of strong diffusion limitation, $\phi^0 \to \infty$, leads to a distorted version of the true deactivation function:

$$\frac{r_A}{r_A^{\,0}} \to \sqrt{1 - \sigma C_{P_s}} \tag{5.2.b-10}$$

Notice also that in this case $r_A/r_A^{\,0}$ decreases less rapidly with $C_{P_s}$, owing to a better utilization of the catalyst surface as the reaction is more poisoned.

### 5.2.c Shell Progressive Poisoning

A similar model that specifically considers the poison deposition in a catalyst pellet was presented by Olson [5] and Carberry and Gorring [6]. Here the poison is assumed to deposit in the catalyst as a moving boundary of a poisoned shell surrounding an unpoisoned core, as in an adsorption situation. These types of models are also often used for noncatalytic heterogeneous reactions, which was discussed in detail in Chapter 4. The pseudo-steady-state assumption is made that the boundary moves rather slowly compared to the poison diffusion or reaction rates. Then, steady-state diffusion results can be used for the shell, and the total mass transfer resistance consists of the usual external interfacial, pore diffusion, and boundary chemical reaction steps in series.

The mathematical statement of the rate of poison deposition is as follows:

$$\tfrac{4}{3}\pi R^3 r_P \rho_s = \frac{d}{dt}\left[C_{Pl\infty}\cdot\tfrac{4}{3}\pi(R^3 - r_c^3)\right]\rho_s \tag{5.2.c-1}$$

$$= 4\pi R^2 k_{gP}(C_P - C_{P_s}^{\,s}) \tag{5.2.c-2}$$

(external interfacial step)

$$= \frac{4\pi D_{eP}}{\left(\dfrac{1}{r_c} - \dfrac{1}{R}\right)}\,(C_{P_s}^{\,s} - C_{P_s}^{\,c}) \tag{5.2.c-3}$$

(steady diffusion through
a spherical shell)

$$= 4\pi r_c^2 k_{sP}\,\sigma_P\,C_{P_s}^{\,c} \tag{5.2.c-4}$$

(deposition rate at
boundary)

where $R$ = radius of particle, $r_c$ = radius of unpoisoned core, $C_P$, $C_{P_s}^{\,s}$, $C_{P_s}^{\,c}$ = bulk fluid, solid surface, core boundary concentrations of poison, $C_{Pl\infty}$ = solid concentration of poison at saturation, $k_{gP}$ = external interfacial mass transfer coefficient, $k_{sP}$ = core surface reaction rate coefficient for poison, $D_{eP}$ = effective pore diffusivity of poison, and $\sigma_P$ = sorption distribution coefficient. The pellet average poison concentration can be denoted by $\langle C_{Pl} \rangle$, and is related to the unpoisoned core radius by

$$\frac{\langle C_{Pl} \rangle}{C_{Pl\infty}} = 1 - \left(\frac{r_c}{R}\right)^3 = 1 - \xi^3 \qquad (5.2.\text{c-}5)$$

If the intermediate concentrations, $C_{P_s}{}^s$ and $C_{P_s}{}^c$, are eliminated in the usual way from Eqs. 5.2.c-1 to 4, one obtains

$$\frac{d}{dt'}\left(\frac{\langle C_{Pl} \rangle}{C_{Pl\infty}}\right) = \frac{N_s C_P / C_{P,\,\text{ref}}}{\dfrac{1}{\text{Sh}'_P} + \dfrac{1 - \xi}{\xi} + \dfrac{1}{\text{Da}\,\xi^2}} \qquad (5.2.\text{c-}6)$$

where the new dimensionless groups are:

$\text{Sh}'_P = k_{gP} R / D_{eP}$ = modified Sherwood number for poison
$\text{Da} = \sigma_P k_{sP} R / D_{eP}$ = Damköhler number
$N_s = 3 D_{eP} t_\text{ref} C_{P,\,\text{ref}} / R^2 \rho_s C_{Pl\infty}$
$t' = t / t_\text{ref}$

The reference time, $t_\text{ref}$ and concentration, $C_{P,\,\text{ref}}$, are chosen for a specific application (e.g., in a flow reactor, the mean residence time and feed concentration, respectively). Equation 5.2.c-6 now permits a solution for the amount of poison, $\langle C_{Pl} \rangle / C_{Pl\infty}$, to be obtained as a function of the bulk concentration, $C_P$, and the physicochemical parameters. In a packed bed tubular reactor, $C_P$ varies along the longitudinal direction, and so Eq. 5.2.c-6 would then be a partial differential equation coupled to the flowing fluid phase mass balance equation—these applications will be considered in Part Two—Chapter 11.

Equation 5.2.c-6 can easily be solved for the case of $C_P$ = constant:

$$N_s \frac{C_P}{C_{P,\,\text{ref}}} t' = \int_0^{(\langle C_{Pl} \rangle / C_{Pl\infty})} \left[\frac{1}{\text{Sh}'_P} + \frac{1 - (1 - \chi)^{1/3}}{(1 - \chi)^{1/3}} + \frac{1}{\text{Da}(1 - \chi)^{2/3}}\right] d\chi$$

$$= \left(\frac{1}{\text{Sh}'_P} - 1\right)\left(\frac{\langle C_{Pl} \rangle}{C_{Pl\infty}}\right) - \frac{3}{2}\left[1 - \left(\frac{\langle C_{Pl} \rangle}{C_{Pl\infty}}\right)\right]^{2/3}$$

$$+ \frac{3}{2} - \frac{3}{\text{Da}}\left[1 - \left(\frac{\langle C_{Pl} \rangle}{C_{Pl\infty}}\right)\right]^{1/3} + \frac{3}{\text{Da}} \qquad (5.2.\text{c-}7)$$

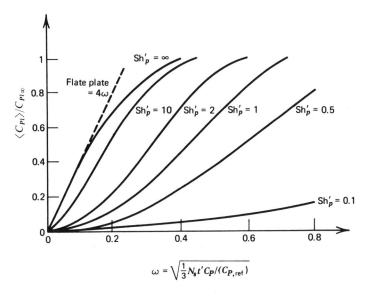

Figure 5.2.c-1 *Fraction of spherical catalyst poisoned versus dimensionless time.* $Da = \infty$ *(after Carberry and Gorring* [6]).

This is an implicit solution for $\langle C_{Pl}\rangle/C_{Pl\infty}$, and is shown in Fig. 5.2.c-1. These results could be used to predict the poison deposition as a function of time and the physicochemical parameters.

Now that the poison concentration is known, the effect on a chemical reaction occurring must be derived. Again, this is based on the assumption that the poisoned shell is completely inactive, and so, for a first-order reaction occurring only in the unpoisoned core of the catalyst, the following mathematical problem must be solved:

$$D'_{eA}\frac{1}{r^2}\frac{d}{dr}\left(r^2\frac{dC'_{As}}{dr}\right) = 0 \qquad r_c < r \leq R \qquad (5.2.c\text{-}8)$$

$$D_{eA}\frac{1}{r^2}\frac{d}{dr}\left(r^2\frac{dC_{As}}{dr}\right) = k_{rA}C_{As} \qquad 0 < r < r_c \qquad (5.2.c\text{-}9)$$

where

$D'_{eA}$ = effective diffusivity of $A$ in poisoned shell
$D_{eA}$ = effective diffusivity of $A$ in unpoisoned core
$k_{rA}$ = rate coefficient for the main chemical reaction

The boundary conditions are

$$D'_{eA} \frac{dC'_{As}}{dr} = k_{gA}(C_A - C_{As}{}^s) \qquad\qquad r = R \qquad (5.2.\text{c-}10a)$$

$$C_{As} = C'_{As} \text{ and } D_{eA} \frac{dC_{As}}{dr} = D'_{eA} \frac{dC'_{As}}{dr} \qquad r = r_c \qquad (5.2.\text{c-}10b)$$

$$C_{As} = \text{finite value} \qquad\qquad r = 0 \qquad (5.2.\text{c-}10c)$$

Note that these equations are again based on a pseudo-steady-state approximation such that the deactivation rate must be much slower than the diffusion or chemical reaction rates. These equations can be easily solved, as in Chapter 3, and the result substituted into the definition of the effectiveness factor, with the following results:

$$\eta\left(\frac{\langle C_{Pl}\rangle}{C_{Pl\infty}}\right) = \frac{4\pi R^2 D'_{eA} \left.\dfrac{dC'_{As}}{dr}\right|_{r=R}}{\dfrac{4}{3}\pi R^3 k_{rA} C_A}$$

$$\eta\left(\frac{\langle C_{Pl}\rangle}{C_{Pl\infty}}\right) = \frac{1/3(\phi')^2}{\dfrac{1}{\text{Sh}'_A} + \dfrac{1-\xi}{\xi} + \left(\dfrac{\phi}{\phi'}\right)^2 \dfrac{1}{\xi[(3\phi\xi)\coth(3\phi\xi) - 1]}} \qquad (5.2.\text{c-}11)$$

$$= \frac{1/3\phi^2}{\dfrac{1}{\text{Sh}'_A} + \dfrac{1-\xi}{\xi} + \dfrac{1}{\xi[(3\phi\xi)\coth(3\phi\xi) - 1]}} \qquad (5.2.\text{c-}12)$$

the latter result being true for $D'_{eA} = D_{eA}$, and so $\phi' = \phi$. Also, the dimensionless parameters are

$$\xi = \frac{r_c}{R} = \left[1 - \left(\frac{\langle C_{Pl}\rangle}{C_{Pl\infty}}\right)\right]^{1/3}$$

$\text{Sh}'_A = k_{rA} R/D'_{eA} = $ modified Sherwood number for main reaction

$$\phi = \frac{R}{3}\sqrt{k_{rA}/D_{eA}} = \text{modulus}$$

$$\phi' = \frac{R}{3}\sqrt{k_{rA}/D'_{eA}} = \text{modulus}$$

Finally, the ratio of the rate at a poison level $\langle C_{Pl} \rangle$ to that at zero poison content, taken at identical $C_A$-values is obtained from:

$$\frac{r_A}{r_A{}^0} = \frac{\eta\left(\dfrac{\langle C_{Pl} \rangle}{C_{Pl\infty}}\right)}{\eta(0)} \qquad (5.2.c\text{-}13)$$

where $\eta(0)$ is the effectiveness factor for the unpoisoned catalyst, and can be found from Eq. 5.2.c-11 or 12 with $\langle C_{Pl} \rangle = 0$. The limiting form of $r_A/r_A{}^0$ for $\phi \to 0$ is

$$r_A/r_A{}^0 = \xi^3 = 1 - \langle C_{Pl} \rangle / C_{Pl\infty}$$

This is just the deactivation function for the shell-progressive model.

To summarize, Eq. 5.2.c-13 gives a theoretical expression for the ratio of rate with to that without poisoning in terms of the reaction physicochemical parameters and the amount of poison ($\langle C_{Pl} \rangle / C_{Pl\infty}$). The amount of poison, in turn, is found from Eq. 5.2.c-6 with the poisoning physicochemical parameters and the fluid phase bulk concentration, $C_P$, at a point in the reactor. It is the only such complete case at the present time, since all other treatments require at least some empirical formulas.

The ratios of rates, $r_A/r_A{}^0$, from Sections 5.2.b and c are illustrated in Fig. 5.2.c-2. We see that the pore mouth poisoning model gives a very rapid decline, especially for strong diffusional limitations. Balder and Petersen [7] presented an interesting experimental technique where both the decrease in overall reaction rate and the centerline reactant concentration in a single particle are measured. The results of the above theories can be replotted as $r_A/r_A{}^0$ versus centerline concentration by eliminating the $\langle C_{Pl} \rangle / C_{Pl\infty}$ algebraically. Thus, the poisoning phenomena can be studied without detailed knowledge of the poison concentrations.

W. H. Ray [8] has considered the case with a nonisothermal particle, which could show instability in a certain narrow range of conditions.

### 5.2.d Effect of Shell Progressive Poisoning on the Selectivity of Complex Reactions

Further extensions of these catalyst poisoning models to complex reactions have been made by Sada and Wen [9]. The poison deposition was described as in Eq. 5.2.c-7, but for very rapid poisoning, $Da \to \infty$, and the results were expressed in terms of the dimensionless position of the poison boundary, $\xi = r_c/R$. Then, the profiles are:

$$N_S \frac{C_P}{C_{P,\,\text{ref}}} t' = 3(1 - \xi)\left(\frac{1 - \xi}{2} + \frac{1}{\text{Sh}'_P}\right) \text{(slab)} \qquad (5.2.d\text{-}1)$$

$$= \tfrac{1}{2}(1 - \xi)^2(1 + 2\xi) + \frac{1 - \xi^3}{\text{Sh}'_P} \text{(sphere)} \qquad (5.2.d\text{-}2)$$

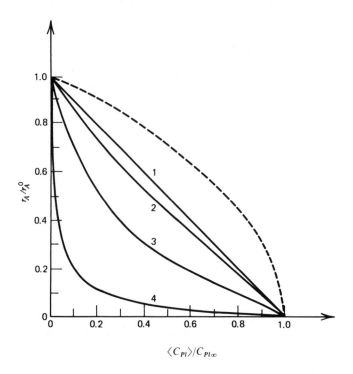

*Figure 5.2.c-2 $r_A/r_A^0$ in terms of amount of poison for homogeneous (Eq. 5.2.b-10 and shell progressive (Eq. 5.2.c-13 models. ($Sh'_A \rightarrow \infty$.)*

---: *uniform poisoning*
—: *shell progressive*
1: $\phi = 0$; $\eta(0) = 1$
2:  3;  0.67
3:  10;  0.27
4:  100;  0.03

The three basic selectivity problems were then solved, for various cases of one or both reactions poisoned. We present only a brief selection of results here—see Sada and Wen [9] for further details.

For independent parallel first-order reactions:

$$A \xrightarrow{\ 1\ } R$$

$$B \xrightarrow{\ 2\ } S$$

The diffusion-reaction problem is:

in the poisoned shell,

$$D_{eA}\nabla^2 C'_{As} = 0$$

$$D_{eB}\nabla^2 C'_{Bs} = k_2 C'_{Bs} \text{ (reaction 2 not poisoned)}$$

$$= 0 \qquad \text{(reaction 2 poisoned)}$$

and in the reactive core,

$$D_{eA}\nabla^2 C_{As} = k_1 C_{As} = -D_{eR}\nabla^2 C_{Rs}$$

$$D_{eB}\nabla^2 C_{Bs} = -D_{eS}\nabla^2 C_{Ss}$$

The boundary conditions at the poison boundary are

$$C_{is} = C'_{is} \qquad \nabla C_{is} = \nabla C'_{is}$$

and these are used together with the usual ones for the external surface and center of the pellet. Note that the effective diffusivities have been assumed constant, and also equal in both the shell and core regions. The solutions of these equations are then used in the definition of selectivity, with the results:

$$\left(\frac{r_R}{r_S}\right)_{bulk} = \frac{D_{eR}\nabla C'_{Rs}}{D_{eS}\nabla C'_{Ss}}\bigg|_{surface}$$

$$= \left(\frac{C_A D_{eA}}{C_B D_{eB}}\right)\frac{\dfrac{1}{\phi_2 \tanh \phi_2} + \dfrac{1}{\text{Sh}'_B}}{\dfrac{1}{\phi_1 \tanh(\phi_1 \xi)} + (1 - \xi) + \dfrac{1}{\text{Sh}'_A}} \qquad (5.2.\text{d-}3)$$

for an infinite slab and for only reaction 1 poisoned, and

$$\left(\frac{r_R}{r_S}\right)_{bulk} = \left(\frac{C_A D_{eA}}{C_B D_{eB}}\right)\frac{\dfrac{1}{\phi_2 \tanh(\phi_2 \xi)} + (1 - \xi) + \dfrac{1}{\text{Sh}'_B}}{\dfrac{1}{\phi_1 \tanh(\phi_1 \xi)} + (1 - \xi) + \dfrac{1}{\text{Sh}'_A}} \qquad (5.2.\text{d-}4)$$

for both reactions poisoned. (Sada and Wen also present solutions for infinite cylinders and for spheres.) In Eq. 5.2.d-3,4

$$\phi_1 = L\sqrt{k_1/D_{eA}} \qquad \phi_2 = L\sqrt{k_2/D_{eB}}$$

An example of the results from Eq. 5.2.d-3 is shown in Fig. 5.2.d-1:
From Eq. 5.2.d-3, for $\phi \to 0$ (and Sh' $\to \infty$)

$$\left(\frac{r_R}{r_S}\right)_{bulk} \simeq \left(\frac{C_A k_1}{C_B k_2}\right)\xi$$

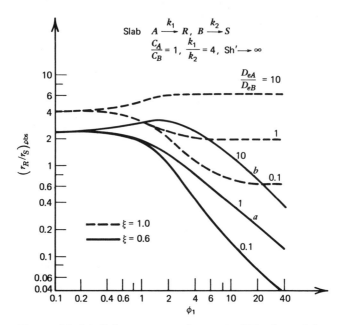

*Figure 5.2.d-1 Selectivity as a function of Thiele modulus $\phi_1$ for independent reactions indicating two types of variations (a and b) (from Sada and Wen [9]).*

which shows that the selectivity is proportional to the unpoisoned fraction of the catalyst volume. Whether the selectivity curve has a maximum or not depends on the values of $\xi$, $\phi$, and $\sqrt{k_{rB}D_{eA}/k_{rA}D_{eB}}$. In the asymptotic region, where $\phi \to \infty$:

$$\left(\frac{r_R}{r_S}\right)_{bulk} \sim \frac{C_A}{C_B}\sqrt{\frac{k_1 D_{cA}}{k_2 D_{cB}}}\,\frac{1}{1 + \phi_1(1 - \xi)} \qquad \phi_1 \gg 1$$

Thus, again a square root change is the dominant factor, as discussed in Chapter 3, but there is an additional change caused by the catalyst poisoning.

For consecutive first-order reactions:

$$A \xrightarrow{\ 1\ } R \xrightarrow{\ 2\ } S$$

The solutions of the appropriate diffusion-reaction equations are used to obtain the selectivity:

$$\left(\frac{r_R}{r_S}\right)_{bulk} = \left.\frac{D_{eR}\nabla C_R}{D_{eS}\nabla C_S}\right|_{surface}$$

$$= \left[\varepsilon\left(\frac{\phi_2}{\phi_1} - \frac{\phi_1}{\phi_2}\right) - 1\right]^{-1} \tag{5.2.d-5}$$

CHEMICAL ENGINEERING KINETICS

where

$$\varepsilon = \frac{\left[1 - \dfrac{1}{Sh'_B}\phi_2 \tanh \phi_2\right]\tanh(\phi_1\xi)\cosh \phi_2}{\tanh(\phi_2\xi) - \dfrac{\phi_1}{\phi_2}\tanh(\phi_1\xi) - \Theta}$$ (5.2.d-6)

$$\Theta = \frac{C_R}{C_A}\frac{\dfrac{k_1}{D_{eA}} - \dfrac{k_2}{D_{eR}}}{\dfrac{k_1}{D_{eR}}}\sinh \phi_2\left\{1 + \phi_1\left[(1-\xi) + \frac{1}{Sh'_A}\right]\tanh(\phi_1\xi)\right\}$$

for only reaction 1 poisoned, and

$$\varepsilon = \frac{\left\{1 + \left[1 - \xi + \dfrac{1}{Sh'_B}\right]\phi_2 \tanh(\phi_2\xi)\right\}\tanh(\phi_1\xi)}{\tanh(\phi_2\xi) - \dfrac{\phi_1}{\phi_2}\tanh(\phi_1\xi) + \Theta}$$ (5.2.d-7)

$$\Theta = \frac{C_R}{C_A}\cdot\frac{k_1/D_{eA} - k_2/D_{eR}}{k_1/D_{eR}}\tanh(\phi_2\xi)\left\{1 + \left[(1-\xi) + \frac{1}{Sh'_A}\right]\phi_1 \tanh(\phi_1\xi)\right\}$$

for both reactions poisoned.

Finally, the case of parallel reactions was considered

for both first-order reactions and the results were as follows:

$$\left(\frac{r_R}{r_S}\right)_{\text{bulk}} = \frac{D_{eR}\nabla C_R}{D_{es}\nabla C_S}\bigg|_{\text{surface}}$$

$$= \left[\left(1 + \frac{k_2}{k_1}\right)\varepsilon' - 1\right]^{-1}$$ (5.2.d-8)

where

$$\varepsilon' = \cosh \phi_1(1 - \xi) + \frac{\phi'_2 \sinh \phi'_2(1 - \xi)}{\phi_{12}\tanh \phi_{12}\xi}$$ (5.2.d-9)

for only reaction 1 poisoned, and

$$\varepsilon' = 1$$ (5.2.d-10)

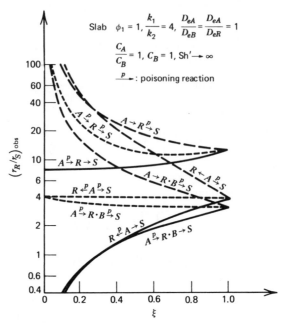

*Figure 5.2.d-2 Selectivities in multiple reactions for three types of poisoning (from Sada and Wen [9]).*

for both reactions poisoned. In Eq. (5.2.d-9)

$$\phi'_2 = L\sqrt{k_2/D_{eA}} \qquad \phi_{12} = L\sqrt{(k_2 + k_1)/D_{eA}}$$

The results of Eq. 5.2.d-10 indicate the obvious result that when both first-order parallel reactions are equally poisoned, the selectivity is not affected, although the conversion would be. The more interesting case of non-first-order parallel reactions would be much more difficult to solve. Figure 5.2.d-2 illustrates the results for several types of poisoning situations:

Many other combinations are also possible, but the method of analyzing these problems should now be clear.

## 5.3 Kinetics of Catalyst Deactivation by Coking

### 5.3.a Introduction

Many petroleum refining and petrochemical processes, such as the catalytic cracking of gas oil, catalytic reforming of naphtha, and dehydrogenation of ethyl-

CHEMICAL ENGINEERING KINETICS

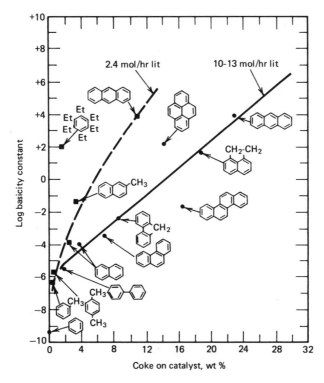

*Figure 5.3.a-1 Coke formation in catalytic cracking and hydrocarbon basicity (from Appleby, et al. [10]).*

benzene and butene hydrofining are accompanied by the formation of carbonaceous deposits, which are strongly adsorbed on the surface, somehow blocking the active sites. Appleby, Gibson, and Good [10] made a detailed study of the coking tendency of various aromatic feeds on silica-alumina catalysts. Figure 5.3.a-1 shows some of their results. Olefins can also readily polymerize to form coke. This "coke" causes a decrease in activity of the catalyst, which is reflected in a drop of conversion to the product(s) of interest. To maintain the production rates within the desired limits, the catalyst has to be regenerated, intermittently or, preferably, continuously. Around 1940, entirely new techniques, such as fluidized or moving bed operation, were developed for the purpose of continuous catalyst regeneration. In what follows the effect of coking on the rates of reaction is expressed quantitatively. Generally, only empirical correlations have been used for this purpose. What is needed, however, for a rational design, accounting for the effect of the coking on the reactor behavior, is a quantitative formulation of the rate of coke deposition. Such a kinetic equation is by no means easy to develop.

The empirical Voorhies correlation for coking in the catalytic cracking of gas oil [11]

$$C_c = At^n \quad \text{with} \quad 0.5 < n < 1$$

has been widely accepted and generalized beyond the scope of the original contribution. Yet, such an equation completely ignores the origin of the coke. Obviously, coke is formed from the reaction mixture itself, so that it must result in some way or other from the reactants, the products or some intermediates. Therefore, the rate of coking must depend on the composition of the reaction mixture, the temperature, and the catalyst activity and it is not justified to treat its rate of formation separately from that of the main reaction. Froment and Bischoff were the first to relate these factors quantitatively to the rate of coking and to draw the conclusions from it as far as kinetics and reactor behavior are concerned [12, 13]. They considered the coke to be formed either by a reaction path parallel to the main reaction

or by a reaction path consecutive to the main:

$$A \longrightarrow R \; — \; \text{intermediates} \longrightarrow C$$

Actually, this can be generalized in case one deals with a main reaction that consists of a sequence of steps itself. Consider the isomerization of $n$-pentane on a dual function catalyst:

$$nC_5 \underset{\longleftarrow}{\overset{Pt}{\longrightarrow}} nC_5^= \underset{\longleftarrow}{\overset{Al_2O_3}{\longrightarrow}} iC_5^= \underset{\longleftarrow}{\overset{Pt}{\longrightarrow}} iC_5$$

Hosten and Froment showed [14] that the rate-determining step for this reaction carried out on a catalyst with a high platinum content is the adsorption of $n$-pentene. In this case, any carbon formation starting from a component situated in this scheme before the rate-determining step would give rise to a characteristic behavior analogous with the parallel scheme given above, even if this component is not the feed component itself. In the example discussed here De Pauw and Froment [15] showed this component to be $n$-pentene. Any coking originating from a component situated in the reaction sequence after the rate determining step could be considered to occur according to the consecutive scheme given above, as if the coke were formed from the reaction product. Indeed, in this case all the components formed after the rate-determining step are in quasi-equilibrium with the final product.

## 5.3.b Kinetics of Coking

Consider a simple reaction $A \rightleftarrows B$ with the conversion of adsorbed $A$ into adsorbed $B$ on a single site as the rate-determining step. The steps may be written:

$$A + l \;\rightleftharpoons\; Al \qquad \text{with } C_{Al} = K_A C_A C_l \qquad (5.3.b\text{-}1)$$

$$Al \;\longrightarrow\; Bl \qquad\qquad\qquad\qquad\qquad (5.3.b\text{-}2)$$

$$Bl \;\rightleftharpoons\; B + l \qquad \text{with } C_{Bl} = K_B C_B C_l \qquad (5.3.b\text{-}3)$$

and since Eq. 5.3.b-2 is the rate-determining step:

$$r_A = k_{sr}\left(C_{Al} - \frac{C_{Bl}}{K_{sr}}\right)$$

or

$$r_A = k_{sr} K_A C_l\left(C_A - \frac{C_B}{K}\right) \qquad (5.3.b\text{-}4)$$

Suppose now some component that will ultimately lead to coke is also adsorbed and competes for active sites:

$$C + l \;\rightleftharpoons\; Cl \qquad\qquad\qquad (5.3.b\text{-}5)$$

so that

$$C_t = C_l + C_{Al} + C_{Bl} + C_{Cl} \qquad (5.3.b\text{-}6)$$

$C_{Al}$ and $C_{Bl}$ may be eliminated from Eq. 5.3.b-6 by means of Eq. 5.3.b-1 and Eq. 5.3.b-3, but not $C_{Cl}$. This coke precursor is generally strongly adsorbed and not found in the gas phase, so that $C_{Cl}$ cannot be referred to a measurable quantity in the gas phase. Then there are two possibilities, starting from Eq. 5.3.b-6 to eliminate $C_l$ from Eq. 5.3.b-4. The first is to write Eq. 5.3.b-6 as follows:

$$C_t = C_l\left(1 + K_A C_A + K_B C_B + \frac{C_{Cl}}{C_l}\right)$$

where $C_{Al}$ and $C_{Bl}$ were eliminated by means of Eq. 5.3.b-1 and Eq. 5.3.b-3. Eq. 5.3.b-4 now becomes

$$r_A = \frac{k_{sr} C_t K_A\left(C_A - \dfrac{C_B}{K}\right)}{1 + K_A C_A + K_B C_B + \dfrac{C_{Cl}}{C_l}} \qquad (5.3.b\text{-}7)$$

Since neither $C_{Cl}$ or $C_l$ can be measured, some empirical correlation for $C_{Cl}/C_l$ has to be substituted into Eq. 5.3.b-7 to express the decline of $r_A$ in terms of the deactivation. The ratio $C_{Cl}/C_l$ could be replaced by some function of a measurable quantity, (e.g., coke) or of less direct factors such as the ratio of total amount of $A$ fed to the amount of catalyst or even process time.

The second possibility is to write Eq. 5.3.b-6 as

$$C_t - C_{Cl} = C_l(1 + k_A C_A + K_B C_B) \tag{5.3.b-8}$$

Substitution of $C_l$ into Eq. 5.3.b-4 leads to:

$$r_A = \frac{k_{sr}{}^0 C_t K_A \Phi_A \left( C_A - \dfrac{C_B}{K} \right)}{1 + K_A C_A + K_B C_B} \tag{5.3.b-9}$$

where $\Phi_A = (C_t - C_{Cl})/C_t$ is the fraction of active sites remaining active. In what follows it will be called the deactivation function. Now $k_{sr} C_t \Phi_A$ can be written as $k = k^0 \Phi_A$. In the absence of information on the coverage of active sites by coke there is no other possibility than to relate $\Phi_A$ empirically to the deactivation. The most direct measure of $C_{Cl}$ and therefore of $\Phi_A$ is the coke content of the catalyst: $\Phi_A = f(C_c)$. On the basis of experimental observations, Froment and Bischoff [12, 13] proposed the following forms:

$$\Phi_A = \exp(-\alpha C_c)$$

$$\Phi_A = \frac{1}{1 + \alpha C_c}$$

The first approach, leading to Eq. 5.3.b-7, was followed in the early work of Johanson and Watson [16] and Rudershausen and Watson [17]. In the terminology of Szepe and Levenspiel [18], Eq. 5.3.b-7 would correspond to a deactivation that is not separable, but Eq. 5.3.b-9 to a separable rate equation.

Equation 5.3.b-5 does not account for the origin of the fouling component. Yet, as previously mentioned, this is an absolute requirement if a rate equation for the deactivation, in other words for the coking, is to be developed. Let the coke precursor be formed by a reaction parallel to the main reaction: ·

$$Al \longrightarrow \, \dot{C}l \tag{5.3.b-10}$$

The coke precursor is an irreversibly adsorbed component whose rate of formation is the rate-determining step in the sequence ultimately leading to coke. Then its rate of formation is given by

$$\frac{dC_{Cl}}{dt} = k'_C C_{Al}$$

Expressing its concentration in terms of coke, which is how it is ultimately determined, and introducing Eq. 5.3.b-1 leads to:

$$r_C = \frac{dC_c}{dt} = k_C K_A C_A C_l$$

and, from Eq. 5.3.b-8

$$r_C = \frac{k_C{}^0 C_t K_A \Phi_c C_A}{1 + K_A C_A + K_B C_B} \tag{5.3.b-11}$$

with $\Phi_c = (C_t - C_{Cl})/C_t$.

Note that even when only one and the same type of active site is involved in the main and coking reaction the deactivation function need not necessarily be identical. Different $\Phi_c$ would result if the rate-determining step in the coking sequence would involve a number of active sites different from that in the main reaction or if the coking sequence would comprise more than one rate-determining step. If the coking would occur exclusively on completely different sites it would only deactivate itself, of course. An example of a complex reaction with more than one deactivation function will be discussed later. A unique deactivation function for both the main and the coking side reactions was experimentally observed by Dumez and Froment [19] in butene dehydrogenation.

If the coke precursor would be formed from a reaction product (i.e., by a consecutive reaction scheme)

$$Bl \rightleftharpoons Cl$$

its rate of formation could be written

$$r_C = \frac{k_C{}^0 C_t K_B \Phi_c C_B}{1 + K_A C_A + K_B C_B} \tag{5.3.b-12}$$

Equations 5.3.b-9 and 5.3.b-11 or 5.3.b-12 form a set of simultaneous equations that clearly shows that the coking of the catalyst not only depends on the mechanism of coking, but also on the composition of the reaction mixture. Consequently, even under isothermal conditions, the coke is not uniformly deposited in a reactor or inside a catalyst particle whenever there are gradients in concentration of reactants and products. This important conclusion will be quantitatively developed in a later section.

The approach followed in deactivation studies is often different from the one outlined here. The starting point of the divergence is the empirical expression for $\Phi_c$, also called "activity." The above approach sets $\Phi_c = f(C_c)$, whereas the alternate approach sets $\Phi = f(t)$. The expressions shown in Table 5.3.b-1 were used to relate $\Phi$, through the ratio of rates or rate coefficients of the main reaction, to time (see Szepe and Levenspiel [18] and Wojchiechowski [20]).

The right-hand side gives the corresponding rates of change of the activity and defines a so-called order of deactivation, from which it has been attempted to get some insight into the mechanism of deactivation—an attempt doomed to fail if not coupled with direct information on the deactivating agent itself.

*Table 5.3.b-1    Activity functions for catalyst deactivation.*

| $\Phi = 1 - \alpha t$ | $-\dfrac{d\Phi}{dt} = \alpha$ |
|---|---|
| $\Phi = \exp(-\alpha t)$ | $= \alpha\Phi$ |
| $\Phi = \dfrac{1}{1 + \alpha t}$ | $= \alpha\Phi^2$ |
| $\Phi = \alpha t^{-0.5}$ | $= \dfrac{\Phi^3}{2\alpha}$ |
| $\Phi = (1 + \alpha t)^{-N}$ | $= \alpha N\Phi^{1 + (1/N)}$ |

At first sight, using $\Phi = f(t)$ instead of $\Phi_c = f(C_c)$ presents definite advantages. An equation like

$$r_A = \frac{k^0 K_A e^{-\alpha t}\left(C_A - \dfrac{C_B}{K}\right)}{1 + K_A C_A + K_B C_B}$$

which has to be compared with Eq. 5.3.b-9, expresses $r_A$ directly in terms of time and therefore suffices in itself to predict the deactivation at any process time, whereas the approach that bases $\Phi_c$ on the coke content of the catalyst leads to an equation for $r_A$ containing the coke content, not time. Consequently, the latter approach requires an additional rate equation for the coke formation to introduce process time. Furthermore, the deactivation function with respect to time is definitely easier to arrive at than the one with respect to coke. However, using the deactivation function with respect to time is far more restricted and it presents several drawbacks.

First, it follows from the definition of $\Phi_c$ and Eq. 5.3.b-11 that

$$-\frac{1}{C_t}\frac{dC_{cl}}{dt} = \frac{d\Phi_c}{dt} = -\frac{k_C K_A \Phi_c C_A}{1 + K_A C_A + K_B C_B}$$

so that

$$\Phi_c = \exp\left[-\int_0^t \frac{k_C K_A C_A}{1 + K_A C_A + K_B C_B}\,dt\right]$$

It is obvious that $\Phi_c$ cannot be a simple function of time, of the type shown in Table 5.3.b-1, except if the coke formation does not depend on the concentrations of the reacting species. Also, in $\Phi = f(\alpha, t)$ the "constant" $\alpha$ is really a function of the operating conditions determining the coke deposition, so that the application of $\Phi = f(t)$ is strictly limited to the conditions prevailing during its determination. With the other approach $\alpha$ is a true constant related to the deactivating event itself, since the effect of the operating variables on the deactivation is explicitly accounted for through the coking rate equation.

Furthermore, when the coke itself is not determined, only one deactivation function can be derived, from the decay with time of the main reaction. The model may then be biased. There is more, however. Since $\Phi = f(t)$ does not contain the coke content, which is related to the local concentration of the reacting species, it predicts a deactivation independent of concentration; that is, the approach predicts a *uniform* deactivation in a pellet or a tubular reactor (e.g., for isothermal conditions at least). In reality, nonuniformity in deactivation, because of coke profiles, does occur in pellets (or tubular reactors), as will be shown in the next section. The consequences of neglecting coke profiles in kinetic studies, in catalyst regeneration, or in design calculations may be serious (see Froment and Bischoff [12, 13]).

### 5.3.c Influence of Coking on the Selectivity

Coking may alter the selectivity when the different reactions have different deactivation functions (see Froment and Bischoff [13] and Weekman [21, 22, 23]). Weekman and Nace [24] represented the catalytic cracking of gasoil ($A$) into gasoline ($Q$), dry gas and coke ($S$) by the following equations:

(Recall Ex.1.4.2.) With rate equations of the power law type, the rates of reaction were written as

$$r_1 = k_1{}^0\Phi_1 y_A{}^2$$

($r_1$ in kg gas oil/kg total · hr)

$$r_2 = k_2{}^0\Phi_2 y_Q$$
$$r_3 = k_3{}^0\Phi_3 y_A{}^2$$

where $\Phi = e^{-\alpha t}$. The selectivity for gasoline may be written:

$$\frac{dy_Q}{dy_A} = -\frac{r_1 - r_2}{r_1 + r_3} = -\frac{1}{1 + (k_3{}^0\Phi_3/k_1{}^0\Phi_1)} + \frac{k_2{}^0\Phi_2}{k_1{}^0\Phi_1 + k_3{}^0\Phi_3}\frac{y_Q}{y_A{}^2}$$

and this relation is readily integrated to yield $y_Q = f(y_A, t)$. Figure 5.3.c-1 shows experimental results of Weekman and Nace [24] from which it follows that the instantaneous gasoline yield is not affected by process time (i.e., by the coke content

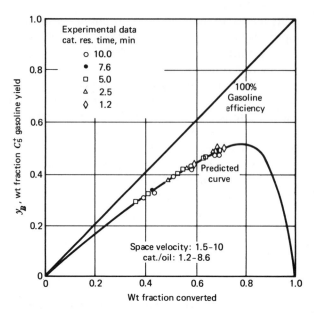

*Figure 5.3.c-1 Catalytic cracking of gasoil. Instantaneous gasoline yield curve (from Weekman and Nace [24]).*

of the catalyst). It may be concluded from this that $\Phi_1 = \Phi_2 = \Phi_3$. When samples are collected over a certain time at the exit of a fixed bed reactor, the time averaged yield will be different from the instantaneous, because the total conversion does vary with time.

In their study of *n*-pentane isomerization on a $Pt/Al_2O_3$ catalyst, to be discussed in more detail later, De Pauw and Froment [15] found the main reaction to be accompanied by hydrocracking and coking. The latter two reactions were shown to occur on sites different from those involved in the main reaction. The three rates decayed through coking, but at different rates, so that the selectivity varied with time as shown in Fig. 5.3.c-2.

### 5.3.d Coking Inside a Catalyst Particle

In the preceding sections, (5.3.b,c) no attention was given to situations where the reaction components encounter important transport resistances inside the catalyst particle. In Chapter 3 it was shown how concentration gradients then build up in the particle, even when the latter is isothermal. In the steady state a feed component *A* then has a descending concentration profile from the surface towards the center

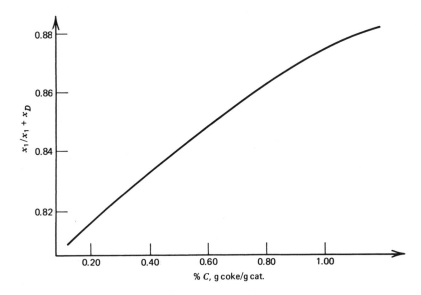

*Figure 5.3.c-2 Isomerization of n-pentane on dual function catalyst. Isomerization selectivity as a function of coke-content.*

and a reaction product $R$ an ascending profile. In such a case it is intuitively clear that the coke will not be uniformly deposited in the particle, but according to a profile, depending on the mechanism of coke formation.

Masamune and Smith [25] applied the approach used by Froment and Bischoff [12] to the situation discussed here. If the rate of coke formation is small compared to the rate of the main reaction, a pseudo steady state may be assumed and the following continuity equations for $A$ and the coke may be written, provided the reaction is irreversible and of the first order and the particle is isothermal:

$$\text{for } A: D_{eA}\nabla_r^2 C_A - \Phi_A r_A{}^0 \rho_s = 0 \qquad (5.3.d\text{-}1)$$

for coking by a parallel mechanism:

$$\frac{\partial C_c}{\partial t} = \Phi_c r_C{}^0(C_A) \qquad (5.3.d\text{-}2)$$

for coking by a consecutive mechanism:

$$\frac{\partial C_c}{\partial t} = \Phi_c r_C{}^0(C_R) \qquad (5.3.d\text{-}3)$$

In Eq. 5.3.d-1 it has again been assumed, in accordance with the pseudo-steady-state hypothesis, that the amount of $A$ involved in the coking reaction is small. Also, the effective diffusivity is presumed to be unaffected by the coke formation. $\Phi_c$ is the deactivation function, assumed to be described by

$$\Phi_c = 1 - C_c/C_{c\infty}$$

where $C_c$ is the instantaneous and local coke content and $(C_c)_\infty$ the value corresponding to complete deactivation. Masamune and Smith numerically integrated Equations 5.3.d-1 and 5.3.d-2 or 5.3.d-3. It was found that with a parallel coking mechanism the coke is deposited according to a descending profile in the particle, whereas with consecutive coke formation the coke profile in the particle is ascending and maximum in the center of the particle. When the diffusivity of the reactants is decreased by the coke deposition, as was verified experimentally by Suga, Morita, Kunugita, and Otake [26] the coke profile would tend to flatten out, however. Also see Butt [37].

Murakami et al. [27] considered very rapid coking so that Eq. 5.3.d-1 had to be completed with an additional term for the coke formation. The pseudo-steady-state approach used above is then no longer valid. With strong diffusion control of the main reaction, both the $A$ and $R$ profiles in the particle are decreasing toward the center during the early part of the transient period. During this period, therefore, the coke profile will be descending toward the center, no matter what the coking mechanism is. In practical situations, however, this early transient period would be brief with respect to the process length and the situation studied by Masamune and Smith [25] would be found. If this were not the case, the catalyst could not be considered as interesting for industrial use.

### Example 5.3.d-1 Coking in the Dehydrogenation of 1-Butene into Butadiene on a Chromia-Alumina Catalyst

(See Dumez and Froment [19].) In the catalytic dehydrogenation of 1-butene into butadiene, which will be described in detail in a later example, coke is observed to be formed from both butene and butadiene, while hydrogen depresses its formation. Figure 1 shows the partial pressure profiles for zero coking and for 0.25 hr, respectively, and the coke profile after 0.25 hr inside the catalyst particle. The solid lines correspond to the results obtained by numerical integration using a Runge-Kutta-Gill routine. The circles represent the partial pressures calculated by means of the collocation method, with constant effective diffusivities (see Villadsen [28]). The rather uniform coke distribution is a result from the parallel-consecutive nature of the coke formation, combined with the inhibiting effect of hydrogen.

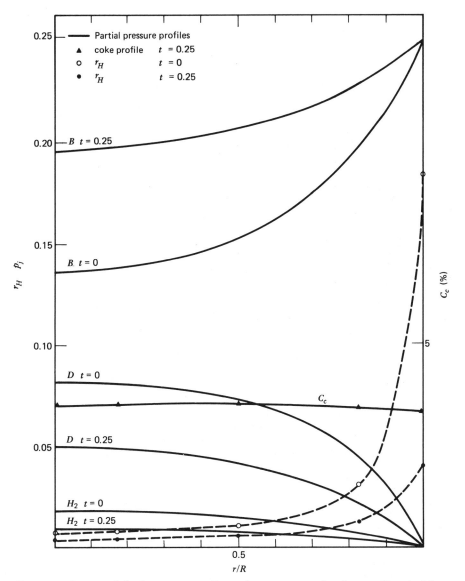

*Figure 1 Butene dehydrogenation. Partial pressure and coke profiles inside a catalyst particle. Parallel-consecutive coking mechanism and inhibition by hydrogen.*

## 5.3.e Determination of the Kinetics of Processes Subject to Coking

The preceding has shown that with processes deactivated by coke deposition the kinetic study should not be limited to the main reaction(s), but also include an investigation of the rate of coke deposition. The kinetic study of the main reaction is in itself seriously complicated by the deactivation, however. Generally, the data are extrapolated to zero process time, when no carbon has been yet deposited. This procedure can be hazardous with very fast coking, of course. In their study of butene dehydrogenation Dumez and Froment [19] were able to take samples of stabilized operation of the fixed bed reactor after two minutes, whereas the total process time lasted about 30 min. In some cases, uncoupling of the main and coking side reactions is possible. When the isomerization of n-pentane is carried out under high hydrogen partial pressure, the coking rate is negligible, so that the kinetics of the main reaction can be conveniently studied (see De Pauw and Froment [15]). The coking kinetics are subsequently obtained from experiments at low hydrogen partial pressure, making use of the known kinetics of the main reaction.

Levenspiel [29] has presented a conceptual discussion of the derivation of rate equations for deactivation from experiments in appropriate equipment. Weekman [23] has also rated various types of laboratory equipment for its adequacy for coking studies. The most useful equipment for coking rate studies is undoubtedly the microbalance used by Takeuchi et al. [30] in the dehydrogenation of isobutene for which they derived a hyperbolic deactivation function $\Phi_c = 1/(1 + \alpha C_c)$, by Ozawa and Bischoff [31] in their investigation of coking associated with ethylene cracking/hydrogenation, by De Pauw and Froment [15] and by Dumez and Froment [19] among others.

Hegedus and Petersen [32, 33] used a single pellet reactor in the hydrogenation of cyclopropane on a Pt-Al$_2$O$_3$ catalyst. They showed how a plot of the ratio of the main reaction rate at any time to that at zero time versus the normalized center-plane concentration of $A$ permits discriminating between coking mechanisms—called self-poisoning mechanisms. The success of this method is strongly dependent on the accuracy with which the center plane concentration can be measured—Thiele moduli in the range one to five are required.

When butene is dehydrogenated around 600°C on a chromia-alumina catalyst (see Dumez and Froment [19]), coke is found to be formed from both butene and butadiene. The rates of coking from both components were studied on a microbalance, which is in fact a differential reactor. For both coking reactions the deactivation function $\Phi_c$ was found to be an exponential function of the coke content of the catalyst, $\Phi_c = \exp(-\alpha C_c)$ and $\alpha$ was identical for both coking reactions. The coking was found to be slowed down by hydrogen availability. The deactivation function of the main reaction was also studied on the electrobalance, by combining the weight variation of the catalyst and conversion measurements.

The same deactivation function was derived, with identical $\alpha$, indicating that the main and the coking reactions occur on the same type of sites.

The kinetics of the main reaction were studied in a classical differential reactor, on the basis of conversions extrapolated to zero time. Since the rate of the main reaction was diffusion controlled, several catalyst sizes had to be investigated. The conversion and coke profiles in a catalyst particle of industrial size were shown already in Fig. 1, Ex. 5.3.d-1.

There are a few recent examples of kinetic studies of deactivating systems in fixed bed reactors. Campbell and Wojciechowski [34] and Pachovsky et al. [35] extensively investigated the catalytic cracking of gas oil into gasoline, associated dry gas and coke on the basis of a triangular mechanism related to that proposed by Weekman, et al. and mentioned in Sec. 5.3.c. The model contained six parameters that were determined by nonlinear regression. As previously mentioned, De Pauw and Froment [15] studied the isomerization of pentane on a platinum-reforming catalyst under coking conditions in a tubular fixed bed reactor. The way in which they derived the kinetics of the main reactions and of the coking side reactions from these experiments is explained in detail in Chapter 11.

## *Example 5.3.e-1 Dehydrogenation of 1-Butene into Butadiene*

Dumez and Froment [19] studied the dehydrogenation of 1-butene into butadiene in the temperature range 480 to 630°C on a chromia-alumina catalyst containing 20 wt% $Cr_2O_3$ and having a surface area of 57 m$^2$/g. The investigation concerned the kinetics of both the main reaction and of the coking.

The kinetics of the main reaction were determined in a differential reactor. The rates in the absence of coke deposition, $r_H^0$, were obtained by extrapolation to zero time. Accurate extrapolation was possible: the reactor was stabilized in less than two minutes after introduction of the butene, whereas the measurements of the rates $r_H$ extended to on stream times of more than 30 minutes.

Fifteen possible rate equations of the Hougen–Watson type were derived from various dehydrogenation schemes and rate-determining steps. The discrimination between these models was achieved by means of sequentially designed experiments, according to the method outlined in Chapter 2. At 525°C, for example, 14 experiments, 7 of which were preliminary, sufficed for the discrimination. The following rate equation, corresponding to molecular dehydrogenation and surface reaction on dual sites as a rate-determining step, was retained:

$$
r_H^0 = \frac{k_H^0 K_B \left( p_B - \dfrac{p_H p_D}{K} \right)}{(1 + K_B p_B + K_H p_H + K_D p_D)^2} \tag{a}
$$

where $K_B$, $K_H$, and $K_D$ and $p_B$, $p_H$, and $p_D$ are adsorption equilibrium constants and partial pressures of butene, hydrogen, and butadiene, respectively.

The kinetics of the coking and the deactivation functions for coking were determined by means of a microbalance. The catalyst was placed in a stainless steel basket suspended at one balance arm. The temperature was measured in two positions by thermocouples placed just below the basket and between the basket and the quartz tube surrounding it. The temperature in the coking experiments ranged from 480 to 630°C, the butene pressure from 0.02 to 0.25 atm, the butadiene pressure from 0.02 to 0.15 atm. Individual components as well as mixtures of butene and butadiene, butene and hydrogen, and butadiene and hydrogen were fed. The hydrogen pressure range was 0 to 0.15 atm. Coke deposition on the basket itself was always negligible.

The deactivation function for coking was determined from the experimental coke versus time curves as described below. Coke was shown to be deposited from both butene and butadiene, while hydrogen exerted an inhibiting effect. An example of the coke content of the catalyst as a function of time is given in Fig. 1. Since the microbalance is a differential reactor, operating at point values of the partial pressures and the temperature, the decrease in the rate of coking observed with increasing coke content reflects the deactivating effect of coke. The rate equation for coke formation therefore has to include a deactivation function, multiplying the rate in the absence of coke:

$$\frac{dC_c}{dt} = r_C{}^0 \Phi_c \qquad\qquad (b)$$

$r_C{}^0$ is the initial coking rate, a function of the partial pressures and temperature that reduces to a constant for a given experiment in the microbalance. Several

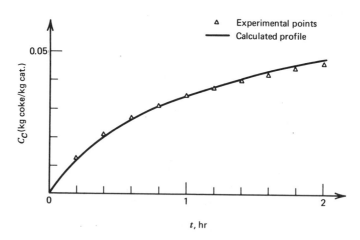

*Figure 1  Butene dehydrogenation. Coke content of catalyst as a function of time in thermobalance experiment.*

expressions were tried for $\Phi_c$:

$$\Phi_c = \exp(-\alpha C_c)$$
$$\Phi_c = 1 - \alpha C_c$$
$$\Phi_c = (1 - \alpha C_c)^2$$
$$\Phi_c = 1/(1 + \alpha C_c)$$
$$\Phi_c = 1/(1 + \alpha C_c)^2$$

Note that the deactivation function is expressed in terms of the coke content of the catalyst, not in terms of time as has been done frequently; indeed, time is not the true variable for the deactivation, as discussed earlier. Substitution of the deactivation function into Eq. (b) and integration with respect to time yields, respectively,

$$C_c = \frac{1}{\alpha} \ln(1 + \alpha r_C^0 t)$$

$$C_c = \frac{1}{\alpha} [1 - \exp(-\alpha r_C^0 t)]$$

$$C_c = \frac{1}{\alpha} \left( 1 - \frac{1}{1 + \alpha r_C^0 t} \right) \qquad \text{(c)}$$

$$C_c = \frac{1}{\alpha} (\sqrt{2\alpha r_C^0 t + 1} - 1)$$

$$C_c = \frac{1}{\alpha} (\sqrt[3]{3\alpha r_C^0 t + 1} - 1)$$

$\alpha$ and $r_C^0$ were determined by fitting of the experimental data by means of a least squares criterion.

For the majority of the 50 experiments $\Phi_c = \exp(-\alpha C_c)$ turned out to give the best fit. An explanation based on a pore blocking mechanism has been attempted (Beeckman and Froment, to be published). The parameter $\alpha$ was found to be identical for coking from either butene or butadiene and independent of the operating variables, as was concluded from the partial correlation coefficients between $\alpha$ and $T$, $p_B$, $p_H$, and $p_D$, respectively, and the $t$-test values for the zero hypothesis for the partial correlation coefficient.

The determination of the complete rate equation for coke deposition required the simultaneous treatment of all experiments, so that $p_B$, $p_H$, $p_D$, and $T$ were varied. The exponential deactivation function was substituted into the rate equation for coking. After integration of the latter, the parameters were determined by minimization of:

$$\mathscr{F} = \sum_{i=1}^{n} (C_c - \hat{C}_c)_i^2 \qquad \text{(d)}$$

where $n$ is the total number of experiments. Several rate equations, either empirical or based on the Hougen-Watson concept were tested. The best global fit was obtained with the following equation:

$$\frac{dC_c}{dt} = \frac{k_{CB}{}^0 p_B{}^{n_{CB}} + k_{CD}{}^0 p_D{}^{n_{CD}}}{(1 + K_{CH}\sqrt{p_H})^2} \exp(-\alpha C_c) \tag{e}$$

with $k_{CB}{}^0 = A_{CB}{}^0 \exp(-E_{CB}/RT)$

$$k_{CD}{}^0 = A_{CD}{}^0 \exp(-E_{CD}/RT) \tag{f}$$

and $K_{CH}$ independent of temperature. The integrated equation used in the objective function (d) was:

$$C_c = \frac{1}{\alpha} \ln \left\{ 1 + \alpha \left[ \frac{k_{CB}{}^0 p_B{}^{n_{CB}} + k_{CD}{}^0 p_D{}^{n_{CD}}}{(1 + K_{CH}\sqrt{p_H})^2} \right] t \right\} \tag{g}$$

The deactivation function for the dehydrogenation was also determined by means of the microbalance, by measuring simultaneously the coke content and the composition of the exit gases as functions of time. To eliminate the effect of by-passing, the conversions were all referred to the first value measured. Figure 2 shows the relation $r_H/r_H{}^0 = \Phi_H$ versus the coke content, easily derived from the measurements $r_H/r_H{}^0 = \Phi_H$ versus time and coke content versus time. Although there is a certain spread of the data, no systematic trend with respect to the temperature or the partial pressures could be detected. The temperature ranged from 520 to 616°C, the butene pressure from 0.036 to 0.16 atm. Again, the best fit was obtained with an exponential function: $\Phi_H = \exp(-\alpha C_c)$. A value of 32.12 was determined for $\alpha$. The agreement with the value found for the deactivation parameter for the two coking reactions is remarkable (compare Eq. i). It may be concluded that the main reaction and the coking reactions occur on the same sites.

The set of rate equations may now be written:

$$r_H = \frac{1.826 \ 10^7 \exp(-29236/RT)\left(p_B - \dfrac{p_H p_D}{K}\right)}{(1 + 1.727 p_B + 3.593 p_H + 38.028 p_D)^2} \exp(-42.12 C_c) \tag{h}$$

$$r_C = \frac{1.5588 \ 10^8 \exp(-32860/RT)p_B{}^{0.743} + 5.108 \ 10^5 \exp(-21042/RT)p_D{}^{0.853}}{(1 + 1.695\sqrt{p_H})^2}$$
$$\times \exp(-45.53 C_c) \tag{i}$$

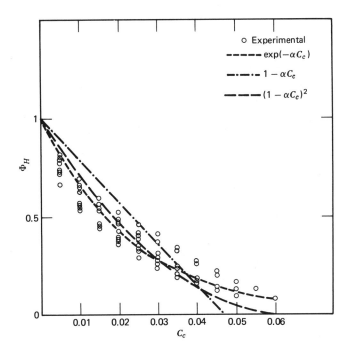

*Figure 2 Butene dehydrogenation. Deactivation function for the main reaction, $\Phi_H$, versus time.*

# Problems

5.1 For shell progressive poisoning, the "shrinking core" model of Chapter 4 was utilized to derive the time rate of change of poison deposition, Eq. 5.2.c-6; complete the steps leading to this result.

5.2 The effect on the reaction rate for shell progressive poisoning is based on Eqs. 5.2.c-8, 9, and 10. Use these to derive the effectiveness factor relation, Eq. 5.2.c-11.

5.3 The amount of poison deposited is given as a function of the dimensionless process time by Fig. 5.2.c-1. Also, the deactivation function for given poison levels is in Fig. 5.2.c-2. Combine these in a figure for the deactivation function as a function of dimensionless time for the shell progressive mechanism.

5.4 Derive Eqs. 5.2.d-5 and 7 for poisoning effects with consecutive reactions.

5.5 Derive Eqs. 5.2.d-8 and 9 for poisoning effects with parallel reactions.

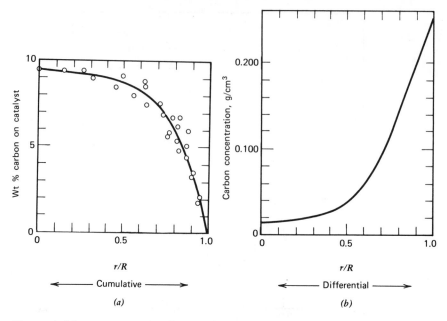

*Figure 1 Measurement of carbon profiles. (a) Experimental data. (b) Calculated profile. Parallel fouling mechanism (from Richardson [36]).*

5.6 Coke profiles in catalyst pellets have been measured by Richardson [36], as shown in Figure 1. Use these to determine the parameters in an appropriate deactivation model:
(a) Which coking mechanism prevailed?
(b) Utilizing reasonable assumptions, which should be stated, complete the analysis with a deactivation model.

# References

[1] Butt, J. B. *Adv. in Chem. Ser.*, **108**, 259, A. C. S. Washington (1972).

[2] Maxted, E. B. *Adv. in Catal.*, **3**, 129 (1951), Maxted, E. B., Evans, H. C. *J. Chem. Soc.* **603**, 1004 (1937).

[3] Mills, G. A., Boedekker, E. R., and Oblad, A. G. *J. Am. Chem. Soc.*, **72**, 1554 (1950).

[4] Wheeler, A. *Catalysis*, P. H. Emmett, ed. Vol. II, Reinhold, New York (1955).

[5] Olson, J. H. *Ind. Eng. Chem. Fund.*, **7**, 185 (1968).

[6] Carberry, J. J., and Gorring, R. L. *J. Catal.*, **5**, 529 (1966).

[7] Balder, J., Petersen, E. E., *Chem. Eng. Sci.* **23**, 1287 (1968).

[8] Ray, W. H. *Chem. Eng. Sci.*, **27**, 489 (1972).

[9] Sada, E. and Wen, C. Y. *Chem. Eng. Sci.*, **22**, 559 (1967).

[10] Appleby, W. G., Gibson, J. W., and Good, G. M. *Ind. Eng. Chem. Proc. Des. Devpt.*, **1**, 102 (1962).

[11] Voorhies, A. *Ind. Eng. Chem.*, **37**, 318 (1945).

[12] Froment, G. F. and Bischoff, K. B. *Chem. Eng. Sci.*, **16**, 189 (1961).

[13] Froment, G. F. and Bischoff, K. B. *Chem. Eng. Sci.*, **17**, 105 (1962).

[14] Hosten, L. H. and Froment, G. F. *Ind. Eng. Chem. Proc. Des. Devpt.*, **10**, 280 (1971).

[15] De Pauw, R. and Froment, G. F. *Chem. Eng. Sci.*, **30**, 789 (1975).

[16] Johanson, L. N. and Watson, K. M. *Natl. Petr. News-Techn. Sect.* (August 1946).

[17] Rudershausen, C. G. and Watson, C. C. *Chem. Eng. Sci.*, **3**, 110 (1954).

[18] Szepe, S. and Levenspiel, O. *Proc. 4th Eur. Symp. Chem. Reaction Engng.*, Brussels (1968), Pergamon Press, London (1971).

[19] Dumez, F. J. and Froment, G. F. *Ind. Eng. Chem. Proc. Des. Devpt.*, **15**, 291 (1976).

[20] Wojchiechowski, B. W. *Can. J. Chem. Eng.*, **46**, 48 (1968).

[21] Weekman, V. W. *Ind. Eng. Chem. Proc. Des. Devpt.*, **7**, 90 (1968).

[22] Weekman, V. W. *Ind. Eng. Chem. Proc. Des. Devpt.*, **8**, 385 (1969).

[23] Weekman, V. W. *A.I.Ch.E. J.*, **20**, 833 (1974).

[24] Weekman, V. W. and Nace, D. M. *A.I.Ch.E. J.*, **16**, 397 (1970).

[25] Masamune, S. and Smith, J. M. *A.I.Ch.E. J.*, **12**, 384 (1966).

[26] Suga, K., Morita, Y., Kunugita, E., and Otake, T. *Int. Chem. Engng.*, **7**, 742 (1967).

[27] Murakami, Y., Kobayashi, T., Hattori, T., and Masuda, M. *Ind. Eng. Chem. Fund.*, **7**, 599 (1968).

[28] Villadsen, J. Selected Approximation Methods for Chemical Engineering Problems, Danmarks Tekniske Højskole (1970).

[29] Levenspiel, O. *J. Catal.*, **25**, 265 (1972).

[30] Takeuchi, M., Ishige, T., Fukumuro, T., Kubota, H., and Shindo, M. *Kag. Kog.* (Engl. Ed.), **4**, 387 (1966).

[31] Ozawa, Y. and Bischoff, K. B. *Ind. Eng. Chem. Proc. Des. Devpt.*, **7**, 67 (1968).

[32] Hegedus, L. and Petersen, E. E. *J. Catal.*, **28**, 150 (1973).

[33] Hegedus, L. and Petersen, E. E. *Chem. Eng. Sci.*, **28**, 69 (1973).

[34] Campbell, D. R. and Wojciechowski, B. W. *Can. J. Chem. Eng.*, **47**, 413 (1969).

[35] Pachovsky, R. A. and Wojciechowski, B. W. *A.I.Ch.E. J.*, **19**, 802 (1973).

[36] Richardson, J. T., *Ind. Eng. Chem. Proc. Des. Devt.* **11**, 8 (1972).

[37] Butt, J. B., *J. Catal.*, **41**, 190 (1976).

# 6

## GAS-LIQUID
## REACTIONS

## 6.1 Introduction

There are many examples of reactions between gases and liquids in industry. They belong to two categories. The first category groups the gas purification processes like removal of $CO_2$ from synthesis gas by means of aqueous solutions of hot potassium carbonate or ethanolamines, or the removal of $H_2S$ and $CO_2$ from hydrocarbon cracking gas by means of ethanolamines or sodium hydroxyde. The second category groups the production processes like the reaction between a gaseous $CO_2$ stream and an aqueous ammonia solution to give ammonium carbonate, air oxidation of acetaldehyde and higher aldehydes to give the corresponding acids, oxidation of cyclohexane to give adipic acid—one of the steps of nylon 66 synthesis. Other production processes are chlorination of benzene and other hydrocarbons, absorption of $NO_2$ in water to give nitric acid, absorption of $SO_3$ in $H_2SO_4$ to give oleum, air oxidation of cumene to cumenehydroperoxide—one of the steps of the Hercules-Distillers phenol-processes.

These processes are carried out in a variety of equipment ranging from a bubbling absorber to a packed tower or plate column. The design of the adsorber itself requires models characterizing the operation of the process equipment and this is discussed in Chapter 14. The present chapter is concerned only with the rate of reaction between a component of a gas and a component of a liquid—it considers only a point in the reactor where the partial pressure of the reactant $A$ in the gas phase is $p_A$ and the concentration of $A$ in the liquid is $C_A$, that of $B$, $C_B$. Setting up rate equations for such a heterogeneous reaction will again require consideration of mass and eventually heat transfer rates in addition to the true chemical kinetics. Therefore we first discuss models for transport from a gas to a liquid phase.

## 6.2 Models for Transfer at a Gas-Liquid Interface

Several models have been proposed to describe the phenomena occurring when a gas phase is brought into contact with a liquid phase. The model that has been used most so far is the two-film theory proposed by Whitman [1] and by Lewis

305

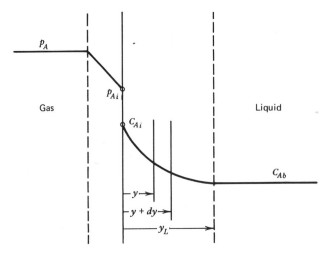

*Figure 6.2-1 Two-film concept for mass transfer be-tween a gas and a liquid.*

and Whitman [2]. In this theory a stagnant layer is supposed to exist in both phases along the interface. In the gas phase the component $A$ experiences a resistance to its transfer to the interface which is entirely concentrated in the film. At the interface itself there is no resistance so that Henry's law is satisfied:

$$p_{Ai} = HC_{Ai} \qquad (6.2-1)$$

where $H$ has the dimension $[m^3 atm/kmol]$.

The resistance to transfer of $A$ from the interface to the bulk liquid is supposed to be entirely located in the liquid film. Beyond that film the turbulence is sufficient to eliminate concentration gradients. This concept is illustrated in Fig. 6.2-1.

The two-film theory originated from the picture adopted for heat transfer between a fluid and a solid surface along which the fluid is flowing in turbulent motion. In that case also it is assumed that at each point along the surface heat is transferred from the fluid to the solid through a laminar boundary layer only by conduction. The entire temperature gradient is limited to this film, since the turbulence is sufficient to eliminate any gradient outside the film. Applying Fourier's law to the conduction through the film in the direction perpendicular to the flow leads to

$$q = \frac{\lambda}{y_h} (T_b - T_s)$$

where $y_h$ is the liquid film thickness for heat transfer, $\lambda$ the conductivity, and $T_b$ and $T_s$ the bulk and surface temperature, respectively. Since the film thickness is

CHEMICAL ENGINEERING KINETICS

not measurable, a convection heat transfer coefficient $\alpha$ is introduced:

$$\alpha = \frac{\lambda}{y_h}$$

The same concept has been applied to mass transfer in the gas and liquid phase, for which one can write, in the absence of reaction:

$$N_{AG} = \frac{D_{AG}}{y_G} (p_A - p_{Ai}) \tag{6.2-2}$$

$$N_{AL} = \frac{D_{AL}}{y_L} (C_{Ai} - C_A) \tag{6.2-3}$$

Again the absence of information on both $y_G$ and $y_L$ leads to the introduction of mass transfer coefficients for the gas and liquid phase, $k_G$ and $k_L$, respectively,

$$k_G = \frac{D_{AG}}{y_G} \quad \text{and} \quad k_L = \frac{D_{AL}}{y_L} \tag{6.2-4}$$

The two-film theory is an essentially steady-state theory. It assumes that the steady-state profiles corresponding to the given $p_A$ and $C_A$ are instantaneously realized. This requires that the capacity of the films be negligible. The two-film theory certainly lacks reality in postulating the existence of films near the interface between the gas and liquid, yet it contains the essential features of the phenomenon, that is, the dissolution and the diffusion of the gas, prior to transfer to the turbulent bulk of the liquid. Nevertheless the theory has enabled consistent correlation of data obtained in equipment in which the postulates are hard to accept completely.

These considerations have led to other models, called "penetration" or "surface renewal" models. In these models the surface at any point is considered to consist of a mosaic of elements with different ages at the surface. An element remains at the surface for a certain time and is exposed to the gas. The element has a volume capacity for mass, is quiescent during its stay at the surface, and is infinitely deep according to some investigators, limited to a certain depth according to others. While at the surface each of the elements is absorbing at a different rate, depending on its age, and therefore also on the concentration profile that has been established. After a certain time of exposure the element is replaced by an element coming from the bulk of the liquid. The mechanism of this replacement is not relevant at this point: it may be due to turbulence or to the flow characteristics in the equipment; for example, think of a packed bed absorber in which the liquid may flow over the particles in laminar flow but is mixed at contact points between particles and in voids, bringing fresh, unexposed elements to the liquid surface. The surface renewal models, in dropping the zero capacity restriction on the film, have to consider the establishment of the profiles with the age of the element at the surface.

Consequently, they are essentially nonsteady state in nature. Furthermore, they have to assume an age distribution function for the elements of the surface, $\Psi(t)$. The average rate of absorption of the surface at the point considered is then:

$$N_A = \int_0^\infty N_A(t)\Psi(t)dt \tag{6.2-5}$$

where $N_A(t)$ is the rate of absorption in an element of the mosaic constituting the surface having an age $t$.

The models discussed above will now be applied to the situation of transfer accompanied by reaction. We first use the two-film theory, then the surface renewal theory. The literature on the subject is overwhelming, and no attempt is made to be complete. Instead, the general concepts are synthesized and illustrated. More extensive coverage can be found in several textbooks more oriented toward gas absorption [39, 40, 41, 42, 43].

## 6.3 Two-Film Theory

### 6.3.a Single Irreversible Reaction with General Kinetics

First consider the case of a chemical reaction that is very slow with respect to the mass transfer, so that the amount of $A$ that reacts during its transfer through the liquid film is negligible. The rate of transfer of $A$ from the liquid interface to the bulk may then be written:

$$N_A A_v = k_L A_v(C_{Ai} - C_{Ab})$$

where $A_v$ represents the interfacial area per liquid volume $(m_i^2/m_L^3)$ while the reaction then occurs completely in the bulk at a rate $r_A = r_A(C_{Ab}, C_{Bb})$. When the two phenomena are purely in series, as assumed here, the resistances may be added, as was shown in Chapter 3 for the simple example of a reaction between a gas and a nonporous solid, to yield the resistance or the rate coefficient for the overall phenomenon. As mentioned previously, in this chapter we only consider a "point" in a reactor, for instance, a volume $dV$ at a certain height in a packed column, with uniform concentrations in a cross section. To arrive at $C_{Ab}$ and $C_{Bb}$ at that point in the reactor requires consideration of the complete reactor with its typical flow pattern and type of operation. This problem is discussed in Chapter 14.

When the rate of reaction cannot be neglected with respect to the mass transfer, the amount reacted in the film has to be accounted for in an explicit way. Let $A$ be the component of the gas phase reacting with a non-volatile component $B$ in the liquid phase and let the film be isothermal. The reaction considered is:

$$aA + bB \longrightarrow qQ + pP$$

and is confined to the liquid phase. Consider only the liquid phase first. Since concentration gradients are limited to the film, a mass balance on $A$ in a slice of thickness $dy$ and unit cross section in the liquid film is set up (Fig. 6.2-1). Since the two-film theory implies steady state, the balance may be written

$$D_A \frac{d^2 C_A}{dy^2} = r_A \tag{6.3.a-1}$$

and, of course,

$$D_B \frac{d^2 C_B}{dy^2} = \frac{b}{a} r_A \tag{6.3.a-2}$$

where

$$r_A = f(C_A, C_B; T)$$

and with $BC$:

$$y = 0 \qquad C_A = C_{Ai} \qquad C_B = C_{Bi} \tag{6.3.a-3}$$

$$y = y_L \qquad C_A = C_{Ab} \qquad C_B = C_{Bb} \tag{6.3.a-4}$$

where $C_{Ab}$ is the bulk concentration of unreacted species.

The bulk concentrations must be determined from an equation for the mass flux through the film-bulk boundary:

$$A_v N_A|_{y=y_L} = (1 - A_v y_L) r_{Ab} + \left( \begin{array}{l} \text{net amount of } A \text{ transported} \\ \text{into corresponding} \\ \text{differential bulk volume by} \\ \text{various mechanisms—for} \\ \text{example, flow} \end{array} \right) \tag{6.3.a-5}$$

The last term of Eq. 6.3.a-5 arises from the fact that the element of bulk fluid considered here is not isolated from its surroundings. When $C_{Ab}$ is not zero, $A$ is transported through liquid flow and diffusion mechanisms into and out of the element, as is discussed in detail in Chapter 14 on gas-liquid reactors. Some past work has ignored this term, presumably to obtain general results relating $C_{Ab}$ to the reactor conditions at the given point but thereby introducing important errors in $C_{Ab}$. For very rapid reactions, for which $C_A$ attains the bulk value (e.g., zero or an equilibrium value $C_{Aeq}$ for a reversible reaction) at $y \leq y_L$, Eq. 6.3.a-4 does not apply, of course. A different approach is given later for this situation.

Integrating Eq. 6.3.a-1 or Eq. 6.3.a-2 with the given boundary conditions and rate equation leads to the concentration profiles of $A$ and $B$ in the liquid film. The rate of the overall phenomenon, as seen from the interface, then follows from the application of Fick's law:

$$N_A = -D_A \frac{dC_A}{dy} \bigg|_{y=0} \tag{6.3.a-6}$$

In general, Eq. 6.3a-1 cannot be integrated analytically. This is only feasible for some special cases of rate equations. We limit ourselves first to those cases in order to illustrate the specific features of gas-liquid reactions.

### 6.3.b First-Order and Pseudo-First-Order Irreversible Reaction

In this case Eq. 6.3.a-1 becomes

$$D_A \frac{d^2 C_A}{dy^2} = k C_A \tag{6.3.b-1}$$

where $k = a k' C_{Bb}$ for a pseudo-first-order reaction. The integral of Eq. 6.3.b-1 may be written as

$$C_A = A_1 \cosh \gamma \frac{y}{y_L} + A_2 \sinh \gamma \frac{y}{y_L}$$

where

$$\gamma = y_L \sqrt{\frac{k}{D_A}} = \frac{\sqrt{k D_A}}{k_L}$$

since $k_L = D_A / y_L$. $\gamma$ is sometimes called the Hatta number and is very similar to the modulus used in the effectiveness factor approach of Chapter 3. Accounting for the BC Eq. 6.3.a-3 and Eq. 6.3.a-4 permits the determination of the integration constants $A_1$ and $A_2$. The solution is

$$C_A = \frac{C_{Ai} \sinh \gamma \left(1 - \dfrac{y}{y_L}\right) + C_{Ab} \sinh \gamma \dfrac{y}{y_L}}{\sinh \gamma} \tag{6.3.b-2}$$

from which it is found, applying Eq. 6.3.a-6, that

$$N_A|_{y=0} = \frac{\gamma D_A}{y_L} \frac{C_{Ai} \cosh \gamma - C_{Ab}}{\sinh \gamma}$$

This equation is easily rewritten into

$$N_A = \frac{\gamma}{\tanh \gamma} \left(1 - \frac{C_{Ab}}{C_{Ai}} \frac{1}{\cosh \gamma}\right) k_L C_{Ai} \tag{6.3.b-3}$$

As mentioned already, this equation has to be combined with a mass balance in the bulk to define $C_{Ab}$ and also $C_{Bb}$, which enters through $\gamma$. The concentration $C_{Bb}$ is constant in any one horizontal slice, but not necessarily over all heights of the equipment. The mass balances yielding $C_{Ab}$ and $C_{Bb}$ is given in Chapter 14.

Let this flux $N_A$ now be compared with that obtained when there is no resistance to mass transfer in the liquid, that is, when the concentration of $A$ in the liquid is

$C_{Ai}$ throughout. From the analogy with the effectiveness factor concept a liquid utilization factor, $\eta_L$, will be defined as follows:

$$\eta_L = \frac{N_A A_v}{k C_{Ai}} = \frac{1}{\gamma \, \mathrm{Sh}_m \tanh \gamma} \left( 1 - \frac{C_{Ab}}{C_{Ai}} \frac{1}{\cosh \gamma} \right) \qquad (6.3.b\text{-}4)$$

where $\mathrm{Sh}_m = (k_L / A_v D_A)$ is a modified Sherwood number.

For very rapid reactions (i.e., when $\gamma$ exceeds 3) $\cosh \gamma > 10$, and since $C_{Ab}/C_{Ai} \leq 1$, Eq. 6.3.b-4 becomes

$$\eta_L = \frac{1}{\gamma \, \mathrm{Sh}_m \tanh \gamma}$$

and for $\gamma > 5$, the only meaningful situation when $C_{Ab} = 0$, the utilization factor reduces to

$$\eta_L = \frac{1}{\gamma \, \mathrm{Sh}_m} = A_v \sqrt{\frac{D_A}{k}}$$

which means that in a plot of log $\eta_L$ versus log $\mathrm{Sh}_m \gamma$, a straight line with slope $-1$ is obtained.

So far the gas film resistance has not been included. This is easily done by eliminating $C_{Ai}$ from Eq. 6.3.b-3 by means of the gas film flux expression $N_A = k_G(p_{Ab} - p_{Ai})$, together with Henry's law and by accounting for the fact that the resistances to transport through the gas and liquid film are purely in series. The following result is obtained:

$$N_A = \frac{p_{Ab} - \dfrac{H C_{Ab}}{\cosh \gamma}}{\dfrac{1}{k_G} + \dfrac{H}{k_L} \dfrac{\tanh \gamma}{\gamma}} \qquad (6.3.b\text{-}5)$$

Note that when $\gamma \to 0$ the equation for physical mass transfer is recovered. When $\gamma \to \infty$ ($\gamma > 5$) Eq. 6.3.b-5 leads to

$$N_A = \frac{p_{Ab}}{\dfrac{1}{k_G} + \dfrac{H}{k_L \gamma}}$$

which is the equation derived when $C_{Ab} = 0$, through a simplified approach that can be found in the literature. Indeed, when $\gamma$ is large the reaction is completed in the film.

It is also possible to base a utilization factor on the bulk gas phase composition, much in the same way as was done already with the $\eta_G$-concept for reaction and transport around and inside a catalyst particle. Let this global utilization factor, based on the bulk gas phase composition, $\eta_G$, be defined as the ratio of the actual rate per unit volume of liquid to the rate that would occur at a liquid concentration equivalent to the bulk gas phase partial pressure of the reactant $A$:

$$\eta_G = \frac{N_A \cdot A_v H}{k p_{Ab}} \qquad (6.3.\text{b-6})$$

Since $N_A \cdot A_v = \eta_L k C_{Ai}$ and $N_A = k_G(p_{Ab} - p_{Ai})$, using $p_{Ai} = HC_{Ai}$ leads to

$$N_A = \frac{p_{Ab}}{\dfrac{1}{k_G} + \dfrac{HA_v}{k\eta_L}} \qquad (6.3.\text{b-7})$$

Combining Eq. 6.3.b-6 and Eq. 6.3.b-7 leads to

$$\frac{1}{\eta_G} = \frac{1}{\eta_L} + \frac{k}{Hk_G A_v}$$

or, in terms of the modified Sherwood number for the liquid phase mass transfer,

$$\frac{1}{\eta_G} = \frac{1}{\eta_L} + \left(\frac{k_L}{Hk_G}\right)\gamma^2 \text{Sh}_m \qquad (6.3.\text{b-8})$$

For large $\gamma$:

$$\frac{1}{\eta_G} = \gamma\,\text{Sh}_m + \left(\frac{k_L}{Hk_G}\right)\gamma^2 \text{Sh}_m$$

In the region of extremely rapid reaction the utilization factor approach, which refers the observed rate to the maximum possible chemical rate, has the drawback of requiring accurate values of the rate coefficient, $k$. An alternate way is to refer to the physical liquid phase mass transfer rate, which is increased by the chemical reaction. This then leads to the definition of an enhancement factor, $F_A$:

$$F_A = \frac{N_A}{k_L(C_{Ai} - C_{Ab})} \qquad (6.3.\text{b-9})$$

and substituting $N_A$ from Eq. 6.3.b-3 leads to

$$F_A = \frac{\gamma}{\tanh \gamma}\left(1 - \frac{C_{Ab}}{C_{Ai}}\frac{1}{\cosh \gamma}\right) \qquad (6.3.\text{b-10})$$

In the literature it is often assumed that in the presence of chemical reaction the concentration of $A$ in the bulk is essentially zero. Starting from Eq. 6.3.a-1 with the BC at $y = y_L$, $C_A = C_{Ab} = 0$. The solution for this situation is easily found in the next section.

$$N_A = F_A k_L C_{Ai}$$

where

$$F_A = \frac{\gamma}{\tanh \gamma} \tag{6.3.b-11}$$

Note that $F_A$ equals $\gamma$ for very large $\gamma$. The reaction is essentially completed in the film when $\gamma > 3$, whereas it takes place mainly in the bulk when $\gamma < 0.3$.

At this point a diagram can be constructed showing $F_A$ as a function of $\gamma$, as first given by Van Krevelen and Hoftijzer [3], but only for the case of no reaction in the bulk (Fig. 6.3.b-1). The other curves in the diagram pertain to second-order and instantaneous reactions and their derivation and discussion are given in the next section.

The enhancement factor approach, like the utilization factor approach, permits accounting for gas phase resistance. Again the gas phase flux equation, Henry's law, the liquid phase flux equation, and the equality of fluxes through both phases can be combined to eliminate $C_{Ai}$, with the result that

$$N_A = \frac{p_{Ab}}{\dfrac{1}{k_G} + \dfrac{H}{F_A k_L}} \tag{6.3.b-12}$$

where $F_A$ is given by Eq. 6.3.b-11.

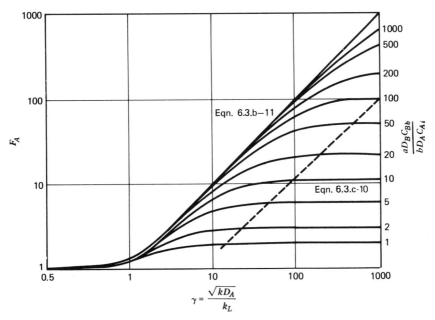

Figure 6.3.b-1 Enhancement factor diagram for $C_{Ab} = 0$.

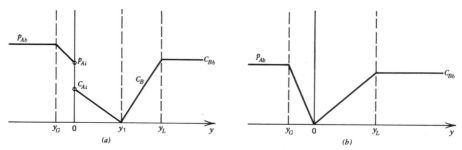

Figure 6.3.c-1 Absorption and infinitely fast reaction. Concentration profiles for A and B. When (a) $C_{Bb} < C'_{Bb}$ and (b) $C_{Bb} = C'_{Bb}$, respectively. See Eq. 6.3.c-7.

### 6.3.c  Single, Instantaneous, and Irreversible Reaction

When the reaction is infinitely fast the thickness of the reaction zone will be reduced to that of a plane situated at a distance $y_1$ from the interface as illustrated in Fig. 6.3.c-1. In the zone of the liquid film between the interface and the reaction plane at $y_1$, $C_A$ varies between $C_{Ai}$ and zero and there is no more $B$ as shown in Fig. 6.3.c-1. In the zone between $y_1$ and $y_L$ there is no more $A$, only $B$, which varies between zero and $C_{Bb}$. The location of the reaction plane is dictated by the concentrations $C_{Ai}$ and $C_{Bb}$, but also by the diffusion rates.

The concentration profile for $A$ in the zone $y = 0$ to $y = y_1$ is obtained from Eq. 6.3.a-1 in which $r_A$ is set equal to zero, since there is no reaction in that zone. A first integration leads to

$$D_A \frac{dC_A}{dy} = \text{constant} = -N_A$$

Notice that in the absence of reaction the concentration profile is linear, as drawn in Fig. 6.3.c-1. A second integration leads to:

$$N_A = \frac{D_A}{y} (C_{Ai} - C_A) \tag{6.3.c-1}$$

Applied to the zone $y = 0$ to $y_1$, where $C_A = 0$, Eq. 6.3.c-1 becomes:

$$N_A = \frac{D_A}{y_1} C_{Ai}$$

The flux of $B$ is obtained in a similar way, leading to:

$$-N_B = \frac{D_B}{y_L - y_1} C_{Bb} \tag{6.3.c-2}$$

314

where

$$\frac{N_A}{a} = -\frac{N_B}{b}$$

Consequently,

$$y_1 N_A = D_A C_{Ai}$$

and from Eq. 6.3.c-2,

$$(y_L - y_1)N_A = \frac{a}{b} D_B C_{Bb}$$

By summing up the two last expressions one obtains

$$N_A = k_L C_{Ai}\left(1 + \frac{a}{b}\frac{D_B}{D_A}\frac{C_{Bb}}{C_{Ai}}\right) \tag{6.3.c-3}$$

where, as before,

$$k_L = \frac{D_A}{y_L}$$

Again define a utilization factor as

$$\eta_L = \frac{N_A A_v}{k C_{Ai}} = \frac{k_L\left(1 + \dfrac{a}{b}\dfrac{D_B}{D_A}\dfrac{C_{Bb}}{C_{Ai}}\right)A_v}{k} \tag{6.3.c-4}$$

or, in terms of the modified Sherwood number,

$$\eta_L = \frac{1}{\gamma^2 \mathrm{Sh}_m}\left(1 + \frac{a}{b}\frac{D_B}{D_A}\frac{C_{Bb}}{C_{Ai}}\right) \tag{6.3.c-5}$$

This is a utilization factor that considers the liquid phase only and represents the slowing-down effect of the mass transfer on the maximum possible chemical rate, which would occur for the interfacial concentration of $A$, $C_{Ai}$, and the bulk concentration of $B$, $C_{Bb}$.

When the gas phase resistance is important an overall utilization factor $\eta_G$ can be derived that is identical to that given in Eq. 6.3.b-8 for the pseudo-first-order case. The value of $\eta_L$ is determined from Eq. 6.3.c-5, which could also be written in terms of the bulk gas phase partial pressure of $A$ as

$$\eta_L = \frac{1}{\gamma^2 \mathrm{Sh}_m}\frac{1 + \dfrac{a}{b}\dfrac{D_B}{D_A}\dfrac{C_{Bb}H}{p_{Ab}}}{1 - \dfrac{a}{b}\dfrac{D_B}{D_A}\dfrac{C_{Bb}}{p_{Ab}}\dfrac{k_L}{k_G}} \tag{6.3.c-6}$$

It can be seen from Eq. 6.3.c-5 or Eq. 6.3.c-6 that the utilization factor (i.e., the rate of the overall phenomenon) is increased by raising the concentration of the liquid phase reactant $C_{Bb}$. This is only true up to a certain point, however. Indeed, as $C_{Bb}$ is increased, the plane of reaction, located at $y = y_1$, moves toward the interface. The reaction takes place at the interface itself when $y_1 = 0$, that is, when

$$y_L = \frac{a}{b} \frac{D_B C_{Bb}}{N_A}$$

The corresponding concentration of $B$ is denoted $C'_{Bb}$:

$$C'_{Bb} = \frac{b}{a} \frac{y_L N_A}{D_B} = \frac{b}{a} \frac{D_A}{D_B} \frac{N_A}{k_L} \qquad (6.3.\text{c-}7)$$

For the value of $C'_{Bb}$ given by Eq. 6.3.c-7 both $C_{Ai}$ and $C_{Bi}$ become zero at the interface, as shown in Fig. 6.3.c-1. Beyond that value no further acceleration of the overall rate is possible by increasing $C_{Bb}$, since the rate is determined completely by the transfer rate in the gas phase. In addition, $p_A$ drops to zero at the interface and the overall rate equation reduces to

$$N_A = k_G p_{Ab} \qquad (6.3.\text{c-}8)$$

so that

$$C'_{Bb} = \frac{b}{a} \frac{D_A}{D_B} \frac{k_G}{k_L} p_{Ab} \qquad (6.3.\text{c-}9)$$

This relationship also shows that $\eta_L$ as determined from Eq. 6.3.c-6 is always a positive quantity.

As mentioned above, in the region of extremely rapid reaction the utilization factor approach, which refers the observed rate to the maximum possible chemical rate, has the drawback of requiring accurate values of the rate coefficient, $k$. An alternate approach is given by the enhancement factor concept. From the definition of $F_A$ given in Eq. 6.3.b-9 and from Eq. 6.3.c-3 it follows that

$$F_A = \frac{k_L C_{Ai}\left(1 + \dfrac{a}{b} \dfrac{D_B}{D_A} \dfrac{C_{Bb}}{C_{Ai}}\right)}{k_L(C_{Ai} - 0)} = 1 + \frac{a}{b} \frac{D_B}{D_A} \frac{C_{Bb}}{C_{Ai}} \qquad (6.3.\text{c-}10)$$

Obviously $F_A \geq 1$, so that the mass transfer rate is "enhanced" by the chemical reaction. As $C_{Bb}$ is increased, Eq. 6.3.c-10 indicates that the enhancement factor, $F_A$, increases, but only until the critical value $C'_{Bb}$ is attained, Eq. 6.3.c-9.

Equation 6.3.c-10 is also represented in Fig. 6.3.b-1. Since $F_A$ is independent of $\gamma$ in the present case, a set of horizontal lines with $(a/b)(D_B/D_A)(C_{Bb}/C_{Ai})$ as a parameter is obtained. The curves in the central part that connect the lines for infinitely fast reactions to the curve for a pseudo-first-order reaction correspond to moderately fast second-order reactions. They were calculated by Van Krevelen and Hoftijzer [3] under the assumption that $B$ is only weakly depleted near the interface. For moderately fast reactions this assumption was reasonably confirmed by more rigorous computations.

When there is appreciable gas phase resistance, again the gas phase flux equation, Henry's law, the liquid phase flux equation and the equality of the fluxes through both phases can be combined to eliminate $C_{Ai}$ with the result:

$$N_A = \frac{p_{Ab} + \dfrac{a}{b}\dfrac{D_B}{D_A} H C_{Bb}}{\dfrac{1}{k_G} + \dfrac{H}{k_L}}$$

which again illustrates the rule of addition of resistances. This equation may also be written in terms of $F_A$ and yields

$$N_A = \frac{p_{Ab}}{\dfrac{1}{k_G} + \dfrac{H}{F_A k_L}} \qquad (6.3.c\text{-}11)$$

This equation is inconvenient to use as such because $F_A$ still contains the interfacial concentration $C_{Ai}$. The enhancement factor $F_A$ can also be expressed explicitly in terms of observables to give:

$$F_A = \frac{1 + \dfrac{a}{b}\dfrac{D_B}{D_A}\dfrac{H C_{Bb}}{p_{Ab}}}{1 - \dfrac{a}{b}\dfrac{D_B}{D_A}\dfrac{C_{Bb}}{p_{Ab}}\dfrac{k_L}{k_G}}$$

So far no attention has been given in this chapter on the effect of the diffusivities. Often instantaneous reactions involve ionic species. Care has to be taken in such case to account for the influence of ionic strength on the rate coefficient, but also on the mobility of the ions. For example, the absorption of HCl into NaOH, which can be represented by $H^+ + OH^- \rightarrow H_2O$. This is an instantaneous irreversible reaction. When the ionic diffusivities are equal the diffusivities may be calculated from Fick's law. But, $H^+$ and $OH^-$ have much greater mobilities than the other ionic species and the results may be greatly in error if based solely on molecular diffusivities. This is illustrated by Fig. 6.3.c-2, adapted from Sherwood and Wei's [4] work on the absorption of HCl and NaOH by Danckwerts. The enhancement factor may be low by a factor of 2 if only molecular diffusion is accounted for in the mobility of the species. Important differences would also occur in the system HAc–NaOH. When $CO_2$ is absorbed in dilute aqueous NaOH the "effective" diffusivity of $OH^-$ is about twice that of $CO_2$.

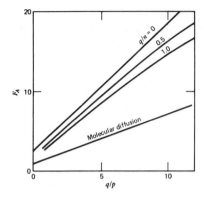

*Figure 6.3.c-2 Enhancement factor for absorption of HCl in aqueous NaOH. $q =$ concentration of $OH^-$ in liquid bulk; $p =$ concentration of $H^+$ at interface; $n =$ concentration of $Na^+$ in liquid bulk (after Danckwerts [43]).*

### 6.3.d Some Remarks on Boundary Conditions and on Utilization and Enhancement Factors

The literature on gas-liquid reactions has mainly dealt with gas-absorption processes, in which the reaction is applied as a means of accelerating the absorption. The reactions used in these absorption processes are very fast, as can be seen from some typical $k$-values, selected from a paper by Sharma and Danckwerts [5] given in Table 6.3.d-1. With such fast reactions $\gamma$ is large and it is often justified to consider the reaction to be completed in the film. But from Table 6.3.d-2 (Barona [6]) which gives characteristic parameters of important industrial gas-liquid reactions, it follows that quite often $\gamma$ is much smaller than one.

*Table 6.3.d-1 Rate coefficients at 25°C of reactions between $CO_2$ or COS and various solutions (after Sharma and Danckwerts [5]).*

|  | Solution | Ionic Strength (kg/m³) | $k$(m³/kmol hr) |
|---|---|---|---|
| $CO_2$ | Monoethanolamine | — | $27.10^6$ |
|  | Diethanolamine | — | $5.4.10^6$ |
|  | Carbonate buffer + arsenite | 1–5 | $3600$–$36.10^6$ |
|  | Morpholine | — | $65.10^6$ |
|  | Aqueous NaOH or KOH | 0.5–6 | $18.10^6$–$36.10^6$ |
|  | Aqueous $Na_2S$ | 1.5–9 | $18.10^6$–$22.10^6$ |
| COS | Monoethanolamine | — | 57,600 |
|  | Diethanolamine | — | 39,600 |
|  | Diisopropylamine | — | 21,600 |
|  | Morpholine | — | 792,000 |
|  | NaOH (1M) | — | 43,200 |

Table 6.3.d-2 Characteristic parameters of some industrial gas-liquid reactions (from Barona [6]).

| Reactions | $T$ (°C) | $C_{A0}$ (kmol/m³) | $C_{B0}$ (kmol/m³) | Catalyst | Cat. conc. (kmol/m³) | $D_A$ (m³/m·hr) | $k_L$ (m³/m² hr) | $k$ | $\gamma$ |
|---|---|---|---|---|---|---|---|---|---|
| *Chlorinations* | | | | | | | | | |
| B + Cl₂ → CB + HCl | 80 | | 10.45 | FeCl₃ | | $2.027 \times 10^{-5}$ | 1.303 | 4.143 m³/kmol hr | 0.0227 |
| id. | 20 | 0.1245 | 11.22 | SnCl₄ | 0.049 | $1.059 \times 10^{-5}$ | 0.716 | 43.09 m³/kmol hr | 0.0999 |
| TCE + Cl₂ → C₂H₄Cl₄ + HCl | 70 | | 10.26 | | | $8.856 \times 10^{-6}$ | 0.576 | 4.619 m³/kmol hr | 0.0357 |
| 1 PB + Cl₂ → MC + HCl | 20 | 0.1750 | 7.183 | SnCl₄ | 0.012 | $1.099 \times 10^{-5}$ | 0.734 | 850.6 m³/kmol hr | 0.353 |
| EB + Cl₂ → MC + HCl | 20 | 0.1060 | 8.179 | SnCl₄ | 0.00208 | $1.234 \times 10^{-5}$ | 0.828 | 2087 m³/kmol hr | 0.554 |
| T + Cl₂ → MC + HCl | 20 | 0.1135 | 9.457 | SnCl₄ | 0.00036 | $1.309 \times 10^{-5}$ | 0.828 | 3468 m³/kmol hr | 0.791 |
| p-X + Cl₂ → MC + HCl | 20 | 0.0685 | 8.066 | SnCl₄ | 0.00066 | $1.234 \times 10^{-5}$ | 0.698 | 14450 m³/kmol hr | 1.718 |
| o-X + Cl₂ → MC + HCl | 20 | 0.1100 | 8.311 | SnCl₄ | 0.00066 | $1.018 \times 10^{-5}$ | 0.796 | 16050 m³/kmol hr | 1.464 |
| *Oxidations* | | | | | | | | | |
| THF + O₂ → HP | 65 | | 12.35 | ADBN | 0.06 | $2.131 \times 10^{-5}$ | 1.145 | 0.0138 hr⁻¹ | 0.00047 |
| EB + O₂ → HP | 80 | | | Cuᴵᴵ-Stearate | $1.62 \times 10^{-5}$ | $3.197 \times 10^{-5}$ | 1.498 | 0.000375 hr⁻¹ | 0.00073 |
| id. | 80 | | 7.736 | Cuᴵᴵ-Stearate | 0.056 | $3.197 \times 10^{-5}$ | 1.498 | 2.627 m³/kmol hr | 0.0170 |
| o-X + O₂ → o-TA | 160 | | | | | $5.389 \times 10^{-5}$ | 0.929 | 0.1025 m³/kmol hr | 0.258 |

B: benzene; MCB: monochlorobenzene; TCE: 1,1,2-trichloroethane; 1 PB: 1-propylbenzene; EB: ethylbenzene; T: toluene; p-X: p-Xylene; o-X: o-xylene; MC: monochloride of 1 PB, EB, T, p-X, and o-X; THF: tetrahydrofurane; HP: hydroperoxide; o-TA: o-toluic acid; ADBN: azodiiso butyronitrile.

Consequently, take care not to resort immediately to the mathematical solutions often encountered in the literature, mainly oriented toward fast processes.

The approach followed in the preceding sections was to start from the most general situation, retaining the possibility of reaction in the bulk. Two approaches have been used throughout to characterize the interaction between mass transfer and chemical reaction between components of a gas and a liquid: one expressing the slowing down of the reaction rate by the mass transfer and leading to the utilization factor and a second expressing the enhancement effect of the reaction on the physical mass transfer and leading to the older concept of the enhancement factor. The much wider acceptance of the enhancement factor approach may again be explained by the historical development of the field, mainly determined by gas-absorption processes. What are the relative merits of the two concepts? It would seem that each approach has its well defined optimum field of application, depending on the process and its rate of reaction. Of course, when the reaction rate is very slow and there is no conversion in the film, the simple series approach for mass transfer and reaction, outlined in Sec. 6.3.a, is logical and there is no need for either the $\eta_L$ or the $F_A$ concept.

For intermediate reaction rates the use of the enhancement factor is not consistent with the standard approach of diffusional limitations in reactor design and may be somewhat confusing. Furthermore, there are cases where there simply is no purely physical mass transfer process to refer to. For example, the chlorination of decane, which is dealt with in the coming Sec. 6.3.f on complex reactions or the oxidation of $o$-xylene in the liquid phase. Since those processes do not involve a diluent there is no corresponding mass transfer process to be referred to. This contrasts with gas-absorption processes like $CO_2$-absorption in aqueous alkaline solutions for which a comparison with $CO_2$-absorption in water is possible. The utilization factor approach for pseudo-first-order reactions leads to $N_A A_v = \eta_L k C_{Ai}$ and, for these cases, refers to known concentrations $C_{Ai}$ and $C_{Bb}$. For very fast reactions, however, the utilization factor approach is less convenient, since the reaction rate coefficient frequently is not accurately known. The enhancement factor is based on the readily determined $k_L$ and in this case there is no problem with the driving force, since $C_{Ab} = 0$. Note also that both factors $\eta_L$ and $F_A$ are closely related. Indeed, from Eqs. 6.3.c-5 and 6.3.c-10 for instantaneous reactions:

$$\frac{\eta_L}{F_A} = \frac{1}{\gamma^2 Sh_m}$$

From Eqs. 6.3.b-4 and 6.3.b-10 for pseudo-first-order reactions the same relation is found.

Finally, the question may be raised if there is any advantage at all in the use of $\eta_L$ and $F_A$. As for the effectiveness factor for solid catalyzed gas phase reactions, the advantage lies in the possibility of characterizing the interaction between mass transfer and reaction by means of a single number, varying between zero and one for the utilization factor. The $N_A$ equation in itself is much less explicit in this respect, of course. As will be evidenced in Chapter 14 there is no advantage or even no need for the explicit use of $\eta_L$ or $F_A$ in design calculations, since the mass flux equations can be directly used.

## 6.3.e Extension to Reactions with Higher Orders

So far, only pseudo-first-order and instantaneous second-order reactions were discussed. In between there is the range of truly second-order behavior for which the continuity equations for $A$ (Eq. 6.3.a-1) or $B$ (Eq. 6.3.a-2), cannot be solved analytically, only numerically. To obtain an approximate analytical solution, Van Krevelen and Hoftijzer [3] dealt with this situation in a way analogous to that applied to pseudo-first-order kinetics, namely by assuming that the concentration of $B$ remains approximately constant close to the interface. They mainly considered very fast reactions encountered in gas absorption so that they could set $C_{Ab} = 0$, that is, the reaction is completed in the film. Their development is in terms of the enhancement factor, $F_A$. The approximate equation for $F_A$ is entirely analogous with that obtained for a pseudo-first-order reaction Eq. 6.3.b-11, but with $\gamma$ replaced by $\gamma'$, where

$$\gamma' = \gamma \sqrt{1 - (F_A - 1)\frac{b\, D_A\, C_{Ai}}{a\, D_B\, C_{Bb}}} \qquad (6.3.e-1)$$

This approximate solution is valid to within 10 percent of the numerical solution. Obviously when $C_{Bb} \gg C_{Ai}$ then $\gamma' = \gamma$ and the enhancement factor equals that for pseudo-first-order. When this is not the case $F_A$ is now obtained from an implicit equation. Van Krevelen and Hoftijzer solved Eq. 6.3.e-1 and plotted $F_A$ versus $\gamma$ in the diagram of Fig. 6.3.c-2, given in Sec. 6.3.c connecting the results for pseudo-first-order and instantaneous second-order reactions.

Porter [7] and also Kishinevskii et al. [8] derived approximate equations for the enhancement factor that were found by Alper [9] to be in excellent agreement with the Van Krevelen and Hoftijzer equation (for Porter's equation when $\gamma \geq 2$) and which are explicit. Porter's equation is:

$$F_A = 1 + \frac{a\, D_B}{b\, D_A}\frac{C_{Bb}}{C_{Ai}}\left[1 - \exp -\left(\frac{\gamma - 1}{\dfrac{a\, D_B}{b\, D_A}\dfrac{C_{Bb}}{C_{Ai}}}\right)\right]$$

Kishinevskii's equation is:

$$F_A = 1 + \frac{\gamma}{\alpha}[1 - \exp(-0.65\gamma\sqrt{\alpha})]$$

where

$$\alpha = \frac{\gamma}{\dfrac{a\,D_B\,C_{Bb}}{b\,D_A\,C_{Ai}}} + \exp\left(\frac{0.68}{\gamma} - \frac{0.45\gamma}{\dfrac{a\,D_B\,C_{Bb}}{b\,D_A\,C_{Ai}}}\right)$$

For an irreversible reaction of global order $m + n$ ($m$ with respect to $A$, $n$ with respect to $B$), the approach followed by Hikita and Asai [10] was very similar to that of Van Krevelen and Hoftijzer. The rate of reaction was written as:

$$r_A = kC_A{}^m C_B{}^n$$

Furthermore, $C_B$ was considered to be nearly constant in the film, while $C_{Ab}$ was again set zero. Hikita and Asai again cast the results into the form of a physical absorption rate multiplied by an enhancement factor

$$F_A = \frac{\gamma''}{\tanh \gamma''}$$

where

$$\gamma'' = \frac{\sqrt{\dfrac{2}{m+1}\,kC_{Ai}{}^{m-1}C_{Bi}{}^n D_A}}{k_L} = \gamma\sqrt{\frac{2}{m+1}C_{Ai}{}^{m-1}C_{Bi}{}^{n-1}}$$

$\gamma''$ evidently reduces to $\gamma$ when $n = 1$ and $m = 1$. Reversible first-order reactions have been considered by Danckwerts and Kennedy and by Huang and Kuo [11, 12]. The latter found for the enhancement factor for the case of a rapid pseudo-first-order reversible reaction (i.e., equilibrium in the liquid bulk) the following expression:

$$F_A = \frac{1 + \dfrac{D_A}{KD_R}}{\dfrac{D_A}{KD_R} + \dfrac{\tanh\left(1 + \dfrac{D_A}{KD_R}\right)\gamma}{\left(1 + \dfrac{D_A}{KD_R}\right)\gamma}}$$

It can be seen from this equation that the reversibility of the reaction can have an important effect on the enhancement factor compared to the corresponding irreversible case with the same $\gamma$-value. Instantaneous reversible reactions were studied by Olander [13].

### 6.3.f Complex Reactions

Complex reactions have also been dealt with. To date, a fairly complete catalog of solutions is available for various reactions, both simple and complex and with fairly general kinetics as long as no solid catalyst is involved. With complex reactions the selectivity is of course crucial and an important question is whether or not the transport limitations alter the selectivities obtained when the chemical reaction is rate controlling.

The following types of complex reactions are the most likely to be encountered:

$$\text{Type 1: } A(g) + B(l) \longrightarrow \text{product}$$
$$A(g) + C(l) \longrightarrow \text{product}$$

$$\text{Type 2: } A(g) + B(l) \longrightarrow \text{product}$$
$$D(g) + B(l) \longrightarrow \text{product}$$

$$\text{Type 3: } A(g) + B(l) \xrightarrow{\ 1\ } R(l)$$

$$A(g) + R(l) \xrightarrow{\ 2\ } S(l)$$

With type 1 reaction systems the concentration profiles of $B$ and $C$ both decrease from the bulk to the interface and no marked selectivity effects can be expected when the transport properties are not greatly dissimilar. The same is true for type 2 reactions. The simultaneous absorption of two gases has been worked out and presented graphically by Goettler and Pigford [14]. Astarita and Gioia analyzed the simultaneous absorption of $H_2S$ and $CO_2$ in NaOH solutions [15].

For type 3 systems $B$ and $R$ have opposite trends, as shown in Fig. 6.3.f-1. In that case the ratio $C_B/C_R$ could change markedly over the film, even for moderate changes in the transport of each species and the selectivity $r_1/r_2 = k_1 C_B/k_2 C_R$ could differ quite a bit from that obtained when the chemical reaction rate is

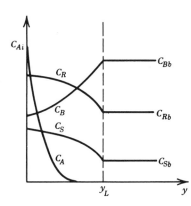

*Figure 6.3.f-1 Type 3 reaction. Typical B and R profiles in the film and bulk.*

controlling. Van de Vusse [16, 17] has discussed the selectivities of type 3 systems with rates $r_1 = k_1 C_A C_B$ and $r_2 = k_2 C_A C_R$, but for fast reactions completed in the film.

The continuity equations for $A$, $B$, and $R$, respectively, may be written, for steady state,

$$D_A \frac{d^2 C_A}{dy^2} = r_1 + r_2$$

$$D_B \frac{d^2 C_B}{dy^2} = r_1$$

$$D_R \frac{d^2 C_R}{dy^2} = -r_1 + r_2$$

with $BC$:

$$C_A = C_{Ai} \qquad \frac{dC_B}{dy} = 0 \qquad \text{and} \qquad \frac{dC_R}{dy} = 0 \text{ at } y = 0$$

$$\frac{dC_A}{dy} = 0 \qquad C_B = C_{Bb} \qquad \text{and} \qquad C_R = C_{Rb} \text{ at } y = y_L$$

The discussion is again in terms of the group $\gamma = \sqrt{k_1 C_{Bb} D}/k_L$ and $C_{Bb}/C_{Ai}$ (Van de Vusse assumed the diffusivities to be equal). When $\gamma$ exceeds 2 (i.e., when the reaction is very fast), gradients of $B$ and $R$ occur in the film when $C_{Bb}/C_{Ai} < \gamma$. Then an effect of mass transfer will be detected, not only on the rate of the global phenomenon, but also on the selectivity. When $\gamma < 0.5$ and $k_1 C_{Bb} < k_L A_v$, the

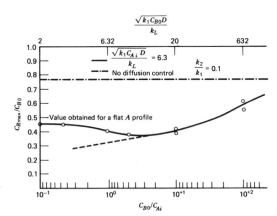

*Figure 6.3.f-2 Type 3 reaction. Influence of $C_{B0}/C_{Ai}$ on selectivity (from Van de Vusse [16]).*

324

CHEMICAL ENGINEERING KINETICS

rate of the global phenomenon corresponds to the true chemical reaction rate; when $k_1 C_{Bb} > k_L A_v$ the rate of the global phenomenon is $k_L A_v C_{Ai}$, which is the rate of mass transfer and in that case there are no gradients of $B$ and $R$. In both these cases there is no change in selectivity with respect to that observed in a homogeneous reaction and determined entirely by the chemical kinetics. Figure 6.3.f-2 illustrates these conclusions quantitatively for certain values of the determining groups.

The values of $k_1$ and $k_2$ are such that in the absence of diffusional limitations the maximum value of $C_R/C_{Bb}$ would be 0.77. This value is found as follows. In a semibatch the ultimate selectivity is an integral value of the instantaneous. For a given $C_{Bb}$ the latter is given by

$$\frac{dC_{Rb}}{dC_{Bb}} = \frac{k_2}{k_1} \frac{C_{Rb}}{C_{Bb}} - 1$$

with boundary conditions $C_{Rb} = 0$ and $C_{Bb} = C_{B0}$ at $t = 0$. Integration of this relation leads to the integral or ultimate selectivity:

$$\left(1 - \frac{k_2}{k_1}\right) \frac{C_{Rb}}{C_{B0}} = \left(\frac{C_{Bb}}{C_{B0}}\right)^{k_2/k_1} - \frac{C_{Bb}}{C_{B0}} \qquad (6.3.f\text{-}1)$$

When $C_{Rb}/C_{B0}$ is plotted versus $C_{B0}$, a maximum is observed, as shown in Fig. 6.3.f-2. The value of $C_{Rb}/C_{B0}$ at this maximum is 0.77. It is seen that for $\sqrt{k C_{Ai} D}/k_L$ = 6.3 the ratio $C_{Rb}/C_{B0}$ is substantially lower for all $C_{B0}/C_{Ai}$ and exhibits a minimum. Only for $C_{B0}/C_{Ai} \gg \gamma$ is the value of 0.77 approached. Extrapolation of the curve to extremely low values of $C_{B0}/C_{Ai}$ is somewhat hazardous, because the boundary condition used by Van de Vusse, $C_{Ab} = 0$, no longer holds for these conditions.

Van de Vusse [16, 17] also performed experiments on the chlorination of $n$-decane, a reaction system of the type considered here, in a semibatch reactor. In such a reactor the chlorine gas is bubbled continuously through a batch of $n$-decane. In some experiments the $n$-decane was pure, in others it was diluted with dichlorobenzene. In some experiments the batch was stirred, in others not. The experimental results could be explained in terms of the above considerations. In all experiments $\gamma \geq 1$ (from 150 to 500), hence the rate of the process was limited by diffusion, but the selectivity was only affected when $C_{B0}/C_{Ai} < \gamma$. This condition was only fulfilled for the experiments in which $n$-decane ($B$) was diluted. For only these experiments were the selectivities in nonstirred conditions found to differ from those with stirring.

Hashimoto et al. [18] considered the same type of reaction, but also accounted for the possibility of reaction in the bulk by setting the boundary conditions at $y = y_L$ as follows: $C_A = C_{Ab}$; $C_B = C_{Bb}$ and $C_R = C_{Rb}$. The order with respect to $A$, the gaseous component, was taken to be zero, that with respect to $B$ and $R$ 1.

This could be encountered in high-pressure oxidation reactions, for example. From typical profiles shown in Fig. 6.3.f-1, it follows that when there are $B$ and $R$ profiles the $R$ selectivity in the film is lower than that in the bulk. In such a case, higher selectivity can be expected when the amount reacting in the bulk is large as compared to that reacting in the film.

The selectivity of $R$ can be written as:

$$\frac{dC_R}{dC_B} = \frac{N_{Rf} + N_{Rb}}{N_{Bf} + N_{Bb}}$$

and with the above boundary conditions this selectivity has to be calculated in two steps. The fluxes in the film are obtained from:

$$N_{Bf} = -D_B \frac{dC_B}{dy}\bigg|_{y=y_L}$$

$$N_{Rf} = -D_R \frac{dC_R}{dy}\bigg|_{y=y_L}$$

and those in the bulk from:

$$N_{Bb} A_v = r_1|_{y=y_L} (1 - A_v y_L)$$
$$N_{Rb} A_v = (-r_1 + r_2)|_{y=y_L} (1 - A_v y_L)$$

The values of $C_{Bb}$ are obtained from the reactor mass balances, as will be shown in Chapter 14 on the design of gas-liquid reactors. Figure 6.3.f-3 shows the effect of the group $(D_A/k_L)\sqrt{k/D_B}$ on the $R$ yield as a function of the conversion of $B$ in a semibatch reactor. When this group is zero (i.e., $k_L \gg k$) the purely chemical yield is obtained. Hashimoto et al. also presented their results in a diagram like that of Fig. 6.3.f-2. Since they accounted for reaction in the bulk, they could

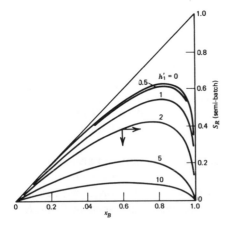

Figure 6.3.f-3 Selectivity for $R$ as a function of conversion of $B$. $S_R = C_{Rb}/C_{B0}$, $h_1' = Parameter$ group $D_A\sqrt{k/D_B}/k_L$ (from Hashimoto, et al. [18]).

CHEMICAL ENGINEERING KINETICS

accurately determine the yield at very low values of $C_{Bb}/C_{Ai}$, in contrast with Van de Vusse.

Derivations were given for reversible, consecutive and parallel reactions with any order by Onda et al. [19, 20, 21, 22]. Onda et al. assumed that the concentrations at $y = y_L$ are the equilibrium values corresponding to the reversible reaction in the bulk. The development was analogous to that of Van Krevelen and Hoftijzer [3] and Hikita and Asai [10]. This led to approximate expressions for the enhancement factor giving values in close agreement with those obtained by numerical integration.

## 6.4 Surface Renewal Theory

In surface renewal models the liquid surface is assumed to consist of a mosaic of elements with different age at the surface. The rate of absorption at the surface is then an average of the rates of absorption in each element, weighted with respect to a distribution function $\psi(t)$—see Eq. 6.2-5. Under this heading of surface renewal theory we will also occasionally mention results of Higbie's [23] so-called "penetration-theory," which can be considered as a special case in which every element is exposed to the gas for the same length of time before being replaced. The main emphasis of this section is on the Danckwerts [24] approach using the distribution function for completely random replacement of surface elements:

$$\psi(t) = se^{-st} \qquad (6.4\text{-}1)$$

By definition of a distribution function it follows that the fraction of the surface having been exposed to the gas for a time between $t$ and $t + dt$ is given by $\psi(t)dt = se^{-st}dt$. Also, since we are dealing with fractions, the distribution is normalized, so that

$$\int_0^\infty \psi(t)dt = 1 \qquad (6.4\text{-}2)$$

Such an age-distribution function would arise when the rate of replacement of the surface elements is proportional to the number of elements or fraction of the surface having that age, $t$:

$$-\frac{d\psi(t)}{dt} = s\psi(t) \qquad (6.4\text{-}3)$$

Integration of Eq. 6.4-3, taking Eq. 6.4-2 into account, leads to Eq. 6.4-1. $s$ is a rate coefficient for surface element replacement and is the parameter of the model. Consequently, with this expression for $\psi(t)$, the average rate of absorption (Eq. 6.2-5) becomes:

$$N_A = \int_0^\infty N_A(t)\psi(t)dt = s \int_0^\infty N_A(t)e^{-st} \, dt \qquad (6.4\text{-}4)$$

Again, as for the two-film theory, analytical integration of the equations is only possible for a few particular cases, especially since the equations are now partial differential equations with respect to position and time. In contrast with what was done for the film theory the instantaneous reaction will be discussed prior to the pseudo-first-order reaction—a more logical sequence to introduce further developments.

### 6.4.a Single Instantaneous Reaction

In contrast with the two-film model the reaction plane is not fixed in space: since the element at the surface is considered to have a finite capacity, transients have to be considered. In the zone between the interface at $y = 0$ and the location of the reaction plane at $y_1(t)$ a non-steady-state balance on $A$ leads to:

$$\frac{\partial C_A}{\partial t} = D_A \frac{\partial^2 C_A}{\partial y^2} \tag{6.4.a-1}$$

In the zone between $y_1(t)$ and infinity:

$$\frac{\partial C_B}{dt} = D_B \frac{\partial^2 C_B}{\partial y^2} \tag{6.4.a-2}$$

with the boundary conditions:

$$
\begin{aligned}
&\text{for } A: t = 0 \quad && y > 0 && C_A = C_{A0} = 0 \text{ in the case considered here} \\
& \quad\quad\quad t > 0 \quad && y = 0 && C_A = C_{Ai} \\
& \quad\quad\quad && y = \infty && C_A = C_{Ab} = C_{A0} = 0 \tag{6.4.a-3} \\
&\text{for } B: t = 0 \quad && y \geq 0 && C_B = C_{B0} \\
& \quad\quad\quad t > 0 \quad && y = \infty && C_B = C_{Bb} = C_{B0}
\end{aligned}
$$

The solution of these equations is well known. It may be obtained by the Laplace transform as

$$C_A = A_1 + A_2 \, \text{erf}\left(\frac{y}{2\sqrt{D_A t}}\right) \tag{6.4.a-4}$$

$$C_B = B_1 + B_2 \, \text{erf}\left(\frac{y}{2\sqrt{D_B t}}\right) \tag{6.4.a-5}$$

where

$$\text{erf}(x) = \frac{2}{\sqrt{\pi}} \int_0^x e^{-\eta^2} \, d\eta$$

Before determining the integration constants by applying the boundary conditions, an inspection of Eq. 6.4.a-4 permits relating the position of the reaction plane, $y_1$, to time. Indeed, in that plane $C_A = 0$ so that necessarily:

$$y_1(t) = 2\beta\sqrt{t} \qquad (6.4.a-6)$$

where $\beta$ is a constant that remains to be determined. Accounting for the boundary conditions, together with Eq. 6.4.a-6 leads to the following expressions for $C_A$ and $C_B$:

$$\frac{C_A}{C_{Ai}} = \frac{\mathrm{erfc}\left(\dfrac{y}{2\sqrt{D_A t}}\right) - \mathrm{erfc}\left(\dfrac{\beta}{\sqrt{D_A}}\right)}{\mathrm{erf}\left(\dfrac{\beta}{\sqrt{D_A}}\right)} \qquad (6.4.a-7)$$

where $\mathrm{erfc}(\eta) = 1 - \mathrm{erf}(\eta)$; for $0 < y < 2\beta\sqrt{t}$

$$\frac{C_B}{C_{B0}} = \frac{\mathrm{erf}\left(\dfrac{y}{2\sqrt{D_B t}}\right) - \mathrm{erf}\left(\dfrac{\beta}{\sqrt{D_A}}\right)}{\mathrm{erfc}\left(\dfrac{\beta}{\sqrt{D_B}}\right)}$$

In the reaction plane $y_1(t) = 2\beta\sqrt{t}$ the stoichiometry requires that $N_A/a = -(N_B/b)$. Writing the fluxes in terms of Fick's law leads to an additional relation that enables $\beta$ to be determined. The result is:

$$e^{\beta^2/D_B}\,\mathrm{erfc}\left(\frac{\beta}{\sqrt{D_B}}\right) = \frac{b}{a}\frac{C_{Bb}}{C_{Ai}}\sqrt{\frac{D_B}{D_A}}\,e^{\beta^2/D_A}\,\mathrm{erf}\left(\frac{\beta}{\sqrt{D_A}}\right) \qquad (6.4.a-8)$$

An example of the evolution of the profiles with time is given in Fig. 6.4.a-1.

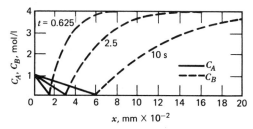

*Figure 6.4.a-1 Location of the reaction plane with time (from Perry and Pigford [44]).*

The flux of $A$ at the interface at any time is obtained by differentiating Eq. 6.4.a-7, as indicated by Eq. 6.3.a-6:

$$N_A(t) = \frac{C_{Ai}}{\text{erf}\left(\frac{\beta}{\sqrt{D_A}}\right)} \sqrt{\frac{D_A}{\pi t}} \tag{6.4.a-9}$$

The average rate of absorption at the surface is, with Higbie's uniform age, $\bar{t}$:

$$N_A = \frac{1}{\bar{t}} \int_0^{\bar{t}} N_A(t)dt = \frac{2C_{Ai}}{\text{erf}\left(\frac{\beta}{\sqrt{D_A}}\right)} \sqrt{\frac{D_A}{\pi \bar{t}}} = \frac{k_L C_{Ai}}{\text{erf}\left(\sqrt{\frac{\beta}{D_A}}\right)} \tag{6.4.a-10}$$

since for purely physical absorption $N_A = 2C_{Ai}\sqrt{D_A/\pi\bar{t}}$, but also $N_A = k_L C_{Ai}$. With Danckwerts' age distribution function Eq. 6.4-1:

$$N_A = \int_0^\infty N_A(t)\Psi(t)dt = \frac{\sqrt{D_A s}C_{Ai}}{\text{erf}\left(\frac{\beta}{\sqrt{D_A}}\right)} = \frac{k_L C_{Ai}}{\text{erf}\left(\frac{\beta}{\sqrt{D_A}}\right)} \tag{6.4.a-11}$$

since for purely physical absorption $N_A = \sqrt{D_A s}C_{Ai} = k_L C_{Ai}$. In this case both results are identical.

Again the results can be expressed in terms of a utilization factor, $\eta_L$ or an enhancement factor, $F_A$. From Eq. 6.4.a-11 it follows immediately that

$$F_A = \frac{1}{\text{erf}\left(\frac{\beta}{\sqrt{D_A}}\right)}$$

When $D_A = D_B$ the enhancement factor is of the same form for both the film and surface renewal models. Indeed, in the film model

$$F_A = 1 + \frac{b}{a}\frac{C_{Bb}}{C_{Ai}}.$$

From Eq. 6.4.a-8 it can be shown that an identical result is obtained for the surface renewal model. The agreement is not surprising for this special case: when $\beta$ is small the rate of displacement of the reaction plane is small, so that steady state is practically realized, as in the film theory. Even when $D_A \neq D_B$ the difference between the film and surface renewal models amounts to only a few percent.

From the definition Eq. 6.3.b-4:

$$\eta_L = \frac{N_A A_v}{kC_{Ai}} = \frac{\sqrt{D_A s}A_v}{\text{erf}\left(\frac{\beta}{\sqrt{D_A}}\right)k} = \frac{k_L A_v}{\text{erf}\left(\frac{\beta}{\sqrt{D_A}}\right)k}, \qquad (D_A = D_B)$$

which can be reduced to the same form as Eq. 6.3.c-4. Finally it should be noted that the calculation of $N_A$ as carried out here required an expression for the concentration profile Eq. 6.4.a-7. With the surface age distribution adopted in the surface renewal model a shortcut may be taken as illustrated in the next section.

### 6.4.b Single Irreversible (Pseudo) First-Order Reaction

The equation governing diffusion, reaction and accumulation of $A$ in a unit volume element of the liquid may be written:

$$\frac{\partial C_A}{\partial t} = D_A \frac{\partial^2 C_A}{\partial y^2} - kC_A \qquad (6.4.b\text{-}1)$$

with boundary conditions:

$$
\begin{array}{lll}
y = 0 & t > 0 & C_A = C_{Ai} \\
y = \infty & t > 0 & C_A = 0 \\
t = 0 & y > 0 & C_A = 0
\end{array}
$$

The first condition expresses that, from a certain time onward, a gas, in which the component $A$ has a partial pressure $p_A$, is brought into contact with the liquid, so that a concentration $C_{Ai}$ is obtained at the interface. The initial concentration of $A$ in the liquid is considered to be zero. Since the exposure time of the element at the surface is rather brief and since its capacity is not considered to be zero the concentration front of $A$ is not likely to extend to the inner edge of the element. This is expressed by the *BC*: for $t > 0$ $C_A = 0$ at $y = \infty$. In the case of a pseudo-first-order reaction $k = ak'C_{Bb}$, of course.

The equation is conveniently integrated by means of Laplace transforms. Transforming with respect to time leads to

$$s\overline{C_A} - C_A(t = 0) = D_A \frac{d^2\overline{C_A}}{dy^2} - k\overline{C_A}$$

or

$$\frac{d^2\overline{C_A}}{dy^2} - \left(\frac{k + s}{D_A}\right)\overline{C_A} = 0$$

The integral of this differential equation is

$$\overline{C_A} = A_1 \exp\left[\sqrt{\frac{k + s}{D_A}}\, y\right] + A_2 \exp\left[-\sqrt{\frac{k + s}{D_A}}\, y\right]$$

The boundary condition $C_A = 0$ for $y = \infty$ requires $A_1$ to be zero. $A_2$ is determined from the boundary condition at $y = 0$:

$$A_2 = \frac{C_{Ai}}{s}$$

so that:

$$\overline{C_A} = \frac{C_{Ai}}{s} \exp\left[-\sqrt{\frac{k+s}{D_A}}\, y\right] \qquad (6.4.b\text{-}2)$$

Finally, $C_A(y, t)$ is obtained by an inverse transformation of Eq. 6.4.b-2, leading to:

$$\frac{C_A}{C_{Ai}} = \frac{1}{2} \exp\left(-y\sqrt{\frac{k}{D_A}}\right) \mathrm{erfc}\left(\frac{y}{2\sqrt{D_A t}} - \sqrt{kt}\right)$$

$$+ \frac{1}{2} \exp\left(y\sqrt{\frac{k}{D_A}}\right) \mathrm{erfc}\left(\frac{y}{2\sqrt{D_A t}} + \sqrt{kt}\right) \qquad (6.4.b\text{-}3)$$

For this solution, see Carslaw and Jaeger [25]. For large values of $kt$:

$$\frac{C_A}{C_{Ai}} = \exp\left[-y\sqrt{\frac{k}{D_A}}\right]$$

since erf $x$ tends to 1 for large $x$ and tends to zero with $x$. Consequently, for sufficiently large times the concentration profiles do not change any more—they have attained the steady state.

At time $t$, the instantaneous rate of absorption $N_A(t)$ in an element having a surface age $t$ is given by

$$N_A(t) = \sqrt{kD_A}\, C_{Ai}\left[\mathrm{erf}(\sqrt{kt}) + \frac{e^{-kt}}{\sqrt{\pi kt}}\right] \qquad (6.4.b\text{-}4)$$

The elements have a distribution of residence times at the surface. The rate that would be observed, at any instant, over a unit surface would be an average

$$N_A = \int_0^\infty N_A(t)\psi(t)dt$$

With Higbie's distribution function all elements at the surface have the same age. Such a situation could be encountered with a quiescent liquid or with completely laminar flow. In that case $N_A$ is simply given by Eq. 6.4.b-4 in which $t$ takes a definite value $\bar{t}$, the uniform time of exposure. With Danckwert's age distribution function Eq. 6.4-1 the average rate of absorption per unit surface, $N_A$, is given by:

$$N_A = \int_0^\infty N_A(t)se^{-st}\, dt = \sqrt{D_A(k+s)}\, C_{Ai} \qquad (6.4.b\text{-}5)$$

Note that the rate of absorption is proportional to $\sqrt{D_A}$, whereas the film theory predicts $N_A \sim D_A$. Equations 6.4.b-5 and 6.4.a-11 for instantaneous reaction were first derived by Danckwerts [26].

The parameter $s$ can be related to the transfer coefficient $k_L$ used in the film model and to the diffusivity in the following way:

In the absence of reaction Eq. 6.4.b-5 reduces to

$$N_A = \sqrt{D_A s}\, C_{Ai}$$

In terms of a transfer coefficient $N_A = k_L C_{Ai}$, so that

$$k_L{}^2 = D_A s \qquad (6.4.\text{b-6})$$

Equation 6.4.b-5 now becomes:

$$N_A = \sqrt{D_A\left(k + \frac{k_L{}^2}{D_A}\right)}\, C_{Ai} = k_L C_{Ai}\sqrt{1 + D_A\frac{k}{k_L{}^2}}$$

or

$$N_A = F_A k_L C_{Ai} \quad \text{where} \quad F_A = \sqrt{1 + D_A\frac{k}{k_L{}^2}} = \sqrt{1 + \gamma^2} \qquad (6.4.\text{b-7})$$

Again the rate of absorption has been expressed as the product of the physical absorption rate and an enhancement factor, $F_A$. The enhancement factor derived from Higbie's result Eq. 6.4.b-5 is easily found to be:

$$F_A = \gamma\left[\left(1 + \frac{\pi}{8\gamma^2}\right)\text{erf}\left(\frac{2}{\sqrt{\pi}}\gamma\right) + \frac{1}{2\gamma}\exp\left(-\frac{4}{\pi}\gamma^2\right)\right] \qquad (6.4.\text{b-8})$$

The three expressions Eq. 6.4.b-7, Eq. 6.4.b-8, and the corresponding Eq. 6.3.b-11 for the film theory look quite different. Yet they lead to identical results when $\gamma \to 0$ and $\gamma \to \infty$, while they differ only by a few percent for intermediate values of $\gamma$. This is illustrated in Table 6.4.b-1 (see Beek [27]) for the film and surface

Table 6.4.b-1 Comparison of model prediction for pseudo-first-order reaction (after Beek [27]).

| $\gamma$ | Film | Surface renewal | Penetration |
|---|---|---|---|
|  | $F_A$ for $C_{Ab} = 0$ | | |
|  | $\dfrac{\gamma}{\tanh \gamma}$ | $\sqrt{1+\gamma^2}$ | $\gamma\left(1 + \dfrac{\pi}{8\gamma^2}\right)\text{erf}\left(\dfrac{2}{\sqrt{\pi}}\gamma\right) + \dfrac{1}{2\gamma}\exp\left(-\dfrac{4}{\pi}\gamma^2\right)$ |
| 0.01 | 1.00 | 1.00 | 0.94 |
| 0.1 | 1.00 | 1.00 | 1.005 |
| 0.3 | 1.04 | 1.04 | 1.035 |
| 1 | 1.31 | 1.41 | 1.37 |
| 3 | 3.02 | 3.16 | 3.12 |
| 10 | 10.00 | 10.05 | 10.39 |

renewal theory of Danckwerts. The utilization factor is given by:

$$\eta_L = \frac{A_v\sqrt{D_A(k + s)}}{k}$$

which is very similar to $\eta_L = A_v\sqrt{D_A/k}$ derived from the film theory when $\gamma$ is large.

In general, for practical applications one is less interested in the concentration profiles near the interface and the rate of absorption in an element having a surface age $t$, $N_A(t)$. What matters primarily is the flux over the total surface, $N_A$. As mentioned already in Sec. 6.4.a a short cut can be taken to obtain $N_A$ when the Danckwerts distribution is adopted, which avoids the difficult inversion of the transform. Indeed,

$$N_A = s \int_0^\infty N_A(t)e^{-st}\, dt = s \int_0^\infty \left[ -D_A \frac{\partial C_A}{\partial y} \right]_{y=0} e^{-st}\, dt$$

$$= -D_A s \frac{d}{dy} \left[ \int_0^\infty C_A e^{-st}\, dt \right]_{y=0} = -D_A s \frac{d\overline{C}_A}{dy} \bigg|_{y=0}$$

where $\overline{C}_A$ is the Laplace transform of $C_A$. Therefore, $N_A$ can be obtained directly by differentiating the Laplace transform with respect to time of the original differential equation—in this case Eq. 6.4.b-1.

The surface renewal models only consider the liquid phase. In Sec. 6.3 on the film model the resistances of both gas and liquid phase were combined into one single expression like Eq. 6.3.b-5. The same can be done here: Danckwerts [24] has shown that in most cases the surface renewal models combined with a gas side resistance lead to the same rules for the addition of resistances as the two-film theory.

### 6.4.c Surface Renewal Models with Surface Elements of Limited Thickness

One feature of the surface renewal model that may not be realistic is that the elements at the surface extend to infinity, as expressed by the boundary condition

$$t > 0 \qquad y = \infty \qquad C_A = 0 \qquad \text{or} \qquad \frac{\partial C_A}{\partial y} = 0$$

As previously mentioned, this arises from the consideration that the residence time of a surface element at the interface is very short, so that it is likely that $A$ has never penetrated to the inner edge of the element before it is replaced. Models

_____

that limit the depth of the surface element have also been proposed, and applied to purely physical mass transfer first—such as the surface rejuvenation model of Danckwerts [28] and the film-penetration model of Toor and Marchello [29]. These were later extended to mass transfer with reaction. More recently Harriott [30] and Bullin and Dukler [31] extended these models by assuming that eddies arriving at random times come to within random distances from the interface. This leads to a stochastic formulation of the surface renewal.

The price that is paid for the greater generality of the models is twofold, however. First, there is the need for two parameters: one expressing the surface renewal and one expressing the thickness of the element. Second, there is the mathematical complexity of the expression for the flux, $N_A$. Is the price worth paying? This question can be partly answered by means of Huang and Kuo's application of the film-penetration model to first-order reactions, both irreversible and reversible [32, 12].

The differential equation is that of Eq. 6.4.b-1, but the boundary condition at $y = y_L$ is now as follows:

$$\text{at } y = y_L \qquad t > 0 \qquad C_A = C_{Ab} \qquad C_B = C_{Bb}$$

For first-order irreversible reactions and Danckwerts' residence time distribution Huang and Kuo derived two solutions: one for long exposure times that expresses the concentration gradients in trigonometric function series and the following solution for rather short exposure times, obtained by Laplace transforms:

$$N_A = \sqrt{D_A(k + s)}C_{Ai}\left[\left(1 - \frac{s\dfrac{C_{Ab}}{C_{Ai}}}{k + s}\right)\coth\sqrt{\frac{k + s}{D_A}}y_L - \frac{C_{Ab}}{C_{Ai}}\frac{k}{k + s}\operatorname{csch}\sqrt{\frac{k + s}{D_A}}y_L\right]$$

$$(6.4.c\text{-}1)$$

The difference in the numerical values predicted from Eq. 6.4.c-1 and the film and the simple surface renewal model turns out to be negligible.

Huang and Kuo also solved two equations for a rapid first-order reversible reaction (i.e., equilibrium in the bulk liquid). The solutions are extremely lengthy and will not be given here. From a comparison of the film, surface renewal, and intermediate film-penetration theories it was found that for irreversible and reversible reactions with equal diffusivities of reactant and product, the enhancement factor was insensitive to the mass transfer model. For reversible reactions with product diffusivity smaller than that of the reactant, the enhancement factor can differ by a factor of two between the extremes of film and surface renewal theory. To conclude, it would seem that the choice of the model matters little for design calculations: the predicted differences are negligible with respect to the uncertainties of prediction of some of the model or operation parameters.

## 6.5 Experimental Determination of the Kinetics of Gas-Liquid Reactions

The approach to be followed in the determination of rates or detailed kinetics of the reaction in a liquid phase between a component of a gas and a component of the liquid is, in principle, the same as that outlined in Chapter 2 for gas-phase reactions on a solid catalyst. In general the experiments are carried out in flow reactors of the integral type. The data may be analyzed by the integral or the differential method of kinetic analysis. The continuity equations for the components, which contain the rate equations, of course depend on the type of reactor used in the experimental study. These continuity equations will be discussed in detail in the appropriate chapters, in particular Chapter 14 on multiphase flow reactors. Consider for the time being, by way of example, a tubular type of reactor with the gas and liquid in a perfectly ordered flow, called plug flow. The steady-state continuity equation for the component $A$ of the gas, written in terms of partial pressure over a volume element $dV$ and neglecting any variation in the total molar flow rate of the gas is as follows:

$$-\frac{F}{p_t}dp_A = N_A A_v \, dV \tag{6.5-1}$$

or, after integration,

$$-\int_{(p_A)_{in}}^{(p_A)_{out}} \frac{dp_A}{N_A A_v} = \frac{p_t V}{F} \tag{6.5-2}$$

where $N_A$ is the rate of the global phenomenon consisting of mass transfer and chemical reaction and $V$ the total reactor volume. In the case of a pseudo-first-order reaction, for example, $N_A$ is given by Eq. 6.3.b-5 when the film theory is adopted.

The integral method of kinetic analysis can be conveniently used when the expression for $N_A$ can be analytically integrated. When the differential method is applied, $N_A A_v$ is obtained as the slope of a curve giving $(p_A)_{in} - (p_A)_{out}$ as a function of $p_t V/F$, arrived at by measuring the amount of $A$ absorbed at different gas flow rates.

From the preceding sections it follows that the global rate of reaction contains several parameters: $k$, $k_L$, $k_G$, and $D_A$, while in many cases, $A_v$, which depends on the equipment and the operating conditions, also has to be determined. As advised already for gas-phase reactions catalyzed by solids, when the true chemical rate is to be measured efforts should be undertaken to eliminate mass transfer limitations and vice versa. If this turns out to be impossible the dependence of the global rate on the factors determining the mass transfer—the liquid and gas flow rates, or the agitation—has to be investigated over a sufficient range, since these are the elements that will vary when extrapolating to other sizes or types of equipment. Except when reliable correlations are available or when use is made of special equipment, to be discussed below, special attention has to be given to the specific interfaçial area, $A_v$. Physical absorption experiments only allow the products $k_G A_v$ and $k_L A_v$ to be measured. Experiments involving both mass transfer and reaction permit $A_v$ to be determined separately. Use is made for this purpose of a fast pseudo-first-order reaction with known kinetics and that makes $N_A$ independent of $k_L$, such as the reaction between $CO_2$ and a carbonate-bicarbonate buffer containing arsenite (see Sharma and Danckwerts and Roberts and Danckwerts [33, 34]) or between $CO_2$ and aqueous amines (see Sharma [35]). If $\gamma$ exceeds 3, $N_A A_v = A_v C_{Ai} \sqrt{kD_A}$ as obtained in Sec. 6.3.b, so that the measurement of $N_A$ for known $C_{Ai}$ and $kD_A$ yields $A_v$. The experiments are devised in such way that there is no gas side resistance (e.g., by using pure $CO_2$, or having sufficient turbulence) and a large excess of $A$ so that the gas phase composition is practically unchanged and $C_{Ai}$ is constant.

When a physical mass transfer experiment is carried out in the same equipment $k_L A_v$ is obtained, so that both $k_L$ and $A_v$ are known. For this purpose it is often preferable to exclusively use experiments involving mass transfer and reaction. This eliminates the problems associated with coming close to gas-liquid equilibrium and with nonideal flow patterns. $k_L A_v$ can be obtained by using an instantaneous reaction in the liquid so that, according to the film theory,

$$N_A A_v = k_L A_v C_{Ai}\left(1 + \frac{a}{b}\frac{D_B}{D_A}\frac{C_{Bb}}{C_{Ai}}\right)$$

Instantaneous reactions include the absorption of $NH_3$ in $H_2SO_4$, of $SO_2$ or $Cl_2$ or HCl in alkali-solutions and of $H_2S$ and HCl in amine solutions. Again gas side resistance is eliminated, generally by using undiluted gas and $C_{Ai}$ is kept constant.

Another possibility is to use a pseudo-first-order reaction, rather slow so that little $A$ reacts in the film, yet sufficiently fast to make $C_{Ab}$ zero. This approach has been used by Danckwerts et al. [36] who interpreted their results in terms of the surface renewal theory. The system they investigated was $CO_2$ absorption in $CO_3^{--}/HCO_3^{-}$-buffers of different compositions. This is a pseudo-first-order reaction for which, the surface renewal model leads to the following rate of absorption, Eq. 6.4.b-5: $N_A A_v = A_v\sqrt{D_A(k + s)}C_{Ai}$. Danckwerts et al. plotted $(N_A A_v/C_{Ai})^2$ versus the different values of $k$ corresponding to the different compositions of the buffer. This led to a straight line with slope $D_A A_v^2$ and intercept $D_A s A_v^2$, from which $A_v$ and $s$ were obtained or $A_v$ and $k_L$ since $k_L = \sqrt{D_A s}$.

If $k_G A_v$ is needed, an instantaneous reaction is convenient. As shown in Sec. 6.3.c when $C_{Bb} > C'_{Bb}$ the reaction is confined to the interface and $N_A A_v = k_G A_v p_{Ab}$. $k_G$ and $A_v$ can be determined separately by means of a rapid reaction, so that $C_{Ab} = 0$. Then, as shown in Sec. 6.3.b:

$$N_A A_v = \frac{p_{Ab}}{\dfrac{1}{k_G A_v} + \dfrac{H}{k_L A_v \gamma}}$$

By plotting

$$\frac{p_{Ab}}{N_A A_v} \quad \text{versus} \quad \frac{H}{\gamma k_L} \quad \text{or} \quad \frac{H}{\sqrt{kD_A}}$$

the intercept is $1/k_G A_v$ and the slope $A_v$. Sharma and Danckwerts [5] have discussed the above methods—and others—and provide valuable quantitative information on the different chemical systems.

As previously mentioned, when the rate coefficient of the reaction has to be determined it is recommended to eliminate mass transfer effects as much as possible. Also, to get rid of the problem of the interfacial area, specific equipment with known $A_v$ has been devised. The wetted wall column was used in early studies to determine the kinetics of the reaction itself. Care has to be taken to have a laminar film (Re < 250–400) and to avoid ripples that increase the interfacial area. In a film flowing down a vertical tube of diameter $d_t$ the velocity $u$ at any depth $y$ from the interface is given by:

$$u = \frac{3}{2}\left(\frac{L}{\pi d_t}\right)^{2/3}\left(\frac{g\rho_L}{3\mu}\right)^{1/3}\left[1 - y^2\left(\frac{\pi g d_t \rho_L}{3\mu L}\right)^{2/3}\right]$$

where $L$ is the liquid flow rate (m$^3$/hr). Since at the wall $u = 0$, the film thickness is $\delta = (3\mu L/\pi g d_t \rho_L)^{1/3}$ and the liquid velocity at the surface equals

$$\frac{3}{2}\left(\frac{L}{\pi d_t}\right)^{2/3}\left(\frac{g\rho_L}{3\mu}\right)^{1/3}$$

In classical versions both the gas and the liquid generally flow, countercurrently. Equations 6.5-1 or 6.5-2 may then serve for the data treatment. In modern versions such as those shown in Fig. 6.5-1, only the liquid flow and the amount of gas absorbed as a function of time is followed by means of a gas buret and a soap film meter. From the lowering of the soap meniscus the amount of $A$ that is absorbed may be calculated. Dividing this amount by the elapsed time yields the rate of absorption. Looking now at the jet, with its known $A_v$, and contact time between gas and liquid, $\bar{t}$, an amount is absorbed: $Q = \int_0^{\bar{t}} N_A(t)dt$. The average rate of absorption is $Q/\bar{t}$. This is exactly the quantity measured by the gas buret and soap film meter, so that $N_A = (1/\bar{t}) \int_0^{\bar{t}} N_A(t)dt$ is known. The contact time is calculated from $u_s$ and the height, $Z$. This equipment was used by Roberts and Danckwerts [34].

Another equipment frequently used for rapid reactions is the laminar jet in which the liquid velocity is uniform, so that the contact time is nothing but the height/velocity. The contact time can be varied from 0.001 sec to 0.1 sec by varying

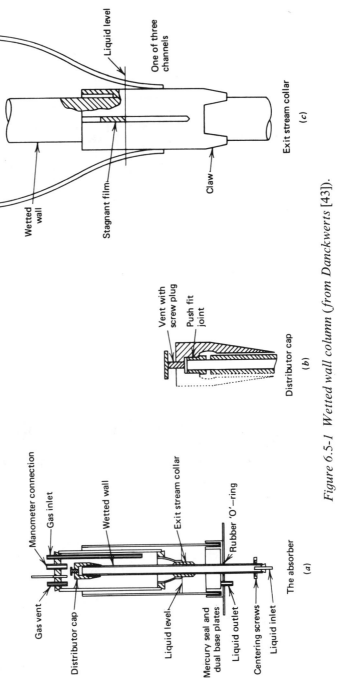

*Figure 6.5-1  Wetted wall column (from Danckwerts [43]).*

**(c)**

Liquid level

One of three channels

Wetted wall

Stagnant film

Exit stream collar

Claw

**(b)**

Vent with screw plug

Push fit joint

Distributor cap

**(a)**

Manometer connection

Gas inlet

Wetted wall

Exit stream collar

Rubber 'O'—ring

Gas vent

Distributor cap

Liquid level

Mercury seal and dual base plates

Liquid outlet

Centering screws

Liquid inlet

The absorber

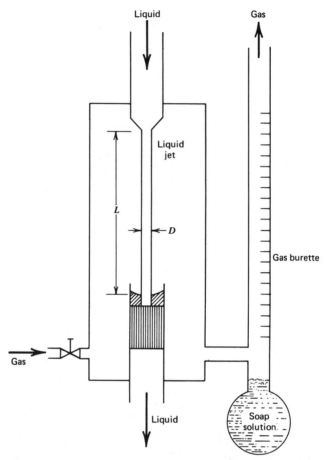

*Figure 6.5-2 Laminar jet with soap-film gas flow meter (after Beek [27]).*

the liquid rate from the jet. Such equipment, an example of which is shown in Fig. 6.5-2, has been used by Nysing et al. [37] and Sharma and Danckwerts [33].

Danckwerts and Kennedy [38] have used the rotating drum shown schematically in Fig. 6.5-3. It has been devised to expose a liquid flowing over a known surface of the rotating drum for a given time to a gas. The contact times can be varied between 0.01 and 0.25 sec. The construction is more complicated than that of the wetted wall and jet equipment.

Danckwerts and co-workers have interpreted the data in terms of contact or exposure time and Higbie's penetration theory as follows. For a pseudo-first-order reaction $N_A(t)$ is given by Eq. 6.4.b-4, the amount absorbed during the

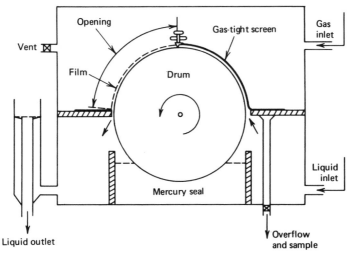

*Figure 6.5-3 Rotating drum (after Danckwerts and Kennedy [38], from Danckwerts [43]).*

contact time $\bar{t}$ per unit surface by:

$$Q = \int_0^{\bar{t}} N_A(t)dt = C_{Ai}\sqrt{\frac{D_A}{k'C_{Bb}}}\left[\left(k'C_{Bb}\bar{t} + \frac{1}{2}\right)\text{erf}(\sqrt{k'C_{Bb}\bar{t}})\right.$$

$$\left. + \sqrt{\frac{k'C_{Bb}\bar{t}}{\pi}}\,\exp(-k'C_{Bb}\bar{t})\right] \tag{6.5-3}$$

where the contact surface is known in this case. The average rate of absorption is $Q/\bar{t} = \int_0^{\bar{t}} N_A(t)dt/\bar{t}$. For short contact times ($k'C_{Bb}\bar{t} < 0.5$) expansion of erf and exp and neglecting higher orders of $k'\bar{t}$ leads to:

$$N_A = \frac{Q}{\bar{t}} = 2C_{Ai}\sqrt{\frac{D_A}{\pi\bar{t}}}\left(1 + \frac{k'C_{Bb}\bar{t}}{3}\right) \tag{6.5-4}$$

For long times $k'C_{Bb}\bar{t} > 2$ the error function goes to one so that

$$N_A = \frac{Q}{t} = C_{Ai}\sqrt{D_A k'C_{Bb}}\left(1 + \frac{1}{2k'C_{Bb}\bar{t}}\right) \tag{6.5-5}$$

By plotting $N_A\sqrt{\bar{t}}$ versus $\bar{t}$, as is obvious from Eq. 6.5-4, $2C_{Ai}\sqrt{D_A/\pi}$ is obtained as intercept. Plotting $N_A$ versus $1/2k'C_{Bb}\bar{t}$ yields $C_{Ai}\sqrt{D_A k'C_{Bb}}$ as an intercept, so that $k'$ and $C_{Ai}$ are obtained. An illustration of this method is given in Sharma and Danckwerts' study of $CO_2$ absorption in a liquid jet [33].

GAS-LIQUID REACTIONS ———————————————— **341**

# Problems

6.1 Derive the rate equation for a reversible first-order gas-liquid reaction

$$A \; \rightleftharpoons \; P$$

using the film theory ($D_A = D_P$).

$$\text{B.C.:} \quad y = 0 \quad C_A = C_{Ai}$$

$$\frac{dC_P}{dy} = 0$$

$$y = y_1 \quad C_A = C_{Ab}$$

$$C_P = KC_{Ab}$$

Show that

$$N_A = \frac{k_L(C_{Ai} - C_{Ab})(1 + K)}{1 + K \dfrac{\tanh \gamma'}{\gamma'}}$$

where

$$\gamma' = \sqrt{\frac{k_1(1 + K)}{KD_A}}$$

6.2 Derive the integral selectivity equation (6.3.f-1).

6.3 A gas is being absorbed into a liquid in which the concentration of the reactive component $B$ is 0.1 $M$. The reaction between the gaseous component $A$ and the component $B$ is extremely fast. The conditions are such that $C_{Ai} = 0.1\ M$. Furthermore, $D_A = 10^{-5}\ cm^2/s$. Compare the enhancement factors based on the film theory and the surface renewal theory for the cases that (a) $D_B = D_A$, (b) $D_B = \frac{1}{2}D_A$, and (c) $D_B = 2D_A$.

6.4 Consider the absorption of gaseous $CO_2$ by a NaOH solution. The stoichiometry is as follows:

$$CO_2 + 2\,NaOH = Na_2CO_3 + H_2O$$

Consider the solubility of $CO_2$ to be independent of the NaOH concentration and let the diffusivities of $CO_2$ and NaOH in the liquid be approximately equal.
(a) Can the reaction be considered as being of the pseudo-first-order when the gas-liquid contact-time is 0.01 s and when
  (i) the partial pressure of $CO_2$ is 0.1 bar and the concentration of NaOH 1 mol/l?
  (ii) the partial pressure of $CO_2$ is 1 bar and the concentration of NaOH 1 mol/l?
(b) When the gas-liquid contact time is 0.1 s and the NaOH concentration is 3 mol/l, what is the partial pressure of $CO_2$ above which the reaction is no longer pseudo-first-order?
Take $k' = 10^7\ cm^3/mol$ s and $H = 25.10^{-3}\ cm^3$ bar/mol.

6.5 $CO_2$ is absorbed at 25°C into a 2.5 $M$ monoethanolamine solution in a rotating drum type of absorber. The contacting surface is 188.5 $cm^2$ and the contact time 0.2 s. The

342 

partial pressure of $CO_2$ in the gas phase is 0.1 atm. The reaction is as follows:

$$CO_2 + 2R_2NH \longrightarrow R_2NCOO^- + R_2NH_2^+$$

The rate of absorption at these conditions is found to be $3.26 \times 10^{-4}$ mol/s. What is the value of the rate coefficient neglecting the gas phase resistance and considering the reaction to obey pseudo-first-order behavior?

Additional data are $D_A = 1.4 \times 10^{-5}$ cm$^2$/s; $D_B = 0.77 \times 10^{-5}$ cm$^2$/s; Henry's constant, $H = 29.8 \times 10^3$ atm cm$^3$/mol.

# References

[1] Whitman, W. G. *Chem. & Met. Eng.*, **29**, 147 (1923).

[2] Lewis, W. K. and Whitman, W. G. *Ind. Eng. Chem.*, **16**, 1215 (1924).

[3] Van Krevelen, D. W. and Holtijzer, P. J. *Rec. Trav. Chim. Pays-Bas*, **67**, 563 (1948).

[4] Sherwood, T. K. and Wei, J. *A.I.Ch.E. J.*, **1**, 522 (1955).

[5] Sharma, M. M. and Danckwerts, P. V. *Brit. Chem. Eng.*, **15**, 522 (1970).

[6] Barona, N. *Proc. 20th Anniv. Dept. Chem. Eng.*, University of Houston (1973).

[7] Porter, K. E. *Trans. Instn. Chem. Engrs.*, **44**, T25 (1966).

[8] Kishinevskii, M. K., Kormenko, T. S., and Popat, T. M. *Theor. Found. Chem. Engng.*, **4**, 641 (1971).

[9] Alper, E. *Chem. Eng. Sci.*, **28**, 2092 (1973).

[10] Hikita, H. and Asai, S. *Int. Chem. Engng.*, **4**, 332 (1964).

[11] Danckwerts, P. V. and Kennedy, A. M. *Trans. Instn. Chem. Engrs.*, **32**, S49 (1954).

[12] Huang, C. J. and Kuo, C. H. *A.I.Ch.E. J.*, **11**, 901 (1965).

[13] Olander, D. R. *A.I.Ch.E. J.*, **6**, 233 (1960).

[14] Goettler, L. A. and Pigford, R. L. *Paper 25e*, 57th Ann. Meeting of A.I.Ch.E. (1964).

[15] Astarita, G. and Gioia, F. *Ind. Eng. Chem. Fund.*, **4**, 317 (1965).

[16] Van de Vusse, J. G. *Chem. Eng. Sci.*, **21**, 631 (1966).

[17] Van de Vusse, J. G. *Chem. Eng. Sci.*, **21**, 645 (1966).

[18] Hashimoto, K., Teramoto, M., Nagayasu, T., and Nagata, S. *J. Chem. Eng. Japan*, **1**, 132 (1968).

[19] Onda, K., Sada, E., Kobayashi, T., and Fujine, M. *Chem. Eng. Sci.*, **25**, 753 (1970).

[20] Ibid. *Chem. Eng. Sci.*, **25**, 761 (1970).

[21] Ibid. *Chem. Eng. Sci.*, **25**, 1023 (1970).

[22] Ibid. *Chem. Eng. Sci.*, **27**, 247 (1972).

[23] Higbie, R. *Trans. Am. Instn. Chem. Engrs.*, **31**, 365 (1935).

[24] Danckwerts, P. V. *Ind. Eng. Chem.*, **43**, 1460 (1951).

[25] Carslaw, H. S. and Jaeger, J. C. *Conduction of Heat in Solids*, Oxford University Press, 2nd ed., London (1959).

[26] Danckwerts, P. V. *Trans. Farad. Soc.*, **46**, 300 (1950).

[27] Beek, W. J. *Stofoverdracht met en zonder Chemische Reaktie.* Notes, University of Delft (1968).

[28] Danckwerts, P. V. *A.I.Ch.E. J.*, **1**, 456 (1955).

[29] Toor, H. L. and Marchello, J. M. *A.I.Ch.E. J.*, **4**, 98 (1958).

[30] Harriott, P. *Chem. Eng. Sci.*, **17**, 149 (1962).

[31] Bullin, J. A. and Dukler, A. E. *Chem. Eng. Sci.*, **27**, 439 (1972).

[32] Huang, C. J. and Kuo, C. H. *A.I.Ch.E. J.*, **9**, 161 (1963).

[33] Sharma, M. M. and Danckwerts, P. V. *Chem. Eng. Sci.*, **18**, 729 (1963).

[34] Roberts, D. and Danckwerts, P. V. *Chem. Eng. Sci.*, **17**, 961 (1962).

[35] Sharma, M. M. *Trans. Far. Soc.*, **61**, 681 (1965).

[36] Danckwerts, P. V., Kennedy, A. M., and Roberts, D. *Chem. Eng. Sci.*, **18**, 63 (1963).

[37] Nysing, R. A. T. O., Hendricksz, R. H., and Kramers, H. *Chem. Eng. Sci.*, **10**, 88 (1959).

[38] Danckwerts, P. V. and Kennedy, A. M. *Trans. Inst. Chem. Engrs.*, **32**, S53 (1954).

[39] Sherwood, T. K. and Pigford, R. L. *Absorption and Extraction*, McGraw-Hill, New York (1952).

[40] Ramm, T. *Absorptionsprozesse in der Chemischen Technik*, VEB Verlag, Berlin (1953).

[41] Astarita, G. *Mass Transfer with Chemical Reaction*, Elsevier, Amsterdam (1967).

[42] Kramers, H. and Westerterp, K. R. *Elements of Chemical Reactor Design and Operation*, Academic Press, New York (1963).

[43] Danckwerts, P. V., *Gas-Liquid Reactions*, McGraw-Hill, New York (1970).

[44] Perry, R. H. and Pigford, R. L., *Ind. Eng. Chem.*, **45**, 1247 (1953).

# Part Two

ANALYSIS
AND
DESIGN
OF
CHEMICAL
REACTORS

# 7

## THE FUNDAMENTAL MASS, ENERGY, AND MOMENTUM BALANCE EQUATIONS

## 7.1 Introduction

The number of types of reactors is very large in the chemical industry. Even for the same operation, such as nitration of toluene, different types are used: the batch reactor, the continuous stirred tank, and a cascade of stirred tanks. Flow reactors of the tubular type are used for such widely different processes as the nitration of glycerine, the sulfonation of aromatics, or gas phase reactions like thermal cracking or the nitration of paraffins. Flow reactors with fixed bed of catalyst particles are used in the ammonia or methanol syntheses and in the oxidation of xylene into phthalic anhydride. A series of such fixed bed reactors is used in $SO_3$ synthesis or in hydrocarbon reforming. Reactors with fluidized or moving beds are used for cracking hydrocarbons, for oxidizing naphthalene or for oxychlorinating ethylene.

The modeling of chemical reactors, as it is conceived in the following chapters, is not based on the external form of the apparatus nor on the reaction taking place in it, nor even on the nature of the medium—homogeneous or not. Focusing on the *phenomena* taking place in the reactor reduces the apparent diversity into a small number of models or basic reactor types. The phenomena occurring in a reactor may be broken down to reaction, transfer of mass, heat, and momentum. The modeling and design of reactors is therefore based on the equations describing these phenomena: the reaction rate equation, the continuity, energy, and momentum equations. The form and complexity of these equations will now be discussed, for introductory and orienting purposes, in general terms. The equations themselves are derived in later sections of this chapter.

## 7.1.a The Continuity Equations

The first step toward the answer to what the conversion of $A$ in the reactor will be consists of applying the law of conservation of mass on a volume-element of the reactor, fixed in space:

$$\begin{bmatrix} \text{Amount of } A \\ \text{introduced} \\ \text{per unit time} \end{bmatrix} - \begin{bmatrix} \text{Amount of } A \\ \text{leaving per} \\ \text{unit time} \end{bmatrix} - \begin{bmatrix} \text{Amount of } A \\ \text{converted} \\ \text{per unit time} \end{bmatrix} = \begin{bmatrix} \text{Amount of } A \\ \text{accumulated} \\ \text{per unit time} \end{bmatrix}$$
$$\quad\;\; \text{I} \qquad\qquad\qquad \text{II} \qquad\qquad\qquad \text{III} \qquad\qquad\qquad \text{IV}$$

$$(7.1.a-1)$$

In mathematical terms Eq. 7.1.a-1 is nothing but the so-called continuity equation for $A$. If $A$ reacts in more than one phase then such an equation is needed for each of these phases.

The mechanisms by which $A$ can enter or leave the volume element considered are: flow and—for those cases where the concentration is not uniform in the reactor —molecular diffusion, in practice generally of minor importance, however. The motion of a fluid, even through empty pipes, is seldom ordered and is difficult to describe. Even if the true detailed flow pattern were known the continuity equation would be so complicated that its integration would be impossible. The crossing of different streamlines, and mixing of fluid elements with different characteristics that result from this crossing, are difficult points in the design of chemical reactors. It is therefore natural to consider, for a first approach, two extreme cases: a first where there is *no* mixing of the streamlines, a second where the mixing is complete. These two extremes may be visualized with sufficient approximation by the tubular reactor with plug flow and continuous flow stirred tank with complete mixing.

In a plug flow reactor all fluid elements move along parallel streamlines with equal velocity. The plug flow is the only mechanism for mass transport and there is no mixing between fluid elements. The reaction therefore only leads to a concentration gradient in the axial flow direction. For steady-state conditions, for which the term IV is zero the continuity equation is a first-order, ordinary differential equation with the axial coordinate as variable. For non-steady-state conditions the continuity equation is a partial differential equation with axial coordinate and time as variables. Narrow and long tubular reactors closely satisfy the conditions for plug flow when the viscosity of the fluid is low.

Reactors with complete mixing may be subdivided into batch and continuous types. In a batch type reactor with complete mixing the composition is uniform throughout the reactor. Consequently, the continuity equation may be written for the entire contents, not only over a volume element. The composition varies with time, however, so that a first-order ordinary differential equation is obtained, with time as variable. The form of this equation is analogous with that for the plug flow case. In the continuous flow type, an entering fluid element is instantaneously mixed with the contents of the reactor so that it loses its identity. This type also operates at a uniform concentration level. In the steady state, the continuity equations are algebraic equations.

Both types of continuous reactors that were considered here are idealized cases. They are important cases, however, since they are easy to calculate and they give the extreme values of the conversions between which those realized in a real reactor will occur—provided there is no bypassing in this reactor. The design of a real reactor, with its intermediate level of mixing, requires information about this mixing. The mixing manifests itself at the outlet of the reactor by a spread or distribution in residence-time (the length of time spent in the reactor) between fluid elements. Such a distribution is relatively easy to measure. The resulting information may then be used as such in the design or used with a model for the real behavior of the reactor. The design of nonideal cases along both lines of approach is discussed in Chapter 12.

### 7.1.b The Energy Equation

In an energy balance over a volume element of a chemical reactor, kinetic, potential, and work terms may usually be neglected relative to the heat of reaction and other heat transfer terms so that the balance reduces to:

$$
\begin{bmatrix} \text{Amount of heat} \\ \text{added} \\ \text{per unit time} \end{bmatrix} - \begin{bmatrix} \text{Amount of heat} \\ \text{out} \\ \text{per unit time} \end{bmatrix} - \begin{bmatrix} \text{Heat effect of} \\ \text{the reaction} \\ \text{per unit time} \end{bmatrix} = \begin{bmatrix} \text{Variation of} \\ \text{heat content} \\ \text{per unit time} \end{bmatrix}
$$

$$\text{I} \qquad\qquad\qquad \text{II} \qquad\qquad\qquad \text{III} \qquad\qquad\qquad \text{IV}$$

$$(7.1.b\text{-}1)$$

The mathematical expression for Eq. 7.1.b-1 is generally called the energy equation, and its integrated form the heat balance. The form of these equations results from considerations closely related to those for the different types of continuity equations. When the mixing is so intense that the concentration is uniform over the reactor, it may be accepted that the temperature is also uniform. When plug flow is postulated, it is natural to accept that heat is also only transferred by that mechanism. When molecular diffusion is neglected, the same may be done for heat conduction. When the concentration in a section perpendicular to flow is considered to be uniform then it is natural to also consider the temperature to be uniform in this section. It follows that when heat is exchanged with the surroundings, the temperature gradient has to be situated entirely in a thin "film" along the wall. This also implies that the resistance to heat transfer in the central core is zero in a direction perpendicular to the flow. This condition is not always fulfilled, especially for fixed bed catalytic reactors—besides heat transfer by convective flow, other mechanisms often have to be introduced in such cases. Even here it is necessary, in order to keep the mathematics tractable, to use simplified models, to be discussed in later chapters.

### 7.1.c The Momentum Equation

This balance is obtained by application of Newton's second law on a moving fluid element. In chemical reactors only pressure drops and friction forces have to be considered in most cases. A number of pressure drop equations are discussed in the chapters on tubular plug flow and on fixed bed catalytic reactors.

## 7.2 The Fundamental Equations ──────────────────────────

### 7.2.a The Continuity Equations

The derivation of differential mass balances or continuity equations for the components of an element of fluid flowing in a reactor is considered in detail in texts on transport processes (e.g., Bird et al. [1]). These authors showed that a fairly general form of the continuity equation for a chemical species $j$ reacting in a flowing fluid with varying density, temperature, and composition is:

$$\frac{\partial C_j}{\partial t} + \nabla \cdot (C_j \mathbf{u}) + \nabla \cdot \mathbf{J}_j = R_j \qquad (7.2.a\text{-}1)$$

If species $j$ occurs in more than one phase such a continuity equation has to be written for each phase. These equations are linked by the boundary conditions and generally also by a term expressing the transfer of $j$ between the phases. Such a term is not included by Eq. 7.2.a-1 since the following discussion is centered on the various forms the continuity equations can take in single phase or "homogeneous" or, by extension, in "pseudo-homogeneous" reactors as a consequence of the flow pattern. Specific modeling aspects resulting from the presence of more than one phase, solid, or fluid is illustrated in detail in Chapter 11 on fixed bed reactors, Chapter 13 on fluidized bed reactors, and Chapter 14 on multiphase reactors.

The terms and symbols used in this equation have the following meaning. $C_j$ is the molar concentration of species $j$ ($kmol/m^3$ fluid), so that $\partial C_j/\partial t$ is the non-steady-state term expressing accumulation or depletion. $\nabla$ is the "nabla" or "del" operator. In a rectangular coordinate system, $x$, $y$, $z$ with unit vectors $\boldsymbol{\delta}_x$, $\boldsymbol{\delta}_y$, and $\boldsymbol{\delta}_z$ the gradient of a scalar function $f$ is represented by $\nabla f$ and the divergence of a vector function $\mathbf{v}$ by $\nabla \cdot \mathbf{v}$. More explicitly:

$$\nabla f = \frac{\partial f}{\partial x} \boldsymbol{\delta}_x + \frac{\partial f}{\partial y} \boldsymbol{\delta}_y + \frac{\partial f}{\partial z} \boldsymbol{\delta}_z$$

$$\nabla \cdot \mathbf{v} = \frac{\partial v_x}{\partial x} + \frac{\partial v_y}{\partial y} + \frac{\partial v_z}{\partial z}$$

$\mathbf{u}$ is the three-dimensional mass average velocity vector, defined by

$$\mathbf{u} = \sum_{j=1}^{N} \frac{M_j C_j}{\rho_f} \mathbf{u}_j \quad \text{(m/s)}$$

where $\rho_f$ is the density of the mixture and $\mathbf{u}_j$ represents the velocity of molecules of species $j$. The term $\nabla \cdot (C_j \mathbf{u})$ thus accounts for the transport of mass by convective flow.

$\mathbf{J}_j$ is the molar flux vector for species $j$ with respect to the mass average velocity (kmol/m$^2$s). When the flow is laminar or perfectly ordered the term $\nabla \cdot \mathbf{J}_j$ results from molecular diffusion only. It can be written more explicitly as an extension, already encountered in Chapter 3, of Fick's law for diffusion in binary systems, as

$$\mathbf{J}_j = -\rho_f D_{jm} \nabla \left( \frac{C_j}{\rho_f} \right) \tag{7.2.a-2}$$

where $D_{jm}$ is the effective binary diffusivity for the diffusion of $j$ in the multicomponent mixture. Of course, appropriate multicomponent diffusion laws could also be used—for ideal gases the Stefan–Maxwell equation, as was done in Sec. 2.c of Chapter 3. In Eq. 7.2.a-2 the driving force has been taken as moles of $j$ per total mass of fluid, for the sake of generality [1]. The term $\nabla \cdot \mathbf{J}_j$ can also account for the flux resulting from deviations of perfectly ordered flow, as encountered with turbulent flow or with flow through a bed of solid particles for example, but this will be discussed further below.

$R_j$ is the total rate of change of the amount of $j$ because of reaction—as defined in Chapter 1, that is, $\alpha_j r$ for a single reaction and $\sum_{i=1}^{M} \alpha_{ij} r_i$ for multiple reactions. The $\alpha_{ij}$ are negative for reactants and positive for reaction products. The units of $R_j$ depend on the nature of the reaction. If the reaction is homogeneous the units could be kmol/m$^3$s but for a reaction catalyzed by a solid preference would be given to kmol/kg cat s, multiplied by the catalyst bulk density in the reactor.

From the definitions given it is clear that $\sum_j M_j \mathbf{J}_j = \sum_j M_j C_j (\mathbf{u}_j - \mathbf{u}) = 0$, while $\sum_j M_j R_j = 0$, due to the conservation of mass in a reacting system. So, if each term of Eq. 7.2.a-1 is multiplied by the molecular weight $M_j$, and the equation is then summed over the total number of species $N$, accounting for the relation $\rho_f = \sum_j M_j C_j$, the total continuity equation is obtained:

$$\frac{\partial \rho_f}{\partial t} + \nabla \cdot (\rho_f \mathbf{u}) = 0 \tag{7.2.a-3}$$

Thus, note that the usual continuity equation of fluid mechanics is also true for a reacting mixture. Equation 7.2.a-3 can be used to rewrite 7.2.a-1 in a form that is sometimes more convenient for reactor calculations. The first two terms can be rearranged as follows:

$$\frac{\partial C_j}{\partial t} + \nabla \cdot (C_j \mathbf{u}) = \frac{\partial}{\partial t}\left(\rho_f \frac{C_j}{\rho_f}\right) + \nabla \cdot \left(\rho_f \frac{C_j}{\rho_f} \mathbf{u}\right)$$

$$= \rho_f \left[\frac{\partial}{\partial t}\left(\frac{C_j}{\rho_f}\right) + \mathbf{u} \cdot \nabla\left(\frac{C_j}{\rho_f}\right)\right] + 0$$

where the last zero term results from the total continuity Eq. 7.2.a-3. This result suggests that $(C_j/\rho_f)$, moles $j$ per unit mass of mixture, is a convenient and natural variable. This is indeed the case, since $(C_j/\rho_f)$ is simply related to the conversion (or extent), a variable frequently used in reactor design:

$$\frac{C_j}{\rho_f} = \frac{(C_j/\rho_f)}{(C_j/\rho_f)_0}(C_j/\rho_f)_0 = \frac{N_j}{N_{j0}}(C_j/\rho_f)_0$$

$$= (1 - x_j)(C_j/\rho_f)_0 \qquad (7.2.a\text{-}4)$$

where $N_j$ is the total number of moles of $j$ present in the reactor and the index zero refers to reactor-inlet values.

Combining these latter results with Eq. 7.2.a-1 and Eq. 7.2.a-2 leads to an equation in terms of conversions:

$$\rho_f \left(\frac{\partial x_j}{\partial t} + \mathbf{u} \cdot \nabla x_j\right) - \nabla \cdot (\rho_f D_{jm} \nabla x_j) = -\left(\frac{\rho_f}{C_j}\right)_0 R_j \qquad (7.2.a\text{-}5)$$

Equations 7.2.a-1 and 7.2.a-5 are in fact extensions of the continuity equations used in previous chapters, where the flow terms were normally not present. These somewhat detailed derivations have been used to carefully illustrate the development of the equations of transport processes into forms needed to describe chemical reactors. It is seldom that the full equations have to be utilized, and normally only the most important terms will be retained in practical situations. However Eqs. 7.2.a-1 or 5 are useful to have available as a fundamental basis.

Equation 7.2.a-5 implicitly assumes perfectly ordered flow in that $\nabla \cdot (\rho_f D_{jm} \nabla x_j)$ is specific for molecular diffusion. Deviations from perfectly ordered flow, as encountered with turbulent flow, lead to a flux that is also expressed as if it arose from a diffusion-like phenomenon, in order to avoid too complex mathematical equations. The proportionality factor between the flux and the concentration gradient is then called the turbulent or "eddy" diffusivity. Since this transport mechanism is considered to have the same driving force as molecular diffusion, the two mechanisms are summed and the resulting proportionality factor is called "effective" diffusivity, $D_e$. In highly turbulent flow the contribution of molecular diffusion is usually negligible, so that $D_e$ is then practically identical for all the species of the mixture. Through its turbulent contribution, the effective diffusion is not isotropic, however. For more details refer to Hinze [2].

Equation 7.2.a-5 now becomes:

$$\rho_f\left(\frac{\partial x_j}{\partial t} + \mathbf{u} \cdot \nabla x_j\right) = \frac{\partial}{\partial x}\left(\rho_f D_{e,x} \frac{\partial x_j}{\partial x}\right) + \frac{\partial}{\partial y}\left(\rho_f D_{e,y} \frac{\partial x_j}{\partial y}\right)$$
$$+ \frac{\partial}{\partial z}\left(\rho_f D_{e,z} \frac{\partial x_j}{\partial z}\right) - \left(\frac{\rho_f}{C_j}\right)_0 R_j \qquad (7.2.a\text{-}6)$$

When the reactor contains a solid catalyst the flow pattern is strongly determined by the presence of the solid. It would be impossible to rigorously express the influence of the packing but again the flux of $j$ resulting from the mixing effect caused by its presence is expressed in the form of Fick's law. Consequently, the form of Eq. 7.2.a-6 is not altered, but the effective diffusivity now also contains the effect of the packing. This topic is dealt with extensively in Chapter 11 on fixed bed catalytic reactors. For further explanation of the effective transport coefficients see Himmelblau and Bischoff [3] and Slattery [4].

## 7.2.b Simplified Forms of the "General" Continuity Equation

As already mentioned, the form of the fundamental continuity equations is usually too complex to be conveniently solved for practical application to reactor design. If one or more terms are dropped from Eq. 7.2.a-6 and or integral averages over the spatial directions are considered, the continuity equation for each component reduces to that of an ideal, basic reactor type, as outlined in the introduction. In these cases, it is often easier to apply Eq. 7.1.a-1 directly to a volume element of the reactor. This will be done in the next chapters, dealing with basic or specific reactor types. In the present chapter, however, it will be shown how the simplified equations can be obtained from the fundamental ones.

It is very common in reactors to have flow predominantly in one direction, say $z$ (e.g., think of tubular reactors). The major gradients then occur in that direction, under isothermal conditions at least. For many cases then, the cross-sectional average values of concentration (or conversion) and temperature might be used in the equations instead of radial point values. The former are obtained from:

$$\langle \zeta \rangle \equiv \frac{1}{\Omega} \iint_\Omega \zeta \, d\Omega$$

where $\zeta$ represents any variable, and $\Omega$ is the cross section inside the rigid boundary and $d\Omega = dx\, dy$. We can see that virtually all the terms contain products of dependent variables, and the first approximation that must be made is that the average of the product is close to the product of the averages; for example,

$$\left\langle \rho_f u_z \frac{\partial x_j}{\partial z} \right\rangle \simeq \langle \rho_f u_z \rangle \frac{\partial \langle x_j \rangle}{\partial z}$$

In this case, the approximation would clearly be best for highly turbulent flow, for which the velocity profiles are relatively flat. The discrepancies actually enter into the effective transport coefficients, which have to be empirically measured in any event. Another approximation concerns the reaction rate term:

$$\langle R_j(C_j, T) \rangle \simeq R_j(\langle C_j \rangle, \langle T \rangle)$$

Thus, Eq. 7.2.a-6 becomes after integration over the cross section:

$$\langle \rho_f \rangle \frac{\partial \langle x_j \rangle}{\partial t} + \langle \rho_f u \rangle \frac{\partial \langle x_j \rangle}{\partial z} = \langle \rho_f D_{e,z} \rangle \frac{\partial^2 \langle x_j \rangle}{\partial z^2} - \left( \frac{\rho_f}{C_j} \right)_0 R_j \qquad (7.2.b\text{-}1)$$

$$\quad\quad (1) \quad\quad\quad\quad\quad (2) \quad\quad\quad\quad\quad (3) \quad\quad\quad\quad (4)$$

where the velocity in the flow direction is represented by $u$. In the presence of packing a distinction would have to be made between the true local fluid velocity, called the interstitial velocity (m/s) and the velocity considered over the whole cross section, as if there were no solid, called the superficial velocity (m³ fluid/m² cross section s). A so called "one-dimensional model" is now obtained. If the convective transport is completely dominant over any diffusive transport, in particular that in the flow direction—that is, the fluid moves like a "plug"—the term (3) may be neglected. Assuming steady state conditions, the term (1) also drops out, so that the simplified Eq. 7.2.b-1 becomes (leaving out the brackets for simplicity):

$$\rho_f u \frac{dx_j}{dz} = - \left( \frac{\rho_f}{C_j} \right)_0 R_j$$

while the continuity Eq. 7.2.a-3 reduces to:

$$\frac{d}{dz} (\rho_f u) = 0$$

This last equation is simply integrated to give:

$$(\rho_f u) = (\rho_f u)_0 = \text{constant} = G(\text{kg/m}^2 \text{ s})$$

where $G$ is usually termed the "mass flow velocity." This result is then combined with the continuity equation for species $j$, giving

$$u_0 \frac{dx_j}{dz} = - \frac{1}{C_{j0}} R_j(x_j) \qquad (7.2.b\text{-}2)$$

354

One modification is normally made before performing the final integration step:

$$\frac{dz}{u_0} = \frac{d(\Omega z)}{\Omega u_0} = \frac{dV}{F_0'}$$

where $F_0'$ is the volumetric flow rate of the feed ($m^3/s$) and $dV$ is a differential element of reactor volume. Integration now gives,

$$\frac{V}{F_0'} = -C_{j0} \int \frac{dx_j}{R_j(x_j)} \tag{7.2.b-3}$$

More often this equation is written in the form

$$\frac{V}{F_{j0}} = -\int \frac{dx_j}{R_j(x_j)} \tag{7.2.b-4}$$

whereby $F_{j0} = F_0' C_{j0}$ is the molar feed rate of species $j$ (kmol/s). The last equation is used to describe the *plug flow reactor*.

Other simplified forms result when the entire reactor may be considered to be uniform—operating under conditions of complete mixing, the idealized picture of a well-mixed vessel. Here, one averages over all the spatial directions so that Eq. 7.2.b-1 can be further integrated over $z$:

$$\bar{x}_j \equiv \frac{\Omega}{V} \int \langle x_j \rangle dz$$

$$= \frac{1}{V} \iiint x_j \, dx \, dy \, dz$$

(For simplicity again, the overlines referring to mean values, will from now be left out.) Moreover, because of the assumption of complete uniformity, no effective transport terms need to be considered. Note that the final coordinate direction here refers to the fluid, which could be expanding, in contrast to the rigid boundary assumed for $x$ and $y$. A more general and more rigorous derivation using the transport theorems of vector/tensor analysis has been given by Bird [5]. In the batch case, when no fluid is entering or leaving the reactor, except at the time of loading or unloading, Eq. 7.2.b-1, with the terms (2) and (3) zero, can be integrated to yield:

$$\frac{d}{dt}(V \rho_f x_j) = -\left(\frac{\rho_f}{C_j}\right)_0 R_j V \tag{7.2.b-5}$$

or,

$$\frac{d}{dt}(N_{j0} x_j) = -R_j V$$

since $V(\rho_f/\rho_{fo})C_{jo} = V_0 C_{jo} = N_{jo}$, the total number of moles of $j$ initially present. $N_j$ is related to $N_{jo}$ by $N_j = N_{jo}(1 - x_j)$, so that finally one obtains:

$$\frac{dN_j}{dt} = R_j V \tag{7.2.b-6}$$

or, in integral form:

$$\theta = \int \frac{dN_j}{R_j V} \tag{7.2.b-7}$$

This is the mass balance equation for the *batch reactor*. The symbol $t$ for "clock time" is replaced here by the more usual symbol $\theta$ for "batch residence time."

For the continuous, completely mixed reactor, it is useful to start from the reduced continuity equation in terms of concentrations, analogous to Eq. 7.2.b-1 (but with no diffusion term):

$$\frac{\partial \langle C_j \rangle}{\partial t} + \frac{\partial \langle u C_j \rangle}{\partial z} = R_j \tag{7.2.b-8}$$

which yields, after integration over $z$ and multiplication by $\Omega$:

$$\frac{d}{dt}(V C_j) + \int \frac{d \langle F_j \rangle}{dz} \cdot dz = R_j V \tag{7.2.b-9}$$

since

$$\langle u C_j \rangle = \frac{\langle F_j \rangle}{\Omega} = \frac{\langle F' C_j \rangle}{\Omega}$$

where $F'$ is the volumetric flow rate, m³/s. If $F_{j,0}$ and $F_{j,e}$ represent, respectively, the inlet and outlet flow rates of species $j$ the following equation is obtained:

$$\frac{dN_j}{dt} = F_{j,0} - F_{j,e} + R_j V \tag{7.2.b-10}$$

Again, Bird [5] presents a more rigorous derivation, with the identical result. Under steady-state conditions Eq. 7.2.b-10 reduces to an algebraic equation:

$$F_{j,e} - F_{j,0} = R_j V \tag{7.2.b-11}$$

which is the mass balance for the *continuous flow stirred tank reactor* (CSTR).

If Eq. 7.2.b-10 is multiplied by the molecular weight $M_j$, and summed on $j$, a total mass balance is obtained:

$$\frac{dm_t}{dt} = F'_0 \rho_{f,0} - F'_e \rho_{f,e} \tag{7.2.b-12a}$$

$$= \dot{m}_0 - \dot{m}_e \tag{7.2.b-12b}$$

_____

where $m_t = \sum_j M_j N_j$ is the total mass, and $\dot{m}$ is the mass flow rate (kg/s). Equation 7.2.b-12 could also be obtained by integrating Eq. 7.2.a-3 over the volume. For liquids, the density is approximately constant, and if the volume is fixed, Eq. 7.2.b-12 shows that the inlet and exit flows must be the same.

## 7.2.c The Energy Equation

Again reference is made to Bird et al. [1] for the rigorous derivation, in various co-ordinate systems, of the fundamental energy equation. The following form, with respect to a rectangular coordinate system, contains the phenomena that are of importance in reactors:

$$\sum_j M_j C_j c_{pj} \underbrace{\left(\frac{\partial T}{\partial t} + \mathbf{u} \cdot \nabla T\right)}_{(1) \quad (2)} = \underbrace{\sum_i (-\Delta H_i) r_i}_{(3)} + \underbrace{\nabla \cdot (\lambda \nabla T)}_{(4)} - \underbrace{\sum_j \mathbf{J}_j \nabla H_j}_{(5)} + \underbrace{Q_{\text{rad}}}_{(6)}$$

$$(7.2.c\text{-}1)$$

where $c_{pj}$ is the specific heat of species $j$ (kcal/kg°C or kJ/kg K), $\lambda$ is the thermal conductivity of the mixture (kcal/m.hr°C or kJ/m.s. K) and the $H_j$ are partial molar enthalpies (kcal/kmol or kJ/kmol). The respective terms arise from: (1) change of heat content with time, (2) convective flow, (3) heat effect of the chemical reactions, (4) heat transport by conduction, (5) energy flux by molecular diffusion, and (6) radiation heat flux.

Other energy terms encountered with particular flow conditions are work of expansion or viscous dissipation terms, primarily important in high speed flow; external field effects, mechanical or electrical, can also occur. Since they usually are of much less importance they will not be considered here. Heat radiation in the reactor is often neglected, except in the case of fixed bed catalytic reactors operating at high temperatures, but then it is generally lumped with the heat conduction and a few more heat transport mechanisms into an "effective" heat conduction having the form of term (4) in Eq. 7.2.c-1. When this is done in Eq. 7.2.c-1 and the diffusion term (5) is neglected the result is:

$$\sum_j M_j C_j c_{pj} \left(\frac{\partial T}{\partial t} + \mathbf{u} \cdot \nabla T\right) = \sum_i (-\Delta H_i) r_i + \frac{\partial}{\partial x}\left(\lambda_{e,x} \frac{\partial T}{\partial x}\right)$$
$$+ \frac{\partial}{\partial y}\left(\lambda_{e,y} \frac{\partial T}{\partial y}\right) + \frac{\partial}{\partial z}\left(\lambda_{e,z} \frac{\partial T}{\partial z}\right) \quad (7.2.c\text{-}2)$$

where $\lambda_e$ is an effective thermal conductivity. Again, when there is more than one phase, more than one energy equation has to be written and a transfer term has to be introduced. For the same reasons as mentioned in Sec. (7.2.a), this has not been done here and will be delayed to the specific cases discussed in the following chapters.

### 7.2.d Simplified Forms of the "General" Energy Equation

The "general" energy equation can be simplified in the same way as the continuity equation, since the approximations introduced there are assumed to be equally applicable here. But, whereas mass is generally not diffusing through the wall, heat frequently is. In deriving the one-dimensional model by averaging over the cross section, a boundary condition for heat transfer at the reactor wall has to be introduced for this reason. This boundary condition is commonly written as:

$$\left(\lambda_{e,n} \frac{\partial T}{\partial n}\right)_w = \alpha_w(T_w - T_R) \tag{7.2.d-1}$$

Here $n$ represents the direction normal to the wall, $\alpha_w$ is a convective heat transfer coefficient, $T_w$ is the temperature of the wall, and $T_R$ is the fluid temperature in the immediate vicinity of the wall. The right-hand side of Eq. 7.2.d-1 would be zero for an adiabatic reactor. Equation 7.2.c-2 then becomes, when averaged over cylindrical geometry, with diameter $d_r$

$$\sum_j \langle M_j C_j c_{pj}\rangle\left(\frac{\partial\langle T\rangle}{\partial t} + \langle u\rangle \frac{\partial\langle T\rangle}{\partial z}\right) = \sum_i \langle -\Delta H_i\rangle\langle r_i\rangle + \frac{\partial}{\partial z}\left(\langle \lambda_{e,z}\rangle \frac{\partial\langle T\rangle}{\partial z}\right)$$

$$+ \frac{4}{d_r}\left(\lambda_{e,n} \frac{\partial T}{\partial n}\right)_w \tag{7.2.d-2}$$

An important point is that the $z$ component of the condition term retains its identity, in terms of averaged variables, but the $x$ and $y$ components are integrated out with the wall boundary condition, Eq. 7.2.d-1, which is now written:

$$\frac{4}{d_r}\left(\lambda_{e,n} \frac{\partial T}{\partial n}\right)_w = \frac{4\alpha_w}{d_r}(T_w - T_R)$$

$$\equiv \frac{4U}{d_r}(T_r - \langle T\rangle)$$

where $T_r$ is the temperature of the surroundings and $U$ is an overall heat transfer coefficient. The latter approximation actually locates the heat transfer with the wall in a thin film. For the tubular reactor considered here, the heat conduction in the $z$-direction is usually much smaller than the heat transported by convection, and also it drops out for the complete mixing case.

Thus, the resulting equation is:

$$\sum_j \langle M_j C_j c_{pj}\rangle\left(\frac{\partial\langle T\rangle}{\partial t} + \langle u\rangle \frac{\partial\langle T\rangle}{\partial z}\right) = \sum_i \langle -\Delta H_i\rangle\langle r_i\rangle + \frac{4U}{d_r}(T_r - \langle T\rangle)$$

$$\tag{7.2.d-3}$$

For steady-state conditions Eq. 7.2.d-3 becomes, after multiplying by $\Omega = \pi d_r^2/4$ (and omitting the brackets):

$$\sum_j \dot{m}_j c_{pj} \frac{dT}{dz} - \Omega \sum_i (-\Delta H_i) r_i - \pi d_r U(T_r - T) = 0 \qquad (7.2.\text{d-4})$$

This is the energy equation for a single-phase tubular reactor with plug flow. Note that Eq. 7.2.d-4 is coupled with the continuity equation, mainly by the reaction term, but also through the heat capacity term on the left-hand side. The latter is sometimes written in terms of a specific heat that is averaged with respect to temperature and composition, that is, $\sum_j \dot{m}_j c_{pj} \equiv \dot{m} \bar{c}_p$.

A rigorous macroscopic energy balance is found by integrating over the entire reactor volume:

$$\sum_j m_j c_{pj} \frac{dT}{dt} = \sum_j F_{j,0}(H_{j,0} - H_{j,e}) + V \sum_i (-\Delta H_i) r_i + \int \pi d_t U(T_r - T)\, dz$$

$$(7.2.\text{d-5})$$

which can also be found by a careful integration of Eq. 7.2.d-3 over the reactor (see Bird [5]). Representing the internal heat exchange surface of the reactor by $A_k$, Eq. 7.2.d-5 reduces to:

$$m_t c_p \frac{dT}{dt} = V \sum_i (-\Delta H_i) r_i + A_k U(T_r - T) \qquad (7.2.\text{d-6})$$

for the batch reactor, or with $F_{j,e} = F_{j,0} + R_j V$

$$\sum_j F_{j,e} H_{j,e} - \sum_j F_{j,0} H_{j,0} = A_k U(T_r - T) \qquad (7.2.\text{d-7})$$

for the continuous flow stirred tank reactor.

The following chapters deal in detail with ideal reactor types like the batch reactor (Chapter 8), the tubular reactor with plug flow (Chapter 9), and the continuous flow reactor with complete mixing (Chapter 10). Deviations from plug flow will be encountered in Chapter 11 on fixed bed catalytic reactors and several degrees of sophistication will be considered there. The problem of modeling nonideal and multiphase reactors will be developed in Chapter 12, while important specific cases of fluidized bed reactors and of gas-liquid-solid reactors will be discussed in Chapters 13 and 14, respectively. Each of these chapters starts from the basic equations developed here or from combinations of these; correlations are given for the mass and heat transfer parameters for each specific case; the operational characteristics of the reactors are derived from the solution of the basic equations; and the performance of reactors in several industrial processes will be simulated and investigated.

# Problems

7.1 Write Eq. 7.2.a-6 in terms of $\xi_i'$, the extent of the $i$th reaction per unit mass of the reaction mixture, defined by

$$\xi_i' = \frac{\xi_i}{N_{j0}} \left(\frac{C_j}{\rho_f}\right)_0 \text{ (kmol/kg)}$$

7.2 Derive the steady-state continuity and energy equations and appropriate boundary conditions for the tubular reactor with turbulent flow, corresponding to the various situations represented in the following diagram (from Himmelblau and Bischoff [3]).

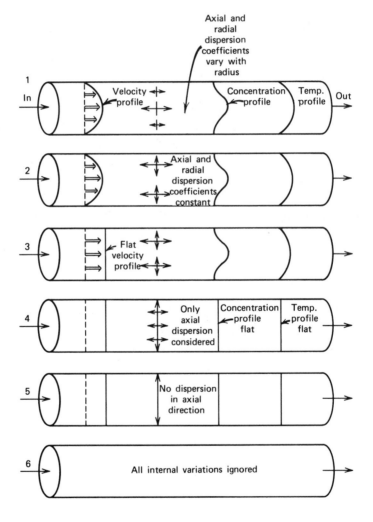

CHEMICAL REACTOR DESIGN

The continuity equation for the first case is given by (in cylindrical coordinates):

$$u(r)\frac{\partial C_j}{\partial z} = D_{e,z}(r)\frac{\partial^2 C_j}{\partial z^2} + \frac{1}{r}\left[\frac{\partial}{\partial r}\,rD_{e,r}(r)\frac{\partial C_j}{\partial r}\right] + R_j$$

with boundary conditions

$$u(r)C_{j0} = u(r)C_j(0, r) - D_{e,z}(r)\frac{\partial C_j(0, r)}{\partial z}$$

$$z = L \qquad \frac{\partial C_j}{\partial z} = 0, \qquad \text{all } r$$

$$r = 0 \qquad \frac{\partial C_j}{\partial r} = 0, \qquad \text{all } z$$

$$r = R_t \qquad \frac{\partial C_j}{\partial r} = 0, \qquad \text{all } z$$

7.3 Write all the above equations in dimensionless form.

# References

[1] Bird, R. B., Stewart, W. E., and Lightfoot, E. N. *Transport Phenomena*, Wiley, New York (1960).

[2] Hinze, J. O. *Turbulence*, McGraw-Hill, New York (1959).

[3] Himmelblau, D. M. and Bischoff, K. B. *Process Analysis and Simulation*, Wiley, New York (1968).

[4] Slattery, J. *Momentum, Energy and Mass Transfer in Continua*, McGraw-Hill, New York (1972).

[5] Bird, R. B. *Chem. Eng. Sci.*, **6**, 123 (1957).

# 8

## THE
## BATCH
## REACTOR

The usual definition of a batch reactor is one in which the only chemical and thermal changes are with respect to time—in other words, the reactor is spatially uniform. We will retain this meaning, and thus the simplified balances from Section 7.3 can be used. Batch reactors are most often used for low production capacities and for short-term productions where the cost of labor and other aspects of the operations are less than capital cost of new equipment, and a small fraction of the unit cost of the product.

## 8.1 The Isothermal Batch Reactor

Because of the uniformity of concentration, the continuity equation for the key reacting component may be written for the entire reactor volume:

$$\frac{dN_A}{d\theta} = -Vr_A(C_A) \qquad (8.1\text{-}1)[1]$$

where $\theta$ = residence time in the reactor. It is convenient to specifically represent this residence time in the reactor by a special symbol—for completely batch systems it is the same as "clock" time, $t$, but in other applications the distinction will be useful. For a general set of reactions, Eq. 8.1-1 can be extended to:

$$\frac{dN_j}{d\theta} = V \sum_{i=1}^{R} \alpha_{ij} r_i \equiv V R_j \qquad (8.1\text{-}2)[1]$$

These mass balances are often written in terms of conversions:

$$\frac{dx_A}{d\theta} = \frac{V}{N_{A0}} r_A \qquad (8.1\text{-}3)$$

[1] Note: Eq. 7.2.b-6 is written for arbitrary species $j$; for species $A$ being taken as a reactant, $\alpha_A = -a = -1$ for a single reaction with stoichiometry referred to $A$, thus leading to Eq. 8.1-1.

362

Then, Eq. 8.1-3 is readily put into integral form:

$$\theta = N_{A0} \int_{x_{A0}}^{x_{Af}} \frac{dx_A}{Vr_A} \tag{8.1-4}$$

Note that the batch residence time, $\theta$, can be interpreted as the area from $x_{A0}$ to $x_{Af}$ under the curve of $N_{A0}/Vr_A(x_A)$ versus $x_A$. The volume of reaction mixture can change because of two reasons: (1) external means (e.g., filling a reaction vessel or adding a second reactant) and (2) changes in densities of reactants or products (e.g., molal expansion of gases). The first possibility is often termed "semibatch" operation, since some sort of flow is involved, and this will be discussed later. The second is usually not very important for liquids, and is neglected. We will derive the proper formulation for gases, although it should be stated that batch gas-phase reactors are not commonly used in industry because of the small mass capacity; however, a gas phase could be part of the reaction mixture, and also laboratory gas-phase reactors have been utilized.

With no expansion, as for liquids, Eq. 8.1-4 becomes

$$\theta = C_{A0} \int_{x_{A0}}^{x_{Af}} \frac{dx_A}{r_A(x_A)} \tag{8.1-5}$$

$$= - \int_{C_{A0}}^{C_{Af}} \frac{dC_A}{r_A(C_A)} \tag{8.1-5a}$$

and for simple rate forms can be easily integrated analytically, as illustrated in Sec. 1.3.

For reactions with the reaction stoichiometry

$$aA + bB \cdots \; \rightleftharpoons \; qQ + sS \cdots$$

the following mole balance can be made at a given extent of reaction based on conversion of $A$:

$$N_A = N_{A0} - N_{A0}x_A$$

$$N_B = N_{B0} - \frac{b}{a} N_{A0}x_A$$

$$N_Q = N_{Q0} + \frac{q}{a} N_{A0}x_A$$

$$N_S = N_{S0} + \frac{s}{a} N_{A0}x_A$$

$$\underline{N_I = N_{I0} \text{ (inert)}}$$

$$N_t = N_{t0} + N_{A0} \frac{q + s - a - b \cdots}{a} x_A$$

Therefore, the total number of moles is given by

$$N_t = N_{t0} + N_{A0}\delta_A x_A$$

from which:

$$\frac{N_t}{N_{t0}} = 1 + (y_{A0}\delta_A)x_A \equiv 1 + \varepsilon_A x_A \tag{8.1-6}$$

Now for gases, let us use the equation of state, for example:

$$p_t V = Z N_t R T$$

Then,

$$\frac{V}{V_0} = \left(\frac{Z}{Z_0}\frac{T}{T_0}\frac{p_{t0}}{p_t}\right)\frac{N_t}{N_{t0}}$$

$$= \left(\frac{Z}{Z_0}\frac{T}{T_0}\frac{p_{t0}}{p_t}\right)(1 + \varepsilon_A x_A) \tag{8.1-7}$$

For constant $(T, p_t)$, this reduces to the special case defined by Levenspiel [1]. Next, the concentrations, for substitution into the rate formula, can be expressed as

$$C_A = \frac{N_A}{V} = \frac{N_{A0}}{V_0}\frac{(1 - x_A)}{(1 + \varepsilon_A x_A)}\left(\frac{Z_0}{Z}\frac{T_0}{T}\frac{p_t}{p_{t0}}\right) \tag{8.1-8}$$

With partial pressures:

$$p_A = p_t \frac{N_A}{N_t} = p_t \frac{N_{A0}}{N_{t0}}\frac{(1 - x_A)}{(1 + \varepsilon_A x_A)}$$

$$= p_{A0}\frac{1 - x_A}{1 + \varepsilon_A x_A}\left(\frac{p_t}{p_{t0}}\right) \tag{8.1-9}$$

As an illustration, for an $n$th-order reaction:

$$Vr_A = VkC_A{}^n = k\frac{N_{A0}{}^n}{V_0{}^{n-1}}\frac{(1 - x_A)^n}{(1 + \varepsilon_A x_A)^{n-1}} \quad \text{(const. } T, p_t)$$

and Eq. 8.1-4 becomes

$$\theta = \frac{1}{kC_{A0}{}^{n-1}}\int_{x_{A0}}^{x_{Af}}\frac{(1 + \varepsilon_A x_A)^{n-1}}{(1 - x_A)^n}dx_A \quad \text{(const. } T, p_t)$$

which for no molar expansion, $\varepsilon_A = 0$, is the same as Eq. 8.1-5, of course.

## Example 8.1-1 Example of Derivation of a Kinetic Equation by Means of Batch Data

The reaction $A + B \rightarrow Q + S$ is carried out in the liquid phase at constant temperature. It is believed the reaction is elementary and, since it is biomolecular, it is natural to first try second order kinetics. The density may be considered constant.

Let $B$ be the component with highest concentration, while the most convenient way to follow the reaction is by titration of $A$.

A batch type experiment led to the following data in Table 1:

Table 1  Concentration
versus time data

| $C_{B0} = 0.585$ kmol/m³ | |
| $C_{A0} = 0.307$ kmol/m³ | |
| Time (hr) | $C_A$(kmol/m³) |
| --- | --- |
| 0 | 0.307 |
| 1.15 | 0.211 |
| 2.90 | 0.130 |
| 5.35 | 0.073 |
| 8.70 | 0.038 |

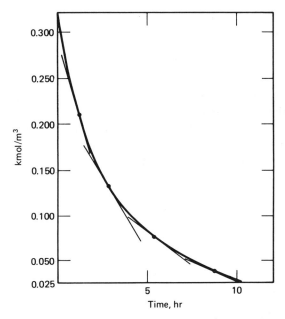

Figure 1 Graphical representation of concentration versus time data. Determination of reaction rates.

*Table 2 Comparison of* k *determined by integral and differential method*

| Time | $C_A$ | $C_B$ | $r_A$ | $k$ | |
|------|-------|-------|-------|-----|---|
| | $\dfrac{\text{kmol}}{\text{m}^3}$ | | $\text{kmol/m}^3\ \text{hr}$ | \multicolumn{2}{c}{$\text{m}^3/\text{kmol hr}$} |
| | | | | from Eq. b | from Eq. e |
| 1.15 | 0.211 | 0.489 | $6.52 \times 10^{-2}$ | $63.1 \times 10^{-2}$ | $61.2 \times 10^{-2}$ |
| 2.90 | 0.130 | 0.408 | $3.32 \times 10^{-2}$ | $62.5 \times 10^{-2}$ | $61.9 \times 10^{-2}$ |
| 5.35 | 0.073 | 0.351 | $1.49 \times 10^{-2}$ | $58.4 \times 10^{-2}$ | $62.0 \times 10^{-2}$ |
| 8.70 | 0.038 | 0.316 | $0.72 \times 10^{-2}$ | $59.6 \times 10^{-2}$ | $60.8 \times 10^{-2}$ |

If the hypotheses of second order is correct, the following relation between the rate and the concentrations of $A$ and $B$ will be valid, for any time, and therefore any composition:

$$r_A = r_B = kC_A C_B \tag{a}$$

and $k$ has to have the same value for all levels of $C_A$ and $C_B$. When the differential method is used Eq. a is the starting point. By substituting the rate equation (a) in the material balances: Eq. 8.1-1 with $C_A = N_A/V$:

$$r_A = -\frac{dC_A}{d\theta} = kC_A C_B \tag{b}$$

This means $r_A$ may be obtained as tangent to the curve $C_A - \theta$ (Fig. 1).

Substituting the corresponding $C_A$ and $C_B$ leads to $k$. The values of $C_B$ follow from $C_B = C_{B0} - (C_{A0} - C_A)$. Table 2 gives the values of $k$ obtained in this way. The variation of $k$ is small and does not invalidate the second-order hypothesis, especially as the precision of the method is getting smaller as the reaction proceeds. A value of $61 \times 10^{-2}$ m$^3$/hr kmol may be used for $k$.

The integral method is based on Eq. 8.1-5. Before integration is possible, $C_A$ and $C_B$ must be expressed as a function of one variable, the fractional conversion, $x_A$. In this case

$$C_A = C_{A0}(1 - x_A)$$

$$C_B = C_{B0}\left(1 - \frac{x_A}{M}\right) \quad \text{where } M = \frac{C_{B0}}{C_{A0}}$$

Eq. 8.1-5a becomes:

$$\theta = \frac{C_{A0}}{kC_{A0}C_{B0}} \int_0^{x_A} \frac{dx_A}{(1 - x_A)(1 - x_A/M)} \tag{c}$$

or

$$C_{BO} k\theta = \frac{M}{1 - M} \ln \frac{M(1 - x_A)}{M - x_A} \tag{d}$$

or with concentrations:

$$k\theta = \frac{1}{C_{AO} - C_{BO}} \ln \frac{C_{BO} \cdot C_A}{C_{AO} \cdot C_B} \tag{e}$$

These equations also lead to a constant value for $k$, which confirms that the reaction has second-order kinetices. Peterson [2] has discussed further aspects of differential versus integral fitting of data from batch reactor experiments.

## 8.2 The Nonisothermal Batch Reactor

In practice, it is not always possible, or even desirable, to carry out a reaction under isothermal conditions. In this situation, both the energy and mass balances must be solved simultaneously:

$$\frac{dx_A}{d\theta} = \frac{V}{N_{AO}} r_A(x_A, T) \tag{8.1-3}$$

$$m_t c_p \frac{dT}{d\theta} = V(-\Delta H) r_A(x_A, T) + q A_k \tag{8.2.1}$$

where Eq. 8.2-1 is the appropriate simplified heat balance and $A_k$ is the heat exchange surface from Section 7.2-d. The term $q A_k$ represents any addition or removal of heat from the reactor. For adiabatic systems, $q = 0$, while for a heat exchange coil it would have the form

$$q = U(T_r - T) \tag{8.2-2}$$

where $T_r$ = temperature of heating or cooling medium. Eq. 8.2-1 can be combined with Eq. 8.1-3 to yield:

$$m_t c_p \frac{dT}{d\theta} - (-\Delta H) N_{AO} \frac{dx_A}{d\theta} = q A_k \tag{8.2-3}$$

$$= 0, \text{ adiabatic}$$

Thus,

$$m_t c_p (T - T_0) - (-\Delta H) N_{AO}(x_A - x_{AO}) = \int_0^\theta q A_k \, d\theta \tag{8.2-4a}$$

$$= q A_k \theta, q = \text{const.} \tag{8.2-4b}$$

$$= 0, \text{ adiabatic} \tag{8.2-4c}$$

For the latter adiabatic situation, the adiabatic temperature change, for a certain conversion level is:

$$T - T_0 = \frac{(-\Delta H)N_{A0}}{m_t c_p}(x_A - x_{A0}) = \frac{1}{\lambda}(x_A - x_{A0}) \tag{8.2-5}$$

Therefore, in this case $T$ can be substituted from Eq. 8.2-5 into Eq. 8.1-3, which then becomes a single differential equation in $x_A$ (or $x_A$ can be substituted into Eq. 8.2-1). This is done by utilizing Eq. 8.1-4, where the integral is evaluated by choosing increments of $x_A$ and the corresponding $T(x_A)$ from Eq. 8.2-5. Again, the reactor residence time, $\theta$, can be represented by the area under the curve

$$\frac{N_{A0}}{V r_A(x_A, T(x_A))} \qquad \text{versus} \qquad x_A$$

Some analytical solutions are even possible for simple-order rate forms—they are given for the analogous situation for plug flow reactors in Chapter 9. Finally, the maximum adiabatic temperature change is found for $x_A = 1.0$, and then (for $x_{A0} = 0$):

$$(\Delta T)_{ad} = T_{ad} - T_0 = \frac{(-\Delta H)N_{A0}}{m_t c_p} \tag{8.2-6}$$

Eq. 8.2-4c can be written in the alternate form:

$$T = T_0 + (\Delta T)_{ad} x_A \tag{8.2-7}$$

More general situations require numerical solutions of the combined mass and heat balances.

Several situations can occur:

1. The temperature is constant or a prescribed function of time, $T(\theta)$—here the mass balance Eq. 8.1-3 can be solved alone as a differential equation:

$$\frac{dx_A}{d\theta} = \frac{V}{N_{A0}} r_A(x_A, T(\theta))$$

Also, Eq. 8.2-1 or 3 can then be solved to find the heating requirements:

$$q(\theta)A_k = m_t c_p \frac{dT(\theta)}{d\theta} - (-\Delta H)V r_A(x_A(\theta), T(\theta))$$

2. Heat exchange is zero, constant, or a prescribed function of time. First Eq. 8.2-4 is used to compute $T = T(x_A, \theta)$ and then substituted into the mass balance Eq. (8.1-3), which can then be integrated:

$$\frac{dx_A}{d\theta} = \frac{V}{N_{A0}} r_A(x_A, T(x_A, \theta))$$

The temperature variation can then be found, if desired, by using the computed values of $x_A(\theta)$:

$$T(\theta) = T(x_A(\theta), \theta)$$

Alternatively, the combined Eqs. 8.1-3, 8.2-1 can be simultaneously solved as coupled differential equations.

3. Heat exchange is given by $q = U(T_r - T)$—direct numerical solution of the coupled mass and heat balances is used.

If convergence problems arise in the numerical solutions, especially for hand calculations, it is often useful to use conversion as the independent variable. Thus, increments of conversion give increments of time from the mass balance, and these give increments of temperature from the heat balance; iterations on the evaluations of the rates are also often required.

For case 3 above, values of the heat transfer coefficient are required. The factor $U$, appearing in Eq. 8.2-2, is a heat transfer coefficient, defined as follows:

$$\frac{1}{U} = \frac{1}{\alpha_k} + \frac{d}{\lambda}\frac{A_k}{A_m} + \frac{1}{\alpha_r}\cdot\frac{A_k}{A_r} \tag{8.2-8}$$

where:

$\alpha_k, A_k$: respectively heat transfer coefficient (kcal/m² hr °C) and heat transfer surface (m²) on the side of the reaction mixture

$\alpha_r, A_r$: the same, but on the side of the heat transfer medium

$A_m$: logarithmic mean of $A_k$ and $A_r$

$\lambda$: conductivity of the wall through which heat is transferred (kcal/m hr°C)

$d$: wall thickness (m)

The literature data concerning $\alpha_k$ and $\alpha_r$ are not always in accordance. As a guide the following relations are given.

For reactors in which heat is transferred through a wall, $\alpha_k$ may be obtained from the following dimensionless equation for stirred vessels:

$$\frac{\alpha_k d_r}{\lambda}\left(\frac{\mu_w}{\mu}\right)^{0.14} = 0.36\left(\frac{d_s^2 N \rho_L}{\mu}\right)^{0.66}\left(\frac{c_p \mu}{\lambda}\right)^{0.33} \tag{8.2-9}$$

where $d_r$ = reactor diameter (m)

$d_s$ = propeller diameter (m)

$\mu_w$ = viscosity of the reaction mixture at the temperature of the wall (kg/m hr)

$\mu$ = viscosity of the reaction mixture at the temperature of the reaction mixture

$\lambda$ = heat conductivity of the reaction mixture (kcal/m hr °C)

$N$ = revolutions per hour (hr$^{-1}$)

$\rho_L$ = density of reaction mixture (kg/m³)

(Chilton, Drew, and Jebens [3].) More extensive work by Chapman, Dallenbach and Holland [4] on a batch reactor with baffles and taking into account the liquid height $(H_L)$ and the propeller position above the bottom $(H_s)$ led to the following equation:

$$\frac{\alpha_k d_r}{\lambda}\left(\frac{\mu_w}{\mu}\right)^{0.24} = 1.15\left(\frac{d_s^2 N \rho_L}{\mu}\right)^{0.65}\left(\frac{c_p \mu}{\lambda}\right)^{0.33}\left(\frac{H_s}{d_s}\right)^{0.4}\left(\frac{H_L}{d_s}\right)^{-0.56} \quad (8.2\text{-}10)$$

Further work on this subject has been done by Strek [5]. For $\alpha_r$ several cases are possible. When the reaction vessel is heated (e.g., with steam) the Nusselt-equation may be applied, provided film condensation is prevailing. Refer to heat transfer texts for this topic. For heat transfer through a coil, $\alpha_k$ may be calculated from an equation such as Eq. 8.2-9, but with a larger coefficient due to the effect of the coil on the turbulence. According to Chilton, Drew and Jebens this coefficient would be 0.87. It is likely to depend also on the mixing intensity; other literature also mentions a value of 1.01.

$\alpha_r$ may be obtained from the following equation, valid for turbulent conditions:

$$\frac{\alpha_r d_t}{\lambda} = 0.023\left(\frac{d_t G}{\mu}\right)^{0.8}\left(\frac{c_p \mu_s}{\lambda}\right)^{0.4}\Phi \quad (8.2\text{-}11)$$

where $\Phi = 1 + 3.5\dfrac{d_t}{d_c}$

and $d_t$ = inner diameter of the pipe (m)
  $d_c$ = coil diameter (m)
  $\mu_s$ = viscosity of the reaction mixture at the surface of the coil (kg/m hr)

Equation 8.2-11 is an adaptation of the classical Dittus and Boelter equation for straight pipes. Further information on this topic can be found in Holland and Chapman [6].

## Example 8.2-1  Decomposition of Acetylated Castor Oil Ester

This example has been adapted from Smith [7] and Cooper and Jeffreys [8]. The overall reaction for the manufacture of drying oil is

$$\left(\begin{array}{c}\text{acetylated}\\\text{castor oil}\\\text{ester}\end{array}\right)(l) \longrightarrow \left(\begin{array}{c}\text{drying}\\\text{oil}\end{array}\right)(l) + CH_3COOH(g)$$

The charge of oil to the batch reactor is 227 kg, and has a composition such that complete hydrolysis gives 0.156 kg acid/kg ester; the initial temperature is $T_0 = 613$ K. The physiochemical properties are: $c_p = 0.6$ kcal/kg °C = 2.51 kJ/kg K,

$M_A = 60 \text{ kg/kmol}, (-\Delta H) = -15 \text{ kcal/mol} = -62.8 \times 10^3 \text{ kJ/kmol}$. The rate is first order (Grummitt and Fleming [9]):

$$r_A = \left(\frac{1}{60}\right) C_A \exp\left(35.2 - \frac{22450}{T}\right) \frac{\text{kg acid}}{\text{m}^3 \text{ s}} \tag{a}$$

$$\text{with } C_A[=]\text{ kg/m}^3,\ T[=]\text{ K}$$

A constant heat supply is provided by an electrical heater, and a final conversion of 70 percent is desired.

This is an example of case 2 discussed above, and so Eq. 8.2-4 is utilized. First, the adiabatic situation is computed, using Eq. 8.2-5. The adiabatic curve is linear in conversion, and has as a slope the adiabatic temperature change from Eq. 8.2-6:

$$(\Delta T)_{ad} = \frac{(-62.8 \times 10^3 \text{ kJ/kmol})(0.156 \text{ kg/kg})(227 \text{ kg})}{(227 \text{ kg})(2.51 \text{ kJ/kg K})(60 \text{ kg/kmol})} \tag{b}$$

$$= -65 \text{ K}$$

Thus,

$$T = 613 - 65x_A, \text{ K} \tag{c}$$

and is shown in Fig. 1. It is seen that for this endothermic reaction, the temperature drops drastically with adiabatic operation, and heating needs to be considered.

Temperature-conversion curves for other heat inputs were calculated by Cooper and Jeffreys, using Eq. 8.2-4b to obtain $T = T(x_A, \theta)$:

$$T = 613 - 65x_A + \frac{qA_k}{m_t c_p}\theta \tag{d}$$

For the $qA_k = 52.8$ kW curve,

$$\frac{qA_k}{m_t c_p} = \frac{(52.8 \text{ kW})}{(227 \text{ kg})(2.51 \text{ kJ/kg K})} = 0.0927 \ K/s$$

Finally, this is substituted into Eq. (8.1-3) to be integrated:

$$\frac{dx_A}{d\theta} = \frac{1}{60C_{A0}} C_{A0}(1 - x_A)\exp\left(35.2 - \frac{22450}{T}\right) \tag{e}$$

$$= \frac{(1 - x_A)}{60} \exp\left(35.2 - \frac{22450}{613 - 65x_A + 0.0927\theta}\right) \tag{f}$$

Figure 1 shows temperature histories for various amounts of heat input. It is seen that the heat input of 52.8 kW or 0.233 kW/kg (200 kcal/kg-hr) is just sufficient to overcome the endothermic cooling past 40 percent conversion, where the reactor temperature begins to rise.

It is also instructive to look at the conversion-time profile (Fig. 2). For the $qA_k = 52.8$ kW results, the first $\frac{1}{2}(0.7) = 0.35$ of the final conversion is reached in about 2.5 min, and the second half requires the remaining 5 min of the total

THE BATCH REACTOR _____

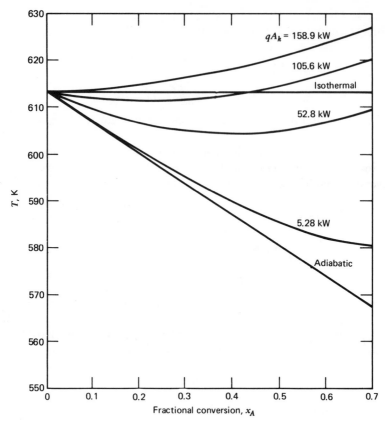

*Figure 1 Temperature-conversion progress for various rates of heat input (from Cooper and Jeffreys [8]).*

batch residence time. This longer time to reach higher conversion is especially severe for the adiabatic case, of course, with its rapid drop in temperature. The total times required for 70 percent conversion are as follows:

| Heat input rate, kW | $\theta_f$ for 70% conversion, min |
|---|---|
| Isothermal, $T = 613$ K | 4.97 |
| Adiabatic, $q = 0$ | 38.25 |
| 5.28 | 23.64 |
| 52.8 | 7.48 |
| 105.6 | 4.72 |
| 158.9 | 3.55 |

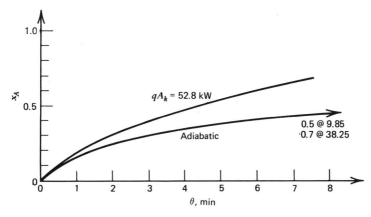

*Figure 2 Conversion versus time curve for adiabatic operation and a heat input rate of 52.8 kW.*

Again notice the large increase in residence time for the smallest heat addition and/or the adiabatic case, caused by the endothermic temperature decrease.

One could also choose the proper heater size to have 70 percent conversion in some chosen time—say 20 min. Here, Eqs. (d) and (e) would have to be solved iteratively for the unknown value of $q$. Actually, after the above range of simulations were available, a simple interpolation is possible; the result is $qA_k = 8$ kW.

## 8.3 Optimal Operation Policies and Control Strategies ____

Two main types of situations are considered:

**1.** Optimal batch operation time for the sequence of operations in a given reactor.

**2.** Optimal temperature (or other variable) variations during the course of the reaction, to minimize the reactor size.

The principles of each of these will be discussed—more extensive details are given in Aris [10]. To simplify the mathematical details, we primarily consider constant volume reactors, but recall from earlier discussion that most practical situations are in this category.

### 8.3.a Optimal Batch Operation Time

The discussion follows that of Aris [11, 18]. The price per kmole of chemical species $A_j$ is $w_j$, and so the net increase in worth of the reacting mixture is

$$W(\theta) = \sum_{j=1}^{N} w_j(N_j - N_{j0}) \tag{8.3.a-1}$$

$$= \sum_{j=1}^{N} w_j \sum_{i=1}^{M} \alpha_{ij} \xi_i$$

$$= \sum_{i=1}^{M} (\Delta W)_i \xi_i$$

where

$$(\Delta W)_i \equiv \sum_{j=1}^{N} \alpha_{ij} w_j \tag{8.3.a-2}$$

which is constant for a given stoichiometry and chemical costs. For a single reaction, it is more common to introduce the conversion of the key species, $A$, into Eq. 8.3.a-1

$$W(\theta) = (\Delta W) \frac{N_{A0} x_A}{|\alpha_A|} \tag{8.3.a-3}$$

The cost of operation is usually based on four steps:

**1.** Preparation and reactor charging time $\theta_P$, with cost per unit time, of $W_P$.

**2.** Reaction time $\theta_R$, with $W_R$.

**3.** Reactor discharge time $\theta_Q$, with $W_Q$.

**4.** Idle or "down" time $\theta_0$, with $W_0$.

The total operation cost then is:

$$W_T = \theta_0 W_0 + \theta_P W_P + \theta_Q W_Q + \theta_R W_R \tag{8.3.a-4}$$

Since our interest is in the reactor operation, all the other times will be taken to be constant, and the main question is to determine the optimal reaction time, with its corresponding conversion. The net profit is

$$W(\theta_R) - W_T \tag{8.3.a-5}$$

and the optimum value of $\theta_R$ is found from

$$\frac{d}{d\theta_R}[W(\theta_R) - W_T] = 0 \tag{8.3.a-6}$$

or

$$\frac{dW(\theta_R)}{d\theta_R} = W_R \qquad (8.3.a\text{-}7)$$

From Eq. 8.3.a-3,

$$\frac{dW(\theta_R)}{d\theta_R} = (\Delta W)\frac{N_{A0}}{|\alpha_A|}\frac{dx_A}{d\theta_R}$$

$$= (\Delta W)V\left.\frac{r_A}{|\alpha_A|}\right|_{\theta_R} \qquad (8.3.a\text{-}8)$$

where the last step used Eq. 8.1-3. Thus, the optimum occurs when

$$\frac{V}{|\alpha_A|}r_A(x_A)|_{\theta_R} = \frac{W_R}{\Delta W} \qquad (8.3.a\text{-}9)$$

The actual optimum reaction time, $\theta_R$, must still, of course, be found from Eq. 8.1-4 evaluated at $x_{AR} = x_A(\theta_R)$ found from Eq. 8.3.a-9:

$$\theta_R = \frac{N_{A0}}{V}\int_0^{x_{AR}}\frac{dx_A}{r(x_A)} \qquad (8.3.a\text{-}10)$$

Instead of the maximum net profit Eq. 8.3.a-5, the maximum of the net profit per unit time may be desired:

$$\frac{W(\theta_R) - W_T}{\theta_T} \qquad (8.3.a\text{-}11)$$

where $\theta_T \equiv \theta_0 + \theta_P + \theta_Q + \theta_R$. Then, the optimum $\theta_R$ is found from:

$$\frac{d}{d\theta_R}\left[\frac{W(\theta_R) - W_T}{\theta_T}\right] = 0 \qquad (8.3.a\text{-}12)$$

or

$$\frac{dW(\theta_R)}{d\theta_R} - W_R = \frac{W(\theta_R) - W_T}{\theta_T} \qquad (8.3.a\text{-}13)$$

Aris [11] has provided a convenient graphical procedure for solving Eq. 8.3.a-13 for the optimal value of $\theta_R$. Figure 8.3.a-1 illustrates a typical curve for net profit. If it is recognized that the right-hand-side of Eq. 8.3.a-13 is precisely the slope of the tangent line $\overline{OL}$, from Eq. 8.3.a-13,

$$\frac{W(\theta_R) - W_T}{\theta_T} = \frac{d}{d\theta_R}[W(\theta_R) - W_T]$$

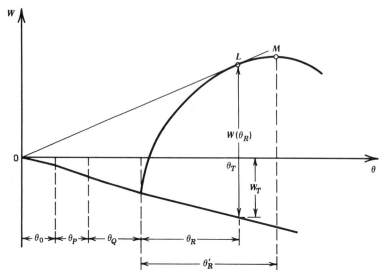

*Figure 8.3.a-1  Net profit curve (from Aris [11]).*

it is seen that the $\theta_R$ indicated in the figure is the one that does satisfy Eq. 8.3.a-13, and is the optimum value for maximum net profit per unit time. The point, $M$, and corresponding $\theta_R'$ gives the optimum for maximum net profit, from Eq. 8.3.a-6.

### Example 8.3.a-1  Optimum Conversion and Maximum Profit for a First-Order Reaction

For a simple first-order reaction, Eq. 8.1-4 gives

$$\theta = C_{A0} \int_0^{x_A} \frac{dx}{kC_{A0}(1 - X_A)} = \frac{-1}{k} \ln(1 - x_A) \tag{a}$$

or

$$x_A = 1 - e^{-k\theta} \tag{b}$$

Thus, from Eq. 8.3.a-3,

$$W(\theta_R) = (\Delta W)N_{A0}(1 - e^{-k\theta_R}) \tag{c}$$

and the value of $\theta_R$ for maximum profit is found from Eq. 8.3.a-7:

$$W_R = (\Delta W)N_{A0}(ke^{-k\theta_R}) \tag{d}$$

or

$$(\theta_R)_{\text{opt}} = \frac{1}{k} \ln\left[\frac{(\Delta W)kN_{A0}}{W_R}\right] \tag{e}$$

CHEMICAL REACTOR DESIGN

The optimum conversion is

$$(x_A)_{opt} = 1 - \left[ \frac{W_R}{(\Delta W)kN_{A0}} \right]$$ (f)

It should be noted that if the result Eq. (f) is substituted into the first-order rate form:

$$V(r_A)_{opt,\theta_R} = VkC_{A0}(1 - x_A) = \frac{W_R}{(\Delta W)}$$

which is Eq. 8.3.a-9 for this situation.

## 8.3.b Optimal Temperature Policies

This section considers two questions: (1) What is the best single temperature of operation? (2) What is the best temperature progression during the reaction time or (as it is sometimes called) the best trajectory? The answers will depend on whether single or complex reaction sequences are of interest. For single reactions, the results are relatively straightforward. If the reaction is *irreversible*, and if the usual situation of the rate increasing with temperature is true, then the optimal temperature for either maximum conversion from a given reactor operation, or minimum time for a desired conversion, is the highest temperature possible. This highest temperature, $T_{max}$, is defined by other considerations such as reactor materials, catalyst physical properties, and the like. Similarly, for reversible *endothermic* reactions where the equilibrium conversion increases with temperature ($E_{for.} > E_{rev.}$), the highest allowable temperature is the best policy.

The case of reversible *exothermic* reactions is more complicated, because even though the rate may increase with temperature, as the equilibrium conversion is reached, higher temperatures have an adverse effect of decreased equilibrium conversion. Thus, there is an optimum intermediate temperature where reasonably rapid rates are obtained together with a sufficiently large equilibrium conversion. The precise value of the optimal temperature can be found with use of Eq. 8.1-4 at the final conversion, $x_{Af}$:

$$\theta_f = C_{A0} \int_0^{x_{Af}} \frac{dx_A}{r(x_A, T)}$$ (8.3.b-1)

This can always—in principle, and usually in practice—be integrated for a constant value of temperature, and then the best temperature found for a given conversion, $x_{Af}$. It can be shown that this is exactly equivalent to the problem of choosing the optimal temperature for the maximum conversion for a given reaction time, $\theta_f$.

## Example 8.3.b-1 Optimal Temperature Trajectories for First Order Reversible Reactions

For a first-order reversible reaction, the reaction rate is:

$$A \underset{2}{\overset{1}{\rightleftharpoons}} S$$

$$r_A = k_1 C_A - k_2 C_S \tag{a}$$
$$= k_1 C_{A0}(1 - x_A) - k_2 C_{A0} x_A \qquad (C_{S0} = 0)$$

It is convenient to use dimensionless variables (e.g., Millman and Katz [12]): with

$$k_1 = A_1 e^{-E_1/RT} \rightarrow u = A_1 \theta_f e^{-E_1/RT} \tag{b}$$

$$k_2 = A_2 e^{-E_2/RT} \rightarrow \beta u^\alpha,$$

where

$$\beta = \frac{\theta_f A_2}{(\theta_f A_1)^\alpha} \tag{c}$$

and

$$\alpha = E_2/E_1$$
$$\tau = \theta/\theta_f$$

Then the mass balance Eq. 8.1-3 becomes

$$\frac{dx_A}{d\tau} = u(1 - x_A) - \beta u^\alpha x_A \tag{d}$$

After the optimum value of $u$ is found, the actual temperature is

$$T = \frac{E_1/R}{\ln(\theta_f A_1/u)} \tag{e}$$

For a given $u$. Eq. (d) can be easily integrated:

$$\tau = \frac{x_{Aeq}}{u} \ln \frac{1 - x_{A0}/x_{Aeq}}{1 - x_A/x_{Aeq}} \tag{f}$$

where the equilibrium conversion is given by

$$x_{Aeq}(u) = (1 + \beta u^{\alpha-1})^{-1} \tag{g}$$

Equation f can be rearranged to

$$x_A = x_{Aeq}[1 - (1 - x_{A0}/x_{Aeq})e^{-u\tau/x_{Aeq}}] \tag{h}$$

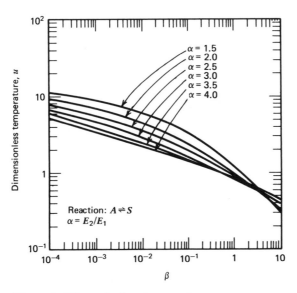

*Figure 1 Dimensionless temperature versus parameter $\beta$ (from Fournier and Groves [13]).*

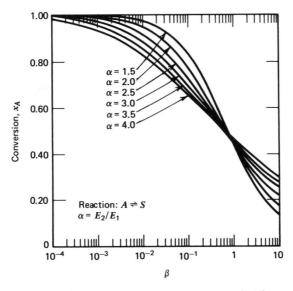

*Figure 2 Conversion versus parameter $\beta$ (from Fournier and Groves [13]).*

379

The value of the optimum $T$, or equivalently $u$, for maximum $x_{Af}$ with a given $\theta_f$ can now be readily found from

$$\left(\frac{\partial x_{Af}}{\partial u}\right)_{\tau=1} = 0 \qquad\qquad \text{(i)}$$

It can be shown that for a single reaction this result is equivalent to the problem of the optimum $u$ for a minimum $\theta_f$ with a given $x_{Af}$; Aris [10] has summarized the results of F. M. J. Horn and others.

Fournier and Groves [13] have provided useful charts based on Eqs. h and i. With Figs. 1 and 2, both equivalent problems can be readily solved by beginning with the known quantities: either $\alpha$ and $\beta(\theta_f)$ or $\alpha$ and $x_{Af}$. Other kinetic schemes have also been evaluated by Fournier and Groves [13].

---

Even better results for the reversible exothermic reaction can be obtained by choosing an optimal temperature variation with time. This type of operation is also feasible in practice, especially with modern automatic control techniques. Qualitative reasoning indicates that a high temperature at the beginning would be best, since this increases the rate constant, and the equilibrium limitations are usually not particularly important at this point. As the reaction progresses, and approaches equilibrium, it is important to have lower temperatures that favor higher equilibrium conversions. Thus, the optimum temperature trajectory would be expected to decrease with time. Also, the maximum overall rate, made up of the cumulative sum of the instantaneous point rates, will be largest if each of the point rates is maximized. This reasoning cannot be extended to multiple reactions, however, since the overall optimum will be made up of the interactions of several rates; this is considered later.

For the single reaction, the condition of optimality to be fulfilled in each point is

$$\frac{\partial r_A}{\partial T} = 0 \qquad\qquad \text{(8.3.b-2)}$$

(A proof is given by, e.g., Aris [11].) This equation can be used for a numerical solution, or in simple cases, it will provide analytical solutions. For the reaction

$$A + B \underset{2}{\overset{1}{\rightleftharpoons}} Q + S$$

the rate is

$$
\begin{aligned}
r_A &= k_1 C_A C_B - k_2 C_Q C_S \\
&= A_1 e^{-E_1/RT} C_{A0}{}^2(1 - x_A)(M - x_A) - A_2 e^{-E_2/RT} C_{A0}{}^2 x_A{}^2 \quad \text{(8.3.b-3)}
\end{aligned}
$$

where

$$M = C_{B0}/C_{A0} \qquad \text{and} \qquad C_{Q0} = 0 = C_{S0}$$

Then, the optimum temperature at each point is found from Eq. 8.3.b-2, with the results

$$
\begin{aligned}
T_{opt} &= \left\{ \left( \frac{-R}{E_1 - E_2} \right) \ln \left[ \left( \frac{A_2 E_2}{A_1 E_1} \right) \left( \frac{x_A{}^2}{(1 - x_A)(M - x_A)} \right) \right] \right\}^{-1} \\
&\equiv \left\{ \frac{1}{(-B_1)} \ln[B_2 B_3] \right\}^{-1}
\end{aligned}
\tag{8.3.b-4}
$$

with

$$
B_1 = (E_1 - E_2)/R \quad \text{and} \quad B_2 = A_2 E_2/A_1 E_1
$$

Fournier and Groves [14] have provided solutions for several other reaction types; the definitions of $B_1$ and $B_2$ are the same for any single reversible reaction, but $B_3$ depends on the reaction type:

| Reaction | $B_3$ |
|---|---|
| $A \rightleftharpoons S$ | $\dfrac{x_A}{1 - x_A}$ |
| $A \rightleftharpoons Q + S$ | $\dfrac{C_{A0} x_A{}^2}{1 - x_A}$ |
| $A + B \rightleftharpoons S$ | $\dfrac{x_A}{C_{A0}(1 - x_A)(M - x_A)}$ |

Other kinetic forms can be similarly handled. The calculation procedure is then as follows. First the result of utilizing Eq. 8.3.b-2, such as Eq. 8.3.b-4, is used to determine $T_{opt}(x_A)$. Then these values are used in the integration of the mass balance Eq. 8.1-4 for $\theta(x_A)$:

$$
\theta = C_{A0} \int_{x_{A0}}^{x_{Af}} \frac{dx_A}{r_A(x_A, T_{opt}(x_A))}
\tag{8.3.b-5}
$$

One complication that occurs can be seen from Eq. 8.3.b-4: for low conversions, $B_3$ may have a sufficiently small value that $B_2 B_3 \leq 1.0$. Then, Eq. 8.3.b-4 gives a value $T_{opt} \to \infty$ (or negative). In practice of course, the temperature will have to be limited to some value $T_{max} < T_{opt}$ over a range of conversion going from zero to some critical value $x_{Ac}$. This critical conversion, $x_{Ac} > 0$, can be found by first using $T_{max}$ in Eq. 8.3.b-5.

A more general consideration of these problems involves the optimization of some sort of "objective function," which usually depends on outlet conversions and total residence time (equipment cost). It is usually difficult to include all possible costs (e.g., safety) and so a simpler compromise quantity, such as selectivity, is often used instead.

Denbigh and Turner [15] consider two major categories:

1. Output problems. These are concerned with the attainment of the maximum output—the amount of reaction product(s) per unit time and reactor volume.

2. Yield problems. These are concerned with maximizing the yield—the fraction of reactant converted to *desired* product.

The first type is most important for simple reactions with no side products and/or very expensive reactors, catalysts, and so on. The second type occurs with complex (usually organic) reactions where the production of undesired products is wasteful. Output problems are somewhat easier to solve in general since their simpler reaction schemes involve less mathematical details.

The above case of single reversible exothermic reactions was an example of an output problem. Intuitive logic led to the qualitative conclusion that the optimum temperature profile was the one that maximized the rate at each point. This was also the quantitative solution, and led to the design techniques presented. For yield problems, if the kinetics are not too complex, the proper qualitative trends of the optimal temperature profiles can also often be deduced by reasoning. However, the quantitative aspects must usually be determined by formal mathematical optimization methods. Simple policies, such as choosing the temperature for maximum local pointwise selectivity, rarely lead to the maximum final overall selectivity because of the complex interactions between the various rates.

A few examples of this qualitative reasoning are worth discussing. Consider the scheme

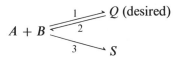

Now, if $E_3 > E_1, E_2 > E_1$, the optimum temperature trajectory is *decreasing* in time, as for the simple reversible, exothermic case, but not quite as high in the beginning to avoid excessive side reaction. If $E_3 > E_1 > E_2$, the reversible reaction is endothermic and so a high temperature level is desirable, but if too high, especially where $C_A$ and $C_B$ are large, too much side reaction occurs. Thus, the optimum trajectory here is *increasing*. If $E_2 > E_1 > E_3$, a decreasing trajectory is again best. Horn (see Denbigh and Turner [15] for references) has worked out the mathematical details of these.

Another example is the familiar

$$A + B \xrightarrow{\ 1\ } Q \xrightarrow{\ 2\ } S \ (Q \text{ desired})$$

If $E_2 > E_1$, the initial temperature should be large for a rapid first reaction but the temperature should be diminished as $Q$ accumulates to preferentially slow down the degradation reaction 2. Here again, a *decreasing* trajectory is best. An example that has two answers depending on whether it is looked at from the output or yield viewpoint is the following:

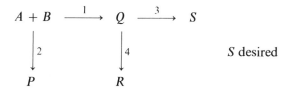

$$A + B \overset{1}{\underset{2}{\diagdown}} \begin{matrix} Q \text{ (desired)} \\ \\ S \end{matrix} \quad , E_2 > E_1$$

From the *output* point of view, the optimum trajectory is an *increasing* one. At the beginning of the reaction, the temperature should be low in order to promote formation of $Q$ rather than $S$, but at the end of the reaction time, the temperature should be high to offset the otherwise low conversion rate—this gives more $Q$ even though it also results in more $S$. If the reactor cost is not important, as in the *yield* problem, the temperature should be *as low as possible* throughout the reaction time. This gives, relatively, the most $Q$ but requires a very large reactor for significant conversion.

As a final example, consider the now classical scheme of Denbigh:

$$A + B \overset{1}{\longrightarrow} Q \overset{3}{\longrightarrow} S$$

$$\downarrow 2 \qquad\qquad \downarrow 4 \qquad\qquad S \text{ desired}$$

$$P \qquad\qquad R$$

Here, the product is formed through an intermediate and both the feed reactants and the intermediates can undergo side reactions. The four possible cases here from the yield viewpoint are:

$$E_1 > E_2, E_3 > E_4 \qquad \text{uniform high temperature}$$
$$E_1 < E_2, E_3 < E_4 \qquad \text{uniform low temperature}$$
$$E_1 < E_2, E_3 > E_4 \qquad \text{increasing trajectory}$$
$$E_1 > E_2, E_3 < E_4 \qquad \text{decreasing trajectory}$$

Denbigh gave some figures for example values of activation energies and showed that for a highest yield of 25 percent under isothermal conditions the optimum temperature trajectory gave over 60 percent; thus, more than double the best isothermal yield was possible.

### *Example 8.3.b-2 Optimum Temperature Policies for Consecutive and Parallel Reactions*

The two basic complex reaction schemes, consecutive and parallel, were considered in an interesting and useful simple way by Millman and Katz [12], and illustrates the computation of optimum temperature trajectories. The details are expressed in dimensionless form, as above:

for consecutive reactions,

$$A \xrightarrow{\;1\;} Q \xrightarrow{\;2\;} S \qquad (Q \text{ desired})$$

$$\frac{dx_A}{d\tau} = u(1 - x_A) \qquad \text{and} \qquad \frac{dx_Q}{d\tau} = u(1 - x_A) - \beta u^\alpha x_Q$$

for parallel reactions,

$$A \underset{2}{\overset{1}{\rightleftarrows}} \begin{matrix} Q \\ S \end{matrix} \qquad (Q \text{ desired})$$

$$\frac{dx_A}{d\tau} = (u + \beta u^\alpha)(1 - x_A) \qquad \text{and} \qquad \frac{dx_Q}{d\tau} = u(1 - x_A)$$

where

$$x_A = \frac{C_{A0} - C_A}{C_{A0}} \qquad x_Q = \frac{C_Q}{C_{A0}} \qquad \tau = \theta/\theta_f$$

$$u = \theta_f A_1 e^{-E_1/RT} \qquad \alpha = E_2/E_1 \qquad \beta = \theta_f A_2/(\theta_f A_1)^\alpha$$

The rigorous optimization could be done with several mathematical techniques— see Beveridge and Schechter [16], and for a concise discussion of the Pontryagin maximum principle see Ray and Szekely [17]; also see Aris [10] for specific chemical reactor examples. Millman and Katz found that the formal optimization techniques were rather sensitive during the calculations and devised a simpler technique whose results appeared to be very close to the rigorous values; it should have further possibilities for practical calculations.

The basic idea was to assume that the temperature trajectory to be determined could be approximated as a linear function of the desired product concentration to be maximized; specifically:

$$u(\tau) \simeq c_0 + c_1 x_Q(\tau)$$

Then the two parameters, $c_0$ and $c_1$, are determined for the optimal condition: $\max\{x_Q(\tau)\}$. This still requires a search technique to obtain the values of $c_0$ and $c_1$, but it was found that these computations were much simpler than the completely rigorous optimization. Actually, further terms in $dx_Q/d\tau$ and $\int x_Q d\tau$ gave better results than the linear function, and are based on standard three-mode process controller actions; however, we will not pursue this further here.

Two typical results are shown in Figs. 1 and 2 and it is seen that the best proportional (simple linear) results are close to the true optimal values. Note that

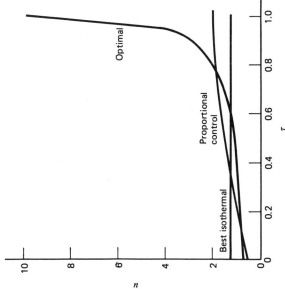

*Figure 2 Temperature histories for parallel reaction:*
$\alpha = 2, \beta = \frac{1}{2}$. *Yields: best isothermal, 0.535; best*
*proportional, 0.559; optimal, 0.575 (from Millman*
*and Katz [12]).*

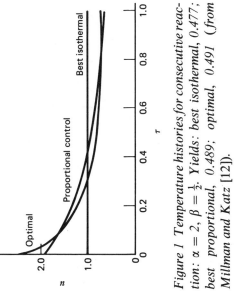

*Figure 1 Temperature histories for consecutive reac-*
*tion:* $\alpha = 2, \beta = \frac{1}{2}$. *Yields: best isothermal, 0.477;*
*best proportional, 0.489; optimal, 0.491 (from*
*Millman and Katz [12]).*

385

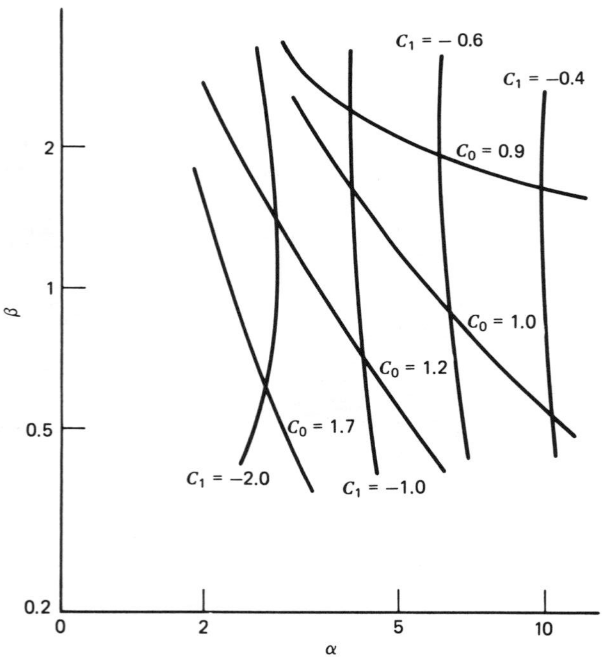

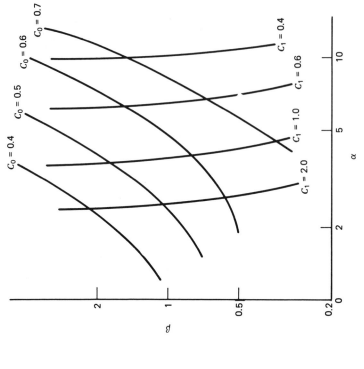

Figure 3 Approximate contour plots of optimal proportional controller settings for consecutive reactions (from Millman and Katz [13]).

Figure 4 Approximate contour plots of optimal proportional controller settings for parallel reactions (from Millman and Katz [12]).

this appears to be true, even though the temperature curves, $u(\tau)$, have some differences between them—apparently the final yield is not particularly sensitive to all the details of the curves.

Also shown in Figs. 3 and 4 are approximate contour plots of the optimal values of the $(c_0, c_1)$ for the two basic reaction types. It can be seen that the $c_0$ are functions of both $\alpha$ and $\beta$, but that $c_1$ primarily depends on the ratio of activation energies, $\alpha = E_2/E_1$. These are only approximate values, and Millman and Katz recommend that they be used as starting values for more detailed calculations.

# Problems

8.1 The esterification of butanol with acetic acid, using sulfuric acid as a catalyst, was studied in a batch reactor:

$$C_4H_9OH + CH_3COOH \xrightarrow{H_2SO_4} C_4H_9-O-\overset{\overset{\displaystyle O}{\displaystyle \|}}{C}-CH_3 + H_2O$$

The reaction was carried out with an excess of butanol. The following data were collected [C. E. Lejes and D. F. Othmer, *I & E.C.* **36**, 968 (1945)].

| Time (hr) | Acetic acid concentration (moles/l) |
|---|---|
| 0 | 2.327 |
| 1 | 0.7749 |
| 2 | 0.4514 |
| 3 | 0.3152 |
| 4 | 0.2605 |

Set up a suitable kinetic model of the homogeneous type.

8.2 The following data on the conversion of hydroxyvaleric acid into valerolactone were collected.

| Time (min) | 0 | 48 | 76 | 124 | 204 | 238 | 289 |
|---|---|---|---|---|---|---|---|
| Acid concentration (mol/l) | 19.04 | 17.6 | 16.9 | 15.8 | 14.41 | 13.94 | 13.37 |

Determine a suitable kinetic model by means of both the differential and integral method of kinetic analysis.

8.3 The batch saponification of ethyl acetate,

$$CH_3COOC_2H_5 + NaOH \rightleftharpoons CH_3COONa + C_2H_5OH,$$

was carried out in a 200-ml reactor at 26°C. The initial concentrations of both reactants were 0.051 N.

(a) From the following time versus concentration data, determine the specific rate and tabulate as a function of composition of the reacting mixture.

| Time, s | NaOH mol/l |
|---------|-----------|
| 30 | 0.0429 |
| 90 | 0.0340 |
| 150 | 0.0282 |
| 210 | 0.0240 |
| 270 | 0.0209 |
| 390 | 0.0164 |
| 630 | 0.0118 |
| 1110 | 0.0067 |

(b) Determine a suitable reaction rate model for this system.

8.4 A daily production of 50,000 kg (50 tons metric) of ethyl acetate is to be produced in a batch reactor from ethanol and acetic acid:

$$\underset{(A)}{C_2H_5OH} + \underset{(B)}{CH_3COOH} \rightleftharpoons \underset{(R)}{CH_3COOC_2H_5} + \underset{(S)}{H_2O}$$

The reaction rate in the liquid phase is given by:

$$r_A = k(C_A C_B - C_R C_S/K)$$

At 100°C:

$$k = 7.93 \times 10^{-6} \text{ m}^3/\text{kmol sec}$$
$$K = 2.93$$

A feed of 23 percent by weight of acid, 46 percent alcohol, and no ester is to be used, with a 35 percent conversion of acid. The density is essentially constant at 1020 kg/m³. The reactor will be operated 24 hours per day, and the time for filling, emptying, and the like, is 1 hour total for reactors in the contemplated size range. What reactor volume is required?

8.5 A gas-phase decomposition $A \rightarrow R + S$ is carried out with initial conditions of: $T_0 = 300$ K, $p_t = 5$ atm, and $V_0 = 0.5$ m³. The heat of reaction is $-1500$ kcal/kmol, and the heat capacity of A, R, and S are 30, 25, and 20 kcal/kmol K. The rate coefficient is

$$k = 10^{14} \exp(-10,000/T), \text{ hr}^{-1}$$

(a) Compute the conversion-time profile for isothermal conditions. Also determine the heat exchange rates required to maintain isothermal conditions.
(b) Compute the conversion-time profile for adiabatic conditions.

388 _____ CHEMICAL REACTOR DESIGN

**8.6** A desired product, $P$, is made according to the following reaction scheme:

$$A \xrightarrow{\;\;1\;\;} P \underset{3}{\overset{2}{\rightleftarrows}} \begin{matrix} X \\ \\ Y \end{matrix} \qquad k_i = A_i e^{-E_i/RT}$$

Discuss qualitatively the optimum temperature profile for the two cases:
(a) $E_2 > E_3 > E_1$
(b) $E_2 > E_1 > E_3$
Describe your reasoning *carefully*.

**8.7** One method of decreasing the large initial heat release in a batch reaction is to utilize "semibatch" operation. Here, the reactor initially contains no reactant, and is filled up with the reacting liquid—thus, there is an inflow but no outflow, and the reacting volume continuously changes. The mass balances are:

Total:
$$\frac{dV}{dt} = F_0'$$

Reactant $A$:
$$\frac{d}{dt}(VC_A) = F_0' C_{A0} - kVC_A$$

(a) Show that the reactant concentration at any time is, with isothermal operation,

$$\frac{C_A}{C_{A0}} = \frac{F_0'(1 - e^{-kt})}{k(V_0 + F_0' t)}$$

where $V_0 =$ initial volume.
(b) Derive an expression for the rate of heat release, and sketch the curve.

**8.8** In a batch reactor having a volume $V = 5\,\mathrm{m}^3$, an exothermic reaction $A \rightarrow P$ is carried out in the liquid phase. The rate equation is

$$r_A = kC_A$$

with

$$k = 4 \times 10^6 \exp(-7900/T), \; S^{-1}$$

The initial temperature, $T_0$, of the reaction mixture is 20°C and the maximum allowable reaction temperature is 95°C. The reactor contains a heat exchanger with area $A_k = 3.3\,\mathrm{m}^2$ and it can be operated with steam ($T_r = 120$°C, $U = 1360\,\mathrm{W/m^2°C}$) or with cooling water ($T_r = 15$°C, $U = 1180\,\mathrm{W/m^2\,°C}$). The times required for filling and emptying the reactor are 10 and 15 min, respectively. Other physicochemical data are: $\Delta H = -1670\,\mathrm{kJ/kg}$; $\rho c_p = 4.2 \times 10^6\,\mathrm{J/m^3°C}$; $M_A = 100\,\mathrm{kg/kmol}$; $C_{A0} = 1\,\mathrm{kmol/m^3}$.

The desired conversion is $x_{Af} \geq 0.9$, and the batch reaction and complete reaction cycle times along with steam and water consumption rates are to be determined for the following policies of operation:
(a) Preheat to 55°C, let the reaction proceed adiabatically, start cooling when either $T = 95$°C or $x_A = 0.9$ occurs, and cool down to 45°C.

(b) Heat to 95°C, let the reaction proceed isothermally until $x_A = 0.9$ occurs, cool down to 45°C. [See H. Kramers and K. R. Westerterp *Elements of Chemical Reactor Design and Operation*, Academic Press, New York (1963).]

8.9 The reversible reaction $A \underset{2}{\overset{1}{\rightleftharpoons}} R$ has the following rate coefficient parameters:

$$A_1 = 7\,s^{-1} \qquad E_1 = 10{,}000 \text{ kcal/kmol}$$
$$A_2 = 5000\,s^{-1} \qquad E_2 = 20{,}000 \text{ kcal/kmol}$$

The reaction is to be carried out in a batch reactor with a maximum allowed temperature of $T_{max} = 800$ K. For a conversion of $x_{Af} = 0.8$:
(a) Determine the optimum isothermal operating temperature, and the resulting batch holding time. Also determine the heat exchange rate required.
(b) If an optimum temperature profile is to be utilized, determine this as a function of conversion and a function of processing time.
(c) Determine the heat exchange rates required for part (b).

*Additional data*

Density of liquid $= 1000$ kg/m$^3$
Heat capacity $= 1$ kcal/kg°C
Initial mole fraction of reactant $A = 0.5$
Molecular weights $= 100$ for $A$
$\qquad\qquad\qquad\quad = 20$ for solvent

8.10 In Example 8.3.b-2 the dimensionless equations for a parallel reaction were derived:

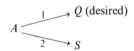

The initial conditions are $x_A = x_Q = 0$ at $\tau = 0$.
(a) Derive an expression for the optimal *single* temperature for $\max_u \{x_Q(\tau = 1)\}$.
(b) For the parameters $\alpha = 2$, $\beta = \frac{1}{2}$, what is $u_{opt}$? If $E_1 = 20{,}000$ kcal/mol, what is $T_{opt}$?

8.11 An endothermic third-order reaction $3A \rightarrow 2B + C$ is carried out in a batch reactor. The reaction mixture is heated up till 400°C. The reaction then proceeds adiabatically. During the heating up period, 10 mol percent of $A$ is converted. From this instant on, what is the time required to reach a conversion of 70 percent?

$$(-\Delta H) = -25000 \text{ kcal/kmol}$$
$$V = 1 \text{ m}^3 = \text{constant}$$
$$c_{pm} = 0.59 \text{ kcal/kg K}$$
$$m_t = 950 \text{ kg}$$
$$N_{A0} = 10.2 \text{ kmol}$$
$$\ln k = -\frac{10000}{RT} + 5 [k \text{ in.(m}^3/\text{kmol } A)^2/s]$$

# References

[1] Levenspiel, O. *Chemical Reaction Engineering*, Wiley, New York (1962).

[2] Peterson, T. I. *Chem. Eng. Sci.*, **17**, 203 (1962).

[3] Chilton, T. H., Drew, T. B., and Jebens, R. H. *Ind. Eng. Chem.*, **36**, 510 (1944).

[4] Chapman, F. S., Dallenbach, H., and Holland, F. A. *Trans. Instn. Chem. Engrs.*, **42**, T398 (1964).

[5] Strek, F. *Int. Chem. Eng.*, **3**, 533 (1963).

[6] Holland, F. A. and Chapman, F. S. *Liquid Mixing and Processing in Stirred Tanks*, Reinhold Publishing Co., New York (1966).

[7] Smith, J. M. *Chemical Engineering Kinetics*, McGraw-Hill, New York (1970).

[8] Cooper, A. R. and Jeffreys, G. V. *Chemical Kinetics and Reactor Design*, Prentice-Hall, Englewood Cliffs, N.J. (1971).

[9] Grummitt, O. and Fleming, F. *Ind. Eng. Chem.*, **37**, 4851 (1945).

[10] Aris, R. *The Optimal Design of Chemical Reactors*, Academic Press, New York (1960).

[11] Aris, R. *Introduction to the Analysis of Chemical Reactors*, Prentice-Hall, Englewood Cliffs, N.J. (1965).

[12] Millman, M. C. and Katz, S. *Ind. Eng. Chem. Proc. Des. Devpt.*, **6**, 447 (1967).

[13] Fournier, C. D. and Groves, F. R. *Chem. Eng.*, **77**, No. 3, 121 (1970).

[14] Fournier, C. D. and Groves, F. R. *Chem. Eng.*, **77**, No. 13, 157 (1970).

[15] Denbigh, K. G. and Turner, J. C. R. *Chemical Reactor Theory*, 2nd ed., Cambridge University Press, London (1971).

[16] Beveridge, G. S. G. and Schechter, R. S. *Optimization Theory and Practice*, McGraw-Hill, New York (1970).

[17] Ray, W. H. and Szekely, J. *Process Optimization*, Wiley, New York (1973).

[18] Aris, R. *Elementary Chemical Reactor Analysis*, Prentice-Hall, Englewood Cliffs, N.J. (1969).

# 9

## THE
## PLUG
## FLOW
## REACTOR

## 9.1 The Continuity, Energy, and Momentum Equations _____

Plug flow is a simplified and idealized picture of the motion of a fluid, whereby all the fluid elements move with a uniform velocity along parallel streamlines. This perfectly ordered flow is the only transport mechanism accounted for in the plug flow reactor model. Because of the uniformity of conditions in a cross section the steady-state continuity equation is a very simple ordinary differential equation. Indeed, the mass balance over a differential volume element for a reactant $A$ involved in a single reaction may be written:

$$F_A - (F_A + dF_A) = r_A \, dV$$

By definition of the conversion

$$F_A = F_{A0}(1 - x_A)$$

so that the continuity equation for $A$ becomes

$$\frac{dx_A}{d\left(\dfrac{V}{F_{A0}}\right)} = r_A \tag{9.1-1}$$

or, in its integrated form:

$$\frac{V}{F_{A0}} = \int_{x_{A0}}^{x_A} \frac{dx_A}{r_A} \tag{9.1-2}$$

Equations 9.1-1 or 9.1-2 are of course, easily derived also from Eqs. 7.2.b-2 or 7.2.b-4 given in Chapter 7. For a single reaction and taking the reactant $A$ as a reference component $\alpha_A = -1$, so that $R_A = -r_A$. Equation 7.2.b-4 then directly yields Eq. 9.1-2.

When the volume of the reactor, $V$, and the molar flow rate of $A$ at the inlet are given, Eq. 9.1-1 permits one to calculate the rate of reaction $r_A$ at conversion $x_A$. For a set of values $(r_A, x_A)$ a rate equation may then be worked out. This outlines how Eqs. 9.1-1 or 9.1-2 may be used for a kinetic analysis, and will be discussed in more detail below.

When the rate of reaction is given and a feed $F_{A0}$ is to be converted to a value of, say $x_A$, Eq. 9.1-2 permits the required reactor volume $V$ to be determined. This is one of the design problems that can be solved by means of Eq. 9.1-2. Both aspects—kinetic analysis and design calculations—are illustrated further in this chapter. Note that Eq. 9.1-2 does not contain the residence time explicitly, in contrast with the corresponding equation for the batch reactor. $V/F_{A0}$, as expressed here in $hr \cdot m^3/kmol\ A$—often called space time—is a true measure of the residence time only when there is no expansion or contraction due to a change in number of moles or other conditions. Using residence time as a variable offers no advantage since it is not directly measurable—in contrast with $V/F_{A0}$.

If there is expansion or contraction the residence time $\theta$ has to be considered first over a differential volume element and is given by

$$d\theta = \frac{dV}{F_A/C_A} \tag{9.1-3}$$

where $F_A$ is the (average) molar flow rate of $A$ in that element. For constant temperature and pressure $C_A$ may be written, as explained already in Chapter 8, Eq. 8.1-8:

$$C_A = \frac{N_A}{V} = \frac{N_{A0}(1 - x_A)}{V_0(1 + \varepsilon_A x_A)} = C_{A0}\frac{1 - x_A}{1 + \varepsilon_A x_A}$$

where $\varepsilon_A$ is the expansion factor, $\varepsilon_A = y_{A0}[(r + s \cdots) - (a + b \cdots)/a]$. Equation 9.1-3 becomes, after formal integration,

$$\theta = \int_0^V \frac{dV}{\dfrac{F_{A0}}{C_{A0}}(1 + \varepsilon_A x_A)} \tag{9.1-4}$$

What remains to be done before the integration is performed is to relate $x_A$ and $V$. This is done by means of Eq. 9.1-1 so that finally:

$$\theta = C_{A0}\int_{x_{A0}}^{x_A} \frac{dx_A}{(1 + \varepsilon_A x_A)r_A} \tag{9.1-5}$$

Equation 9.1-5 shows that the calculation of $\theta$ requires the knowledge of the function $r_A = f(x_A)$. But, establishing such a relation is precisely the objective of a kinetic investigation. The use of $\theta$ is a superfluous intermediate step, since the test of a rate equation may be based directly on Eq. 9.1-2. Note also that when there is no change in number of moles due to the reaction—and only then—is there complete correspondence between Eq. 9.1-5 and the batch reactor equation 8.1-5.

Thus, it is seen that the most direct measure of the reactor's capability for carrying out the conversion is the space time, $V/F_{A0}$, which is the result of making a rigorous mass balance in the steady-state plug flow system. In industrial practice, the reciprocal is commonly used—termed "space velocity." Specifically, using

$$\frac{F_{A0}}{V} = \left(\frac{F'_0}{V}\right)C_{A0}$$

the group $(F'_0/V)$ with units volume of feed (measured at some reference conditions) per unit time, per unit volume of reactor, is the space velocity. One must be careful concerning the choice of the reference conditions, since several customs are in use— see Hougen and Watson [1]. For example, if a liquid feed is metered, and then vaporized before entry into the reactor, it is common to use the liquid volumetric rate rather than the actual gas rate at reactor conditions, which is implied in $F'_0$ corresponding to $C_{A0}$. Use of the molar flow rate, $F_{A0}$, obviates these difficulties, and is the choice in this book. However, the space velocity customs need to be known in order to properly interpret existing literature data.

Take a first-order rate equation $r_A = kC_A$ or $kC_{A0}(1 - x_A)$. Substitution of the latter expression in Eq. 9.1-2 leads to

$$k\frac{V}{F_{A0}/C_{A0}} = \ln\frac{1}{1 - x_A}$$

or in terms of the concentration of $A$:

$$k\frac{V}{F_{A0}/C_{A0}} = \ln\frac{C_{A0}}{C_A}$$

Since for constant temperature, pressure, and total number of moles $V/(F_{A0}/C_{A0})$ is nothing but the residence time, these results are identical to the integrated forms given for a first order reaction in Table 1.3-1 of Chapter 1. All the other reactions considered in that table and those of more complex nature dealt with in the rest of that chapter will lead to the same integrated equations as those given here, provided $\theta$ is replaced by $V/(F_{A0}/C_{A0})$. This will not be so when these reactions are carried out in the flow reactor with complete mixing as will be shown in Chapter 10.

Equations 9.1-1 or 9.1-2 can serve as basic equation for the analysis or design of isothermal empty tubular reactors or of packed catalytic reactors of the tubular type. The applicability of these equations is limited only by the question of how well plug flow is approximated in the real case. For empty tubular reactors this is generally so with turbulent flow conditions and sufficiently high ratio length/diameter so that entrance effects can be neglected, such as in tubular reactors for thermal cracking. Deviations from the ideal plug flow pattern will be discussed in detail in Chapter 12.

394 _____

For fixed bed catalytic reactors the idealized flow pattern is generally well approximated when the packing diameter, $d_p$, is small enough with respect to the tube diameter, $d_t$, to have an essentially uniform void fraction over the cross section of the tube, at least till the immediate vicinity of the wall. According to a rule of thumb the ratio $d_t/d_p$ should be at least 10. This may cause some problems when investigating a catalyst in small size laboratory equipment. The application of the plug flow model to the design of fixed bed catalytic reactors will be dealt with extensively in Chapter 11. For this reason the examples of this chapter deal exclusively with empty tubular reactors.

Tubular reactors do not necessarily operate under isothermal conditions in industry, be it for reasons of chemical equilibrium or of selectivity, of profit optimization, or simply because it is not economically or technically feasible. It then becomes necessary to consider also the energy equation, that is, a heat balance on a differential volume element of the reactor. For reasons of analogy with the derivation of Eq. 9.1-1 assume that convection is the only mechanism of heat transfer. Moreover, this convection is considered to occur by plug flow and the temperature is completely uniform in a cross section. If heat is exchanged through the wall the entire temperature difference with the wall is located in a very thin film close to the wall. The energy equation then becomes, in the steady state:

$$\sum_j \dot{m}_j c_{pj}\, dT + U\pi\, d_t(T - T_r)\, dz - r_A(-\Delta H)\, dV = 0 \qquad (9.1\text{-}6)$$

where $\dot{m}_j$:    mass flow rate of the component $j$ (kg/hr)

    $c_{pj}$:    specific heat of $j$ (kcal/kg°C)

    $T, T_r$:   temperature of the fluid, respectively the surroundings (°C)

    $U$:     overall heat transfer coefficient (kcal/m² hr °C), based on the inside diameter of the tube. The formula for $U$ and a correlation for the inside heat transfer coefficient have been given already in Chapter 8.

    $z$:     length coordinate of the reactor ($m$)

Note that Eq. 9.1-6 is nothing but Eq. 7.2.d-4 of Chapter 7, obtained by simplifying the "general" energy Eq. 7.2.c-1, provided that $d_r$, the reactor diameter, is replaced by $d_t$, the tube diameter. The benefit of using the general equations of Chapter 7 is that the precise assumptions required to use a given form of the balances is clear, and also the route required to improve an inadequate model has then been outlined.

Equations 9.1-1 and 9.1-6 are coupled through the rate of reaction. The integration of this system of ordinary differential equations generally requires numerical techniques. Note that the group $\sum_j \dot{m}_j c_{pj}$ has to be adjusted for each increment. It is often justified to use a value of $c_p$ averaged over the variations of temperature and compositions so that $\sum_j \dot{m}_j c_{pj}$ may be replaced by $\dot{m}c_p$, where $\dot{m}$ is the total mass flow rate. $(-\Delta H)$ is frequently also averaged over the temperature interval in the reactor.

Introducing Eq. 9.1-1 into Eq. 9.1-6 leads to

$$\frac{dT}{dz} = \frac{F_{A0}(-\Delta H)}{\sum_j \dot{m}_j c_{pj}} \frac{dx_A}{dz} - \frac{U\pi d_t}{\sum_j \dot{m}_j c_{pj}} (T - T_r) \tag{9.1-7}$$

It follows that for an adiabatic reactor, for which the second term on the right hand side is zero, there is a direct relation between $\Delta x$ and $\Delta T$.

Equations 9.1-1 and 9.1-6 or 9.1-7 are applicable to both empty tubular reactors or fixed bed tubular reactors provided the assumptions involved in the derivation are fulfilled. Again, the application to the latter case is discussed in detail in Chapter 11.

Sometimes the pressure drop in the reactor is sufficiently large to be necessary to account for it, instead of using an average value. For an empty tube the Fanning equation may be used in the usual Bernoulli equation:

$$-\frac{dp_t}{dz} = 2f\alpha \frac{\rho_g u^2}{d_t} + \alpha \rho_g u \frac{du}{dz} \tag{9.1-8}$$

(assuming no significant effects of elevation changes). The value of the conversion factor, $\alpha$, depends on the dimensions of the total pressure, $p_t$ and the flow velocity, $u$. Some values are listed in Table 9.1-1.

Table 9.1-1 Values of $\alpha$, conversion factor in the Fanning pressure drop equation

| $p_t$ | $\alpha$ | |
|---|---|---|
| | $u$ (m/s) | $u$ (m/hr) |
| N/m$^2$ | 1 | $7.72 \times 10^{-8}$ |
| bar | $10^{-5}$ | $7.72 \times 10^{-13}$ |
| atm | $9.87 \times 10^{-6}$ | $7.62 \times 10^{-19}$ |
| kgf/m$^2$ | $\dfrac{1}{g_c} = \dfrac{1}{9.81} = 0.102$ | $7.78 \times 10^{-9}$ |

The Fanning friction factor, $f$, equals 16/Re for laminar flow in empty tubes. An expression that is satisfactory for Reynolds numbers between 5000 and 200,000 (i.e., for turbulent flow) is

$$f = 0.046\, Re^{-0.2}$$

CHEMICAL REACTOR DESIGN

When the density of the reaction mixture varies with the conversion, $\rho_g$ in Eq. 9.1-8 has to also account for this. This is illustrated in the example on the thermal cracking of ethane, later in this chapter. Pressure drop equations for packed beds will be discussed in Chapter 11.

## Example 9.1-1 Derivation of a Kinetic Equation from Experiments in an Isothermal Tubular Reactor with Plug Flow. Thermal Cracking of Propane

The thermal cracking of propane was studied at atmospheric pressure and 800°C in a tubular reactor of the integral type. The experimental results are given in Table 1.

Table 1 Thermal cracking of propane. Conversion versus space time data

| $V/F_{A0}$ (m³ s/kmol) | $x_A = \dfrac{F_{A0} - F_A}{F_{A0}}$ |
|---|---|
| 32 | 0.488 |
| 50 | 0.630 |
| 59 | 0.685 |
| 64 | 0.714 |
| 75 | 0.760 |
| 82 | 0.782 |

The global reaction propane → products is considered to be irreversible. When first order is assumed, the rate equation may be written:

$$r_A = kC_A$$

The kinetic analysis starts from either Eqs. 9.1-1 or 9.1-2.

This reaction is carried out in the presence of a diluent, steam. The diluent ratio is $\kappa$ (moles diluent/moles hydrocarbon). Furthermore, 1 mole of propane leads to 2 moles of products, in other words the molar expansion $\delta_A = 1$. The relation between the propane concentration and the conversion has to account for the dilution and expansion and is obtained as follows. For a feed of $F_{A0}$ moles of propane per second, the flow rates in the reactor at a certain distance where a conversion $x_A$ has been reached may be written

$$
\begin{array}{ll}
\text{Propane:} & F_{A0}(1 - x_A)(\text{kmol/s}) \\
\text{Products:} & F_{A0}(1 + \delta_A)x_A \\
\text{Diluent:} & F_{A0}\kappa
\end{array}
$$

---

$$
\text{Total:} \qquad F_{A0}[1 - x_A + (1 + \delta_A)x_A + \kappa]
$$

so that, for the feed rate $F_{A0}$:

$$
F_t = F_{A0}(1 + \delta_A x_A + \kappa)
$$

while the mole fraction of propane consequently equals $(1 - x)/(1 + \delta_A x_A + \kappa)$ and the concentration $[(1 - x_A)/(1 + \delta_A x_A + \kappa)]C_t$. The diluent ratio, $\kappa$, is often used in industrial practice, although exactly the same end results are found with the use of

$$
\varepsilon_A = y_{A0}\delta_A = \frac{1}{1 + \kappa}\,\delta_A.
$$

The rate equation that has to be introduced in Eqs. 9.1-1 or 9.1-2 now becomes

$$
r_A = k\,\frac{1 - x_A}{1 + \delta_A x_A + \kappa}\,C_t
$$

*Integral Method of Kinetic Analysis*

Substituting the rate equation in Eq. 9.1-2 leads to:

$$
\frac{V}{F_{A0}} = \frac{1}{kC_t}\int_0^{x_A}\frac{1 + \delta_A x_A + \kappa}{1 - x_A}\,dx_A
$$

from which, with $\delta_A = 1$,

$$
k = -\frac{F_{A0}}{C_t V}\left[(2 + \kappa)\ln(1 - x_A) + x_A\right]
$$

$k$ can be calculated for a set of experimental conditions, remembering that $C_t = p_t/RT$. For $V/F_{A0} = 32$ and $x_A = 0.488$ a value of $4.14\text{ s}^{-1}$ is obtained for $k$. When $k$ takes the same value for all the sets of $x_A$ versus $V/F_{A0}$ data the asumption of first order is correct. We can see from Table 2 that this condition is indeed fulfilled.

*Differential Method of Kinetic Analysis*

The slope of the tangent at the curve $x_A$ versus $V/F_{A0}$ is the rate of reaction of $A$ at the conversion $x_A$, from Eq. 9.1-1. The rates are shown in Table 2.

Table 2 Thermal cracking of propane. Rate
versus conversion. k-values from the integral
and differential method of kinetic analysis

| | $r_A$ | $k$ $(s^{-1})$ | |
| $x_A$ | $(kmol/m^3\ s)$ | Integral | Differential |
|---|---|---|---|
| 0.488 | 0.00978 | 4.14 | 4.15 |
| 0.630 | 0.00663 | 4.11 | 4.11 |
| 0.685 | 0.00543 | 4.11 | 4.04 |
| 0.714 | 0.00492 | 4.15 | 4.08 |
| 0.760 | 0.00409 | 4.09 | 4.11 |
| 0.782 | 0.00354 | 4.03 | 3.94 |

If the order of reaction were $n$, the rate equation would have to be written

$$r_A = kC_A{}^n$$

which becomes, after taking the logarithm,

$$\log r_A = \log k + n \log C_t + n \log\left(\frac{1 - x_A}{1 + \delta_A x_A + \kappa}\right)$$

A straight line is obtained in a plot of $\log r_A$ versus $\log[(1 - x_A)/(1 + \delta_A x_A + \kappa)]$. The slope is the order, while the intercept on the ordinate yields $k$.

If an order of one is assumed, the formula permits checking the constancy of $k$. For $x_A = 0.488$ and $r_A = 0.00978$ kmol/m³ s a $k$ value of 4.15 s⁻¹ is obtained, for $x_A = 0.714$ and $r_A = 0.00492$ kmol/m³ s a value of 4.08 s⁻¹ is obtained. The assumption of first order is verified.

## 9.2 Kinetic Analysis of Nonisothermal Data

The above example deals with a simple isothermal situation. In Chapters 1 and 2 it was suggested to operate reactors for kinetic studies, whenever possible, in an isothermal way. There are cases, however, in which isothermal operation is impossible, in spite of all precautions, for example, a homogeneous reaction like thermal cracking of hydrocarbons. In such a case it is inevitable that part of the reactor is used to bring the feed to the desired temperature. In contrast with catalytic reactors there is no clear-cut separation between preheat and reaction section in such a case. If the rate is to be determined at a reference temperature, say $T_1$, and if the reaction volume is counted from the point where $T_1$ is reached, then the conversion in the preheat section that cannot be avoided is not accounted for. Similarly at the outlet there is a section where the conversion continues to some extent while the reaction mixture is being cooled.

Such a situation can be dealt with in two ways. The first way is to analyze the data as such. The temperature dependence of the rate parameters is then directly included into the continuity equation and the resulting equation is numerically integrated along the tube with estimates for the parameters. If the gas temperature profile itself is not available or insufficiently defined, the energy equation has to be coupled to the continuity equation. To determine both the form of the rate equation and the temperature dependence of the parameters directly from nonisothermal data would require excessive computations.

Recently, however, apparently successful attempts have been reported of the derivation of rate parameters from the direct treatment of nonisothermal data, given the form of the rate equation. (See Emig, Hofmann, and Friedrich [2]; Lambrecht, Nussey, and Froment [3]; and Van Damme, Narayanan, and G. F. Froment [4].) The work of Van Damme et al. [4] on the kinetic analysis of the cracking of propane is taken as an example. In this work the gas temperature rose from 600°C at the inlet of the cracking section to 850°C at the exit, to simulate industrial operation. Since the gas temperature profile was given, the Arrhenius temperature dependence was directly accounted for in the continuity equation for propane, but no energy equation had to be coupled to it. The pressure profile was also directly accounted for. The resulting continuity equation was numerically integrated assuming a power law rate equation and with estimated values for the order with respect to propane, $n$, for the frequency factor, $A_0$, and for the activation energy, $E$. The calculated exit propane conversions were compared with the experimental. The sum of squares of deviations between calculated and experimental propane conversions was used as an objective function; the latter was minimized by nonlinear regression using Marquardt's routine.

The strong correlation between $A_0$ and $E$ necessitated a reparameterization. Setting

$$A_0' = A_0 \exp\left(-\frac{E}{R\overline{T}}\right)$$

where $\overline{T}$ represents the average of all the measured temperatures, the continuity equation for propane

$$F_{A0} \, dx = k\left(\frac{1-x}{1+\kappa+\delta x}\right)^n \left(\frac{p_t}{RT}\right)^n dV$$

became

$$F_{A0} \, dx = A_0' \exp\left(\frac{E}{R\overline{T}} - \frac{E}{RT}\right)\left[\frac{1-x}{1+\kappa+\delta x}\right]^n \left(\frac{p_t}{RT}\right)^n dV$$

For 1.4 atm abs (1.37 bar) and a steam dilution of 0.4 kg steam/kg propane, Van Damme et al. obtained $n = 1.005$, $E = 51167$ kcal/kmol $= 214226$ kJ/kmol, and $A_0 = 1.7.10^{11}$ as compared with $n = 1$, $E = 51000$ kcal/kmol $= 213500$ kJ/kmol, and $A_0 = 1.08 \times 10^{11}$ by a pseudo-isothermal analysis using the equivalent reactor volume concept to be described next.

The equivalent reactor volume concept, introduced by Hougen and Watson [1] allows for a second way of dealing with nonisothermal data: it first reduces the data to isothermality and determines the temperature dependence of the rate parameters in the second stage only. The equivalent reactor volume has been defined as that volume $V_R$, which, at the reference temperature $T_1$ and the reference total pressure $p_{t1}$, would give the same conversion as the actual reactor, with its temperature and pressure profiles. It follows that

$$r_{T_1, p_{t1}} \, dV_R = r_{T, p_t} \, dV$$

so that, for a reaction with order $n$,

$$V_R = \int_0^V \left(\frac{p_t T_1}{p_{t1} T}\right)^n \exp\left[\frac{E}{R}\left(\frac{1}{T_1} - \frac{1}{T}\right)\right] dV$$

Once $V_R$ has been derived, the calculation of the rate is straightforward, as for isothermal experiments, and is based solely on the continuity equation. Calculating $V_R$ requires the knowledge of the temperature and pressure profiles along the tube and of the activation energy, $E$. Note also that where several reactions are occurring simultaneously the dependence of $V_R$ on $E$ leads to different $V_R$ for each of the reactions considered.

In a kinetic study the activation energy is generally not known a priori, or only with insufficient accuracy. The use of the equivalent reactor volume concept therefore leads to a trial-and-error procedure: a value of $E$ is guessed and with this value and the measured temperature profile $V_R$ is calculated by graphical or numerical integration. Then, for the rate model chosen, the kinetic constant is derived. This procedure is carried out at several temperature levels and from the temperature dependence of the rate coefficient, expressed by Arrhenius' formula, a value of $E$ is obtained. If this value is not in accordance with that used in the calculation of $V_R$ the whole procedure has to be repeated with a better approximation for $E$.

Froment et al. [5,6] proposed a short-cut method for the first estimate of $E$, which turned out to be extremely efficient. Consider two experiments carried out in an isobaric flow reactor, one at a reference temperature $T_1$, the other at the reference temperature, $T_2$ and let the conditions be such that the temperature difference $\Delta T = T_1 - T_2$ is the same over the whole length of the reactor. The reaction taking place is homogeneous and of the type $A \rightarrow B$. If the feed rates are adjusted so that equal conversions are obtained then the conversion $x$ or the $p_A$ versus $V$

profiles are identical in both cases. Then in all points:

$$\frac{dx_1}{dV} = \frac{dx_2}{dV}$$

$$(p_A)_1 = (p_A)_2$$

while $\Delta T$ is independent of $V$.

From the continuity equation

$$F_{A0}\, dx = A_0 \exp\left(-\frac{E}{RT}\right)(p_A)^n\, dV$$

applied to both experiments it follows that

$$\frac{(F_{A0})_1}{(F_{A0})_2} = \exp\left[-\frac{E}{R}\left(\frac{1}{T_1} - \frac{1}{T_2}\right)\right] = \exp\left(\frac{E}{R} \cdot \frac{\Delta T}{T_1 \cdot T_2}\right)$$

from which

$$E = 2.3R\, \frac{T_1 \cdot T_2}{\Delta T} \log \frac{(F_{A0})_1}{(F_{A0})_2}$$

This means that the activation energies may be obtained from two experiments at different temperatures, without even knowing the rate constants, provided the conversions are the same and the temperature profiles, plotted versus $V$, are parallel.

The latter condition is not always fulfilled in practice. It requires that the heat effect of the reaction is negligible or entirely compensated for by the heat flux from or to the surroundings or (and) that the specific heat of the gases is very large. If the reactant $A$ is consumed by more than one reaction than, at equal conversion to the product of interest, $B$, the partial pressure $p_A$ is only equal in both experiments when the activation energies of the parallel reactions are equal. If not, the approximation is better the more the principal reaction prevails over the side reaction(s). Froment et al. [6] applied the $V_R$ concept and the short-cut method for estimation of $E$ to their data on the thermal cracking of acetone. Since then it was also successfully applied by Buekens and Froment to the thermal cracking of propane and isobutane [7,8] and by Van Damme et al. [3] to the thermal cracking of propane and propane-propylene mixtures.

## *Example 9.2-1 Derivation of a Rate Equation for the Thermal Cracking of Acetone from Nonisothermal Data*

When submitted to thermal cracking conditions acetone decomposes according to the overall reaction:

$$CH_3COCH_3 \longrightarrow CH_2{=}CO + CH_4$$

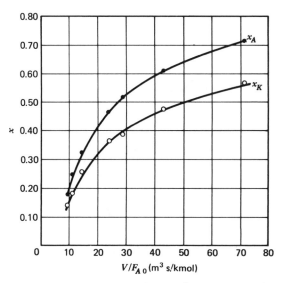

*Figure 1 Acetone cracking. Conversion versus space-time diagram at 750°C (from Froment, et al. [5, 6]).*

that may be considered irreversible in the range of practical interest (700 to 750°C). Ketene and methane are not the only products, however. In the range considered, ethylene, carbon monoxide and dioxide, hydrogen, and carbon are also obtained, probably according to the overall reactions:

$$2CH_2{=}CO \longrightarrow CH_2{=}CH_2 + 2CO$$
$$CH_3COCH_3 \longrightarrow 3H_2 + CO + 2C$$

From isobaric experiments at atmospheric pressure in a laboratory flow reactor with 6 mm inside diameter and 1.20 m length, Froment et al. [5] obtained at 750°C the $x$ versus $V/F_{A0}$ diagram of Fig. 1.

$$\text{whereby } x_A = \frac{\text{moles of acetone decomposed}}{\text{moles of acetone fed}}$$

$$x_K = \frac{\text{moles of ketene formed}}{\text{moles of acetone fed}}$$

We see how the curves do not extrapolate through the origin. This results from the fact that not all of the volume accounted for is at the reference temperature considered. The equivalent reactor volume concept will be used to reduce the data to "isothermal" conditions.

THE PLUG FLOW REACTOR ——————————————— **403**

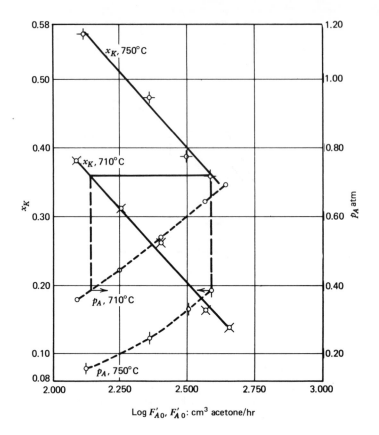

Figure 2 *Acetone cracking. Short-cut method for estimation of activation energy (from Froment, et al. [5, 6]).*

First the short-cut method is used to estimate the activation energy. In Fig. 2 $x_A$ and $p_A$ are plotted versus log $F'_{A0}$ for two series of experiments, one at 750°C, the other at 710°C. The two $x_K$ lines are parallel, while although side reactions do occur, equal values of $p_A$ correspond very nearly to equal $x_K$. The conditions for a satisfactory estimate of $E$ are fulfilled. The horizontal distance between the two parallel $x_K$ lines leads to a value of 51,800 kcal/kmol (216,900 kJ/kmol). This value looks quite plausible. $E$ for the cracking of diethylether is 53,500 kcal/kmol (223,000 kJ/kmol), for dimethylether 47,000 kcal/kmol (196,800 kJ/kmol).

With this value of $E$ and the temperature profiles the equivalent reactor volumes may be obtained as shown in Fig. 3. The curve $x_K$ versus $V_R/F_{A0}$ is shown in Fig. 4. The curves now extrapolate through the origin. With such a diagram the derivation of a rate equation may now be undertaken.

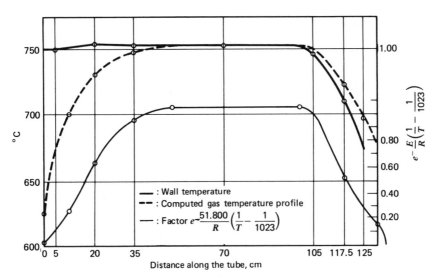

*Figure 3  Acetone cracking. Calculation of equivalent reactor volume (from Froment, et al. [5, 6]).*

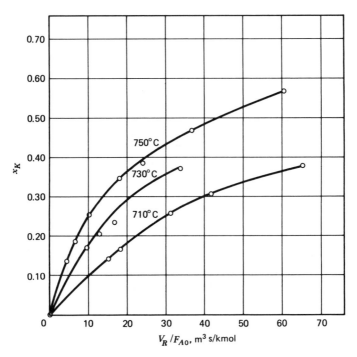

*Figure 4  Acetone cracking. Corrected conversion versus space-time curves (from Froment, et al. [5, 6]).*

405

When a rate equation of the form

$$r_A = k p_A^n \qquad \text{(a)}$$

is postulated the continuity equation for acetone reacting into ketene becomes:

$$\frac{V_R}{F_{AO}} = \frac{1}{k} \int_0^x \frac{dx_k}{p_A^n} \qquad \text{(b)}$$

The differential method is based on Eq. (a), the integral on Eq. (b).

*Differential Method of Kinetic Analysis*

The rate, $r_A$, has to be derived from a $p_A$ versus $V_R/F_{AO}$ plot by graphical differentiation or by fitting a mathematical function to the experimental points first and then differentiating analytically. The values of $k$ and $n$ are then obtained from a log plot of Eq. (a) by means of a least-square fit of the points to a straight line. The results are shown in Table 1.

*Table 1 Thermal cracking of acetone. Rate coefficients and order by the integral and differential methods of kinetic analysis*

| | Differential | | Integral | |
|---|---|---|---|---|
| °C | k | n | k | n |
| 750 | 0.041 | 1.48 | 0.040 | 1.50 |
| 730 | 0.023 | 1.37 | 0.024 | 1.50 |
| 710 | 0.014 | 1.54 | 0.013[a] | — |

[a] When $n = 1.50$.

*Integral Method of Kinetic Analysis*

Before the integral in Eq. (b) may be worked out it is necessary to express $p_A$ as a function of $x_k$. A rigorous expression would only be possible if all reactions taking place were exactly known. Therefore an empirical fit of this function was undertaken. The function was found to be, for the temperature range investigated of 710 to 750°C:

$$p_A = (1 - 1.05 x_K)^2$$

CHEMICAL REACTOR DESIGN

Equation (b) then becomes

$$\text{for all } n \neq \tfrac{1}{2} \qquad k = \frac{1}{1.05 \dfrac{V_R}{F_{AO}}(2n-1)}\left[\frac{1}{(1-1.05x_K)^{2n-1}} - 1\right] \qquad \text{(c)}$$

$$\text{for } n = \tfrac{1}{2} \qquad k = \frac{1}{1.05 \dfrac{V_R}{F_{AO}}} \ln \frac{1}{1-1.05x_K} \qquad \text{(d)}$$

It follows that, when the values of $x_K$ and $V_R/F_{AO}$ corresponding to the different experiments are substituted in Eq. (c) or Eq. (d) $k$ becomes a function of $n$ only for each experiment. The point of intersection of the $k$ versus $n$ curves should give the value of $k$ at the temperature considered and the unique value of $n$. This is shown in Fig. 5 for 750°C. The order is found to be 1.5, also at 730° and 710°C. This order is quite plausible on the basis of radical mechanisms for the reaction. The values of $k$ are given in Table 1. The Arrhenius diagram for $k$ is shown in Fig. 6.

A value for $E = 52,900$ kcal/kmol (221,500 kJ/kmol) is obtained, very close to that obtained by the short-cut method 51,800 kcal/kmol (216,900 kJ/kmol), so that no iteration is required.

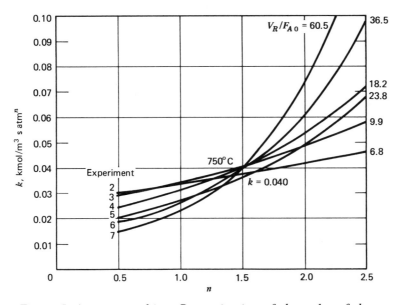

*Figure 5  Acetone cracking. Determination of the order of the re-action at 750°C (from Froment et al. [5, 6]).*

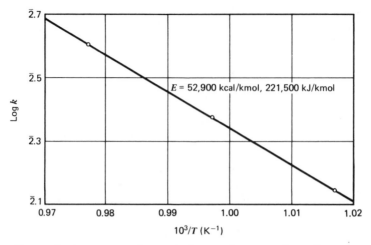

*Figure 6 Acetone cracking. Arrhenius diagram (from Froment, et al. [5, 6]).*

The rate equation for the ketene formation from acetone may therefore be written:

$$r_A = \exp\left(22.780 - \frac{26,600}{T}\right)p_A^{1.5}$$

## 9.3 Design of Tubular Reactors with Plug Flow

It is clear from the preceding that the kinetic analysis of a process based upon non-isothermal data may be a demanding problem from the computational point of view. The reverse problem: designing a reactor when the kinetics are known is generally much more straightforward. In this section two examples of the design of a nonisothermal tubular reactor with plug flow are given. The first example deals with a very simple situation allowing (semi) analytical integration. The second example deals with a reactor for the thermal cracking of hydrocarbons.

### *Example 9.3-1 An Adiabatic Reactor with Plug Flow Conditions*

For simple irreversible reactions a (semi) analytical solution of the continuity and energy equations is possible. Douglas and Eagleton [9] published solutions for zero-, first-, and second-order reactions, both with a constant and varying number of moles. For a first-order reaction with constant density the integration proceeds as follows:

Continuity equation for $A$:

$$F_{A0}\, dx = r_A\, dV \tag{a}$$

With $r_A = kp_{A0}(1 - x)$ Eq. (a) may be written:

$$dx = kp_{A0}(1 - x)\frac{dV}{F_{A0}}$$

Energy equation:

$$F_{A0}\, dx(-\Delta H) = \dot{m}c_P\, dT$$

or

$$dx = \frac{\dot{m}c_p}{F_{A0}(-\Delta H)}\, dT \tag{b}$$

and after integration:

$$x - x_0 = \frac{\dot{m}c_p}{F_{A0}(-\Delta H)}\,(T - T_0)$$

or

$$x = x_0 + \lambda(T - T_0) \tag{c}$$

where

$$\lambda = \frac{\dot{m}c_p}{F_{A0}(-\Delta H)}$$

Note the simple relation between the conversion and the temperature variation in adiabatic situations: the variation in temperature is a measure of conversion and vice versa.

Formal integration of Eq. (a) leads to:

$$\frac{V}{F_{A0}} = \int_{x_0}^{x}\frac{dx}{kp_{A0}(1 - x)}$$

After substituting $dx$ with its expression based on Eq. (c) and of $k$ by its Arrhenius expression, we obtain

$$\frac{V}{F_{A0}} = \frac{1}{A_0}\int_{T_0}^{T}\frac{e^{E/RT}}{p_{A0}}\frac{\lambda\, dT}{\{1 - [x_0 + \lambda(T - T_0)]\}}$$

$$= \frac{\lambda}{A_0 p_{A0}}\int_{T0}^{T}\frac{e^{E/RT}\, dT}{(1 - x_0 + \lambda T_0) - \lambda T}$$

$$= \frac{\lambda}{A_0 p_{A0}}\int_{T0}^{T}\frac{e^{E/RT}\, dT}{\lambda T\left[\dfrac{1 - x_0 + \lambda T_0}{\lambda T} - 1\right]} \tag{d}$$

Let

$$u = \frac{E}{RT} \qquad \text{then } T = \frac{E}{Ru} \qquad \text{and} \qquad dT = -\frac{E\,du}{Ru^2}$$

Eq. (d) then becomes:

$$= -\frac{1}{A_0 p_{A0}} \int_{E/RT_0}^{E/RT} \frac{e^u\,du}{u\left[\frac{R}{\lambda E}(1 - x_0 + \lambda T_0)u - 1\right]}$$

$$= \frac{1}{A_0 p_{A0}} \int_{E/RT_0}^{E/RT} \frac{e^u\,du}{u} - \frac{1}{A_0 p_{A0}} \int_{E/RT_0}^{E/RT} \frac{e^u\,du}{u - \dfrac{\lambda E}{R(1 - x_0 + \lambda T_0)}}$$

Let

$$Z = u - \frac{E}{R(1 - x_0 + \lambda T_0)} \equiv u - \alpha, \, dZ = du$$

$$Z_0 = \frac{E}{RT_0} - \frac{E}{R(1 - x_0 - \lambda T_0)}$$

Then

$$\frac{V}{F_{A0}} = \frac{1}{A_0 p_{A0}} \int_{E/RT_0}^{E/RT} \frac{e^u\,du}{u} - \frac{e^\alpha}{A_0 p_{A0}} \int_{Z_0}^{Z} \frac{e^Z\,dZ}{Z}$$

$$\frac{V}{F_{A0}} = \frac{1}{A_0 p_{A0}}\left[Ei\left(\frac{E}{RT}\right) - Ei\left(\frac{E}{RT_0}\right)\right] - \frac{e^\alpha}{A_0 p_{A0}}[Ei(Z) - Ei(Z_0)] \qquad (e)$$

For given feed conditions, Eq. (e) permits the calculation of the $V/F_{A0}$, which limits the outlet temperature and therefore the outlet conversion to a set value. Obviously for a given $V/F_{A0}$ one can calculate the corresponding outlet conditions, but the expression is implicit with respect to $T$.

For more complicated rate equations semianalytical integration is no longer possible.

---

### Example 9.3-2  Design of a Nonisothermal Tubular Reactor for Thermal Cracking of Ethane

The thermal cracking of hydrocarbons is carried out in long coils that are horizontally or vertically placed inside a gas-fired furnace. The burners are located on both sides of the tubes. The furnace consists (1) of a convection section in which the hydrocarbon feed and the steam diluent are preheated and (2) of a radiant section in which the reaction takes place. A given conversion per pass has to be achieved in the cracking coil, together with an optimum product distribution. If the conversion is too low, the product distribution may not meet the specifications; if it is too high, unwanted side reactions lead to strong coke formation and frequent shutdowns of the furnace.

Figure 1 schematically represents an ethane cracker with horizontal coils. Two coils are running in parallel through the furnace. The coil length in the radiant section is 95 m. The length of the straight portions of the coil is 8.85 m, the length of the bends 0.55 m. The radius of the latter is 0.178 m. The internal diameter of the tube is 0.108 m. The ethane feed per coil is 68.68 kg/m²·s. The ethane is 98.2 mol percent pure, the impurities being $C_2H_4$ (1 mol percent) and $C_3H_6$ (0.8 mol percent). The steam dilution amounts to 0.4 kg of steam per kilogram of ethane. The inlet pressure is 2.99 atm abs (2.93 bars) and the outlet pressure 1.2 atm abs (1.18 bars). The temperature is measured in three locations: inlet, 680°C; 80 percent of coil length, 820°C; exit, 835°C. The ethane conversion at the exit is 60 percent. The products of the cracking are hydrogen, methane, acetylene, ethylene, propadiene, propylene, propane, butenes, butadiene, and small amounts of benzene—all building blocks of the petrochemical industry. The yearly ethylene production capacity is of the order of 10,000 tons/coil.

In early work the simulation of such a furnace was attempted on the basis of the overall rate of disappearance of the hydrocarbon feed (see Buekens and Froment [10]). The advantage is that only one continuity equation has to be used for ethane in the present example, but this approach does not generate the product distribution. Knowing the exit conversion the product distribution can be obtained, however, from yield versus conversion diagrams. For example, the ethylene yield is defined as the number of kilograms of ethylene produced per kilogram of ethane fed. The product distribution obtained in this way is only correct when the yield-conversion relation is independent of temperature. Fortunately, this is very nearly so for the thermal cracking of paraffins, at least in the usual range covered by industrial operation. Another difficulty is the heat of reaction, which has to be substituted in the energy equation Eq. 9.1-6. Since the reaction consists of many parallel and consecutive steps it is not possible to assign a single fixed value for the $(-\Delta H)$ of the overall reaction. Global $(-\Delta H)$ values will not lead to a satisfactory fit of the temperature profile without distorting the correct kinetic parameters. In propane cracking Buekens and Froment calculated a $(-\Delta H)$ of $-24,800$ kcal/kmol (103,800 kJ/kmol) from an approximation of the true reaction scheme by the greatly simplified scheme:

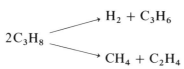

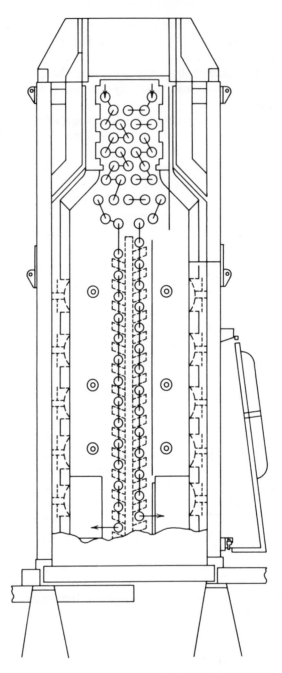

*Figure 1  Configuration of ethane furnace.*

412

whereby the two reactions are approximately of equal importance at zero conversion—but not at higher conversions. To be more rigorous in the design requires a detailed reaction scheme. This leads to a set of continuity equations instead of only one, but in this way the product distribution is directly predicted and the effect of the temperature level is correctly accounted for. Furthermore, the $(-\Delta H)$ is correctly calculated from the $(-\Delta H_i)$ of the individual reactions, at all stages of conversion. There are a few examples of this approach (Myers and Watson, Snow and Shutt, Shah, Petryschuk and Johnson, Fair and Rase, and Lichtenstein [11,12,13,14, 15,16]).

Simulating the furnace described above requires the following set of continuity equations for the components to be integrated, together with the energy equation and the pressure drop equation:

$$\frac{dF_j}{dz} = R_j \frac{\pi d_t^2}{4} = \left(\sum_i \alpha_{ij} r_i\right) \frac{\pi d_t^2}{4} \tag{a}$$

$j = 1 \cdots 8, i = 1 \cdots 7$

$$\frac{dT}{dz} = \frac{1}{\sum_{j=1}^{9} F_j c_{pj}} \left[ q(z)\pi d_t + \frac{\pi d_t^2}{4} \sum_i (-\Delta H_i) r_i \right] \tag{b}$$

$$-\frac{dp_t}{dz} = \alpha \left[ \frac{2f}{d_t} + \frac{\zeta}{\pi r_b} \right] \rho_g u^2 + \alpha \rho_g u \frac{du}{dz} \tag{c}$$

with initial conditions: $F_j = F_{j0}$, $T = T_0$ and $p_t = p_{t0}$ at $z = 0$. In Eq. (a) $R_j$ is the total rate of change of the amount of the component $j$ and $r_i$ is the rate of the $i$th reaction. This rate can be expressed as

$$r_i = k_i \Pi C_j^{a'_j}$$

with

$$C_j = \frac{F_j}{\sum_{j=1}^{9} F_j} \frac{p_t}{RT}$$

and whereby the product is taken over all the reactants of the $i$th reaction.

The radical reaction schemes for thermal cracking mentioned in Chapter 1 have not been used so far in design. They lead to a set of continuity equations for the reacting components that are mathematically stiff in nature, because of the orders of magnitude of difference between the concentrations of molecular and radical species. Only recently have satisfactory numerical integration routines for sets of stiff differential equations been worked out (see Gear [17]). In addition, the rate parameters of radical reactions are frequently not known with sufficient precision, so far. The radical scheme has therefore been approximated by a set of reactions containing only molecular species.

*Table 1 Molecular reaction scheme and kinetic parameters for the thermal cracking of ethane*

| Reaction | Order | $A_0(s^{-1})$ or $(m^3/kmol\ s)^+$ | $E(kcal/kmol)$ | $E(kJ/kmol)$ |
|---|---|---|---|---|
| 1. $C_2H_6 \rightarrow C_2H_4 + H_2$ | 1 | $4.65 \times 10^{13}$ | 65,210 | 273,020 |
| 2. $C_2H_4 + H_2 \rightarrow C_2H_6$ | 2 | $8.75 \times 10^{8+}$ | 32,690 | 136,870 |
| 3. $2C_2H_6 \rightarrow C_3H_8 + CH_4$ | 1 | $3.85 \times 10^{11}$ | 65,250 | 273,190 |
| 4. $C_3H_6 \rightarrow C_2H_2 + CH_4$ | 1 | $9.81 \times 10^8$ | 36,920 | 154,580 |
| 5. $C_2H_2 + CH_4 \rightarrow C_3H_6$ | 2 | $5.87 \times 10^{4+}$ | 7,040 | 29,480 |
| 6. $C_2H_2 + C_2H_4 \rightarrow C_4H_6$ | 2 | $1.03 \times 10^{12+}$ | 41,260 | 172,750 |
| 7. $C_2H_4 + C_2H_6 \rightarrow C_3H_6 + CH_4$ | 2 | $7.08 \times 10^{13+}$ | 60,430 | 253,010 |

The kinetic model used here has been developed by Sundaram and Froment [18] by a rigorous screening between several plausible molecular reaction schemes on the basis of thermodynamic considerations and statistical tests on the kinetic parameters. The scheme, together with the kinetic parameters, is given in Table 1. It should be added that the kinetic parameters for the reverse reactions (2) and (5) have been obtained from equilibrium data.

Table 2 shows the matrix of the stoichiometric coefficients $\alpha_{ij}$ for this set of reactions, according to:

$$\sum_{j=1}^{N} \alpha_{ij} A_j = 0 \qquad i = 1, 2, \ldots, 7$$

*Table 2 Matrix of stoichiometric coefficients*

| | $CH_4$ | $C_2H_2$ | $C_2H_4$ | $C_2H_6$ | $C_3H_6$ | $C_3H_8$ | $C_4H_6$ | $H_2$ | $H_2O$ |
|---|---|---|---|---|---|---|---|---|---|
| $i \setminus j$ | 1 | 2 | 3 | 4 | 5 | 6 | 7 | 8 | 9 |
| 1 | | | 1 | −1 | | | | 1 | |
| 2 | | | −1 | 1 | | | | −1 | |
| 3 | 1 | | | −2 | | 1 | | | |
| 4 | 1 | 1 | | | −1 | | | | |
| 5 | −1 | −1 | | | 1 | | | | |
| 6 | | −1 | −1 | | | | 1 | | |
| 7 | 1 | | −1 | −1 | 1 | | | | |

The specific heat in Eq. (b) is calculated from Rihani and Doraiswamy's formula given in Reid and Sherwood's book [19]. The specific heat of the mixture follows from

$$c_p = \sum_j \frac{F_j}{F_t} c_{pj}$$

The heat of reaction is the algebraic sum of heats of formation of reactants and products:

$$-\Delta H_i = -\sum_j \alpha_{ij} \Delta H_{fj}$$

where $\alpha_{ij}$ are the stoichiometric coefficients of the reaction and

$$\Delta H_{fj} = \Delta H_{fj}{}^0 + \int_{T_{ref}}^{T} c_{pj} \, dT$$

$\Delta H_{fj}{}^0$ is calculated from group contributions at the reference temperature.

The pressure drop equation Eq. (c) not only accounts for friction losses in the straight portions and in the bends of the coil but also for changes in momentum. The first term in the brackets on the right-hand side arises from the Fanning equation, the second from Nekrasov's equation [20] for the additional pressure drop resulting from the curvature in the bends. Furthermore, since

$$u = \frac{M_m F_t}{\rho_g \dfrac{\pi d_t^2}{4}} = \frac{G}{M_m} \frac{RT}{p_t}$$

$$\frac{du}{dz} = \frac{GR}{p_t} \left[ T \frac{d\left(\dfrac{1}{M_m}\right)}{dz} + \frac{1}{M_m} \frac{dT}{dz} \right] - \frac{G}{M_m} \frac{RT}{p_t^2} \frac{dp_t}{dz}$$

so that Eq. (c) finally becomes:

$$\frac{dp_t}{dz} = \frac{\dfrac{d}{dz}\left(\dfrac{1}{M_m}\right) + \dfrac{1}{M_m}\left[\dfrac{1}{T}\dfrac{dT}{dz} + \left(\dfrac{2f}{d_t} + \dfrac{\zeta}{\pi r_b}\right)\right]}{\dfrac{1}{M_m p_t} - \dfrac{p_t}{\alpha G^2 RT}}$$

with

$$\frac{d}{dz}\left(\frac{1}{M_m}\right) = \frac{d}{dz}\left(\frac{\sum_j F_j}{G\Omega}\right) = \frac{\sum_j \dfrac{dF_j}{dz}}{G\Omega}$$

The friction factor for straight tubes is taken from Knutzen and Katz [21].

$$f = 0.046 \, \text{Re}^{-0.2} \quad \text{when} \quad \text{Re} = \frac{d_t G}{\mu}$$

The factor $\zeta$ used in the equation for the supplementary pressure drop in the bends is given by Nekrasov:

$$\zeta = \left(0.7 + \frac{\Lambda}{90°} 0.35\right)\zeta'$$

where $\zeta' = \left(0.051 + 0.19\frac{d_t}{r_b}\right)$

$\Lambda$ = angle described by the bend, here 180°
$r_b$ = radius of the bend

If the value of the viscosity is not found in the literature the corresponding state equations can be used, as illustrated already in Chapter 3.

The following heat flux profile was generated from independent simulations of the heat transfer in the firebox. First tube: 23 kcal/m² s (96 kJ/m² s); second tube: 20 (84); third tube: 19 (80); fourth tube: 17 (71); fifth tube: 15 (63); sixth, seventh, eighth, ninth, and tenth tubes, 14 (59). With this heat flux profile, the conversion, temperature and total pressure profile of Fig. 2 was obtained. The agreement with the industrial data is really excellent. Also, the product distribution is in complete agreement as can be seen from Fig. 3: the simulated yields for ethylene, hydrogen, and methane, for example, are, respectively, 47.92, 3.79, and 3.49; the

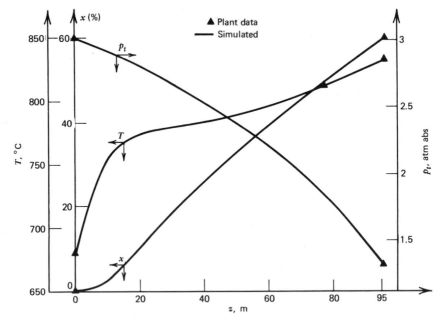

*Figure 2 Ethane cracking. Reactor simulation.*

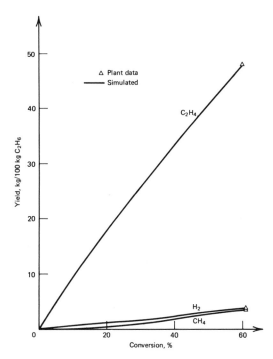

*Figure 3 Ethane cracking. Product distribution.*

industrial 48.7, 3.65, and 3.4. The ethane conversion is seen to be limited to approximately 60 percent to avoid too much coking.

Instead of using a given heat flux profile [i.e., an energy equation like Eq. (b)], the calculation could be started from the furnace gas temperature, which would then involve an energy equation like Eqs. 9.1-6 or 9.1-7. A mean furnace temperature may be calculated by the method of Lobo and Evans [22]. A further refinement would be to consider the temperature distribution in the furnace, which in addition requires taking into account the geometrical configuration and the location of the burners. This is a rather involved procedure, called the zoning method, which has been developed by Hottel and Cohen [23] and recently refined by Vercammen and Froment [24].

## Problems

9.1 A gas phase reaction, $A \rightarrow 2R$, is carried out in a tubular plug flow reactor at $T = 60°C$ and $p_t = 4.75$ atm. The feed consists of 50 mol percent $A$ and 50 mol percent inert at a rate of 4000 kg/hr. The molecular weights of $A$ and inert are 40 and 20, respectively, and the rate coefficient is $k = 2000$ hr$^{-1}$.

Determine the reactor size for 35 percent conversion of $A$.

9.2 The process $A \xrightarrow{k_1} B \xrightarrow{k_2} C$ is carried out in a tubular reactor with plug flow. Both reactions are of first order. The feed consists of pure $A$. Given the following data

$$C_{A0} = 0.05 \text{ kmol/m}^3$$
$$F_0' = 0.15 \text{ m}^3/\text{hr}$$
$$\Omega = 2.10^{-3} \text{ m}^2$$
$$k_1 = 172.5 \text{ hr}^{-1}$$

Calculate the length of the tube to maximize the yield of $B$ in the cases (a) $k_2 = k_1/2$, (b) $k_2 = k_1$.

What are the exit concentrations of $A$, $B$, and $C$ in both cases?

9.3 (a) Repeat the derivations of Example 9.3-1, but for a zeroth-order reaction
    (b) Given the data

| | |
|---|---|
| $F_{A0} = 20$ kmol/hr | $T_0 = 300$ K |
| $c_p = 0.5$ kcal/kg°C | $x_0 = 0; x = 0.9$ |
| $(-\Delta H) = 10000$ kcal/kmol | $k_0 = 3.27 \ 10^6$ kmol/m$^3$·hr |
| $E = 9000$ kcal/kmol | $\dot{m} = 1000$ kg/hr |

Calculate the reactor volume and exit temperature when the reaction is of zeroth order.

(c) Compare with the volume required when the reaction is carried out isothermally
(a) at $T = T_0$
(b) at $T = (T_0 + T_e)/2$
where $T_e$ is the exit temperature of the adiabatic reactor.

9.4 Prove that the curve $x$ versus $(V/F_{t0})$, where $x$ is the conversion, $V$ the total reactor volume, and $F_{t0}$ the total molar inlet flow rate of reactant plus inert diluent, is independent of the dilution ratio for a reversible reaction where both forward and reverse reactions are of first order only.

9.5 Consider the following data for the enzymatic hydrolysis of $n$-benzoyl 1-arginine ethyl ester (BAEE) by trypsin bound to particles of porous glass in a fixed bed reactor:

$$C_{A0} = 0.5 \text{ mM}$$

| $X_A$ | $V/F_0'$, min |
|---|---|
| 0.438 | $5.90 \times 10^{-2}$ |
| 0.590 | 8.03 |
| 0.670 | 9.58 |
| 0.687 | 9.46 |
| 0.910 | 14.72 |
| 0.972 | 18.00 |

(a) For Michaelis–Menten kinetics, show that a plot of $(1/C_{A0}x_A)\ln(1 - x_A)^{-1}$ versus $V/F_{A0}x_A$ should give a straight line, from which the constants can be determined.
(b) Compute values for the constants.

# References

[1] Hougen, O. A. and Watson, K. M. *Chemical Process Principles*, Vol. III, Wiley, New York (1947).

[2] Emig, G., Hofmann, H., and Friedrich, H. *Proc. 2nd Intl. Symp. Chem. React. Engng.*, p. B5–23, Elsevier Publishing Co., Amsterdam (1972).

[3] Lambrecht, G., Nussey, C., and Froment, G. F. *Proc. 2nd Intl. Symp. Chem. React. Engng.*, p. B2–19, Elsevier Publishing Co., Amsterdam (1972).

[4] Van Damme, P. S., Narayanan, S., and Froment, G. F. *A.I.Ch.E. J.*, **21**, 1065 (1975).

[5] Froment, G. F., Pijcke, H., and Goethals, G. *Chem. Eng. Sci.*, **13**, 173 (1961).

[6] Froment, G. F., Pijcke, H., and Goethals, G. *Chem. Eng. Sci.*, **13**, 180 (1961).

[7] Buekens, A. G. and Froment, G. F. *Ind. Eng. Chem. Proc. Des. Devpt.*, **7**, 435 (1968).

[8] Buekens, A. G. and Froment, G. F. *Ind. Eng. Chem. Proc. Des. Devpt.*, **10**, 309 (1971).

[9] Douglas, J. M. and Eagleton, L. C. *Ind. Eng. Chem. Fund.*, **1**, 116 (1962).

[10] Buekens, A. G. and Froment, G. F. *Proc. 4th Eur. Symp. Chem. React. Engng. 1968*, Pergamon Press, London (1971).

[11] Myers, P. F. and Watson, K. M. *Nat. Petrol. News*, **18**, 388 (1946).

[12] Snow, R. H. and Shutt, H. C. *Chem. Eng. Prog.*, **53**, No. 3, 133 (1957).

[13] Shah, M. J. *Ind. Eng. Chem.*, **59**, 70 (1967).

[14] Petryschuk, W. E. and Johnson, A. I. *Can. J. Chem. Eng.*, **46**, 172 (1968).

[15] Fair, J. R. and Rase, H. F. *Chem. Eng. Prog.*, **50**, No. 8, 415 (1954).

[16] Lichtenstein, T. *Chem. Eng. Prog.*, **12**, 64 (1964).

[17] Gear, C. W. *Numerical Initial Value Problems in Ordinary Differential Equations*, Prentice-Hall, Englewood Cliffs, N.J. (1971).

[18] Sundaram, K. M. and Froment, G. F. *Chem. Eng. Sci.*, **32**, 601 (1977).

[19] Reid, R. C. and Sherwood, T. K. *The Properties of Gases and Liquids*, 2nd ed., McGraw-Hill, New York (1966).

[20] Nekrasov, B. B. *Hydraulics*, Peace Publishers, Moscow (1969).

[21] Knudsen, J. G. and Katz, D. L. *Fluid Dynamics and Heat Transfer*, McGraw-Hill, New York (1958).

[22] Lobo, W. E. and Evans, J. E. *Trans. A.I.Ch.E.*, **35**, 743 (1939).

[23] Hottel, H. C. and Cohen, E. S. *A.I.Ch.E. J.*, **4**, 3 (1958).

[24] Vercammen, H. and Froment, G. F. *Proc. 5th Intl. Symp. Chem. React. Engng.*, p. 271, ACS Symp. Ser. 65, Amer. Chem. Soc., Washington, D.C. (1978).

# 10

## THE PERFECTLY MIXED FLOW REACTOR

### 10.1 Introduction

This reactor type is the opposite extreme from the plug flow reactor considered in Chapter 9. The essential feature is the assumption of complete uniformity of concentration and temperature throughout the reactor, as contrasted with the assumption of no intermixing of successive fluid elements entering a plug flow vessel. Therefore, in the perfectly mixed flow reactor, the conversion takes place at a unique concentration (and temperature) level which, of course, is also the concentration of the effluent. In order to approach this ideal mixing pattern, it is necessary that the feed be intimately mixed with the contents of the reactor in a time interval that is very small compared to the mean residence time of the fluid flowing through the vessel. Further discussion of deviations from these ideal flow patterns are given in Chapter 12; in this chapter, we assume that perfect mixing has been achieved.

The stirred flow reactor is frequently chosen when temperature control is a critical aspect, as in the nitration of aromatic hydrocarbons or glycerine (Biazzi-process). The stirred flow reactor is also chosen when the conversion must take place at a constant composition, as in the copolymerization of butadiene and styrene, or when a reaction between two phases has to be carried out, or when a catalyst must be kept in suspension as in the polymerization of ethylene with Ziegler catalyst, the hydrogenation of $\alpha$-methylstyrene to cumene, and the air oxidation of cumene to acetone and phenol (Hercules-Distillers process).

Finally, several alternate names have been used for what here is called the "perfectly mixed flow reactor." One of the earliest was "continuous stirred tank reactor," or CSTR, which some have modified to "continuous flow stirred tank reactor," or CFSTR. Other names are "backmix reactor," "mixed flow reactor," and "ideal stirred tank reactor." All of these terms appear in the literature, and must be recognized.

420

## 10.2 Mass and Energy Balances _____

### 10.2.a  Basic Equations

Since the reactor contents are completely uniform with perfect mixing, the reactor-integrated balances from Chapter 7 are used. From Eq. 7.2.b-12,

$$\frac{dm_t}{dt} = F_0' \rho_{f0} - F_e' \rho_{fe} \qquad (10.2.a\text{-}1)$$

from Eq. 7.2.b-10,

$$\frac{dN_j}{dt} = F_{j,0} - F_{j,e} + VR_j \qquad (10.2.a\text{-}2)$$

$$= F_0' C_{j,0} - F_e' C_{j,e} + VR_j \qquad (10.2.a\text{-}2a)$$

and from Eq. 7.2.d-5,

$$m_t c_p \frac{dT}{dt} = \sum_j F_{j,0}(H_{j,0} - H_{j,e}) + V \sum_i (-\Delta H_i) r_i + Q(T) \qquad (10.2.a\text{-}3)$$

where $Q(T)$ represents external heat addition or removal from the reactor [e.g., $A_k U(T_r - T)$].

For single reactions, it is useful to write Eq. 10.2.a-2 in terms of conversion of reactant $A$.

$$F_A = F_{A0}(1 - x_A)$$

leading to:

$$N_{A0} \frac{dx_A}{dt} = F_{A0} x_{A0} - F_{A0} x_A - V r_A \qquad (10.2.a\text{-}4)$$

$$= -F_{A0} x_A - V r_A \qquad (10.2.a\text{-}4a)$$

where the latter equation is when the inlet conversion is taken to be zero. Aris [1] has discussed the reductions possible for the general set of reactions $0 = \sum_j \alpha_{ij} A_j$. For arbitrary feed and/or initial compositions, which may not have stoichiometrically interrelated compositions, the mass balance can be written in terms of an extent for each independent reaction, plus variables related to the incompatibility of the feed and initial compositions. For constant feed, this single latter variable is related to the "washout" of the initial contents. In these general situations, it is probably just as easy to directly integrate the Eq. 10.2.a-2.

In the energy balance, mean specific heats are generally used, so that Eq. 10.2.a-3 reduces to

$$V \rho_f c_p \frac{dT}{dt} = F_0' \rho_f c_p (T_0 - T) + V \sum_i (-\Delta H_i) r_i + Q(T) \qquad (10.2.a\text{-}5)$$

Finally, since most reactions carried out in stirred tank reactors are in the liquid phase, with constant density, the special cases for constant volume and total mass are useful. From Eq. (10.2.a-1), it is seen then that $F_0' = F_e' \equiv F' = $ constant, and Eq. (10.2.a-2a) can be written

$$V \frac{dC_j}{dt} = F'(C_{j0} - C_j) + VR_j \qquad (10.2.a\text{-}6)$$

## 10.2.b Steady-State Reactor Design

As a consequence of the complete mixing, a continuous flow stirred tank reactor also operates isothermally. Therefore, in the steady state it is not necessary to consider the mass and energy balances simultaneously. Optimum conditions may be computed on the basis of the material balance alone, and then afterwards the energy balance is used, in principle (see Sec. 10.4), to determine the external conditions required to maintain the desired temperature.

Thus, the design equation, from Eq. 10.2.a-4, is either

$$x_A - x_{A0} = \frac{V}{F_{A0}} r_A \qquad (10.2.b\text{-}1)$$

or for constant densities,

$$C_{A0} - C_A = \frac{V}{F'} r_A = \tau r_A \qquad (10.2.b\text{-}2)$$

where $\tau = V/F' = C_{A0}V/F_{A0}$ is called the mean residence (or holding) time. Given an expression for $r_A$, the above equations can then be readily solved for $x_A$ as a function of the system parameters. For a first-order reaction,

$$r_A = kC_A = kC_{A0}(1 - x_A)$$

so that, with

$$x_{A0} = 0$$

$$x_A = \frac{kC_{A0}V/F_{A0}}{1 + kC_{A0}V/F_{A0}} = 1 - \frac{1}{1 + kC_{A0}V/F_{A0}} \qquad (10.2.b\text{-}3)$$

For constant densities, the result is usually written as:

$$\frac{C_A}{C_{A0}} = \frac{1}{1 + kV/F'} = \frac{1}{1 + k\tau} \qquad (10.2.b\text{-}4)$$

When two perfectly mixed reactors are connected in series, the mass balance for the second reactor is:

$$x_{A2} - x_{A1} = \frac{V}{F_{A0}} r_A(x_{A2}) = \frac{V}{F_{A0}} kC_{A0}(1 - x_{A2}) \qquad (10.2.b\text{-}5)$$

When $x_{A1}$ is eliminated by means of Eq. 10.2.b-3, so that the final conversion is written solely in terms of the conditions at the inlet ($x_{A0} = 0$), the following equation is obtained:

$$x_{A2} = 1 - \frac{1}{(1 + kC_{A0}V/F_{A0})^2} \qquad (10.2.b-6)$$

Note that $V$ is the volume of *one* reactor. For $n$ reactors in series

$$x_{An} = 1 - \frac{1}{(1 + kC_{A0}V/F_{A0})^n} \qquad (10.2.b-7)$$

These formulas may be used for the study of the kinetics of a first-order reaction by measuring $x_A$, $C_{A0}$, $F_{A0}$, and $V$ and then determining $k$. Alternately, for a given reaction, they can be used for determining the volume required to achieve a certain production.

For second-order reactions, Eq. (10.2.b-1) becomes

$$F_{A0}x_A = VkC_A C_B \qquad (10.2.b-8)$$

With the irreversible reaction $A + B \rightarrow$, when equimolar quantities of $A$ and $B$ are fed to the reactor, the following equation is obtained:

$$x_A = 1 + \frac{F_{A0}}{2kC_{A0}^2 V} - \left[\left(\frac{F_{A0}}{2kC_{A0}^2 V}\right)^2 + \frac{F_{A0}}{kC_{A0}^2 V}\right]^{1/2} \qquad (10.2.b-9)$$

The conversion at the exit of a second reactor of equal volume, and placed in series with the first is

$$x_A = 1 + \frac{F_{A0}}{2kC_{A0}^2 V} - \left\{\left(\frac{F_{A0}}{2kC_{A0}^2 V}\right)^2 + \frac{F_{A0}}{kC_{A0}^2 V}\left[\left(\frac{F_{A0}}{2kC_{A0}^2 V}\right)^2 + \frac{F_{A0}}{kC_{A0}^2 V}\right]^{1/2}\right\}^{1/2}$$

$$(10.2.b-10)$$

The results of Eqs. 10.2.b-7 and 10.2.b-10 have been represented by Schoenemann and Hofmann [2] in convenient diagrams; Fig. 10.2.b-1 is the diagram for first-order reactions with constant density. The conversion for more complex kinetic forms must often be obtained numerically, by solving the algebraic Eq. 10.2.b-1 with the appropriate rate function on the right-hand side.

Note that Eqs. 10.2.b-1, 7, and 10 do not explicitly contain the residence time, just as was the case for the continuity equations for the plug flow reactor in Chapter 9. They could be reformulated (e.g., see Levenspiel [3]) but, again as in Chapter 9, there is no advantage, and it is simpler to just use the directly manipulated variables $F_{A0}$, $V$, and $C_{A0}$. It is only in constant density systems that the residence time, $V/F'$, directly appears, as is illustrated by Eq. 10.2.b-4.

With the perfectly mixed flow reactor, the actual residence times of individual fluid elements is a continuous spectrum: by the completely random mixing, some fluid elements immediately reach the exit after their introduction, while some remain in the reactor for a very long time. The above results did not specifically

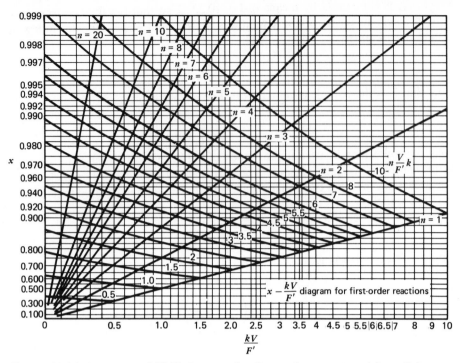

*Figure 10.2.b-1 x versus kV/F′ diagram for first-order reactions (from Schoene-mann and Hofmann [2]).*

consider this spread in residence times; the reason is that the assumption of *perfect* mixing implies that each fluid element instantaneously loses its identity. In principle, this means that the molecular environment is also completely uniform for the reacting species. By implicitly defining the molecular environment, the perfect mixing model only requires the conservation of mass to predict the overall conversion. If the intensity of actual mixing is not so intense that the molecular environment is made uniform before significant reaction occurs, then specific account must be taken of the spectrum, or distribution, of fluid residence times. These so-called "nonideal flow patterns" are considered in Chapter 12.

### Example 10.2.b-1 Single Irreversible Reaction in a Stirred Flow Reactor

Kermode and Stevens [4] studied the reaction of ammonia and formaldehyde to make hexamine, a classical chemical process:

$$4\,NH_3 + 6\,HCHO \longrightarrow (CH_2)_6N_4 + 6\,H_2O$$
$$(A) \qquad\quad (B)$$

The continuous flow reactor was a 490 cm$^3$ baffled stainless steel tank vigorously stirred at 1800 rpm, with several precautions to ensure almost perfect mixing. The kinetics were separately studied, and the overall reaction had a rate

$$r_A = kC_A C_B^2 \qquad \text{mol } A/\text{l s} \tag{a}$$

with

$$k = 1.42 \times 10^3 e^{-3090/T} \tag{b}$$

The reactants were each fed to the reactor in streams 1.50 cm$^3$/s with the ammonia concentration 4.06 mol/l and the formaldehyde concentration 6.32 mol/l. The average reaction temperature was 36°C.

In this constant density system, the mass balance equations 10.2.b-2 could be used for each reactant:

$$F'C_{A0} - F'C_A = r_A V \tag{c}$$

$$F'C_{B0} - F'C_B = \tfrac{6}{4} r_A V \tag{d}$$

as was essentially done by Kermode and Stevens. The total volumetric feed rate $F'$ is 3.0 (cm)$^3$/s, and the inlet concentrations are

$$C_{A0} = \frac{1.5(4.06)}{3.0} = 2.03 \text{ mol/l}$$

$$C_{B0} = \frac{1.5(6.32)}{3.0} = 3.16 \text{ mol/l}$$

The concentrations of $A$ and $B$ can be interrelated through extent or conversion; choosing the latter:

$$C_A = C_{A0}(1 - x_A) \tag{e}$$

$$C_B = C_{B0} - \tfrac{3}{2} C_{A0} x_A = C_{A0}\left(\frac{C_{B0}}{C_{A0}} - \frac{3}{2} x_A\right) \tag{f}$$

It is actually simpler to use the mass balance based on conversion, Eq. 10.2.b-1, which here becomes

$$x_A = \frac{V}{F'C_{A0}} kC_{A0}{}^3(1 - x_A)\left(\frac{C_{B0}}{C_{A0}} - \frac{3}{2} x_A\right)^2 \tag{g}$$

$$= \frac{(490)(0.065)(2.03)^2}{3.0} (1 - x_A)\left(\frac{3.16}{2.03} - \frac{3}{2} x_A\right)^2$$

$$= 43.8(1 - x_A)(1.557 - 1.5 x_A)^2 \tag{h}$$

Solving Eq. h gives $x_A = 0.82$, which in turn, leads to the concentrations

$$C_A = 0.36 \text{ mol/l} \qquad C_B = 0.66 \text{ mol/l}$$

The combined mass and heat balances were solved by Kermode and Stevens by means of an analog computer to obtain $C_B = 0.637$ mol/l and $T = 37.3°C$. The average experimental values were $C_B = 0.64$ mol/l and $T = 36°C$. (The actual purpose of their study was to investigate the *transient* behavior; see Sec. 10.4).

Turning to some design considerations, we now utilize a simple first-order irreversible reaction, with $kV/F' = 2.0$, and the conversion will be $x_A = 0.667$ from Eq. 10.2.b-3 or 4 or from Fig. 10.2.b-1 (the ordinate corresponding to the intersection of $kV/F' = 2.0$ and the $n = 1$ line). In a plug flow or batch reactor, the conversion would be

$$x_A = 1 - e^{-kV/F'} = 0.865$$

If this conversion were desired in a perfectly mixed flow reactor, Fig. 10.2.b-1 gives $(kV/F') = 6.5$ (abcissa of the intersection of the ordinate level of 0.865 and the $n = 1$ line); that is, for the given $k$, the reactor volume would have to be 6.5 times the flow rate rather than only twice, as with plug flow. This example clearly illustrates that results obtained in a batch or plug flow tubular reactor cannot be directly extrapolated to a continuous flow stirred tank reactor—there may be large differences in conversion levels.

It also follows from the above discussion that it is difficult to obtain high conversions in a continuous flow stirred tank reactor (at least for first-order kinetics) without resorting to large volumes, in which perfect mixing may not be easily achieved. Therefore, it is often preferable to connect two or more smaller reactors in series, which will be shown to also reduce the total volume required to achieve a given conversion. Indeed, from Fig. 10.2.b-1, it is seen that a conversion of 0.865 can be obtained with $kV/F' = 1.75$. This means that the volume of *each* of the two tanks has to be 1.75 times the flow rate, for a total volume ratio of 3.5 instead of 6.5—a savings of almost a factor of 2.

When the *total* volume $(= nV)$ is kept constant, the subdivision of the reactor permits one to increase the overall conversion. Consider again the value $kV/F' = 2.0$ for a single tank reactor, and determine the conversions when the total volume is such that $nVk/F' = 2.0$ while increasing $n$. The results can easily be found from Fig. 10.2.b-1 by following the $nVk/F'$ curve as it intersects the $n = 1, 2, 3, \ldots$ curves, and reading the ordinate values:

| $n$ | $nVk/F'$ | $x_A$ |
|-----|----------|-------|
| 1 | 2 | 0.67 |
| 2 | 2 | 0.75 |
| 3 | 2 | 0.78 |
| 5 | 2 | 0.81 |
| $\infty$ | 2 | 0.87 (plug flow) |

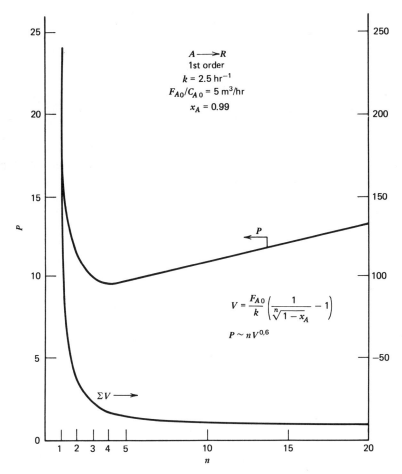

*Figure 10.2.b-2 Economic optimum choice of number of perfectly mixed reactors in series.*

Alternatively, a given conversion may be reached with either a single large reactor volume or with a series of smaller reactors. The ultimate choice is based on economic factors, as illustrated in Fig. 10.2.b-2. The total reactor volume required decreases with more subdivision (larger $n$), but with the cost per reactor proportional to $V^{0.6}$, the total cost proportional to $nV^{0.6}$ shows a definite minimum—in this case at about $n = 4$. Plant operational difficulties may also increase with $n$, and the optimum choice is usually a relatively small number of reactors in series, especially since most of the savings in total volume occur for $n < 5$. Exceptions are in multistage contacting devices, but this is a more complicated situation.

For reaction orders other than 1, the best choice is not two equal size tanks in series. Several situations have been analyzed (see Levenspiel [5] for a clear discussion); Luss [6] has provided a simple analytical procedure for determining the optimum size ratio. For second-order reactions in two tanks in series, this ratio is about 1 : 1 for low conversions and 1 : 2 for high conversions. However, the overall advantage of the variable-sized multistage system is rather small compared to equal sizes, and this, plus the above comments, usually dictates only considering equal size reactors in series.

The result that for a given conversion the perfectly mixed flow reactor requires a larger volume than the plug flow reactor is only valid for reaction rate expressions such that the rate monotonically decreases with decreasing reactant concentration (e.g., simple orders greater than zero). For these reactions, it is clearly advantageous to operate a reactor at the highest average concentration level possible. In a perfectly mixed flow reactor, the conversion takes place at the concentration level of the effluent, which is low, while the plug flow reactor takes advantage of the higher concentrations at the entrance. The subdivision of the total volume by a series of stirred tanks is an intermediate situation, which approaches the continuous concentration profile of the plug flow reactor, and therefore yields a higher conversion compared to that in a single tank.

These conclusions can be readily quantitatively visualized as shown in Fig. 10.2.b-3, which is based on the geometric nature of the plug flow or batch reactor design equation versus that for the perfectly mixed flow reactor.

For a plug flow or batch reactor:

$$\frac{V}{F_{A0}} = \int_{x_{A0}}^{x_{Ae}} \frac{dx_A}{r_A(x_A)}$$

For a perfectly mixed flow reactor:

$$\frac{V}{F_{A0}} = \frac{x_{Ae} - x_{A0}}{r_A(x_{Ae})}$$

Thus, the reactor size for plug flow is given by the area under a curve ($1/r_A$) versus $x_A$ — area $1$ in Fig. 10.2.b-3a. For perfectly mixed flow, on the other hand, the size is given by the area of the rectangle with ordinate $1/r_A(x_{Ae})$ (i.e., evaluated at the exit conversion), which is the sum of areas $1 + 2$. Clearly, for this case where the rate monotonically increases with concentration, the plug flow reactor will always have the smaller area, and thus a smaller size. However, Fig. 10.2.b-3b is a plot for another type of rate form, which could result from an autocatalytic reaction, a dual site catalytic mechanism, "negative order," or any other form where the rate has a maximum in the concentration range. Here, we can see that

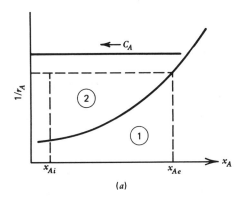

(a)

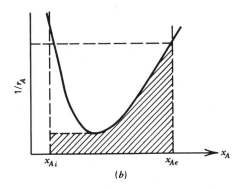

(b)

*Figure 10.2.b-3 Comparison of plug flow and perfectly mixed flow reactor volumes.*

the optimum arrangement is a perfectly mixed reactor followed by a plug flow reactor, and that a combined volume could result significantly smaller than that of *either* type of single reactor. More detailed analyses are given by Levenspiel [5]; Bischoff [7] has treated the case of Michaelis–Menten kinetics important in enzyme and fermentation reactions. The latter reference shows for a typical case that the total volume is reduced by a factor of 2.77 with the optimal design. Finally, adiabatic systems of exothermic reversible reactions have the same type of characteristics, and Aris [8] has considered this case; in combustion systems, it is often beneficial to begin with a mixing region (which may also have other benefits), followed by a plug flow region.

In many cases, however, the reactor volume is not the main factor in the choice of the reactor type. Most reactions of industrial importance are actually complex reactions. In such cases, the selectivity is far more important than the reactor size. Therefore it is important that a judicious choice of reactor type permits one to influence the selectivity, which may depend on the concentration levels and therefore on the degree of mixing in the reactor. This is discussed in the next section.

## 10.3 Design for Optimum Selectivity in Complex Reactions ___

### 10.3.a General Considerations

The effects of concentration levels on the selectivity of complex reactions can most readily be seen by considering a few examples. We begin with the two basic cases: parallel and consecutive reactions. For the parallel reactions

where $Q$ is the desired product, the rate equations for the formation of $Q$ and $S$ are:

$$r_Q = k_1 C_A^{a_1'}$$
$$r_S = k_2 C_A^{a_2'}$$

from which

$$\frac{r_Q}{r_S} = \frac{k_1}{k_2} C_A^{a_1' - a_2'} \qquad (10.3.a\text{-}1)$$

The relative rates of formation then depend on the ratio of the rate coefficients, $k_1/k_2$, and the difference in orders, $a_1' - a_2'$. If both rates have the same order, then it is clear that the selectivity will not depend on the concentration level (although the conversion will). For given $k_1/k_2$, and $a_1' \neq a_2'$, the selectivity can be altered by the concentration environment, and this should then be chosen to maximize the desired product, $Q$. When $a_1' < a_2'$, $r_Q/r_S$ is small when $C_A$ is large. In the batch and plug flow reactors, part of the conversion is occurring at the high initial concentrations. In the perfectly mixed flow reactor, the feed concentration is immediately reduced to that of the outlet, which is low. Therefore, it is clear that the selectivity (to $Q$) will be higher in the perfectly mixed flow reactor. Similar reasoning indicates that the opposite would be true for $a_1' > a_2'$. The former case is illustrated by the calculated results presented in Fig. 10.3.a-1, which compares the conversions to $Q$, $x_Q \equiv C_Q/C_{A0}$, in batch or plug flow reactors with a cascade of perfectly mixed reactors. It is seen, as expected, that a single stirred tank would give the highest conversion to $x_Q$ and thus the highest selectivity for $Q$.
Next consider consecutive reactions:

$$A \xrightarrow{\ 1\ } Q \xrightarrow{\ 2\ } S$$

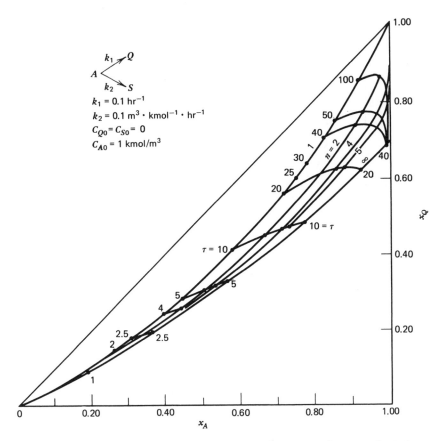

*Figure 10.3.a-1 Conversions and selectivities with various degrees of mixing as a function of the mean residence times $\tau = V/F$.*

where $Q$ is the desired product. For both reactions first order, the resulting rate equations have been integrated in Chapter 1 for the batch or plug flow reactors, and example curves are shown in Fig. 10.3.a-2. Also from Chapter 1, the maximum value of $C_Q$ is:

$$\left(\frac{C_Q}{C_{A0}}\right)_{max} = \frac{k_1}{k_2 - k_1}(e^{-k_1/k_{lm}} - e^{-k_2/k_{lm}}) \qquad (10.3.a-2)$$

which occurs at the particular holding time:

$$\theta_{max} = \left(C_{A0}\frac{V}{F_{A0}}\right)_{max} = \frac{1}{k_{lm}} \equiv \frac{\ln k_2/k_1}{k_2 - k_1}$$

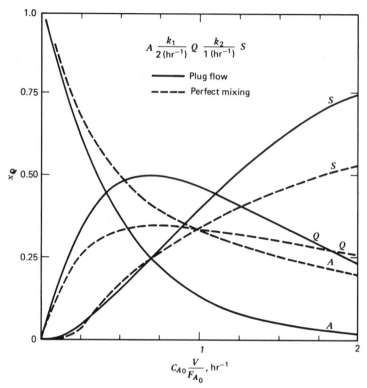

*Figure 10.3.a-2 Consecutive reaction species profiles in batch or plug flow and perfectly mixed reactors.*

For the perfectly mixed flow reactor, the mass balances Eq. 10.2.b-2 lead to:

$$C_A = \frac{C_{A0}}{1 + k_1 C_{A0} V/F_{A0}}$$

$$C_Q = \frac{k_1 C_{A0}^2 V/F_{A0}}{(1 + k_1 C_{A0} V/F_{A0})(1 + k_2 C_{A0} V/F_{A0})}$$

$$C_S = C_{A0} - C_A - C_Q$$

Then, it can easily be shown that the maximum value of $C_Q$ is

$$\left(\frac{C_Q}{C_{A0}}\right)_{max} = \left[\left(\frac{k_2}{k_1}\right)^{1/2} + 1\right]^{-2} \tag{10.3.a-3}$$

which occurs at

$$\left(C_{A0}\frac{V}{F_{A0}}\right)_{max} = (k_1 k_2)^{-1/2}$$

    CHEMICAL REACTOR DESIGN

Comparing Eqs. 10.3.a-2 and 10.3.a-3 shows that again there are differences between the yields in the reactor types. A specific example is shown in Fig. 10.3.a-2, where it is seen that the batch or plug flow reactor has greater selectivity for $Q$ relative to the perfectly mixed flow reactor. For complex first-order reaction systems Wei [9] has shown that the convexity of reaction paths is decreased from plug flow to mixed reactors, because of the intermingling of fluid elements with different extents of reaction, and so the relative selectivities will decrease. Also, if the orders of the two reactions are different, this can additionally affect the relative rates of the reactions in different reactor types. Thus, the broader distribution of residence times of the fluid elements in a perfectly mixed flow reactor will cause a broader maximum in the intermediate species concentrations.

For more complicated reaction networks, it is not always completely obvious how to apply the above concepts, as is seen from consideration of the example of van de Vusse [10]:

$$A \xrightarrow{\ 1\ } Q \xrightarrow{\ 2\ } S$$

$$A + A \xrightarrow{\ 3\ } R$$

where $Q$ is the desired product. Here the rates of reaction are,

$$r_A = k_1 C_A + k_3 C_A^2 \tag{10.3.a-4}$$

$$r_Q = k_1 C_A - k_2 C_Q \tag{10.3.a-5}$$

and the yield $C_Q/C_{A0}$, or the selectivity $C_Q/(C_{A0} - C_A)$ can be found from the relative rates:

$$\frac{r_Q}{r_A} = \frac{k_1}{k_1 + k_3 C_A} - \frac{k_2 C_Q}{C_A(k_1 + k_3 C_A)}$$

$$= \frac{1}{1 + a_1(1 - x_A)} - \frac{a_2 x_Q}{(1 - x_A)[1 + a_1(1 - x_A)]} \tag{10.3.a-6}$$

We see that the results will depend on the two parameter groups:

$$a_1 \equiv k_3 C_{A0}/k_1 \quad \text{and} \quad a_2 \equiv k_2/k_1 \tag{10.3.a-7}$$

Now for $k_3 C_{A0} \gg k_2$, or $a_1 \gg a_2$, it seems reasonable to expect that the parallel reaction is more critical than the consecutive step in decreasing the yield of $Q$, and based on the above paragraphs the optimum choice would be a perfectly mixed reactor rather than a plug flow reactor—this will be verified by calculations. Also, for $k_3 C_{A0} < k_2$, or $a_1 < a_2$, the consecutive reaction should dominate, and the plug flow reactor should be best. However, for $a_1 \simeq a_2$, it is not so clear which is the optimum reactor type.

Van de Vusse [10] performed computations to determine the proper choices. By using the ratio of the mass balances for perfectly mixed reactors, Eq. 10.2.b-1 or 2, we obtain:

$$\frac{r_Q}{r_A} = \frac{0 - C_Q}{C_{A0} - C_A} = \frac{-x_Q}{x_A}$$

which, with Eq. 10.3-7, has the solution

$$x_Q = \frac{x_A(1 - x_A)}{a_1(1 - x_A)^2 + (1 - a_2)(1 - x_A) + a_2} \tag{10.3.a-8}$$

For plug flow, the relationship is:

$$-\frac{dx_Q}{dx_A} = \frac{r_Q}{r_A}$$

which with Eq. 10.3.a-7 gives

$$x_Q = \left[\frac{1 - x_A}{1 + a_1(1 - x_A)}\right]^{a_2} \int_0^{x_A} \frac{[1 + a_1(1 - x)]^{a_2 - 1}}{(1 - x)^{a_2}} \, dx \tag{10.3.a-9}$$

$$= \frac{(1 - x_A)}{1 + a_1(1 - x_A)} \ln\left(\frac{1}{1 - x_A}\right), \text{ for } a_2 = 1 \tag{10.3.a-9a}$$

(certain other cases can also be analytically integrated). From these results, the maximum yield and selectivity can be found by the equations:

$$\frac{dx_Q}{dx_A} = 0 \quad \text{or} \quad \frac{d}{dx_A}\left(\frac{x_Q}{x_A}\right) = 0$$

Results of such computations were summarized by van de Vusse in Fig. 10.3.a-3 and Table 10.3.a-1. We see that the conjectures concerning the optimum reactor type in the extreme regions of $a_1$ and $a_2$ are indeed verified, but also the more complicated middle region is clarified.

Further consideration of the van de Vusse reaction sequence leads to the conclusion that even better results might be obtained with a combination of reactor types or with a reactor of intermediate mixing level. At the beginning of the conversion, when $C_A$ is high, and very little $Q$ has been formed, it is most important to suppress the parallel reaction, and so a perfectly mixed flow reactor is advantageous. However, at higher conversion when $C_A$ is relatively low, and an appreciable amount of $Q$ has been formed, the loss of yield by the consecutive step dominates. To minimize this, plug flow is required. Thus, the optimum configuration is a perfectly mixed followed by a plug flow reactor. Using a theoretical model of intermediate mixing levels, allowing for adjustment of the levels along the reactor length, Paynter and Haskins [11] were able to formally optimize the intermediate

CHEMICAL REACTOR DESIGN

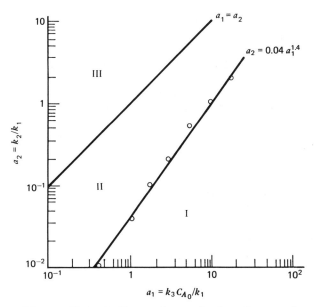

*Figure 10.3.a-3 Comparison of plug flow and perfectly mixed reactors (yield and selectivity). The points correspond to pairs of values of $a_1$ and $a_2$ for which both types of reactor give the same maximum yield. The upper line corresponds to equal selectivity for both types of reactor (at zero conversion) (from van de Vusse [10]).*

| Region | Highest yield of Q | Highest selectivity of Q |
|--------|--------------------|--------------------------|
| I | *perfectly mixed* | *perfectly mixed* |
| II | *plug flow* | *perfectly mixed* |
| III | *plug flow* | *perfectly mixed* = *plug flow* (*zero conversion*) |

mixing levels for complex reaction systems, including the one under discussion. An alternate procedure was utilized by Gillespie and Carberry [12] and van de Vusse [13], who considered a recycle reactor (either actual or as a model) where a portion of the product stream from a plug flow reactor is returned to the entrance. For zero recycle one obviously has a plug flow reactor, and it can be visualized that for infinite recycle the system in some sense behaves as a perfectly

*Table 10.3.a-1 Optimum reactor choice for consecutive and parallel reactions.*

| Region | Highest yield of $Q$ | Highest selectivity of $Q$ |
|--------|----------------------|----------------------------|
| I | perfectly mixed | perfectly mixed |
| II | plug flow | perfectly mixed |
| III | plug flow | perfectly mixed = plug flow (zero conversion) |

mixed reactor because of the large "feedback" of material. Gillespie and Carberry [12] showed that for some values of $a_1$ and $a_2$, an intermediate recycle rate indeed provided the best performance.

Many other examples of optimizing the chemical environment have been discussed in the literature. For example, van de Vusse and Voetter [14] have considered the parallel second-order reactions:

$$A + B \xrightarrow{\;\;1\;\;} Q$$

$$A + A \xrightarrow{\;\;2\;\;} S$$

where $Q$ is the desired product. Here, the best results would be obtained by keeping $C_A$ low, throughout the reactor. The suggested way to do this was to have a plug flow reactor with an entrance feed of $B$ and some $A$, together with side feed of $A$ along the length of the reactor. The purpose was to always keep $C_A$ low, by continually converting it, but also provide sufficient $A$ to convert the $B$ fed to the reactor. A more practical system, of course, would be a series of stirred tank reactors with intermediate feeds of $A$. Refer to Kramers and Westerterp [15] and Denbigh and Turner [16] for further details concerning these problems.

If one is interested in achieving a specified product distribution, rather than just maximizing a yield, the problem is naturally more complicated. Usually numerical simulations with the reactor design equations is necessary, often combined with formal optimization procedures. A study of choice of reactor type, together with separation and recycle systems, was presented by Russell and Buzzelli [17] for the important class of reactions

$$A + B \xrightarrow{\;\;1\;\;} P_1$$

$$P_1 + B \xrightarrow{\;\;2\;\;} P_2$$

$$\vdots$$

$$P_{n-1} + B \xrightarrow{\;\;n\;\;} P_n$$

which are encountered in industrial processes such as the production of mono-, di-, and tri-ethylene glycol from ethylene oxide and water; mono-, di-, and tri-ethanolamine from ethylene oxide and ammonia; mono-, di-, and tri-glycol ethers from ethylene oxide and alcohols; mono-, di-, and tri-chlorobenzenes from benzene and chlorine; and methylchloride, di- and tri-chloromethane from methane and chlorine. In these cases, usually the lower members of the product spectrum, $P_1$ or $P_2$, are primarily desired, and the proper reactor design is crucial to success of the operation. Except for a few general categories such as this, most cases must be handled on an individual basis by the above methods.

## 10.3.b Polymerization Reactions

One of the most important areas for application of concepts discussed in the previous section is the selection of polymerization reactors. The properties of polymers depend on their molecular weight distribution (MWD) and so the design should ultimately use this as its basis. The subject is a vast one, and so only the basic concepts will be briefly discussed. Several excellent reviews now exist, covering various aspects of the area from a chemical reaction engineering viewpoint: see Shinnar and Katz, Keane, and Gerrens, [18, 19, 20]. The latter presents a masterful survey of the effects of the choice of reactor type.

The quite different results that may be obtained by performing polymerization reactions in batch or plug flow versus perfectly mixed flow reactors were described early by Denbigh [21]. The key point concerns the relative lifetimes of the active propagating polymer species. If this is long relative to the mean holding time of the fluid in the reactor, the rules in Sec. 10.3.a apply, and so the product distribution (the MWD) is narrow in a batch reactor (BR)/plug flow reactor (PFR) and broader in a perfectly mixed flow reactor (PMFR), just as in the earlier examples. Recall that the reason was the broader distribution of residence times in the PMFR. However, if the active propagating polymer lifetimes are much shorter than the mean holding time, the residence time of almost all the fluid elements is approaching infinity compared to the local reaction speed. In this case, the constant availability of monomer tends to produce a more uniform product, and so the PMFR produces a narrower MWD than the BR/PFR. Figure 10.3.b-1 shows results computed by Denbigh [21] for a free radical polymerization as considered in Example 1.4-6, and illustrates the striking differences that may be obtained. Also, for the copolymerization of two monomers, the uniform concentrations of a PMFR tend to produce a product of more uniform composition than a BR/PFR. Excellent summaries of the mathematical modeling of polymerization reactors are provided by Ray [22] and Min and Ray [23].

Table 10.3.b-1 shows a summary by Gerrens [20] of the MWD results from the main reactor types for simple polyreactions:

**1** Monomer coupling with termination (e.g., radical polymerization).

Initiation $\qquad I \xrightarrow{\;k_i\;} 2R^{\cdot}$

$$R^{\cdot} + M_1 \xrightarrow{\;k_i'\;} P_1$$

Propagation $\qquad P_{n-1} + M_1 \xrightarrow{\;k_{pr}\;} P_n$

Termination $\qquad P_n + P_m$
$$\begin{cases} \xrightarrow{\;k_{t,d}\;} M_n + M_m \text{ (disproportionation)} \\ \xrightarrow{\;k_{t,c}\;} M_{n+m} \qquad \text{ (combination)} \end{cases}$$

**2** Monomer coupling without termination (e.g., living polymerization).

Initiation $\qquad I + M_1 \xrightarrow{\;k_i\;} P_1$

Propagation $\qquad P_{n-1} + M_1 \xrightarrow{\;k_{pr}\;} P_n$

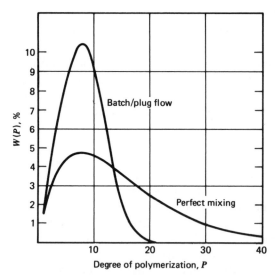

*Figure 10.3.b-1a Molecular weight distribution when active propagating polymer lifetime is long compared to reactor mean holding time (after Denbigh [21], from Levenspiel [3]).*

CHEMICAL REACTOR DESIGN

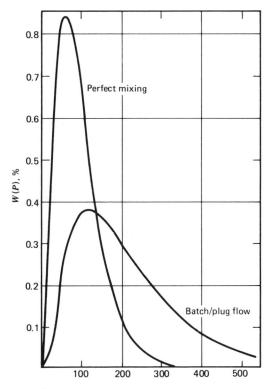

*Figure 10.3.b-1b Molecular weight distribution when active propagating polymer lifetime is short compared to reactor mean holding time (after Denbigh [21], from Levenspiel [3]).*

**3** Polymer coupling (e.g., polycondensation)

Propagation $\quad M_n + M_m \xrightarrow{\quad k \quad} M_{n+m}$

The third column in Table 10.3.b-1 gives results for a reactor with a special state of mixing, where the fluid elements are randomly distributed in the reactor, but also retain their individual identities—called "segregated flow"—which will be considered in more detail in Chapter 12. This situation is considered to be characteristic for very viscous fluids (see Nauman [24]). The entries 1.1, 1.2 and 2.1, 2.2 refer to short and long active propagating polymer relative lifetimes, respectively.

For the free radical polymerization considered in Ex. 1.4-6, Fig. 10.3.b-2 indicates how the MWD evolves in a stirred reactor as the conversion proceeds

Table 10.3.b-1 Molecular weight distributions resulting from polyreactions in various reactor types (from Gerrens [20])

| Reaction \ Reactor | BR or PFR | HCSTR[a] | SCSTR[b] |
|---|---|---|---|
| Monomer coupling with termination | Broader than Schultz–Flory (1.1) | Schultz–Flory distribution (1.2) | Broader than 1.1 (1.3) |
| Monomer coupling without termination | Narrower than Schultz–Flory (Poisson) (2.1) | Schultz–Flory distribution (2.2) | Between 2.1 and 2.2 (2.3) |
| Polymer coupling | Schultz–Flory distribution (3.1) | Much broader than Schultz–Flory (3.2) | Between 3.1 and 3.2 (3.3) |

[a] HCSTR = homogeneous continuous stirred tank reactor.
       = perfectly mixed flow reactor of this chapter.
[b] SCSTR = segregated continuous stirred tank reactor.

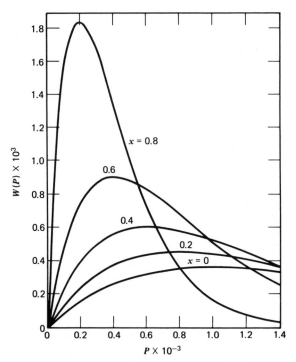

Figure 10.3.b-2 Weight distribution in HCSTR; parameter is conversion, $x$; $\bar{P}_N$ at zero conversion = 1000 (from Gerrens [20]).

440

(the quasi-steady-state approximation for various reactor types was discussed by Ray [25]). Recall that the number average chain length is:

$$\bar{P}_N = \frac{k_{pr}}{\sqrt{k_i k_t}} \frac{M_{10}}{I^{1/2}} \left(\frac{M_1}{M_{10}}\right) = \bar{P}_{N0}(1-x) \tag{10.3.b-1}$$

where

$M_{10}$ is the initial feed monomer concentration

$x = (M_{10} - M_1)/M_{10}$, the monomer conversion

The weight polymer distribution is

$$W(P) = (1-p)^2 Pp^{P-1} \tag{10.3.b-2}$$

with

$$p \simeq 1 - \frac{(k_i k_t)^{1/2}}{k_{pr}} \frac{I^{1/2}}{M_1} = 1 - \frac{1}{\bar{P}_{N0}(1-x)} \tag{10.3.b-3}$$

The conversion, $x$, would be found from the usual relation, Eq. 10.2.b-2:

$$M_1 - M_{10} = -k_{pr}\left(\frac{k_i I}{k_t}\right)^{1/2} M_1 \frac{V}{F'}$$

or

$$1 - x = \frac{1}{1 + k_{pr}\left(\dfrac{k_i I}{k_t}\right)^{1/2} \tau} \tag{10.3.b-4}$$

When real systems are to be described, several practical complications must also be accounted for. Gerrens [20] lists eight of these for radical polymerization:

- Thermal initiation $\qquad\qquad M_1 + M_1 \rightarrow 2R^{\cdot}$

- Decrease of $R^{\cdot}$ during polymerization

- Chain transfer to monomer $\qquad P_n + M_1 \rightarrow P_1 + M_n$

- Chain transfer to solvent, etc. $\qquad P_n + S \rightarrow S^{\cdot} + M_n$

- Chain transfer to polymer $\qquad P_n + M_m \rightarrow P_m + M_n$

- Diffusion control of propagation (glass effect) $\qquad k_{pr} = f(x, \bar{P}_W, \ldots)$

- Diffusion control of termination (Trommsdorff effect) $\qquad k_t = f(x, \bar{P}_W, \ldots)$

- Copolymerization $\qquad\qquad P_1 + M_2 \rightarrow P_2$
  $\qquad\qquad\qquad\qquad\qquad P_2 + M_1 \rightarrow P_1$

The chain transfer, or branching, steps are very important for the polymer properties, but also because as the second step in a series of consecutive reactions, they are especially sensitive to mixing effects.

Nagasubramanian and Graessley [26] have provided a detailed study of these effects for vinyl-acetate polymerization. Here, the strong branching phenomena can reverse the conclusions reached above as to which reactor type will have the narrowest MWD. This is true because through the effect on the branching, the residence time distribution of fluid elements again is the predominant factor. Figure 10.3.b-3 shows that the MWD-breadth, $\bar{P}_W$, is larger when changing from a BR/PFR to a PMFR in this rapid chain, but branching, reaction system—the opposite of Fig. 10.3.b-2. Experimental results obtained by Nagasubramanian and Graessley [26] are shown in Fig. 10.3.b-4, where the theoretical predictions are verified, including the fact that at larger conversions the higher viscosity reacting fluid appears to be better represented by the segregated flow condition (Chapter 12). Also see Hyun, Graessley, and Bankoff [27].

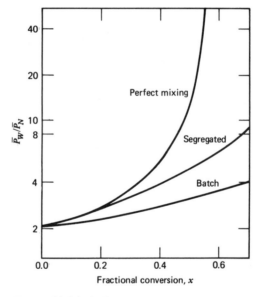

*Figure 10.3.b-3  Dispersion ratio versus conversion, calculated for the three reactor types with typical parameter values for vinyl acetate polymerization (from Nagasubramanian and Graessley [26]).*

442

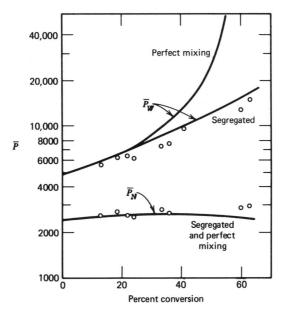

*Figure 10.3.b-4 Degrees of polymerization for continuous flow stirred tank reactors (from Nagasubramanian and Graessley [26]).*

## 10.4 Stability of Operation and Transient Behavior _____

### 10.4.a Stability of Operation

At the beginning of Sec. 10.2.b, it was stated that after the solution to the mass balance is used to decide the reactor operating conditions for optimum conversion (or selectivity), then the energy balance is utilized to determine the external conditions required to maintain the desired temperature. Thus, Eq. 10.2.b-1 is solved together with the steady-state form of Eq. 10.2.a-5:

$$x_A - x_{A0} = \tau \frac{r_A}{C_{A0}} \qquad (10.4.a-1)$$

and

$$T - T_0 = \frac{\tau}{\lambda} \frac{r_A}{C_{A0}} - Q_r(T) \qquad (10.4.a-2)$$

with

$$\tau = C_{A0} V/F_{A0} = V/F'_0$$

$$\lambda = \rho_f c_p/C_{A0}(-\Delta H) > 0, \text{exothermic}$$

$$Q_r(T) = (-Q)/\rho_f c_p F'_0$$

Note that $Q_r(T)$ is proportional to the heat removal rate by external heat exchange.

To illustrate the procedure, consider an irreversible first-order reaction at constant volume, where

$$r_A = kC_A = kC_{A0}(1 - x_A)$$

so that Eq. 10.4.a-1 gives Eq. 10.2.b-3:

$$x_A = 1 - \frac{1}{1 + k\tau}$$

This is then substituted into Eq. 10.4.a-2, which, when rearranged, becomes

$$\frac{1}{\lambda} \frac{A_0 \exp(-E/RT)}{1 + A_0 \exp(-E/RT)} = \frac{T - T_0}{\tau} + \frac{A_k U}{\rho_p c_p F'_0 \tau}(T - T_r)$$

$$= \left(\frac{1}{\tau} + \frac{UA_k}{\rho_f c_p V}\right) T - \left(\frac{T_0}{\tau} + \frac{UA_k}{\rho_f c_p V} T_r\right) \quad (10.4.a\text{-}3)$$

where the simplest expression was used for the heat removal rate,

$$Q_r = \frac{UA_k}{\rho_f c_p F'_0}(T - T_r)$$

Equation 10.4.a-3 is a nonlinear algebraic equation to be solved for $T$, given values for all of the parameters. For the general case, similar manipulations would lead to

$$\frac{1}{\lambda C_{A0}} r_A(x_A, T) = \frac{1}{\tau}[T - T_0 + Q_r(T)] \quad (10.4.a\text{-}4)$$

and $x_A$ is found from Eq. 10.4.a-1 for given $\tau$ and kinetic parameters, the latter depending on the temperature, of course.

Each of Eqs. 10.4.a-3 and 4 have been arranged in such a way that the left-hand side represents the rate of heat generated per total heat capacity of the reactor, $Q_G(T)$, and the right-hand side represents the net heat removed, $Q_R$, by both flow and external heat exchange. The heat balance just states, then, that at a steady-state operating point, these must be equal: $Q_G(T) = Q_R(T)$. The solution(s) to Eq. 10.4.a-4 can be profitably visualized by plotting both $Q_G(T)$ and $Q_R(T)$ against $T$, and noting the intersection(s) of the curves, as illustrated in Fig. 10.4.a-1 for exothermic reactions.

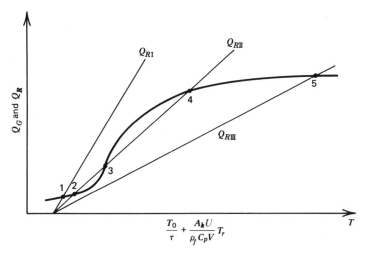

*Figure 10.4.a-1  Heat generation and removal rates.*

The S shape of the curve results from the Arrhenius dependent rate coefficients, while it follows from Eq. 10.4.a-3 that $Q_R$ leads to (essentially) straight lines. Any of the points 1, 2, 3, 4, and 5 represent possible steady states; that is, solutions of the combined mass and energy balances for a particular design. Since the slope of $Q_R$ is the specific heat removal rate, the steepest line $Q_{RI}$ is for high heat removal, and consequently steady-state point 1 means that the reactor will operate at a low temperature, low-heat generation rate, and, consequently, low conversion.

Point 5 is just the converse-low specific heat removal, high temperature, high conversion. $Q_R$-lines falling in between can lead to three intersections, therefore, three solutions. This multiplicity of steady states is caused by the highly nonlinear nature of the heat generation and by the (internal) feedback associated with the complete mixing. The classical discussion of instabilities resulting from this multiplicity was published by Van Heerden [28]. (A similar discussion was previously published by Liljenroth in 1922 [28a]). Van Heerden reasoned that a small increase in temperature from point 2 would lead to the heat removal, $Q_R$, increasing more than the heat generation, $Q_G$; thus, it would seem that the system would tend to decrease in temperature and return to the operating point 2. The same is true for points 1 or 5. However, a small increase (or decrease) about point 3 would tend to be accentuated, and the system thus migrate to the new operating point 4 (or 2). Operating point 3 is called an unstable steady state, and would not be maintained in a real reactor (without automatic control). Which intersection occurs would appear to be based on the slopes of the heat generation versus heat removal curves. These results will be made more specific below.

Another interesting aspect concerns continuous changes in the operating conditions. If the reactor is operating at point 1 on Fig. 10.4.a-1 and the heat removal is decreased to $Q_{RII}$, the temperature will increase, and operating point 2 will be reached. If the heat removal is further decreased to $Q_{RIII}$, the reactor can only operate at point 5, and a large jump in temperature will be generated—this is termed ignition. Then if the heat removal is increased back to $Q_{RII}$ and $Q_{RI}$, the reactor will operate at points 4 and jump to 1, respectively; the latter is called quenching or extinguishing. In addition, it is seen that different paths are followed for increasing versus decreasing the heat removal, and so a hysteresis phenomenon occurs. Further detailed discussion is given by Aris [1].

### Example 10.4.a-1 Multiplicity and Stability in an Adiabatic Stirred Tank Reactor

Experimental verification of the above phenomena was provided by Vejtasa and Schmitz [29] for the exothermic reaction between sodium thiosulfate and hydrogen peroxide—a well-characterized test reaction. It is useful for an adiabatic reactor to use an altered rearrangement of Eqs. 10.4.a-1, 2 whereby the rate term is eliminated to give

$$x_A - x_{A0} = \lambda(T - T_0) \tag{a}$$

Thus, this adiabatic operating relation can be used to eliminate the conversion in terms of the temperature and the thermal properties of the fluid, so that the rate of reaction can be written $r_A(x_A(T), T) \rightarrow r_A(T)$. Then Eq. 10.4.a-4 becomes

$$\frac{1}{\lambda C_{A0}} r_A(T) = \frac{1}{\tau}(T - T_0) \tag{b}$$

and the mean holding time only appears in the right-hand side for heat removal. The heat generation can be plotted for the reaction, and the effect of changing flow rates, for example, only alters the straight lines for $Q_R$. Figure 1 shows this data (also note that a mathematical expression for the reaction rate is not even really needed).

We see in Fig. 1 that the holding times $\tau = 6.8$ and 17.8 sec should be the values between which multiple steady-state and hysteresis phenomena should occur. By starting up the experimental reactor in various ways, and then altering operating conditions, Vejtasa and Schmitz obtained data illustrating this, as shown in Fig. 2.

We see that good agreements with the above predictions were obtained for the steady-state results. Simulations [29] of the complete transient changes, however, were much more sensitive to the details of the models, especially thermal capacity parameters.

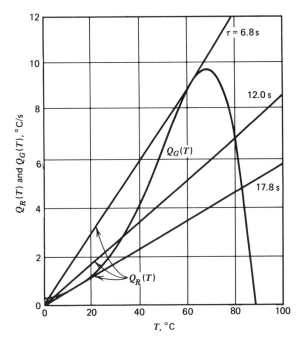

*Figure 1 Heat generation and removal functions for feed mixture of 0.8M $Na_2S_2O_3$ and 1.2M $H_2O_2$ at 0°C (from Vejtassa and Schmitz [29]).*

Heat is added for endothermic reactions so that the straight lines $Q_R(T)$ have negative slopes and only one intersection is possible.

Reversible exothermic reactions have an ultimate decrease of rate with temperature, and so the heat generation curve turns down (as in Example 10.4.a-1); however, the qualitative features remain the same. The heat generation curve for complex reactions can have more than one "hump," and thus more than three steady states are possible for a given operating condition. The humps also tend to be smaller, leading to more readily obtained transitions between steady states, and so on—Westerterp [30]. Also, other types of multiple steady states and instabilities can occur. For example, with certain forms of rate expressions highly nonlinear in concentration, just the mass balance Eq. 10.4.a-1 may have more than one solution. This is summarized in Perlmutter [31] (as well as many other techniques).

These considerations can be put in analytical form, following the reasoning of van Heerden [28] given above. The slopes of the heat removal and generation

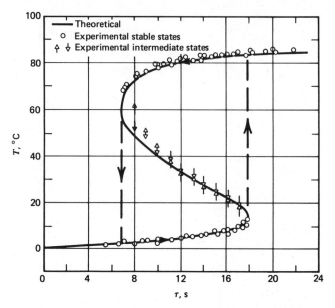

*Figure 2  Steady-state results (from Vejtassa and Schmitz [29]).*

rates in Eq. 10.4.a-4 are found as follows:

$$\frac{dQ_R}{dT} = \frac{d}{dT}\left\{\frac{1}{\tau}\left[T - T_0 + Q_r(T)\right]\right\} = \frac{1}{\tau}\left(1 + \frac{dQ_r}{dT}\right) \qquad (10.4.a\text{-}5)$$

$$\frac{dQ_G}{dT} = \frac{d}{dT}\left[\frac{1}{\lambda C_{A0}} r_A(x_A, T)\right]$$

Now

$$\frac{dr_A}{dT} = \frac{\partial r_A}{\partial x_A}\frac{dx_A}{dT} + \frac{\partial r_A}{\partial T}$$

$$= \frac{\partial r_A}{\partial x_A}\cdot\frac{\tau}{C_{A0}}\frac{dr_A}{dT} + \frac{\partial r_A}{dT}$$

where the last line utilized the mass balance, Eq. 10.4.a-1. The total change of the heat generation rate with temperature is:

$$\frac{dQ_G}{dT} = \frac{\dfrac{1}{\lambda C_{A0}}\dfrac{\partial r_A}{dT}}{1 - \dfrac{\tau}{C_{A0}}\dfrac{\partial r_A}{\partial x_A}} \qquad (10.4.a\text{-}6)$$

448      CHEMICAL REACTOR DESIGN

Then, the reactor *can* be stable if the heat removal slope is greater than the heat generation slope at the steady-state operating point,

$$\frac{dQ_R}{dT} > \frac{dQ_G}{dT}$$

leading to:

$$\left(1 - \frac{\tau}{C_{A0}}\frac{\partial r_A}{\partial x_A}\right)\left(1 + \frac{dQ_r}{dT}\right) > \frac{\tau}{\lambda C_{A0}}\frac{\partial r_A}{\partial T} \qquad (10.4.a\text{-}7)$$

For the case of a first-order irreversible reaction in a reactor with simple heat exchange, as in Eq. 10.4.a-3, this criterion becomes:

$$\lambda(1 + k\tau)\left(1 + \frac{UA_k}{\rho_f c_p F_0'}\right) > \frac{E}{RT^2}\frac{k\tau}{1 + k\tau} \qquad (10.4.a\text{-}8)$$

Equation 10.4.a-7 is a *necessary* but not sufficient condition for stability. In other words, if the criterion is satisfied, the reactor may be stable; if it is violated, the reactor will be unstable. (Aris [1] prefers to use the reverse inequality as a sufficient condition for instability.) The reason is that in deriving Eq. 10.4.a-7, it was implicitly assumed that only the special perturbations in conversion and temperature related by the steady-state heat generation curve were allowed. To be a general criterion giving both necessary and sufficient conditions, arbitrary perturbations in both conversion and temperature must be considered. Van Heerden's reasoning actually implied a sense of time ("tends to move..."), and so the proper criteria can only be clarified and deduced by considering the complete transient mass and energy balances.

### 10.4.b Transient Behavior

The time-dependent mass and energy balances are given by Eqs. 10.2.a-4 and 5:

$$\tau\frac{dx_A}{dt} = x_{A0} - x_A + \tau\frac{r_A}{C_{A0}} \qquad (10.4.b\text{-}1)$$

$$\tau\frac{dT}{dt} = T_0 - T + \frac{\tau}{\lambda}\frac{r_A}{C_{A0}} - Q_r(T) \qquad (10.4.b\text{-}2)$$

Analytical solution of this system of differential equations is not possible. Therefore Aris and Amundson [32] linearized it by a Taylor expansion, about the steady-state operating points. Consider the small perturbations:

$$x = x_A - x_{A,s}$$
$$y = T - T_s$$

where the subscript $s$ refers to a steady-state solution. Then, substracting Eq. 10.4.a-1, 2 from Eq. 10.4.b-1, 2 gives:

$$\tau \frac{dx}{dt} = -x + \frac{\tau}{C_{A0}}(r_A - r_{A,s}) \qquad (10.4.b\text{-}3)$$

$$\tau \frac{dy}{dt} = -y + \frac{\tau}{\lambda C_{A0}}(r_A - r_{A,s}) - [Q_r(T) - Q_r(T_s)] \qquad (10.4.b\text{-}4)$$

Expanding $r_A$ and $Q_r(T)$ in Taylor series and neglecting second-order terms leads to:

$$r_A = r_{A,s} + \left(\frac{\partial r_A}{\partial x_A}\right)_s x + \left(\frac{\partial r_A}{\partial T}\right)_s y$$

$$Q_r(T) = Q_r(T_s) + \left(\frac{dQ_r}{dT}\right)_s y$$

Substituting into Eq. 10.4.b-3, 4 yields

$$\tau \frac{dx}{dt} = -\left[1 - \frac{\tau}{C_{A0}}\left(\frac{\partial r_A}{\partial x_A}\right)_s\right]x + \left[\frac{\tau}{C_{A0}}\left(\frac{\partial r_A}{\partial T}\right)_s\right]y \qquad (10.4.b\text{-}5)$$

$$\tau \frac{dy}{dt} = \left[\frac{\tau}{\lambda C_{A0}}\left(\frac{\partial r_A}{\partial x_A}\right)_s\right]x - \left[1 - \frac{\tau}{\lambda C_{A0}}\left(\frac{\partial r_A}{\partial T}\right)_s + \left(\frac{dQ_r}{dT}\right)_s\right]y \qquad (10.4.b\text{-}6)$$

These equations (10.4.b-5, 6) are linear differential equations, whose solutions are combinations of exponentials of the form $\exp[mt/\tau]$, where the values of $m$ are solutions of the characteristic equation:

$$m^2 + a_1 m + a_0 = 0 \qquad (10.4.b\text{-}7)$$

where

$$a_1 = \left[1 - \frac{\tau}{C_{A0}}\left(\frac{\partial r_A}{\partial x_A}\right)_s\right] + \left[1 + \left(\frac{dQ_r}{dT}\right)_s\right] - \left[\frac{\tau}{\lambda C_{A0}}\left(\frac{\partial r_A}{\partial T}\right)_s\right]$$

$$a_0 = \left[1 - \frac{\tau}{C_{A0}}\left(\frac{\partial r_A}{\partial x_A}\right)_s\right]\left[1 + \left(\frac{dQ_r}{dT}\right)_s\right] - \left[\frac{\tau}{\lambda C_{A0}}\left(\frac{\partial r_A}{\partial T}\right)_s\right]$$

The solutions will only go to zero as $t \to \infty$ when the real parts of the roots are negative (e.g., see Himmelblau and Bischoff [33]).

The solution of Eq. 10.4.b-7 is:

$$m = \tfrac{1}{2}(-a_1 \pm \sqrt{a_1{}^2 - 4a_0})$$

and we see that this stability condition is only always met when

$$a_1 > 0 \qquad\qquad (10.4.\text{b-}8)$$

and

$$a_0 > 0 \qquad\qquad (10.4.\text{b-}9)$$

If $a_0 < 0$, at least one of the roots will be positive, and the solution will diverge for $t \to \infty$. If $a_1 = 0$ and $a_0 > 0$, the roots will be purely imaginary numbers, with oscillatory solutions for $x$ and $y$. Thus, the necessary and sufficient conditions for stability (i.e., $x_A$ and $T$ return to the steady state after removal of the perturbation or $x$ and $y \to 0$ as $t \to \infty$) are Eqs. 10.4.b-8 and 9. In terms of the physical variables those equations can be written as follows:

$$\left[1 - \frac{\tau}{C_{AO}}\left(\frac{\partial r_A}{\partial x_A}\right)_s\right] + \left[1 + \left(\frac{dQ_r}{dT}\right)_s\right] > \left[\frac{\tau}{\lambda C_{AO}}\left(\frac{\partial r_A}{\partial T}\right)_s\right] \quad (10.4.\text{b-}10)$$

and

$$\left[1 - \frac{\tau}{C_{AO}}\left(\frac{\partial r_A}{\partial x_A}\right)_s\right]\left[1 + \left(\frac{dQ_r}{dT}\right)_s\right] > \left[\frac{\tau}{\lambda C_{AO}}\left(\frac{\partial r_A}{\partial T}\right)_s\right] \quad (10.4.\text{b-}11)$$

Comparing Eq. 10.4.b-11 with Eq. 10.4.a-7, we see that they are identical, and the above discussion shows that the "slope" criterion, $a_0 > 0$, is indeed a necessary condition for stability. We also see that $a_0 > 0$ is not sufficient, for if $a_1 = 0$, the oscillations are not stable, in that $x_A$ and $T$ do not return to their steady-state values. Thus, the second criterion, Eq. 10.4.b-10, seems to be related to oscillatory behavior—a discussion is given by Gilles and Hofmann [34].

For the case of a first-order irreversible reaction with simple heat exchange, as in Eq. 10.4.a-3, the second ("dynamic") criterion Eq. 10.4.b-10 becomes

$$(1 + k\tau) + \left(1 + \frac{UA_k}{\rho_f c_p F'_0}\right) > \frac{1}{\lambda}\frac{E}{RT^2}\frac{k\tau}{1 + k\tau} \quad (10.4.\text{b-}12)$$

Also note that for an adiabatic reactor, $Q_r \equiv 0$, the "slope" criterion Eq. 10.4.b-11 implies the other Eq. 10.4.b-10, and so the slope criterion is both necessary and sufficient for this case.

*Example 10.4.b-1 Temperature Oscillations in a Mixed Reactor for the Vapor Phase Chlorination of Methyl Chloride*

This system was studied by Bush [35]; the reaction was:

$$CH_3Cl \xrightarrow{Cl_2} CH_2Cl_2 \xrightarrow{Cl_2} CHCl_3 \xrightarrow{Cl_2} CCl_4$$

Experimental measurements were made of the several relevant variables so that an evaluation of the above criteria could be made. First, the steady-state heat generation and removal rates were determined as shown in Fig. 1.

We see that the necessary "slope" criterion is satisfied over the entire range of conditions:

$$\frac{dQ_R}{dT} > \frac{dQ_G}{dT}$$

Thus, there may be a unique steady-state reactor temperature for a given bath temperature, $T_r$.

However, it was found that the reactor showed oscillatory behavior in certain ranges (see Bush [36]). Therefore, the second "dynamic" criterion, Eq. 10.4.b-10

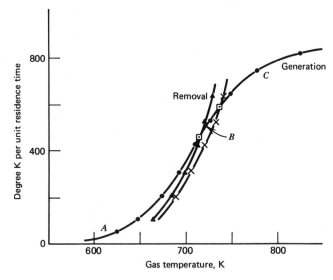

*Figure 1 Steady-state temperature:* ●, *heat evolution;* ×, *heat removal for* $T_r = 400°C$; △, *heat removal for* $T_r = 390°C$; ☐, *steady-state temperatures* $T_s$ *(from Bush [35]).*

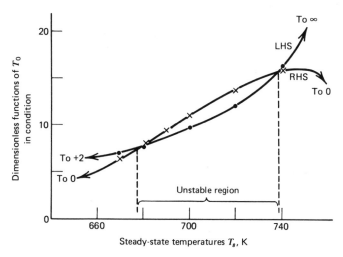

*Figure 2 Stable and unstable reaction temperatures:* ●, *LHS;* ×, *RHS (from Bush [35]).*

was also checked with relationships similar to Eq. 10.4.b-12, but also accounting for some of the additional complexities in the real (gas phase) experimental system. The results are shown in Fig. 2, where the left-hand side (LHS) and right-hand side (RHS) of the criterion are plotted. We see that a central region exists where the criterion is violated. This is verified by the experimentally observed large temperature excursions:

| Bath $T_r$°C | Reactor $T_s$°C | Frequency Hz | Amplitude, °C Nominal measured | Computed | % High Temp. Product |
|---|---|---|---|---|---|
| <375 | <400 | ← | —————— No oscillations —————— | → | <0.1 |
| 392 | 445 | 1 | 150 | 280 | 24 |
| 397 | 453 | 1.12 | 110 | 170 | 10 |
| >410 | >466 | ← | —————— No oscillations —————— | → | <1 |

Good agreement was obtained for the temperature range of oscillatory phenomena, along with rough comparison of the excursion amplitudes.

Studies where the reactor was deliberately allowed to oscillate have been reported by Baccaro, Gaitonde, and Douglas [37] and by Chang and Schmitz [38]; also see the review by Bailey [39].

THE PERFECTLY MIXED FLOW REACTOR _____ **453**

Many other aspects of this basic system have also been developed. For example, if the heat exchange is controlled by an automatic device such that additional rates are proportional to the reactor temperature perturbations, $Q_r$ in Eq. 10.4.b-2 is modified to:

$$Q_r + [\beta + \mu(T - T_s)]$$

Then the corresponding term in the linearized Eq. 10.4.b-6 is changed to,

$$\left(\frac{dQ_r}{dT}\right)_s \rightarrow \left(\frac{dQ_r}{dT}\right)_s + \mu$$

The two criteria Eq. 10.4.b-10, 11 then become

$$\left[1 - \frac{\tau}{C_{A0}}\left(\frac{\partial r_A}{\partial x_A}\right)_s\right] + \left[1 + \left(\frac{dQ_r}{dT}\right)_s + \mu\right] > \left[\frac{\tau}{\lambda C_{A0}}\left(\frac{\partial r_A}{\partial T}\right)_s\right] \quad (10.4.b\text{-}13)$$

and

$$\left[1 - \frac{\tau}{C_{A0}}\left(\frac{\partial r_A}{\partial x_A}\right)_s\right]\left[1 + \left(\frac{dQ_r}{dT}\right)_s + \mu\right] > \left[\frac{\tau}{\lambda C_{A0}}\left(\frac{\partial r_A}{\partial T}\right)_s\right] \quad (10.4.b\text{-}14)$$

We see that the criteria will be easier to satisfy for large $\mu$ (control), since the LHS is larger; in fact, an inherently unstable reactor can be made stable in this way. However, the above is only a very simple consideration of a control system, and real-life complications can modify the results—see Aris [1].

All of the above conclusions were based on the linearized equations for small perturbations about the steady state. A theorem of differential equations states that if the linearized calculations show stability, then the nonlinear equations will also be stable for sufficiently small perturbations. For larger excursions, the linearizations are no longer valid, and the only recourse is to (numerically) solve the complete equations. A definitive study was performed by Uppal, Ray, and Poore [40] where extensive calculations formed the basis for a detailed mathematical classification of the many various behavior patterns possible; refer to the original work for the extremely complex results. The evolution of multiple steady states when the mean holding time is varied leads to even more bizarre possible behavior (see Uppal, Ray, and Poore [41]. Further aspects can be found in the comprehensive review of Schmitz [42] and in Aris [1], Perlmutter [31], and Denn [43].

## Problems

10.1 A perfectly mixed flow reactor is to be used to carry out the reaction $A \rightarrow R$. The rate is given by

$$r_A = kC_A\left(\frac{\text{kmol}}{\text{m}^3\text{ s}}\right)$$

with

$$k = 4 \times 10^6 \exp[-8000/T(K)]\text{s}^{-1}$$

Other physicochemical data is:

$$\Delta H = -40{,}000 \text{ kcal/kmol} \qquad \rho c_p = 1000 \text{ kcal/m}^3 \,^\circ\text{C}$$
$$M_A = 100 \text{ kg/kmol} \qquad C_{A0} = 1 \text{ kmol/m}^3$$

At a temperature of 100°C and a desired production rate of 0.4 kg/s, determine:
(a) the reactor volume required at a conversion of 70 percent
(b) the heat exchange requirement

10.2 The first-order reversible reaction

$$A \underset{2}{\overset{1}{\rightleftharpoons}} R$$

is carried out in a constant volume perfectly mixed flow reactor. The feed contains only $A$, at a concentration of $C_{A0}$, and all initial concentrations are zero.
(a) Show that the concentration of $A$ is given by

$$\frac{C_A}{C_{A0}} = \frac{1 + k_2 \tau}{1 + k_1 \tau + k_2 \tau} - \frac{k_2}{k_1 + k_2} e^{-t/\tau} - \frac{k_1 e^{-(1 + k_1\tau + k_2\tau)t/\tau}}{(k_1 + k_2)(1 + k_1 \tau + k_2 \tau)}$$

where $\tau = V/F' = $ mean residence time.
(b) Find $C_A/C_{A0}$ at steady state, and also show that for very rapid reactions, $(k_1, k_2) \to \infty$, the equilibrium concentration is

$$\frac{C_A}{C_{A0}} = \frac{1}{1 + K} \qquad K \equiv k_1/k_2$$

(c) For very rapid reactions, $(k_1, k_2) \to \infty$, show that, in general,

$$\frac{C_A}{C_{A0}} = \frac{1}{1 + K} - \frac{1}{1 + K} e^{-t/\tau}$$

and explain how this can be physically interpreted as the final steady-state equilibrium minus the equilibrium "washout."

10.3 For a first-order reaction, the conversion to be expected in a series of $n$-stirred tanks can be formed from Fig. 10.2.b-1. Alternatively, at a given conversion level, and for a given rate coefficient and mean residence time, $k\tau$, the total volume required to carry out the reaction can be determined.
(a) With this basis, plot $V_{total}/V_{plug\ flow}$ versus the fraction of unreacted reactant, $1 - x_A$, for various values of $n = 1, 2, 5, 10, 40$. Study the effect of utilizing several stirred tank reactors in series compared to a plug flow reactor.
(b) Add further lines of constant values of the dimensionless group $k\tau_{total}$ to the plot— these are convenient for reactor design calculations.

10.4 (a) For the reversible consecutive reactions

$$A \underset{2}{\overset{1}{\rightleftharpoons}} R \underset{4}{\overset{3}{\rightleftharpoons}} S$$

taking place in a steady state, constant volume perfectly mixed reactor, show that the concentration of $R$, when the feed contains only $A$ at concentration $C_{A0}$, is:

$$\frac{C_R}{C_{A0}} = \frac{\dfrac{k_1 \tau}{1 + k_1 \tau}\left(1 + \dfrac{k_2 \tau}{K_b}\right)}{\left(\dfrac{k_1 \tau}{1 + k_1 \tau}\dfrac{1}{K_a} + 1\right)\left(1 + \dfrac{k_2 \tau}{K_b}\right) + k_2 \tau}$$

where

$$K_a = k_1/k_2 = \text{equilibrium constant for the first reaction}$$
$$K_b = k_3/k_4$$

(b) For both reactions irreversible, show that the results of part (a) reduce to the equation given in Sec. 10.3.

(c) If the first reaction is very rapid, it is always close to its equilibrium as $R$ is reacting further to $S$. Explain how this can be represented by $k_1 \to \infty$ but $K_a = $ finite, and find the expression for $C_R/C_{A0}$ by appropriately reducing the result of part (a). This is similar to a rate-determining step situation, and is more simply derived by taking the first reaction to always be in instantaneous equilibrium, $C_A \simeq C_R/K_a$. Show that a new derivation of the mass balances with this basis leads to the same result as above. Note that this is a useful technique in more complex situations of this type, when the general expression may not be possible to derive.

10.5 Consider the startup of a perfectly mixed flow reactor containing a suspended solid catalyst. For a first-order reaction, $r_A = kC_A$, and assuming constant volume, show that the outlet concentration of reactant $A$ is

$$\frac{C_A(t)}{C_{A0}} = \frac{1}{\left[1 + \dfrac{(1-\varepsilon)V}{F'}k\right]} + \left\{\frac{C_A(0)}{C_{A0}} - \frac{1}{\left[1 + \dfrac{(1-\varepsilon)V}{F'}k\right]}\right\}$$

$$\times \exp\left\{-\left[1 + \frac{(1-\varepsilon)V}{F'}k\right]\left(\frac{F'}{\varepsilon V}t\right)\right\}$$

where

$C_A(0) = $ initial concentration
$C_{Ai} = $ feed concentration
$\varepsilon = $ void fraction, not occupied by solids
$\varepsilon V = $ fluid volume

Note that the steady-state $(t \to \infty)$ result depends only on the group $(1 - \varepsilon)Vk/F'$, the *solid catalyst* inverse space velocity-rate coefficient group, but the transient effects also require knowledge of $(F'/\varepsilon V)^{-1}$, or the *fluid* mean residence time.

CHEMICAL REACTOR DESIGN

10.6 In a process to make compound $R$, the following reactions occur:

$$A + B \xrightarrow{\quad 1 \quad} 2R$$

$$A + A \xrightarrow{\quad 2 \quad} 2S$$

(a) Based on the text discussion, explain why the optimum chemical environment would be high $B$ and low $A$ concentrations.

(b) An idealized reactor configuration to achieve this is a reactor with side stream feeds of $A$:

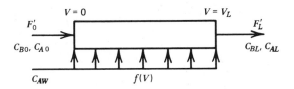

where $f(V)$ (m³ side feed/hr − m³ reactor volume) is the distribution of side feed additions along the reactor length (volume), to be determined. Assuming the reactor to be plug flow, derive the following mass balances:

Total:
$$\frac{dF'}{dV} = f(V)$$

$A$:
$$\frac{d}{dV}(F'C_A) = C_{AW} f(V) - k_1 C_A C_B - k_2 C_A{}^2$$

$B$:
$$\frac{d}{dV}(F'C_B) = -k_1 C_A C_B$$

(c) As an approximate optimal design, the condition will be used that the side feed be adjusted to maintain $C_A = $ constant (i.e., $C_A = C_{A0} = C_{AL}$). Also, a high conversion of $A$ is desired, and to simplify the calculations, it will be assumed that the side feed concentration is high, $C_{AW} \gg C_A = C_{A0} = C_{AL}$. For these special conditions, show that the three mass balances become

$$F' \simeq \text{constant} = F'_0$$

$$0 = C_{AW} f(V) - k_1 C_{AL} C_B - k_2 C_{AL}{}^2$$

$$F' \frac{dC_B}{dV} = -k_1 C_{AL} C_B$$

(d) Using the simplified balances, determine the total reactor volume required as a function of $F'_0$, $C_{AL}$, $C_{B0}$, $C_{BL}$.

(e) Show that the side feed distribution as a function of reactor length, to maintain the above condition of constant $C_A$ is given by

$$f(V) = \frac{C_{AL}}{C_{AW}} (k_2 C_{AL} + k_1 C_{BO} e^{-k_1 C_{AL} V/F_0'})$$

(f) As a final condition, equal stoichiometric feeds of $A$ and $B$ are to be used:

$$F_0' C_{BO} = F_0' C_{AO} + C_{AW} \int_0^{V_L} f(V) dV$$

Show for this case that the relationship between the outlet levels of $A$ and $B$ is:

$$C_{AL} = \frac{C_{BL}}{1 - \frac{k_2}{k_1} \ln \frac{C_{BL}}{C_{BO}}}$$

(g) A useful measure is the reactor yield of the desired $R$:

$$Y \equiv \frac{\text{total } R \text{ formed}}{\text{total } A \text{ fed}}$$

For $k_2/k_1 = 1$, compare the yield as a function of conversion with that found in a single perfectly mixed reactor and with a single plug flow reactor without side feeds.

*Note:*

This problem was first solved by van de Vusse and Voetter [14], who also considered more general cases, and a true mathematically optimal profile, $f(V)$. These latter results were rather close to the approximately optimal basis of $C_A = $ constant. Finally, such an ideal scheme might be implemented in practice by using a series of stirred tank reactors with intermediate feed additions of $A$.

10.7 A perfectly mixed reactor is to be used for the hydrogenation of olefins, and will be operated isothermally. The reactor is 10 m$^3$ in size, and the feed rate is 0.2 m$^3$/s, with a concentration of $C_{AO} = 13$ kmol/m$^3$. For the conditions in the reactor, the rate expression is:

$$r_A = \frac{C_A}{(1 + C_A)^2} \frac{\text{kmol}}{\text{m}^3 \cdot \text{s}}$$

It is suspected that this nonlinear rate form, that has a maximum value, may cause certain regions of unstable operation with multiple steady states.
(a) From the reactor mass balance Eq. 10.2.b-2 determine if this is the case by plotting $r_A$ and $[(1/\tau)(C_{AO} - C_A)]$ on the same graph.
(b) To what concentration(s) should the feed be changed to avoid this problem?

*Note:*

This problem was investigated by Matsuura and Kato [*Chem. Eng. Sci.*, **22**, 17 (1967)], and general stability criteria are provided by Luss [*Chem. Eng. Sci.*, **26**, 1713 (1970)].

10.8 Using the expressions for the necessary and sufficient conditions for stability of a stirred tank chemical reactor as derived in Sec. 10.4:

(a) Show that for a single endothermic reaction the steady state is always stable.

(b) Show that for an adiabatic reactor, the slope condition

$$\left[1 - \frac{\tau}{C_{AO}} \left(\frac{\partial r_A}{\partial x_A}\right)_s\right]\left[1 + \left(\frac{dQ_r}{dT}\right)_s\right] > \left[\frac{\tau}{\lambda C_{AO}} \left(\frac{\partial r_A}{\partial T}\right)_s\right]$$

is sufficient, as well as necessary.

(c) If the reactor is controlled on concentration,

$$Q_r(x, y) = Q_r(y) + vx$$

show that it is not always possible to get control of an unstable steady state. Note here that $Q_r = Q_r(x, y)$, and be careful of the criteria that you use.

10.9 Show that recycling the effluent of a perfectly mixed reactor has no effect on the conversion.

10.10 Consider two perfectly mixed reactors in series. For a given total volume, determine optimal distribution of the sub-volumes for (a) first-order reaction, (b) second-order reaction.

# References

[1] Aris, R. *Introduction to the Analysis of Chemical Reactors*, Prentice-Hall, Englewood Cliffs, N.J. (1965).

[2] Schoenemann, K. *Dechema Monographien*, **21**, 203 (1952).

[3] Levenspiel, O. *Chemical Reaction Engineering*, 1st ed., Wiley, New York (1962).

[4] Kermode, R. I. and Stevens, W. F. *Can J. Chem. Eng.*, **43**, 68 (1965).

[5] Levenspiel, O. *Chemical Reaction Engineering*, 2nd ed., Wiley, New York (1972).

[6] Luss, D. *Chem. Eng. Sci.*, **20**, 17 (1965).

[7] Bischoff, K. B. *Can. J. Chem. Eng.*, **44**, 281 (1966).

[8] Aris, R. *Can. J. Chem. Eng.*, **40**, 87 (1962).

[9] Wei, J. *Can J. Chem. Eng.*, **44**, 31 (1966).

[10] van de Vusse, J. G. *Chem. Eng. Sci.*, **19**, 994 (1964).

[11] Paynter, J. D. and Haskins, D. E. *Chem. Eng. Sci.*, **25**, 1415 (1970).

[12] Gillespie, B. M. and Carberry, J. J. *Chem. Eng. Sci.*, **21**, 472 (1966).

[13] van de Vusse, J. G. *Chem. Eng. Sci.*, **21**, 611 (1966).

[14] van de Vusse, J. G. and Voetter, H. *Chem. Eng. Sci.*, **14**, 90 (1961).

[15]  Kramers, H. and Westerterp, K. R. *Elements of Chemical Reactor Design and Operation*, Academic Press, New York (1963).

[16]  Denbigh, K. and Turner, J. C. R. *Chemical Reactor Theory*, 2nd ed., Cambridge University Press, London (1971).

[17]  Russell, T. W. F. and Buzzelli, D. T. *I&EC Proc. Des. Devt.*, **8**, 2 (1969).

[18]  Shinnar, R. and Katz, S. *Proc. 1st Intl. Symp. Chem. Reac. Eng.*, Am. Chem. Soc. Adv. Chem. Ser. 109, Washington, D.C. (1972).

[19]  Keane, T. R. *Proc. 2nd Intl. Symp. Reac. Eng.*, Elsevier, Amsterdam (1972).

[20]  Gerrens, H. *Proc. 4th Intl. Symp. Chem. Reac. Eng.*, DECHEMA (1976).

[21]  Denbigh, K. G. *Trans. Farad. Soc.*, **40**, 352 (1944); **43**, 648 (1947); *J. Appl. Chem.*, **1**, 227 (1951).

[22]  Ray, W. H. *J. Macromolec. Sci.—Rev. Macromolec. Chem.*, **C8**, 1 (1972).

[23]  Min, K. W. and Ray, W. H. *J. Macromolec. Sci.—Rev. Macromolec. Chem.*, **C11**, 177 (1974).

[24]  Nauman, E. B. *J. Macromolec. Sci.—Rev. Macromolec. Chem.*, **C10**, 75 (1974).

[25]  Ray, W. H. *Can. J. Chem. Eng.*, **47**, 503 (1969).

[26]  Nagasubramanian K. and Graessley, W. W. *Chem. Eng. Sci.*, **25**, 1549, 1559 (1970).

[27]  Hyun, J. C., Graessley, W. W., and Bankoff, S. G. *Chem. Eng. Sci.*, **31**, 945 (1976).

[28]  van Heerden, C. *Ind. Eng. Chem.*, **45**, 1245 (1953).

[29]  Vejtassa, S. A. and Schmitz, R. A. *A.I.Ch.E. J.*, **16**, 410 (1970).

[30]  Westerterp, K. R. *Chem. Eng. Sci.*, **17**, 423 (1969).

[31]  Perlmutter, D. D. *Stability of Chemical Reactors*, Prentice-Hall, Englewood Cliffs, N. J. (1972).

[32]  Aris, R. and Amundson, N. R. *Chem. Eng. Sci.*, **7**, 121 (1958).

[33]  Himmelblau, D. M. and Bischoff, K. B. *Process Analysis and Simulation*, Wiley, New York (1968).

[34]  Gilles, E. D. and Hofmann H. *Chem. Eng. Sci.*, **15**, 328 (1961).

[35]  Bush, S. F. *Proc. Roy. Soc.*, **A309**, 1 (1969).

[36]  Bush, S. F. *Proc. 1st Intl. Symp. Chem. Reac. Eng.*, Amer. Chem. Soc. Adv. Chem. Ser. No. 109, p. 610, Washington, D. C. (1972).

[37]  Baccaro, G. P., Gaitonde, N.Y., and Douglas, J. M. *A.I.Ch. E. J.*, **16**, 249 (1970).

[38]  Chang, M. and Schmitz, R. A. *Chem. Eng. Sci.*, **30**, 21 (1975).

[39]   Bailey, J. E. *Chem. Eng. Commun.*, **1**, 111 (1973).

[40]   Uppal, A., Ray, W. H., and Poore, A. B. *Chem. Eng. Sci.*, **29**, 967 (1974).

[41]   Uppal, A., Ray, W. H., and Poore, A. B. *Chem. Eng. Sci.*, **31**, 205 (1976).

[42]   Schmitz, R. A. *Chem. Reac. Eng. Rev.* (3rd Intl. Symp.), Am. Chem. Soc. Adv. Chem. Ser. 148, Washington, D.C. (1975).

[43]   Denn, M. M. *Stability of Reaction and Transport Processes*, Prentice-Hall, Englewood Cliffs, N.J. (1975).

[44]   Liljenroth, F. G. *Chem. Metal. Eng.*, **19**, 287 (1922).

# 11

## FIXED
## BED
## CATALYTIC
## REACTORS

### Part One

### Introduction

### 11.1 The Importance and Scale of Fixed Bed Catalytic Processes

The discovery of solid catalysts and their application to chemical processes in the early years of this century has led to the breakthrough of chemical industry. Since these days, this industry has diversified and grown in a spectacular way, through the development of new or the rejuvenation of established processes, mostly based on the use of solid catalysts.

The major part of these catalytic processes is carried out in fixed bed reactors. Some of the main fixed bed catalytic processes are listed in Table 11.1-1. Except for the catalytic cracking of gas oil, which is carried out in a fluidized bed to enable the continuous regeneration of the catalyst, the main solid catalyzed processes of today's chemical and petroleum refining industry appear in Table 11.1-1. However, there are also fluidized bed alternatives for phthalic anhydride— and ethylene dichloride synthesis. Furthermore, Table 11.1-1 is limited to fixed bed processes with only one fluid phase; trickle bed processes (e.g., encountered in the hydrodesulfurization of heavier petroleum fractions) are not included in the present discussion. Finally, important processes like ammonia oxidation for nitric acid production or hydrogen cyanide synthesis, in which the catalyst is used in the form of a few layers of gauze are also omitted from Table 11.1-1.

Todays fixed bed reactors are mainly large capacity units. Figure 11.1-1 shows growth curves of reactor capacity for ammonia-synthesis and phthalic-anhydride synthesis on German catalysts. Such a spectacular rise in reactor capacity is evidently tied to the growing market demand, but its realization undoubtedly also reflects progress in both technological and fundamental areas, pressed by the booming construction activity of the last years.

462

*Table 11.1-1  Main fixed bed catalytic processes*

| Basic chemical industry | Petrochemical industry |
|---|---|
| Steam reforming { Primary / Secondary | Ethylene oxide |
| | Ethylene dichloride |
| Carbon monoxide conversion | Vinylacetate |
| Carbon monoxide methanation | Butadiene |
| Ammonia ⎫ | Maleic anhydride |
| Sulfuric acid ⎪ synthesis | Phthalic anhydride |
| Methanol ⎪ | Cyclohexane |
| Oxo ⎭ | Styrene |
| | Hydrodealkylation |

| Petroleum refining | |
|---|---|
| Catalytic reforming | Polymerization |
| Isomerization | (Hydro)desulfurization |
| | Hydrocracking |

From Froment [148]

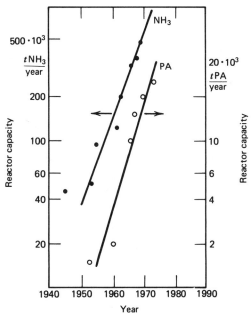

*Figure 11.1-1 Growth curves of reactor capacity in ammonia and phthalic anhydride synthesis (from Froment [148]; data from [1] and [6]).*

463

## 11.2 Factors of Progress: Technological Innovations and Increased Fundamental Insight

Among the many technological innovations of recent years are the following examples:

- The introduction of better materials of construction (e.g., in steam reforming) where the use of centrifugal cast 25% Cr-20% Ni steel tubes has enabled increasing the operating temperature and consequently the throughput.

- Better design of reactor internals (e.g., in phthalic anhydride synthesis), improving the rate and uniformity of heat removal by molten salts.

- More adequate shop techniques and increased shipping clearance, permitting the construction of multitubular reactors of large diameters containing up to 20,000 tubes.

- Modification of auxiliary equipment (e.g., the introduction of centrifugal compressors) boosted the capacity of well-established processes like ammonia and methanol synthesis.

- Modification of flow pattern (e.g., the use of radial flow reactors in catalytic reforming and ammonia synthesis) to reduce the pressure drop and thus enhance the recycle compressor capacity.

- The use of small catalyst particles in regions where heat transfer matters and larger particles in other zones to limit the pressure drop, as in primary steam reformers.

- The design of improved control schemes.

Examples of progress that may be termed fundamental are:

- The development of new catalysts or the modification of existing ones. Major recent achievements concerning fixed bed processes were the addition of rhenium and other rare metals to platinum-$Al_2O_3$ catalysts for catalytic reforming, to increase stability; the formulation of a stable low-pressure methanol synthesis catalyst; the introduction of a low-temperature CO shift catalyst, permitting operation under thermodynamically more favorable conditions; and a $V_2O_5$ catalyst allowing high throughputs at relatively low temperatures in phthalic anhydride synthesis.

- Advances in fundamental data. Intensive research has led to more extensive and more reliable physicochemical data; heat transfer in packed beds has been studied more carefully. Large companies are now well aware of the importance of reliable kinetic data as a basis for design and kinetic studies have benefited from more systematic methods for the design of experiments and improved methods for analysis of the data.

- The use of reactor models as a basis for design, associated with the ever-increasing possibilities of computers. This is an aspect that will be dealt with extensively further in this chapter. To place this aspect in the right perspective, earlier stages of design in which decisions are taken on the basis of sound judgment and semiquantitative considerations will be discussed first.

## 11.3 Factors Involved in the Preliminary Design of Fixed Bed Reactors

When a reactor has to be scaled up from its bench scale version, a certain number of questions arise as to its ultimate type and operation. In general, several alternatives are possible. These may be retained up to a certain degree of progress of the project, but a choice will have to be made in as early a stage as possible, on the basis of qualitative or semiquantitative considerations, before considerable effort is invested into the detailed design.

The first and most elementary type of reactor to be considered is the adiabatic. In this case, the reactor is simply a vessel of relatively large diameter. Such a simple solution is not always applicable, however. Indeed, if the reaction is very endothermic, the temperature drop may be such as to extinguish the reaction before the desired conversion is attained—this would be the case with catalytic reforming of naphtha or with ethylbenzene dehydrogenation into styrene. Strongly exothermic reactions lead to a temperature rise that may be prohibitive for several reasons: for its unfavorable influence on the equilibrium conversion, as in ammonia, methanol, and $SO_3$ synthesis, or on the selectivity, as in maleic anhydride or

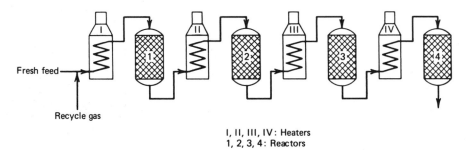

I, II, III, IV: Heaters
1, 2, 3, 4: Reactors

*Figure 11.3-1 Multibed adiabatic reactor for catalytic reforming (from Smith [5]).*

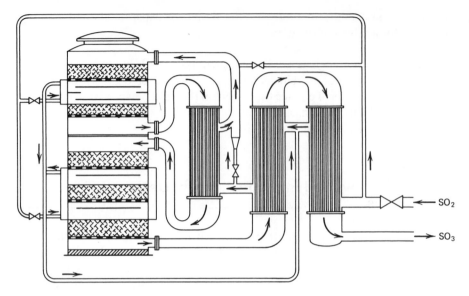

*Figure 11.3-2 Multibed adiabatic reactor for SO$_3$ synthesis (after Winnacker and Kuechler [2], from Froment [148]).*

ethylene oxide synthesis, or on the catalyst stability, or simply because it would lead to unsafe operation. A solution that can be applied to endothermic reactions, although it is not without drawbacks, is to dilute the reactant with a heat carrier. More often, however, the reactor is subdivided into several stages, with intermediate heat exchange. An example of such a multibed adiabatic reactor is shown in Figure 11.3-1 for an endothermic process, catalytic reforming.

The exothermic process of SO$_3$ synthesis is carried out in reactors as illustrated in Figure 11.3-2, and exothermic NH$_3$ synthesis as in Figure 11.3-3. In ammonia or SO$_3$ synthesis the intermediate cooling may be achieved by means of heat exchangers or by injection of cold feed. With SO$_3$ synthesis the heat exchangers are generally located outside the reactor. Special care has to be taken to provide homogeneous distribution of the quench or the flow coming from an intermediate heat exchanger over the bed underneath.

The temperature-composition relation in such a multibed adiabatic reactor is illustrated in Fig. 11.3-4 for ammonia synthesis [3]. The $\Gamma_e$ curve in this diagram represents the equilibrium relation between composition and temperature. The maximum ammonia content that could be obtained in a single adiabatic bed with inlet conditions corresponding to $A$ would be 14 mole % as indicated by point $B'$, and this would theoretically require an infinite amount of catalyst. The five-bed quench converter corresponding to the reaction path $ABCDEFGHIJ$ permits

_____ CHEMICAL REACTOR DESIGN

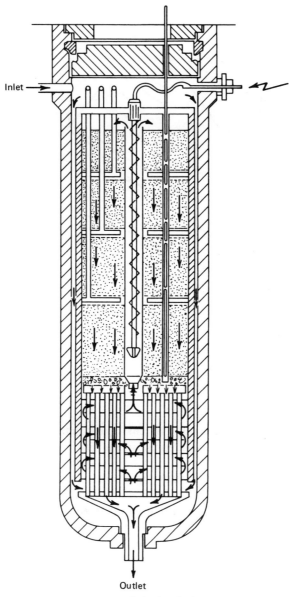

Inlet

Outlet

*Figure 11.3-3 Multibed adiabatic reactor for NH₃ synthesis (after Winnacker and Kuechler [2], from Froment [148]).*

467

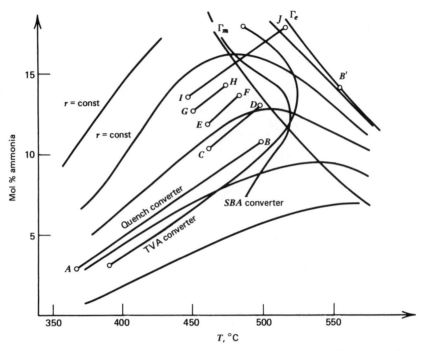

*Figure 11.3-4 Mole percent ammonia versus temperature diagram (after Shipman and Hickman* [3]).

attaining a much higher ammonia content. The reaction path *ABCDEFGHIJ* evolves around the curve $\Gamma_m$, which represents the relation conversion-temperature that ensures maximum reaction rate in each point of the reactor. Clearly, for each bed the question is how close the adiabatic outlet condition will be allowed to approach equilibrium and how far the reaction mixture will have to be cooled in the heat exchanger before proceeding to the next stage. This is a problem of optimization, requiring a more quantitative approach. For the specific case considered here, another possibility is to depart from the adiabatic stages in order to follow more closely the curve of optimum reaction rates, $\Gamma_m$. The continuous removal of excess heat implied by this is only possible in a multitubular reactor. The way in which this is achieved in ammonia synthesis is shown in Fig. 11.3-5. We can see that use is made of the feed stream to remove the heat from the reaction section [4]. How well the objective is met by a proposed design (i.e., how well the actual trajectory approximates the $\Gamma_m$ curve) can only be found by a more quantitative approach involving modeling, discussed further in Parts Two and Three.

468  CHEMICAL REACTOR DESIGN

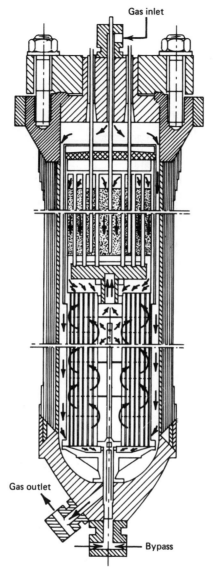

Gas inlet

Gas outlet

Bypass

*Figure 11.3-5 Ammonia synthesis reactor with tubular heat exchanger ( from Vancini* [4]).

469

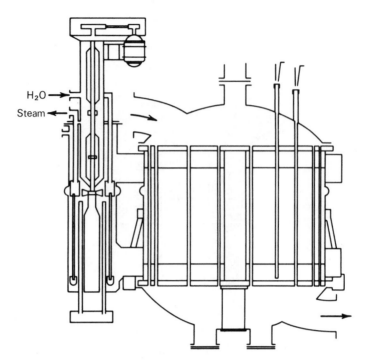

*Figure 11.3-6 Multitubular reactor for phthalic anhydride synthesis by o-xylene oxidation (from Suter [6]).*

With other very exothermic reactions, such as air oxidation of aromatic hydro-carbons, the number of beds would have to be uneconomically large to limit the temperature increase per bed, so that the multitubular reactor is definitely preferred. Cooling the reactor with the incoming reactant would be insufficient, however, and require too much heat exchanging surface. Such reactors are therefore cooled by means of circulating molten salts which in turn give off their heat to a boiler. The phthalic anhydride synthesis reactor shown in Fig. 11.3-6 [6] may contain up to 10,000 tubes of 2.5 cm inside diameter. The tube diameter has to be limited to such a small value to avoid excessive overtemperatures on the axis, a feature that is discussed later in this chapter.

A different type of multitubular reactor has to be used in natural gas or naphtha reforming into hydrogen or synthesis gas, an endothermic reaction (Fig. 11.3-7). In this case, the gases are gradually heated from 500 to 850°C. To obtain the highest possible capacity for a given amount of catalyst, heat fluxes of 65,000 kcal/m$^2$ hr (75.6 kJ/m$^2$ s) are applied to tubes of 10 cm inner diameter. The tubes, 10 m long, are suspended in two rows in a furnace that may contain as many as 300 tubes.

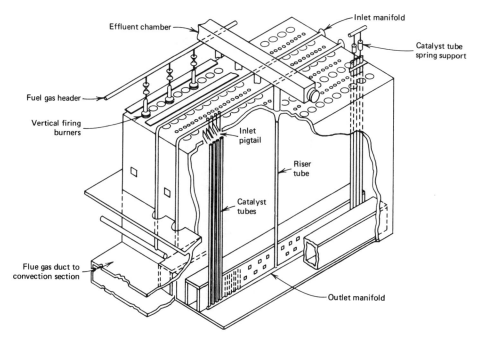

*Figure 11.3-7 Multitubular steam reformer with furnace (after* [7], *from Froment* [148]).

In several cases effluent gases have to be recycled (e.g., in catalytic reforming—hydrogen and light hydrocarbons; in ammonia synthesis—the noncondensed fraction of the effluent, because of equilibrium limitations on the conversion per pass). To limit the cost of recycling and get a maximum capacity out of the centrifugal recycle compressor, the pressure drop over the catalyst bed has to be kept as low as possible. This requires limiting the bed depth, which means, in conventional reactors at least, the diameter would have to be increased. This is no longer possible for the giant ammonia synthesis converters, so that other solutions had to be sought. Figure 11.3-8 shows three different ways of increasing the flow area without increasing the bed depth [8]. Note that radial flow has been applied for quite a number of years in catalytic reforming. Clearly, in all the decisions related to the above discussion, the following elements had to be considered all the time: technology in all its various aspects, the rate of reaction, reaction scheme, equilibrium, catalyst composition and properties, heat transfer, pressure drop, with constant reference to safety, reliability, and economics.

The same factors will, of course, have to be considered in the next stage of design, only more quantitatively and in a way accounting for their interaction. This stage requires some degree of mathematical modeling of the reactor.

FIXED BED CATALYTIC REACTORS _____ **471**

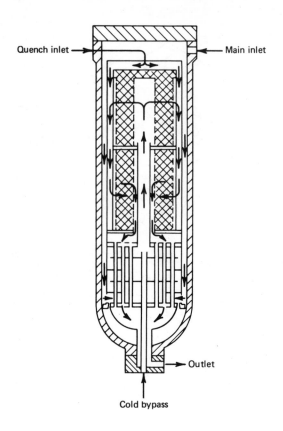

*Figure 11.3-8 Modern ammonia synthesis reactors. (a) Radial H. Topsoe converter (from [8]).*

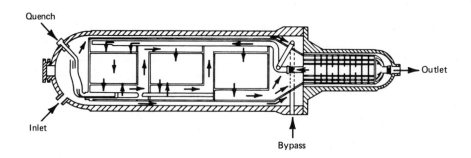

*Figure 11.3-8 (b) Horizontal multibed Kellogg reactor (from [8]).*

472

Gas outlet

Gas inlet

Catalyst
discharge port

*Figure 11.3-8 (c) ICI reactor
(from [8]).*

473

## 11.4 Modeling of Fixed Bed Reactors

In this chapter it is not possible to concentrate on specific cases and processes. Instead, we discuss general models and principles involved in the design and analysis of any type of fixed bed reactor, no matter what the process.

The development in recent years of chemical reaction engineering as a recognized discipline and the increasing possibilities of computers have led to extensive exploration of reactor design and performance, both in the steady and nonsteady state. Models now range from the very simple ones that could be handled before 1960, to some very sophisticated ones presented in the last two or three years.

Reactor design and analysis groups are continuously confronted with the degree of sophistication that can be justified. This is a question that cannot be answered in a general manner: the required degree of sophistication depends in the first place on the process, that is, on the reaction scheme and on its sensitivity to perturbations in the operating conditions. Of equal importance, however, is the degree of accuracy with which the kinetic and transport parameters are known. To establish a better insight into the models a classification is proposed in Table 11.4-1 [9, 10]. In this table the models have been grouped in two broad categories: pseudo-homogeneous and heterogeneous. Pseudo-homogeneous models do not account explicitly for the presence of catalyst, in contrast with heterogeneous models, which lead to separate conservation equations for fluid and catalyst. Within each category the models are classified in an order of growing complexity. The basic model, used in most of the studies until now, is the pseudo-homogeneous one-dimensional model, which only considers transport by plug flow in the axial direction Sec. 11.5. Some type of mixing in the axial direction may be superposed on the plug flow so as to account for non ideal flow conditions Sec. 11.6. If radial gradients have to be accounted for, the model becomes two dimensional Sec. 11.7. The basic model of the heterogeneous category considers only transport by plug flow again, but distinguishes between conditions in the fluid and on the solid Sec. 11.8. The next step towards complexity is to take the gradients inside the catalyst into account Sec. 11.9. Finally, the most general models used today,

*Table 11.4-1 Classification of fixed bed reactor models*

|  | Pseudo-homogeneous models $T = T_s; C = C_s$ | Heterogeneous models $T \neq T_s; C \neq C_s$ |
|---|---|---|
| One dimensional | Sec. 11.5 basic, ideal | Sec. 11.8 + interfacial gradients |
|  | Sec. 11.6 + axial mixing | Sec. 11.9 + intraparticle gradients |
| Two dimensional | Sec. 11.7 + radial mixing | Sec. 11.10 + radial mixing |

CHEMICAL REACTOR DESIGN

(the two dimensional heterogeneous models) are discussed in Sec. 11.10. In the following sections, the specific features of these models and their adequacy with respect to industrial practice are discussed.

## Part Two

### Pseudo-homogeneous Models

## 11.5 The Basic One-Dimensional Model

### 11.5.a Model Equations

The basic or ideal model assumes that concentration and temperature gradients only occur in the axial direction. The only transport mechanism operating in this direction is the overall flow itself and this is considered to be of the plug flow type. The conservation equations may be written for the steady state and a single reaction carried out in a cylindrical tube:

$$-u_s \frac{dC_A}{dz} = \rho_B r_A \qquad (11.5.a\text{-}1)$$

$$u_s \rho_g c_p \frac{dT}{dz} = (-\Delta H)\rho_B r_A - 4\frac{U}{d_t}(T - T_r) \qquad (11.5.a\text{-}2)$$

$$-\frac{dp_t}{dz} = \frac{2f\rho_g u_s^2}{gd_p} \qquad (11.5.a\text{-}3)$$

With initial conditions: at $z = 0$; $C_A = C_{A0}$; $T = T_0$; $p_t = p_{t0}$. The integration of the system Eq. 11.5.a-1, 2, 3 is a straightforward matter, either on a digital or an analog computer. This permits a simulation of the reactor. Questions that can be answered by such simulation and that are important in fixed bed reactor design are: What is the tube length required to reach a given conversion? What will the tube diameter have to be? Or the wall temperature? Before considering such problems, however, we will discuss some features of the system of differential Eqs. 11.5.a-1, 2, 3.

Equation 11.5.a-1 is obtained from a material balance on a reference component, say $A$, over an elementary cross section of the tubular reactor, containing an amount of catalyst $dW$. Indeed, as previously mentioned, rate equations for heterogeneously catalyzed reactions are generally referred to unit catalyst weight, rather than reactor volume, in order to eliminate the bed density. Obviously, different packing densities between the laboratory reactor in which kinetic data were determined and the industrial reactor, calculated on the basis of these data would lead to different results.

When use is made of conversion the material balance for $A$ over an elementary weight of catalyst may be written:

$$r_A \, dW = F_{A0} \, dx_A \tag{11.5.a-4}$$

where $F_{A0}$ is the molar feed rate of $A$ or

$$r_A \rho_B \Omega \, dz = F_{A0} \, dx_A \tag{11.5.a-5}$$

from which Eq. 11.5.1-1 is easily obtained. $U$ in Eq. 11.5.a-1 is an overall heat transfer coefficient defined by:

$$\frac{1}{U} = \frac{1}{\alpha_i} + \frac{d}{\lambda} \frac{A_b}{A_m} + \frac{1}{\alpha_u} \frac{A_b}{A_u} \tag{11.5.a-6}$$

where $\alpha_i$ = heat transfer coefficient on the bed side (kcal/m$^2$ hr °C)
$\quad \alpha_u$ = heat transfer coefficient, heat transfer medium side (kcal/m$^2$ hr °C)
$\quad A_b$ = heat exchanging surface, bed side (m$^2$)
$\quad \lambda$ = heat conductivity of the wall (kcal/m hr °C)
$\quad A_u$ = heat exchanging surface, heat transfer medium side (m$^2$)
$\quad A_m$ = log mean of $A_b$ and $A_u$ (m$^2$)

In general, the thickness of the wall, $d$, is small, so that the ratio of surfaces is close to 1. $\alpha_u$ is found from classical correlations in books on heat transfer. $\alpha_i$ may be based on Leva's correlation [12] for heating up the reaction mixture:

$$\frac{\alpha_i d_t}{\lambda_g} = 0.813 \left( \frac{d_p G}{\mu} \right)^{0.9} e^{-6 d_p / d_t}$$

for cooling: $\tag{11.5.a-7}$

$$\frac{\alpha_i d_t}{\lambda_g} = 3.50 \left( \frac{d_p G}{\mu} \right)^{0.7} e^{-4.6 d_p / d_t}$$

where $d_t$ = tube diameter (m)
$\quad d_p$ = equivalent particle diameter (m)

Further correlations of this type were published by Maeda [17] and Verschoor and Schuit [18].

De Wasch and Froment, on the other hand, found a linear relation between the Nusselt and the Reynolds numbers [19]:

$$\frac{\alpha_i d_p}{\lambda_g} = \frac{\alpha_i^0 d_p}{\lambda_g} + 0.033 \left( \frac{c_p \mu}{\lambda_g} \right) \left( \frac{d_p G}{\mu} \right) \tag{11.5.a-8}$$

The influence of the tube diameter and of the catalyst properties enter the correlation through $\alpha_i^0$, the so-called static contribution,

$$\alpha_i^0 = \frac{2.44 \lambda_e^0}{d_t^{4/3}}$$

$\lambda_e^0$ is the static contribution to the effective thermal conductivity of the bed and will be discussed in detail in Sec. 11.10. The friction factor $f$ now remains to be specified in the pressure drop equation.

Some well-known equations for the friction factor for flow in packed beds are: Ergun's equation [14]

$$f = \frac{1 - \varepsilon}{\varepsilon^3} \left( 1.75 + 150 \frac{1 - \varepsilon}{\text{Re}'} \right) \qquad (11.5.a\text{-}9)$$

where $f$ is the friction factor, defined by $f = [(-\Delta p_t)g\rho_g\psi\, d_p]/(LG^2)$, provided that $p_t$ is in kgf/m². Otherwise, the acceleration of gravity, $g$, should be replaced by a conversion factor. Re' is a modified Reynolds number: $\text{Re}' = (\psi\, d_p G)/\mu$ in which $d_p$ is the equivalent diameter of a sphere with a volume equal to that of the actual particle:

$$d_p = \left( \frac{6}{\pi} \text{ volume of particle} \right)^{1/3}$$

and $\psi$ is the shape factor or sphericity of the particle, defined by:

$$\psi = \frac{S_s}{S_x} = \frac{\pi}{S_x} \left( \frac{6}{\pi} V_p \right)^{2/3}$$

In this equation $S_x$ and $V_p$ are the external surface area and the volume of the particle and $S_s$ is the surface of the equivalent volume sphere ($\psi = 1$ for spheres, 0.874 for cylinders with height equal to the diameter, 0.39 for Raschig-rings, 0.37 for Berl saddles). $\psi$ extends the correlation to particles of arbitrary shape. The product $\psi\, d_p$ is sometimes written as a diameter $d_p'$:

$$d_p' = \psi d_p = 6V_p/S_x$$

Handley and Hegg's equation is [110]:

$$f = \frac{1 - \varepsilon}{\varepsilon^3} \left( 1.24 + 368 \frac{1 - \varepsilon}{\text{Re}'} \right) \qquad (11.5a\text{-}10)$$

Hicks reviewed several pressure drop equations [15]. It may be concluded from his work that the Ergun equation is limited to $\text{Re}/(1 - \varepsilon) < 500$ and Handley and Hegg's equation to $1000 < \text{Re}/(1 - \varepsilon) < 5000$. Hicks proposes an equation for spheres that can be written

$$f = 6.8 \frac{(1 - \varepsilon)^{1.2}}{\varepsilon^3} \text{Re}^{-0.2} \qquad (11.5a\text{-}11)$$

which fits Ergun's, Handley, and Hegg's data and the results of Wentz and Thodos obtained at very high Reynolds numbers [111].

Leva did extensive work on the pressure drop in packed beds of particles with various shapes [16]. He suggests the following equation for laminar flow through packed beds:

$$\frac{dp_t}{dz} + 200G\frac{\mu(1-\varepsilon)^2}{d_p^2\psi^2\rho_g g\varepsilon^3} = 0 \qquad (11.5.a\text{-}12)$$

For turbulent flow Leva proposed the following equation:

$$\frac{dp_t}{dz} + 3.50\frac{G^{1.9}\mu^{0.1}(1-\varepsilon)^{1.1}}{d_p^{1.1}\psi^{1.1}\rho_g g\varepsilon^3} = 0 \qquad (11.5.a\text{-}13)$$

If the density varies $\rho_g$ has to be replaced by

$$\rho_g = \frac{M_m}{V} = \frac{p_t M_m}{RT} = \frac{p_t M_{m0}}{(1+\varepsilon_A x_A)RT}$$

$M_{m0}$ = initial molecular weight of reaction mixture
$M_m$ = mean molecular weight at conversion $x$
$\varepsilon_A$ = expansion factor

Leva also proposed an equation valid for both laminar and turbulent flow conditions [12]

$$\frac{dp_t}{dz} + \frac{2f_m G^2(1-\varepsilon)^{3-n}}{d_p g\rho_g\psi^{3-n}\varepsilon^3} = 0 \qquad (11.5.a\text{-}14)$$

The friction factor $f_m$ and the power $n$ follow from Fig. 11.5.a-1.

Brownell, Dombrowsky, and Dickey [13] correlated the results of several authors on the basis of the $f$-Re diagram for empty pipes. In order to make the results for packed tubes coincide with those for empty pipes, the characteristic length in $f$ and Re is taken to be the particle diameter. This is not sufficient: one has to account for the true fluid velocity and true path length. Brownell et al. introduced two correction factors, one for the Reynolds number, $F_{Re}$ and one for the friction factor, $F_f$. These were determined as functions of $\varepsilon$ and $\psi$. The results are shown in Figs. 11.5.a-2, 11.5.a-3 and 11.5.a-4.

The abscissa that has to be substituted in the $f$-Re plot for empty pipes is:

$$\frac{d_p G}{\mu}xF_{Re}$$

The ordinate is $f$, where

$$fF_f = 2\frac{(-\Delta p_t)}{L}\frac{g\rho_g d_p}{G^2} \qquad (11.5.a\text{-}15)$$

The pressure drop is then calculated from Eq. 11.5.a-15. Note that the definition of $f$ is different from that given in Eq. 11.5.a-3: it is *not* the Fanning definition.

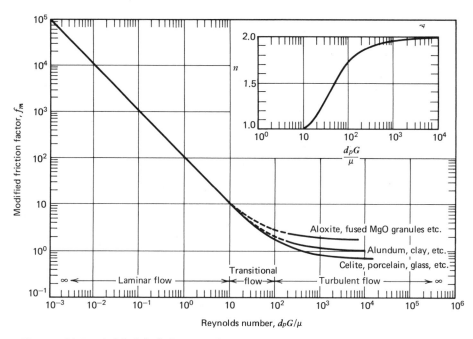

Figure 11.5.a-1 Modified friction factor versus Reynolds number (from Leva [12]).

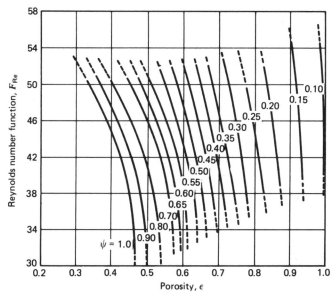

Figure 11.5.a-2 Reynolds number function versus porosity with parameters of sphericity (from Brownell, et al. [13]).

479

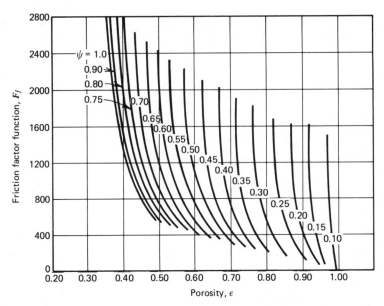

*Figure 11.5.a-3 Friction factor function versus porosity with parameters of sphericity (from Brownell, et al. [13]).*

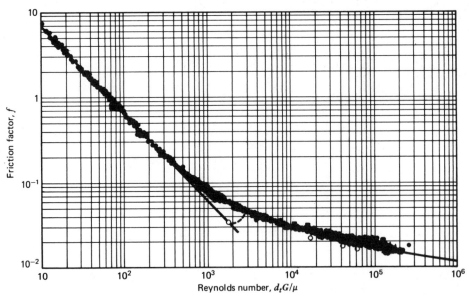

*Figure 11.5.a-4 Correlation of data for porous media to friction factor plot for smooth pipe (from Brownell, et al. [13]).*

480

Further pressure drop data may be found in a recent paper by Reichelt [112]. In most cases the pressure drop in a fixed bed reactor is relatively small, so that it is frequently justified to use an average pressure in the calculations.

## Example 11.5.a-1 Calculation of Pressure drop in Packed Beds

A tube 2.5 m long and having an inside diameter of 0.025 m is packed with $Al_2O_3$ cylinders with $d_p = 0.003$ m. The bulk density of the bed is 1300 kg/m³ and the void fraction $\varepsilon = 0.38$. Air flows through the tube at 372°C with a superficial mass flow velocity of 4,684 kg/m² hr.

Calculate the pressure drop in the bed.

Viscosity of air at 372°C = 0.031 cps = 0.1116 kg/m hr; specific weight of air at 372°C: 0.9487 kg/m³; sphericity, $\psi = 0.874$.

*Solution According to Ergun*

Since Re' = 110, Ergun's equation should be chosen and not Handley and Hegg's. One obtains:

$$-\Delta p_t = \frac{2.5 \times (4684)^2 \times 0.62}{(0.38)^3 \times 9.81 \times (3600)^2 \times 10^4 \times 0.9487 \times 0.003 \times 0.874}$$

$$\times \left( \frac{150 \times 0.62 \times 0.1116}{4684 \times 0.003 \times 0.874} + 1.75 \right)$$

$$-\Delta p_t = 0.509 \text{ kg/cm}^2 = 0.499 \text{ bar.}$$

*Solution According to Hicks*

If it is assumed that Hicks' equation also applies to beds of nonspherical particles (e.g. cylinders, provided $d$ is replaced by $\psi d_p$), substitution of the numerical values into Eq. 11.5.a-11 yields:

$$-\Delta p_t = \frac{2.5 \times (4684)^2}{9.81 \times (3600)^2 \times 10^4 \times 0.9487 \times 0.003 \times 0.874}$$

$$\times \left[ 6.8 \times \frac{(0.62)^{1.2}}{(0.38)^3} \times \left( \frac{0.874 \times 0.003 \times 4684}{0.1116} \right)^{-0.2} \right]$$

$$-\Delta p_t = 0.473 \text{ kg/cm}^2 = 0.464 \text{ bar}$$

*Solution According to Max Leva*

From Fig. 11.5.a-1: $f_m = 2$ and $n = 1.73$. From Eq. 11.5.a-14

$$-\Delta p_t = \frac{2.5 \times 2 \times 2 \times (4684)^2 \times (0.62)^{1.27}}{0.003 \times 9.81 \times (3600)^2 \times 10^4 \times 0.9487 \times (0.874)^{1.27} \times (0.38)^3}$$

$$-\Delta p_t = 0.714 \text{ kg/cm}^2 = 0.700 \text{ bar}$$

*Table 1 Influence of the flow velocity and of the packing diameter on the pressure drop*

| Re | $d_p$ (cm) | $-\Delta p_t$ (kg/cm$^2$) | $-\Delta p_t$ (bars) |
|------|------|------|------|
| 55.5 | 0.3 | 0.239 | 0.234 |
| 126 | 0.15 | 1.910 | 1.873 |
| 126 | 0.3 | 0.743 | 0.729 |
| 126 | 0.6 | 0.318 | 0.312 |
| 242 | 0.3 | 3.819 | 3.745 |

*Solution According to Brownell et al.*

From Figs. 11.5.a-2 and 11.5.a-3 it follows that for $\varepsilon = 0.38$ and $\psi = 0.874$ the correction factors are: $F_{Re} = 51$ and $F_f = 2800$. The value of the abcissa in Fig. 11.5.a-4 follows from:

$$Re'' = Re\, F_{Re} = 126 \times 51 = 6426$$

From Fig. 11.5.a-4 one obtains $f = 3.5.10^{-2}$ and from Eq. 11.5.a-15:

$$-\Delta p_t = \frac{2.5 \times 3.5.10^{-2} \times (4684)^2 \times 2800}{2 \times 9.81 \times (3600)^2 \times 10^4 \times 0.003 \times 0.9487}$$

$$-\Delta p_t = 0.743 \text{ kg/cm}^2 = 0.729 \text{ bar.}$$

The difference between the correlations of Ergun and Hicks on one hand and those of Leva and Brownell on the other hand is important.

Analogous calculations, based on Brownell's correlation lead to the results given in Table 1, which illustrates the influence of the flow velocity and of the packing diameter on the pressure drop.

## 11.5.b  Design of a Fixed Bed Reactor According to the One-Dimensional Pseudo-Homogeneous Model

This design example is suggested from hydrocarbon oxidation processes such as benzene oxidation into maleic anhydride or the synthesis of phthalic anhydride from *o*-xylene. Such strongly exothermic processes are carried out in multi-tubular reactors, cooled by molten salt that is circulating around the tubes and that exchanges heat to an internal or external boiler. The length of the tubes is 3 m and their internal diameter 2.54 cm. One reactor may contain 2500 tubes in parallel and even up to 10,000 in the latest versions. In German processes the catalyst is $V_2O_5$ on promoted silica gel and the operating temperature range is 335 to 415°C. The particle diameter is 3 mm and the bulk density is 1300 kg/m$^3$. The hydrocarbon is vaporized and mixed with air before entering the reactor. Its concentration is kept below 1 mole %, in order to stay under the explosion limit.

The operating pressure is nearly atmospheric. The phthalic anhydride production from such a reactor with 2500 tubes is 1650 tons/yr. It follows that with this catalyst a typical mass flow velocity of the gas mixture is 4684 kg/m² hr. With a mean fluid density of 1293 kg/m³, this leads to a superficial fluid velocity of 3600 m/hr. A typical heat of reaction is 307,000 kcal/kmol and the specific heat is 0.237 kcal/kg°C (0.992 kJ/kg K). In this example the kinetic equation for the hydrocarbon conversion will be considered in first approximation to be pseudo first order, due to the large excess of oxygen.

$$r_A = kp_B^0 p$$

where $p_B^0 = 0.208$ atm $= 0.211$ bar represents the partial pressure of oxygen. Let $k$ be given by

$$\ln k = 19.837 - \frac{13,636}{T}$$

More complex rate equations for this type of reaction will be used in a later example given in Sec. 11.7.

The continuity equation for the hydrocarbon, in terms of partial pressures and the energy equation, may be written, for constant density

$$u_s \frac{dp}{dz} + \frac{M_m p_t \rho_B}{\rho_g} kp_B^0 p = 0 \tag{11.5.b-1}$$

$$u_s \rho_g c_p \frac{dT}{dz} - (-\Delta H)\rho_B kp_B^0 p + \frac{4U}{d_t}(T - T_r) = 0 \tag{11.5.b-2}$$

with $p = p_0$      at     $z = 0$

$T = T_0 = T_r$    at     $z = 0$

The total pressure is considered to be constant and equal to 1 atm. The overall heat transfer coefficient $U$ may be calculated from the correlations given above to be 82.7 kcal/m² hr (0.096 kJ/m² s). $T_r$ is chosen to be 352°C.

The figures reveal a "hot spot" in the bed, which is typical for strongly exothermic processes. The magnitude of this hot spot depends, of course, on the heat effect of the reaction, the rate of reaction, the heat transfer coefficient and transfer areas as shown by Bilous and Amundson [21]. Its location depends on the flow velocity. It is also observed that the profiles become sensitive to the parameters from certain values onward. If the partial pressure of the hydrocarbon were 0.018 atm an increase of 0.0002 atm would raise the hot spot temperature beyond permissible limits. Such a phenomenon is called runaway. Note that for the upper part of the curves with $p_0 = 0.0181$, 0.0182, and 0.019 (Figs. 11.5.b-1 and 2) the model used here is not longer entirely adequate: heat and mass transfer effects would have to be taken into account. There is no doubt however as to the validity of the lower part indicating excessive sensitivity in this region.

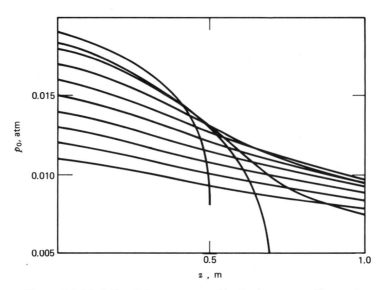

*Figure 11.5.b-1 Partial pressure profiles in the reactor illustrating the sensitivity with respect to the inlet partial pressure (from van Welsenaere and Froment [20]).*

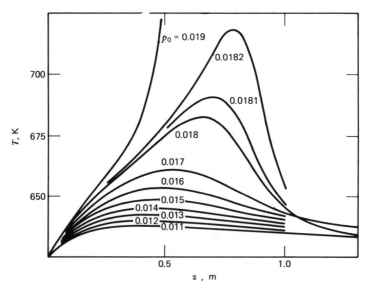

*Figure 11.5.b-2 Temperature profiles corresponding to partial pressure profiles of Fig. 11.5.b-1 (from van Welsenaere and Froment [20]).*

484

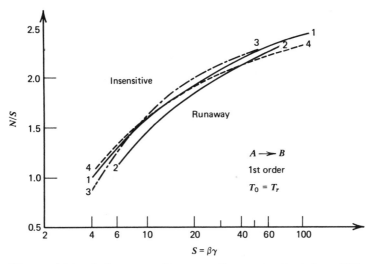

*Figure 11.5.c-1 Runaway diagram. Curve 1: Barkelew [22]; curve 2: Dente and Collina [23]; curve 3: Hlavacek et al. [24]; curve 4: Van Welsenaere and Froment [20].*

### 11.5.c Runaway Criteria

In the above example it was shown how hot spots develop in fixed bed reactors for exothermic reactions. An important problem associated with this is how to limit the hot spot in the reactor and how to avoid excessive sensitivity to variations in the parameters. Several approaches have been attempted to derive simple criteria that would permit a selection of operating conditions and reactor dimensions prior to any calculation on the computer. Such criteria are represented in Fig. 11.5.c-1. In this figure the abscissa is $S = \beta\gamma$ (i.e., the product of the dimensionless adiabatic temperature rise)

$$\frac{T_{ad} - T_0}{T_0} = \frac{(-\Delta H)p_0}{M_m p_t c_p T_0} = \frac{(-\Delta H)C_0}{\rho_g c_p T_0}$$

and the dimensionless activation energy $E/RT_0$, two groups characterizing the reaction properties and the operating conditions. The ordinate $N/S$ is the ratio of the rate of heat transfer per unit reactor volume at $f_r = 1$, where $f_r = (E/RT_r^2)(T - T_r)$, to the rate of heat generation per unit volume at $f_r = 0$ and zero conversion (i.e., at the reactor inlet). Specifically, using the volumetric rate coefficient of Chapter 3:

$$N = \frac{2}{R_t} \frac{U}{\rho c_p k_v}$$

and with the rate form of Sec. 11.5.b,

$$N = \frac{2U}{R_t c_p M_m} \frac{1}{\rho_B k p_B^0 p_t}$$

Also,

$$\frac{N}{S} = \frac{2U(T - T_r)}{R_t} \frac{1}{k_v C_{A0}(-\Delta H)} \frac{R T_r^2}{E(T - T_r)}$$

as stated above in physical terms. Further details are given in Ex. 11.5.c-1. The curves 1, 2, 3, and 4 define a band that bounds two regions. If the operating conditions are such that they lead to a point in the diagram above the curves, the reactor is insensitive to small fluctuations, but if it is situated under the curves runaway is likely.

Barkelew arrived at curve 1 by inspecting a very large number of numerical integrations of the system (Eqs. 11.5.a-1 to 11.5.a-3) for a wide variation of the parameter values, but used a simplified temperature dependence of the reaction rate [22]. Dente and Collina came to essentially the same curve with less effort by taking advantage of the observation that in drastic conditions the temperature profile through the reactor has two inflection points before the maximum, which coincide in critical situations [23]. Hlavacek et al. [24] and Van Welsenaere and Froment [20] independently utilized two properties of the $T$-$z$ curve to derive criteria without any of the integrations involved in the approach of Barkelew and with the Arrhenius temperature dependence for the rate coefficient. From an inspection of the temperature and partial pressure profiles in the reactor they concluded that extreme parametric sensitivity and runaway is possible (1) when the hot spot exceeds a certain value and (2) when the temperature profile develops inflection points before the maximum, as noticed already by Dente and Collina. Van Welsenaere and Froment transposed the peak temperature and the conditions at the inflection points into the $p - T$ phase plane, a diagram often used in the study of the dynamic behavior of a reactor.

In Fig. 11.5.c-2 the locus of the partial pressure and temperature in the maximum of the temperature profile and the locus of the inflection points before the hot spot are shown as $p_m$ and $(p_i)_1$, respectively. Two criteria were derived from this. The first criterion is based on the observation that extreme sensitivity is found for trajectories—the $p$-$T$ relations in the reactor—intersecting the maxima curve $p_m$ beyond its maximum. Therefore, the trajectory going through the maximum of the $p_m$-curve is considered as critical. This is a criterion for runaway based on an intrinsic property of the system, not on an arbitrarily limited temperature increase. The second criterion states that runaway will occur when a trajectory intersects $(p_i)_1$, which is the locus of inflection points arising before the maximum. Therefore, the critical trajectory is tangent to the $(p_i)_1$-curve. A more convenient version of this criterion is based on an approximation for this locus represented by $p_s$ in

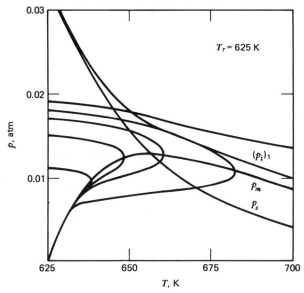

*Figure 11.5.c-2 p-T phase plane, showing trajectories, maxima-curve, loci of inflexion points and the "simplified" curve $p_s$ (from van Welsenaere and Froment [20]).*

Fig. 11.5.c-2. Representation of the trajectories in the $p$-$T$ plane requires numerical integration, but the critical points involved in the criteria—the maximum of the maxima curve and the point of tangency of the critical trajectory with $p_s$ are easily located by means of elementary formulas. Two simple extrapolations from these points to the reactor inlet conditions lead to upper and lower limits for the inlet partial pressures.

The formulas used in the first criterion are easily derived as follows. Considering again the case of a pseudo-first-order reaction treated under Sec. 11.5.b and dividing Eq. 11.5.b-2 by Eq. 11.5.b-1 leads to

$$\frac{dT}{dp} = -\frac{B}{A} + \frac{C}{A} \frac{T - T_r}{p \exp\left(-\dfrac{E}{RT} + b\right)} \qquad (11.5.c\text{-}1)$$

where

$$A = \frac{M_m p_t \rho_B}{\rho_g} p_B^0 \qquad B = \frac{(-\Delta H)\rho_B}{c_p \rho_g} p_B^0 \qquad C = \frac{4U}{c_p d_t \rho_g}$$

Trajectories in the $p$-$T$ diagram may be obtained from this equation by numerical integration. The locus of the $p$ and $T$ values in the maximum of the temperature profile in the reactor is obtained by setting $dT/dz = 0$ in Eq. 11.5.b-2 or $dT/dp = 0$ in Eq. 11.5.c-1. This leads to

$$p_m = \frac{T_m - T_r}{\dfrac{B}{C} \exp\left(-\dfrac{E}{RT_m} + b\right)} \tag{11.5.c-2}$$

This curve is called the maxima curve. It can be seen from Fig. 11.5.c-2 that it has a maximum. The temperature corresponding to this maximum, $T_M$, is obtained by differentiating Eq. 11.5.c-2 with respect to $T_m$ and setting the result equal to zero:

$$T_M = \frac{1}{2}\left[\frac{E}{R} - \sqrt{\frac{E}{R}\left(\frac{E}{R} - 4T_r\right)}\right]$$

or, in dimensionless form,

$$f_M = \frac{E}{RT_M{}^2}(T_M - T_r) = 1 \tag{11.5.c-3}$$

Notice the slightly different definition of $f_M$ in this formula, compared to that of $f_r$ used in conjunction with Fig. 11.5.c-1.

What remains to be done is to find the inlet conditions leading to the critical situations. Rigorously this requires numerical back integration. Approximate values for the critical inlet conditions may be obtained by simple extrapolations, however. Two ways of extrapolation were retained to define an upper and a lower limit for the approximated critical inlet conditions. The lower limit is based on the property of the trajectories to start in the $p$-$T$ plane with an adiabatic slope for $T_0 = T_r$ and to bend under this line, due to heat exchange through the wall. Therefore, an adiabatic line starting from a point on a critical trajectory leads to a lower limit for the critical inlet conditions. Indeed, the critical trajectory through the critical point starts from inlet partial pressures that are higher than those of the adiabatic lines. This extrapolation defines a lower limit for $p_0$, which is entirely safe. The upper limit is based on the observation that tangents to the trajectories taken at a given $T$ between $T_r$ and $T_M$ all intersect the ordinate at $T_r$ at values for $p$ that are higher than those intersected by the trajectories themselves. The intercepts of these tangents are determined by two opposing effects: the higher the trajectory is situated the smaller the value of the slope of the tangent. One of the trajectories will therefore lead to a minimum intercept on the ordinate through $T_r$. The corresponding inlet partial pressure, which is higher than that of the critical trajectory, will be the best possible approximation and is considered as an upper limit for the inlet partial pressure. The value of the abscissa at which the tangents are drawn is the critical temperature, ($T_M$ for the first criterion). This extrapolation defines an upper limit above which runaway will certainly occur.

The following formulas are easily derived:

From the first criterion:

Lower limit

$$p_0{}^l = \frac{A}{B}(T_M - T_r)(1 + Q^2)$$

or

$$\frac{\Delta T_{ad}}{\Delta T} = 1 + Q^2$$

where

$$Q = \frac{1}{\sqrt{\dfrac{A}{C} \exp\left(-\dfrac{E}{RT_M} + b\right)}} \qquad (11.5.c\text{-}4)$$

Upper limit

$$p_0{}^u = \frac{A}{B}(T_M - T_r)(1 + Q)^2$$

or

$$\frac{\Delta T_{ad}}{\Delta T} = (1 + Q)^2 \qquad (11.5.c\text{-}5)$$

A very accurate approximation of the true critical value is given by the mean

$$\frac{\Delta T_{ad}}{\Delta T} = 1 + Q + Q^2 \qquad (11.5.c\text{-}6)$$

which is represented in Fig. 11.5.c-1 as curve 5.

From the second criterion, Van Welsenaere and Froment derived the following formulas:

Lower limit:

$$\frac{\Delta T_{ad}}{\Delta T} = 1 + \frac{Q^2}{f_c}$$

where

$$f_c = \frac{E}{RT_c^2}(T_c - T_r)$$

$T_c$ is the critical temperature derived from the second criterion

Upper limit:

$$\frac{\Delta T_{ad}}{\Delta T} = (1 + Q)^2$$

Mean:

$$\frac{\Delta T_{ad}}{\Delta T} = 1 + Q + \frac{Q^2}{2}\left(1 + \frac{1}{f_c}\right) \qquad (11.5.c\text{-}7)$$

These formulas lead to values that are in close agreement with those based on the first criterion.

As previously mentioned the methods discussed here are helpful in first stages of design to set limits on the operating conditions, but cannot answer questions related to the length of the reactor—these require integration of the set of Eqs. 11.5.b-1 to 11.5.b-2. Also, Fig. (11.5.c-1) is limited to single reactions, except if some meaningful lumping could be applied to the reaction system, a topic investigated by Luss and co-workers [25].

### Example 11.5.c-1 Application of the First Runaway criterion of Van Welsenaere and Froment

The reaction and operating variables are those considered in Sec. 11.5.b, so that $A = 6,150$, $B = 257.10^6$, and $E/R = 13,636$.

1. Calculation of the permissible inlet partial pressure for a given wall and inlet temperature and given tube radius. Let $T_r = T_0 = 635$ K and $R_t = 0.0125$ m. According to the first criterion, the critical temperature is $T_M$, the maximum of the maxima curve and $f_M = (E/RT_M^2)(T_M - T_r) = 1$ from which $T_M = 667.69$ K so that to avoid runaway the maximum $\Delta T$ in the reactor is 32.7°. From 11.5.c-4 it follows that $Q = 2.9203$. Once $Q$ is known $\Delta T_{ad}$ can be calculated and from $\Delta T_{ad} = (B/A)p_0$ the inlet partial pressure $p_0$ is obtained. The results are given in Table 1, where use had been made of (11.5.c-4), (11.5.c-5), and (11.5.c-6), respectively, for calculating $\Delta T_{ad}$. If $T_r = T_0 = 625$ K and $R_t = 0.0125$ m, $T_M$ is 656.6 K and $\Delta T$ has to be limited to 31.6 K. Then

*Table 1*

|  | $\Delta T_{ad}$ |  | $p_0$ (atm) | $p_0$ (bar) |
|---|---|---|---|---|
| Lower limit | $\Delta T(1 + Q^2)$ | $= 310°$ | 0.0074 | 0.0075 |
| Upper limit | $\Delta T(1 + Q)^2$ | $= 504°$ | 0.012 | 0.0012 |
| Mean | $\Delta T(1 + Q + Q^2)$ | $= 407°$ | 0.00965 | 0.00978 |

## Table 2

| | $\Delta T_{ad}$ | | $p_0$ (atm) | $p_0$ (bar) |
|---|---|---|---|---|
| Lower limit | $\Delta T(1 + Q^2)$ | $= 411.5$ | 0.01353 | 0.01371 |
| Upper limit | $\Delta T(1 + Q)^2$ | $= 521.1$ | 0.01976 | 0.02002 |
| Mean | $\Delta T(1 + Q + Q^2)$ | $= 466.3$ | 0.01665 | 0.01687 |

$Q = 3.4675$. The $p_0$ values are given in Table 2. Numerical back integration from the critical point onward leads to a critical inlet value for $p_0$ of 0.01651 atm which is in excellent agreement with the mean.

2. Calculation of the critical radius. Given $p_0 = 0.0125$ atm (0.0127 bar) and $T_r = T_0 = 625$ K. What would be the radius leading to critical conditions? From Eq. 11.5.c-3 $T_M$ is found to be 656.6 K, so that $\Delta T = 31.6°$. $\Delta T_{ad}$ amounts to 521.09 K. From Eq. 11.5.c-4 it follows that $Q = 3.4675$. From

$$C = Q^2 A \exp\left(-\frac{E}{RT} + b\right)$$

the radius $R_t$ is found to be 0.0175 m.

3. Subcritical conditions. Given a radius $R_t = 0.0125$ m and $p_0 = 0.0075$ atm $= 0.0076$ bar determine the wall temperature that limits the hot spot to 675 K. For this maximum to be critical the wall temperature would have to be, from Eq. 11.5.c-3:641 K. The lower limit for the inlet partial pressure would be, from Eq. 11.5.c-4: 0.0086 atm, the upper limit from Eq. 11.5.c-5: 0.0136 atm. Therefore, the maximum is definitely subcritical. With $p = 0.0075$ atm it follows from $B/A\, p_0 = \Delta T_{ad} = 312.6$ K. $Q$ is found to be 3.094. Equation 11.5.c-6 then leads to $\Delta T = 22.9°$, so that $T_r = 652.2$ K. A numerical integration starting from $p_0 = 0.0075$ atm and $T_0 = T_r = 652.2$ K yields a maximum temperature of 677 K.

Figure 11.5.c-1 also permits a check on the criticality of the conditions. Therefore we require the numerical values of the groups:

$$S = \frac{(-\Delta H)p_0}{c_p M_m p_t}\left(\frac{E}{RT_r^2}\right) \quad \text{and} \quad N = \frac{2h}{k_B(T_r)R_t c_p}$$

where Barkelew's symbols and units were used in the group $N$. Thus, $k_B(T_r)$ is Barkelew's rate constant. Take care when translating this formula into the groups used here. Indeed, Barkelew expressed the rate as follows:

$$r = k_B \cdot C$$

where $C =$ mole fraction of key reacting component $A$ and $k_B$ has the dimensions [mol fluid/cm$^3$ bed s].

Van Welsenaere and Froment used a pseudo-first-order rate law:

$$r = k_F p_B{}^0 p$$

with $r$ in kmol $A$/kg cat.hr. It follows that

$$k_B = \frac{k_F \cdot p_t p_B{}^0 \rho_B}{3.6 \cdot 10^6},$$

so that with the symbols and units used here, $N$ becomes

$$N = \frac{2U}{3.6 \cdot 10^6 R_t c_p M} \times \frac{3.6 \cdot 10^6}{p_t p_B{}^0 \rho_B} \times \frac{1}{k_F(T_r)} = \frac{C}{A} \frac{1}{\exp\left(-\dfrac{a}{T_r} + b\right)}$$

Furthermore,

$$S = \frac{B}{A} p_0 \cdot \frac{E}{RT_r^2} = \frac{\Delta T_{ad}}{T_r} \cdot \frac{E}{RT_r} = \beta\gamma$$

For $R_t = 0.0125$ m; $p_0 = 0.0075$ atm, and $T_r = 652.2$ K it has been calculated that $(B/A)p_0 = \Delta T_{ad} = 312.6$ K so that

$$S = \beta\gamma = 312.6 \times \frac{13{,}636}{(652.2)^2} = 10$$

Since

$$C = \frac{2U}{R_t c_p \rho_g} = \frac{2 \times 82.7}{0.0125 \times 0.323} = 41{,}000$$

and

$$A = \frac{M p_t p_B{}^0 \rho_B}{\rho_g} = \frac{29.48 \times 1 \times 0.208 \times 1300}{1.293} = 6{,}150$$

$$\frac{N}{S} = \frac{C}{A} \frac{1}{\exp\left(-\dfrac{a}{T_r} + b\right)} \cdot \frac{1}{\beta\gamma} = \frac{41{,}000}{6{,}150 \times 0.348} \times \frac{1}{10} = 1.92$$

The point (1.92; 10) falls well above curves 1, 2, 4, and 5, in Fig. 11.5-c-1, so that the reactor is insensitive.

## 11.5.d The Multibed Adiabatic Reactor

In discussing the preliminary design of fixed bed reactors in Sec. 11.3 we mentioned that adiabatic operation is frequently considered in industrial operation because of the simplicity of construction of the reactor. It was also mentioned why straight adiabatic operation may not always be feasible and examples of multibed adiabatic reactors were given. With such reactors the question is how the beds should be sized. Should they be designed to have equal $\Delta T$'s or is there some optimum in the $\Delta T$'s, therefore in the number of beds and catalyst distribution? In Section 11.3. this problem was already discussed in a qualitative way. It is taken up in detail on the basis of an example drawn from $SO_2$ oxidation, an exothermic reversible reaction. To simplify somewhat it will be assumed, however, that no internal gradients occur inside the catalyst so that the effectiveness factor is one.

A very convenient diagram for visualizing the problem of optimizing a multibed adiabatic reactor is the conversion versus temperature plot already encountered in Sec. 11.3, and drawn in Fig. 11.5.d-1 for the $SO_2$ oxidation based on the rate equation of Collina, Corbetta, and Cappelli [113] with an effectiveness factor of 1. (For further reading on this subject see [114] and [115].) This equation is

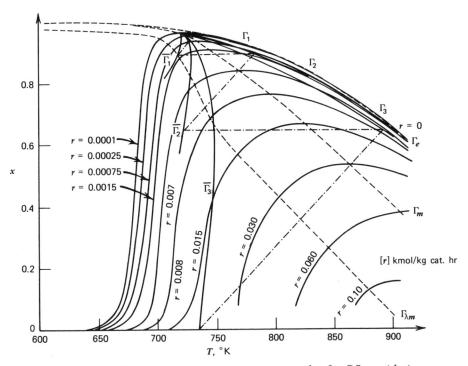

*Figure 11.5.d-1 Conversion versus temperature plot for $SO_2$ oxidation.*

FIXED BED CATALYTIC REACTORS ⸺⸺⸺⸺⸺⸺ **493**

based on the Langmuir–Hinshelwood concept and on the observation that the reaction between adsorbed $SO_2$ and oxygen from the gas phase is the rate-controlling step:

$$r = \frac{k_1 p_{O_2} p_{SO_2}\left(1 - \dfrac{p_{SO_3}}{p_{SO_2} \cdot p_{O_2}^{1/2} K_p}\right)}{22.414(1 + K_2 p_{SO_2} + K_3 p_{SO_3})^2} \qquad (11.5.d\text{-}1)$$

where $r = $ kmol $SO_2$/kg cat hr
$k_1 = \exp(12.160 - 5473/T)$
$K_2 = \exp(-9.953 + 8619/T)$
$K_3 = \exp(-71.745 + 52596/T)$
$K_p = \exp(11300/T - 10.68)$

The coefficients $k_1$, $K_2$, and $K_3$ were determined by nonlinear regression on 59 experiments carried out in a temperature range 420 to 590°C. The partial pressures are converted into conversions by means of the formulas: For 1 mol $SO_2$ fed per hour, the molar flow rates in a section where the conversion is $x$ is given in the left-hand column. The partial pressures are given in the right-hand column.

| | | |
|---|---|---|
| mol $SO_2$ | $1 - x$ | $\dfrac{1 - x}{(1 + a + b) - \frac{1}{2}x} p_t$ |
| $a$ mol $O_2$ | $a - \frac{1}{2}x$ | $\dfrac{a - \frac{1}{2}x}{(1 + a + b) - \frac{1}{2}x} p_t$ |
| $b$ mol $N_2$ | $b$ | |
| $SO_3$ | $x$ | $\dfrac{x}{(1 + a + b) - \frac{1}{2}x} p_t$ |

Total molar flow:  $(1 + a + b) - \frac{1}{2}x$

Figure 11.5.d-1 contains curves of equal reaction rate ("rate contours") $r(x, T)$ = const. These are obtained by finding the root of $r(x, T) - C = 0$ for a given temperature value. The shape of these contours is intuitively clear: at a constant conversion, the rate first increases with temperature but then decreases as the influence of the equilibrium is more strongly felt. The figure also contains the $\Gamma_e$ curve. This is the locus of equilibria conditions. The $\Gamma_m$ curve is the locus of the points in which the rate is maximum, by the appropriate selection of the temperature (i.e., $\partial r / \partial T = 0$). This locus is found by determining the root of $\partial r(x, T)/\partial T = 0$ for given values of the temperature. The curve $\Gamma_{\lambda m}$, also shown, is the locus of the points in which the rate is maximum when the reaction is carried out adiabatically. This locus is found by determining along the adiabatic line, starting from the inlet temperature and by means of a search method, the position where the rate is maximum. The $\Gamma_{\lambda m}$-curve is also the locus of the contact points between the adiabatic lines, which are straight lines with a slope

$$\lambda = \frac{\dot{m}c_p}{F_{A0}(-\Delta H)}$$

and the rate contours.

The figure has been calculated for the following feed composition: 7.8 mole % $SO_2$; 10.8 mole % $O_2$; 81.4 mole % inerts, atmospheric pression, a feed temperature of 37°C, a mean specific heat of 0.221 kcal/kg °C (0.925 kJ/kg K) and a $(-\Delta H)$ of 21,400 kcal/kmol (89,600 kJ/kmol). Cooling by means of a heat exchanger is represented by a parallel to the abscissa in this diagram.

If only the amount of catalyst is considered in an optimization of the reactor, the curve $\Gamma_m$ would have to be followed. If, however, in addition it is attempted to realize this by adiabatic operation the curve $\Gamma_{\lambda m}$ would have to be followed as closely as possible. This is realized by the zigzag line shown in the figure and corresponding to multibed adiabatic operation. The more beds there are the better $\Gamma_{\lambda m}$ is approximated. However, when the cost of equipment, supervision, control, and the like is also taken into account there is an optimum in the number of beds. Accounting in the optimization for the profit resulting from the conversion will, of course, also affect the location of the optimal zigzag line. The choice of the inlet temperature to a bed and the conversion realized in it determine the amount of catalyst required in that bed and also the heat exchanger. With $N$ beds $2N$ decisions have to be taken. The simultaneous variation of $2N$ variables to find the optimum policy leads to an enormous amount of computation, that rapidly becomes prohibitive, even for fast computers. There are methods for systematizing the search for the optimum and for reducing the amount of computation. A technique that is very well suited for stagewise processes is the technique of "dynamic programming," which allows one to reduce a $2N$-dimensional problem to a sequence of two-dimensional problems. The method introduced by Bellman [116] has been discussed in detail in books by Aris [30] and by Roberts [117]. Only a brief discussion, oriented toward direct application, is given here.

The calculations do not necessarily proceed according to the direction of the process flow. This is only so for a final-value problem (i.e., when the conditions at the exit of the reactor are fixed). For an "initial-value" problem, whereby the inlet conditions are fixed, the direction of computation for the optimization is opposite to that of the process stream. In what follows an initial-value problem is treated. First consider the last bed. No matter what the policy is before this bed the complete policy cannot be optimal when the last bed is not operating optimally for its feed. The specifications of the feed of the last bed are not known yet. Therefore, the optimal policy of the last bed has to be calculated for a whole set of possible inlet conditions of that bed.

Next, consider the last two beds. There exists an optimal policy for the two beds as a whole. In this optimal policy the first of the two (considered in the direction of the process flow) does not operate necessarily in the conditions which would be optimal if it were alone. The second has to be optimal for the feed coming from the first, however, or the combined policy would not be optimal. So that

$$
\begin{bmatrix} \text{Maximum profit} \\ \text{from two beds} \end{bmatrix} = \begin{array}{c} \text{maximum} \\ \text{of} \end{array} \left[ \begin{array}{c} \text{profit of} \\ \text{first bed} \end{array} + \left( \begin{array}{c} \text{maximum profit} \\ \text{of second bed} \\ \text{with feed} \\ \text{from first} \end{array} \right) \right] \qquad (11.5.\text{d-}2)
$$

To find this maximum it suffices to choose the conditions in the first of the two beds, since the optimal policy of the second has been calculated already. Equation (11.5.d-2) is Bellman's "optimum principle." Consider now the last three beds. These can be decomposed into a first bed (in the direction of process flow) and a pseudo stage consisting of the last two beds, for which the optimal policy has already been calculated for a series of inlet conditions. The procedure is continued in the same way towards the inlet of the multibed reactor.

Finally, all stages are done again in the direction of the process stream to determine "the" optimal policy, corresponding to the given feed to the whole reactor, among all available data. Dynamic programming is a so-called imbedding technique. Optimal policies are computed for all possible feed conditions of which ultimately only one is retained—that corresponding to the given feed conditions. Nevertheless, dynamic programming permits an enormous saving in computation time, because the conditions are only varied step by step (i.e., sequentially and not simultaneously over the $N$ stages).

The optimization procedure is illustrated for a particular case. The case considered is that of an exothermic, reversible reaction. The cooling between the beds is realized by means of heat exchangers. With $N$ stages $2N$ decisions have to be taken: $N$ inlet temperatures to the beds and $N$ conversions at the exit of the beds. The beds are numbered in the opposite direction of the process flow and the computation proceeds backward since the case considered is an initial value problem. The symbols are shown in Fig. 11.5.d-2. $x_{N+1}$ and $T_{N+1}$ are given. The conversion is not affected by the heat exchanger so that $\bar{x}_j = x_{j+1}$. The choice of the inlet temperature to bed $j$ together with the exit temperature of bed $j + 1$ determines the heat exchanger between $j + 1$ and $j$; the choice of $x_j$ the amount of catalyst in $j$.

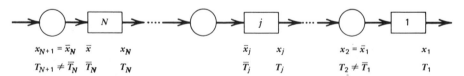

Figure 11.5.d-2 Definition of symbols used in multibed adiabatic reactor optimization by dynamic programming.

CHEMICAL REACTOR DESIGN

These decisions have to be optimal with respect to a certain objective or profit function. Such a profit function contains the profit resulting from the conversion of $A$ (e.g., $SO_2$) into the product $P$ (e.g., $SO_3$), but also the costs (catalyst, construction, control ...). If the costs were not taken into account it would follow from the computations that the conversions should proceed to the equilibrium values, and this would require an infinite amount of catalyst. Let $\alpha$ represent the profit resulting from the conversion of 1 kmol of $A$ into $P$. (e.g., U.S. \$/kmol). Per stage the value of the reaction mixture increases by an amount (in \$ per hr):

$$\alpha F_{A0}(x_j - \bar{x}_j)$$

where $F_{A0}$ is the molar feed rate of $A$ (kmol/hr). The only negative item considered in this example is the cost of the catalyst.

The cost of cooling is not considered here. For a detailed example taking this into account see Lee and Aris [32]. For a given purchase prize and life the cost of 1 kg of the catalyst can be expressed per hour, say $\beta$/hr. If the conversion in bed $j$ requires $W_j$ kg catalyst then the cost of this stage is $\beta W_j$ and the net profit is

$$\alpha F_{A0}(x_j - \bar{x}_j) - \beta W_j$$

Summing up over all the beds the total profit, $P_N$, becomes:

$$P_N = \sum_1^N [\alpha F_{A0}(x_j - \bar{x}_j) - \beta W_j]$$

or

$$P_N = \alpha F_{A0} \sum_1^N \left[(x_j - \bar{x}_j) - v\frac{W_j}{F_{A0}}\right] = \alpha F_{A0} \sum_1^N p_j \qquad (11.5.d\text{-}3)$$

where

$$v = \beta/\alpha \quad \text{and} \quad p_j = (x_j - \bar{x}_j) - v\frac{W_j}{F_{A0}}$$

Since $\alpha F_{A0}$ is a fixed amount it suffices to optimize the quantity in the straight brackets: the maximum profit is obtained subsequently by multiplying by $\alpha F_{A0}$.

The problem is now to optimize Eq. 11.5.d-3, that is, to find

$$\text{Max} \sum_1^N p_j = g_N(N + 1)$$

by the proper choice of $\bar{T}_N, x_N; \bar{T}_{N-1}, \ldots, \bar{T}_1, x_1$. For the bed numbered 1:

$$g_1(x_2) = \text{Max } p_1 = \text{Max}\left(x_1 - x_2 - v\frac{W_1}{F_{A0}}\right) = \text{Max} \int_{x_2}^{x_1} \left(1 - \frac{v}{r_1}\right) dx \qquad (11.5.d\text{-}4)$$

Therefore, $\bar{T}_1$ and $x_1$ have to be chosen such that:

$$\frac{\partial p_1}{\partial x_1} = 1 - \frac{v}{r_1} = 0 \tag{11.5.d-5}$$

and

$$\frac{\partial p_1}{\partial \bar{T}_1} = v \int_{x_2}^{x_1} \frac{1}{r_1^2} \frac{\partial r_1}{\partial \bar{T}_1} \, dx = 0 \tag{11.5.d-6}$$

Equation 11.5.d-5 means that the reaction has to be stopped when the rate has reached a value of $v$. Beyond that point the increase in cost outweighs the increase in profit resulting from the conversion. It is clear that this point is situated on that part of the adiabatic reaction path that is beyond $\Gamma_{\lambda m}$ and $\Gamma_m$. That part of the rate contour that has a value $v$ and that is to the right of $\Gamma_m$ is represented by $\Gamma_1$.

The second condition in Eq. (11.5.d-6) is satisfied only when $\partial r/\partial \bar{T}_1$, the partial derivative of the rate with respect to the temperature is partly positive and partly negative. Substituting into this partial derivative the relation between $x$ and $T$ along an adiabatic reaction path starting from $\Gamma_1$ (condition Eq. 11.5.d-5) turns $\partial p/\partial \bar{T}_1 = f_1(x, T_1)$ into a function $\partial p/\partial \bar{T}_1 = f_2(\bar{T}_1)$. The root of this equation is easily found by a one-dimensional search procedure and is the optimum inlet temperature leading to the exit conditions represented by the point chosen on $\Gamma_1$. This procedure is repeated for a certain number of points on $\Gamma_1$, to obtain the locus of optimum inlet conditions for bed 1, represented by $\bar{\Gamma}_1$ in Fig. 11.5.d-1. It follows from Eq. 11.5.d-6 that $\Gamma_1$ and $\bar{\Gamma}_1$ intersect on $\Gamma_m$, not on $\Gamma_{\lambda m}$.

Consider now two steps, the last two of the multibed adiabatic reactor. From Bellman's maximum principle it follows that the optimal policy of bed 1 is preserved. This time $x_2$ and $\bar{T}_2$ have to be chosen in an optimal way to arrive at

$$g_2(x_3) = \text{Max}(p_2 + p_1) = \text{Max} \int_{x_3}^{x_2} \left[ \left(1 - \frac{v}{r_2}\right) dx + g_1(x_2) \right]$$

To do so the following conditions have to be fulfilled:

$$\frac{\partial}{\partial x_2} (p_2 + p_1) = 0 \tag{11.5.d-7}$$

$$\frac{\partial}{\partial \bar{T}_2} (p_2 + p_1) = 0 \tag{11.5.d-8}$$

$x_2$ is the upper limit of the integral but appears also in $g_1(x_2)$, so that it is necessary to calculate $dg_1/dx_2$. Since $g_1 = f(x_2, x_1, \bar{T}_1)$:

$$\frac{dg_1}{dx_2} = -\left[1 - \frac{v}{r_1(x_2, \bar{T}_1)}\right] + \left[1 - \frac{v}{r_1(x_1, \bar{T}_1)}\right] \frac{dx_1}{dx_2} + \left[\int_{x_2}^{x_1} \frac{v}{r_1^2} \frac{\partial r_1}{\partial \bar{T}_1} \, dx\right] \frac{dT}{dx_2}$$

whereby $dT/dx_2$ has to be taken along an adiabatic path, so that

$$T = \overline{T}_1 + \frac{1}{\lambda}(x - x_2)$$

It follows that

$$\frac{dT}{dx_2} = \frac{d\overline{T}_1}{dx_2} - \frac{1}{\lambda}$$

Stage 1 has been determined in such a way that the parts between the brackets of the last two terms are zero, and since neither $dx_1/dx_2$ nor $dT/dx_2$ are infinite

$$\frac{dg_1}{dx_2} = -\left[1 - \frac{v}{r_1(x_2, \overline{T}_1)}\right]$$

and

$$\frac{\partial}{\partial x_2}(p_2 + p_1) = \left[1 - \frac{v}{r_2(x_2)}\right] - \left[1 - \frac{v}{r_1(x_2)}\right] = 0$$

The optimal policy, therefore, requires that

$$r_2(x_2, T_2) = r_1(x_2, \overline{T}_1) \qquad (11.5.d\text{-}9)$$

which means that the rate at the exit of bed 2 must equal that at the inlet of bed 1. This determines the heat exchanger: it should change the temperature in such way that Eq. 11.5.d-8 is fulfilled. Figure 11.5.d-3 illustrates how the curve representing

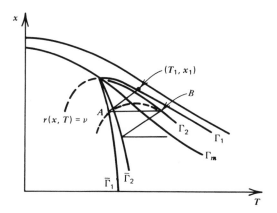

*Figure 11.5.d-3 Optimal reaction paths in a multibed adiabatic reactor according to dynamic programming.*

optimal exit conditions for bed 2, (i.e., $\bar{\Gamma}_2$), may be obtained from curve $\Gamma_1$—the heat exchanger does not modify the conversion.

The second condition (11.5.d-7) leads to:

$$\frac{\partial}{\partial \bar{T}_2}(p_2 + p_1) = \frac{\partial p_2}{\partial \bar{T}_2} = v \int_{x_3}^{x_2} \frac{1}{r_2^2} \frac{\partial r}{\partial \bar{T}_2} dx = 0 \qquad (11.5.d\text{-}10)$$

Equation 11.5.d-10 is completely analogous to Eq. 11.5.d-6 and the locus of optimal inlet temperatures to bed 2, represented by $\bar{\Gamma}_2$, is derived from $\Gamma_2$ in the same way as $\bar{\Gamma}_1$ from $\Gamma_1$. The procedure outlined above may be continued for further beds. Figure 11.5.d-1 shows $\Gamma$ and $\bar{\Gamma}$-curves for a three-bed $SO_2$ oxidation reactor. For a given feed represented by the point $A$ on the abscissa the optimal policy is determined as shown: first preheat to $B$, then adiabatic reaction in bed 3 until curve $\Gamma_3$ and so on.

The above discussion was based on a graphical representation. In reality the computations are performed on a computer and the $x_j$, $T_j$, $\bar{x}_j$, and $\bar{T}_j$ are stored. The above has been applied to a three-bed adiabatic reactor for $SO_2$ oxidation, using Collina, Corbetta, and Cappelli's [113] rate equation. The pressure is considered constant. There are no $\Delta T$ and $\Delta p$ over the film surrounding the catalyst. Also, to simplify the treatment the effectiveness factor is considered to be one in this illustrative example. The objective function to be optimized consists of two parts:

1. The profit resulting from the conversion.

2. The cost of the catalyst and the reactor, $v$.

From an example treated by Lee and Aris $\alpha = 2.5$ \$/kmol $SO_2$ converted; $\beta = 0.0017$ \$/kg cat hr. The amount of gas fed was 55,000 kg/hr [32].

The feed composition is that mentioned already in Eq. (10.5.d-1). The results are represented graphically in Fig. 11.5.d-1. This figure shows, besides the reaction contours and $\Gamma_e$, $\Gamma_m$, and $\Gamma_{\lambda m}$ curves the $\Gamma$ and $\bar{\Gamma}$ curves and the optimum reaction

Table 11.5.d-1

| Bed | $\bar{x}_j$ | $x_j$ | $\bar{T}_j$ (°K) | $T_j$ (°K) | (Profit) \$/kg gas | Total profit \$/kg gas | Catalyst weight kg |
|---|---|---|---|---|---|---|---|
| Optimal three-bed reactor for $SO_2$ synthesis (Cappelli's rate equation) | | | | | | | |
| 1 | 0 | 0.65 | 734.2 | 891.2 | 0.00405 | 0.00405 | 3248 |
| 2 | 0.65 | 0.90 | 723.3 | 783.3 | 0.00152 | 0.00557 | 7819 |
| 3 | 0.90 | 0.96 | 718.5 | 733.6 | 0.00029 | 0.00586 | 14.427 |

path. Notice that Eqs. 11.5.d-6 and 11.5.d-10 require the reaction path to lie on both sides of $\Gamma_m$, but not necessarily on both sides of $\Gamma_{\lambda m}$. $\Gamma_{\lambda m}$ may differ significantly from $\Gamma_m$ for rate contours of the type encountered with the rate equations considered here. Table 11.5.d-1 contains the weights of catalyst in each bed for both rate equations. Notice the large conversion in the first bed obtained with relatively little catalyst and the large amount of catalyst required in the third bed. In practice, intermediate cooling is also realized by cold shot cooling. The optimization of such a reactor has been discussed in detail by Lee and Aris [32]. Further work on the optimization of $SO_2$ oxidation has been published by Paynter et al. [33] and by Burkhardt [34].

### 11.5.e Fixed Bed Reactors with Heat Exchange between the Feed and Effluent or between the Feed and Reacting Gas
### "Autothermic Operation"

In industrial operation it is necessary, for economic reasons, to recover as much as possible the heat produced by exothermic reactions. One obvious way of doing this, mentioned earlier in Section 11.3, is to preheat the feed by means of the reacting fluid and/or the effluent. When the heat of reaction is sufficient to raise the temperature of the feed to such a value that the desired conversion is realized in the reactor without further addition of heat, the operation is called "autothermic." Some of the most important industrial reactions like ammonia and methanol synthesis, $SO_2$ oxidation, and phthalic anhydride synthesis, the water gas shift reaction can be carried out in an autothermic way. Coupling the reactor with a heat exchanger for the feed and the reacting fluid or the effluent leads to some special features that require detailed discussion.

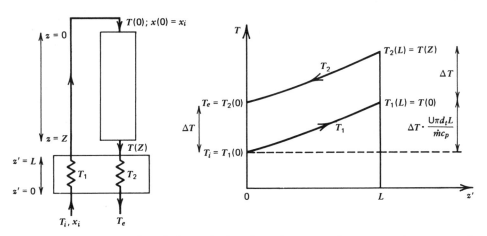

*Figure 11.5.e-1 Single adiabatic bed with preheating of reactants by means of effluent gases.*

Consider, as an example, a modern large-size ammonia-synthesis process. In such a process, producing 1000 T/day of ammonia in a single converter, the feed is preheated by the effluent in a heat exchanger. In the catalytic bed itself the reaction is carried out adiabatically. For reasons discussed in Sec. 11.3 the reactor is subdivided into several beds with intermediate cold shot cooling. The principles are discussed first on a simplified scheme consisting of only one adiabatic bed with given amount of catalyst and one heat exchanger with given exchange surface shown in Fig. 11.5.e-1.

Consider a single reaction and let the pressure drop over the reactor be small so that a mean value may be used with sufficient accuracy. Then the continuity equation for the key component and the energy equations for the reactor and the heat exchanger, respectively, may be written (in the steady state):

$$\frac{dx}{dz} = \frac{\pi d_t^2}{4F_{A0}} \rho_B r_A \tag{11.5.e-1}$$

$$\frac{dT}{dz} = \frac{\pi d_t^2}{4} \frac{\rho_B(-\Delta H) r_A}{\dot{m} c_p} \tag{11.5.e-2}$$

$$\frac{dT_1}{dz'} = \frac{U \pi d_t}{(\dot{m} c_p)_1} (T_2 - T_1) \tag{11.5.e.3}$$

$$(\dot{m} c_p)_1 dT = -(\dot{m} c_p)_2 dT_2$$

with boundary conditions:

Reactor:

$$x(0) = x_i$$

$$T(0) = T_1(L)$$

Heat exchanger, with reference to the complete system inlet and outlet temperatures:

$$T_1(0) = T_i$$

$$T_2(0) = T_e, \text{ unknown}$$

or with reference to the reactor inlet and outlet temperatures:

$$T_1(L) = T(0),$$

$$T_2(L) = T(Z), \text{ unknown}$$

For the situation represented in Fig. 11.5.e-1, $(\dot{m} cp)_1 = (\dot{m} c_p)_2$. It is seen how the reactor and heat exchanger are coupled through the boundary or initial conditions. Even more, the problem is a so-called two-point boundary value problem. Indeed, the inlet temperature to the reactor $T(0)$ is not known, since $T_1(L)$ depends on the

CHEMICAL REACTOR DESIGN

outlet temperature of the reactor. For the heat exchanger $T_1(0) = T_i$ is given, but not $T_2(0)$, which depends on the reactor outlet-temperature. Solving the problem, therefore, involves trial and error. One procedure assumes a value for $T(0) = T_1(L)$ and simultaneously integrates the differential equations describing the reactor Eqs. 11.5.e-1 and 11.5.e-2, yielding $T(Z) = T_2(L)$. Then, the heat exchanger Eq. 11.5.e-3 may be integrated, yielding $T_1(0)$. This value has to be compared with the given inlet temperature $T_i$. If it corresponds the assumed value of $T(0)$ is correct and the calculated values are the final ones, if not $T(0)$ has to be improved. Problems of this type will be encountered later, but for a better insight the one under discussion will be approached in a somewhat less formal way along the lines set by Van Heerden [28] by decomposing it into two parts. First, consider the adiabatic reactor. In the formal procedure described above the integration of Eqs. 11.5.e-1 and 11.5.e-2 was performed for various values of $T(0)$. What are then the possible outlet conditions for the reactor? After integration of the ratio of Eqs. 11.5.e-1 and 11.5.e-2 from the inlet to the outlet,

$$\Delta x = \frac{\dot{m} c_p}{F_{A0}(-\Delta H)} \Delta T = \lambda \Delta T = \lambda [T(Z) - T(0)] \qquad (11.5.e-4)$$

and Eq. 11.5.e-1 to:

$$\frac{W}{F_{A0}} = \int_{x(0)}^{x(Z)} \frac{dx}{r_A}$$

Let the reaction be reversible $A \rightleftarrows B$, and first order in both directions, so that the rate can be written:

$$r_A = k\left(p_A - \frac{p_B}{K}\right)$$

or

$$r_A = k p_t\left[(1 - x) - \frac{x}{K}\right]$$

or

$$r_A = A_0 p_t\left[\exp\left(-\frac{E}{RT}\right)\right]\left[(1 - x) - \frac{x}{K}\right]$$

in which $T$ is to be substituted by $T(0) + \Delta x/\lambda$. This means that, for a given reaction [given $A_0$, $E$, $K$, $c_p$, $(-\Delta H)$, feed $x(0)$ and feed rate $F_{A0}$ and a given amount of catalyst, $W$]:

$$x(Z) - x(0) = f[T(0)] \qquad \text{and also} \qquad T(Z) = g[T(0)] \quad (11.5.e-5)$$

The shape of this relation between the outlet and inlet conditions is shown in Fig. 11.5.e-2 as curve ⓐ. The rising, sigmoidal-shaped part of curve ⓐ stems

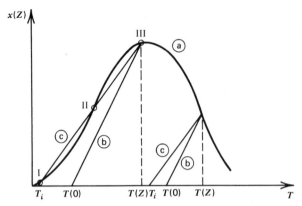

*Figure 11.5.e-2 Possible steady-state operating points in reactor-heat exchanger system, for two inlet temperatures $T_i$.*

from the Arrhenius temperature dependence of the rate, the descending part from the unfavorable influence of the equilibrium. With such a bell-shaped curve there is an optimum region for $T(0)$ if a maximum conversion is to be reached. The location of the curve ⓐ obviously depends on the factors that determine the kinetics of the reaction: total pressure, the reactant concentrations (in ammonia synthesis—the presence of inerts and catalyst activity. Curve ⓐ can also be considered as a measure of the amount of heat produced by the reaction.

In this diagram of possible outlet conditions for various inlet conditions, the adiabatic reaction path corresponding to given $x(0)$ and $T(0)$ is represented by the straight line ⓑ having a slope $\lambda$ and ending in a point of ⓐ. Curve ⓐ and the line ⓑ have only one point in common. The second step in the formal procedure is to calculate the temperatures in the heat exchanger and to check whether or not the assumed $T(0)$ leads to $T_1(0) = T_i$. The coupling with a heat exchanger will obviously impose a restriction on the $T(0)$.

Simplifying somewhat and considering the difference $T_2 - T_1 = \Delta T$ to be constant over the total length of the heat exchanger, $L$, the equation for the heat exchanger Eq. 11.5.e-3 becomes, after integration

$$\int_T^{T(0)} \frac{dT_1}{\Delta T} = \frac{U\pi d_t}{\dot{m}c_p} L$$

or

$$T(0) - T_i = \frac{U\pi d_t}{\dot{m}c_p} L\Delta T \qquad (11.5.e-6)$$

and by adding $T(Z)$ to both sides:

$$T(Z) - T_i = \Delta T\left(1 + \frac{U\pi d_t}{\dot{m}c_p}L\right)$$

since

$$\Delta T = T_e - T_i = T_2 - T_1 = T(Z) - T(0)$$

Now, $\Delta T = T(Z) - T(0)$ is the adiabatic temperature rise in the reactor, which is given by Eq. 11.5.e-4. So, combining the reactor and the heat exchanger leads to:

$$x(Z) - x(0) = \frac{\lambda[T(Z) - T_i]}{1 + \frac{U\pi d_t}{\dot{m}c_p}L} \qquad (11.5.e\text{-}7)$$

which reduces to Eq. 11.5.e-4 when $L = 0$ [i.e., when there is no heat exchanger, since $T_i$ then equals $T(0)$].

Equation 11.5.e-7 is represented in the $x - T$ diagram by a straight line ©, starting from $T_i$, with a slope

$$\frac{\lambda}{1 + \frac{U\pi d_t}{\dot{m}c_p}L}$$

smaller than $\lambda$, and ending on a point of ⓐ. The line © may be considered to be representative of the amount of heat exchanged.

The steady state of the complete system—reactor and heat exchanger—has to satisfy both Eqs. 11.5.e-5 and 11.5.e-7. But it is easily seen that, depending on the location of $T_i$ and on the slope, the straight line © can have up to three intersections with ⓐ (i.e., three steady-state operating conditions are possible for the system reactor + heat exchanger, whereas the reactor on itself has a unique steady state). The multiplicity of steady states in the complete system is a consequence of the thermal feedback realized in the heat exchanger between the feed and the effluent.

The operating point represented by I is of no practical interest: the conversion achieved under these conditions is far too low. The operating point corresponding to II is a naturally unstable point (i.e., extremely sensitive to perturbations in the operating conditions). Indeed, for these conditions the slightest increase in $T(Z)$ has a much larger effect on the heat produced than on the heat exchanged (curve ⓐ has a much larger slope than the line ©) and the operating point will shift to III). The reverse would happen for a decrease in $T(Z)$: the operating point would shift to I and the reaction would practically extinguish. By the same reasoning it can be shown that III represents intrinsically stable operating conditions.

The conditions represented by III are not optimal, however, since the point is beyond the maximum of ⓐ, which means that the rate is influenced considerably

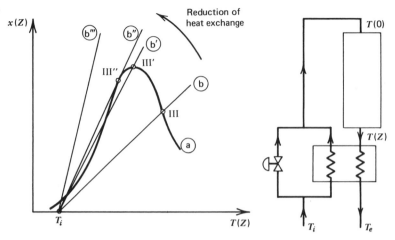

*Figure 11.5.e-3 Modification of the location of the steady-state operating conditions as the amount of heat exchanged is reduced.*

already by the equilibrium. Operation corresponding to the straight line ⓑ" in Fig. 11.5.e-3, tangent to the curve ⓐ would realize an optimum conversion for a given catalyst weight. This can be achieved by suitable design of the heat exchanger, more precisely by decreasing the heat exchanging surface or $L$ or by partly bypassing the heat exchanger as shown also in Fig. 11.5.e-3. This is effectively done in ammonia-synthesis. Operation in III′ is not so easy, however, since it corresponds to the limit of stability—the reactor easily extinguishes. Obviously, with a heat exchanging line ⓑ‴ no intersection with ⓐ is possible: the only intersection with ⓐ is the low conversion point, which is of no practical interest. The sensitivity of an ammonia-synthesis reactor is well known to its operators. Thus point III′ is a reasonable compromise.

The bypass illustrated in Fig. 11.5.e-3 is important also for compensating for a decrease in catalyst activity. In the beginning of its life the catalyst is very active, but due to poisons, temperature variations, and other operational vices inducing structural changes the activity gradually decreases. Such a situation is represented in Fig. 11.5.e-4. A decrease in activity of the catalyst has to be compensated for by higher operating temperatures in the reactor, which means that curve ⓐ shifts to the right in the $x$-$T$ diagram. The only intersection left between ⓐ and ⓒ would be the low conversion point. The slope of ⓒ therefore has to be reduced, which means increasing the heat exchanging surface (or $L$) or increasing $\dot{m}$ (i.e., the amount of gas flowing through the heat exchanger or decreasing the amount bypassed). We can see from Fig. 11.5.e-4 how $T(0)$ and $T(Z)$ correspondingly increase so that the desired conversion is maintained.

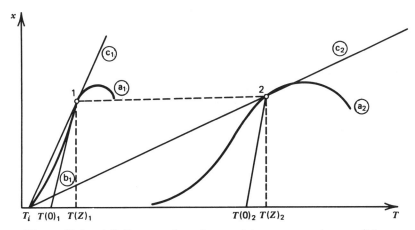

*Figure 11.5.e-4 Influence of catalyst activity on operating conditions.*

It is stressed again that the occurrence of multiple steady states is due to the feedback of heat. Tubular reactors that are not coupled with a heat exchanger generally do not exhibit this feature—except in very particular situations, as will be shown later. This does not mean that perturbations in the operating conditions cannot give rise to drastic changes in the conversion and temperature profiles, but this is then caused by parametric sensitivity, a feature already discussed in Sec. 11.5.b. The preceding discussion is illustrated quantitatively by the results of a simulation study by Shah [29], who numerically integrated the system of differential equations describing an ammonia-synthesis reactor of the type represented schematically in Fig. 11.5.e-5 and including an intermediate quench by means of cold feed. Figure 11.5.e-6 shows the hydrogen-conversion and outlet temperature as a function of the percentage of the feed being preheated. When the fraction of the feed being preheated exceeds 0.7, two exit conditions are possible;

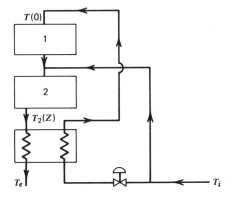

*Figure 11.5.e-5 Schematic representation of two-stage adiabatic reactor.*

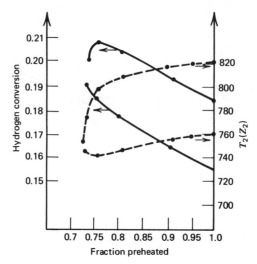

*Figure 11.5.e-6 Two-stage adiabatic re-actor. Hydrogen conversion as a function of fraction of feed being preheated (from Shah [29]).*

for instance, when all the feed is preheated—18.5 percent and 15.5 percent for the conversion and 547°C and 487°C for the temperature. If the amount of "cold split" becomes too important no solution is found (i.e., no autothermic operation is possible).

The same situation is reflected in Fig. 11.5.e-7, which shows the effect of inlet temperature on the exit conditions; again two steady states are found. Also, when

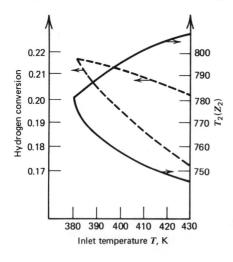

*Figure 11.5.e-7 Two-stage adiabatic reaction. Hydrogen conversion vs inlet temperature (from Shah [29]).*

the inlet temperature is decreased below 107°C the reactor is extinguished. It also follows from Fig. 11.5.e-7 that the maximum conversion is obtained close to the conditions leading to extinction, as shown already in the preceding discussion. The simulation also predicts that the reactor would extinguish when the pressure is decreased from 240 to 160 atm or the inerts content increased from 9 to 18 percent. Therefore, the inlet temperature $T_i$ should be kept sufficiently above the blowout temperature (20 to 25°C) to avoid the possibility that an increase in inerts content or of the feed rate may cause instability. The question of which steady state will be attained depends on the initial conditions and cannot be answered by steady-state calculations; transients have to be considered.

The scheme illustrated by Fig. 11.5.e-5 is not the only one possible for autothermic operation. Another possibility is the multitubular arrangement with internal heat exchanger, represented schematically in Fig. 11.5.e-8, together with the temperature profiles in the catalyst bed and in the heat exchanger tubes. For constant total pressure the simulation of such a reactor with built-in heat exchanger requires the simultaneous integration of the continuity equation(s) for the key component(s) and of two energy equations, one for the effluent gas in the tubes and one for the reacting gas in the catalyst bed.

The steady-state continuity equation for the key component may be written

$$\frac{dx}{dz} = r_A \rho_B \frac{\Omega}{F_{A0}}$$

Provided the heat capacities of the feed and the reacting gas are constant the energy equations may be written

$$\frac{dT_1}{dz} = -\frac{U A_k}{\dot{m} c_p}(T_2 - T_1)$$

for the reacting gas in the catalyst bed

$$\frac{dT_2}{dz} = \frac{F_{A0}(-\Delta H)}{\dot{m} c_p}\frac{dx}{dz} - \frac{U A_k}{\dot{m} c_p}(T_2 - T_1)$$

The boundary or initial conditions are

$$T_1 = T_2 \quad \text{at} \quad z = 0$$
$$T_1 = T_i \quad \quad z = Z$$
$$x = x_i \quad \quad z = 0$$

Again, this is a two-point boundary value problem and again three steady states are possible, the outer two of which are stable, at least to small perturbations, the intermediate being unstable. A figure completely analogous to Fig. 11.5.e-2 may be constructed, with two types of curves: the first, bell shaped for reversible reactions, which is a measure of the heat generated, and the second, which is a

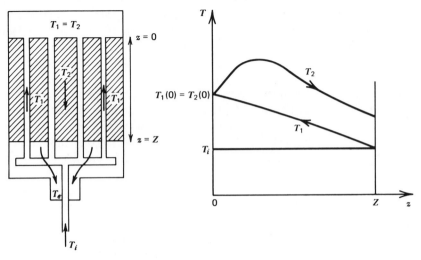

*Figure 11.5.e-8 Multitubular reactor with internal heat exchange.*

measure of the heat exchanged and which is a straight line. Again, for a certain range of operating variables, more than one intersection is possible.

The NEC (Nitrogen Engineering Co.) and TVA (Tennessee Valley Authority) —ammonia synthesis reactors are practical realizations of the above principles. Figure 11.3-5 of Sec. 11.3 schematically represents a TVA reactor. The corresponding temperature profiles inside the tubes and in the catalyst bed section, calculated by Baddour, Brian, Logeais, and Eymery [26] are shown in Fig. 11.5.e-9. Reactor dimensions for the TVA converter simulated by Baddour et al. and also by Murase, Roberts and Converse [27] are

### Catalyst bed

| | |
|---|---|
| Total catalyst volume | 4.07 m$^3$ |
| Reactor length | 5.18 m |
| Reactor basket diameter | 1.1 m |
| Reactor basket cross-sectional area | 0.95 m$^2$ |
| Catalyst bed cross-sectional area | 0.78 m$^2$ |

### Cooling tubes

| | |
|---|---|
| Number | 84 |
| Tube outside diameter | 50.8 mm |
| Tube inside diameter | 38.1 mm |
| Tube heat exchange area (outer) | 69.4 m$^2$ |
| Tube heat exchange area (inner) | 52.0 m$^2$ |

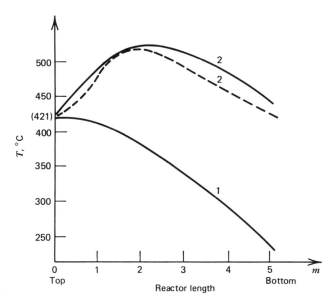

*Figure 11.5.e-9 Temperature profiles inside TVA. ammonia-synthesis reactor. 1 = gas in heat exchanger tubes; 2 = gas in catalyst bed; full curve 2 = simulated; dashed curve 2 = plant (from Baddour, et al. [26]).*

Typical operating conditions are

|                              |                           |
|------------------------------|---------------------------|
| Production capacity          | 120 T $NH_3$/day          |
| $H_2$ mole fraction in feed  | 0.65                      |
| $N_2$ mole fraction in feed  | 0.219                     |
| $NH_3$ mole fraction in feed | 0.052                     |
| Inert                        | 0.079                     |
| Mass flow rate               | 26,400 kg/hr              |
| Space velocity               | 13,800 hr$^{-1}$          |
| Pressure                     | 286 atm (290 bar)         |
| Top temperature              | 421°C                     |

The rate equations used in these simulations is that proposed by Temkin and Pyzhev [118]:

$$r_A \rho_B = f\left( k_1 \frac{p_{N_2} \cdot p_{H_2}^{1.5}}{p_{NH_3}} - k_2 \frac{p_{NH_3}}{p_{H_2}^{1.5}} \right)$$

where $r_A$ is the rate of reaction of nitrogen (kmol/kg cat hr) and $f$ is the catalyst activity (one, at zero process time).

$$k_1 = 1.79 \times 10^4 \exp\left(-\frac{20,800}{RT}\right) \qquad k_2 = 2.57 \times 10^{16} \exp\left(-\frac{47,400}{RT}\right)$$

$$(-\Delta H) = 26,600 \text{ kcal/kmol } N_2 \text{ reacted}$$
$$= 111,370 \text{ kJ/kmol}$$

Baddour [26] retained the above model equations after checking for the influence of heat and mass transfer effects. The maximum temperature difference between gas and catalyst was computed to be 2.3°C at the top of the reactor, where the rate is a maximum. The difference at the outlet is 0.4°C. This confirms previous calculations by Kjaer [120]. The inclusion of axial dispersion, which will be discussed in a later section, altered the steady-state temperature profile by less than 0.5°C. Internal transport effects would only have to be accounted for with particles having a diameter larger than 6 mm, which are used in some high-capacity modern converters to keep the pressure drop low. Dyson and Simon [121] have published expressions for the effectiveness factor as a function of the pressure, temperature and conversion, using Nielsen's experimental data for the true rate of reaction [119]. At 300 atm and 480°C the effectiveness factor would be 0.44 at a conversion of 10 percent and 0.80 at a conversion of 50 percent.

Figure 11.5.e-10 shows the relation between the inlet temperature and the top temperature for the TVA reactor simulated by Baddour et al. for the conditions given above. The curve given corresponds to a space velocity of 13,800 m³/m³ cat. hr. The space velocity, often used in the technical literature, is the total volumetric feed rate under normal conditions, $F'_0$(Nm³/hr) per unit catalyst volume (m³), that is, $\rho_B F'_0/W$. It is related to the inverse of the space time $W/F_{A0}$ used in this text (with $W$ in kg cat. and $F_{A0}$ in kmol $A$/hr). It is seen that, for the nominal space velocity of 13,800 (m³/m³ cat. hr) and inlet temperatures between 224 and 274°C, two top temperatures correspond to one inlet temperature. Below 224°C no autothermal operation is possible. This is the blowout temperature. By the same reasoning used in relation with Fig. 11.5.e-2 it can be seen that points on the left branch of the curve correspond to the unstable, those on the right branch to the upper stable steady state. The optimum top temperature (425°C), leading to a maximum conversion for the given amount of catalyst, is marked with a cross. The difference between the optimum operating top temperature and the blowout temperature is only 5°C, so that severe control of perturbations is required. Baddour et al. also studied the dynamic behavior, starting from the transient continuity and energy equations [26]. The dynamic behavior was shown to be linear for perturbations in the inlet temperature smaller than 5°C, around the conditions of maximum production. Use of approximate transfer functions was very successful in the description of the dynamic behavior.

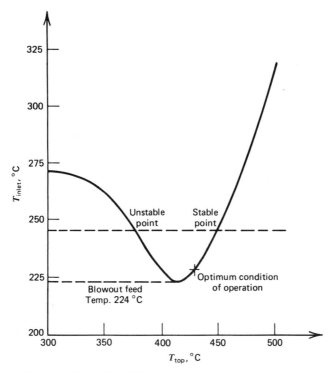

*Figure 11.5.e-10 TVA ammonia synthesis reactor. Relation between inlet and top temperature (from Baddour, et al. [26]).*

In the preceding section (11.5.d) on multibed adiabatic reactors, the optimization of the reactor was discussed in detail and, for example, one way of doing this rigorously according to dynamic programming was worked out. The multitubular reactor with feed-effluent heat exchange considered in this section has also been the object of optimization, first by sound judgment, more recently by a more systematic and rigorous approach. Again the problem is best illustrated by means of an $x$-$T$ diagram or a mole $\% - T$ diagram like that of Fig. 11.3-4 of Section 11.3, which shows the mole $\% - T$ diagram of a TVA reactor compared with that of a five-bed quench converter of the same capacity [3]. Murase et al. [27] optimized the profit of the countercurrent TVA-$NH_3$-synthesis converter by optimizing the temperature profile. This was done by means of Pontryagin's maximum principle [31], which is the method best suited to systems with continuous variables. The countercurrent flow in the reactor-heat exchanger systems tends to lower the temperature in the first catalyst layers. It follows from Murase's calculation that in the TVA-reactor this effect would have to be enhanced. The

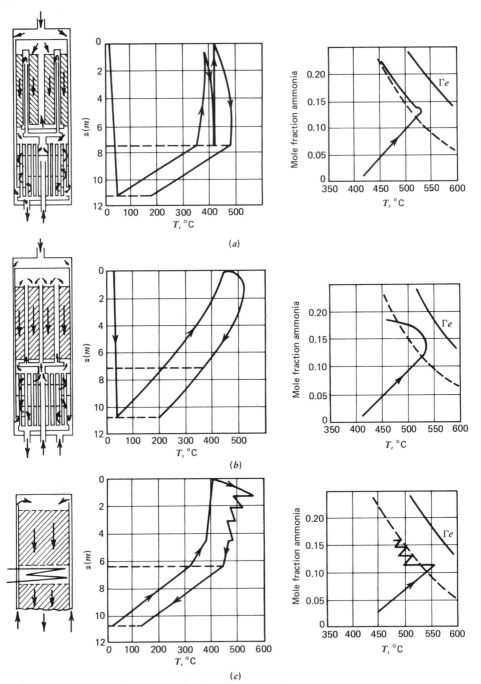

*Figure 11.5.e-11 Operating diagrams for various types of ammonia synthesis reactors. (a) Multitubular reactor with cocurrent flow. (b) Multitubular reactor with countercurrent flow. (c) Multibed adiabatic reactor (from Fodor [122]).*

production of the considered TVA-reactor could have been increased by 5.4 percent if it could have been designed with a continuously varying heat transfer coefficient along the bed. In practice this is not so simple: it means that, without changing the number of tubes, the tubes would have to be finned in the top zone. It would then be almost impossible to pack the catalyst homogeneously. The SBA reactor, which also has countercurrent flow, has a much larger number of tubes (900 for production of 200 T $NH_3$/day) so that the temperature in the first layers is more optimized. However, this large exchange surface would also lower too much the temperature toward the end of the bed and lower the reaction rate too much, in spite of the more favorable equilibrium. The remedy is then to decrease the heat exchange in the second half of the catalyst bed. SBA chose to do this by shielding the tubes by concentric tubes. The trajectory for such a 200-tons/day reactor is also shown in Fig. 11.3-4 of Sec. 11.3.

There are also cocurrent flow-type reactors as shown in Fig. 11.5.e-11. They permit a closer approach to the curve of maximum rate, at the expense of a more complicated construction [122].

### 11.5.f Non-Steady Behavior of Fixed Bed Catalytic Reactors Due to Catalyst Deactivation

In Chapter 5, rate equations were set up for several types of catalyst deactivation. In this section we discuss the consequences of catalyst deactivation on fixed bed reactor performance. Clearly, when the catalyst deactivates in a point in the reactor the conversion in that point is affected. Consequently, the conversion profile and the temperature profile will be modified with time, in other words the reactor is operating in non-steady-state conditions. The way the profiles are shifted and the rate at which this happens depends on the mechanism of deactivation, of course. This shift is well known in industrial practice. In an ammonia-synthesis reactor, for example, the hot spot is known to migrate slowly through the reactor, due to sintering of the catalyst: if the first layers the feed contacts are becoming less active more catalyst will be required to reach a given conversion and the hot spot moves down the bed. If no precautions are taken this would mean a decrease of production of the reactor. What is done in this case is to oversize the reactor so that sufficient catalyst is available to compensate for loss in activity until it has to be replaced for other, more imperative reasons, such as excessive pressure drop due to powder formation. Another way to compensate for loss in activity is to increase the inlet temperature, as discussed already. Another example is catalytic reforming of naphtha, where the catalyst is deactivated by coke deposition. In this case the deactivation is compensated for by increasing the operating temperature so that the conversion, measured here by the octane number of the reformate, is kept constant. There is a limit to this temperature increase, of course, since it causes a higher production of light gases (i.e., decreases the selectivity) so that the coke has to be burned off and the catalyst regenerated. It is clear that it is important to predict the behavior of reactors subject to deactivation. This requires setting up a mathematical model. This model consists of the set of continuity and energy equations we have set up already, but considering the transient nature of the process and the variable catalyst activity, reflected in a rate equation that contains a deactivation function.

We will illustrate this by means of an example of the effect of fouling by coke deposition. We will simplify somewhat by considering only isothermal operation. The continuity equation for the reactant, $A$, may be written in terms of mole fractions, assuming that both the density and the number of moles remains constant (see Froment and Bischoff [35]):

$$\frac{\partial y_A}{\partial t} + \frac{M_M F_T}{\varepsilon \rho_g \Omega} \frac{\partial y_A}{\partial z} = - \frac{M_m \rho_B}{\varepsilon \rho_g} r_A \qquad (11.5.f\text{-}1)$$

When the following dimensionless variables are introduced:

$$z' = \frac{z}{d_p} \qquad t' = \frac{M_m F_t}{\varepsilon \rho_g \Omega d_p} t$$

Eq. (11.5.f-1) becomes

$$\frac{\partial y_A}{\partial t'} + \frac{\partial y_A}{\partial z'} = - \frac{\Omega \rho_B d_p}{F_t} r_A \qquad (11.5.f\text{-}2)$$

The continuity equation for the catalyst coking compound is

$$\frac{\partial C_c}{\partial t} = r_C \qquad (11.5.f\text{-}3)$$

In this equation $C_c$ is really written in terms of amount of carbon per unit weight of catalyst, since the amount of carbonaceous compound is usually measured as carbon. When the dimensionless variables defined above are introduced, Eq. 11.5.f-3 becomes

$$\frac{\partial C_c}{\partial t} = \frac{\varepsilon \rho_g \Omega d_p}{M_m F_t} r_C \qquad (11.5.f\text{-}4)$$

Now the rate terms $r_A$ and $r_C$ remain to be specified. Then if it is assumed, for simplicity, that both the main reaction and the coke deposition are of first order, $r_A$ and $r_C$ may be written

For a parallel coking mechanism:

$$r_A = k_A p_t y_A + \frac{k_C p_t y_A}{\psi_A M_A}$$

$$r_C = k_C p_t y_A$$

CHEMICAL REACTOR DESIGN

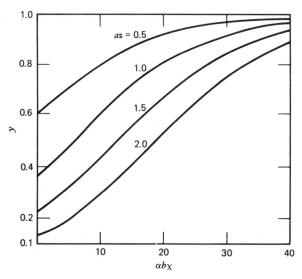

*Figure 11.5.f-1 Reactant mole fraction versus time group for parallel reaction mechanism with exponential activity function (from Froment and Bischoff [35]).*

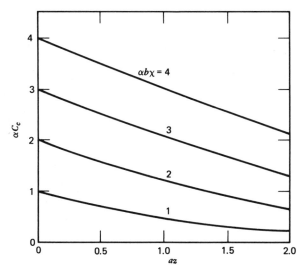

*Figure 11.5.f-2 Coke profiles for parallel reaction mechanism with exponential activity function (from Froment and Bischoff [35]).*

517

with $k_C$ in kg coke/kg cat hr atm and the conversion factor $\psi_A$ in kg coke/kg $A$. For a consecutive coking mechanism:

$$r_A = k_A p_t y_A$$
$$r_C = k_C p_t (1 - y_A)$$

Next, the fouling function has to be introduced. Let this function be of the exponential type, then:

$$k_A = k_A{}^0 e^{-\alpha_A C_c}$$
$$k_C = k_C{}^0 e^{-\alpha_C C_c}$$

In order to permit analytical integration of the system (Eqs. 11.5.f-2 to 11.5.f-4), Froment and Bischoff considered $\alpha_2 = 0$, that is, the amount of coke already deposited influences the rate of the main reaction, but not that of the parallel reaction leading to coke because it is of thermal nature, rather than catalytic. Integrating the system of equations with suitable boundary conditions leads to the results represented in Figs. 11.5.f-1 and 11.5.f-2. In Figs. 11.5.f-1 and 2,

$$a = \frac{\Omega \rho_B d_p p_t}{F_t} k_1{}^0$$

$$b = \frac{\Omega \rho_g \varepsilon d_p p_t}{F_t} k_2{}^0$$

$$\chi = t' - z'$$

For all practical cases $\chi \cong t'$.

It is clear from Fig. 11.5.f-1 that the mole fraction of the reactant $A$ increases with time at a given bed depth, in other words, the conversion decreases. From Fig. 11.5.f-2 we see that the carbon is not deposited uniformly along the bed, but according to a descending profile. This is intuitively clear: when the carbonaceous compound is deposited by a reaction parallel to the main, its rate of formation is maximum at the inlet of the reactor, where the mole fraction of the reactant is maximum.

Froment and Bischoff also treated the consecutive coking mechanism along the lines given above and obtained the diagrams of Figs. 11.5.f-3 and 11.5.f-4. The difference with the parallel coking case is striking, particularly in the carbon content of the catalyst, which again is not uniform, but increases with bed length. It follows from the preceding that equations that try to relate the activity of the bed with time (Voorhies formula) can only be approximate. The Voorhies formula can at best be valid only for a *given* bed length. By plotting the *average* carbon content of the bed versus time in a double logarithmic plot—the way Voorhies and others did (Chapter 5)—Froment and Bischoff obtained a power of 1 for the Voorhies formula with the parallel coking mechanism with exponential activity function. For a consecutive coking mechanism they obtained a power of 1 at low

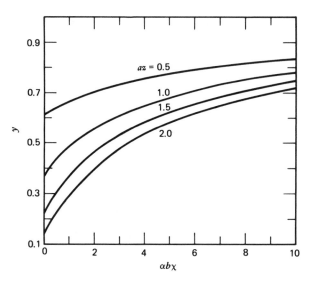

*Figure 11.5.f-3 Reactant mole fraction versus time group for consecutive reaction mechanism with exponential deactivation function (from Froment and Bischoff [35]).*

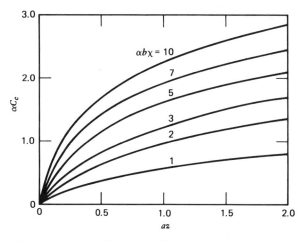

*Figure 11.5.f-4 Coke profiles for consecutive reaction mechanism with exponential deactivation function (from Froment and Bischoff [35]).*

process times and 0.5 at higher process times. The same is true for a parallel fouling mechanism with a hyperbolic activity function. These values are in the same range as those of the experimental studies quoted in Chapter 5. In their study of the dehydrogenation of butene into butadiene, Dumez and Froment [141] observed coke formation from both butene (parallel mechanism) and butadiene (consecutive mechanism), while hydrogen inhibited the coking. The power in the Voorhies relation decreased from 0.55 to 0.35. In the catalytic cracking of light East Texas gasoil, Eberly et al. [142] found a power varying from 0.77 to 0.55 as process time increased.

It further follows from Froment and Bischoff's study that, for a given bed length. both the point and the average carbon content increase with increasing space time (or decrease with space velocity) for the consecutive reaction mechanism, but decrease for the parallel mechanism (increase in terms of space velocity). Eberly's data [142] also indicate that the power in the Voorhies relation depends on the space time or on the liquid hourly space velocity. Another consequence of this analysis is shown in Fig. 11.5.f-5 for a parallel reaction. In the absence of fouling, and for isothermal conditions, the maximum rate of reaction $A \rightarrow R$ is always at the reactor entrance. This is not necessarily true when the catalyst is fouled

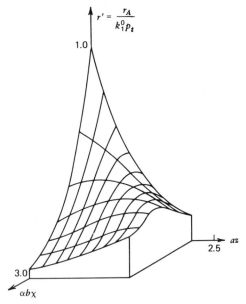

$$r' = \frac{r_A}{k_1^0 p_t}$$

*Figure 11.5.f-5 Rate surface for parallel reaction mechanism with exponential deactivation function (from Froment and Bischoff [35]).*

according to a parallel mechanism. Indeed, in that case the inlet is then deactivated to a greater extent than other portions of the reactor so that a ridge develops in the rate surface. A maximum in the reaction rate is developed that travels down the bed as time progresses. If the operation is not isothermal this activity wave will be reflected in a temperature wave, which may complicate the control of the reactor. Such a behavior was observed by Menon and Sreeramamurthy [38] in their study of hydrogen sulfide air oxidation into water and sulfur on a charcoal catalyst. The sulfur progressively covers the catalyst surface, deactivating the catalyst in this way. An activity profile results, which is revealed by a temperature peak traveling through the bed. In the case of an ascending carbon profile, which is obtained when the carbonaceous deposit results from a consecutive fouling mechanism, the rate is continuously decreasing with time in all points of the reactor, except at $z = 0$, where coke is not deposited yet. A gradually decreasing part of the reactor will then be effective in the conversion to the main product.

Descending coke profiles were observed experimentally by Van Zoonen in the hydroisomerization of olefins [39]. In butene dehydrogenation into butadiene on a chromium-aluminium-oxide catalyst at 595°C, Dumez and Froment [141] measured nearly uniform coke contents in the integral reactor. It was shown, by experiments on a thermobalance, that coke originated from both butene and butadiene, while hydrogen inhibited its formation. The absence of a pronounced coke profile results from a balance between the three phenomena. In the isothermal isomerization of $n$-pentane under hydrogen pressure on a platina-alumina reforming catalyst, DePauw and Froment [123] observed ascending coke profiles. The coke content of the catalyst was not zero at the inlet of the reactor, however, so that a parallel–consecutive coking mechanism—confirmed by independent measurements on a thermobalance—was adopted. Since, in addition, some hydrocracking had to be accounted for, the following set of continuity equations was considered:

For $n$-pentane, in terms of partial pressures:

$$\frac{\partial p_A}{\partial t} + \frac{M_m F_t}{\Omega \rho_g} \frac{\partial p_A}{\partial z} = - \frac{M_m p_t \rho_B}{\Omega \rho_g} r_A$$

For the lumped hydrocracking products (methane, ethane, propane, butane):

$$\frac{\partial p_D}{\partial t} + \frac{M_m F_t}{\Omega \rho_g} \frac{\partial p_D}{\partial z} = \frac{M_m p_t \rho_B}{\Omega \rho_g} r_D$$

for the coke:

$$\frac{\partial C_c}{\partial t} = r_c$$

The rate of isomerization, $r_I$, was found to be controlled by the surface reaction of $n$-pentene on $Al_2O_3$, instead of the adsorption on $n$-pentene on these sites, as found by Hosten and Froment for a slightly different catalyst and with continuous chlorine injection (see Chapter 2)

$$r_I = \frac{(k^I)\left(p_A - \dfrac{p_B}{K}\right)}{p_{H_2} + K_{AB}{}^I(p_A + p_B) + K_D{}^I p_D}$$

in which the adsorption equilibrium constants of $n$- and $i$-pentene are taken to be identical, so that $K_{AB}{}^I = K_A{}^I = K_B{}^I$. The total rate of disappearance of $n$-pentane is given by

$$r_A = r_I + 0.5 r_D$$

since two moles of hydrocracked products are formed from one mole of pentane and since the rate of coke formation, $r_C$, is small compared with $r_I$ and $r_D$. Hydrocracking was shown to originate from both $n$- and $i$-pentene. The rate of hydrocracking was found to be given by

$$r_D = \frac{k^D(p_A + 0.714 p_B)}{\left(1 + K_{AB}{}^D \dfrac{p_A + p_B}{p_{H_2}} + K_D{}^D \dfrac{p_D}{p_{H_2}}\right)^2}$$

The coke also originated from $n$- and $i$-pentene and its rate of formation was described by

$$r_C = \frac{k_C{}^p p_A}{p_{H_2}{}^2} + \frac{k_C{}^c p_B}{p_{H_2}{}^2}$$

Each rate coefficient in the above equations contains the corresponding deactivation function. This was shown, by experiments on a thermobalance, to be of the exponential type. Furthermore, it was shown that coking and hydrocracking occurred on the same sites so that

$$\phi_C = \phi_D = e^{-\alpha_D C_C} \qquad \text{whereas} \qquad \phi_I = e^{-\alpha_I C_C}$$

and

$$k^I = (k^I)^0 \phi_I \qquad k^D = (k^D)^0 \phi_D \qquad k_C = (k_C)^0 \phi_C$$

The rate constants, the adsorption equilibrium constants, and the deactivation parameters $\alpha_I$ and $\alpha_D$ were determined from the measurement of $p_A$ and $p_D$ as a function of time and position (i.e., $W/F_{A0}$) in the bed, through a special sampling device. In addition, the coke profile was measured at the end of the run. The parameters were found to be statistically significant, and the rate coefficients obeyed the Arrhenius temperature dependence.

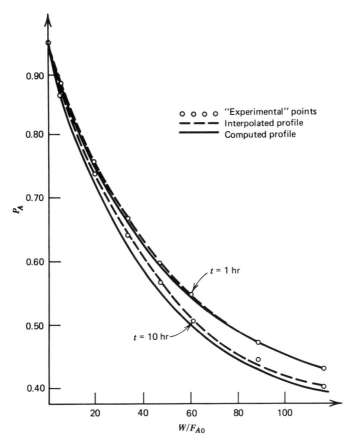

*Figure 11.5.f-6 Pentane isomerization. Partial pressure profiles of n-pentane versus $W/F_{A0}$.*

Figures 11.5.f-6, 11.5.f-7, and 11.5.f-8 compare experimental data with results obtained from a simulation on the basis of the above equations and the best set of parameters. The agreement is quite satisfactory and the approach appears to be promising for the characterization of catalysts deactivated by coke deposition.

An important aspect of coking is its influence on the selectivity. As the product distribution or the selectivity depends on the ratios of the various rate coefficients it is evident that the selectivity may also be affected by changes in catalyst activity, when the different reactions are not influenced in the same way by the catalyst activity decline. Froment and Bischoff [37] worked out the theory for such a situation. Figure 11.5.f-9 shows the results for a complex reaction with parallel coking scheme. The variation of the selectivity for the isomerization of *n*-pentane

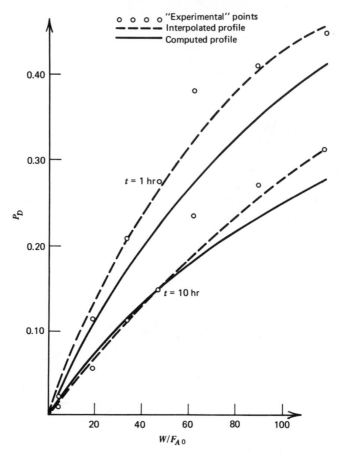

*Figure 11.5.f-7 Pentane isomerisation. Partial pressure profiles of lumped hydrocracked products versus $W/F_{A0}$.*

with the coke content of the catalyst [123] has been given in Fig. 5.3.c-2 of Chapter 5. Since the hydrocracking is more affected by the coke content of the catalyst, the isomerization selectivity increases under those conditions. Ultimately, however, the decrease in the isomerization rate would become too important. To compensate for this, before regeneration of the catalyst is resorted to, the temperature could be increased. Thereby, hydrocracking and coking would be more promoted than the isomerization and the selectivity would seriously decrease. In the catalytic cracking of gasoil on silica-alumina catalysts, on the other hand, the selectivity for gasoline was found to be independent of the coke content of the catalyst (see Weekman and Nace [12]).

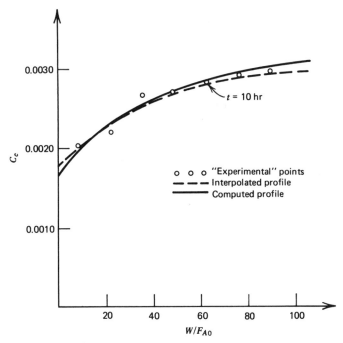

*Figure 11.5.f-8 Pentane isomerization. Coke profiles after 10 hours.*

Recently, considerable attention has been given to the problem of maintaining the conversion of a reactor constant by adapting the temperature level. Butt has studied this problem for simple and for bifunctional catalysts, such as used in reforming [36]. Optimization techniques have been applied to this problem by Jackson [124], Chou, Ray, and Aris [125], and Ogunye and Ray [126].

## 11.6 One-Dimensional Model with Axial Mixing

In Sec. 11.5 the one-dimensional pseudohomogeneous model was discussed in considerable detail. Several aspects like runaway, optimization, and transient behavior due to catalyst coking, which are in fact entirely general, were analyzed under this model. This is sound justification for doing so. Most of the practical reactor design work so far has been based on this model, sometimes because it was considered sufficiently representative, more often because it was more convenient to use. Yet, several assumptions involved in the model are subject to criticism. It may be argued that the flow in a packed bed reactor deviates from the ideal pattern because of radial variations in flow velocity and mixing effects due to the presence of packing. Furthermore, it is an oversimplification also to assume

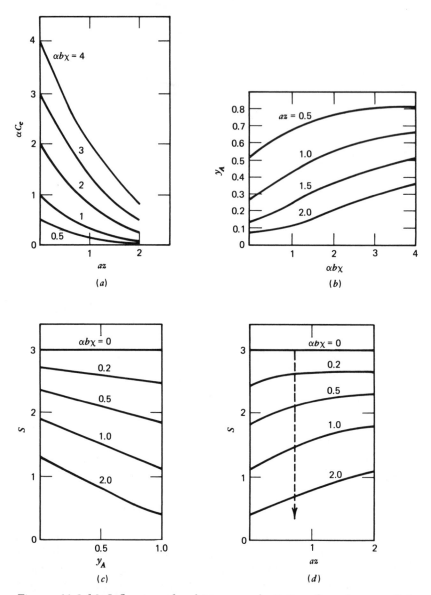

Figure 11.5.f-9 Influence of coking on selectivity. Complex parallel reaction scheme. (a) Coke profiles. (b) Mole fraction of reactant versus process time group. (c) Selectivity versus mole fraction of reactant. (d) Selectivity versus dimensionless position (from Froment and Bischoff [37]).

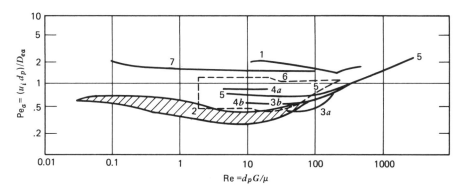

*Figure 11.6-1 Peclet number for axial effective diffusion, based on particle diameter, versus Reynolds number. 1: McHenry and Wilhelm; 2: Ebach and White; 3: Carberry and Bretton; 4: Strang and Geankoplis; 5: Cairns and Prausnitz; 6: Hiby; 7: Hiby, without wall effect (from Froment [76]).*

that the temperature is uniform in a cross section. The first objection led to a development that will be discussed in the present section, the second to a model discussed in Sec. 11.7.

Accounting for the velocity profile is practically never done, because it immediately complicates the computation in a serious way. In addition, very few data are available to date and no general correlation could be set up for the velocity profile (Schwartz and Smith [85], Schertz and Bischoff [40], Cairns and Prausnitz [41], and Mickley et al. [42]).

The mixing in axial direction, which is due to the turbulence and the presence of packing, is accounted for by superposing an "effective" transport mechanism on the overall transport by plug flow. The flux due to this mechanism is described by a formula analogous to Fick's law for mass transfer or Fourier's law for heat transfer by conduction. The proportionality constants are "effective" diffusivities and conductivities. Because of the assumptions involved in their derivation they implicitly contain the effect of the velocity profile. This whole field has been reviewed and organized by Levenspiel and Bischoff [43]. The principal experimental results concerning the effective diffusivity in axial direction are shown in Fig. 11.6-1 [44, 45, 46, 47, 48]. For design purposes $Pe_a$, based on $d_p$, may be considered to lie between 1 and 2. Little information is available on $\lambda_{ea}$. Yagi, Kunii, and Wakao [49] determined $\lambda_{ea}$ experimentally, while Bischoff derived it from the analogy between heat and mass transfer in packed beds [50].

The steady-state continuity equation for a component $A$ may be written:

$$\varepsilon D_{ea} \frac{d^2 C_A}{dz^2} - u_s \frac{dC_A}{dz} - r_A \rho_B = 0 \qquad (11.6-1)$$

and the energy equation:

$$\lambda_{ea}\frac{d^2T}{dz^2} - \rho_g u_s c_p \frac{dT}{dz} + (-\Delta H)r_A \rho_B - \frac{4U}{d_t}(T - T_r) = 0 \qquad (11.6\text{-}2)$$

Axial mixing smoothens axial gradients of concentration and temperature so that it decreases the conversion obtained in a given reactor, in principle at least.

The boundary conditions have given rise to extensive discussion [52, 53, 54, 55, 56]. Those generally used are:

$$u_s(C_{A0} - C_A) = -\varepsilon D_{ea}\frac{dC_A}{dz} \qquad \text{for } z = 0$$

$$\rho_g u_s c_p(T_0 - T) = -\lambda_{ea}\frac{dT}{dz}$$

$$\frac{dC_A}{dz} = \frac{dT}{dz} = 0 \qquad\qquad \text{for } z = L$$

We see that this leads to a two-point boundary value problem, requiring trial and error in the integration. It has been shown several times that for the flow velocities used in industrial practice the effect of axial dispersion of heat and mass on conversion is negligible when the bed depth exceeds about 50 particle diameters [51]. In spite of this, the model has received great attention recently, more particularly the adiabatic version. The reason is that the introduction of axial mixing terms into the basic equations leads to an entirely new feature, that is, the possibility of more than one steady-state profile through the reactor [57].

Indeed, for *a certain range of operating conditions* three steady-state profiles are possible with the same feed conditions, as is shown in Fig. 11.6-2. The outer two of these steady state profiles are stable, at least to small perturbations, while

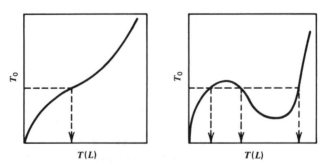

*Figure 11.6-2 Relations between $T_0$ and $T(L)$, which lead to a unique and to three steady-state profiles, respectively (after Raymond and Amundson [57]).*

the middle one is unstable. Which steady-state profile will be predicted by steady-state computations depends on the initial guesses of $C_A$ and $T$ involved in the integration of this two-point boundary value problem. Physically, this means that the steady state actually experienced depends on the initial profile in the reactor. For all situations where the initial values are different from the feed conditions transient equations have to be considered in order to make sure the correct steady state profile is predicted. In order to avoid those transient computations when they are unnecessary it is useful to know a priori if more than one steady-state profile is possible. From Fig. 11.6-2 we see that a necessary and sufficient condition for uniqueness of the steady state profile in an adiabatic reactor is that the curve $T_0 = f[T(L)]$ has no hump. Mathematically this means that the equation

$$\frac{d^2 T}{dz'^2} - Pe_a' \frac{dT}{dz'} + f(T) = 0$$

where

$$z' = \frac{z}{L} \qquad Pe_a' = \frac{u_i L}{D_{ea}}$$

and

$$f(T) = \frac{k_0 L^2}{\varepsilon D_{ea}} \rho_B (T_{ad} - T) \exp \frac{E}{R T_0} \left( 1 - \frac{T_0}{T} \right)$$

has no bifurcation point, whatever the length of the reactor. This led Luss and Amundson [58] to the following conditions:

$$\text{Sup}\left[ f'(T) - \frac{Pe_a'^2}{4} \right] \le 0 \qquad (11.6\text{-}4)$$

$$T_0 \le T \le T_{ad}$$

which can be satisfied by diluting the reaction mixture. Another way of realizing a unique profile is to limit the length of the adiabatic reactor so that

$$\text{Sup}\left[ f'(T) - \frac{Pe_a'^2}{4} \right] \le \mu_1 \qquad (11.6\text{-}5)$$

$$T_0 \le T \le T_{ad}$$

where $\mu_1$ is the smallest positive eigenvalue of

$$\Delta v + \mu v = 0 \qquad (11.6\text{-}6)$$

and where $v(z)$ is the difference between two solutions $T_1(z)$ and $T_2(z)$. Uniqueness is guaranteed only if the only solution of Eq. 11.6-6 is $v(z) = 0$.

When applied to a first-order irreversible reaction carried out in an adiabatic reactor, these conditions lead to Eqs. 11.6-7 and 11.6-8, respectively:

$$\frac{k_0 L^2}{\varepsilon D_{ea}} \rho_B \left[ \frac{E}{R} \left( \frac{T_{ad} - T_0}{T_p^2} \right) - 1 \right] \exp\left[ \frac{E}{RT_0} \left( 1 - \frac{T_0}{T_p} \right) \right] - \frac{u_i^2 L^2}{4D_{ea}^2} \leq 0 \quad (11.6\text{-}7)$$

or

$$< \mu_1 \quad (11.6\text{-}8)$$

where $T_{ad} - T_0 = [(-\Delta H)/(\rho_g c_p)]C_{A0}$ is the adiabatic temperature rise and $T_p$ is the value for $T$ for which $f'(T)$ assumes its supremum. A sufficient, but not necessary, condition for Eq. 11.6-7 is that

$$\frac{E}{RT_0} \cdot \frac{T_{ad} - T_0}{T_0} \leq 1 \quad (11.6\text{-}9)$$

or

$$\gamma\beta \leq 1$$

Luss later refined these conditions and arrived at [59]

$$(T - T_0) \frac{d \ln f(T)}{dT} \leq 1 \quad (11.6\text{-}10)$$

The magnitude of the axial effective diffusivity determines which of the two conditions, Eq. 11.6-4 or 11.6-10, is stronger. For a first-order irreversible reaction carried out in an adiabatic reactor, Eq. 11.6-10 leads to

$$\frac{E}{RT_0} \cdot \frac{T_{ad} - T_0}{T_0} \leq 4 \frac{T_{ad}}{T_0} \quad \text{or} \quad \gamma\beta \leq 4 \frac{T_{ad}}{T_0} \quad (11.6\text{-}11)$$

which is far less conservative than Eq. 11.6-9, based on Eq. 11.6-5. Hlavacek and Hofmann [60] derived the following form, which is identical to the Luss criterion (Eq. 11.6-11):

$$\frac{E}{RT_0} \frac{T_{ad} - T_0}{T_0} < \frac{4}{1 - \frac{4RT_0}{E}}$$

Hlavacek and Hofmann also defined necessary and sufficient conditions for multiplicity, for a simplified rate law of the type Barkelew used (Eq. 11.5-c) and equality of the Peclet numbers for heat and mass transfer. The necessary and sufficient conditions for multiplicity, which have to be fulfilled simultaneously are

1. The group $(E/RT_0)[(-\Delta H)C_{A0}/(\rho_g c_p T_0)] = \gamma\beta$ has to exceed a certain value.
2. The group $Lk_0/u_i = Da$ has to lie within a given interval.
3. The Peclet number based on reactor length $u_i L/D_{ea}$ has to be lower than a certain value.

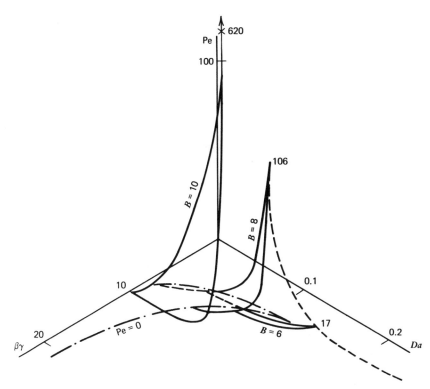

*Figure 11.6-3 Region of multiple steady states. Relation between Peclet, Damköhler, and βγ group (after Hlavacek and Hofmann [60]).*

From a numerical study Hlavacek and Hofmann derived the results represented in Fig. 11.6-3.

This figure clearly illustrates that the range within which multiple steady states can occur is very narrow. It is true that, as Hlavacek and Hofmann calculated, the adiabatic temperature rise is sufficiently high in ammonia, methanol and oxo-synthesis and in ethylene, naphthalene, and o-xylene oxidation. None of the reactions are carried out in adiabatic reactors, however, although multibed adiabatic reactors are sometimes used. According to Beskov (mentioned in Hlavacek and Hofmann) in methanol synthesis the effect of axial mixing would have to be taken into account when $Pe'_a < 30$. In industrial methanol synthesis reactors $Pe'_a$ is of the order of 600 and more. In ethylene oxidation $Pe'_a$ would have to be smaller than 200 for axial effective transport to be of some importance, but in industrial practice $Pe'_a$ exceeds 2500. Baddour et al. in their simulation of the TVA ammonia synthesis converter found that the axial diffusion of heat altered the steady-state temperature profile by less than 0.6°C. Therefore, the length of industrial fixed bed reactors removes the need for reactor models including axial diffusion and the risks involved with multiple steady states, except perhaps for very shallow beds. In practice, shallow catalytic beds are only encountered in the first stage of multibed adiabatic reactors. One may question if very shallow beds can be described by effective transport models, in any event. The question remains if shallow beds really exhibit multiple steady states. The answer to this question probably requires a completely different approach, based on better knowledge of the hydrodynamics of shallow beds.

In summary, in our opinion there is no real need for further detailed study of the axial transport model: there are several other effects, more important than axial mixing, which have to be accounted for.

## 11.7 Two-Dimensional Pseudo-Homogeneous Models _____

### 11.7.a The Effective Transport Concept

The one-dimensional models discussed so far neglect the resistance to heat and mass transfer in the radial direction and therefore predict uniform temperatures and conversions in a cross section. This is obviously a serious simplification when reactions with a pronounced heat effect are involved. For such cases there is a need for a model that predicts the detailed temperature and conversion pattern in the reactor, so that the design would be directed towards avoiding eventual detrimental overtemperatures in the axis. This then leads to two-dimensional models.

The model discussed here uses the effective transport concept, this time to formulate the flux of heat or mass in the radial direction. This flux is superposed on the transport by overall convection, which is of the plug flow type. Since the effective diffusivity is mainly determined by the flow characteristics, packed beds are not isotropic for effective diffusion, so that the radial component is different from the axial mentioned in Sec. 11.6.b. Experimental results concerning $D_{er}$ are shown in Fig. 11.7.a-1 [61, 62, 63]. For practical purposes $Pe_{mr}$ may be considered to lie between 8 and 10. When the effective conductivity, $\lambda_{er}$, is determined from heat transfer experiments in packed beds, it is observed that $\lambda_{er}$ decreases strongly in the vicinity of the wall. It is as if a supplementary resistance is experienced near the wall, which is probably due to variations in the packing density and flow velocity. Two alternatives are possible: either use a mean $\lambda_{er}$ or consider $\lambda_{er}$ to be constant in the central core and introduce a new coefficient accounting for the heat transfer near the wall, $\alpha_w$, defined by:

$$\alpha_w(T_R - T_w) = -\lambda_{er}\left(\frac{\partial T}{\partial r}\right)_w \qquad (11.7.a-1)$$

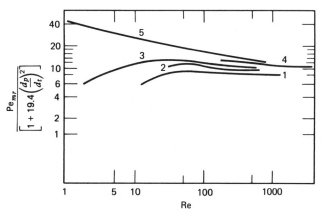

*Figure 11.7.a-1 Peclet number for radial effective diffusion, based on particle diameter, versus Reynolds number. 1: Fahien and Smith, 2: Bernard and Wilhelm, 3: Dorweiler and Fahien, 4: Plautz and Johnstone, 5: Hiby.*

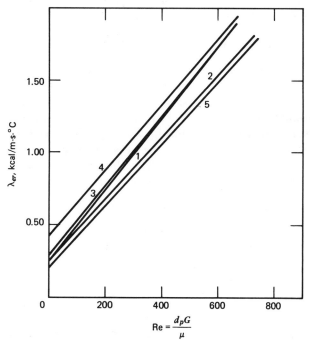

*Figure 11.7.a-2 Effective radial thermal conductivity versus Reynolds number. 1: Coberly and Marshall, 2: Campbell and Huntington, 3: Calderbank and Pogorsky, 4: Kwong and Smith, 5: Kunii and Smith.*

533

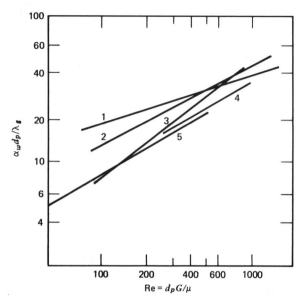

*Figure 11.7.a-3 Nusselt number for wall heat transfer coefficient versus Reynolds number. 1: Coberly and Marshall, 2: Hanratty (cylinders), 3: Hanratty (spheres), 4: Yagi and Wakao, 5: Yagi and Kunii.*

When it is important to predict point values of the temperature with the greatest possible accuracy the second approach is preferred, so that two parameters are involved to account for heat transfer in radial direction. Figure 11.7.a-2 and 11.7.a-3 show some experimental results for $\lambda_{er}$ and $\alpha_w$ [64, 65, 66, 67, 69, 76].

The data for $\alpha_w$ are very scattered. Recently De Wasch and Froment [19] published data that are believed to have the high degree of precision required for the accurate prediction of severe situations in reactors. The correlations for air are of the form:

$$\lambda_{er} = \lambda_{er}{}^0 + \frac{0.0025}{1 + 46\left(\dfrac{d_p}{d_t}\right)^2} \text{Re} \qquad (11.7.a-2)$$

$$\alpha_w = \alpha_w{}^0 + \frac{0.0115\, d_t}{d_p} \text{Re} \qquad (11.7.a-3)$$

where $\lambda_{er}{}^0$ and $\alpha_w{}^0$ are static contributions, dependent on the type and size of the catalyst. The correlation for $\alpha_w$ is of an entirely different form of those published until now, but confirms Yagi and Kunii's theoretical predictions [72].

       CHEMICAL REACTOR DESIGN

Since both solid and fluid are involved in heat transfer $\lambda_{er}$ is usually based on the total cross section and therefore on the superficial velocity, in contrast with $D_{er}$. This is reflected in Eq. (11.7.b-1). Yagi and Kunii [70, 72] and Kunii and Smith [68] and later Schlünder [74, 75] have set up models for calculating $\lambda_{er}$ and $\alpha_w$. In these models the flux by effective conduction is considered to consist of two contributions, the first dynamic (i.e., dependent on the flow conditions) and the second static so that

$$\lambda_{er} = \lambda_{er}{}^0 + \lambda_{er}{}^t$$

*The Static Contribution*

In the absence of flow the following mechanisms contribute to the effective conduction, according to Kunii and Smith [68].

**1.** Transport through the fluid in the voids.
  (a) By conduction.
  (b) By radiation between neighboring voids.

**2.** Transport in which the solid phase is involved.
  (a) Conduction through the contact surface between particles.
  (b) Conduction in the stagnating film in the vicinity of the contact surface.
  (c) Radiation from particle to particle.
  (d) Conduction through the particles.

Except in high vacuum the contribution 2(a) may be neglected. Figure 11.7.a-4 represents this model by means of an electrical network. By expressing each of these contributions by means of the basic formulas for heat transfer and combining

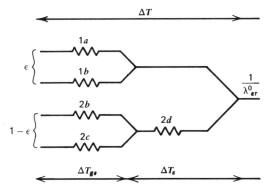

*Figure 11.7.a-4 Model for heat transfer in packed bed according to Yagi and Kunii [70].*

them in the appropriate way, depending on whether they operate in series or parallel, the following equation is obtained:

$$\frac{\lambda_{er}^{\,0}}{\lambda_g} = \varepsilon\left(1 + \beta\,\frac{d_p \alpha_{rv}}{\lambda_g}\right) + \frac{\beta(1 - \varepsilon)}{\dfrac{1}{\dfrac{1}{\gamma} + \dfrac{\alpha_{rs} d_p}{\lambda_g}} + \gamma\,\dfrac{\lambda_g}{\lambda_s}}$$  (11.7.a-4)

where

$\lambda_g, \lambda_s$ = conductivities of fluid and solid, respectively
$\quad\varepsilon$ = void fraction
$\quad\alpha_{rv}$ = radiation coefficient from void to void, used when the expression for heat transfer by radiation is based on a temperature difference $T_1 - T_2$, in view of combining it with transport by convection or conduction

$$\alpha_{rv} = \frac{0.1952}{1 + \dfrac{\varepsilon}{2(1 - \varepsilon)} \cdot \dfrac{1 - p}{p}}\,\frac{(T + 273)^3}{100}$$

where $p$ is the emissivity of the solid and $T$ is in °C.
$\quad\alpha_{rs}$ = radiation coefficient for the solid

$$\alpha_{rs} = 0.1952\,\frac{p}{2 - p}\,\frac{(T + 273)^3}{100}$$

$\beta$ = a coefficient that depends on the particle geometry and the packing density, comprised between 0.9 and 1.0
$\gamma = \frac{2}{3}$
$\phi$ = depends on the packing density

$\phi$ may be calculated when $\phi_1$ and $\phi_2$ are known. $\phi_1$ is the value of $\phi$ for the loosest possible packing ($\varepsilon = 0.476$). $\phi_2$ is the value of $\phi$ for the densest packing ($\varepsilon = 0.260$). $\phi_1$ and $\phi_2$ may be calculated from the knowledge of $\lambda_s/\lambda_g$. These functions are plotted in Fig. 11.7.a-5. When $\varepsilon$ is comprised between its two extreme values $\phi$ is calculated according to:

$$\phi = \phi_2 + (\phi_1 - \phi_2)\,\frac{\varepsilon - 0.260}{0.476 - 0.260}$$

Zehner and Schlünder [74, 75] arrived at the following formula for the static contribution:

$$\frac{\lambda_{er}^{\,0}}{\lambda_g} = (1 - \sqrt{1 - \varepsilon})\left(1 + \varepsilon\,\frac{\lambda_{rs} d_p}{\lambda_g}\right) + \frac{2\sqrt{1 - \varepsilon}}{1 + \left(\dfrac{\lambda_{rs} d_p}{\lambda_g} - B\right)\dfrac{\lambda_g}{\lambda_s}}\,\Theta$$

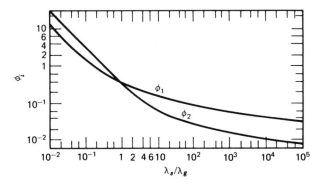

*Figure 11.7.a-5 Curves $\phi_1$ and $\phi_2$ versus ratio of solid to gas conductivity, $\lambda_s/\lambda_g$ (after Kunii and Smith [68]).*

with

$$
\Theta = \left\{ \frac{\left[ 1 + \left( \frac{\alpha_{rs} d_p}{\lambda_g} - 1 \right) \frac{\lambda_g}{\lambda_s} \right] B}{\left[ 1 + \left( \frac{\alpha_{rs} d_p}{\lambda_g} - B \right) \frac{\lambda_g}{\lambda_s} \right]^2} \ln \frac{1 + \frac{\alpha_{rs} d_p}{\lambda_s}}{B \frac{\lambda_g}{\lambda_s}} - \frac{B - 1}{1 + \left( \frac{\alpha_{rs} d_p}{\lambda_g} - B \right) \frac{\lambda_g}{\lambda_s}} + \frac{B + 1}{2B} \left( \frac{\alpha_{rs} d_p}{\lambda_g} - B \right) \right\}
$$

where $B = 2.5[(1 - \varepsilon)/\varepsilon]^{10/9}$ for cylinders.

*The Dynamic Contribution*

This contribution arises exclusively from the transport in the fluid and corresponds to the mixing that is described by the effective diffusion in radial direction, $D_{er}$. When the analogy between heat and mass transfer is complete the following relation may be written:

$$
\lambda_{er}^t = \varepsilon \rho_g c_p D_{er}
$$

from which

$$
\frac{\lambda_{er}^t}{\lambda_g} = \Psi \, \mathrm{Pr} \, \mathrm{Re}
$$

where

$$
\Psi = \frac{1}{\mathrm{Pe}_{mr}}
$$

FIXED BED CATALYTIC REACTORS _____ 537

For $Pe_{mr} = 10$, $\Psi = 0.1$. Yagi and Kunii [70] have derived $\Psi$ from experimental data on $\lambda_{er}$ and obtained a value of $\Psi$ for spherical and cylindrical packing between 0.10 and 0.14.

De Wasch and Froment [19] obtained the following equation:

$$\Psi = \frac{0.14}{1 + 46\left(\dfrac{d_p}{d_t}\right)^2}$$

The wall heat transfer coefficient can be predicted by a model that is analogous to that outlined here for $\lambda_{er}$. [72, 73]. It should be stressed here that $\alpha_w$ is intrinsically different from the "global" coefficients discussed in Sec. 11.5.a. Indeed, the latter are obtained when the experimental heat transfer data are analyzed on the basis of a one-dimensional model that does not consider radial gradients in the core of the bed. This comes down to localizing the resistance to heat transfer in radial direction completely in the film along the wall.

## 11.7.b Continuity and Energy Equations

The continuity equation for the key reacting component, $A$, and the energy equation can now be written as follows, for a single reaction and steady state

$$\left\{\begin{array}{l} \varepsilon D_{er}\left(\dfrac{\partial^2 C}{\partial r^2} + \dfrac{1}{r}\dfrac{\partial C}{\partial r}\right) - u_s\dfrac{\partial C}{\partial z} - \rho_B r_A = 0 \\[3mm] \lambda_{er}\left(\dfrac{\partial^2 T}{\partial r^2} + \dfrac{1}{r}\dfrac{\partial T}{\partial r}\right) - u_s\rho_g c_p\dfrac{\partial T}{\partial z} + \rho_B(-\Delta H)r_A = 0 \end{array}\right. \qquad (11.7.b\text{-}1)$$

with boundary conditions:

$$\begin{array}{llll} C = C_0 & & & \\ T = T_0 & \text{at} & z = 0 & 0 \leq r \leq R_t \\ \partial C/\partial r = 0 & \text{at} & r = 0 & \text{and} \quad r = R_t \\ \partial T/\partial r = 0 & \text{at} & r = 0 & \text{all } z \\ \partial T/\partial r = -\dfrac{\alpha_w}{\lambda_{er}}(T_R - T_w) & \text{at} & r = R_t & \end{array}$$

Note that the term accounting for effective transport in the axial direction has been neglected in this model, for the reasons given already in Sec. 11.6. This system of nonlinear second order partial differential equations was integrated by Froment using a Crank-Nicolson procedure [76, 77], to simulate a multitubular fixed bed reactor for a reaction involving yield problems.

Mihail and Iordache [145] compared the performance of some numerical techniques for integrating the system (11.7.b-1): Liu's average explicit scheme with a five-point grid [129], the Crank-Nicholson implicit scheme [76, 77], and orthogonal collocation (Finlayson [130]). The reactor was an *o*-xylene oxidation reactor and the reaction scheme that of Froment [76], discussed in the next section. The computation time was of the same order of magnitude with the Crank-Nicolson and Liu's scheme, but orthogonal collocation only required two thirds of this time. Liu's explicit scheme was very sensitive to step size and led to problems of stability and convergence for severe operating conditions leading to important hot spots.

### 11.7.c Design or Simulation of a Fixed Bed Reactor for Catalytic Hydrocarbon Oxidation

In this example the design or simulation of a multitubular reactor for catalytic hydrocarbon oxidation is discussed (Froment [76]). The case considered here is of a rather complex nature, that is,

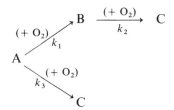

This reaction model is fairly representative of the gas phase air oxidation of *o*-xylene into phthalic anhydride on $V_2O_5$ catalysts. *A* represents *o*-xylene, *B* phthalic anhydride, and *C* the final oxidation products CO and $CO_2$, lumped together. The process conditions were already described in Sec. 11.5.b. The purpose of this example is mainly to check whether or not serious radial temperature gradients occur in such a reactor. For a better approximation of reality a reaction model is chosen that is closer to the true model than the one used in Sec. 11.5.b. In addition, it illustrates a yield problem, such as is often encountered in industrial practice.

Due to the very large excess of oxygen the rate equations will be considered to be of the pseudo-first-order type, so that, at atmospheric pressure,

$$r_A = (k_1 + k_3)y_{A0}y_0(1 - x_A)$$
$$r_B = y_{A0}y_0[k_1(1 - x_A) - k_2 x_B]$$
$$r_C = y_{A0}y_0[k_2 x_B + k_3(1 - x_A)]$$

where $y_0$ represents the mole fraction of oxygen, $x_A$ is the total conversion of *o*-xylene, and $x_B$ the conversion of *o*-xylene into phthalic anhydride. When $x_C$

represents the conversion into CO and $CO_2$ then $x_A = x_B + x_C$. The rate coefficients are given by the following expressions:

$$\ln k_1 = -\frac{27{,}000}{1.987(T' + T_0)} + 19.837$$

$$\ln k_2 = -\frac{31{,}400}{1.987(T' + T_0)} + 20.86$$

$$\ln k_3 = -\frac{28{,}600}{1.987(T' + T_0)} + 18.97$$

where $T_0$ is the inlet temperature to the reactor and $T' = T - T_0$. When this reduced temperature $T'$ and the following dimensionless variables are used,

$$z = \frac{z'}{d_p} \qquad r' = \frac{r}{d_p} \qquad R_t' = \frac{R_t}{d_p}$$

the steady-state continuity and energy equations may be written, in cylindrical coordinates and in terms of conversion,

$$\frac{\partial x_B}{\partial z} = a_1 \left( \frac{\partial^2 x_B}{\partial r'^2} + \frac{1}{r'} \frac{\partial x_B}{\partial r'} \right) + b_1 r_B$$

$$\frac{\partial x_C}{\partial z'} = a_1 \left( \frac{\partial^2 x_C}{\partial r'^2} + \frac{1}{r'} \frac{\partial x_C}{\partial r'} \right) + b_1 r_C$$

$$\frac{\partial T'}{\partial z'} = a_2 \left( \frac{\partial^2 T'}{\partial r'^2} + \frac{1}{r'} \frac{\partial T'}{\partial r'} \right) + b_2 r_B + b_3 r_C$$

The constants in these equations have the following meaning:

$$a_1 = \frac{D_{er}}{u_i d_p} = \frac{1}{Pe_{mr}} \qquad b_1 = \frac{\rho_B d_p M_m}{G N_{A0}}$$

$$a_2 = \frac{\lambda_{er}}{G c_p d_p} = \frac{1}{Pe_{hr}} \qquad b_2 = \frac{\rho_B d_p (-\Delta H_1)}{G c_p}$$

$$b_3 = \frac{\rho_B d_p (-\Delta H_3)}{G c_p}$$

The boundary conditions are those of the previous section and are the same for $x_C$ as for $x_B$, of course.

Bulk mean values are obtained from

$$\langle \zeta \rangle = 2 \int_0^1 \zeta \frac{r}{R_t} \, d\left( \frac{r}{R_t} \right)$$

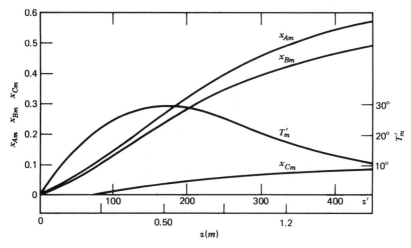

*Figure 11.7.c-1 Radial mean conversions and temperature profile in multitubular o-xylene oxidation reactor (from Froment [76]).*

The following typical data were used in the computations: $y_{A0} = 0.00924$, corresponding to 44 g/m$^3$; $(-\Delta H_1) = 307,000$ kcal/kmol $= 1.285 \times 10^6$ kJ/kmol, and $(-\Delta H_3) = 1,090,000$ kcal/kmol $= 4.564 \times 10^6$ kJ/kmol. All the other data were already given in Ex. 11.5.b. From Kunii and Smith's correlation [68] it follows that at Re $= 121$, $\lambda_{er} = 0.67$ kcal/m hr °C $= 0.78. 10^{-3}$ kJ/m s K and from Yagi and Kunii's equation [72] $\alpha_w = 134$ kcal/m$^2$ hr °C $= 0.156$ kJ/m$^2$ s K so that Pe$_{hr} = 5.25$, whereas Pe$_{mr} = 10$. In all cases the feed inlet temperature equaled that of the salt bath.

Figure 11.7.c-1 shows the results obtained for an inlet temperature of 357°C. The bulk mean conversion and temperature profile is shown. The conversion to phthalic anhydride tends to a maximum, as is typical for consecutive reaction systems, but which is not shown on the figure. Also typical for exothermic systems, as we have seen already, is the hot spot, where $T'_m$ equals about 30°C. Even for this case, which is not particularly drastic, and with a small tube diameter of only 2.54 cm, the radial temperature gradients are severe, as seen from Fig. 11.7.c-2. The temperature in the axis is well above the mean.

Notice from Fig. 11.7.c-1 that a length of 3 m is insufficient to reach the maximum in phthalic anhydride concentration. What happens when the inlet temperature is raised by only 3°C to overcome this is shown in Fig. 11.7.c-3. Again we have a case here of parametric sensitivity. Hot spots as experienced in such cases, even less dramatic than that experienced with $T_0 = 360$°C, may be detrimental for the catalyst. Even if it were not, important hot spots would be unacceptable for reasons of selectivity. Indeed, the kinetic equations are such that

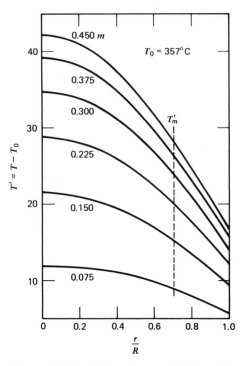

Figure 11.7.c-2 O-xylene oxidation. Radial temperature profiles at various bed depths (*from Froment* [76]).

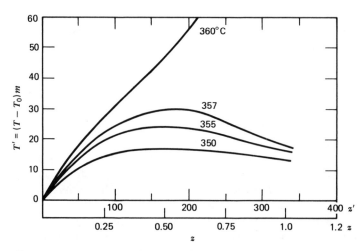

Figure 11.7.c-3 O-xylene oxidation parametric sensitivity. Influence of inlet temperature (*from Froment* [76]).

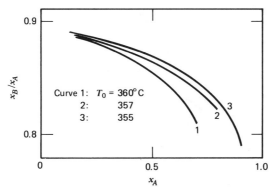

*Figure 11.7.c-4 O-xylene oxidation. Effect of hot spot on phthalic anhydride yield (from Froment [76]).*

the side reactions are favored by increasing the temperature. The effect of the hot spot on the yield is shown in Fig. 11.7.c-4 in which the yield is plotted as a function of total conversion for several inlet temperatures. A few percent more in yield, due to judicious design and operation, are important in high tonnage productions.

As illustrated in Sec. 10.1.a. the inlet temperature is not the only parameter determining the runaway temperature. The influence of the hydrocarbon inlet concentration is shown in Fig. 11.7.c-5 which summarizes Fig. 11.7.c-3 obtained with 44 g/m³ and two more diagrams like this, but with 38 and 32 g/m³. Fig. 11.7.c-5 shows how the runaway limit temperature rises with decreasing hydrocarbon inlet concentration, but it is important to note no noticeable gain in safety

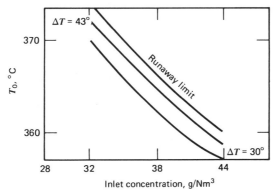

*Figure 11.7.c-5 O-xylene oxidation. Influence of hydrocarbon inlet concentration on critical inlet temperature (from Froment [76]).*

FIXED BED CATALYTIC REACTORS _____ 543

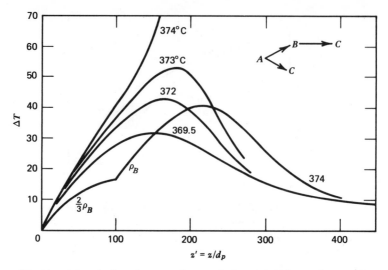

*Figure 11.7.c-6 O-xylene oxidation. Effect of diluting the catalyst bed with inert material.*

margin is obtained by lowering the inlet concentration. Moreover, such a measure would decrease the production capacity and unfavorably influence the economics of the plant. Yet, as designed, the risk of operating the reactor is too large; a safety margin of 3°C is unthinkable. With the given length of 3 m there seems to be only one way out, that is to realize an entirely different type of temperature profile, showing no pronounced hot spot, but leading all together to a higher average temperature. An appropriate dilution of the catalyst with inert packing in the front section of the bed would enable this. This is shown in Fig. 11.7.c-6. The dilution of the catalyst in an optimal way has been discussed by Calderbank et al. [131], by Adler et al. [79], and by Narsimhan [146].

Finally the question rises how well the results predicted by the one-dimensional model of Sec. 11.5.a. correspond with those of the model discussed here. For such a comparison to be valid and reflect only the effect of the model itself the heat transfer coefficient $\alpha_i$ of the one-dimensional model has to be based on $\lambda_{er}$ and $\alpha_w$ according to

$$\frac{1}{\alpha_i} = \frac{1}{\alpha_w} + \frac{R_t}{4\lambda_{er}}$$

as derived by Froment [77, 78]. Slightly different but analogous relations are given by Crider and Foss [82], Marek et al. [83], and Hlavacek [84]. Figures 11.7.c-7 and 11.7.c-3 can be used to compare the predictions based on the two

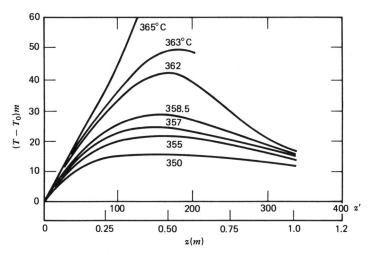

*Figure 11.7.c-7 O-xylene oxidation. Predictions of one-dimensional model for influence of inlet temperature (from Froment [76]).*

models. The two-dimensional model predicts runaway at an inlet temperature of less than 360°C, the one-dimensional at 365°C. The discrepancy between the predictions of both models grows as the conditions become more drastic.

It follows from Froment [76] and from calculations by White and Carberry [80] that the computed results are not very sensitive with respect to $Pe_{mr}$, but very sensitive with respect to $\lambda_{er}$ and $\alpha_w$. Beek [81] and Kjaer [120] have also discussed features of this model.

The present model could be refined by introducing a velocity profile. This was done by Valstar [128], who used the velocity profiles published by Schwartz and Smith [85] that exhibit a maximum at 1.5 $d_p$ of the wall, and also by Lerou and Froment [144]. The latter authors concluded from a simulation of experimental radial temperature profiles that a radial velocity profile inversely proportional to the radial porosity profile led to the best fit. Such a radial velocity profile exhibits more than one peak. It follows from these studies that the influence of radial nonuniformities in the velocity profile are worthwhile accounting for in the simulation of severe operating conditions. Progress in this field will require more extensive basic knowledge of the packing pattern and hydrodynamics of fixed beds.

This discussion of the tubular reactor with radial mixing has been based on a continuum model leading to a system of differential equations with mixing effects expressed in terms of effective diffusion or conduction. There exists a different approach that considers the bed to consist of a two-dimensional network of perfectly mixed cells with two outlets to the subsequent row of cells. Alternate rows are offset half a stage to allow for radial mixing. In the steady state a pair of algebraic equations must be solved for each cell. This model was proposed by Deans and Lapidus [86] and applied by McGuire and Lapidus [87] to non-steady-state cases. Agnew and Potter [88] used it to set up runaway diagrams of the Barkelew type. In fact, the model is not completely analogous to the one discussed above, since it considers heat to be transferred only through the fluid. It is clear already from the correlations for $\lambda_{er}$ given above that this is a serious simplification, as will be illustrated in Sec. 11.10. More elaborate cell models, with a coupling between the particles to account for conduction or radiation, are possible [89, 93], but the computational problems then become overwhelming. The effective transport concept keeps the problem within tractable limits.

To conclude this section, it is believed that the possibilities of present-day computers are such that there is no longer any reason for not using two-dimensional models for steady-state calculations, provided the available reaction rate data are sufficiently accurate. The one-dimensional model of Sec. 11.5.a will continue to be used for on-line computing and process control studies.

## Part Three

### Heterogeneous Models

For very rapid reactions with an important heat effect it may be necessary to distinguish between conditions in the fluid and on the catalyst surface Sec. (11.8) or even inside the catalyst Sec. (11.9). As in Part II the reactor models may be of the one- or two-dimensional type.

## 11.8 One-Dimensional Model Accounting for Interfacial Gradients

### 11.8.a Model Equations

The steady-state equations are, for a single reaction carried out in a cylindrical tube and with the restrictions already mentioned in Sec. 11.5.a for the basic case:

Fluid

$$-u_s \frac{dC}{dz} = k_g a_v (C - C_s^s) \qquad (11.8.a-1)$$

$$u_s \rho_g c_p \frac{dT}{dz} = h_f a_v (T_s^s - T) - 4 \frac{U}{d_t} (T - T_r) \qquad (11.8.a-2)$$

Solid

$$\rho_B r_A = k_g a_v (C - C_s^s) \tag{11.8.a-3}$$

$$(-\Delta H)\rho_B r_A = h_f a_v (T_s^s - T) \tag{11.8.a-4}$$

With boundary conditions:

$$C = C_0 \quad \text{at} \quad z = 0$$

$$T = T_0$$

In this set of equations and in those to follow in this chapter $C$ stands for the concentration of a reactant, $A$. Figure 3.2.a-1 and 3.2.b-1 of Chapter 3 show most of the correlations available to date for $k_g$ and $h_f$ [9]. Except perhaps for the most stringent conditions these parameters are now defined with sufficient precision. Also, for the special case of very fine ($< 100 \ \mu m$) particles, the possible agglomerating tendencies cannot yet be completely defined.

The distinction between conditions in the fluid and on the solid leads to an essential difference with respect to the basic one-dimensional model, that is, the problem of stability, which is associated with multiple steady states. This aspect was studied first independently by Wicke [90] and by Liu and Amundson [89, 91]. They compared the heat produced in the catalyst, which is a sigmoidal curve when plotted as a function of the particle temperature, with the heat removed by the fluid through the film surrounding the particle, which leads to a straight line. The steady state for the particle is given by the intersection of both lines. It turns out that for a certain range of gas—and particle temperatures—three intersections, therefore three steady states are possible.

From a comparison of the slopes of the sigmoid curve and the straight line in these three points it follows that the middle steady state is unstable to any perturbation, but not necessarily to large ones. It follows that when multiple steady states are possible, the steady state the particle actually operates in also depends on its initial temperature. When this is now extended from a particle to an adiabatic reactor it follows that the concentration and temperature profiles are not determined solely by the feed conditions but also by the initial solid temperature profile. If this is not equal to the fluid feed temperature, transients are involved. The design calculations would then have to be based on the system (Eqs. 11.8.a-1 to 11.8.a-4) completed with non-steady-state terms. Figures 11.8.a-1 and 11.8.a-2 illustrate this for an adiabatic reactor [91]. Figure 11.8.a-1 shows a situation with a unique steady state profile. In Fig. 11.8.a-2 the gas is first heated up along the lower steady state and then jumps to the upper steady state as soon as the gas temperature exceeds 480°C. The higher the initial temperature profile the earlier the profile jumps from the lower to the higher steady state. From a comparison with the unique steady-state case of Fig. 11.8.a-1 it follows that the shift from one steady state to another leads to temperature profiles that are much steeper. The reactor

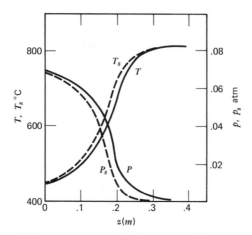

Figure 11.8.a-1 One-dimensional heterogeneous model with interfacial gradients.
Unique steady-state case; $p_0 = 0.007$ atm,
$T_0 = 449°C$ (after Liu and Amundson [91],
from Froment [9]).

of Fig. 11.8.a-2 may be unstable while the reactor of Fig. 11.8.a-1 is stable, which
does not exclude parametric sensitivity and runaway, as discussed under Sec.
11.5.c, however.

Are these multiple steady states possible in practical situations? From an
inspection of Figs. 11.8.a-1 and 11.8.a-2 it is clear that the conditions chosen for
the reaction are rather drastic. It would be interesting to determine the limits on the
operating conditions and reaction parameters within which multiple steady states

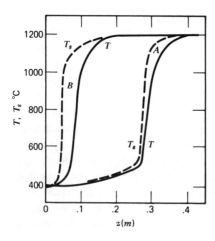

Figure 11.8.a-2 One-dimensional hetero-
geneous model with interfacial gradients.
Nonunique steady-state case; $p_0 = 0.15$ atm,
$T_0 = 393°C$. Initial $T_s$: $A \leq 393°C$, $B =
560°C$ (after Liu and Amundson [91], from
Froment [9]).

CHEMICAL REACTOR DESIGN

could be experienced. These limits will probably be extremely narrow, so that the phenomena discussed here would be limited to very special reactions or to very localized situations in a reactor, which would probably have little effect on its overall behavior. Indeed, in industrial fixed bed reactors the flow velocity is generally so high that the temperature and concentration drop over the film surrounding the film is small, at least in the steady state.

A criterion for detecting the onset of interphase temperature gradients has been proposed by Mears. If the observed rate is to deviate less than 5 percent from the true chemical rate the criterion requires:

$$\frac{(-\Delta H)r_A \rho_B d_p}{2h_f T} < 0.15 \frac{RT}{E}$$

Baddour et al. [26] in their simulation of the TVA ammonia-synthesis converter, already discussed in Sec. 11.5.e, found that in steady-state operation the temperature difference between the gas and the solid at the top, where the rate of reaction is a maximum, amounts to only 2.3°C and decreases as the gas proceeds down the reactor to a value of 0.4°C at the outlet. In the methanol reactor simulated in Sec. 11.9.b the difference between gas and solid temperature is of the order of 1°C. This may not be so with highly exothermic and fast reactions involving a component of the catalyst as encountered in the reoxidation of Fe and Ni catalysts used in ammonia synthesis and steam reforming plants or involving material deposited on the catalyst, coke for example.

Notice that the model discussed here does not provide any axial coupling between the particles. Consequently, heat is transferred in axial direction only through the fluid. Recently, Eigenberger added heat transfer through the solid to the model and this was found to significantly modify the behavior [92]. He also showed the influence of the boundary conditions to be quite pronounced.

## 11.8.b Simulation of the Transient Behavior of a Reactor

The system of equations for the transient state is easily derived from the system Eq. 11.8.b-1 to 4. The following equations are found for a single reaction with constant density:

Fluid:

$$\frac{u_s \rho_g}{M_m p_t} \frac{\partial p}{\partial z} + \frac{\varepsilon \rho_g}{M_m p_t} \frac{\partial p}{\partial t} = k_g a_v (p_s^{\ s} - p)$$

$$u_s \rho_g c_p \frac{\partial T}{\partial z} + \varepsilon \rho_g c_p \frac{\partial T}{\partial t} = h_f a_v (T_s^{\ s} - T) - 4 \frac{U}{d_t} (T - T_r)$$

Solid:

$$\rho_B r_A = k_g a_v(p - p_s^s) - \frac{C_t}{p_t} \varepsilon_s(1 - \varepsilon) \frac{\partial p_s^s}{\partial t}$$

$$(-\Delta H)\rho_B r_A = h_f a_v(T_s^s - T) + c_{ps}\rho_B \frac{\partial T_s^s}{\partial t}$$

B.C: $t = 0$  $\quad z \geq 0$  $\quad T = T_s = T_r$
$\quad\quad\quad\quad\quad\quad\quad\quad\quad\quad p = p_s = 0$
$\quad\quad\quad t > 0$  $\quad z = 0$  $\quad T = T_0$
$\quad\quad\quad\quad\quad\quad\quad\quad\quad\quad p = p_0$

The example considered here is again the hydrocarbon oxidation process with its simplified kinetic scheme used in Sec. 11.5.b. Suppose the reactor is at a temperature of 362°C and let the gas entering the bed be 362°C. How long will it take to

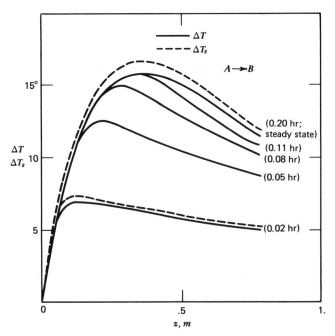

Figure 11.8.b-1 One-dimensional heterogeneous model with interfacial gradients. Start up of reactor, transient temperature profiles. $\Delta T$ = temperature increase of gas phase above feed value; $\Delta T_s$ = increase of solid temperature above initial value.

reach the steady state and what will the difference between gas and solid temperature be? The integration is performed numerically along the characteristics.

The results are shown in Fig. 11.8.b-1. We see how the fluid phase temperature approaches the steady state quite closely within 1 to 2°C already after 0.11 hr. The steady-state profiles are attained, within the accuracy of the computations, after 0.20 hr. The difference in temperature between the gas and solid is really very small and of the order of 1°C. Yet, this is a very exothermic reaction and the operating conditions used in these calculations are realistic.

## Example 11.8.b-1  A Gas-Solid Reaction in a Fixed Bed Reactor

In Chapter 4 some gas solid processes were mentioned and rate equations were derived that permit a quantitative description of the progression with time of the reaction inside the solid. When the solid particles are packed and form a fixed bed reaction, the approach discussed in the present section can be followed to model this reactor. Obviously, the model has to distinguish between the fluid and solid phase—it is "heterogeneous." Furthermore, non-steady-state equations will have to be set up to account for the inherently transient character of the operation, not only in the solid but also in the fluid phase. Indeed, since the fluid phase is depleted in reactant the reaction is confined to a zone that gradually moves through the reactor as the solid reactant is converted. The example that will be worked out in what follows concerns the reaction between an oxygen-containing gas phase on one hand and hydrogen and nickel contained in a steam-reforming catalyst on the other hand. The rate equation used does not explicitly consider the presence of intraparticle gradients. This is the reason why the example is dealt with under this section.

A secondary reformer is an adiabatic reactor which is a part of an ammonia-synthesis gas production line. In this reactor the mixture of $CH_4$, $CO$, $CO_2$, $H_2$, and steam coming from the primary reformer is brought into contact with air to oxidize the remaining $CH_4$ (and also some hydrogen) and to add the required amount of nitrogen for the synthesis of ammonia. The secondary reformer is packed with a NiO on $Al_2O_3$ catalyst, operating in the reduced state. When the reactor has to be opened for inspection or repair, the catalyst, which is very pyrophoric, has to be reoxidized. This has to be done in a controlled way, to avoid an excessive temperature rise. First, the reactor is cooled by means of a flow of steam. At about 250°C the steam is switched off to avoid any condensation which would damage the catalyst and further cooling is achieved by means of a flow of nitrogen containing only a very small fraction of oxygen. The oxygen content is gradually increased as the temperature decreases.

In this example, the temperature rise in an adiabatic bed during reoxidation is investigated by means of simulation on a digital computer. This requires the combination of a rate equation for the gas-solid reaction and of the model equations for the reactor. The first reaction to consider in the present case is the reaction between the oxygen of the gas phase and the hydrogen adsorbed on the catalyst. It is this hydrogen that is mainly responsible for the pyrophoric character of such a catalyst. The reaction between the oxygen and the adsorbed hydrogen will be considered to be infinitely fast. It causes a temperature rise that may initiate the oxidation of the nickel of the catalyst.

Viville studied the rate of the latter process on an electrobalance [132]. Hatcher, Viville, and Froment [147] described its rate by a homogeneous gas-solid model with a Hougen–Watson expression for the reaction rate:

$$r_{Ni} = -\frac{dC_{Ni}}{dt} = \frac{kK_A p C_{Ni}^2}{1 + K_A p} \quad \left(\frac{\text{mol Ni}}{\text{cm}^3 \text{cat. s}}\right) \tag{a}$$

where $C_{Ni}$ is the nickel concentration of the catalyst (mol Ni/cm$^3$ cat.) and $p$ the oxygen partial pressure. The temperature dependencies of the reaction rate coefficient, $k$, and of the oxygen adsorption coefficient, $K_A$, are given by the following Arrhenius relations:

$$k = 21.7e^{-3520/T}$$
$$K_A = 0.305e^{3070/T}$$

From the stoichiometry $r_{O_2} = \frac{1}{2}r_{Ni}$.

The reactor model chosen is one dimensional and heterogeneous with interfacial gradients. Intraparticle gradients are accounted for implicitly through the fit of the experimental data by means of Eq. 1. The operation is adiabatic. Furthermore, as mentioned already, the operation is of a non-steady-state nature, since the oxidation takes place in a zone that gradually moves through the reactor.

Continuity equation for oxygen in the gas phase:

$$\frac{G}{M_m p_t a_v k_g}\left(\frac{\partial p}{\partial z} + \frac{\varepsilon \rho_g}{G}\frac{\partial p}{\partial t}\right) = p_s - p \tag{b}$$

Energy equation for the gas phase:

$$\frac{Gc_p}{a_v h_f}\left(\frac{\partial T}{\partial z} + \frac{\varepsilon \rho_g}{G}\frac{\partial T}{\partial t}\right) = T_s - T \tag{c}$$

Continuity equations for the solid phase:

$$6k_g(p - p_s) = d_p \rho_s r_{O_2} \tag{d}$$

also

$$6k_g(p - p_s) = -\frac{1}{2}d_p \rho_s \frac{\partial C_{Ni}}{\partial t} \tag{e}$$

Energy equation for the solid phase:

$$h_f(T_s - T) + \frac{1}{6}d_p c_{ps} \rho_s \frac{\partial T_s}{\partial t} = \frac{d_p \rho_s(-\Delta H)_1 r_{O_2}}{6} \tag{f}$$

With initial conditions:

$$\begin{aligned} \text{at } t = 0 \quad & z \geq 0 \quad && T_s = T = T_0 \\ & && p_s = p = 0 \\ & && C_{Ni} = C_{Nio} \\ \text{at } t \geq 0 \quad & z = 0 \quad && p = p_0 \text{ and} \\ & && T = T_0 \end{aligned}$$

The oxygen first reacts with the hydrogen. When the latter is completely oxidized the reaction with nickel starts, provided the temperature has risen sufficiently. For the stage during which oxygen reacts with the adsorbed hydrogen, $p_s$ is set equal to zero in Eqs. (b) and (d), since this reaction is considered to be infinitely fast. In that case the right-hand side of Eq. (c) is nothing but $k_g p(-\Delta H)_2$ as can be seen from Eq. (d), whereas Eq. (e) is replaced by

$$6k_g p = -\frac{1}{2}d_p \rho_s \frac{\partial C_{H_2}}{\partial t} \tag{7}$$

The boundary conditions are identical, except that hydrogen replaces nickel.
The following conditions are typical for industrial operation:

$G = 2786.9$ kg/m² hr     Vol % oxygen at the inlet : 3%

$M_m = 28.2$ kg/kmol     $C_p = 0.25$ kcal/kg °C $= 1.05$ kJ/kg K

$p_t = 1.5$ atm $= 1.52$ bar     $T_0 = 250°C$     $d_p = 0.014$ m

$\rho_B = 950$ kg/m³     $\rho_s = 1515$ kg/m³ cat.     $a_v = 352.5$ m²/m³

$C_{ps} = 0.24$ kcal/kg °C $= 1.01$ kJ/kg K

$C_{Nio} = 0.00624$ kmol/kg cat.     $C_{H_2} = 0.00093$ kmol/kg cat.

From the $j_{D_2}$ and $j_H$ correlations, $k_g$ is calculated to be 13.7 kmol/m² hr atm $= 0.00381$ kmol/m² s bar and $h_f$: 83.75 kcal/m² hr °C $= 0.0974$ kJ/m² s bar

Figure 1 shows the hydrogen and nickel concentration profiles through the reactor as a function of time. Notice the steep hydrogen profile, caused by the infinitely fast reaction between hydrogen and oxygen, which raises the tempera-

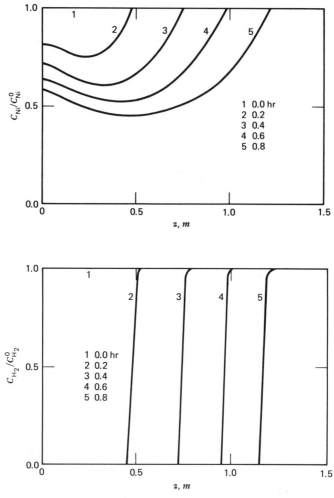

*Figure 1 Reoxidation of secondary steam reforming catalyst bed. Profiles of (a) nickel and (b) hydrogen concentrations as a function of time.*

ture to such values where the nickel also reacts rapidly. The temperature peaks travel through the bed with time as shown in Fig. 2. The peak widens as time progresses. The temperature increase is very important with 3 vol % of oxygen in the nitrogen. Notice also in Fig. 1 that for this catalyst the nickel is not completely reoxidized, due to the very low rate of reoxidation beyond the initial rapid reaction mentioned above. This degree of reoxidation suffices to handle the catalyst

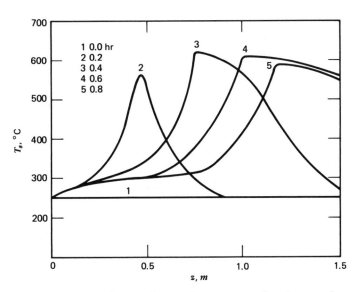

*Figure 2 Reoxidation of secondary steam reforming catalyst bed. Temperature peak as a function of time.*

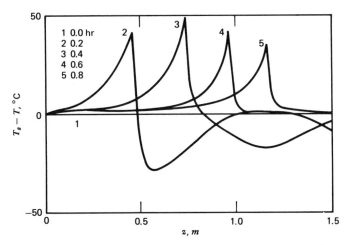

*Figure 3 Reoxidation of secondary steam reforming catalyst bed. Differences in solid and gas temperatures.*

555

without any further risk, however. From Fig. 3 it follows that the temperature difference between the solid and the gas attains about 50 °C. This justifies the use of a heterogeneous reaction model.

## 11.9 One-Dimensional Model Accounting for Interfacial and Intraparticle Gradients

### 11.9.a Model Equations

When the resistance to mass and heat transfer inside the catalyst particle is important the rate of reaction is not uniform throughout the particle. The set of Eqs. 11.8.a-1 to 11.8.a-4 of Sec. 11.8.a then no longer adequately describes the system. It has to be completed with equations describing the concentration and temperature gradients inside the particle, so that the complete set may be written:

Fluid:

$$-u_s \frac{dC}{dz} = k_g a_v (C - C_s^s) \tag{11.9.a-1}$$

$$u_s \rho_g c_p \frac{dT}{dz} = h_f a_v (T_s^s - T) - \frac{4U}{d_t}(T - T_r) \tag{11.9.a-2}$$

Solid:

$$\frac{D_e}{\xi'^2} \frac{d}{d\xi'} \left( \xi'^2 \frac{dC_s}{d\xi} \right) - \rho_s r_A(C_s, T_s) = 0 \tag{11.9.a-3}$$

$$\frac{\lambda_e}{\xi'^2} \frac{d}{d\xi'} \left( \xi'^2 \frac{dT_s}{d\xi'} \right) + \rho_s(-\Delta H) r_A(C_s, T_s) = 0^1 \tag{11.9.a-4}$$

with boundary conditions

$$C = C_0 \qquad \text{at } z = 0 \tag{11.9.a-5}$$
$$T = T_0$$

$$\frac{dC_s}{d\xi'} = \frac{dT_s}{d\xi'} = 0 \qquad \text{at } \xi' = 0 \tag{11.9.a-6}$$

$$k_g(C_s^s - C) = -D_e \frac{dC_s}{d\xi'} \qquad \text{at } \xi' = \frac{d_p}{2} \tag{11.9.a-7}$$

$$h_f(T_s^s - T) = -\lambda_e \frac{dT_s}{d\xi'} \qquad \text{at } \xi' = \frac{d_p}{2} \tag{11.9.a-8}$$

---

[1] The signs in Eqs. 11.9.a-3 and 11.9.a-4 are those obtained when the rate is defined for first order as $r_A = kC_A$. Many books define $r_A = -kC_A$. See Chapter 1 for use of the stoichiometric coefficients.

Numerical values for $D_e$ and $\lambda_e$ are given in Satterfield's book [133] and in Weisz and Hicks' classical paper on this subject [98], as discussed in Chapter 3.

The system of second-order, nonlinear differential equations (Eqs. 11.9.a-1 to 8) has to be integrated in each node of the computational grid used in the integration of the fluid field equations, Eqs. 11.9.a-1 and 11.9.a-2. This is feasible on present-day computers, but still extremely lengthy. Analytical solution is only possible for isothermal situations inside the particle and first-order irreversible reactions. Fortunately, even with strongly exothermic reactions the particle is practically isothermal: the main resistance inside the pellet is to mass transfer and the main resistance in the film surrounding the particle to heat transfer [98, 99]. When gradients occur inside the catalyst particle use is often made of the effectiveness factor, $\eta$ introduced in Chapter 3. In its classical sense $\eta$ is a factor that multiplies the reaction rate at the particle surface conditions to yield the rate that is actually experienced when the conditions inside the particle are different. The use of the effectiveness factor reduces the system (Eqs. 11.9.a-1 to 11.9.a-4) to (Eqs. 11.9.a-1 to 11.9.a-2) with boundary conditions (Eq. 11.9.a-5) and the following algebraic equations:

$$k_g a_v (C - C_s^s) = \eta \rho_B r_A(C_s^s, T_s^s) \qquad (11.9.a-9)$$

$$h_f a_v (T_s^s - T) = \eta \rho_B(-\Delta H) r_A(C_s^s, T_s^s) \qquad (11.9.a-10)$$

with

$$\eta = f(C_s^s, T_s^s)$$

Note that Eqs. 11.9.a-9 and 11.9.a-10 differ only from Eqs. 11.9.a-3 and 11.9.a-4 by the factor $\eta$.

The effectiveness factor depends on the local conditions, which are introduced through the modulus, $\phi$, and therefore has to be computed in each node of the grid. With a first-order reaction and an isothermal spherical particle, the following relation (Eq. 11.9.a-11) between $\eta$ and the determining variables may be obtained by analytical integration of Eqs. 11.9.a-3 and 11.9.a-4:

$$\eta = (3/\phi^2)(\phi \coth \phi - 1) \qquad (11.9.a-11)$$

where $\phi$, the modulus, which contains these variables, is related to the ratio of the reaction rate at surface conditions to the mass transfer rate toward the inside

$$\phi = \frac{d_p}{2} \sqrt{\frac{k(T_s^s)}{D_e}}$$

When an analytical solution is *not* available for $\eta$, there is no gain in the use of $\eta$ from the reactor design and computational viewpoint. The only advantage of $\eta$ then is its possibility of characterizing the situation inside the particle by means of a single number. For reactions with an order different from 1 but isothermal conditions advantage can be taken of approximate expressions for $\eta$ (e.g., through the use of the generalized modulus (Bischoff [96]), which avoids the numerical integration of Eq. (11.9.a-3) and only requires the evaluation of an integral in each point of the grid.

Until recently $\eta$ has always been expressed as a function of the surface conditions, $C_s^s$ and $T_s^s$. It is also possible to refer to bulk fluid conditions in case these are different from those prevailing at the particle surface. With this version of the effectiveness factor, represented by $\eta_G$, the rate $r_A$ has to be expressed as a function of fluid-conditions $(C, T)$ and $\eta_G$ enters into Eq. 11.9.a-9 and Eq. 11.9.a-10. The equivalent of Eq. 11.9.a-11 then is for isothermal conditions in both the film and the particle [104]:

$$\eta_G = \frac{3Sh}{\phi^2} \left[ \frac{\phi \cosh \phi - \sinh \phi}{\phi \cosh \phi + (Sh - 1)\sinh \phi} \right] \qquad (11.9.a-12)$$

where

$$Sh = \frac{k_g \delta}{D_e}$$

and

$$\phi = \frac{d_p}{2} \sqrt{\frac{k(T)}{D_e}}$$

($\delta$ represents the film thickness). The use of $\eta_G$ reduces the system, Eqs. 11.9.a-1 to 11.9.a-4, to

$$-u_s \frac{dC}{dz} = \eta_G \rho_B r_A(C, T)$$

$$u_s \rho_g c_p \frac{dT}{dz} = \eta_G \rho_B r_A(C, T) - 4\frac{U}{d_t}(T - T_r)$$

with

$$\eta_G = f(C, T)$$

Figure 11.9.a-1 shows the relation between the effectiveness factors $\eta$ and $\eta_G$ and the modulus $\phi$ [104, 108]. This relation can only be obtained by numerical integration of the system, Eqs. 11.9.a-1 to 11.9.a-8, except for the cases already mentioned. With isothermal situations $\eta$ tends to a limit of 1 as $\phi$ increases, with nonisothermal situations, however, $\eta$ or $\eta_G$ may exceed 1. Curve 1 corresponds to the $\eta$ concept, curves 2, 3, and 4 to $\eta_G$. The dotted part of curve 4 corresponds to a region of conditions within which multiple steady states inside the catalyst are

CHEMICAL REACTOR DESIGN

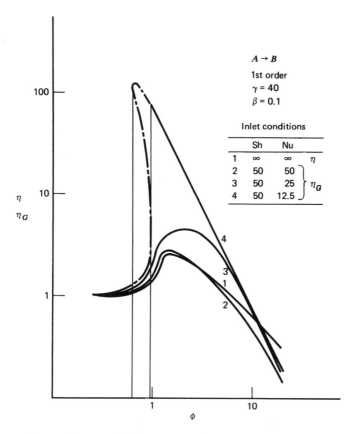

*Figure 11.9.a-1 Effectiveness factor diagram for non-isothermal situations (after McGreavy and Cresswell [108], from Froment [9]).*

possible. But it should be stressed that the range of parameters leading to this situation is very narrow. Again, this may be extended to the reactor, which will then have nonunique profiles. The steady state actually experienced may depend on the initial conditions, so that transients have to be computed. To avoid these when they are unnecessary, considerable effort has gone recently into defining criteria for the uniqueness of the steady state of a particle. We illustrate these for a first-order irreversible reaction treated numerically by Drott and Aris [103] with $\beta = 0.1$ and $\gamma = 50$. It should be noted that $\beta = 0.1$ is an extremely high value. Normal values for $\gamma$ are 20 to 30. According to Luss and Amundson's criterion for uniqueness $\beta\gamma \leq 1$ or Luss' less conservative criterion $\beta\gamma \leq 4(1 + \beta)$ [58, 59] multiplicity is possible, since $\beta\gamma = 5$. Multiplicity may still be avoided by the appropriate choice of the modulus, that is, the size of the particle. According to Weisz and Hicks $\phi$ should be smaller than 0.1 [98], according to Luss and Amundson smaller than 0.4218. But from Fig. 11.9.a-1 it follows that multiplicity can only occur for a certain interval of $\phi$ values. Luss gives approximate values for the bounds on $\phi$ [101]. In this example 0.4191 and 0.4655 are values for the lower and 0.4965 for the upper bound. The values computed by Drott and Aris are 0.4707 and 0.4899, respectively. Once more this confirms multiplicity is possible only for extremely exothermic reactions and over an extremely narrow region of $\phi$.

McGreavy and Thornton [134] included the film surrounding the particle into their analysis of the multiplicity of solutions. With gradients over the film, multiplicity is possible even when the particle is isothermal, of course. As mentioned in Chapter 3, Luss [135] came to the following criterion for uniqueness for a first-order reaction:

$$\beta\gamma < 4\left(\frac{Nu'}{Sh'} + \beta\right)$$

while Kehoe and Butt obtained, for an $n$th-order reaction,

$$\beta\gamma < 4(n + 1)\left(\frac{Nu'}{Sh'} + \beta\right)$$

Even with an isothermal particle and without including the film, multiplicity is theoretically possible when the rate increases with conversion as may occur for some type of Langmuir–Hinshelwood rate equations and some particular parameter values. According to Luss the sufficient condition for uniqueness of the concentration profile in the particle is:

$$(C_s - C_s^s)\frac{d\ln f(C_s)}{dC_s} \le 1$$

where $f(C_s)$ is the rate equation. Luss and Lee [102] developed a method for obtaining stability regions for the various steady states, based on the knowledge of the steady-state profiles. It thus becomes possible to predict toward which steady state a particle will tend starting from given initial conditions.

This whole field of uniqueness and stability has been reviewed recently by Aris [100]. As previously mentioned the possibility of multiple steady states seriously complicates the design of the reactor. Indeed, transient computations have to be performed in order to make sure that the correct steady-state profile throughout the reactor is predicted. Another way would be to check the possibility of multiple steady states on the effectiveness factor chart for every point in the reactor. This would, in principle, require an infinite set of such charts, because $\beta$ and $\gamma$ vary throughout the reactor.

McGreavy and Thornton [134] reformulated the problem in order to enable a single graph to be used for the whole reactor and to reduce the effectiveness factor curve to a single point in the new chart. For this purpose they introduced a new parameter

$$\theta = \frac{d_p}{2} \sqrt{\frac{A_0}{D_e}}$$

replacing the Thiele modulus and based on the frequency factor $A_0$ rather than the rate coefficient itself. Another convenient group is

$$\frac{\beta}{\gamma Nu'} = \frac{(-\Delta H)CD_e R}{d_p h_f E}$$

For a given system this group depends only on the reactant concentration in the fluid. It is, therefore, an implicit function of axial position. Taking advantage of the fact that the particle is generally isothermal, a relatively simple formula for the bounds on the *fluid* temperature within which multiple steady states may occur can be derived. It is represented graphically in Fig. 11.9.a-2. With a reactant concentration corresponding to a $\beta/\gamma$ Nu value of $8.10^{-5}$, for example, multiple steady states are possible when the gas temperature is between 265°C and 320°C. The diagram of Fig. 11.9.a-2 is essentially a $C$-$T$ phase plane and allows a trajectory through the reactor to be plotted. When a trajectory intersects the nonunique region, multiple profiles are possible. The figure shows a trajectory for adiabatic conditions that just intersects the zone. In the corresponding reactor zone the temperature may jump to the higher steady state but as the region is very narrow it is possible that instabilities will be damped. As soon as the conditions are no longer adiabatic only unique profiles are possible in the example considered. Since McGreavy and Thornton really used a two-dimensional model for the reactor in the nonadiabatic case the figure shows longitudinal profiles in the axis and at the wall. The parameter values used appear to be realistic, although rather drastic.

McGreavy and Adderley [105, 136] and in a more detailed way Rajadhyasha et al. [143] studied the parametric sensitivity and instability of reactors with concentration gradients inside the particle and interfacial temperature gradients by extending the treatment of Van Welsenaere and Froment, outlined and applied in Sec. 11.5.c. A runaway line is easily derived for such a model and represented in the $p$-$T$ phase plane. For a given coolant temperature McGreavy and Adderley obtained the critical inlet partial pressure by adiabatic extrapolation toward the ordinate from the intersection of the maxima curve and the runaway line onward. This prediction is rather conservative, however. McGreavy and Adderley also present an estimate of the critical inlet partial pressure that would lead to runaway

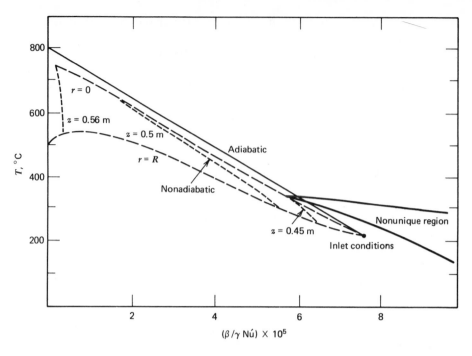

*Figure 11.9.a-2 Tubular reactor with interfacial and intraparticle gradients. C-T phase plane and region of multiple steady states (after McGreavy and Thornton [134], from Froment [9]).*

caused, not by parametric sensitivity but by the occurrence of multiple steady states. This involves an adiabatic extrapolation from the intersection of the runaway line and the lower bound of the nonunique region, shown in Fig. 11.9.a-2, onward. Rajadhyasha et al. derived upper and lower limits for the critical inlet partial pressure. The average of these values was an excellent approximation for the numerical solution, as experienced by Van Welsenaere and Froment with the pseudo-homogeneous model.

### Example 11.9.a-1 Simulation of a Fauser–Montecatini Reactor for High-Pressure Methanol Synthesis

This example is taken from Cappelli, Collina, and Dente [95] but reworked to follow the general approach followed in this book.

Methanol is synthesized from CO and $H_2$ in a reactor represented schematically in Fig.1, having an internal diameter of 580 mm and consisting of four adiabatic

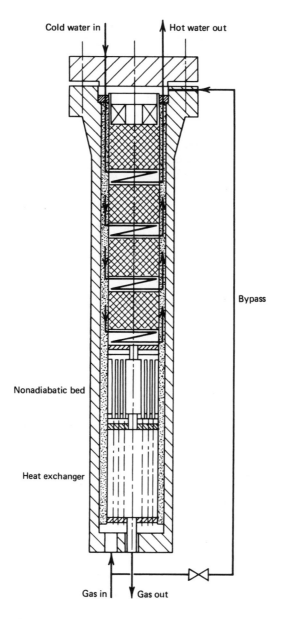

*Figure 1 Fauser–Montecatini methanol synthesis reactor (from Cappelli, et al.* [95]).

563

beds, a nonadiabatic bed, and an internal heat exchanger. Part of the feed bypasses the heat exchanger and non adiabatic layer. Typical operating conditions are given in Table 1. The catalyst characteristics are as follows:

**Composition**

| | |
|---|---|
| ZnO | 75.3 wt % |
| $Cr_2O_3$ | 11.6 wt % |
| True density | 5.36 g/cm³ |
| Particle density | 2.35 g/cm³ |
| Specific surface | 70 m²/g |
| Porosity | |
| <75 Å | 0.0602 cm³/g |
| 75–75,000 Å | 0.1837 cm³/g |
| >75,000 Å | 0.0014 cm³/g |
| Total porosity | 0.2453 cm³/g |

The main reaction can be written

$$CO + 2H_2 \rightleftharpoons CH_3OH$$

Since the synthesis gas contains about 2 percent of $CO_2$ the following side reaction has to be accounted for:

$$CO_2 + H_2 \rightleftharpoons CO + H_2O$$

The mathematical model used by Capelli et al. to simulate the reactor may be classified as a one-dimensional heterogeneous model considering external and internal gradients, but not axial diffusion and conduction. The steady-state model equations are, in terms of the partial pressures

*Gas Phase*

*Continuity equation for methanol:*

$$-\frac{1}{\Omega}\frac{dF_M}{dz} = k_M a_v [p_M - (p_M^s)_s] \tag{a}$$

but

$$F_M = F_t \frac{p_M}{p_t}$$

so that

$$dF_M = \frac{1}{p_t}(p_M\,dF_t + F_t\,dp_M)$$

Eq. (a) therefore becomes:

$$-\frac{p_M}{\Omega p_t}\frac{dF_t}{dz} - \frac{F_t}{\Omega p_t}\frac{dp_M}{dz} = k_M a_v[p_M - (p_M{}^s)_s] \qquad \text{(b)}$$

$F_t$ changes with position according to the stoichiometry of the main reaction, so that

$$\frac{dF_t}{dz} = 2k_M a_v[p_M - (p_M{}^s)_s]$$

A second continuity equation is required to account for the side reaction, for example, for $CO_2$

$$-\frac{p_{CO_2}}{\Omega p_t}\frac{dF_t}{dz} - \frac{F_t}{\Omega p_t}\frac{dp_{CO_2}}{dz} = k_{CO_2} a_v[p_{CO_2} - (p_{CO_2}{}^s)_s] \qquad \text{(c)}$$

*Energy equation*

$$-\frac{F_t c_p}{\Omega}\frac{dT}{dz} - \frac{c_p T}{\Omega}\frac{dF_t}{dz} = k_M a_v[p_M - (p_M{}^s)_s](-\Delta H)_1 + k_{CO_2} a_v[p_{CO_2} - (p_{CO}{}^s)_s]$$

$$\times (-\Delta H)_2 - 4\frac{U}{d_t}(T - T_{ex}) \qquad \text{(d)}$$

In Eq. (d) $c_p$ is a function of temperature and composition. The third term on the right-hand side is zero for adiabatic beds. For the nonadiabatic bed $T_{ex}$ represents the temperature of the feed that is preheated inside the tubes. For this part of the reactor another equation has to be added to those written above:

$$\frac{F_t c_p}{\Omega}\frac{dT_{ex}}{dz} = 4\frac{U}{d_t}(T - T_{ex})$$

where $U$ is an overall heat transfer coefficient. The initial conditions are:

at $z = 0$ $\qquad p_M = (p_M)_0$ $\qquad p_{CO_2} = (p_{CO_2})_0$ $\qquad$ etc; $\qquad T = T_0$ $\qquad F_t = F_{t0}$

*Solid Phase*

*Continuity equation for methanol:*

$$\frac{(D_e)_M}{RT}\frac{1}{\xi'^2}\frac{d}{d\xi'}\left[\xi'^2\frac{d(p_M)_s}{d\xi'}\right] = r_M \rho_s \qquad \text{(e)}$$

*Continuity equation for $CO_2$:*

$$\frac{(D_e)_{CO_2}}{RT}\frac{1}{\xi'^2}\frac{d}{d\xi'}\left[\xi'^2\frac{d(p_{CO_2})_s}{d\xi'}\right] = r_{CO_2} \rho_s \qquad \text{(f)}$$

FIXED BED CATALYTIC REACTORS ⎯⎯⎯⎯⎯⎯⎯⎯⎯⎯⎯ **565**

*Energy equation:*

$$-\frac{\lambda_e}{\xi'^2}\frac{d}{d\xi'}\left(\xi'^2\frac{dT_s}{d\xi'}\right) = r_{M_s}(-\Delta H)_1\rho_s - r_{CO_{2,s}}(-\Delta H)_2\rho_s \qquad (g)$$

*With boundary conditions:*

$$\text{at} \quad \xi' = \frac{d_p}{2}: (p_M)_s = (p_M{}^s)_s \qquad (p_{CO_2})_s = (p_{CO_2}{}^s)_s \qquad T_s = T_s^s$$

$$\text{at} \quad \xi' = 0: \frac{dp_M{}^s}{d\xi'} = 0 \qquad \frac{dp_{CO_2}{}^s}{d\xi'} = 0 \qquad \frac{dT_s}{d\xi'} = 0$$

The kinetic equation for methanol production is based on work by Natta et al. [137] and Cappelli and Dente [138] and accounts for the nonideal behavior of the reacting gases.

$$r_M = \frac{f_{CO}(p_{CO})_s f_{H_2}{}^2(p_{H2}{}^2)_s - f_M\left(\dfrac{p_M}{K_1}\right)^s}{A^3[1 + Bf_{CO}(p_{CO})_s + Cf_{H_2}(p_{M_2})_s + Df_M(p_M)_s + Ef_{CO_2}(p_{CO_2})_s]} \qquad \left[\frac{\text{kmol methanol}}{\text{kg cat hr}}\right] \quad (h)$$

where $A = 2.78 \times 10^5 \exp\left(-\dfrac{8280}{RT_s}\right)$

$B = 1.33 \times 10^{-10} \exp\left(\dfrac{23,850}{RT_s}\right)$

$C = 4.72 \times 10^{-14} \exp\left(\dfrac{30,500}{RT_s}\right)$

$D = 5.05 \times 10^{-12} \exp\left(\dfrac{31,250}{RT_s}\right)$

$E = 3.33 \times 10^{-10} \exp\left(\dfrac{23,850}{RT_s}\right)$

The fugacity coefficients are estimated by adopting Amagat's hypothesis, adapted by Newton [139] and by Clayton and Giaucque [140].

At the temperature of the methanol synthesis (350 to 400°C) the conversion reaction $CO_2 + H_2 \rightleftarrows CO + H_2O$ is very fast, so that equilibrium is reached at the catalyst surface and the fugacities obey

$$K_2 = \frac{f_{CO}(p_{CO})_s f_{H_2O}(p_{H_2O})_s}{f_{CO_2}(p_{CO_2})_s f_{H_2}(p_{H_2})_s} \qquad (i)$$

Note that when the effectiveness factor $\eta$ is calculated explicitly, Equations (e) and (f) may be replaced by:

$$-k_M a_v[p_M - (p_M{}^s)_s] = r_M{}^s \rho_B \eta_M \tag{j}$$

$$k_{CO_2} a_v[p_{CO_2} - (p_{CO_2}{}^s)_s] = r_{CO_2}{}^s \rho_B \eta_{CO_2} \tag{k}$$

Furthermore, the catalyst particle is considered to be isothermal [99]. Then a heat balance on the particle and the surrounding film or, by extension, on the solid phase in a cross section and the surrounding film may be written

$$h_f a_v(T_s^s - T) = r_M{}^s(-\Delta H)_1 \rho_B \eta_M - r_{CO_2}{}^s(-\Delta H)_2 \rho_B \eta_{CO_2} \tag{l}$$

or

$$h_f a_v(T_s^s - T) = -k_M a_v[p_M - (p_M{}^s)_s](-\Delta H)_1 - k_{CO_2} a_v[p_{CO_2} - (p_{CO_2}{}^s)_s](-\Delta H)_2 \tag{m}$$

Alternatively, Eq. (3) may be derived from Eq. (g) in the following way.

A first integration of Eq. 11.9.b-7 leads to

$$-\lambda_e \xi'^2 \frac{dT}{d\xi}\Big|_{\xi'=d_p/2} = \frac{4}{d_p{}^2} \int_0^{d_p/2} \xi'^2 [r_M(-\Delta H)_1 \rho_s - r_{CO_2}(-\Delta H)_2 \rho_s] d\xi'$$

and since by definition (see Chapter 3)

$$\eta_M = \frac{4\pi \int_0^{d_p/2} \xi'^2 r_M(\xi') d\xi'}{\frac{4}{3}\pi \left(\frac{d_p}{2}\right)^3 r_M{}^s} \tag{n}$$

For a single particle:

$$h_f(T_s^s - T) = -\lambda_e \frac{dT}{d\xi'}\Big|_{\xi'=d_p/2}$$

so that

$$h_f(T_s^s - T) = \frac{d_p}{6}[r_M{}^s(-\Delta H)_1 \rho_s \eta_M - r_{CO_2}{}^s(-\Delta H)_2 \rho_s \eta_{CO_2}]$$

or

$$h_f a_v(T_s^s - T) = r_M{}^s(-\Delta H)_1 \rho_B \eta_M - r_{CO_2}{}^s(-\Delta H)_2 \rho_B \eta_{CO_2}$$

which is nothing but Eq. (l).

Since $\eta_M$ cannot be obtained analytically, Eqs. (e) and (f) would have to be integrated in each node of the grid used for the integration of the gas phase equations (b), (c), and (d). However, Cappelli et al. preferred to rewrite Eq. (e) in the form of an integral equation. Two successive integrations of Eq. (e) accounting for the boundary conditions, lead to

$$p_M(\xi') = p_M(0) + \frac{RT\rho_s}{D_e} \int_0^{\xi'} \frac{dy}{y^2} \int_0^y \xi'^2 r_M(p_M) d\xi' \qquad (o)$$

This equation may be solved in an iterative way, calculating the $n + 1$st approximation of the profile from the $n$th approximation:

$$p_M^{(n+1)}(\xi') = p_M^{(n)}(0) + \frac{RT\rho_s}{D_e} \int_0^{\xi'} \frac{dy}{y^2} \int_0^y \xi'^2 r_M(p_M^{(n)}) d\xi' \qquad (p)$$

To start the procedure a uniform value of $x_M$ is chosen, namely

$$p_M(0) = p_M\left(\frac{d_p}{2}\right) = p_M{}^s$$

Then, since $r_M(p_M{}^s)$ is a constant,

$$p_M^{(1)}(\xi') = p_M(0) + \frac{RT\rho_s}{D_e} \frac{r_M(p_M{}^s)}{6} \xi'^2 \qquad (q)$$

and at

$$\xi' = \frac{d_p}{2}$$

$$p_M^{(1)}\left(\frac{d_p}{2}\right) = p_M{}^s + \frac{RT\rho_s}{D_e} \frac{r_M(p_M{}^s)d_p}{24} \qquad (r)$$

The profile Eq. (r) is now substituted into Eq. (p) to give the second approximation $p_M^{(2)}(\xi')$ and so on, until convergence is achieved. The corresponding $(p_M{}^s)_s$ then has to be used in Eqs. (b), (c), (h), (i), (r), and (m).

The calculation of the reactor now proceeds as follows. The gas-phase equations are integrated by means of a Runge–Kutta procedure. To do this, $p_M{}^s$, $p_{CO_2}{}^s$, and $T_s$ have to be calculated first. This cannot be done directly from Eq. (j) since this equation contains $r_M{}^s$ and $\eta_M$ which are functions of $(p_M{}^s)_s$. Assume first a value of $(p_M{}^s)_s$. From the equilibrium relation (Eq. (h)), which replaces Eq. (f), $p_{CO_2}$ is calculated, using the relation between $p_M{}^s$, $p_{CO_2}{}^s$, and the other partial pressures.

Then $T_s^s$ is calculated from Eq. (m) and subsequently $r(T)$. Next the integral equation is solved, starting with the assumed $p_M{}^s$ as first approximation for the profile inside the particle. The $(p_M{}^s)_s$ obtained from Eqs. (o) is compared with the assumed value. If they do not correspond the procedure described here is repeated. If they do, $p_{CO_2}$ and $T_s^s$ are recalculated so that everything is known at $z = 0$. Then using the Runge–Kutta routine, the values of $p_M$, $p_{CO_2}$, and $T$ at the end of the first increment $\Delta z$ are calculated and from this $F_t$. Then the values of $p_M$, $p_{CO_2}{}^s$, and so on have to be computed in the way outlined above. Cappelli et al. performed the simulation on a Univac 1108 computer. One such computation

*Table 1 Methanol synthesis. Comparison of simulation with industrial results*

| | | | | | |
|---|---|---|---|---|---|
| Total feed flow rate (Nm$^3$/hr) | | | | | 26,245 |
| Pressure, atm | | | | | 254 |
| Composition of feed gas (mol %) | | | | | |
| CO | | | | | 11.20 |
| CH$_3$OH | | | | | 0.11 |
| H$_2$ | | | | | 65.46 |
| H$_2$O | | | | | 0.15 |
| CH$_4$ | | | | | 13.75 |
| N$_2$ | | | | | 7.72 |
| CO$_2$ | | | | | 1.60 |
| Production tons/day | | | | | |
| Experimental | | | | | 25.6 |
| Calculated | | | | | 25.7 |
| Catalyst bed | 1 | 2 | 3 | 4 | 5[c] |
| Catalyst volume, m$^3$ | 0.250 | 0.094 | 0.135 | 0.197 | 0.250 |
| Compositions at bed exit, mol % | | | | | |
| CO | | | | | |
| Experimental | 10.85 | 10.46 | 10.05 | 9.64 | 9.45 |
| Calculated | 10.99 | 10.70 | 10.22 | 9.65 | 9.11 |
| CH$_3$OH | | | | | |
| Experimental | 1.11 | 1.61 | 2.14 | 2.55 | 3.14 |
| Calculated | 0.93 | 1.48 | 2.15 | 2.90 | 3.15 |
| CH$_4$ | | | | | |
| Experimental | 13.98 | 14.10 | 14.25 | 14.50 | 14.65 |
| Calculated | 13.98 | 14.13 | 14.31 | 14.52 | 14.59 |
| N$_2$ | | | | | |
| Experimental | 7.85 | 8.04 | 8.02 | 8.11 | 8.20 |
| Calculated | 7.85 | 7.93 | 8.03 | 8.15 | 8.19 |
| CO$_2$ | | | | | |
| Experimental | 1.08 | 1.09 | 1.11 | 1.17 | 1.33 |
| Calculated | 1.20 | 1.09 | 1.07 | 1.07 | 1.43 |
| Experimental inlet temperature (°C) | | | | | |
| | 350 | 369 | 368 | 364 | 368 |
| Calculated outlet temperature (°C) | | | | | |
| | 373 | 385 | 388 | 386 | 294 |
| $\Delta T = T_s - T$ (°C) | $-1.5^a$; $0^b$ | $0.3^a$; $0^b$ | $0.4^a$; $0^b$ | $0.5^a$; $0^b$ | $0.6^a$; $0^b$ |
| $\eta_M$ | $0.34^a$; $0.44^b$ | $0.44^a$; $0.48^b$ | $0.48^a$; $0.52^b$ | $0.53^a$; $0.56^b$ | $0.59^a$; $0.85^b$ |

[a] Inlet.
[b] Outlet.
[c] Nonadiabatic bed.

took about one minute. The program is currently used for checking the operation of reactors in various Montecatini-Edison plants and for design calculations.

Table 1 and Fig. 2 show the results of such a simulation and comparison with industrial results.

The agreement between simulated and experimental results is remarkable, indicating that the rate of expression and the model are excellent. Note that the $\Delta T$ over the gas film is practically zero.

The conservation equations could have been written in terms of conversions as well. The continuity equation for methanol would be written as follows.

$$-\frac{(F_{CO})_0}{\Omega}\frac{dx_M}{dz} = \begin{array}{l}\text{(rate of transport of methanol from the surface to the}\\ \text{bulk in terms of conversions}\end{array} \qquad \text{(s)}$$

The derivation of the conversions has to take the variation in number of moles into account. The way this is done was illustrated already in Chapter 9 for another example of a reaction with a change in number of moles. The molar flow rates and partial pressures in a section where the conversion of CO into methanol is $x_M$ and

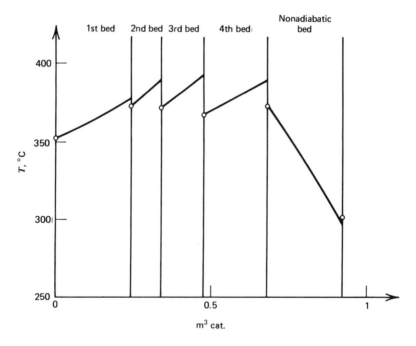

*Figure 2 Simulation of methanol synthesis reactor. Comparison with industrial results (from Cappelli, et al. [95]).*

the conversion of $CO_2$ is $x_{CO_2}$, are as follows, when expressed with respect to a CO-feed of 1 mol/hr

$$CO = 1 - x_M + x_{CO_2} \qquad p_{CO} = \frac{1 - x_M + x_{CO_2}}{8.92 - 2x_M} p_t$$

$$CH_3OH = 0.01 + x_M \qquad p_M = \frac{0.01 + x_M}{8.92 - 2x_M} p_t$$

$$H_2 = 5.85 - 2x_M - x_{CO_2} \qquad \cdot$$

$$CH_4 = 1.22 \qquad \cdot$$

$$N_2 = 0.69 \qquad \cdot$$

$$\underline{CO_2 = 0.14 - x_{CO2}} \qquad \cdot$$

$$8.92 - 2x_M$$

The total flow rate, per mol CO fed, may be related to the flow rate at the inlet as follows:

$$F_t = F_{t0} - 2x_M \qquad \text{or here:} \qquad F_t = 8.92 - 2x_M$$

Equation (s) is now easily derived from Eq. (b) by substituting $F_t$ and $p_M$ by the values given above. The final equation is

$$-\frac{(F_{CO})_0}{\Omega} \frac{dx_M}{dz} = k_M a_v p_t \left[ \frac{0.01 + x_M}{8.92 - 2x_M} - \frac{0.01 + (x_M^s)_s}{8.92 - 2(x_M^s)_s} \right]$$

## Example 11.9.a-2 Simulation of an Industrial Reactor for 1-Butene Dehydrogenation into Butadiene

This is an example, taken from Dumez and Froment [141], combining reactor simulation on the basis of a slightly simplified version of the model considered in this section but accounting for transients resulting from start up and from catalyst deactivation due to coke deposition.

The 1-butene dehydrogenation process considered here is carried out at temperatures of about 600°C in an adiabatic reactor and under reduced pressure to minimize coke formation. The reaction is strongly endothermic and the catalyst bed is diluted with inert particles that provide a heat reservoir that reduces to a certain extent the temperature drop resulting from the endothermicity. Because of catalyst deactivation by coking the time on stream is generally limited to about 15 minutes. Before regeneration the reactor has to be purged. The heat given off by the regeneration restores the original temperature level in the reactor. After purging the bed the butene may be fed again and a new cycle starts. The characteristics of a typical reactor and its operating conditions are given in Table 1.

## Table 1 Characteristics of an Industrial Reactor for 1-Butene Dehydrogenation

| | |
|---|---|
| Length | 0.8 m |
| Cross section | 1 m² |
| Catalyst and inert particle diameter | 0.0046 m |
| Amount of catalyst | 800 kg |
| Catalyst bulk density | 400 kg of cat./m³ of diluted bed |
| Inert bulk density | 900 kg of solid/m³ of diluted bed |
| Catalyst geometrical surface area | 274 m²/m³ of diluted bed |
| Inert surface area | 411 m²/m³ of diluted bed |
| Total pressure | 0.25 atm abs (0.245 bar) |
| Inlet butene pressure | 0.25 atm abs |
| Molar flow rate | 15 kmol/m² hr |
| Feed temperature | 600°C |
| Initial bed temperature | 600°C |

Since the amount of butene and butadiene lost in coking reactions cannot be neglected (i.e., since there are three independent reactions), three continuity equations have to be written for the fluid phase components. Also, since the flow velocity is high, the interfacial concentration and temperature gradients were neglected so that the fluxes at the catalyst surface are directly linked to the variations in the bulk gas-phase composition and enthalpy.

The continuity and energy equations are as follows.

Fluid phase:

$$\frac{\partial(u_s C_B)}{\partial z} + \varepsilon \frac{\partial C_B}{\partial t} = -a_k D_{eB}\left(\frac{\partial C_{B,k}}{\partial \xi'}\right)_{d_p/2} \tag{a}$$

$$\frac{\partial(u_s C_H)}{\partial z} + \varepsilon \frac{\partial C_H}{\partial t} = -a_k D_{eH}\left(\frac{\partial C_{H,k}}{\partial \xi'}\right)_{d_p/2} \tag{b}$$

$$\frac{\partial(u_s C_D)}{\partial z} + \varepsilon \frac{\partial C_D}{\partial t} = -a_k D_{eD}\left(\frac{\partial C_{D,k}}{\partial \xi'}\right)_{d_p/2} \tag{c}$$

$$\frac{\partial(u_s \rho_g c_p T)}{\partial z} + \varepsilon \frac{\partial(\rho_g c_p T)}{\partial t} = -[a_k h_k(T - T_k) + a_i h_i(T - T_i)] \tag{d}$$

in which the indices $k$ and $i$ refer to catalyst and inert material.

Solid phase:

$$\rho_{B,k} c_{p_k} \frac{\partial T_k}{\partial t} = a_k h_k(T - T_k) + a_{ik} h_{ik}(T_i - T_k) + a_k(-\Delta H)D_{eH}\left(\frac{\partial C_{H,k}}{\partial \xi'}\right)_{d_p/2} \tag{e}$$

$$\rho_{B,i} c_{p_i} \frac{\partial T_i}{\partial t} = a_i h_i(T - T_i) - a_{ik} h_{ik}(T_i - T_k) \tag{f}$$

Inside the catalyst particles:

$$\frac{\partial^2 C_{B,k}}{\partial \xi'^2} + \frac{2}{\xi'}\frac{\partial C_{B,k}}{\partial \xi'} - \frac{\varepsilon_s}{D_{eB}}\frac{\partial C_{B,k}}{\partial t} = \frac{\rho_k}{D_{eB}}\left(r_H + \frac{r_{CB}}{\Psi_{CB}M_B}\right) \tag{g}$$

$$\frac{\partial^2 C_{H,k}}{\partial \xi'^2} + \frac{2}{\xi'}\frac{\partial C_{H,k}}{\partial \xi'} - \frac{\varepsilon_s}{D_{eH}}\frac{\partial C_{H,k}}{\partial t} = -\frac{\rho_k}{D_{eH}}r_H \tag{h}$$

$$\frac{\partial^2 C_{D,k}}{\partial \xi'^2} + \frac{2}{\xi'}\frac{\partial C_{D,k}}{\partial \xi'} - \frac{\varepsilon_s}{D_{eD}}\frac{\partial C_{D,k}}{\partial t} = -\frac{\rho_k}{D_{eD}}\left(r_H - \frac{r_{CD}}{\Psi_{CD}M_D}\right) \tag{i}$$

$$\frac{\partial C_c}{\partial t} = r_{CB} + r_{CD} \tag{j}$$

The pressure-drop equation is taken from Leva for turbulent flow. The boundary conditions are:

$z = 0$     all $t$: $C_B = C_B^0$     $C_H = C_H^0$     $C_D = C_D^0$     $u_s = u_s^0$

$t = 0$     all $z$: $T_k = T_k^0$     $T_i = T_i^0$

$\xi' = d_{p/2}$     all $z$ and $t$: $(C_{B,k})_R = C_B$
                                           $(C_{H,k})_R = C_H$
                                           $(C_{D,k})_R = C_D$

$\xi' = 0$     all $z$ and $t$: $\left(\dfrac{\partial C_{B,k}}{\partial \xi'}\right)_{\xi'=0} = \left(\dfrac{\partial C_{H,k}}{\partial \xi'}\right)_{\xi'=0} = \left(\dfrac{\partial C_{D,k}}{\partial \xi'}\right)_{\xi'=0} = 0$

$t = 0$     all $z$ and $\xi$: $C_c = 0$

The continuity and energy equations for the fluid phase in the reactor contain non-steady-state terms. However, since the interstitial flow velocity is 4 m/s and since the bed length is only 0.8 m, the second terms on the left-hand side of Eqs. (a), (b), (c), and (d) can be neglected. The products $(u_s C_B), (u_s C_H) \ldots$ are kept under the differential in these equations to account for the important change in number of moles in the gas phase owing to the dehydrogenation and to a certain extent to the coking. The right-hand side in Eq. (d) expresses the amount of heat exchanged between the gas and the solid particles, both catalytic and inert. The second term in the right-hand side of Eqs. (e) and (f) expresses the amount of heat exchanged between the catalyst and the inert particles by conduction and radiation. Of course, the non-steady-state terms have to be kept in Eqs. (e) and (f). In the particle equations (Eqs. (g), (h), and (i)) again the non-steady-state terms can be neglected. No energy equation is written for the catalyst particle, which may be considered as isothermal for the reasons explained in Chapter 3.

The rate equations were determined by Dumez and Froment by means of sequentially designed experimental programs for model discrimination and parameter estimation, discussed and illustrated in Chapter 2. The following equations were found:

- For the rate of formation of hydrogen in the absence of coking, $r_H^0$:

$$r_H^0 = \frac{1.826 \times 10^7 \exp(-29236/RT)\left(p_B - \frac{p_H p_D}{K}\right)}{(1 + 18727 p_B + 3.593 p_H + 38028 p_D)^2} \tag{k}$$

- For the rate of disappearance of butene into butadiene and coke on fresh catalyst, $r_B^0$:

$$r_B^0 = r_H^0 + \frac{r_{CB}^0}{\Psi_{CB} M_B} \tag{l}$$

where $r_{CB}^0$ is the rate of coke formation from butene on fresh catalyst. Since $r_B^0$ is in kmol butene/kg cat. hr and $r_{CB}^0$ in kg coke/kg cat. hr the conversion factor $\Psi_{CB}$, expressed in kg coke/kg butene, is required. $M_B$ is the molecular weight of butene.

- For the net rate of production of butadiene, which also reacts further into coke

$$r_D^0 = r_H^0 - \frac{r_{CD}^0}{\Psi_{CD} M_D} \tag{m}$$

where $\Psi_{CD}$ is the corresponding conversion factor and $M_D$ the molecular weight of butadiene.

The influence of coke on the rate of these reactions is accounted for by exponential deactivation functions based upon the coke content, as advocated in Chapter 5:

$$\begin{aligned} r_H &= r_H^0 \exp(-42.12 C_c) \\ r_{CB} &= r_{CB}^0 \exp(-45.53 C_c) \\ r_{CD} &= r_{CD}^0 \exp(-45.53 C_c) \end{aligned} \tag{n}$$

The form of the deactivation function and the numerical value of the deactivation parameter were determined by means of an electrobalance. The deactivation constant is essentially identical for the three reactions in this case. The total rate of coke formation, $r_C = r_{CB} + r_{CD}$ has been given in Chapter 5 as Eq. (i) of Ex. 5.3.e-1.

The effective diffusivities for transport inside the catalyst were determined from experiments with particle radii varying from 0.35 to 2.3 mm. For butene, for example, the effective diffusivity contains the tortuosity, the internal void fraction, and the molecular and Knudsen diffusivity:

$$\frac{1}{D_{eB}} = \frac{\tau}{\varepsilon_s} \left( \frac{1}{D_{B,m}} + \frac{1}{D_{K,B}} \right) \tag{o}$$

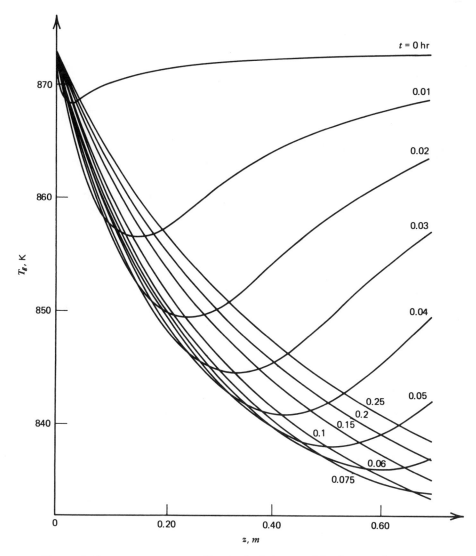

*Figure 1 Gas temperature profiles in the 1-butene dehydrogenation reactor.*

The molecular diffusivities were calculated from a weighted average of the binary diffusion resistances which in turn were estimated from the formula of Fuller et al. The calculation of the Knudsen diffusivities requires information about the pore size. The catalyst had a bimodal pore distribution. Furthermore, via electron microscopy, it was found that the catalyst consisted of crystallites of about 5 $\mu$, separated by voids of about 1 $\mu$. Consequently, the maximum length of the micropores with average diameter 70 Å cannot exceed 5 $\mu$. It is easily calculated that this is too short to develop any significant concentration gradients. Therefore, only the macropores with a pore volume of 0.155 cm$^3$/g and with average pore diameter of 10000 Å were considered in the evaluation of the Knudsen diffusivity.

The only unknown parameter left in Eq. (o) is the tortuosity factor, $\tau$. This factor was determined from a comparison between the experimental rate at zero coke content, measured in a differential reactor and the surface fluxes. The latter were calculated using Fick's law and for a given $\tau$ from the concentration profiles obtained by numerical integration of the system (Eqs. (g), (h), (i), and (j)). A value of $\tau = 5$ led to the best fit of all six experiments. This is the generally accepted value for the tortuosity factor in a catalyst of the type used in this work. It was also possible to calculate an effectiveness factor from these results. A value of 0.20 was obtained for a particle radius of 2.3 mm at 550°C. The Bischoff general modulus approach, presented in Chapter 3, leads to a value of 0.28. Finally, the heat transfer coefficients were calculated from the $j_H$ correlation of Handley and Heggs, mentioned in Chapter 3.

With all this information the design calculations can now be performed. The particle equations were solved by means of collocation, the continuity equations for the fluid phase by means of a Runge–Kutta–Merson routine and the energy equations in a semianalytical manner. The results for one particular operation are shown in Figs. 1, 2, and 3. In Fig. 1 the temperature is seen to drop rapidly

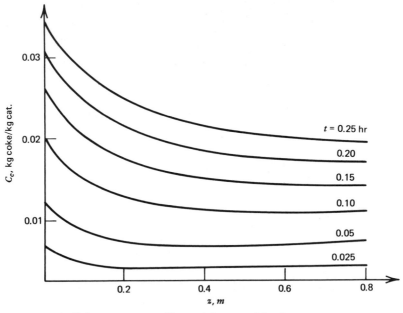

*Figure 2  Coke content profiles in 1-butene dehydrogenation reactor.*

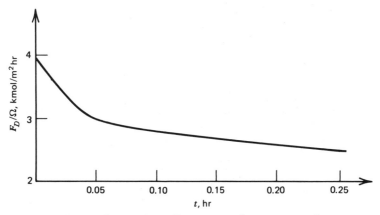

*Figure 3 Butadiene flow rate at the reactor outlet.*

under the initial value, due to the endothermic nature of the reaction. A temperature wave rapidly travels through the reactor as the reaction gradually extends to increasing depths. In the presence of coking no true steady state is reached, however: the temperature profile is slowly translated upward, because of the decrease in reaction rate caused by the catalyst deactivation. The corresponding coke profiles are shown in Fig. 2. The profile is decreasing, mainly because of the higher temperature at the inlet and of the inhibition of the coke formation due to the hydrogen concentration near the outlet. Figure 3 illustrates the butadiene flow rate at the outlet as a function of time. The rapid initial decrease corresponds to the initial temperature drop. Beyond this initial period the decrease in production is much slower. The catalyst is deactivated by the coking but as the dehydrogenation rate is lowered the bed temperature slowly rises, thus favoring the butadiene production. Further aspects of the problem and an optimization of the on stream time can be found in Dumez and Froment [141].

## 11.10 Two-Dimensional Heterogeneous Models

Finally, in recent years, attempts have been made to develop two-dimensional heterogeneous models. McGreavy and Cresswell [108] proceeded by adding to the one-dimensional model accounting for interfacial and intraparticle gradients discussed in Sec. 11.9 the terms accounting for radial heat and mass transfer in the bed. From an inspection of the equations it is clear, however, that it is implicitly assumed heat transfer in the radial direction only occurs through the fluid phase. Figure 11.7.a-2 shows that even for typical industrial flow rates the solid and stagnant films contribute for at least 25 percent in the radial heat flux (i.e., $\lambda_{er}^0/\lambda_{er}$). With regard to the extreme sensitivity of the profiles to $\lambda_{er}$ [76; 78] the model used by McGreavy and Cresswell can only be considered as a rough approximation. The model set up by Carberry and White [80] is hybrid in this sense that it distinguishes between conditions in the gas and on the solid, but makes use of the $\lambda_{er}$ and $\alpha_w$ concept of Yagi and Kunii [70] and Kunii and Smith [68], which lumps gas and solid, as is explained in Sec. 11.7. In order to account for heat transfer through the solid correctly, the equations concerning the solid should not be limited to a *single* particle as is generally done, but extended to the complete cross section occupied by the catalyst. This was done for the one-dimensional model (11.8.a-3 and 4 of 11.8.a), but without accounting for eventual radial temperature gradients, of course. In addition, one has to distinguish between the effective thermal conductivity for the fluid phase, $\lambda_{er}^f$ and that for the solid phase $\lambda_{er}^s$ [106]. Strangely enough, this concept of $\lambda_{er}^f$ and $\lambda_{er}^s$ was introduced as far back as 1953 by Singer and Wilhelm [107]. All subsequent work in this field made use of the global $\lambda_{er}$ concept, briefly reviewed in Sec. 11.7, however.

The preceding considerations led De Wasch and Froment [106] to the following mathematical model:

$$
\left\{
\begin{aligned}
&u_s \frac{\partial C}{\partial z} = \varepsilon D_{er}\left(\frac{\partial^2 C}{\partial r^2} + \frac{1}{r}\frac{\partial C}{\partial r}\right) - k_g a_v(C - C_s^s) \\
&u_s \rho_g c_p \frac{\partial T}{\partial z} = \lambda_{er}^f\left(\frac{\partial^2 T}{\partial r^2} + \frac{1}{r}\frac{\partial T}{\partial r}\right) + h_f a_v(T_s^s - T) \\
&k_g a_v(C - C_s^s) = \eta \rho_B r_A \\
&h_f a_v(T_s^s - T) = \eta \rho_B(-\Delta H)r_A + \lambda_{er}^s\left(\frac{\partial^2 T_s}{\partial r^2} + \frac{1}{r}\frac{\partial T_s}{\partial r}\right)
\end{aligned}
\right.
\qquad (11.10\text{-}1)
$$

with boundary conditions

$$C = C_0 \qquad \text{at } z = 0$$

$$T = T_0$$

$$\frac{\partial C}{\partial r} = 0$$

$$\frac{\partial T}{\partial r} = \frac{\partial T_s}{\partial r} = 0 \qquad \text{at } r = 0 \qquad \text{all } z$$

$$\frac{\partial C}{\partial r} = 0$$

$$\alpha_w^f(T_w - T) = \lambda_{er}^f\frac{\partial T}{\partial r} \qquad \text{at } r = R \qquad \text{all } z$$

$$\alpha_w^s(T_w - T_s) = \lambda_{er}^s\frac{\partial T_s}{\partial r}$$

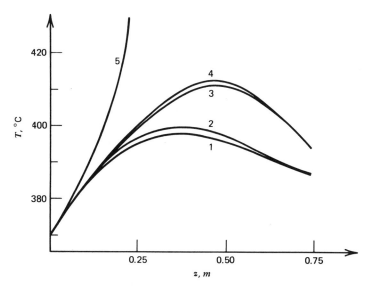

*Figure 11.10-1 Radial mean temperature as a function of bed length. Comparison of model predictions. 1: Basic pseudo-homogeneous one-dimensional model; 2: Heterogeneous model with interfacial gradients; 3: Pseudo-homogeneous two-dimensional model; 4: Two-dimensional heterogeneous model with boundary conditions given in model Equation (11.10-1); 5: Two-dimensional heterogeneous model with no heat transfer through solid.*

Note that the distinction between solid and fluid also appears in the boundary conditions for heat transfer at the wall. There are several possibilities for the boundary condition for the "solid" phase at the wall. The simplest is to set the temperature of the solid equal to that of the wall itself. A better approximation is to consider the temperature profile in the "solid" phase to be linear near the wall $(\partial^2 T_s/\partial r^2 = 0)$. Still another possibility is to use a boundary condition for the "solid" analogous to that for the fluid, as shown in the above equations. These different possibilities and the numerical values to be given to the parameters are discussed by De Wasch and Froment. Figure 11.10-1 shows radial mean temperature profiles through a reactor for the boundary conditions shown above and no intraparticle resistance ($\eta = 1$). In this figure curves 1 and 2 represent the results obtained by means of the one-dimensional models, 1 without, 2 with interparticle gradients. Curve 3 is obtained by means of the two-dimensional pseudo-homogeneous model, curve 4 by means of the two-dimensional heterogeneous model discussed in this section and with the heat transfer boundary condition

$$\alpha_w{}^s(T_w - T_s) = \lambda_{er}{}^s \frac{\partial T_s}{\partial r}$$

Curve 5 corresponds to no-heat transfer through the solid and this predicts a hot spot that is far too important. Such a model is no improvement at all with respect to the two-dimensional pseudo-homogeneous model of Sec. 11.7. It is interesting also to note that, for the conditions used in these calculations, the solid temperature only exceeds the gas temperature by 1 or 2°C. This is generally so in industrial reactors. Finally, the radial mean temperatures of the two-dimensional models are significantly different from the temperature predicted by the one-dimensional models. Provided the physical data are available the two-dimensional models would definitely have to be preferred for the simulation of this reactor.

## Problems

11.1 Show that Leva's pressure drop equation for packed beds, (Eq. 11.5.a-4), yields a parabolic pressure profile when the fluid density is constant.

11.2 (a) The kinetics of the catalytic reaction $A \rightleftharpoons R + S$ are given by:

$$r = \frac{dx}{d\dfrac{W}{F_{A0}}} = \frac{kK_A(p_A - p_R \cdot p_S/K)}{(1 + K_A p_A + K_R p_R + K_S p_S)^2} \tag{1}$$

The reaction is carried out isothermally in a packed bed reactor with plug flow at 275°C. The feed contains 0.155 moles water per mole of reactant. Water is not adsorbed on the catalyst and acts purely as an inert diluent. Given the following data:

$F_{t0} = 4.2$ kmol/hr (total flow rate)
$\rho_b = 1500$ kg/m³
$d_t = 0.05$ m $\qquad\qquad\qquad\qquad k = 4.3593$ [kmol/kg cat. hr]
$p_t = 3$ atm $\qquad\qquad\qquad\qquad\quad K_A = 0.43039$ [1/atm]
$K = 0.589$ [atm] $\qquad\qquad\quad K_R + K_S = 2.8951$ [1/atm]

Calculate the length of the reactor required to reach an exit conversion of (i) 40 percent, (ii) 70 percent.

(b) Suppose the reaction is carried out under the same conditions in a multitubular reactor. The tube length is 3 m. The total feed per tube is 4 kmoles/hr. An annual production of 20.000 metric tons of product is required. The molecular weight of the product is 44. One year on stream is equivalent to 8000 hr. Determine the number of tubes required to meet the production.

11.3 Discuss whether the following is correct or wrong.
(a) The relation conversion versus reactor length is linear for a zero-order reaction without heat effect carried out in an adiabatic reactor
(b) In an adiabatic reactor the relation temperature vs conversion is a straight line for first-order reactions only.

(c) Optimization of a multibed adiabatic reactor turns out to be roughly equivalent with distributing the catalyst in equal amounts over the different beds.

11.4 (a) Consider an isothermal fixed bed reactor with axial mixing superposed on plug flow conditions, in which an irreversible first-order reaction takes place. Show that, for a given set of operating variables, the effect of axial diffusion decreases with increasing reactor length.

(b) Given the values

$$u_s = 0.01 \text{ m/s} \qquad d_p = 0.004 \text{ m}$$
$$\varepsilon = 0.4 \qquad \rho_b = 1200 \text{ kg/m}^3$$
$$\text{Pe}_a = 2 \qquad k = 1.\text{E-5 m}^3/\text{kg cat. s}$$

Compute the $C_A$ profiles in tubular reactors with axial mixing as a function of total length and compare with the plug flow profile.

(c) Verify for this case if a bed depth of $50d_p$ is sufficient for eliminating axial mixing effects.

(d) On what basis has the 50 $d_p$ rule been established?

(e) Explain why the concentration profiles under axial mixing conditions never converge to the plug profile, not even in the limiting case $L \to \infty$.

11.5 Calculate the heat transfer parameters of the two-dimensional pseudo homogeneous models for the design of the reactor for hydrocarbonoxidation of Ex. 11.7.c, using the correlations given in Sec. 11.7.a. Compare the value of $\lambda_{er}^{0}$ calculated from the expressions given by (a) Kunii and Smith, (b) Zehner and Schlünder. Determine their sensitivity with respect to the solid conductivity. Additional data:

$$\lambda_s = 1 \text{ kcal/m hr } ^\circ\text{C} = 1.163 \times 10^{-3} \text{ kJ/m s K}$$
$$\lambda_g = 0.0429 \text{ kcal/m hr } ^\circ\text{C} = 4.99 \times 10^{-5} \text{ kJ/m s K}$$
$$p = 0.8$$
$$\beta = 0.95$$
$$d_p = 0.003 \text{ m}$$
$$d_t = 0.0254 \text{ m}$$
$$T_m = 382 ^\circ\text{C}$$
$$\varepsilon = 0.38$$

11.6 Check whether multiple steady states can occur in the case of the hydrocarbon oxidation of Ex. 11.7.c, but considering only the reaction $A \to B$. Required data:

$$-\Delta H = 307,000 \text{ kcal/kmol} = 1,285,350 \text{ kJ/kmol}$$
$$\lambda_s = 1 \text{ kcal/m hr } ^\circ\text{C} = 1.163 \cdot 10^{-3} \text{ kJ/m s K}$$
$$T_f = 655 \text{ K}$$
$$E = 27,000 \text{ kcal/kmol} = 113,044 \text{ kJ/kmol}$$
$$R = 1.98 \text{ kcal/kmol K} = 8.3144 \text{ kJ/kmol K}$$
$$C_f = 0.4145 \cdot 10^{-3} \text{ kmol/m}^3$$
$$D_e = 5 \cdot 10^{-4} \text{ m}^2/\text{hr}$$

11.7* A tubular, fixed bed catalytic reactor is to be used for a highly exothermic reaction, and the preliminary design must consider the possibility of a hot spot. The following

* This problem was contributed by Prof. J. H. Olson, University of Delaware.

parameters have been established in the initial design estimates.

$$\beta = \frac{\Delta T_{ad}}{T_{Ref}} = \frac{600}{400} = 1.5 \qquad \text{dimensionless adiabatic temperature rise}$$

$$\gamma = \frac{E}{R T_{ref}} = \frac{32000}{2 \times 400} = 40 \qquad \text{dimensionless activation energy}$$

$$k_v C_{A0} V/F_{A0} \qquad\qquad = 3 \qquad \text{reaction rate group or the number of reactor units}$$

$$U A/F' \rho c_p \qquad\qquad = 22.5 \quad \text{number of heat transfer units}$$

(a) Will there be an excessive hot spot for the set of parameters given in the preliminary design?

(b) An obvious (if expensive) way to overcome a hot-spot problem is to add diluent to the reactor feed. What must the dilution be to achieve design 10 percent safer than the hot-spot minimum dilution?

(c) Indicate how the two parameters of the hot-spot analysis are altered by the following design choices:

- Decrease inlet (reference) temperature by 10 K.
- Decrease the tube diameter by 20 percent.
- Increase the reactor length by 20 percent.
- Change the catalyst to lower the activation energy by 40 percent.

(d) The reactor designed in part (b) is expected to undergo two changes during extended operation: the catalyst activity will decrease by 30 percent and the heat transfer coefficient will decrease by 20 percent. The loss in catalyst activity will be compensated by raising the reactor temperature enough to keep $R$ constant. Investigate the probability of developing a hot spot in the reactor under the revised conditions.

## 11.8 Fixed bed reactor for styrene production

### (a) Introduction

Styrene is produced by catalytic dehydrogenation of ethylbenzene. The reaction is endothermic and reversible and takes place with an increase in the number of moles. Consequently, the styrene conversion is favored by high temperatures, low pressures, and by dilution of the feed by means of an inert component, like benzene or more generally steam. The steam also serves as a heat carrier, reducing the temperature drop in adiabatic operation.

### (b) Reactor

The following data on the styrene reactor of the Polymer Corporation, Sarnia, Ontario, have been presented by Sheel and Crowe.[2]

*Operating conditions*
Hydrocarbon flow: 4080 kg/h
Steam flow : 8160 kg/h

[2] J. B. P. Sheel and C. M. Crowe, *Can. J. Chem. Engng.*, **47**, 183 (1969).

Temperature of the superheated steam: $T_D = 755°C$
Inlet temperature of the mixture: $T_0 = 650°C$
Temperature at the reactor exit: $T_e = 577°C$
Inlet pressure: 2.37 atm abs

*Reactor dimensions*

Diameter: 1.95 m
Depth of catalyst bed: 1.61 m

*Additional data*

Catalyst equivalent diameter: $d_p = 0.005$ m
Void fraction of the bed: $\varepsilon = 0.455$
Bulk density of the bed: $\rho_B = 1300$ kg/m³

(c) *Kinetics*

Sheel and Crowe considered 10 reacting species. Six linear independent stoichiometric equations are needed to describe the variation of the amounts of these species along the reactor.

*Reactions*

1. $C_6H_5 - C_2H_5 \rightleftharpoons C_6H_5 - C_2H_3 + H_2$
2. $C_6H_5 - C_2H_5 \longrightarrow C_6H_6 + C_2H_4$
3. $C_6H_5 - C_2H_5 + H_2 \longrightarrow C_6H_5 - CH_3 + CH_4$
4. $\frac{1}{2}C_2H_4 + H_2O \longrightarrow CO + 2H_2$
5. $CH_4 + H_2O \longrightarrow CO + 3H_2$
6. $CO + H_2O \longrightarrow CO_2 + H_2$

All reactions were assumed to be catalytic. Only the main reaction 1 was assumed to be reversible. The following rate expressions were proposed for the above reactions.

*Table 1 Rate equations and rate parameters*

| Reaction | Rate equation | Rate parameters $A_i \dfrac{\text{kmol}}{\text{kg s}}$ | $E/R(°K)$ |
|---|---|---|---|
| 1 | $r_1 = k_1(p_E - p_S p_{H_2}/K_p)$ | 1.51286 | 10,925 |
| 2 | $r_2 = k_2 p_E$ | 5.6197 10⁵ | 25,000 |
| 3 | $r_3 = k_3 p_E p_{H_2}$ | 1.3446 | 11,000 |
| 4 | $r_4 = k_4 p_W p_{C_2H_4}$ | $9.3016 \times 10^{-1}$ | 12,500 |
| 5 | $r_5 = k_5 p_W p_{CH_4}$ | $6.3163 \times 10^{-2}$ | 7,900 |
| 6 | $r_6 = k_6(p_t/T^3)p_W p_{CO}$ | $1.6769 \times 10^9$ | 8,850 |

$p_i$ represents the partial pressure of species $i$ and $p_t$ is the total pressure.

The frequency factors and activation energies were determined by Sheel and Crowe from plant data. The equilibrium constant, $K_p$, is taken to be 0.4 atm in the temperature range experienced in the reactor.

*(d)* *Reactor Model*

The reactor is assumed to be adiabatic with plug flow. Axial dispersion can be ignored. Any effect of limitations of mass or heat transfer inside the catalyst pellet is lumped into the rate constants given in Table 1. The catalyst activity is assumed to be constant. Use the conversion of ethylbenzene or water in the set of continuity equations. Use the Ergun equation to describe the pressure drop.

*(e)* *Physicochemical Data*

*(i)* *Reaction Enthalpies*

These are fitted by means of linear equations $(-\Delta H_i) = a_i + b_i T$. The sets of $(a_i, b_i)$ are given in Table 2.

*Table 2 Coefficients in reaction enthalpy equations*

| Reaction | $a_i$ $\dfrac{\text{kcal}}{\text{kmol}}$ | $b_i$ $\dfrac{\text{kcal}}{\text{kmol}}$ |
|:---:|:---:|:---:|
| 1 | $-28843$ | $-1.09$ |
| 2 | $-25992$ | $1.90$ |
| 3 | $12702$ | $3.15$ |
| 4 | $19602$ | $-2.11$ |
| 5 | $50640$ | $-3.96$ |
| 6 | $10802$ | $-2.50$ |

*(ii)* *Heat Capacities*

These are calculated from Reid, Prausnitz, and Sherwood[3]. They are calculated in the form of quadratic functions of $T$. The coefficients are given in Table 3.

*Table 3 Coefficients of the quadratic equation $c_p = A + BT + CT^2 + DT^3$ (kcal/kmol °K)*

| | Species | $A$ | $B$ | $C$ | $D$ | MW |
|:---:|:---|:---:|:---:|:---:|:---:|:---:|
| 1 | $E$ | $-10.294$ | $1.689 \times 10^{-1}$ | $-1.149 \times 10^{-4}$ | $3.107 \times 10^{-8}$ | 106.168 |
| 2 | $S$ | $-6.747$ | $1.471 \times 10^{-1}$ | $-9.609 \times 10^{-5}$ | $2.373 \times 10^{-8}$ | 104.151 |
| 3 | $B$ | $-8.101$ | $1.133 \times 10^{-1}$ | $-7.206 \times 10^{-5}$ | $1.703 \times 10^{-8}$ | 78.114 |
| 4 | Tol | $-5.817$ | $1.224 \times 10^{-1}$ | $-6.605 \times 10^{-5}$ | $1.173 \times 10^{-8}$ | 92.141 |
| 5 | $C_2H_4$ | $0.909$ | $3.740 \times 10^{-2}$ | $-1.994 \times 10^{-5}$ | $4.192 \times 10^{-9}$ | 28.054 |
| 6 | $CH_4$ | $4.598$ | $1.245 \times 10^{-2}$ | $2.860 \times 10^{-6}$ | $-2.703 \times 10^{-9}$ | 16.043 |
| 7 | $H_2O(W)$ | $7.701$ | $4.595 \times 10^{-4}$ | $2.521 \times 10^{-6}$ | $-0.859 \times 10^{-9}$ | 18.015 |
| 8 | CO | $7.373$ | $-0.307 \times 10^{-2}$ | $6.662 \times 10^{-6}$ | $-3.037 \times 10^{-9}$ | 28.010 |
| 9 | $CO_2$ | $4.728$ | $1.754 \times 10^{-2}$ | $-1.338 \times 10^{-5}$ | $4.097 \times 10^{-9}$ | 44.010 |
| 10 | $H_2$ | $6.483$ | $2.215 \times 10^{-3}$ | $-3.298 \times 10^{-6}$ | $1.826 \times 10^{-9}$ | 2.016 |

[3] R. C. Reid, J. M. Prausnitz, and T. K. Sherwood, *The Properties of Gases and Liquids* (3rd ed.), McGraw Hill, New York (1977).

*(iii) Viscosity*
The viscosity can be assumed to be that of steam at the reaction temperature (0.03 cp).

*(iv) Simulation*
Simulate the profiles of conversion into styrene, benzene, and toluene, and the temperature and pressure profiles in the reactor.

# References

[1] Dickinson, N. L., Finneran, J. A. and Solomon, E. *Eur. Chem. News.* "Large Plants," Sept. 29 (1967).

[2] Winnacker, K. and Kuechler, L. *Chemische Technologie*, C. Hanser Verlag, München (1970).

[3] Shipman, L. M. and Hickman, J. B. *Chem. Eng. Progr.*, **64**, No. 5, 59 (1968).

[4] Vancini, C. A. *Synthesis of Ammonia*, Macmillan, London (1971).

[5] Smith, R. B. *Chem. Eng. Progr.*, **55**, No. 6, 76 (1959).

[6] Suter, H. *Phthalsäureanhydrid*, Steinkopf Verlag, Darmstadt (1972).

[7] *High Performance Process Furnaces* M. W. Kellogg Co, New York.

[8] Finneran, J. A., Buividas, L. J. and Walen, N. Hydrocarbon processing **51** (4), 127 (1972). Oct. 11, 70 (1971). Eschenbrenner, G. P. and Wagner, G. A. *Chem. Eng. Progr.* **68**, 66 (1972).

[9] Froment, G. F. *Chemical Reaction Engineering*, Advances in Chemistry Series 109, Am. Chem. Soc. (1972).

[10] Froment, G. F. *Proc. 5th Eur. Symp. Chem. React. Engng.* Amsterdam (1972), Elsevier Publ. Co., New York (1972).

[11] Ray, W. H. *Proc. 5th Eur. Symp. Chem. React. Engng.* Amsterdam (1972), Elsevier Publ. Co., New York (1972).

[12] Leva, M. *Chem. Eng.*, **56**, 115 (1949).

[13] Brownell, L. E., Dombrowsky, H. S. and Dickey, C. A. *Chem. Eng. Progr.*, **46**, 415 (1950).

[14] Ergun, S. *Chem. Eng. Progr.*, **48**(2), 89 (1952).

[15] Hicks, R. E. *Ind. Eng. Chem. Fund.*, **9**, 500 (1970).

[16] Leva, M. *Ind. Eng. Chem.*, **40**, 747 (1948).

[17] Maeda, S. *Techn. Dep. Tohoku Univ.*, **16**, 1 (1952).

[18] Verschoor, H. and Schuit, G. *Appl. Sci. Res.*, **A.2**, 97 (1950).

[19] De Wasch, A. P. and Froment, G. F. *Chem. Eng. Sci.*, **26**, 629 (1971).

[20] Van Welsenaere, R. J. and Froment, G. F. *Chem. Eng. Sci.*, **25**, 1503 (1970).

[21] Bilous, O. and Amundson, N. R. *A.I.Ch.E. J.*, **2**, 117 (1956).

[22] Barkelew, C. R. *Chem. Eng. Progr.*, Symp. Ser., **55**(25), 38 (1959).

[23] Dente, M. and Collina, A. *Chim. et Industrie*, **46**, 752 (1964).

[24] Hlavacek, V., Marek, M. and John, T. M. *Coll. Czechoslov. Chem. Comm.*, **34**, 3868 (1969).

[25] Hutchinson, P. and Luss, D. *Chem. Eng. J.*, **1**, 129 (1970).
Luss, D. and Hutchinson, P. *Chem. Eng. J.*, **2**, 172 (1971).
Golikeri, S. V. and Luss, D. *A.I.Ch.E. J.*, **18**, 277 (1972).

[26] Baddour, R. F., Brian, P. L. T., Logeais, B. A. and Eymery, J. P. *Chem. Eng. Sci.*, **20**, 281 (1965).

[27] Murase, A., Roberts, H. L. and Converse, A. O. *Ind. Eng. Chem. Proc. Des. Devpt.*, **9**, 503 (1970).

[28] Van Heerden, C. *Ind. Eng. Chem.*, **45**, 1242 (1953).

[29] Shah, M. J. *Ind. Eng. Chem.*, **59**, No. 1, 72 (1967).

[30] Aris, R. *The Optimal Design of Chemical Reactors*, Academic Press, New York (1960).

[31] Pontryagin, L. S., Boltryanski, V. G., Gamkrelidze, R. V. and Mishenko, E. F. *Mathematische Theorie Optimalier Prozesse*, Holdenburg Verlag, München (1969).

[32] Lee, K. Y. and Aris, R. *Ind. Eng. Chem. Proc. Des. Devpt.*, **2**, 300 (1963).

[33] Paynter, J. D., Dranoff, J. S. and Bankoff, S. G. *Ind. Eng. Chem. Proc. Des. Devpt.*, **10**, 244 (1971).

[34] Burkhardt, D. B. *Chem. Eng. Progr.*, **61** (1968).

[35] Froment, G. F. and Bischoff, K. B. *Chem. Eng. Sci.*, **16**, 189 (1961).

[36] Butt, J. B. *Chemical Reaction Engineering*, Advances in Chemistry Series 109, Am. Chem. Soc. (1972).

[37] Froment, G. F. and Bischoff, K. B. *Chem. Eng. Sci.*, **17**, 105 (1962).

[38] Menon, P. G. and Sreeramamurthy, R. *J. Catal.*, **8**, 95 (1967).
Menon, P. G., Sreeramamurthy, R. and Murti, P. S. *Chem. Eng. Sci.*, **27**, 641, 1972.

[39] Van Zoonen, D. D. *Proc. III Int. Congr. Catal.*, North Holland Publishing Co., Amsterdam (1965).

[40] Schertz, W. W. and Bischoff, K. B. *A.I.Ch.E. J.*, **15**, 597 (1969).

[41] Cairns, E. J. and Prausnitz, J. M. *Ind. Eng. Chem.*, **51**, 1441 (1959).

[42] Mickley, H. S., Smith, K. A. and Korchack, E. I., *Chem. Eng. Sci.*, **20**, 237 (1965).

[43] Levenspiel, O. and Bischoff, K. B. *Advan. Chem. Eng.* **4**, 95–198 (1963).

[44] McHenry, K. W. and Wilhelm, R. H. *A.I.Ch.E. J.*, **3**, 83 (1957).

[45] Ebach, E. A. and White, R. R. *A.I.Ch.E. J.*, **4**, 161 (1958).

[46] Carberry, J. J. and Bretton, R. H. *A.I.Ch.E.J.*, **4**, 367 (1958).

[47] Strang, D. A. and Geankoplis, C. I. *Ind. Eng. Chem.*, **50**, 1305 (1958).

[48] Hiby, J. W. *Interaction between Fluids and Particles*, Institution of Chemical Engineers, London (1962).

[49] Yagi, S., Kunii, D. and Wakao, N. *A.I.Ch.E. J.*, **6**, 543 (1960).

[50] Bischoff, K. B. *Can. J. Chem. Eng.*, **40**, 161 (1963).

[51] Carberry, J. J. and Wendel, M. *A.I.Ch.E. J.*, **9**, 132 (1963).

[52] Danckwerts, P. V. *Chem. Eng. Sci.*, **2**, 1 (1953).

[53] Wehner, J. F. and Wilhelm, R. H. *Chem. Eng. Sci.*, **6**, 89 (1956).

[54] Pearson, J. R. A. *Chem. Eng. Sci.*, **10**, 28 (1959).

[55] Bischoff, K. B. *Chem. Eng. Sci.*, **16**, 731 (1961).

[56] Van Cauwenberghe, A. R. *Chem. Eng. Sci.*, **21**, 203 (1966).

[57] Raymond, L. R. and Amundson, N. R. *Can. J. Chem. Eng.*, **42**, 173 (1964).

[58] Luss, D and Amundson, N. R. *Chem. Eng. Sci.*, **22**, 253 (1967).

[59] Luss, D. *Chem. Eng. Sci.*, **23**, 1249 (1968).

[60] Hlavacek, V. and Hofmann, H. *Chem. Eng. Sci.*, **25**, 173, 187 (1970).

[61] Bernard, R. A. and Wilhelm, R. H. *Chem. Eng. Progr.*, **46**, 233 (1950).

[62] Dorweiler, V. P. and Fahien, R. W. *A.I.Ch.E. J.*, **5**, 139 (1959).

[63] Fahien, R. W. and Smith, J. M. *A.I.Ch.E. J.*, **1**, 25 (1955).

[64] Calderbank, P. H. and Pogorsky, L. A. *Trans. Inst. Chem. Eng.*, (London) **35**, 195 (1957).

[65] Campbell, T. M. and Huntington, R. L. *Petrol. Refiner*, **31**, 123 (1952).

[66] Coberly, C. A. and Marshall, W. R. *Chem. Eng. Progr.*, **47**, 141 (1951).

[67] Kwong, S. S. and Smith, J. M. *Ind. Eng. Chem.*, **49**, 894 (1957).

[68] Kunii, D. and Smith, J. M. *A.I.Ch.E. J.*, **6**, 71 (1960).

[69] Plautz, D. A. and Johnstone, H. F. *A.I.Ch.E. J.*, **1**, 193 (1955).

[70] Yagi, S. and Kunii, D. *A.I.Ch.E. J.*, **3**, 373 (1957).

[71] Hanratty, T. J. *Chem. Eng. Sci.*, **3**, 209 (1954).

[72] Yagi, S. and Kunii, D. *A.I.Ch.E. J.*, **6**, 97 (1960).

[73] Yagi, S. and Wakao, N. *Chem. Eng. Sci.*, **5**, 79 (1959).

[74] Schlünder, E. U. *Chem. Eng. Techn.*, **43**, 651 (1971).

[75] Zehner, P. and Schlünder, E. U. *Chem. Eng. Techn.*, **42**, 333 (1970), **44**, 1303 (1972).

[76] Froment, G. F. *Ind. Eng. Chem.*, **59**, No. 2, 18 (1967).

[77] Froment, G. F. *Chem. Eng. Sci.*, **7**, 29 (1961).

[78] Froment, G. F. *Periodica Polytechnica* (Budapest), **15**, 219 (1971).

[79] Adler, R., Nagel, G., Hertwig, K. and Henkel, D. K. *Chem. Tech.*, **24**, 600 (1972).

[80] Carberry, J. J. and White, D. *Ind. Eng. Chem.*, **61**, 27 (1969).

[81] Beek, J. *Advan. Chem. Eng.*, **3**, 249 (1962).

[82] Crider, J. E. and Foss, A. S. *A.I.Ch.E. J.*, **11**, 1012 (1965).

[83] Marek, M., Hlavacek, V. and John, M. T. *Coll. Czechosl. Chem. Comm.*, **34**, 3664 (1969).

[84] Hlavacek, V. *Ind. Eng. Chem.*, **62**, 8 (1970).

[85] Schwartz, C. S. and Smith, J. M. *Ind. Eng. Chem.*, **45**, 1209 (1953).

[86] Deans, H. A. and Lapidus, L. *A.I.Ch.E. J.*, **6**, 656 (1960).

[87] McGuire, M. and Lapidus, L. *A.I.Ch.E. J.*, **11**, 85 (1965).

[88] Agnew, J. B. and Potter, O. E. *Trans. Inst. Chem. Eng.* (London), **44**, T216 (1966).

[89] Amundson, N. R. *Ber. Bunsen Gesellschaft*, **74**, 90 (1970).

[90] Wicke, E. *Acta Techn. Chim.*, Acad. Naz. Lincei, Varese (1960).

[91] Liu, S. L. and Amundson, N. R. *Ind. Eng. Chem. Fund.*, **1**, 200 (1962), **2**, 12 (1963).

[92] Eigenberger, G. *Chem. Eng. Sci.*, **27**, 1909 (1972), **27**, 1917 (1972).

[93] Kunii, D. and Furusawa *Chem. Eng. J.*, **4**, 268 (1972).

[94] Nielsen, A. *Catalysis Reviews*, Vol. 4, M. Dekker, New York (1971).

[95] Cappelli, A., Collina, A. and Dente, M. *Ind. Eng. Chem. Proc. Des. Devpt.*, **11**, 184 (1972).

[96] Bischoff, K. B. *Chem. Eng. Sci.*, **22**, 525 (1967).

[97] Petersen, E. E. *Chemical Reaction Analysis*, Prentice-Hall, Englewood Cliffs, N.J. (1965).

[98] Weisz, P. B. and Hicks, J. S. *Chem. Eng. Sci.*, **17**, 265 (1962).

[99] Carberry, J. J. *A.I.Ch.E. J.*, **7**, 350 (1961).

[100] Aris, R. *Chem. Eng. Sci.*, **46**, 343 (1969).

[101] Luss, D. *Chem. Eng. Sci.*, **26**, 1713 (1971).

[102] Luss, D. and Lee, J. C. *Chem. Eng. Sci.* **23**, 1237 (1968). **26**, 1433 (1971).

[103] Drott, D. W. and Aris, R. *Chem. Eng. Sci.*, **24**, 541 (1969).

[104] Kehoe, J. P. G. and Butt, J. B. *Proc. 5th Eur. Symp. Chem. React. Engng.* Amsterdam (1972), Elsevier Publishing Co., Amsterdam (1972).

[105] McGreavy, C. and Adderley, C. I. *Chem. Eng. Sci.*, **28**, 577 (1973).

[106] De Wasch, A. P. and Froment, G. F. *Chem. Eng. Sci.*, **26**, 629 (1971).

[107] Singer, E. and Wilhelm, R. M. *Chem. Eng. Progr.*, **46**, 343 (1950).

[108] McGreavy, C. and Cresswell, D. L., *Can. J. Chem. Eng.*, **47**, 583 (1969).

[109] Dente, M., Biardi, G. and Ranzi, E. *Proc. 5th Eur. Symp. Chem. React. Engng.*, Amsterdam (1972), Elsevier Publishing Co., Amsterdam (1972).

[110] Handley, D. and Heggs, P. J. *Trans. Instn. Chem. Engrs.*, **46**-T251 (1968).

[111] Wentz, C. A. and Thodos, G. *A.I.Ch.E. J.*, **9**, 81 (1963).

[112] Reichelt, W. *Chem. Eng. Techn.*, **44**, 1068 (1972).

[113] Collina, A., Corbetta, D. and Cappelli, A. Eur. Symp. "Use of Computers in the Design of Chemical Plants," Firenze (1971).

[114] Livbjerg, H. and Villadsen, J. *Chem. Eng. Sci.*, **27**, 21 (1972).

[115] Calderbank, P. H. *J. Appl. Chem.*, **2**, 482 (1952).
Calderbank, P. H. *Chem. Eng. Progr.*, **49**, 585 (1953).

[116] Bellman R. *Dynamic Programming*, Princeton University Press, Princeton, N.J. (1957).

[117] Roberts, S. M. *Dynamic Programming in Chemical Engineering and Process Control*, Academic Press, New York (1964).

[118] Temkin, M. I. and Pyzhev, V. *Acta Physicochim* (USSR), **12**, 327 (1960).

[119] Nielsen, A. *J. Catal.*, **3**, 68 (1964).

[120] Kjaer, J. *Measurement and Calculation of Temperature and Conversion in Fixed Bed Catalytic Reactors* Gjellerup, Copenhagen (1958).

[121] Dyson, D. C. and Simon, J. M. *Ind. Eng. Chem. Fund.*, **7**, 605 (1968).

[122] Fodor, L., *Génie Chimique*, **104**, 1002 (1971).

[123] DePauw, R. and Froment, G. F. *Chem. Eng. Sci.*, **30**, 789 (1975).

[124] Jackson, R. *Trans. Instn. Chem. Engrs.*, **45**, 160 (1967).

[125] Chou, A., Ray, H. W. and Aris, R. *Trans. Instn. Chem. Engrs.*, **45**, 153 (1967).

[126] Ogunye, A. F. and Ray, M. W. *A.I.Ch.E. J.*, **17**, 43 (1970).

[127] Weekman, V. W. and Nace, D. M. *A.I.Ch.E. J.*, **16**, 397 (1970).

[128] Valstar, J. "A Study of the Fixed Bed Reactor with Application to the Syntheis of Vinylacetate," Ph.D. Thesis, Delft University, Netherlands (1969).

[129] Liu, S. L. *A.I.Ch.E. J.*, **16**, 501 (1970).

[130] Finlayson, B. A. *Chem. Eng. Sci.*, **28**, 1081 (1971).

[131] Calderbank, P. H., Caldwell, A. and Ross, G. *Proc. 4th Eur. Symp. Chem. React. Eng.*, Pergamon Press, New York (1968).

[132] Viville, L. "De oxydatie, reductie en pyrofoor karakter van ijzer-nikkel katalysatoren," Ph.D. Thesis, Rijksuniversiteit Ghent, Belgium (1975).

[133] Satterfield, C. N. *Mass Transfer in Heterogeneous Catalysis*, M.I.T. Press, Cambridge, Mass. (1970).

[134] McGreavy, C. and Thornton, J. M. *Can. J. Chem. Eng.*, **48**, 187 (1970).

[135] Luss, D. *Proc. 4th Int. Symp. Chem. React. Engng.*, Heidelberg (1976).

[136] McGreavy C. and Adderley, C. I. *Adv. Chem. Ser.*, **133**, 519 (1974).

[137] Natta G., Mazzanti G. and Pasquon, I. *Chim. Ind.*, Milan, **37**, 1015 (1955).

[138] Cappelli, A. and Dente M. *Chim. Ind.*, Milan, **47**, 1068 (1965).

[139] Newton, R. *Ind. Eng. Chem.*, **27**, 302 (1935).

[140] Clayton, J. O. and Giaucque, W. F. *J.A.C.S.*, **54**, 2610 (1932).

[141] Dumez, F. J. and Froment, G. F. *Ind. Eng. Chem. Proc. Des. Devpt.*, **15**, 291 (1976).

[142] Eberly, P. E., Kimberlin, C. N., Miller, W. H. and Drushel, H. V. *Ind. Eng. Chem. Proc. Des. Devpt.*, **5**, 193 (1966).

[143] Rajadhyasksha, R. A., Vasudeva, K. and Doraiswamy, L. K. *Chem. Eng. Sci.*, **30**, 1399 (1975).

[144] Lerou, J. and Froment, G. F. *Chem. Eng. Sci.*, **32**, 853 (1977).

[145] Mihail, R. and Iordache, C. *Chem. Eng. Sci.*, **31**, 83 (1976).

[146] Narsimhan, G. *Ind. Eng. Chem. Proc. Des. Devpt.*, **15**, 302 (1976).

[147] Hatcher, W., Viville, L. and Froment, G. F. *Ind. Eng. Chem. Proc. Des. Devt.* **17**, 491 (1978).

[148] Froment, G. F. *Chem. Ing. Tech.* **46**, 374 (1974).

# 12

## NONIDEAL
## FLOW
## PATTERNS
## AND
## POPULATION
## BALANCE
## MODELS

## 12.1 Introduction

The preceding chapters were almost completely concerned with the analysis, design, and operation of chemical reactors with the ideal flow patterns of plug flow or perfect mixing. The only exceptions were the effective transport models of Chapter 11, which were based on reasonable physical models for that situation. However, there are cases that cannot be handled by these techniques, since the flow patterns are not close to either of the extreme limits. This can be caused by such aspects of real equipment as corners, baffles, and so on, that can lead to stagnant regions, or by nonuniform flow paths that can lead to bypassing of fluid. The problem here is often of a diagnostic nature of devising tests to determine the exact flow nonideality and, hopefully, remedy it. Another important application is to reactors with inherently complex configurations such as bubbling fluidized beds, three-phase trickle beds, bubble column slurry reactors, and so on. For these cases, an adequate model to represent their behavior is often required, and are the subject of Chapters 13 and 14 of this book.

We will see that the techniques that have been developed to handle these questions utilize notions of distributions of properties of the reacting fluids in the vessel, in the sense of probability theory. These properties can be the residence times of elements of flowing fluid(s), catalyst activity of particles, crystal size in a crystallizer, and others. The first type is usually termed a "residence time distribution" (RTD) and the others are handled by "particle population balances" that are more general and can be used for properties other than just residence time. The more widely developed and utilized RTD methods will first be discussed, followed by the general population balances.

592

Since extensive treatments of this subject are available in survey chapters and books on process modeling, only a concise discussion will be given (see Levenspiel and Bischoff, Wen and Fan, Levenspiel, and Himmelblau and Bischoff [1, 2, 3, 4]).

## 12.2 Age-Distribution Functions

The "age" of an element of fluid is defined as the time elapsed since it entered the reactor. The concept of a fluid element or "point" was introduced by Danckwerts [5] to mean a volume small with respect to the reactor vessel size, but still large enough to contain sufficient molecules so that continuous properties such as density and concentration can be defined. In liquids and gases at not too low a pressure, this is probably reasonable, but in vacuum systems the methods of kinetic theory would have been used, just as in transport phenomena.

An experiment can be visualized, and actually performed with tracers, such that at an instant of time all fluid elements entering a process vessel are marked. If the vessel outlet stream is then monitored for these tagged particles, several possibilities could be observed. If the vessel had plug flow, no tagged fluid would be seen until a time elapsed equal to the mean holding or residence time of the vessel, at which point all the tagged elements would leave (see Fig. 12.2-1). The other extreme of perfect mixing would show another behavior, also shown in Fig. 12.2-1. This shape is obtained since the instantaneous mixing of the tracer at time zero gives a certain initial concentration, which is then "washed out" of a vessel by the continued inflow of nontracer fluid. Many vessels, of course, give an intermediate behavior, where very little tracer (fluid) leaves the vessel directly from the inlet, most of the fluid spends about one holding time in the vessel, and a little stays in for a long time.

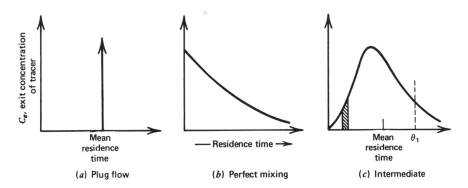

*Figure 12.2-1 Exit concentrations of tracer observed in vessel when input stream is tagged at time zero.*

This type of curve, then, has an ordinate that gives the fraction of fluid that has a certain residence time, which is plotted on the abscissa. In more formal terms, the curve defines the residence time distribution or exit age distribution. The exact definition uses the common symbol $E(\theta)$ for the exit age-distribution frequency function as defined by Danckwerts [6] (see Himmelblau and Bischoff [4] for more details):

$$E(\theta)d\theta = \left( \begin{array}{l} \text{fraction of fluid leaving vessel that has} \\ \text{residence time (exit age) of } (\theta, \theta + d\theta) \end{array} \right) \qquad (12.2\text{-}1)$$

as shown by the striped area in Fig. 12.2-1. Since all the fluid (in the absence of any permanent retention in the vessel) has *some* residence time, the RTD is properly normalized; another way of stating this is that if one waits a sufficiently long time, all the tracer will eventually come out:

$$\int_0^\infty E(\theta)d\theta = 1 \qquad (12.2\text{-}2)$$

The fraction of fluid in the exit stream with age less than $\theta_1$ is

$$\int_0^{\theta_1} E(\theta)d\theta \qquad (12.2\text{-}2a)$$

as shown on Fig. 12.2-1.

Finally, the mean residence time is the centroid of the RTD, and is

$$\tau \equiv \int_0^\infty \theta E(\theta)d\theta \qquad (12.2\text{-}3)$$

$$= \text{(mean holding time)}$$

$$= V/F' \qquad (12.2\text{-}3a)$$

where $V$ = vessel volume and $F'$ = volumetric flow rate, as used in Chapters 9 and 10. Spaulding [7] has shown that this is true for any arbitrary "closed vessel"— one without back diffusion at the flow boundaries.

It is often convenient to use a dimensionless time of $\theta' \equiv \theta/\tau$, and a corresponding version of the RTD, $E(\theta')$. The relation with $E(\theta)$ is found from the basis that both represent the same physical entity, the fraction of exit fluid with age $\theta$:

$$E(\theta)d\theta = E(\theta')d\theta'$$

$$= E(\theta')\frac{d\theta}{\tau}$$

and so:

$$E(\theta') = \tau E(\theta) \qquad (12.2\text{-}4)$$

It can be seen that this relation is also consistent with the normalization, Eq. (12.2-2).

## Example 12.2-1 RTD of a Perfectly Mixed Vessel

The mathematical statement of the above type of experiment is to inject an impulse of tracer into the vessel inlet at time zero. This is represented by the Dirac delta function or perfect unit impulse function:

$$\delta(\theta - \theta_0) \begin{array}{l} = 0,\ \theta \neq \theta_0 \\ \neq 0,\ \theta = \theta_0 \end{array} \tag{a}$$

where $\theta_0$ is a specified value of $\theta$, and

$$\int_{-\infty}^{\infty} \delta(\theta - \theta_0)dt = 1 \tag{b}$$

that is, unit area under the curve.

Another property of use is the "filtering property":

$$\int_{-\infty}^{\infty} \delta(\theta - \theta_0)f(\theta)d\theta = f(\theta_0) \tag{c}$$

This can be heuristically derived by considering the nature of the integrand with Eq. (a), and then using Eq. (b) at $\theta = \theta_0$.

The transient mass balance for a perfectly mixed vessel for which $M$ units of tracer are injected into the inlet stream is:

$$V\frac{dC}{dt} = M\delta(t) - F'C \tag{d}$$

The solution of this simple first-order differential equation, with the initial concentration before injection taken to be zero, $C(0) = 0$, is:

$$C = e^{-(V/F')\theta} \int_0^\theta e^{+(V/F')\theta_1} \frac{M}{V} \delta(\theta_1)d\theta_1$$

$$= \frac{M}{V} e^{-(V/F')\theta} e^0$$

where Eq. (c) has been used. The *fractional* tracer concentration in the outlet stream is, in dimensionless form:

$$\frac{C(\theta)}{(M/V)} = e^{-(V/F')\theta} \tag{e}$$

$$= e^{-\theta/\tau} \tag{e'}$$

$$= e^{-\theta'} \tag{e''}$$

$$= E(\theta') \tag{f}$$

Thus, for a perfect mixer, the RTD is an exponential curve, which is the exact shape sketched in Fig. 12.2-1:

$$E(\theta') = e^{-\theta'} \qquad \text{(g)}$$

or

$$E(\theta) = \frac{1}{\tau} e^{-(\theta/\tau)} \qquad \text{(h)}$$

---

## Example 12.2-2 Determination of RTD from Experimental Tracer Curve

If an impulse of tracer is injected into an arbitrary vessel, the outlet stream tracer concentration would be actually measured in some arbitrary units. The data then have to be properly manipulated to give the distribution function. If $M$ units of tracer are injected into *any* vessel, an overall mass balance gives (for no permanent retention—sometimes an experimental difficulty):

$$M = \int_0^\infty F'C(\theta)d\theta$$

and for constant $F'$,

$$= F' \int_0^\infty C(\theta)d\theta \qquad \text{(a)}$$

As in the previous example, the RTD is given in dimensionless form by:

$$E(\theta') = \frac{C(\theta)}{(M/V)}$$

$$= \frac{V}{F'} \frac{C(\theta)}{\displaystyle\int_0^\infty C(\theta)d\theta} \qquad \text{(b)}$$

or

$$E(\theta) = \frac{C(\theta)}{\displaystyle\int_0^\infty C(\theta)d\theta} \qquad \text{(c)}$$

Thus, $E(\theta)$ is found from the measured outlet concentrations in arbitrary units, and the exact amount of tracer injected doesn't even have to be known. In practice, it's best, however, to also use the mass balance Eq. (a) to check the quality of the experiment. In addition, by knowing $M$ and the integral under the output concentration curve, the flow rate can be found (often used in physiology).

Knowing $E(\theta)$ directly from the experimental data also permits the calculation of the mean residence time:

$$\frac{V}{F'} = \tau = \frac{\displaystyle\int_0^\infty \theta C(\theta)d\theta}{\displaystyle\int_0^\infty C(\theta)d\theta} \tag{d}$$

Equation (d) is often used in two-phase systems, for example, to determine the vessel volume for the flowing phase, $V$ (usually called holdup then), which is difficult to measure by other techniques.

Finally, note that the dimensionless RTD, Eq. 12.2-4, can be obtained directly from experimental tracer data only, by utilizing Eq. (d) to compute $\tau = V/F'$.

Another distribution function of interest in some applications is the "internal age distribution":

$$I(\alpha)d\alpha = \begin{pmatrix} \text{fraction of the fluid inside the vessel} \\ \text{with age } (\alpha, \alpha + d\alpha) \end{pmatrix} \tag{12.2-5}$$

where $\alpha$ is the age, or length of time a fluid element has been in the vessel. It has properties similar to $E(\theta)$:

$$\int_0^\infty I(\alpha)d\alpha = 1 \tag{12.2-6}$$

$$\int_0^{\alpha_1} I(\alpha)d\alpha = (\text{fraction of fluid in vessel younger than age } \alpha_1) \tag{12.2-7}$$

$$\tau_I \equiv \int_0^\infty \alpha I(\alpha)d\alpha = \text{mean internal age} \tag{12.2-8}$$

As might be expected, there is an interrelationship between $I(\alpha)$ and $E(\theta)$ since the fluid entering a vessel at a given time obviously either leaves it or stays inside. This interrelationship is (e.g., 4):

$$\tau I(\alpha) = 1 - \int_0^\alpha E(\theta)d\theta \tag{12.2-9}$$

or

$$E(\theta) = -\tau \frac{d}{d\theta} I(\theta) \tag{12.2-10}$$

For the perfectly mixed flow reactor, the internal age distribution function is found, from Eq. (h) of Ex. 12.2-1,

$$\tau I(\theta) = 1 - \int_0^\theta \frac{1}{\tau} e^{-\theta_1/\tau} \, d\theta_1$$

$$= 1 - (1 - e^{-\theta/\tau})$$

Thus,

$$I(\theta) = \frac{1}{\tau} e^{-\theta/\tau} \qquad\qquad (12.2\text{-}11)$$

$$= E(\theta) \qquad\qquad (12.2\text{-}11a)$$

Equation 12.2-11a simply expresses that in a perfectly mixed vessel the internal and exit conditions are identical.

Equation 12.2-9 is a special case of the general convolution result from linear systems theory, based on the fact that $E(\theta)$ is the impulse response of the flow system:

$$[\text{OUTPUT } (\theta)] = \int_0^\theta E(\theta_1)[\text{INPUT } (\theta - \theta_1)] d\theta_1 \qquad (12.2\text{-}12)$$

This result is valid for *any* input. In particular, for a step function of tracer, INPUT $(\theta) = U(\theta)$ and Eq. 12.2-12 give

$$[\text{OUTPUT } (\theta)] = \int_0^\theta E(\theta_1)[1] d\theta_1$$

$$= 1 - \tau I(\theta)$$

from Eq. 12.2-9; thus, this type of tracer test permits determination of the internal age distribution function.

A comprehensive listing of various tracers and experimental methods is provided by Wen and Fan [2], including flow visualization techniques.

### Example 12.2-3 Calculation of Age-Distribution Functions from Experimental Data from Himmelblau and Bischoff [4]

In this example we use the data obtained by Hull and von Rosenberg [8] as presented in an article entitled "Radiochemical Tracing of Fluid Catalyst Flow." They injected a pulse of radioactive tracer into the catalyst inlet at the bottom of a reactor. The radioactive tracer concentration was then measured at various points in the reactor. With this information, the mixing patterns at the various measuring locations could be determined. Their Run No. 4 gave the following data:

| Time $\theta$, min | Counts/min $\times$ $10^{-3}$ (smoothed to equidistant points) |
|---|---|
| 0 | 0 |
| 0.5 | 5 |
| 1.0 | 22 |
| 1.5 | 27 |
| 2.0 | 26 |
| 2.5 | 22 |
| 3.0 | 19 |
| 3.5 | 15 |
| 4.0 | 10 |
| 4.5 | 7 |
| 5.0 | 4 |
| 5.5 | 3 |
| 6.0 | 3 |
| 6.5 | (1) |
| 7.0 | (0) |
| | 164 |

Each of the age-distribution integrals previously described will be approximately evaluated by summation instead of integration.

$$\int_0^\infty C d\theta \simeq \sum_{t=0}^{7.0} C \Delta\theta = (164 \times 10^{+3})(0.5)$$

$$= 82 \times 10^{+3} (cpm)(min)$$

Then, using Eq. (c) of Ex. 12.2-2:

$$E(\theta) = \frac{C(\theta)}{\int_0^\infty c d\theta} = \frac{C(\theta)}{82 \times 10^{+3}}$$

The calculated values of $E(\theta)$ are

| Time $\theta$, min | $E(\theta)$, (min)$^{-1}$ | $\theta E(\theta)$ |
|---|---|---|
| 0.0 | 0 | 0 |
| 0.5 | 0.061 | 0.030 |
| 1.0 | 0.27 | 0.27 |
| 1.5 | 0.33 | 0.50 |
| 2.0 | 0.32 | 0.64 |
| 2.5 | 0.27 | 0.67 |
| 3.0 | 0.23 | 0.69 |
| 3.5 | 0.18 | 0.63 |
| 4.0 | 0.12 | 0.48 |
| 4.5 | 0.085 | 0.38 |
| 5.0 | 0.049 | 0.25 |
| 5.5 | 0.037 | 0.20 |
| 6.0 | 0.037 | 0.22 |
| 6.5 | 0.012 | 0.08 |
| 7.0 | 0 | 0 |
| | $2.001 = 1/\Delta\theta$ | 5.04 |

We can now find the mean residence time from the distribution curve

$$\tau = \int_0^\infty \theta E(\theta)d\theta \simeq \Sigma\theta E(\theta)\Delta\theta$$

$$= (5.04)(0.5) = 2.5 \text{ min}$$

The article also indicated that the catalyst rate was 340 lb/hr (0.0428 kg/s) and the holdup was 18.4 lb (8.35 kg). This gives

$$\tau_{\text{expt}} = \frac{18.4(60)}{340} = 3.25 \text{ min}$$

There is approximately a 20 percent deviation from the value of $\tau_{\text{expt}}$, which is not too bad considering the approximations involved in the analysis and the model itself.

The actual purpose of these authors was not, of course, to check the mean residence time by using radioactive tracers. They took data for various conditions and used this information qualitatively to improve the reactor operation. They stated that an appraisal of the data "is sufficient to demonstrate the unique ability of the tracer technique to provide vital information concerning the effects of operating conditions and structural designs on solids-mixing patterns in fluidized systems."

The internal age-distribution function, $I(\theta)$, could also be found from the tabulated $E(\theta)$ values. The function $I(\theta)$ would be of value in considering properties of the solid within the reactor, such as the catalyst activity level. Recalculations of $E(\theta)$, $I(\theta)$, are given below from an accurate calculation using a digital computer:

| $\theta'$ | $E(\theta')$ | $I(\theta')$ |
|-----------|--------------|--------------|
| 0.000 | 0.0000 | 1.000 |
| 0.197 | 0.156 | 0.991 |
| 0.394 | 0.686 | 0.914 |
| 0.591 | 0.842 | 0.760 |
| 0.788 | 0.811 | 0.594 |
| 0.986 | 0.686 | 0.447 |
| 1.183 | 0.593 | 0.322 |
| 1.380 | 0.468 | 0.217 |
| 1.577 | 0.312 | 0.139 |
| 1.774 | 0.218 | 0.087 |
| 1.971 | 0.125 | 0.053 |
| 2.168 | 0.094 | 0.032 |
| 2.365 | 0.094 | 0.014 |
| 2.562 | 0.031 | 0.003 |
| 2.759 | 0.000 | 0.000 |
| $\tau = 2.54$ min | $\sigma^2 = 0.262$ | |

Figure 1 shows these results.

We see from the figure that the exit age distribution, $E(\theta')$, is not very close to that for perfect mixing; $e^{-\theta'}$, while the internal age distribution, $I(\theta')$, is much better represented by this approximation. The main region of divergence is for small $\theta' < 1.0$, and this can cause significant differences in various applications to be discussed later.

---

The fluid age distributions defined above for steady flow systems can be extended to nonsteady situations. Nauman [9] illustrated this for vessels with time-varying inflow, outflow, and volume related by

$$\frac{dV}{dt} = F'_0(t) - F'_e(t) \tag{12.2-13}$$

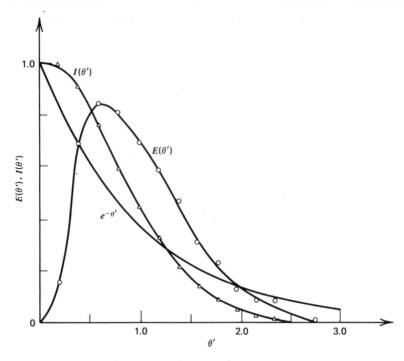

*Figure 1  Age distribution functions for solids in a fluidized bed.*

(e.g., Eq. 7.2.b-12a for constant density). Then, the RTD and the like will be a function of both the residence time and the "clock time," $E(\theta, t)$. For example, Nauman [9] showed that, for a perfect mixer,

$$E(\theta, t) = \frac{F_0'(t - \theta)}{V(t - \theta)} \exp\left[ - \int_{t-\theta}^{t} \frac{F_0'(t_1)}{V(t_1)} dt_1 \right] \qquad (12.2\text{-}14)$$

which reduces to Eq. (h) of Ex. 12.2-1 for $F_0'/V = $ constant.

Other more general situations are discussed in Himmelblau and Bischoff [4], and by Chen [10]; recycle and nonzero initial inventory of fluid have been considered by Chu and Fan [11]; also see Wen and Fan [2].

These tracer techniques can also be utilized to determine local sojourn time distributions; see Zvirin and Shinnar [12]. The interpretations are based on rather sophisticated notions from probability theory combined with interconnected "compartments" representing the internal regions of the system. More detailed information about the important flow patterns can be obtained in this way, but there are as yet no specific applications.

## 12.3 Interpretation of Flow Patterns from Age-Distribution Functions _____

### 12.3.a Measures of the Spectrum of Fluid Residence Times

Now that the physical meaning, definitions, and basis of experimental determination of the various age-distribution functions has been presented, let us briefly discuss how inspection of the curves can be used to infer certain properties of the flow patterns. The most obvious characteristic, perhaps, is the "width" of the curve. For plug flow, $E(\theta)$ is very narrow ("zero" width), while for perfect mixing, the curve is rather wide—refer to Fig. 12.2-1. Thus, the width of the observed curve can be used to decide where the system is in the spectrum of mixing between plug flow and perfect mixing. If a plug flow pattern is required for the particular application (e.g., Chapter 10), this information can be used directly to decide whether the reactor has the proper flow design.

Using the analogy with probability theory, a useful measure of the "width" of the RTD is the variance with respect to the mean residence time:

$$\sigma^2 \equiv \int_0^\infty (\theta' - 1)^2 E(\theta') d\theta'$$

$$= \int_0^\infty (\theta')^2 E(\theta') d\theta' - 1 \tag{12.3.a-1}$$

The dimensionless residence time, $\theta'$, is used so that comparisons of "widths" are for curves all at the same "location"—$\theta = \tau$ or $\theta' = 1$. Higher statistical moments can also be used in principle, but calculating them from the usual rather scattered tracer data is quite difficult in practice.

### *Example 12.3-1 Age-Distribution Functions for a Series of n-Stirred Tanks*

This is a representation for a RTD yielding simple mathematical results. The mass balance for the $i$th tank, in a series of $n$, is as follows:

$$\left(\frac{V_t}{n}\right) \frac{dC_i}{dt} = F'C_{i-1} - F'C_i \tag{a}$$

with

$$C_{i \neq 0} = 0$$

$$\text{at } t = 0 \tag{b}$$

$$C_{i=0} = \frac{M}{F'} \delta(t)$$

The simplest solution method uses Laplace transforms, although the set of differential equations can be solved successively for $i = 1, 2, 3 \ldots n$. The Laplace transform of Eqs. (a) and (b) is:

$$\left(\frac{\tau}{n}s + 1\right)\bar{C}_i = \bar{C}_{i-1}$$

$$\bar{C}_{i=0} = \frac{M}{F'} \tag{c}$$

(note that $\tau$ is here the *total* mean residence time of the $n$ reactors) and so

$$\bar{C}_i = \frac{\bar{C}_{i-1}}{1 + s\tau/n}$$

$$= \frac{\bar{C}_{i-2}}{(1 + s\tau/n)^2}$$

or

$$\vdots$$

$$\bar{C}_n = \frac{M/F'}{(1 + s\tau/n)^n} \tag{d}$$

The inverse transform is directly found from tables:

$$E(\theta') = \frac{C(\theta)}{(M/V)} = \frac{(n)^n}{(n-1)!}(\theta')^{n-1}e^{-n\theta'} \tag{e}$$

These $E$-curves are illustrated in Fig. 1.

By using a table of integrals, one can easily show that

$$\int_0^\infty E(\theta')d\theta' = \frac{(n)^n}{(n-1)!}\int_0^\infty (\theta')^{n-1}e^{-n\theta'}\,d\theta' = 1$$

and

$$\int_0^\infty \theta' E(\theta')d\theta' = \frac{(n)^n}{(n-1)!}\int_0^\infty (\theta')^n e^{-n\theta'}\,d\theta' = 1$$

and that

$$\sigma^2 = \int_0^\infty (\theta')^2 E(\theta')d\theta' - 1 = \frac{(n+1)(n)}{n^2} - 1 = \frac{1}{n} \tag{f}$$

Thus, the curve width, or variance, changes from $\sigma^2 = 1$ for $n = 1$ to $\sigma^2 = 0$ for $n \to \infty$, and indicates that the behavior of $j$-stirred tanks in series ranges from perfect mixing to plug flow. Recall from Chapter 10 that similar behavior was also indicated for chemical reactions occurring in a series of steady-state stirred-tank chemical reactors. One practical implication of this is that if a plug flow

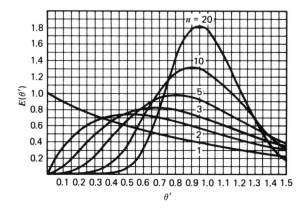

*Figure 1  E curve for tanks-in-series model.*

pattern is desired for the reaction aspects (Chapter 10), but stirring is essential for other reasons (e.g., a suspended slurry), a compromise design would call for a series of stirred tanks. This is often done for polymerization, where stirring is needed for heat transfer and/or to maintain an emulsion, but a plug flow profile might be desired for the molecular weight distribution.

---

## *Example 12.3-2  RTD for Combinations of Noninteracting Regions*

If an impulse input is used at the beginning of a combination of independent regions, such that the RTD of any one region is unaffected by the presence of the others, the combination RTD is given by the convolution formula, Eq. 12.2-12:

$$E(\theta) = \int_0^\theta E_1(\theta_1)E_2(\theta - \theta_1)d\theta_1 \qquad (a)$$

If $E_1(\theta)$ and $E_2(\theta)$ are normalized, so is $E(\theta)$:

$$\int_0^\infty E(\theta)d\theta = \int_0^\infty d\theta \int_0^\theta d\theta_1 \, E_1(\theta_1)E_2(\theta - \theta_1)$$

Changing the order of integration, and defining the new variable $\theta_2 \equiv \theta - \theta_1$, leads to

$$\int_0^\infty E(\theta)d\theta = \int_0^\infty d\theta_1 E_1(\theta_1) \int_0^\infty d\theta_2 E_2(\theta_2)$$

$$= (1)(1) = 1 \qquad (b)$$

NONIDEAL FLOW PATTERNS AND POPULATION BALANCE MODELS _____ **605**

Similarly, the overall mean residence time is found:

$$\tau = \int_0^\infty \theta\, d\theta \int_0^\theta d\theta_1\, E_1(\theta_1) E_2(\theta - \theta_1)$$

$$= \int_0^\infty d\theta_1\, E_1(\theta_1) \left[ \int_0^\infty \theta_2\, d\theta_2\, E_2(\theta_2) + \theta_1 \int_0^\infty d\theta_2\, E_2(\theta_2) \right]$$

$$= \tau_2 + \tau_1 \qquad\qquad\qquad\qquad\qquad\qquad (c)$$

where the same procedure was utilized.

Finally, a similar treatment for the variance gives

$$\sigma_\theta^2 = \int_0^\infty (\theta - \tau)^2 E(\theta)\, d\theta$$

$$= \sigma_{\theta,1}^2 + \sigma_{\theta,2}^2 \qquad\qquad\qquad\qquad (d)$$

This can be put into dimensionless form for consideration of the width of the overall RTD:

$$\sigma^2 = \frac{\sigma_\theta^2}{\tau^2} \qquad\qquad\qquad\qquad\qquad (e)$$

By consecutively repeating these manipulations for multiple region convolutions, the moments would be

$$\tau = \sum_i \tau_i \qquad \sigma_\theta^2 = \sum_i \sigma_{\theta,i}^2 \qquad\qquad (f)$$

These results are useful in computing the moments of complicated flow systems, especially for the flow models in Sec. 12.5.

---

### 12.3.b  Detection of Regions of Fluid Stagnancy from Characteristics of Age Distributions

In addition to use of statistical measures, such as the variance $\sigma^2$, to deduce certain features of the vessel flow patterns from the RTD curves, certain other aspects (usually malfunctions) are also of interest. These include possible significant bypassing of some of the fluid and/or regions of very stagnant fluid, or "dead space." Since these cause either gross under- or overconversion of the reactants, they are usually detrimental to the operation of the reactor. Thus, if a reactor does not appear to be properly behaving, tracer tests can be used to decide if the problem concerns the flow patterns, or if other sources for the problem should be sought. These "troubleshooting" applications can sometimes suggest improvements to be made in the reactor equipment, for example, adding (or eliminating) baffles, repacking, modifying the feed sparger, and so on.

The characteristic shapes of the age-distribution function curves when these flow patterns occur are as follows. If a region of the vessel retains a portion of the fluid for an order of magnitude greater than the mean residence time of the total fluid, then, for all practical purposes, that portion is essentially at rest and the region is wasted space in the vessel. In addition, this material will also have experienced an order of magnitude greater severity in reaction conditions, and often leads to undesired side reactions. When this phenomenon occurs, the $E(\theta)$ curve will have a very long tail, indicating that the fluid is held in the dead space.

The converse situation of bypassing of a significant portion of the fluid produces, in ideal form, two "humps" in the $E(\theta)$ curve—one corresponding to the bypassing fluid and the second for the remainder. This maldistribution is often easier seen on the $I(\theta)$ curve, since a portion of the inventory very rapidly leaves the vessel, followed by the main portion of the fluid.

Further discussion of these techniques is provided in Himmelblau and Bischoff [4] and specific examples illustrated in Bischoff and McCracken [13]. Also see Wen and Fan [2].

## 12.4 Application of Age-Distribution Functions

The direct application of RTD to predict chemical reactor behavior is based on considering the extents of chemical reaction in the fluid elements and then summing over all the elements when they exit from the reactor. In other words, each fluid element is considered as a little batch reactor, and the total reactor conversion is the average over fluid elements. Mathematically, this can be stated as:

$$\begin{bmatrix} \text{Mean concentration} \\ \text{of reactant leaving} \\ \text{the reactor unreacted} \end{bmatrix} = \sum \begin{bmatrix} \text{Concentration of} \\ \text{reactant remain-} \\ \text{ing in an element} \\ \text{of age } (\theta,\, \theta + d\theta) \end{bmatrix} \begin{bmatrix} \text{Fraction of exit} \\ \text{stream that con-} \\ \text{sists of elements} \\ \text{of age } (\theta,\, \theta + d\theta) \end{bmatrix}$$

where the sum includes all the elements in the exit stream. Then,

$$\bar{C}_A = \int_0^\infty C_A(\theta) E(\theta) d\theta \tag{12.4-1}$$

where the concentration in an element $C_A(\theta)$, depends on the residence time of the element according to:

$$-\frac{dC_A}{d\theta} = r_A(C_A(\theta)) \tag{12.4-2}$$

with

$$C_A|_{\theta=0} = C_{A0} = \text{feed concentration}$$

For a first-order reaction,

$$\frac{dC_A}{d\theta} = -kC_A$$

or

$$C_A = C_{A0} e^{-k\theta}$$

Then, Eq. 12.4-1 becomes

$$\frac{\bar{C}_A}{C_{A0}} = \int_0^\infty e^{-k\theta} E(\theta) d\theta \qquad (12.4\text{-}3)$$

Now the exit concentrations can be easily found for various RTD. For a perfectly mixed reactor, the use of Eq. (h) of Ex. 12.2-1 gives

$$\frac{\bar{C}_A}{C_{A0}} = \int_0^\infty \frac{1}{\tau} e^{-[k + (1/\tau)]\theta} d\theta$$

$$= \frac{1}{1 + k\tau}$$

which is identical to that derived in Chapter 10. At this point, it appears that there are two approaches for determining the conversion in a perfectly mixed reactor. In Chapter 10, it was assumed that there was complete mixing in the reactor, and no RTD appeared to be explicity necessary. In this section, the RTD *was* used, and for the first-order reactions gave identical results. Further clarification of the physical meaning of these two approaches will be discussed below. For plug flow,

$$\frac{\bar{C}_A}{C_{A0}} = \int_0^\infty e^{-k\theta} \delta(\theta - \tau) d\theta$$

$$= e^{-k\tau}$$

which is identical to the result in Chapter 9. For *n*-stirred tanks in series, as in Ex. 12.3-1,

$$\frac{\bar{C}_A}{C_{A0}} = \frac{1}{\tau^n} \frac{(n)^n}{(n-1)!} \int_0^\infty \theta^{n-1} e^{-[k + (n/\tau)]\theta} d\theta$$

$$= \frac{1}{[1 + (k\tau/n)]^n} = \frac{1}{[1 + (k\tau_n)]^n}$$

which once more is the same as derived in Chapter 10 (Here, $\tau$ is taken as the *total* holding time of the *n*-reactors.)[1]

---

[1] The above integral is easiest done by recognizing that Eq. 12.4-3 is really the definition of the Laplace transform with respect to $\theta$ of $E(\theta)$, but with $k$ substituted for $s$. Thus, the exit concentration for a first-order reaction is found from the transforms of any $E(\theta)$ by merely substituting $k$ for $s$.

The real utility of Eq. 12.4-1, 2 is not to rederive simple results, of course, but for use with more complex reactions and RTD. It is emphasized that Eqs. 12.4-1, 2 can also be used without any models at all, in that experimental concentrations as a function of residence time, and the measured RTD, can be directly numerically integrated to predict the conversion in the flow reactor. An interesting application to performance criteria for industrial reactors was given by Murphree, Voorhies, and Mayer [14].

Wei [15] has shown, by using the Wei–Prater decomposition scheme discussed in Chapter 1, that the same concept can be used for complex first-order reactions. In terms of the vector of concentrations:

$$\overline{\mathbf{C}}(\tau) = \int_0^\infty E(\theta)[\mathbf{C}(\theta)]d\theta$$

$$= \int_0^\infty E(\theta)\exp[-\mathbf{K}\theta]\mathbf{C}(0)d\theta$$

This assumes that all chemical species have the same residence time distribution, and is very convenient to compute the reaction paths for different contacting patterns. Matsuyama and Miyauchi [16] have also considered some aspects of this. An important conclusion of Wei [15] is that "for a reactor with distribution of residence times, all reactions are slowed down in comparison with those in a plug flow reactor, but the faster reactions are slowed down a great deal more than the slower ones. Consequently, the occurrence of distribution of residence times makes all reaction rates of the characteristic species nearly equal." That is, the differences between the various reaction rates are decreased, thereby decreasing the selectivity. This is similar to the diffusion effects considered in Chapter 3.

### Example 12.4-1 Mean Value of the Rate Constant of a Reaction Carried Out in a Well-Mixed Reactor

The flow pattern of solids in fluidized bed can, as a first approximation, be assumed to be perfectly mixed—Chapter 13 deals with this in detail. Weekman [17] used this concept together with a time variable rate constant to compute its mean value. This result was then used in a comparison of fluidized and fixed bed reactors for catalytic cracking.

When expressing the catalyst activity as a function of the *coke content* of the catalyst, the rate coefficient of the main reaction takes the form (see Chapter 5):

$$k(C_c) = k^0 e^{-\alpha C_c} \tag{a}$$

where $\alpha$ is the deactivation constant. When the deactivation function is identical for the main reaction and for the coking reaction the kinetic equation for the latter can be written:

$$r_C = \frac{dC_c}{d\theta} = k_C^0 f(C_A, C_B, \ldots) e^{-\alpha C_c} \tag{b}$$

whereby $f(C_A, C_B, \ldots)$ is a constant over the reactor for complete mixing conditions. After integration:

$$e^{-\alpha C_c} = \frac{1}{1 + \alpha k_C^0 f(C_A, C_B, \ldots)\theta} \tag{c}$$

and Eq. 1 becomes:

$$k(\theta) = \frac{k^0}{1 + \alpha k_c^0 f(C_A, \ldots)\theta} \tag{d}$$

The mean value of $k$ over the internal age distribution for the solids $I_s(\theta)$ is given by:

$$\bar{k} = \int_0^\infty k(\theta) I_s(\theta) d(\theta) \tag{e}$$

According to Eq. 12.2-11, this internal age distribution for the perfectly mixed solids is:

$$I_s(\theta) = \frac{1}{\tau_s} e^{-\theta/\tau_s} \tag{f}$$

where $\tau_s$ = mean residence time of the *solids*. The mean value of $k$ then becomes

$$\frac{\bar{k}}{k^0} = \int_0^\infty \frac{e^{-(\theta/\tau_s)}}{1 + [\alpha k_c^0 f(C_A, \ldots)\tau_s](\theta/\tau_s)} d(\theta/\tau_s) \tag{g}$$

$$= \frac{e^{1/[\alpha k_c^0 f(C_A, \ldots)\tau_s]}}{[\alpha k_c^0 f(C_A, \ldots)\tau_s]} E_1[1/[\alpha k_c^0 f(C_A, \ldots)\tau_s]] \tag{h}$$

For the typical particular case of $[\alpha k_c^0 f(C_A, \ldots)\tau_s] = 1.0$

$$\frac{\bar{k}}{k^0} = 0.592$$

This value can now be compared with the $\bar{k}$ value calculated from the data from Ex. 12.2-3, to gauge the appropriateness of the perfect solid mixing approximation for a commercial fluidized bed. Retaining Eq. (d) for $k(\theta)$ (i.e., complete mixing for the gas phase), the following integral has to be evaluated from the data:

$$\frac{\bar{k}}{k^0} = \int_0^\infty \frac{I_s(\theta_s')}{1 + \theta_s'} d\theta_s' \tag{i}$$

where $\theta'_s = \theta/\tau_s$. The results are given in Table 1:

Table 1

| $\theta'_s$ | $I_s(\theta'_s)$ | $\dfrac{I_s}{1 + \theta'_s}$ |
|---|---|---|
| 0 | 1.000 | 1.000 |
| 0.197 | 0.991 | 0.828 |
| 0.394 | 0.914 | 0.656 |
| 0.591 | 0.760 | 0.478 |
| 0.788 | 0.594 | 0.332 |
| 0.986 | 0.447 | 0.225 |
| 1.183 | 0.322 | 0.147 |
| 1.380 | 0.217 | 0.091 |
| 1.577 | 0.139 | 0.054 |
| 1.774 | 0.087 | 0.031 |
| 1.971 | 0.053 | 0.018 |
| 2.168 | 0.032 | 0.010 |
| 2.365 | 0.014 | 0.004 |
| 2.562 | 0.003 | 0.001 |
| 2.759 | 0.000 | 0.000 |
| | | 3.875 |

Thus $\dfrac{\bar{k}}{k^0} = \sum \dfrac{I_s(\theta_s)}{1 + \theta_s} \, \Delta\theta_s = 3.875 \times 0.197$

$$= 0.763$$

Thus, in this case, the perfect mixing approximation could be about 25 percent in error. The figure of Ex. 12.2-3 shows that the perfect mixing approximation is in error primarily for small $\theta < 1.0$, and it is seen from Eq. (i) that this is just that region *most heavily weighted by the integration procedure.*

The next example considers the slightly more complicated case of a second-order reaction in a perfectly mixed reactor, and also introduces a subtle assumption that has actually been made in the derivation of Eq. 12.4-1, 2.

## Example 12.4-2  Second-Order Reaction in a Stirred Tank

For a second-order reaction, the solution of Eq. 12.4-2 is (Chapter 1 or 8):

$$\frac{C_A}{C_{A0}} = \frac{1}{1 + kC_{A0}\theta} \tag{a}$$

Then, using Eq. 12.4-1 and $E(\theta)$ for perfect mixing:

$$\frac{\bar{C}_A}{C_{A0}} = \frac{1}{\tau} \int_0^\infty \frac{e^{-\theta/\tau}}{1 + kC_{A0}\theta}\, d\theta \tag{b}$$

$$= \frac{e^{1/kC_{A0}\tau}}{kC_{A0}\tau} E_1\left(\frac{1}{kC_{A0}\tau}\right)$$

$$E_1(x) \equiv \text{exponential integral} \equiv \int_x^\infty \frac{e^{-y}}{y}\, dy$$

However, in Chapter 10 an *alternate* result was found for a second-order reaction in a perfectly mixed reactor by simply writing a mass balance without reference to RTD:

$$\frac{C_A}{C_{A0}} = \frac{-1 + \sqrt{1 + 4kC_{A0}\tau}}{2kC_{A0}\tau} \tag{c}$$

This comparison by Metzner and Pigford [18] indicates that matters are not as straightforward as the above developments would indicate. The next example will illustrate another facet of this problem, and then a discussion will be given.

---

### Example 12.4-3 Reactions in Series Plug Flow and Perfectly Mixed Reactors

Consider a plug flow reactor, with mean residence time $\tau_1$, followed by a perfectly mixed reactor, with mean residence time, $\tau_2$. The overall RTD for this system will merely be that for a perfectly mixed vessel, but with a time delay caused by the plug flow vessel:

$$E(\theta) = \frac{1}{\tau_2} U(\theta - \tau_1)e^{-[(\theta - \tau_1)]/\tau_2} \tag{a}$$

If the order of the two vessels is reversed, exactly the same RTD would be obtained, even though the time delay is now after rather than before the mixing. Thus, a tracer test would not differentiate between the two arrangements.

Now a first-order reaction will give an exit concentration of

$$\frac{C_A}{C_{A0}} = \frac{e^{-k\tau_1}}{1 + k\tau_2} \tag{b}$$

for either case, as can easily be derived by the results of Chapters 9 and 10. Kramers and Westerterp [19] pointed out, however, that for a second-order reaction, two different results are found. When the plug flow reactor is followed by the completely mixed reactor,

$$\frac{C_A}{C_{A0}} = \frac{-1 + \sqrt{1 + 4\left(\dfrac{kC_{A0}\tau_2}{1 + kC_{A0}\tau_1}\right)}}{2kC_{A0}\tau_2} \tag{c}$$

and when the completely mixed reactor comes first,

$$\frac{C_A}{C_{A0}} = \frac{-1 + \sqrt{1 + 4kC_{A0}\tau_2}}{2kC_{A0}\tau_2 + kC_{A0}\tau_1(-1 + \sqrt{1 + 4kC_{A0}\tau_2})} \tag{d}$$

The conversions, $1 - C_A/C_{A0}$, for $kC_{A0}\tau_1 = 1$ and $\tau_2/\tau_1 = 4$, are as follows: For plug flow followed by complete mixing,

$$x = 75 \text{ percent}$$

For complete mixing followed by plug flow,

$$x = 72 \text{ percent}$$

For segregated flow,

$$x = 77 \text{ percent (Eq. 12.4-1)}$$

Let us now discuss the meaning of these results.

---

The most obvious result of the above example is that the linear (first-order) reactions differed in their results from the nonlinear (second-order). This was true in the first example, where more detailed consideration will indicate that result Eq. (b) from Eqs. 12.4-1, 2 was based on the concept of separate, completely segregated fluid elements while the "standard" result Eq. (c) actually assumed completely intimate mixing of *all* the fluid elements right down to the molecular level. Thus, these two extremes of mixing are not detectable by the linear processes of concentration mixing (RTD tracer tests) or first-order reaction, but are different for the second-order reaction. Briefly, the reason for this is that the first-order processes are in principle dependent *only* on the length of time the molecules spend in the vessel, but *not* on exactly where they are located during their sojourn. Nonlinear processes, on the other hand, depend on the encounter of two different sets of molecules (or fluid elements), and so depend both on how long the fluid element was in the system and *also* on what it "saw" while there. Thus, the RTD measures *only* the time that various fractions of fluid reside in the vessel, but gives no information on the mixing details inside the system. The term "macromixing" is used for the former and "micromixing" for the latter. Thus, for the given state of perfect macromixing, two extremes of micromixing are possible: complete segregation and perfect micromixing.

Example 12.4-3 is another example of the same phenomenon in that the difference between the two arrangements is in the earliness of mixing: at the end of the reactor or at the beginning. Also, the state of segregation again is important — that is, the third conversion value. For (macromixing) RTD's (apart from perfect mixing) then, both the degree of segregation as well as the earliness of mixing through the reactor, which now has concentration profiles, needs to be considered. Zwietering [20] developed a general treatment of these micromixing effects using "life expectation" distribution functions, in addition to the age distribution functions already discussed above. He showed that for a given (macro) RTD, two bounds can be put on the reactor behavior: complete segregation and maximum mixing. The latter is perfect mixing only for the case of perfect macromixing — a stirred tank.

For other cases, the following arguments led to the formula to compute the reactor performance with maximum mixing. The complete segregation case, described above in terms of batch fluid particles moving through the reactor, can also be considered as an abstract plug flow reactor with side exit streams in proportion to the actual reactor RTD. Thus, the opposite extreme, maximum mixing, would be defined by an abstract plug flow reactor with side entrance streams in proportion to the actual reactor RTD — this provides for as early a mixing as possible of all the fluid elements with the same life expectations. The final result of Zwietering [20] was that one should solve Eq. 12.4-4 and find the outlet conversion from $C_{Am}$ (at $\lambda = 0$):

$$\frac{dC_{Am}}{d\lambda} = r_A(C_{Am}) + \Lambda(\lambda)(C_{Am} - C_{A0}) \qquad (12.4\text{-}4)$$

with

$$\frac{dC_{Am}(\infty)}{d\lambda} = 0 \qquad (12.4\text{-}4a)$$

where

$C_{Am}$ = concentration involved in maximum mixing calculation
$\lambda$ = life expectation
$\Lambda(\lambda) = E(\lambda)/I(\lambda)$ = life expectation function

In using Eq. 12.4-4a, the boundary condition at "infinity" can be applied at about $\lambda_\infty \simeq 6\tau$. Also, the value of $C_{Am}(\infty)$ can be found from Eq. 12.4-4 using Eq. 12.4-4a.

To summarize, for a given rate equation, $r_A(C_A)$, and RTD, $E(\theta)$, bounds on the reactor performance are obtained by using Eq. 12.4-1, 2 for the complete segregation limit, and Eq. 12.4-4 for the maximum mixing limit. The actual state of micromixing lies, of course, between the two extremes, but further details are extremely difficult to utilize, either theoretically or experimentally. Thus, it is of

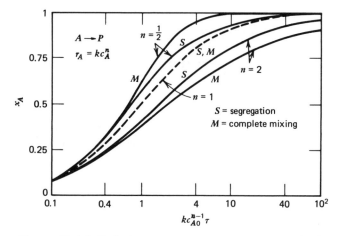

*Figure 12.4-1 Relative conversion in a reactor with the residence time distribution of a tank reactor; a comparison between segregated flow (S) and complete mixing (M). (from Kramers and Westerterp [19]).*

some importance to get an idea of the magnitude of the spread in the bounds under various conditions as follows.

Kramers and Westerterp [19] present results for the extreme case of a perfectly macromixed reactor with various reactions occurring. Figure 12.4-1 illustrates the results, and it can be seen that the differences between the levels of micromixing are small. Certain other kinetic forms, having autocatalytic ranges of approximately "negative order" behavior could cause larger differences. This may also be true with complex kinetics (e.g., see Zoulalian and Villermaux [21]). In particular, the free radical polymerization discussed in Chapter 10 appears to have large differences—computations based on extended forms of Eq. 12.4-1, 2 are given by Nagasubramanian and Graessley [22].

Hofmann [23] has presented a series of calculations giving the two limiting bounds for various widths of RTD and various reaction orders. For a second-order reaction with $kC_{A0}\tau = 10$:

| Per Cent Conversion values for | Number of tanks, $n$ | | | |
|---|---|---|---|---|
| | 1 | 2 | 3 | $\infty$ |
| Complete segregation | 79.8 | 86.0 | 87.8 | 90.9 |
| Maximum mixing | 73.0 | 80.4 | 83.4 | 90.9 |
| Difference in conversion | −6.8 | −5.6 | −4.4 | 0 |

These results show that the magnitude of the difference in bounds decreases for a narrow RTD, and are zero, of course, for plug flow where micromixing differences have no meaning.

These results indicate that, for most cases, the extreme limits of micromixing are about 10 to 20 percent of the conversion level. If it is kept in mind that the actual state lies between the bounds, the possible differences are usually rather small in homogeneous reactors, except for exceptional cases or very high conversion levels. Also, the effects may be larger for product distributions from complex reactions. When nonhomogeneous systems are considered, however, larger effects can occur. For example, in emulsion or suspension polymerization (especially involving free radical mechanisms), the degree of segregation can be quite important for the reactor behavior. In this case, and also for situations where the reaction is occurring inside a flowing solid phase, the actual conditions are probably close to the complete segregation limit.

A final point is that the rule derived by Danckwerts [5] is confirmed: for reaction rate order $> 1$, segregation gives a higher conversion, and conversely for order $< 1$. A rigorous proof is given by Chauhan, Bell, and Adler [24].

To summarize, the effects of micromixing are important in principle, but for "thin" fluids and not extremely sensitive reactions, are of a lesser order of magnitude effect than the macromixing. Therefore, the simple-to-use segregation solution, Eq. 12.4-1, 2, almost always can be utilized to predict conversions without actually knowing the real details of the micromixing. For those cases, such as combustion, where extremely fast reactions occur, some of the population balance methods to be described later can provide a general method of attack.

Several types of models have been proposed for intermediate levels of micromixing with arbitrary macromixing RTD. One category considers the reactor contents to be comprised of two parts—one in a segregated state and one in a maximum mixing state: Weinstein and Adler [25] and see Villermaux and Zoulalian [26]. The fraction in each state is to be fit to actual reactor data and then correlated. Similar models also include mass transfer rates between regions— see Rippin [27]. Other approaches are to consider the fluid elements as periodically colliding, coalescing, and redividing, usually based on the population balance concepts to be presented below. A simplified version that only considers mean micromixing collision effects, but that appears to give similar results to the more complicated population balance equations, is that of Villermaux and Devillon [28] and Aubry and Villermaux [29]. Some experimental results have been interpreted with these models, but there are as yet no general correlations available.

Other descriptions of micromixing are based on the physical bases of the ultimate microturbulent and molecular diffusivities of the fluid elements. For example, Nauman [30, 31] has argued that the ultimate molecular mixing events before reaction should be based on molecular diffusion within the smallest eddy— presumably determined by the mixing and/or turbulence level. Thus the micro time scale would be the diffusion time, and a segregation number can be defined as the ratio of the micro to macro times. Nauman chose the reactor mean residence time $\tau$ as the macrotime but Berty [32] has proposed that the inverse of a characteristic first-order rate constant would be a better choice. Nauman [31] then solved the diffusion equation for "droplets" in a homogeneous perfectly macro-mixed reactor, allowing for mass transfer between "droplets" as if they were suspended in a fluid with essentially the reactor mean concentration.

Similar approaches have been taken by Truong and Methot [33]. Also, Danckwerts [5] defined an index of segregation based on the variances of the ages in the fluid elements and the vessel as a whole; however, there is no direct relation with reactor performance.

Again, no general correlations of these parameters are available. Surveys of various aspects of relating turbulent mixing concepts to micromixing in reactors appear in Brodkey [34]. An overall survey of mixing and chemical reactions is by Olson and Stout [35].

A final important point is that all of the above has only been for isothermal conditions in single-phase systems. Extension to other cases requires the introduction of interactions with the second phase and/or heat exchange walls, and so on, and the age-distribution and micromixing functions depend much more on the details of the system. A formal treatment would use joint probability distribution functions, but this rapidly gets extremely complex. The population balance models can give some insight into the two-phase situation, and will be discussed below.

## 12.5 Flow Models

### 12.5.a Basic Models

The above methods are quite general in that they are not based on any specific physical model. On the other hand, they are limited in presuming that the RTD is available and also by restriction to homogeneous phases. Thus, other techniques are required for predictive puposes and design where the actual reactor is obviously not available for a tracer test. In order to develop general correlations of behavior, the usual engineering tactics, of utilizing mathematical models with parameters to be determined from the experimental data, are useful. These parameters are then correlated as functions of fluid and flow properties, reactor configurations, along with other important features, and can be used in design calculations. The flow models are semiempirical in nature, but hopefully will bear some relation to the actual flow patterns in the vessel.

There are several types of models that have found useful applications, but two are most common—at least when found to be adequate to represent the physical situation. The first is usually termed the "axial dispersion" or "axial dispersed plug flow" model, Levenspiel and Bischoff [1], and takes the form of the one-dimensional diffusion equation with a convective term. The second is a series of perfectly mixed vessels, some features of which have been discussed in Sec. 12.3.

The mass balance for the axial dispersion model is, from Chapter 7,

$$\frac{\partial c}{\partial t} + u\frac{\partial c}{\partial z} = \frac{\partial}{\partial z}D_a\frac{\partial c}{\partial z} + r(c) \tag{12.5a-1}$$

In Eq. 12.5a-1, $u$ is taken to be the mean (plug flow) velocity through the vessel, and $D_a$ is a mixing-dispersion coefficient to be found from experiments with the system of interest. One important application is to fixed beds, as discussed in detail in Chapter 11, and then it is usually termed an effective transport model, with $D_a = D_{ea}$. However, the axial dispersion model can also be used to approximately describe a variety of other reactors.

One of the main benefits of this model is its analogy to the diffusion equation, and the possibility of utilizing all of the classical mathematical solutions that are available (e.g., Carslaw and Jaeger and Crank, [36]). (Of course, it is an exact model for the pure diffusion reaction problem.)

One common approach to determining the model parameter, $D_a$, is to perform a residence time-distribution test on the reactor, and choose the value of $D_a$ so that the model solution and experimental output curve agree [e.g., by least squares techniques (see Sec. 12.5.c)]. Figure 12.5.a-1 shows the $E(\theta')$—curves of the model for an impulse input with "closed" boundaries (here $\theta' = \theta F'/V = \theta u/L$, with $L$ = length of the reactor):

$$\frac{C}{(M/V)} = E(\theta') = e^{Pe_a/2}\sum_{i=1}^{\infty}\frac{(-1)^{i+1}8\alpha_i^2}{4\alpha_i^2 + 4Pe_a + Pe_a^2} \times \exp[-\theta'(Pe_a^2 + 4\alpha_i^2)/4Pe_a]$$

$$\tag{12.5.a-2}$$

with

$$\tan\alpha_i = \frac{4Pe_a\alpha_i}{4\alpha_i - Pe_a^2}$$

or

$$\left(\frac{\alpha_i}{2}\right)\tan\left(\frac{\alpha_i}{2}\right) = \frac{Pe_a}{4}; \quad \left(\frac{\alpha_i}{2}\right)\cot\left(\frac{\alpha_i}{2}\right) = -\frac{Pe_a}{4}$$

(See Carslaw and Jaeger [36].) A thorough evaluation was given by Brenner [37]. It is seen that the axial dispersion model can represent mixing behavior ranging from perfect mixing ($D_a \to \infty$) to plug flow ($D_a \to 0$).

Other types of boundary conditions can also be used when solving Eq. 12.5.a-1 (some details are given below when discussing chemical reaction applications); for example, in an "infinite " pipe, the dispersion characteristics upstream from

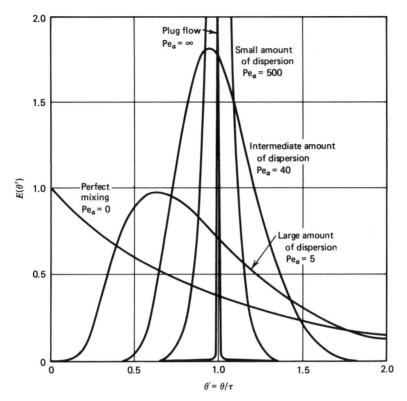

*Figure 12.5.a-1 E curves in closed vessels for various extents of dispersion; $Pe_a = uL/De_a$ (after Levenspiel [3].)*

the tracer injection would be identical to those within the system, and a different mathematical solution results:

$$\frac{C}{(M/V)} = E(\theta') = \frac{1}{2}\left(\frac{Pe_a}{\pi\theta'}\right)^{1/2} \exp\left[-\frac{Pe_a(1-\theta')^2}{4\theta'}\right] \quad (12.5\text{-}3)$$

The $E$ curves from Eq. 12.5-3 are similar to those from Eq. 12.5-2 for reasonably large values of $Pe_a$.

Many of the available solutions are given in Levenspiel and Bischoff [1], Himmelblau and Bischoff [4], and Wen and Fan [2]. As a practical matter, these mathematical complexities should probably not be taken too literally since the precise conditions at the boundaries of real equipment cannot usually be exactly defined in any event. That is, if the boundary condition details give significantly different results, be cautious about the axial dispersion model as being a faithful representation of the flow patterns in the vessel. These differences are only seen for large values of $D_a$ (actually small $Pe_a$), but the physical basis is a diffusionlike situation, which would not be valid for extremely large levels of mixing.

### Example 12.5.a-1 Axial Dispersion Model for Laminar Flow in Round Tubes

This is one situation where the exact convective-diffusion equation can be solved and compared with the one-dimensional dispersion equation. The complete equation is

$$\frac{\partial C}{\partial t} + 2u\left(\frac{R^2 - r^2}{R^2}\right)\frac{\partial C}{\partial z} = \mathscr{D}\left(\frac{1}{r}\frac{\partial}{\partial r} r \frac{\partial C}{\partial r} + \frac{\partial^2 C}{\partial z^2}\right) \tag{a}$$

where $u$ is the mean velocity.

Taylor [38] has discussed this in detail, and a summary of the results is given here. Taylor and Aris [39] showed that the main variable of interest, the mean concentration,

$$\langle C \rangle \equiv \frac{2}{R^2} \int_0^R rC dr$$

could be found by averaging Eq. (a) (for impermeable walls):

$$\frac{\partial\langle C\rangle}{\partial t} + u \frac{\partial\langle C\rangle}{\partial z} = \mathscr{D}\frac{\partial^2\langle C\rangle}{\partial z^2} + \frac{4u}{R^2}\frac{\partial}{\partial z}\int_0^R [C(r) - \langle C\rangle]r^3 dr \tag{b}$$

$$\sim D_a \frac{\partial^2\langle C\rangle}{\partial z^2} \tag{c}$$

with

$$D_a = \mathscr{D} + \frac{1}{48}\frac{U^2 R^2}{\mathscr{D}} \tag{d}$$

(See Hunt [40] for a recent exposition.) The step from Eq. (b) to Eq. (c) is only possible under conditions that the radial concentration profile has a chance to "stabilize" somewhat, or, in practical terms, that the length to diameter ratio be sufficiently large:

$$\frac{L}{d_t} > 0.04 \frac{u d_t}{\mathscr{D}} \tag{e}$$

For this case, these results provide a theoretical basis for an "effective transport model." The dispersion coefficient, $D_a$, is the sum of the ordinary diffusion contribution, and the second term that actually represents the effect of the velocity profile causing some fractions of an initial plane of fluid in the pipe to move faster than others. Note that this velocity profile effect causes a "distribution of residence times" and can also be represented in terms of $E$ curves. It is remarkable that the rather complicated Eq. (a) can often be represented by the much simpler Eq. (c) and this is often called "Taylor diffusion" after the original paper. Extensive numerical comparison by Gill et al. [41] and Wen and Fan [2] have shown that the axial dispersion model results do agree with the full equation for the proper ranges of variables and for situations where an initial distribution of solutes "spreads" with time. However, this agreement is not true for all problems. One easily visualized case is when a solute would be fed in a steady stream at a downstream point—very little back diffusion would occur in the actual laminar flow system, but the symmetrical dispersion model with large $D_a$ would predict extensive back diffusion. Thus, approximate flow models that provide an adequate description for one situation may be completely inadequate for another, even in the same type of vessel. Also, the flow models imply some state of micromixing, which may, or may not, correspond to the real reactor being modeled.

The practical importance of this Taylor diffusion analysis lies in the justification of the effective transport models to take into account complicated velocity and concentration profiles in a simple manner, as well as providing a theoretical framework for the dispersion coefficient, $D_a$. Similar results have been worked out for turbulent flow, packed columns, and other situations. For correlations of the axial dispersion coefficients, see Himmelblau and Bischoff [4] and Wen and Fan [2].

---

As described in Sec. 12.3, the moments of the impulse response can be used to characterize the RTD curve. This technique is also useful here for estimating the model parameter, $D_a$, although better techniques will be described below for the latter. It is found for a closed system that

$$\mu = 1 \tag{12.5.a-4}$$

$$\sigma^2 = \frac{2}{Pe_a} \left[ 1 - \frac{1}{Pe_a} (1 - e^{-Pe_a}) \right] \tag{12.5.a-5}$$

$$\sim \frac{2}{Pe_a} = 2 \frac{D_a}{uL} \tag{12.5.a-5a}$$

Thus, it is seen that the "width" of the axial dispersion model RTD is essentially inversely proportional to the Peclet number (Pe), or directly proportional to the axial dispersion coefficient, $D_a$.

For the so-called "tanks in series" model, Ex. 12.3-1 gives

$$E(\theta') = \frac{(n)^n}{(n-1)!} (\theta')^{n-1} e^{-n\theta'} \tag{12.5a-6}$$

The model parameter need not strictly be an integer for curve-fitting purposes, and then Eq. 12.5a-6 is interpreted as a gamma distribution. The mean is equal to unity, and the variance is

$$\sigma^2 = \frac{1}{n} \tag{12.5a-7}$$

In many ways, the two models are rather similar, although the mathematical details for the tanks in series model is much simpler than for the axial dispersion model. On the other hand, no theoretical justification such as Taylor diffusion is possible in general, nor are theoretical estimates of the model parameter, $n$; that is, $n$ is strictly empirical. The only exceptions to this are the finite stage models for packed bed interstices as briefly discussed in Chapter 11.

There is no one exact way to compare the tanks-in-series and dispersion models, since the responses are never identical. However, a useful relation is obtained from equating the variances for the two models:

$$\sigma^2 = \frac{1}{n} = \frac{2}{\mathrm{Pe}_a} \left[ 1 - \frac{1}{\mathrm{Pe}_a} (1 - e^{-\mathrm{Pe}_a}) \right] \tag{12.5a-8}$$

or

$$n \simeq \frac{\mathrm{Pe}_a}{2} + \frac{1}{2} \qquad \mathrm{Pe}_a > 2 \tag{12.5a-8a}$$

which is similar to the relation pointed out by Kramers and Alberda [42]. This gives consistent results for plug flow, $D_a \to 0$ or $\mathrm{Pe}_a \to \infty$ and $n \to \infty$, and also for perfect mixing, $D_a \to \infty$ or $\mathrm{Pe}_a \to 0$ and $n \to 1$. This type of comparison of the moments of the distribution curve of the two models has a wide applicability. It is based on the notion that two distributions must be similar if the following are equal:

1. The "size" of the curve—normalized.

2. The "location" of the curve—the mean.

3. The "width" of the curve—the variance.

Finer details of the curves can also be matched through use of higher moments, but, especially with experimental data, these higher moments are almost impossible to obtain. For the axial dispersion or tanks-in-series models that have only one (macro) mixing parameter, the higher moments give no further information (beyond model consistency) but more complex flow models can have more parameters.

Even though the two models agree in many respects, one can devise situations where this is not true. For example, if a steady stream of tracer is injected into the reactor halfway to the exit, the axial dispersion model predicts "backmixing" because of the symmetrical nature of diffusion, but the tanks-in-series model obviously has no provision for backwards transport of tracer. Thus, in situations where this phenomena is important, the two models would *not* be similar at all, and only one would be definitely closer to the physical situation.

A closer similarity would be to modify the tanks-in-series model to include backflow streams. This model has mass balances

$$\frac{V}{n}\frac{dC_1}{dt} = F_0' C_0 + F_b' C_2 - (F_0' + F_b')C_1 \tag{12.5.a-9}$$

where $F_0'$ is the throughflow feed, and $F_b'$ is the backflow between stages, Eq. 12.5.a-9 can also be written as:

$$\frac{\tau}{n}\frac{dC_1}{dt} = C_0 + \beta C_2 - (1 + \beta)C_1 \tag{12.5.a-10}$$

where $\tau = V/F_0'$ and $\beta = F_b'/F_0'$. For the general $i$th tank,

$$\frac{\tau}{n}\frac{dC_i}{dt} = (1 + \beta)C_{i-1} - (1 + 2\beta)C_i + \beta C_{i+1} \tag{12.5.a-11}$$

and for the end tank,

$$\frac{\tau}{n}\frac{dC_n}{dt} = (1 + \beta)C_{n-1} - (1 + \beta)C_n \tag{12.5.a-12}$$

Roemer and Durbin [43] gave a thorough treatment of this model. The transfer function for an impulse input is:

$$\frac{\bar{C}_n(s)}{(M/F_0')} = 2\left(\frac{1+\beta}{\beta}\right)^n \frac{a}{\lambda_1{}^n\left(1 + \frac{s}{n} + 2\beta\frac{s}{n} + a\right) - \lambda_2{}^n\left(1 + \frac{s}{n} + 2\beta\frac{s}{n} - a\right)} \tag{12.5.a-13}$$

where

$$\lambda_1, \lambda_2 = (b \pm a)/2\beta$$
$$a = [b^2 - 4\beta(1 + \beta)]^{1/2}$$
$$b = 1 + 2\beta + s/n$$

The moments are readily found:

$$\mu_{t1} = V/F_0' = \tau \tag{12.5.a-14}$$

$$\frac{\sigma_t{}^2}{\mu_{t1}{}^2} = \frac{1 + 2\beta}{n} - \frac{2\beta(1 + \beta)}{n^2}\left[1 - \left(\frac{\beta}{1+\beta}\right)^n\right] \tag{12.5.a-15}$$

$$\sim \frac{1 + 2\beta}{n} \tag{12.5.a-15a}$$

for large values of $n$.

Klinkenberg [44] pointed out that Eq. 12.5.a-15a has a useful physical interpretation: for each stage, the flowing fluid moves $(1 + \beta)$ forward and $(\beta)$ backward; thus, the mean is

$$\mu \propto n(1 + \beta - \beta) = n$$

and the variance about the mean is

$$\sigma^2 \propto n(1 + \beta + \beta) = n(1 + 2\beta)$$

If $\beta = 0$, these results naturally reduce to those for the simple tanks-in-series model. Roemer and Durbin [43] also provide results for intermediate values of $i$, to interpret tracer measurements from internal regions of the vessel. A comprehensive treatment of moments for the model is given in Klinkenberg [45, 46]. General analysis of systems with internal reflux, using this type of model, is presented by Shinnar and Naor [47]. An interesting example of applying the model to a real reactor with internal staged mixing characteristics that should be able to be represented by the model is by Lelli, Magelli, and Sama [48] including correlations of the parameters. Finally, further details of the model, including comparisons with the axial dispersion model, are given in Seinfeld and Lapidus [49].

Once the model parameters are determined and correlated, the flow model can be used to predict chemical reactor behavior. For example, with a first-order reaction occurring in a steady flow reactor with axial dispersion, the mass balance is:

$$D_a \frac{d^2 C_A}{dz^2} - u \frac{dC_A}{dz} - kC_A = 0 \tag{12.5.a-16}$$

The proper boundary conditions to use in solving Eq. 12.5.a-16 have been extensively considered in the literature. For a "closed" reactor, consideration of flux balances at the entrance and exit provide what are usually termed the "Danckwerts boundary conditions" [6]:[2]

$$C_{A0} = C_A(0^+) - \frac{D_a}{u} \frac{dC_A(0^+)}{dz} \tag{12.5.a-17}$$

$$\frac{dC_A(1^-)}{dz} = 0 \tag{12.5.a-18}$$

(These boundary conditions were used to derive the solution Eq. 12.5.a-2 for Eq. 12.5.a-1.) The same results are also found for other situations, as shown by Wehner and Wilhelm [50] and Bischoff [51]; transient cases are more complex—see van Cauwenberghe [52].

[2] As a matter of historic interest, these boundary conditions were utilized by Langmuir [*J. Am. Chem. Soc.*, **30**, 1742 (1908).

However, for specific reactors, the boundary conditions Eq. 12.5.a-17, 18 may not be the most appropriate, since they contain the implication that the local diffusionlike symmetric "backdispersion" is valid at the boundaries.[3] For example, in packed beds, it is known from the work of Hiby [53] that there is essentially no true "backmixing." Thus, Wicke [54] concludes that the only back transport would be by strictly molecular diffusion, which is usually small in magnitude compared to other processes, and so the boundary conditions Eq. 12.5.a-17 would be replaced with

$$C_{A0} = C_A(0^+) \qquad (12.5.a\text{-}17a)$$

Also, the outlet boundary condition Eq. 12.5.a-18 is equivalent to a partially reflecting boundary, and taking the packed bed as a portion of a semiinfinite region may be preferable; this changes the boundary conditions Eq. 12.5.a-18 to:

$$C_A(\infty) \to 0 \qquad (12.5.a\text{-}18a)$$

Again, there is a significant difference between the various solutions only for large amounts of mixing—small $Pe_a$. Other complicating situations are discussed in Young and Finlayson [55] and Zvirin and Shinnar [56], Gill [41], and Wen and Fan [2].

With the Danckwerts boundary conditions the solution of Eq. 12.5.a-16, 17, 18 is

$$\frac{C_A}{C_{A0}} = \frac{4a \exp(\tfrac{1}{2}Pe_a)}{(1 + a)^2 \exp[(a/2)Pe_a] - (1 - a)^2 \exp[-(a/2)Pe_a]} \qquad (12.5.a\text{-}19)$$

with

$$a \equiv [1 + 4k\tau/Pe_a]^{1/2}$$

It is useful to compare this result for small deviations from plug flow—large $Pe_a$—by an expansion about the plug flow reactor conversion; the result is:

$$\frac{C_A}{(C_A)_{PF}} \sim 1 + (k\tau)^2/Pe_a \qquad (12.5.a\text{-}20)$$

$$\simeq 1 + \frac{D_a}{uL}\left[\ln\left(\frac{C_{A0}}{C_A}\right)\right]^2 \qquad (12.5.a\text{-}20a)$$

[3] This point was discussed in a general way by Beek and Miller, *Chem. Eng. Prog. Symp. Ser.*, **55**, No. 25, 23 (1959).

An alternate form is found by equating the actual and plug flow conversion:

$$\frac{k\tau}{(k\tau)_{PF}} \sim 1 + (k\tau)/\text{Pe}_a \qquad (12.5.a-21)$$

$$\simeq 1 + \frac{D_a}{uL} \ln\left(\frac{C_{A0}}{C_A}\right) \qquad (12.5.a-21a)^4$$

These results show that for a given plug flow conversion level, the reactor with axial dispersion will produce essentially the same results for sufficiently large $\text{Pe}_a$.

For a given type of reactor, the axial dispersion coefficient is usually correlated using a characteristic local length, $l$ (e.g., the tube diameter, packing size)—see Himmelblau and Bischoff [4] and Wen and Fan [2]; then

$$\text{Pe}_a = \frac{uL}{D_a} = \left(\frac{ul}{D_a}\right)\left(\frac{L}{l}\right) = \text{Pe}_{l,a}\left(\frac{L}{l}\right) \qquad (12.5.a-22)$$

where $\text{Pe}_{l,a}$ is the *local* Peclet number for mass dispersion. Equation 12.5.a-22 with Eq. 12.5.a-20 or 21 indicates that axial dispersion effects may be neglected for reactors that are sufficiently long with respect to the characteristic dispersion length, $l$. Mears [57] has combined Eq. 12.5.a-21a and 22 to provide a formal criterion:

$$\frac{L}{l} > \frac{1}{e}\frac{1}{\text{Pe}_{l,a}} \ln \frac{C_{A0}}{C_A} \qquad (12.5.a-23)$$

where $e$ is the allowable error (e.g., 5 percent, $e = .05$). This is usually the case for industrial reactors of the simple empty tube or packed bed types. However, it may not be true for laboratory studies, especially in differential reactors, or for reactors with more complicated flows leading to large characteristic dispersion lengths. Results similar to Eq. 12.5.a-20 for more complex rate forms, including adiabatic reactors, have been derived using formal perturbation techniques—see Turian [58].

It was pointed out by Young and Finlayson [55] that for reactors with finite wall heat transfer, the ultimate conditions for a large reactor will be determined by the heat exchange, whereas in isothermal and adiabatic situations the concentrations and temperatures are determined only by the reaction, and so for the latter case, the differences between plug flow and axial dispersion model results always diminish with increasing reactor length. But, differences between the two models at the entrance may persist for reactors with wall heat transfer. They provide alternate criteria for the unimportance of axial dispersion (of both mass and heat) effects. Mears [59] pointed out that these new criteria were not general, and provided alternate ones for equal feed and wall temperatures. The new criteria are

---

[4] This result was also obtained by I. Langmuir, *J. Am. Chem. Soc.*, **30**, 1742 (1908).

usually more stringent than Eq. 12.5.a-20 or 22, but since their exact form is based on the precise boundary conditions at the reactor entrance, the uncertainties discussed below (Eq. 12.5.a-17) should make one cautious in their implementation. For practical purposes, axial dispersion effects can usually be neglected in the common industrial and laboratory reactor if they are sufficiently long.

In addition to use of the axial dispersion model to represent the longitudinal distribution of residence times in a reactor, the transverse or radial dispersion characteristics can also often be adequately modeled with a diffusionlike equation. This is most often done for empty tubes or packed beds; the latter has been thoroughly covered in Chapter 11. Further aspects of the topic can be found in Levenspiel and Bischoff [1] and Wen and Fan [2].

The conversion for the tanks-in-series model was presented in Chapter 10 and is similar to the results from the axial dispersion model for $n \sim Pe_a/2 \gg 1$. In addition, two-dimensional networks of such compartments can be used to describe dispersion—this was briefly discussed in Chapter 11. For a general mathematical treatment, see Wen and Fan [2].

## 12.5.b  Combined Models

Even though the general methods and results discussed above can be used to handle a wide variety of nonideal flow situations, many reactors contain elements that do not satisfy the fundamental basis of a diffusionlike model: the overall mixing should be the result of a large number of small random events. For example, if the mixing is caused by a relatively small number of bubbles rapidly moving through a continuous phase, one would not expect the RTD curves to have the proper shape of a diffusion equation result, nor might it be possible to develop consistent correlations of the dispersion coefficients. In these situations, more complicated models need to be utilized, usually built from combinations of the basic models of Sec. 12.5.a.

One of the most important cases is when there are two (or more) distinct regions within the reactor. This might be a packed bed of porous solids, two fluid phases, partially stagnant regions, or other complicated flows through a vessel that can basically be described by an axial dispersion type model. Transport balances can be made for each phase, per unit reactor volume:

$$\frac{\partial C_1}{\partial t} + u_1 \frac{\partial C_1}{\partial z} = D_{a1} \frac{\partial^2 C_1}{\partial z^2} - \frac{\Delta C}{(V_1/V)} + r_1 \qquad (12.5.b\text{-}1)$$

and

$$\frac{\partial C_2}{\partial t} + u_2 \frac{\partial C_2}{\partial z} = D_{a2} \frac{\partial^2 C_2}{\partial z^2} + \frac{\Delta C}{(V_2/V)} + r_2 \qquad (12.5.b\text{-}2)$$

where

$$u_i = \text{velocity in region } i = F'_i/\Omega_i$$
$$r_i = \text{reaction rate per volume of region } i$$
$$\Delta C = \text{rate of mass (or heat) interchange (kg/m}^3\text{s)}$$
$$= k_{T1} C_1 - k_{T2} C_2$$
$$= k_T a_v (C_1 - C_2/K), \text{ for example (with } K = \text{equilibrium constant)}$$

In Eq. 12.5.b-1, 2, the concentrations, $C_i$, could be mass concentrations, or for heat transfer, $C_i = \rho_i c_{pi} T_i$. Also, the axial dispersion coefficients for heat transfer would be $D_{ai} = \lambda_{ai}/\rho_i c_{pi}$, and the transfer coefficient would be $k = h/\rho_1 c_{p1}$ (for region 1 normally being the fluid, and then $K \equiv \rho_2 c_{p2}/\rho_1 c_{p1}$).

Equations 12.5.b-1, 2 reduce to a large variety of special cases. As written, they are the standard equations for interfacial mass transfer used in Chapters 6 and 14 (also see Pavlica and Olson [60]. They are also the basis of the "cross-flow" models for fluidized beds (see Chapter 13) or other multiphase reactors, and have been used for heat transfer studies.

For packed beds, the void fraction is $\varepsilon = (V_1/V)$ and $(V_2/V) = 1 - \varepsilon$; also $u_2 = 0$ (except for moving beds). Then the balance Eqs. 12.5.b-1, 2 become

$$\frac{\partial C_1}{\partial t} + u_1 \frac{\partial C_1}{\partial z} = D_{a1} \frac{\partial^2 C_1}{\partial z^2} - \frac{k_T a_v}{\varepsilon} (C_1 - C_2/K) + r_1 \quad (12.5.b\text{-}3)$$

and

$$\frac{\partial C_2}{\partial t} = D_{a2} \frac{\partial^2 C_2}{\partial z^2} + \frac{k_T a_v}{1 - \varepsilon} (C_1 - C_2/K) + r_2 \quad (12.5.b\text{-}4)$$

Equations 12.5.b-3, 4 are commonly used in chromotography—see Seinfeld and Lapidus [49]. The transport and reaction within individual particles is often specifically modeled; in this case, the concentrations in Eq. 12.5.b-4 are to be interpreted as averaged values for the particles:

$$C_2 = \varepsilon_s \langle C_s \rangle$$
$$= \varepsilon_s \frac{3}{R^3} \int_0^R C_s r^2 \, dr \quad (12.5.b\text{-}5)$$

for spherical particles.

From Chapter 3, the mass balance for the solid particles is:

$$\varepsilon_s \frac{\partial C_s}{\partial t} = D_e \nabla_r^2 C_s + r_s \quad (12.5.b\text{-}6)$$

with the boundary condition:

$$D_e \frac{\partial C_s}{\partial r}\bigg|_R = k_T(C_1 - C_s/K_s) \text{ at } r = R$$

Averaging Eq. 12.5.b-6 gives:

$$\varepsilon_s \frac{\partial \langle C_s \rangle}{\partial t} = D_e \cdot \frac{3}{R} \frac{\partial C_s}{\partial r}\bigg|_R + \langle r_s \rangle$$

or

$$\frac{\partial C_2}{\partial t} = \frac{a_v}{1 - \varepsilon} k_T \left( C_1 - \frac{C_s}{K_S}\bigg|_R \right) + \langle r_s \rangle \qquad (12.5.b\text{-}7)$$

In order to rigorously determine the terms on the right-hand side of Eq. 12.5.b-7, the complete particle mass balance, Eq. 12.5.b-6, must be solved, and this result then utilized. However, if only the average values are of interest, the same type of approximations discussed in Chapter 7 lead to the result:

$$\frac{\partial C_2}{\partial t} \simeq \frac{k_T a_v}{1 - \varepsilon} \left( C_1 - \frac{C_2}{K} \right) + r_2 \qquad (12.5.b\text{-}8)$$

Equation 12.5.b-8 is the same as Eq. 12.5.b-4 except for the term in the latter with $D_{a2}$, representing direct effective particle-to-particle transport. For several types of situations, this term may be of definite importance: mass transfer in highly porous solids at low Reynolds numbers (Wakao [63]) and the heat transfer situation discussed by Littman and Barile [61], which is analogous to the model for radial heat transfer proposed by De Wasch and Froment [62].

Solutions to Eqs. 12.5.b-1 and 2 are most conveniently obtained with Laplace transforms (for rates linear in concentrations). Thus, with zero initial concentration,

$$s\bar{C}_1 + u_1 \frac{\partial \bar{C}_1}{\partial z} = D_{a1} \frac{\partial^2 \bar{C}_1}{\partial z^2} - \frac{\overline{\Delta C}}{(V_1/V)} - k_1 \bar{C}_1 \qquad (12.5.b\text{-}9)$$

It can be shown that the term $\overline{\Delta C}$ for a wide variety of interregion transport mechanisms will have the form

$$\overline{\Delta C} = H(s)\bar{C}_1 \qquad (12.5.b\text{-}10)$$

where $H(s)$ is particle transfer function with the interregion transport coefficients. This generality was pointed out by Miller and Bailey [64].

For a specified concentration at $z = 0$, and a semiinfinite region, the reactor transfer function is

$$\frac{\bar{C}_1}{\bar{C}_1(z = 0)} = \exp\left[\frac{u_1 z}{2Da_1}\left\{1 - \left[1 + \frac{4D_{a1}}{u_1^2}\left((s + k_1) + \left(\frac{V}{V_1}\right)H(s)\right)\right]^{1/2}\right\}\right]$$

(12.5.b-11)

The moments of the reactor transfer function, Eq. 12.5.b-11, can be readily found by the formula of van der Laan [65]:

$$\Delta\mu_{t,p} = (-1)^p \frac{\partial^p}{\partial s^p}\left[\frac{\bar{C}_1}{\bar{C}_{10}}\right]_{s=0}$$

(12.5.b-12)

Then, one finds:

$$\Delta\mu_{t,0} = \exp\left[\frac{u_1 z}{2D_{a1}}\left\{1 - \left[1 + \frac{4D_{a1}}{u_1^2}\left(k_1 + \left(\frac{V}{V_1}\right)H(0)\right)\right]^{1/2}\right\}\right]$$

(12.5.b-13)

$$= 1$$

(12.5.b-13a)

if $k_1 = 0$ and $H(0) = 0$; the latter is true for reversible solid transport processes. Next,

$$\frac{\Delta\mu_{t,1}}{\Delta\mu_{t,0}} = \frac{z}{u_1}\frac{1 + (V/V_1)(\partial H/\partial s)_{s=0}}{\left[1 + \frac{4D_{a1}}{u_1^2}(k_1 + (V/V_1)H(0))\right]^{1/2}}$$

(12.5.b-14)

$$= \frac{z}{u_1}[1 + (V/V_1)(\partial H/\partial s)_{s=0}]$$

(12.5.b-14a)

if $k_1 = 0$ and $H(0) = 0$. Finally, the variance is,

$$\frac{\Delta\sigma_t^2}{\Delta\mu_{t,0}} = 2\left(\frac{z}{u_1}\right)^2\left\{\frac{D_{a1}}{u_1 z}\frac{[1 + (V/V_1)(\partial H/\partial s)_0]^2}{\left[1 + \frac{4Da_1}{u_1^2}(k_1 + (V/V_1)H(0))\right]^{3/2}}\right.$$

$$\left. - \frac{u_1}{2z}\frac{(V/V_1)(\partial^2 H/\partial s^2)_0}{\left[1 + \frac{4D_{a1}}{u_1}(1 + (V/V_1)H(0))\right]^{1/2}}\right\}$$

(12.5.b-15)

$$= 2\left(\frac{z}{u_1}\right)^2\left\{\frac{D_{a1}}{u_1 z}\left[1 + \left(\frac{V}{V_1}\right)\left(\frac{\partial H}{\partial s}\right)_0\right]^2 - \frac{u_1}{2z}\left(\frac{V}{V_1}\right)\left(\frac{\partial^2 H}{\partial s^2}\right)_0\right\}$$

(12.5.b-15a)

if $k_1 = 0$ and $H(0) = 0$.

## Example 12.5.b-1 Transient Mass Transfer in a Packed Column

For this situation, region 2 is stationary, $u_2 = 0$, and also if no interparticle transfer is assumed, $D_{a2} = 0$ (but recall Wakao [63]). If, in addition there are no reactions occurring, $r_1 = 0 = r_2$. Then, Eq. 12.5.b-4 becomes, when Laplace-transformed:

$$s\bar{C}_2 = \frac{k_T a_v}{1 - \varepsilon} (\bar{C}_1 - \bar{C}_2/K) \tag{a}$$

and

$$H(s) = \frac{\overline{\Delta C}}{\bar{C}_1} = \left( \frac{1}{s(1 - \varepsilon)K} + \frac{1}{k_T a_v} \right)^{-1} \tag{b}$$

Note that for this case, $H(0) = 0$.
Then, the moments become

$$\Delta\mu_{t,1} = \frac{z}{u_1} \left( 1 + \frac{1 - \varepsilon}{\varepsilon} K \right) \tag{c}$$

and

$$\Delta\sigma_t^2 = 2\left( \frac{z}{u_1} \right)^2 \left[ \frac{D_{a1}}{zu_1} \left( 1 + \frac{1 - \varepsilon}{\varepsilon} K \right)^2 + \frac{u_1}{z} \frac{1 - \varepsilon}{\varepsilon} \frac{K^2}{k_T a_v/(1 - \varepsilon)} \right] \tag{d}$$

For spherical particles of radius $R$,

$$a_v = 3(1 - \varepsilon)/R$$

and then Eq. (d) is identical to the results of Schneider and Smith [66] used for chromatographic studies in determining reactor transport properties. Actually, they consider the more complex situation with individual spherical particles with internal pore diffusion resistance and additionally a linear adsorption process. Their complete results could be found by utilizing Eq. 12.5.b-6, with linear $r_s$, in place of Eq. 12.5.b-4 to determine the particle transfer function, $H(s)$. The method is identical to that illustrated here, but the equations are naturally more complex. A collection of results for various types of intraparticle transport was provided by Gunn [67].

It's seen from this example why the use of moments is a popular way to determine the model parameters—Eqs. (c) and (d) are relatively simple compared to the complete solution of Eq. 12.5.b-3, 4 by attempting to invert the transfer function Eq. 12.5.b-11, for example. The method appears to work well in chromatographic columns and has been widely used. However, for a broader range of systems that may be of interest in chemical reaction engineering, moments will often not provide the best parameter estimates. Section 12.5.c will discuss this further.

A similar model has been found useful for modeling the flow patterns in "trickle bed" reactors—see Hochman and Effron [68]. In this situation, some of the liquid readily passes through the packing (region 1 in the model) but some liquid forms into stagnant pockets (region 2) that do have mass transfer with the main flowing fluid. In this case, the mass balance for the flowing liquid is, from Eq. 12.5.b-1,

$$\frac{\partial C_1}{\partial t} + u_1 \frac{\partial C_1}{\partial z} = D_{a1} \frac{\partial^2 C_1}{\partial z^2} - \frac{(k_T a_v)}{(V_1/V)}(C_1 - C_2) \qquad (12.5.b\text{-}16)$$

In region 2, there is no flow and no interstagnant pocket dispersion, and so Eq. 12.5.b-2 becomes

$$\frac{\partial C_2}{\partial t} = \frac{(k_T a_v)}{(V_2/V)}(C_1 - C_2) \qquad (12.5.b\text{-}17)$$

Here, the regions refer to the fluid in the voids between the packing, and one must be careful to account properly for the various volume fractions. The first moment, from Eq. 12.5.b-14a, is found to be similar to Eq. (c) of Ex. 12.5.b-1:

$$\Delta\mu_{t,1} = \frac{z}{u_1}\left(1 + \frac{V_2}{V_1}(1)\right)$$

$$= \frac{z}{u_{s1}}\varepsilon_l,$$

where $u_{s1} = u_1 V_1/V$ = superficial velocity, and $\varepsilon_l$ = void fraction available to all the liquid (total liquid holdup). The variance, Eq. 12.5.b-15, can also be evaluated in a similar fashion.

Villermaux and van Swaaij [69] have provided a thorough analysis of this model. The transfer function and moments were derived, as well as an analytical time-domain solution; the latter is very complex, involving several transcendental functions. (It should be mentioned that they solved the equations for an impulse input to a semiinfinite system, and with these boundary conditions, the moments, and so on, are slightly different from those above; refer to the discussion in Sec. 12.5.a. A summary of the moments for various boundary conditions is given by Arva, Kafarov, and Dorokhov [70]. Alternate forms of these solutions are possible—see Popović and Deckwer [71].) Villermaux and van Swaaij [69] also provided preliminary correlations of the model parameters, such as fraction stagnant fluid, crossflow $(k_T a_v)$, and flowing fluid dispersion coefficients. For the case when the flowing fluid can be taken to be in piston flow, a useful set of correlations that can be applied to industrial operations is in Hochman and Effron [68].

At this point, many of the problems of trickle bed reactor operations have been fairly well studied (see Satterfield [72]). Even if a priori design is not readily done (see Hofmann [73]), many aspects of pilot plant scale-up focus on the fluid-particle mass transfer (see Koros [74]); the details of the liquid axial dispersion are probably not critical, with sufficient radial mixing, especially for industrial scale reactors. Further consideration of the design aspects are presented in Chapter 14.

In the more general case with Eq. 12.5.b-2, it is readily seen that the (Laplace transformed) concentration in region 2 is given by:

$$D_{a2} \frac{\partial^2 \bar{C}_2}{\partial z^2} - u_2 \frac{\partial \bar{C}_2}{\partial z} - \left[ s + \frac{(k_T a_v)}{(V_2/V)K} + k_2 \right] \bar{C}_2 = - \frac{(k_T a_v)}{(V_2/V)} \bar{C}_1 \quad (12.5.b\text{-}18)$$

Equation 12.5.b-18 can be solved, together with Eq. 12.5.b-1, and substituted into $\overline{\Delta C}$, to yield

$$H(s) = \frac{\overline{\Delta C}}{\bar{C}_1} = f(z, s; \text{parameters})$$

This could then be substituted into Eq. 12.5.b-11 in order to derive the overall system transfer function—see Pavlica and Olson [60] for details.

When considering larger-scale mixing flows, for example, in real stirred tanks, other types of models are required. From the viewpoint of macromixing, the existence of significant regions of bypassing or stagnant fluid are particularly important. A basic model that can account for these features has been developed by Cholette et al. [75, 76] (see Fig. 12.5.b-1). The parameters of fraction bypassing and fraction dead (totally stagnant) volume have been correlated for a wide range of operating variables. Also see Levenspiel [3], Himmelblau and Bischoff [4], and Wen and Fan [2] for further aspects of these idealized models.

A model that more closely agrees with the observed mixing patterns in real stirred tanks was given by van de Vusse [77]. The basis is a region of intense shear and mixing around the impeller followed by recirculating flows around the rest of the tank. These "signal-flow" models have the general features as shown in Fig. 12.5.b-2. Each of the elements can represent a flow region with a certain amount of dispersion, or idealized model tanks-in-series. Compact mathematical methods for the solution of the mass balances for these types of models have been given by Gibilaro, Kropholler, and Spikins [78]. Experimental determination of the model parameters was illustrated by Taniyama and Sato [79], and an interesting example for better explaining microbial reactions was by Sinclair and Brown [80].

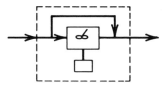

*Figure 12.5.b-1 Model of real stirred tank with bypassing and dead space.*

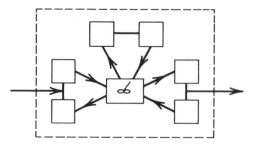

*Figure 12.5.b-2 Signal flow model of real stirred tank.*

A comprehensive review of these and other aspects of mixing and chemical reactions was given by Olson and Stout [35]. One conclusion was that the signal flow models are often overly sensitive to rather ill-defined parameters such as exact inflow conditions. These models also cannot account for mass interchange between internal flow paths that undoubtedly occurs in most highly stirred tanks. A similar model has been used by Khang and Levenspiel [81] to predict mixing times in batch mixing vessels (also see the earlier references given there).

Stirred-tank scale up is usually based on power input per unit volume—a survey of various applications is given by Connolly and Winter [82]. A more detailed consideration of chemical reaction behavior is by Paul and Treybal [83] where they found that "Power per unit volume is ... incapable of correlating the widely different local conditions within a stirred vessel." Instead, a criterion involving both the reaction rate constants and a microtime scale based on turbulent fluctuations was developed. Measurements of these fluctuations have been measured—see Gray [84] and Günkel and Weber [85], including references to prior work. Clearly, these considerations are returning us to the question of micromixing effects, and there is as yet no comprehensive method for the utilization of this information for predicting chemical reactor behavior.

A final topic concerns two phase liquid-liquid reactors, where droplet breaking and coalescence is of great importance. This complicated area cannot be covered here; for a recent useful reference, see Coulaloglou and Tavlarides [86]. The methods are based on the use of population balance techniques, to be discussed in Sec. 12.6.

### Example 12.5.b-2 Recycle Model for Large-Scale Mixing Effects

A variety of reactors have large-scale mixing flow patterns that are primarily in one direction for a portion of the vessel cross section and in the reverse direction elsewhere. Examples are bubble columns or vigorously bubbling fluidized beds, where the center portion consists of rapidly rising fluid, with flow down the wall regions. For these applications, two (or more) region models provide the best description of the physical events.

A sketch of this type of model is as follows:

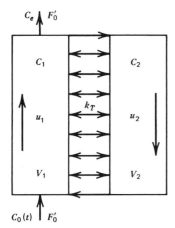

*Figure 1 Recycle model.*

The balance equations are Eq. 12.5.b-1, 2 with $u_2$ having a negative sign because of the direction of the velocity vector. Usually, the dispersion in each phase is neglected, since essentially the same effect is provided by sufficiently large crossflow, $k_T$. This model becomes

$$\frac{\partial C_1}{\partial t} + u_1 \frac{\partial C_1}{\partial z} = -\frac{(k_T a_v)}{(V_1/V)}(C_1 - C_2) + r_1 \tag{a}$$

$$\frac{\partial C_2}{\partial t} - u_2 \frac{\partial C_1}{\partial z} = \frac{(k_T a_v)}{(V_2/V)}(C_1 - C_2) + r_2 \tag{b}$$

Note that the flows are related by

$$\frac{V_1}{L} u_1 = F_0' + \frac{V_2}{L} u_2 \tag{c}$$

or

$$F_1' = F_0' + F_2' \tag{c'}$$

The special feature of the model is the recycle boundary condition:

$$F_1' C_1(0, t) = F_0' C_0(t) + F_2' C_2(0, t)$$

or

$$C_1(0, t) = \frac{1}{1 + r} [C_0(t) + r C_2(0, t)] \tag{d}$$

where

$$r = F_2'/F_0'$$
$$= \text{recycle ratio}$$

The other boundary condition is

$$C_1(L, t) = C_2(L, t) \tag{e}$$

This is a very flexible model, and reduces to several other special cases, as pointed out by Hochman and McCord [87]:

1. As $k_T \to 0$, the ordinary recycle model for reactors, with delay in both streams, is obtained. In addition, if $V_2/F_2' \to 0$, the model used by Carberry [88] to study reaction selectivity behavior is obtained; and if $V_1/F_1' \to 0$, there is instantaneous bypassing of a fraction of the feed.

2. As $r \to \infty$, the model approaches a perfectly mixed vessel.

3. As $r \to 0$, the model is similar to that described above for flow with transfer to a stagnant region, used for trickle bed reactors, for example.

4. As $k_T \to \infty$, the model approaches plug flow, with net flow rate $F_0' = F_1' - F_2'$, because of the instantaneous transverse mixing; this limit can also be seen by adding Eqs. 1 and 2 with $C_2 \to C_1$.

In addition to the above reasoning, the various limits can all be derived mathematically by careful consideration of Eqs. 1 and 2, or the solutions to be presented below.

The solution of Eqs. (a) and (b) for an impulse injection was found by Laplace transforms (with $r_1 = 0 = r_2$) in Hochman and McCord [87]:

$$\bar{C}_1(L, s) = \frac{\rho \exp[(\tau_m - \tau_1)s + \gamma]}{[(2r + 1)\tau_m s]\sinh \rho + \rho \cosh \rho} \tag{f}$$

where

$$\tau_i = L/u_i = V_i/F_i'$$
$$= \text{mean residence time of each region} \tag{g}$$

$$\tau_m = (\tau_1 + \tau_2)/2 \tag{h}$$

$$\rho = [\tau_m^2 s^2 + 2(2r + 1)\gamma\tau_m s + \gamma^2]^{1/2} \tag{i}$$

$$\gamma = \frac{k_T a_v V/F_0'}{2r(r + 1)} \tag{j}$$

Hochman and McCord [87] found several useful approximate time domain solutions from Eq. 6, but it is probably easier to directly solve Eqs. 1 and 2 numerically, as done by Dayan and Levenspiel [89].

However, the moments can be readily determined:

$$\mu_t = \tau_1 + 2r\tau_m \tag{k}$$

$$= \frac{V_1 + V_2}{F_0'} = \frac{V}{F_0'} = \tau \tag{k'}$$

which can be proven using Eqs. (c'), (g), (h). Thus, it can be seen that for a given $\tau$, only two of the three parameters $\tau_1$, $\tau_2$, or $r$ are independent. Also,

$$\sigma_t^2 = 4\tau_m^2 r(r + 1) \cdot \frac{1}{\gamma}\left[1 - \frac{1}{2\gamma}(1 - e^{-2\gamma})\right] \tag{l}$$

The procedure recommended by Hochman and McCord [87] is to use two tracer probes to directly measure the bottom-to-top residence time, $\tau_1$, by the first appearance from an impulse injection at $z = 0$, and the top-to-bottom residence time, $\tau_2$, with an impulse injection at $z = L$. Then, the recycle parameter, $r$, can be found from $\mu_t = V/F_0'$, Eq. (k) and the crossflow parameter, $k_T a_v$, from $\sigma^2$, Eq. (l), using the definition of $\gamma$ (Eq. j). Alternatively, other parameter estimation techniques can be used, as to be described in Sec. 12.5.c.

A practical example of the utilization of this model was provided by Goldstein [90] for the improved design of an industrial stirred-tank reaction with high $L/d_t \simeq 4$. For the rapid reaction of interest, the goal was to increase the overall uniformity of concentrations and temperatures in the reactor by choosing the appropriate impeller type, size, number and stirring speed, as well as effects of reactor throughput rate, feed nozzles, and liquid viscosity. The flow model was found to be very effective in rapidly and efficiently organizing the results from varying the large number of experimental variables.

From simulation of chemical reactor behavior using Eqs. (a) and (b), and the corresponding heat balances, it was found that the most important model parameter was the recycle ratio, $r$, as shown in Fig. 1. Thus, it would appear that the transverse between-regions mixing is not critical, but the longitudinal mixing is only adequate when $r > 40$ (here the flow is actually down in the center and up at the wall). Thus, the various combinations of reactor internals were studied to attempt to obtain high values of $r$ in this range. This was done by performing RTD tests in a scaled laboratory reactor—the detailed results are in the original reference. The optimum combination, when tested in the plant, was able to reduce the undesirable temperature gradients by a factor of 20 (to $\sim 3°C$), with reasonable power inputs. It was also found that the plant reactor experienced a 30 percent increase in catalyst efficiency and a 70 percent increased throughput rate was possible.

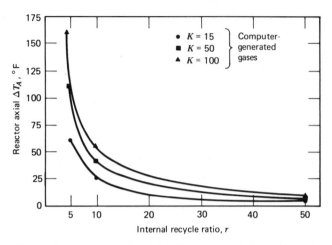

*Figure 2 Variation of axial $\Delta T$ with internal recycle ratio ( from Goldstein [90]).*

The scale-up procedure was based on maintaining constant $r$ values in the plant and laboratory model vessel:

$$r_p = r_m \qquad (m)$$

Relating the measured $r_m$ to the impeller pumping capacity resulted in a correlation

$$r = \frac{F'_2}{F'_0} = (\text{const}) \frac{Nd^2b}{F'_0} \qquad (n)$$

where

$$N = \text{impeller stirring speed}$$
$$d = \text{impeller diameter}$$
$$b = \text{impeller width (or other size characteristic)}$$

From standard correlations of power requirements, it is known that in the turbulent mixing regime:

$$P = (\text{const})\rho N^3 d^4 b \qquad (o)$$

Thus, the scaling relations are

$$\frac{r_p}{r_m} = \left(\frac{N_p}{N_m}\right)\left(\frac{d_p}{d_m}\right)^2\left(\frac{b_p}{b_m}\right)\left(\frac{F'_{0m}}{F'_{0p}}\right) \qquad (p)$$

$$= 1 \text{ (by choice)}$$

and

$$\frac{P_p}{P_m} = \left(\frac{N_p}{N_m}\right)^3 \left(\frac{d_p}{d_m}\right)^4 \left(\frac{b_p}{b_m}\right) \left(\frac{\rho_p}{\rho_m}\right) \tag{q}$$

Combining Eqs. 16 and 17 to eliminate $(N_p/N_m)$:

$$\frac{P_p}{P_m} = \left(\frac{d_m}{d_p}\right)^2 \left(\frac{b_m}{b_p}\right)^2 \left(\frac{F'_{0p}}{F'_{0m}}\right)^3 \left(\frac{\rho_p}{\rho_m}\right) \tag{r}$$

which was used by Goldstein [90]. For complete geometric similarity, $d$ and $b$ would be proportional to the tank diameter, $d_t$, and then Eq. 18 becomes

$$\frac{P_p}{P_m} = \left(\frac{d_{tm}}{d_{tp}}\right)^4 \left(\frac{F'_{0p}}{F'_{0m}}\right)^3 \left(\frac{\rho_p}{\rho_m}\right) \tag{s}$$

Nienow pointed out [91] that for equivalent performance as a chemical reactor, equal holding times must also exist:

$$\tau = \frac{V}{F'_0} \propto \frac{d_t^{\,3}}{F'_0} = \text{constant} \tag{t}$$

Then Eq. 19 becomes

$$\frac{P_p}{P_m} = \left(\frac{d_{tp}}{d_{tm}}\right)^5 \tag{u}$$

(for $\rho_p = \rho_m$).

It is seen that Eq. 21 decrees a scaling factor more than the traditional (power/unit volume):

$$\frac{(P_p/d_{tp}^{\,3})}{(P_m/d_{tm}^{\,3})} = \left(\frac{d_{tp}}{d_{tm}}\right)^2 \tag{v}$$

which may be difficult to accomplish with large scaling factors. Nienow [91] also showed that equivalent scaling factors arise from criteria of equal RTD and approximately equal mixing times in batch vessels.

Extensions to the basic model have been made by Mah [92], who utilized a series of stirred tanks for each region, rather than assuming plug flow. There could be some computational advantage in this, and this more general case also then includes the stirred tanks with backmixing models. However, the cost is another model parameter—the numbers of stirred tanks. Van Deemter [93] has applied the same model to mixing in fluidized beds, and also shown some interrelationships with the axial dispersion models. Gwyn, Moser, and Parker [94] have described a similar three-region model that more completely describes events in fluidized beds.

This example provides many aspects of the application of the recycle-type model. It is probably the most flexible model that has been successfully utilized to help describe large scale mixing patterns in actual chemical reactors.

### 12.5.c Flow Model Parameter Estimation

With multiparameter flow models, the accurate estimation of the parameters can be far from a trivial task. The basic problem is, of course, similar to those considered in Chapters 1 and 2 for kinetic rate coefficients, but since many flow models are partial differential equations, the problems are more severe. The mixing of tracer concentrations is inherently a linear process, and if other diffusion and dispersion steps are also linear, the governing differential equations will then be linear (although the parameters may appear in nonlinear ways), and the methods of systems engineering can be useful. We will only give a brief outline here, focusing on a few of the special problems involved for flow models. An excellent reference to many useful techniques is Seinfeld and Lapidus [49].

Let us first discuss the choice of input signals. The practical aspects of type of tracer for various situations is surveyed in Wen and Fan [2]. Also see Hougen [95] for an extensive discussion. The advantages and disadvantages of various types of signals are as follows.

1. A pulse input is often best in principle, since the output is directly the impulse response of the system; also, the model response is often simplest in this case. In addition, the test only causes a brief disturbance to the process. The main disadvantage is that it is difficult to experimentally generate a perfect impulse input, and especially, small deviations in the tail can cause important deviations from ideal behavior. Finally, in principle, all frequency modes of the system are stimulated, and this may not be optimum.

2. A step input is relatively easy to experimentally generate, and the result is simply related to the impulse response. However, a long time of input stimulation is required to achieve the final "lined out" response; a related problem is that it may be difficult to accurately determine the final value, and thus the normalization factor.

3. A sinusoidal input (frequency response) has the advantage of only requiring measurement of the amplitude ratio and phase lag of the output, and also several waves can be averaged for better accuracy. Some disadvantages are that it is relatively complex to experimentally generate a sine wave in concentration (or temperature). Also, one must wait for the initial transients to die out to achieve the desired stationary response, leading to a long process disturbance. Since tests at several frequencies are required, it is a rather time-consuming method; also the frequencies must be chosen in some optimum way.

**4.** In principle, arbitrary inputs (including "noise") can be utilized by measuring both input and output. The advantages are lack of need to generate any perfect specific input signal, and the results can be related to the general convolution properties of the system (e.g., Eq. 12.2-12). The random noise with cross-correlation technique is usually rather difficult with chemical process systems. Some disadvantages are that two signals must be measured; also, the deconvolution is not always easy to perform numerically (often, in fact, iterative convolutions are the only possibility), and usually necessitates rather sophisticated computational techniques.

Next, let us discuss the various techniques that have been used for the actual parameter estimation, describing their strong and weak points. For the axial dispersion model, this has been done by Böxkes and Hofmann [96], and illustrates the typical problems involved.

**1.** Direct nonlinear regression in the time domain. This would appear to be the most straightforward in principle, since one is comparing the actual data with the model response. Also, the least squares, maximum likelihood, or other criterion is minimizing errors directly with the data, which may often be well approximated as Gaussian even though this may not be true for transformations of the data—recall the discussion in Chapter 2. The primary disadvantage is that the time-domain solution of the model must be available. The alternative of repetitive numerical solutions during an extensive parameter search routine can involve very large amounts of computer time. Also, the objective functions often have local in addition to global minima, and/or long ridges because of correlation between the parameters—both of these cause difficulties. Seinfeld and Lapidus [49] provide a comprehensive discussion of the standard techniques. Michelson [97] describes a method for partially eliminating the problems associated with correlation between parameters (ridges in the objective function), and other approaches are given in Hosten [97] and Hosten and Emig [99]. However, there does not appear to be a completely sure technique at present.

**2.** The second technique is that of using time moments of the data to be compared with those of the model. The advantages are that they can be easily computed by numerical integration of the data, and the integrations tend to somewhat smooth the data. Also, the moments can be readily found from the Laplace transform of the model, without the necessity of a time-domain solution—recall the formula of van der Laan [65] utilized in Sec. 12.5.b. The main disadvantage is that the moments tend to emphasize the data for large values of time, and this data in the tail of the curve is commonly not very accurate. Also it is often difficult to decide just where to truncate the tail of the data curve. In addition, the number of moments must equal the number of parameters, which usually means using the second or third moments, with corresponding heavy weighting of the tails. Finally, one does not know how well the model using the moment-evaluated parameters finally fits the actual data in the time domain. In fact, the agreement should always be checked with this method, and is not too difficult since only a few solutions of the model equations will be required.

A general discussion was given by Bischoff [100]. The basis was the transfer function, which is the ratio of the Laplace transforms of the outlet and inlet concentrations. The model equations can be formally expanded in $s$:

$$\mathcal{T}(s; \alpha) = \mathcal{T}(0; \alpha) + \left[\frac{\partial \mathcal{T}(0; \alpha)}{\partial s}\right]s + \frac{1}{2}\left[\frac{\partial^2 \mathcal{T}(0; \alpha)}{\partial s^2}\right]s^2 + \cdots \quad (12.5.\text{c-}1)$$

$$\equiv 1 - \beta_1(\alpha)s + \tfrac{1}{2}\beta_2(\alpha)s^2 + \cdots \quad (12.5.\text{c-}2)$$

where the $\alpha$ represent the model parameters. Recalling the formula of van der Laan, Eq. 12.5.b-12, the moments of the experimental data can be related to the $\beta_i$:

$$\Delta\mu = \beta_1 \quad (12.5.\text{c-}3)$$

$$\Delta\sigma^2 = (\beta_2 - \beta_1{}^2) \quad (12.5.\text{c-}4)$$

where $\mu$ and $\sigma^2$ are defined as in Eq. 12.3.a-1. For a given flow model transfer function, the $\beta_i$ can readily be found. Equations 12.5.c-3, 4 reduce to the results given earlier for the dispersion and other models. A comparison of this relative to other methods will be given later.

3. The moments can be defined with a weighting factor to decrease the importance of the tail. A common choice of the weighting factor is $e^{-st}$, which then obviously bears a relation to considering the Laplace transform. These weighted moments possess all the same advantages of ordinary moments, without the critical disadvantage of too heavy weighting of the tails. However, the question of truncation still remains, and also numerical values of the weighting function parameter, $s$, must be chosen on some basis. Again, the final results should be checked by a time-domain solution. A useful reference is Andersson and White [101].

4. Since the Laplace transform-space solution for the model is so readily obtained, another method uses fitting (e.g., least squares) in transform space. This, of course, necessitates performing numerical transforming the time domain data:

$$\mathcal{T}_{\text{exp}}(s) = \int_0^\infty e^{-st}C_m(t)dt \bigg/ \int_0^\infty e^{-st}C_0(t)dt$$

There are subclasses depending on how $s$ is chosen (i.e., exactly which type of transform is used), and, once again, the final results should be checked in the time domain.

## A. Laplace Transform

One choice is to choose real values of $s$ for both the numerical transformation of the data as well as for the $s$-domain fitting. The advantages are again decreased emphasis of the tail. Also, the complete model equation in the transform domain is utilized in the fitting, not just $s = 0$ as in ordinary moments, or a particular value(s) as in weighted moments. The disadvantages are that the numerical Laplace transformation of the data must be performed, including the question of the best choices of numerical values of $s$, often chosen around $s = 1/\tau$, and truncation of the tail. For this technique, see Williams, Adler, and Zolner [102].

## B. Fourier Transform

If the transform variable is chosen to be purely imaginary, $s = i\omega$, the Fourier transform is generated. The same advantage as for the Laplace transform are also true here; in addition, by Parseval's theorem, least squares fitting in the Fourier domain is equivalent to least squares fitting in the time domain. The disadvantages are again choosing numerical values of $\omega$. For details of this technique, see Clements [103].

## C. Special Rearrangements

Some of the numerical problems in nonlinear regression seem to be able to be partially avoidable by utilizing special features of a given transfer function, as proposed by Mixon, Whitaker, and Orcutt [104] and extended by Østergaard and Michelsen [105]. This is best illustrated by considering the transfer function of the axial dispersion model with semiinfinite boundary conditions:

$$\mathcal{T}(s; \tau, \text{Pe}_a) = \exp\left[\frac{\text{Pe}_a}{2}\left(1 - \sqrt{1 + \frac{4s\tau}{\text{Pe}_a}}\right)\right] \qquad (12.5.\text{c-}5)$$

Rearrangement of Eq. 12.5.c-5 gives:

$$[-\ln \mathcal{T}(s)]^{-1} = \tau s[-\ln \mathcal{T}(s)]^{-2} - \frac{1}{\text{Pe}_a} \qquad (12.5.\text{c-}6)$$

Thus, a plot of the left-hand side of Eq. 12.5.c-6 versus

$$s[-\ln \mathcal{T}(s)]^{-2}$$

should give a straight line with slope $\tau$ and intercept $(-1/\text{Pe}_a)$.

The quality of the straight line gives an indication of the appropriateness of the chosen model to represent the data. Also, it appears that separating the parameters $\tau$ and $\text{Pe}_a$ into the slope and intercept permits a less equivocal determination of their values. Of course, not all model transfer functions can be rearranged in this way, and Bashi and Gunn [106] have recently described various other approaches such as linearizations of the transfer function.

Numerical comparisons of the various methods by Böxkes and Hofmann [96], Østergaard and Michelsen [105], Bashi and Gunn [106], and others have shown that the use of ordinary moments is the least accurate, often leading even to negative values of the Peclet number. Direct fitting in the transform domain usually leads to valid parameter values, but they are not always the same as would be found by time domain fitting. As emphasized above, the final time domain curve should be checked for agreement with the original data. Also, in models with more than one mixing parameter, there may be problems with correlations between the parameters—see Michelsen [97]. Using the special features of the transfer function seems also to yield good estimates. It does appear, however, that the information contained in RTD experiments can sometimes only be used to determine a certain (small) number of parameters. Unique discrimination will then require different types of experiments, such as downstream injection of tracer with measurement of backmixing. Then a crossplot can indicate which parameters are most sensitive to which type of experiments, and how these combinations can be used to define the appropriate parameter combinations—for example, see Mireur and Bischoff [107] for a cross-flow model. Finally, using first-order test reactions could provide alternate information, as suggested by Glasser, Katz, and Shinnar [108], but the experimental difficulties may be formidable. Clearly, one should use care in evaluating and utilizing flow model and other transport parameters.

## 12.6 Population Balance Models

The notion of using the distribution of ages of fluid elements to characterize flow patterns can be extended to the monitoring of other variables. For example, if the tracer is reversibly adsorbed within the system, the output is determined by both the fluid age distribution plus the time the tracer is adsorbed. Extensions of the standard age distribution formulation to these situations was made by Nauman and Collinge [109], Orth and Schügerl [110], and Glasser et al. [108], and recently to nonisothermal reactors by Nauman [111]. Also see Mann et al. [112] for extensions to circulating systems.

Even more general models are based on probability methods—see Katz and Shinnar [113] and Zvirin and Shinnar [12]. This section considers the specific type of model that is used to determine the evolution of the distribution function measuring the properties of interest of the population of particles. One example would be the distribution function for the ages of fluid elements. However, many other variables could also be of interest: coke content and/or activity levels of catalyst pellets, concentrations within coalescing droplets, changing pore sizes in a reacting solid, biochemical properties of growing cells, sizes of growing crystals, and many others.

A distribution of particles is defined, $\psi(x, y, z, \zeta_1, \zeta_2, \ldots \zeta_m, t)$, where $(x, y, z)$ are the usual spatial coordinates, and the $\zeta_i$ represent the other variables of interest (e.g., fluid element age, catalyst activity). Specifically,

$$\psi(x, y, z, \zeta_1, \zeta_2, \ldots \zeta_m, t)dx\, dy\, dz\, d\zeta_1 d\zeta_2 \ldots d\zeta_m$$

is taken as the fraction of particles in the geometric volume element $(\Delta x, \Delta y, \Delta z)$ with property values in the ranges $(\zeta_i, \zeta_i + \Delta\zeta_i)$. The balance on the distribution of particles can be derived similarly to equations of transport phenomena (e.g., see Himmelblau and Bischoff [4] or Randolph and Larson [114]):

$$\frac{\partial \psi}{\partial t} + \nabla \cdot (\mathbf{u}\psi) + \sum_{i=1}^{n} \frac{\partial}{\partial \zeta_i} (v_i \psi) + D - B = 0 \tag{12.6-1}$$

where $D, B$ represent the destruction and generation (birth) of particles by some processes.

In Eq. 12.6-1, the second term is the physical transport of the particles, and in the third term the $v_i = d\zeta_i/dt$ represent the rate of change of property $\zeta_i$ in the "phase space" of $\zeta$. The equation is very similar to the Boltzmann equation of kinetic theory, where $\psi$ then represents the distribution function of molecular velocities, $\zeta$.

Determination of the detailed spatial dependence of $\psi$ is often very difficult to find, and often only the distribution of properties $\zeta_i$ within the entire system is desired. Then, the volume integral average of Eq. 12.6-1 is needed:

$$\frac{1}{V}\frac{\partial}{\partial t}(V\langle\psi\rangle) + \sum_{i=1}^{n} \frac{\partial}{\partial \zeta_i}(\langle v_i\psi\rangle) + \langle D\rangle - \langle B\rangle$$

$$= \frac{1}{V}[F_0'\psi_0 - F_e'\psi_e] \tag{12.6-2}$$

where the volume average distribution function is

$$\langle\psi(\zeta, t)\rangle \equiv \frac{1}{V}\int \psi(\mathbf{x}, \zeta, t)dx\, dy\, dz \tag{12.6-3}$$

The term on the right-hand side of Eq. 12.6-2 represents any flow of the particles into or out of the macroscopic volume. The derivation is similar to that used in Chapter 7, and is given in Himmelblau and Bischoff [4].

It should be mentioned that $\psi$ actually represents the distribution of the mathematical expected value of the particles, and (as in kinetic theory) higher-order distribution functions are required to rigorously determine the system evolution—see Ramkrishna, Borwanker, and Shah [115, 116, 117]. However, Eq. 12.6-1 or 2 seem to be adequate for a variety of problems.

Using the population balance approach is most readily seen by considering several examples. One could rederive all of the age distribution formalism by choosing $\zeta_1 = \alpha$ (age) in the macroscopic population balance Eq. 12.6-2:

$$\frac{1}{V}\frac{\partial}{\partial t}(V\langle\psi\rangle) + \frac{\partial\langle\psi\rangle}{\partial\alpha} = \frac{1}{V}(F_0'\psi_0 - F_e'\psi_e)$$

where $v_1 = d\alpha/dt = 1$ was used. The correspondence with the functions of Sec. 12.2 is:

$$\langle\psi\rangle = \text{internal age distribution}$$
$$= I(\alpha)$$
$$\psi_e = \text{exit age distribution}$$
$$= E(\alpha)$$
$$\psi_0 = \text{inlet age distribution}$$
$$= \delta(\alpha)$$

Thus

$$\frac{1}{V}\frac{\partial}{\partial t}(VI) + \frac{\partial I}{\partial\alpha} = \frac{1}{V}(F_0'\psi_0 - F_e'\psi_e) \tag{12.6-4}$$

which reduces to many of the earlier results for special cases. More detailed examples of the use of this is given in Himmelblau and Bischoff [4], Chen [10] and Wen and Fan [2].

### Example 12.6-1  Population Balance Model for Micromixing

If a flow system is visualized as consisting of a large number of fluid elements that collide, coalesce, and then reform into two new elements; a population balance model similar to that for immiscible droplet interaction can be formulated. The latter was devised by Curl [118]; also see Rietema [119] for a review of this complex area.

For a perfectly macromixed reactor, Evangelista, Katz, and Shinnar [120] utilized Eq. 12.6-2 for the particle distribution of concentrations; thus,

$$\langle\psi\rangle \equiv \psi(C_A, t)$$

(The averaging symbol will be dropped for simplicity.) With this choice of $\zeta_1 \equiv C_A$,

$$v_1 = \frac{dC_A}{dt} = -r_A = \text{reaction rate}$$

For a constant volume system, Eq. 12.6-2 becomes

$$\frac{\partial\psi}{\partial t} - \frac{\partial}{\partial C}(r\psi) = \frac{1}{\tau}(\psi_0 - \psi) + (B - D) \tag{a}$$

The "birth" of a fluid element of concentration $C$ occurs when two other elements coalesce with concentrations such that $C = (C' + C'')/2$. The "death" term means any collision with another fluid element so that the original concentration is changed. Therefore, the net birth and death rate is given by

$$(B - D) = 2\beta\left\{\iint \psi(c')\psi(c'')\delta\left(\frac{c' + c''}{2} - c\right)dc'\,dc'' - \psi(c)\int \psi(c')dc'\right\}$$ (b)

where the first term in the brackets picks out only those collisions of two elements producing a concentration $C$, and the second term represents all the collisions of elements of concentration $C$. $\beta$ is a measure of the agglomerative mixing of the fluid elements in the reactor. The main variable of interest is the "mean" concentration, which is actually measured (the "expected value" of $\psi(c, t)$. This will be denoted by

$$\bar{C}(t) \equiv \int c\psi(c, t)dc$$ (c)

The variance will also be defined in the usual way:

$$\sigma_c^2(t) = \int (C - \bar{C})^2\psi\, dC = \int C^2\psi\, dC - \bar{C}^2$$ (d)

For the mean, multiply the population balance Eqs. (a, b) by $C$, and integrate; the terms are:

$$\frac{\partial \bar{C}}{\partial t} - \left[cr(c)\psi\Big|_{\text{limits}}^{0} - \int r(c)\psi\, dc\right]$$

$$= \frac{1}{\tau}(\bar{C}_0 - \bar{C}) + 2\beta(\bar{C}^* - \bar{C})$$

Thus, the equation for the mean (measured) concentration is

$$\frac{d\bar{C}}{dt} = -\int r\psi\, dC + \frac{1}{\tau}(\bar{C}_0 - \bar{C})$$ (e)

---

$$^* \int cdc \iint \psi(c')\psi(c'')\delta\left(\frac{c' + c''}{2} - c\right)dc'\,dc''$$

$$= \iint \psi(c')\psi(c'') \int c\delta\left(\frac{c' + c''}{2} - c\right)dc\,dc'\,dc''$$

$$= \tfrac{1}{2}\int c'\psi(c')dc' \int \psi(c'')dc'' + \tfrac{1}{2}\int \psi(c')dc' \int c''\psi(c'')dc''$$

$$= \tfrac{1}{2}(\bar{c})(1) + \tfrac{1}{2}(1)(\bar{c}) = \bar{c}$$

A similar procedure gives an equation for the variance:

$$\frac{d\sigma_c^2}{dt} = -2 \int (C - \bar{C})r\psi \, dC + \frac{1}{\tau}[\sigma_{co}^2 - \sigma_c^2 + (\bar{C}_0 - \bar{C})^2] - \beta\sigma_c^2 \quad \text{(f)}$$

For a pure tracer experiment, $r \equiv 0$, the equation for $\bar{C}$ indicates that the result (measured $\bar{C}$) is independent of $\beta$ (i.e., the micromixing ($\beta$) has no effect, as was reasoned in Sec. 12.4). For a step input in *time*, $\psi_0(c, t) = \delta(C - C_0)$, or $\bar{C}_0 = C_0$, and, with no reaction,

$$\bar{C}(t)|_{\text{step}} = C_0(1 - e^{-t/\tau}), \quad \bar{C}(0) = 0 \quad \text{(g)}$$

as usual. The variance is (with $\sigma_{co}^2 = \int C^2 \delta(C - C_0)dC - C_0^2 = 0$):

$$\sigma_c^2 = C_0^2 e^{-(t/\tau)} \frac{e^{-(t/\tau)} - e^{-\beta\tau(t/\tau)}}{\beta\tau - 1} \quad \text{(h)}$$

Note that $\beta\tau = \infty$ implies perfect mixing—no fluctuations. These results are as indicated in Fig. 1.

The variance about the mean is usually difficult to measure, but Evangelista et al., [120] suggested that the time at maximum variance might be observable:

$$t_{\text{max}} = \tau \frac{\ln\left(\dfrac{\beta\tau + 1}{2}\right)}{\beta\tau - 1}$$

which is one way to obtain values of $\beta$.

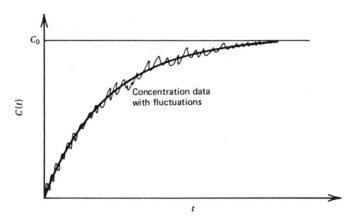

*Figure 1 Sketch of appearance of concentration data, with micromixing fluctuations, in a perfectly mixed flow vessel after a step input.*

_____ CHEMICAL REACTOR DESIGN

In a batch system, $\tau \to \infty$, and with no reaction, the equations are:

$$\frac{d\bar{C}}{dt} = 0 \longrightarrow \bar{C} = \text{const} = \bar{C}(0) \tag{i}$$

$$\frac{d\sigma_c^2}{dt} = -\beta\sigma_c^2 \longrightarrow \sigma_c^2 = \sigma_c^2(0)e^{-\beta t} \tag{j}$$

Thus, in a perfectly macromixed batch system, the mean concentration is constant (obviously), and the variance of the fluctuations decreases exponentially with time—the latter is obverved in turbulent systems, and is also predicted from isotropic, homogeneous turbulence theory (see Fig. 2).

These types of tracer experiments can be used to find the micromixing parameter, $\beta$. Evangelista et al. [120] describe typical values:

$$\beta \simeq (0.1 - 0.5)(\varepsilon/L^2)^{1/3}$$

where

$$\varepsilon = \text{power/mass of fluid } (\propto \text{power/volume})$$
$$\propto N^3L^2, \text{ geometrically similar vessels}$$
$$N = \text{impeller rpm}$$
$$L = \text{characteristic length}$$

For large processing vessels, they estimate

$$\beta \simeq 0.5 \text{ sec}^{-1}$$

These "scale-up" parameters are similar to what would be found from simpler dimensional arguments, of course, but the theory also indicates how this information can be used to predict concentration fluctuations (micromixing). In addition, more rigorous turbulence theories, or experiments, could be used to provide more exact values of $\beta$ for a specific situation.

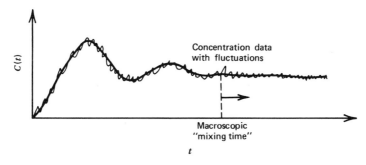

*Figure 2 Sketch of appearance of concentration data, with micromixing fluctuations, in a batch vessel.*

Using these relations with a reaction occurring illustrates the principles discussed in this chapter. For a first-order reaction,

$$r = kC$$

The Eq. (e) for $\bar{C}$ becomes:

$$\frac{d\bar{C}}{dt} = -k\bar{C} + \frac{1}{\tau}(\bar{C}_0 - \bar{C}) \tag{k}$$

which can be directly solved, given the feed $\bar{c}_0$. Thus, the micromixing parameter, $\beta$, has no effect here on the outlet conversion, as reasoned and illustrated earlier. For a steady-state perfectly macromixed reaction,

$$\frac{\bar{C}}{\bar{C}_0} = \frac{1}{1 + k\tau} \tag{l}$$

as is normally found. There is a variance in the outlet concentration, caused by imperfect micromixing:

$$\left(\frac{\sigma_c}{\bar{\bar{C}}}\right)^2 = \frac{(k\tau)^2}{2k\tau + \beta\tau + 1} \tag{m}$$

Again, $\beta\tau = \infty$ implies perfect mixing $-\sigma_c{}^2 = 0$.

The case of a steady-state (or unsteady-state) system with a second-order (or any non-first-order) reaction is more complicated:

$$r = kC^2$$

and from Eq. (e):

$$k \int C^2\psi \, dC = \frac{1}{\tau}(\bar{C}_0 - \bar{C})$$

$$= k(\sigma_c{}^2 + \bar{C}^2) \tag{n}$$

With no fluctuations (perfect micromixing), $\sigma_c{}^2 = 0$, and the usual result is obtained:

$$k\bar{C}^2 = \frac{1}{\tau}(\bar{C}_0 - \bar{C}) \tag{o}$$

However, *with* micromixing effects, the mean concentration cannot be obtained without knowledge of $\sigma_c{}^2$; thus, the result depends on the micromixing, which was reasoned earlier. If the concentration fluctuations are symmetric (third central moment is zero), the variance can be found, with knowledge of $\beta$,

$$4k\bar{C}\tau + \beta\tau + 1)\left(\frac{\sigma_c}{\bar{\bar{C}}}\right)^2 = (k\bar{C}\tau)\left[\left(\frac{\sigma_c}{\bar{\bar{C}}}\right)^2 + 1\right]^2 \tag{p}$$

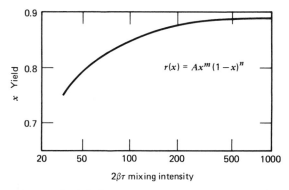

*Figure 3 Adiabatic reactions; x = conversion
to product, yield = x averaged over contents of
vessel. ---- Asymptotic value = 0.897 (from
Evangelista, et al. [120]).*

In general, the details of the micromixing would have to be known. A perturbation technique of Evangelista et al. for large $\beta\tau$ can also be useful.

In standard process vessels with "thin" fluids, the approximate values of $\beta \simeq 0.5 \text{ s}^{-1}$ are such that $\beta\tau$ is large—several hundred +. Thus, micromixing effects are not very critical, as is well known, at least for not too rapid reactions. However, for viscous systems with smaller $\beta$, very fast reactions, or short residence times $\tau$, the value of $\beta\tau$ can be small relative to $k$, and the micromixing effects then become important.

Evangelista et al. [120] and [121] illustrate this for an adiabatic combustion process, where the rate equation in terms of conversion is

$$r(x) = Ax^m(1 - x)$$

(The $x^m$, $m = 5 - 20$, is an approximation to the Arrhenius temperature dependency with the adiabatic temperature rise expressed in terms of conversion from the mass and heat balances.) The figure shows how the conversion $x$ decreases with imperfect micromixing (for $A = 15$, $m = 5$) (see Fig. 3).

It was also found that the variance divided by the conversion could range from about 20 percent, to zero for very high mixing intensity ($\beta\tau \rightarrow \infty$). Figure 4 shows that the stability criteria could also be affected by the micromixing effects; this is even more true for highly exothermic reactions, if $m = 15$. A useful review is by Pratt [122].

These coalescence-redispersion models can also be formulated as a direct numerical Monte Carlo computation, as shown by Spielman and Levenspiel [123]. Similar results are obtained, but, of course, each situation must be individually simulated.

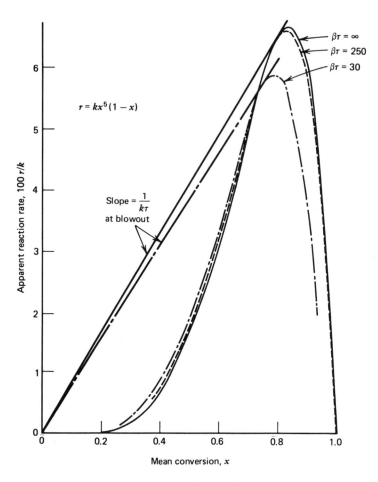

*Figure 4 Effect of mixing intensity on apparent reaction rate (from Evangelista, et al. [121]).*

The Monte Carlo approach was extended to reactors with a plug flow macromixing RTD by Kattan and Adler [124]. Here, the coalescing fluid elements are moved through the reactor at a speed corresponding to the constant mean fluid velocity. Kattan and Adler [124] were able to simulate experimental results of Vassilatos and Toor [125]. The coalescence frequency was found from data for extremely rapid reactions, where the observed rate is essentially completely controlled by the micromixing in this situation with a flat velocity profile. Then, these coalescence rates were used to predict the experimental results for rapid and slow reactions taking place in the same equipment.

A generalization of these population balance methods to reactions with arbitrary RTD was given by Kattan and Adler [126]. They expanded the phase space of the distribution functions to include the life expectation as well as concentration of the individual fluid elements: $\psi(C, \lambda, t)$. The population balance then reduces to all of the previous developments for the various special cases of segregated or micromixed flow, the perfect macromixing coalescence-redispersion model, and can be solved as continuous functions or by discrete Monte Carlo techniques. Goto and Matsubara [127] have combined the coalescence and two-environment models into a general, but very complex, approach that incorporates much of the earlier work.

---

## Example 12.6-2 Surface Reaction-Induced Changes in Pore-Size Distribution

Many types of porous media can have their pore-size distribution changed by reaction of the fluid in the pores with the walls. Examples would be heterogeneous fluid–solid reactions (e.g., ore reductions), catalyst pellet modifications, leaching, and the acidation of oil wells, where an acid is pumped into the well to dissolve the surrounding rock partially and thereby increase its porosity and permeability.

The gradual growth in pore size with treatment can be handled by the usual methods of reaction and diffusion (as in Chapter 4), but if the pores grow into each other a more detailed model is required. This latter effect is especially important in modifying permeabilities, since the new large pore areas have a strong effect on the overall hydraulic resistance, or in decreasing the new macropore diffusional resistance (a common jargon term is "wormholing").

This problem was considered by Schechter and Gidley [128]; an outline of their treatment follows. Define a distribution of pore areas at a point in the porous medium, $\psi(A, t)$ (this assumes that an average pore length can be utilized). Then the population balance will consist of terms for the time evolution of the distribution, the rate of change is a given pore size, and terms accounting for the intersection of separate pores:

$$\frac{\partial \psi}{\partial t} + \frac{\partial}{\partial A}(v\psi) = \int_0^A v(A')\psi(A - A')\psi(A')dA'$$
$$- \int_0^\infty [v(A) + v(A')]\psi(A)\psi(A')dA' \qquad \text{(a)}$$

In Eq. (a), $v(A, C_f(t))$ (pore and reaction properties) is the rate at which the pore walls change because of interaction with the fluid reactant concentration $C_f(t)$. For the terms on the right-hand side, the first term accounts for the "birth" of a new pore of area $A$ from two smaller pores of sizes $A'$ and $(A - A')$. The second term accounts for the losses of pores of area $A$ by their intersecting another pore

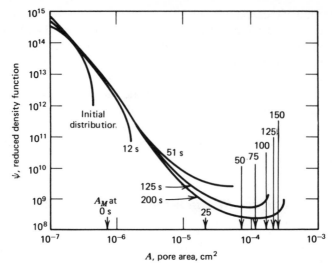

Figure 1 Time change in pore structure (from Schechter and Gidley [128]).

of area $A'$, and by a pore of area $A'$ intersecting the original pore. Once this population balance equation is solved for the distribution function $\psi(A, t)$, the change in porosity can be found from (for average pore length):

$$\varepsilon(t) = \int_0^\infty A\psi(A, t)dA \qquad (b)$$

Schechter and Gidley [128] utilized an approximation to the classical Graetz problem for the wall reaction to obtain $v(A, C_f(t))$, and then solved the population balance Eq. (a) by numerical methods. An example of the evolution of the pore area distribution for typical conditions for well acidation is shown in Fig. 1.

We see that the larger pores increase significantly in amount at the expense of the slightly smaller pores. The smallest pores remain about the same, however, because of the diffusional limitations in the pore wall reaction. The permeability change of Darcy's law can then be found from

$$\frac{k(t)}{k_0} = \frac{\displaystyle\int_0^\infty A^2\psi(A, t)dA}{\displaystyle\int_0^\infty A^2\psi(A, 0)dA} \qquad (c)$$

(Note the strong weighting with respect to area.) The computation showed a many fold increase, which, of course, is the purpose of the process. Further investigation of the details of the wall reactions and other specific aspects is given in Sinex, Schechter, and Silberberg [129] and Swift and Fogler [130].

Applications to catalyst deactivation and activity are given by Rudd [131], and use of these concepts for reactor-regenerator systems (e.g., in catalytic cracking) is analyzed by Petersen [132], Gwyn and Colthart [133], and Jacob [134]; an outline of the developments is given in Chapter 13.

## Problems

12.1 Supply the derivation of Eq. 4 of Ex. 12.3-2.

12.2 Consider the following flow system:

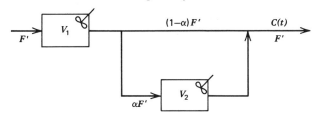

where a fraction $\alpha$ of the flow passes through the second tank, $V_2$.
(a) For an impulse function input to tank $V_1$, derive the RTD for the complete system, that would be measured at $C(t)$.
(b) Show that, for the special flow split,

$$\alpha = \frac{V_2}{V_1 + V_2}$$

The overall RTD becomes as for a *single, perfectly mixed region*!
   Note that this example illustrates that overall perfect mixing, as measured by RTD, need not mean that the specific details of the flow system are a simple perfect mixer.
   Further discussion is provided by T. Fitzgerald [*Chem. Eng. Sci.*, **29**, 1019 (1974)].

12.3 Find the mean and variance for the RTD of Ex. 12.4-3.

12.4 Derive Eqs. 2, 3, and 4 of Ex. 12.4-3.

12.5 Given the reactor configurations

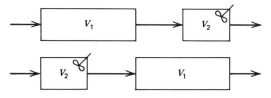

If the total volume $V = V_1 + V_2$ is imposed, what should be the ratio of the volumes $V_1/V_2$ in both cases to obtain the same exit conversion, $X_{Af}$ for a first-order reaction? For a second-order reaction?

12.6 Calculate the effect of recycle on the conversion in tubular reactors with plug flow.

**12.7** Suppose a reactor system consists of many piston-flow reactors in parallel:

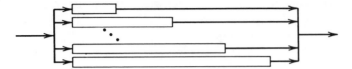

From Chapter 9 (or 1), the exit concentration in any one of them is, with a first-order reaction:

$$\frac{C_A(L)}{C_{A0}} = e^{-\Omega k L/F'}$$

(a) If the reactor lengths are arranged according to the gamma distribution,

$$f(L) = \left(\frac{\alpha + 1}{\bar{L}}\right)^{\alpha + 1} \frac{L^\alpha}{\Gamma(\alpha + 1)} \exp\left[-(\alpha + 1)\frac{L}{\bar{L}}\right]$$

where

$$\bar{L} = \text{mean length}$$

$$= \int_0^\infty L f(L) dL \,(\text{check this})$$

show that the total output of the system is given by:

$$\frac{\bar{C}}{\bar{C}_{A0}} = \int_0^\infty \frac{C_A(L)}{C_{A0}} f(L) dL$$

$$= \left(1 + \frac{\bar{L}}{\alpha + 1} \frac{\Omega k}{F'}\right)^{-(\alpha + 1)}$$

(b) For the special case of $\alpha = 0$, the exponential distribution, the result of part (a) becomes

$$\frac{\bar{C}_A}{C_{A0}} = \left(1 + \frac{\bar{L}\Omega k}{F'}\right)^{-1}$$

which is the result for a single perfectly mixed reaction, Chapter 10. Discuss this result in view of reactor flow models.

**12.8** Compute the second order reaction results for Fig. 12.4-1.

**12.9** Derive Eqs. 12.5.a-13 to 15 and plot a few curves for various values of the parameters.

**12.10** Derive Eq. 12.5.a-19 and plot a few curves for various values of the parameters.

**12.11** Show that for an axial dispersion flow system with variable density, area, velocity, and dispersion coefficients between two adjacent sections, equating total fluxes leads to the boundary condition:

$$\left(\frac{C_1}{\rho_1}\right) - D_1 \frac{d}{dz}\left(\frac{C_1}{\rho_1}\right) = \left(\frac{C_2}{\rho_2}\right) - D_2 \frac{d}{dz}\left(\frac{C_2}{\rho_2}\right)$$

12.12 Derive Eqs. 12.5.b-11, 13–15.

12.13 Plot the dimensionless variances of the recycle model of Ex. 12.5.b-2, Eq. (1), as a function of the parameters $r$ and $(k_T a_v V/F_0')$.

# References

[1] Levenspiel, O. and Bischoff, K. B. *Adv. Chem. Engng.*, Vol. 4, Academic Press, New York (1963).

[2] Wen, C. Y. and Fan, L. T. *Models for Flow Systems and Chemical Reactors*, Marcel Dekker, New York (1975).

[3] Levenspiel, O. *Chemical Reaction Engineering*, 2nd ed., Wiley, New York (1972).

[4] Himmelblau, D. M. and Bischoff, K. B. *Process Analysis and Simulation*, J. Wiley, New York (1968).

[5] Danckwerts, P. V. *Chem. Eng. Sci.*, **8**, 93 (1958).

[6] Danckwerts, P. V. *Chem. Eng. Sci.*, **2**, 1 (1953).

[7] Spaulding, D. B. *Chem. Eng. Sci.*, **9**, 74 (1958).

[8] Hull, D. E. and von Rosenberg, D. U. *Ind. Eng. Chem.*, **52**, 989 (1960).

[9] Nauman, E. B. *Chem. Eng. Sci.*, **24**, 1461 (1969).

[10] Chen, M. S. K. *Chem. Eng. Sci.*, **26**, 17 (1971).

[11] Chu, L. C. and Fan, L. T. *Can. J. Chem. Eng.*, **41**, 60 (1963).

[12] Zvirin, Y. and Shinnar, R. *Int. J. Multiphase Flow*, **2**, 495 (1976).

[13] Bischoff, K. B. and McCracken, E. A. *Ind. Eng. Chem.*, **58**, No. 7, 18 (1966).

[14] Murphree, E. V., Voorhies, A., and Mayer, F. X. *Ind. Eng. Chem. Proc. Des. Devpt.*, **3**, 381 (1964).

[15] Wei, J. *Can. J. Chem. Eng.*, **44**, 31 (1966).

[16] Matsuyama, H. and Miyauchi, T. *J. Chem. Eng. Japan*, **2**, 80 (1969).

[17] Weekman, V. W. *Ind. Eng. Chem. Proc. Des. Devpt.*, **7**, 90 (1968).

[18] Metzner, A. B. and Pigford, R. L. *Scaleup in Practice*, R. Fleming, ed., Reinhold Publishing Corp., New York (1958), p. 16.

[19] Kramers, H. and Westerterp, K. R. *Elements of Chemical Reactor Design and Operation*, Academic Press, New York (1963).

[20] Zwietering, T. N. *Chem. Eng. Sci.*, **11**, 1 (1959).

[21] Zoulalian, A. and Villermaux, J. *Proc. 3rd Intl. Symp. Chem. Reac. Eng.*, *Adv. Chem. Ser. 133*, Am. Chem. Soc. (1974).

[22] Nagasubramanian, K. and Graessley, W. W., *Chem. Eng. Sci.*, **25**, 1549 (1970).

[23] Hofmann, H. "Interaction of Fluid Flow and Chemical Kinetics in Homogeneous Reactions," presented at A.I.Ch.E. National Meeting, Houston, Texas (December 1963).

[24] Chauhan, S. P., Bell, J. P., and Adler, R. J. *Chem. Eng. Sci.*, **27**, 585 (1972).

[25] Weinstein, H. and Adler, R. J. *Chem. Eng. Sci.*, **22**, 65 (1967).

[26] Villermaux, J. and Zoulalian, A. *Chem. Eng. Sci.*, **24**, 1513 (1969).

[27] Rippin, D. W. T. *Chem. Eng. Sci.*, **22**, 247 (1967).

[28] Villermaux, J. and Devillon, J. C. *Proc. Second Intl. Symp. Chem. Reac. Engng.*, Elsevier, Amsterdam (1972).

[29] Aubry, C. and Villermaux, J. *Chem. Eng. Sci.*, **30**, 457 (1975).

[30] Nauman, E. B. *J. Macromol. Sci.—Rev. Macromol. Chem.*, **c10**, 75 (1973).

[31] Nauman, E. B. *Chem. Eng. Sci.*, **30**, 1135 (1975).

[32] Berty, J. M. *J. Macromol. Sci. Chem.*, **A8**, 919 (1974).

[33] Truong, K. T. and Methot, J. C. *Can. J. Chem. Eng.*, **54**, 572 (1976).

[34] Brodkey, R. S., ed. *Turbulence in Mixing Operations*, Academic Press, New York (1975).

[35] Olson, J. H. and Stout, L. E. *Mixing Theory and Practice*, Vol. II, V. W. Uhl and J. B. Gray, eds., Academic Press, New York (1966).

[36] Carslaw, H. S. and Jaeger, J. C. *Conduction of Heat in Solids*, Oxford Universities Press, London (1959). Crank, J. *Mathematics of Diffusion*, Oxford University Press, London (1956).

[37] Brenner, H. *Chem. Eng. Sci.*, **17**, 229 (1962).

[38] Taylor, G. I. *Proc. Roy. Soc.*, **A219**, 186 (1953).

[39] Aris, R. *Proc. Roy. Soc.*, **A235**, 67 (1956).

[40] Hunt, B. *Int. J. Heat Mass Trans.*, **20**, 393 (1977).

[41] Gill, W. N. *Chem. Eng. Sci.*, **30**, 1123 (1975).

[42] Kramers, H. and Alberda, G. *Chem. Eng. Sci.*, **2**, 1731 (1953).

[43] Roemer, M. H. and Durbin, L. D. *Ind. Eng. Chem. Fund.*, **6**, 121 (1967).

[44] Klinkenberg, A. *Ind. Eng. Chem. Fund.*, **5**, 283 (1966).

[45] Klinkenberg, A. *Chem. Eng. Sci.*, **23**, 175 (1968).

[46] Klinkenberg, A. *Chem. Eng. Sci.*, **26**, 1133 (1971).

[47] Shinnar, R. and Naor, P. *Chem. Eng. Sci.*, **22**, 1369 (1967).

[48] Lelli, U., Magelli, F., and Sama, C. *Chem. Eng. Sci.*, **27**, 1109 (1972).

[49] Seinfeld, J. H. and Lapidus, L. *Process Modeling, Estimation, and Identification*, Prentice-Hall, Englewood Cliffs, N.J. (1974).

[50] Wehner, J. F. and Wilhelm, R. H. *Chem. Eng. Sci.*, **6**, 89 (1956).

[51] Bischoff, K. B. *Chem. Eng. Sci.*, **16**, 131 (1961).

[52] van Cauwenberghe, A. R. *Chem. Eng. Sci.*, **21**, 203 (1966).

[53] Hiby, J. W. *Interactions Between Fluids and Solids*, Inst. Chem. Eng., London (1962), p. 312.

[54] Wicke, E. *Chemie Ing. Techn.*, **47**, 547 (1975).

[55] Young, L. C. and Finlayson, B. A. *Ind. Eng. Chem. Fund.*, **12**, 412 (1973).

[56] Zvirin, Y. and Shinnar, R. *Water Research*, **10**, 765 (1976).

[57] Mears, D. E. *Chem. Eng. Sci.*, **26**, 1361 (1971).

[58] Turian, R. M. *Chem. Eng. Sci.*, **28**, 2021 (1973).

[59] Mears, D. E. *Ind. Eng. Chem. Fund.*, **15**, 20 (1976).

[60] Pavlica, R. T. and Olson, J. H. *Ind. Eng. Chem.*, **62**, No. 12, 45 (1970).

[61] Littman, H. and Barile, R. G. *Chem. Eng. Prog. Symp. Ser. No. 67*, **62**, 10 (1966).

[62] DeWasch, A. P. and Froment, G. F. *Chem. Eng. Sci.*, **26**, 629 (1971).

[63] Wakao, N. *Chem. Eng. Sci.*, **31**, 1115 (1976).

[64] Miller, G. A. and Bailey, J. E. *A.I.Ch.E. J.*, **19**, 876 (1973).

[65] van der Laan, E. T., *Chem. Eng. Sci.*, **7**, 187 (1957).

[66] Schneider, P. and Smith, J. M. *A.I.Ch.E. J.*, **14**, 762 (1968).

[67] Gunn, D. J. *Chem. Eng. Sci.*, **25**, 53 (1970).

[68] Hochman, J. M. and Effron, E. *Ind. Eng. Chem. Fund.*, **8**, 63 (1969).

[69] Villermaux, J. and van Swaaij, W. P. M. *Chem. Eng. Sci.*, **24**, 1097, 1083 (1969).

[70] Arva, P., Kafarov, V. V., and Dorokhov, I. N. *Theo. Found. Chem. Eng.*, **3**, 221 (1969).

[71] Popović, M. and Deckwer, W. D. *Chem. Eng. J.*, **11**, 67 (1976).

[72] Satterfield, C. N. *A.I.Ch.E. J.*, **21**, 209 (1975).

[73] Hofmann, H. *Int. Chem. Eng.*, **17**, 19 (1977).

[74] Koros, R. M. *Proc. Fourth Intl. Symp. Chem. Rxn. Engng.*, Heidelberg, Fed. Rep. Germany, DECHEMA, Frankfurt (M), (1976), p. 372.

[75] Cholette, A. et al. *Can. J. Chem. Eng.*, **37**, 105 (1959).

[76] Cholette, A. et al. *Can. J. Chem. Eng.*, **50**, 348 (1972).

[77] van de Vusse, J. G. *Chem. Eng. Sci.*, **17**, 507 (1962).

[78] Gibilaro, L. G., Kropholler, H. W., and Spikins, D. J. *Chem. Eng. Sci.*, **22**, 517 (1967).

[79] Taniyama, I. and Sato, T. *Kagaku Kogaku*, **29**, 709 (1965).

[80] Sinclair, C. G. and Brown, D. E. *Biotech. Bioengng.*, **12**, 1001 (1970).

[81] Khang, S. J. and Levenspiel, O. *Chem. Eng. Sci.*, **31**, 569 (1976).

[82] Connolly, J. R. and Winter, R. L. *Chem. Eng. Progr.*, **65**, No. 8, 70 (1969).

[83] Paul, E. L. and Treybal, R. E. *A.I.Ch.E. J.*, **17**, 718 (1971).

[84] Gray, J. B. *Mixing—Theory and Practice*, V. W. Uhl and J. B. Gray, eds., Vol. I, II, Academic Press, New York (1966).

[85] Günkel, A. E. and Weber, M. E. *A.I.Ch.E. J.*, **21**, 931 (1975).

[86] Coulaloglou, C. A. and Tavlarides, L. L., *A.I.Ch.E. J.*, **22**, 289 (1976).

[87] Hochman, J. M. and McCord, J. R. *Chem. Eng. Sci.*, **25**, 97 (1970).

[88] Carberry, J. J. *Ind. Eng. Chem.*, **58**, No. 10, 40 (1966).

[89] Dayan, J. and Levenspiel, O. *Chem. Eng. Prog. Symp. Ser. No. 101*, **66**, 28 (1970).

[90] Goldstein, A. M. *Chem. Eng. Sci.*, **28**, 1021 (1973).

[91] Nienow, A. W. *Chem. Eng. Sci.*, **29**, 1043 (1974).

[92] Mah, R. S. H. *Chem. Eng. Sci.*, **26**, 201 (1971).

[93] van Deemter, J. J. *Proc. Intl. Symp. Fluidization*, Eindhoven, A. A. H. Drinkenberg, ed., Netherlands Universities Press, Amsterdam (1967).

[94] Gwyn, J. E., Moser, J. H., and Parker, W. A. *Chem. Eng. Prog. Symp. Ser. No. 101*, **66**, 19 (1970).

[95] Hougen, J. O. *Experiences and Experiments with Process Dynamics*, CEP Monograph Ser. No. 4, **60** (1964).

[96] Böxkes, W. and Hofmann, H. *Chemie Ing. Techn.*, **44**, 882 (1972).

[97] Michelsen, M. L. *Chem. Eng. J.*, **4**, 171 (1972).

[98] Hosten, L. H. *Chem. Eng. Sci.*, **29**, 2247 (1974).

[99] Hosten, L. H. and Emig, G. *Chem. Eng. Sci.*, **30**, 1357 (1975).

[100] Bischoff, K. B. *Can. J. Chem. Eng.*, **41**, 129 (1963).

[102] Andersson, A. S. and White, E. T. *Chem. Eng. Sci.*, **26**, 1203 (1971).

[102] Williams, J. A., Adler, R. J., and Zolner, W. J. *Ind. Eng. Chem. Fund.*, **9**, 193 (1970).

[103] Clements, W. C. *Chem. Eng. Sci.*, **24**, 957 (1969).

[104] Mixon, F. O., Whitaker, D. R., and Orcutt, J. C. *A.I.Ch.E. J.*, **13**, 21 (1967).

[105] Østergaard, K. and Michelsen, M. L. *Can. J. Chem. Eng.*, **47**, 107 (1969).

[106] Bashi, H. and Gunn, D. *A.I.Ch.E. J.*, **23**, 40 (1977).

[107] Mireur, J. P. and Bischoff, K. B. *A.I.Ch.E. J.*, **13**, 839 (1967).

[108] Glasser, D., Katz, S., and Shinnar, R. *Ind. Eng. Chem. Fund.*, **12**, 165 (1973).

[109] Nauman, E. B. and Collinge, C. N. *Chem. Eng. Sci.*, **23**, 1309, 1317 (1968).

[110] Orth, P. and Schügerl, K. *Chem. Eng. Sci.*, **27**, 497 (1972).

[111] Nauman, E. B. *Chem. Eng. Sci.*, **32**, 359 (1977).

[112] Mann, U., Crosby, E. J., and Rubinovitch, M. *Chem. Eng. Sci.*, **29**, 761 (1974).

[113] Katz, S. and Shinnar, R. *Ind. Eng. Chem.*, No. 4, **61**, 60 (1969).

[114] Randolph, A. D. and Larson, M. A. *Theory of Particulate Processes*, Academic Press, New York (1971).

[115] Ramkrishna, D., Borwanker, J. D., and Shah, B. H. *Chem. Eng. Sci.*, **28**, 1423 (1973).

[116] Ibid., **29**, 1711 (1974).

[117] Ibid., **31**, 435 (1976).

[118] Curl, R. L. *A.I.Ch.E. J.*, **9**, 175 (1963).

[119] Rietema, K. *Adv. Chem. Eng.*, **5**, 237 (1964).

[120] Evangelista, J. J., Katz, S., and Shinnar, R. *A.I.Ch.E. J.*, **15**, 843 (1969).

[121] Evangelista, J. J., Shinnar, R., and Katz, S. *12th Symp. (Intl.) Comb.* (1969), p. 901.

[122] Pratt, D. T. *15th Symp. (Intl.) on Combustion*, The Combustion Institute, Pittsburgh, Pa. (1975), p. 1339.

[123] Spielman, L. A. and Levenspiel, O. *Chem. Eng. Sci.*, **20**, 247 (1965).

[124] Kattan, A. and Adler, R. J. *A.I.Ch.E. J.*, **13**, 580 (1967).

[125] Vassilatos, G. and Toor, H. L. *A.I.Ch.E. J.*, **11**, 666 (1965).

[126] Kattan, A. and Adler, R. J. *Chem. Eng. Sci.*, **27**, 1013 (1972).

[127] Goto, S. and Matsubara, M. *Chem. Eng. Sci.*, **30**, 61, 71 (1975).

[128] Schechter, R. S. and Gidley, J. L. *A.I.Ch.E. J.*, **15**, 339 (1969).

[129] Sinex, W. E., Schechter, R. S., and Silberberg, I. H. *Ind. Eng. Chem. Fund.*, **11**, 205 (1972).

[130] Swift, S. T. and Fogler, H. S. *Chem. Eng. Sci.*, **32**, 339 (1977).

[131] Rudd, D. F. *Can. J. Chem. Eng.*, **40**, 197 (1962).

[132] Petersen, E. E. *A.I.Ch.E.J.*, **6**, 488 (1960).

[133] Gwyn, J. E. and Colthart, J. D. *A.I.Ch.E. J.*, **15**, 932 (1969).

[134] Jacob, S. M. *Ind. Eng. Chem. Proc. Des. Devpt.*, **9**, 635 (1970).

# 13

## FLUIDIZED BED REACTORS

## 13.1 Introduction _____

The breakthrough of fluidization in 1942 was associated with catalytic cracking of gasoil into gasoline. Before that, catalytic cracking had been carried out in fixed bed reactors. Catalytic cracking deposits carbonaceous products on the catalyst, causing rapid deactivation of the latter. In order to maintain the production capacity, the carbon had to be burned off. This regeneration required switching the reactor out of production. In order to eliminate the switching, attempts were made to circulate the catalyst and burn off the coke in a separate vessel, the regenerator. The first approach was to use a moving catalyst bed: in the reactor the catalyst moved downward, against the fluid stream, in a very compact mass, into the regenerator. From the regenerator the catalyst was taken back to the top of the reactor by a bucket conveyor. Later, the conveyor system was replaced by a gas lift. Finally, both the reactor and the regenerator were operated under transport conditions. The high velocities required for pneumatic transport caused considerable attrition of the catalyst. In order to lower the velocities, very fine solids had to be used. It was then found that the dense mixture of solids and gas behaved in many aspects like a fluid, and this makes the control of streams much more convenient. Since then fluidized beds have been used in other fields where solids have to be handled, like the roasting of ores and in catalytic reactors for highly exothermic reactions. The latter application is based upon another advantage: the high turbulence created in the fluid–solid mixture leads to much higher heat transfer coefficients than those which can be obtained in fixed beds. Therefore, fluidized bed reactors have now also found use in exothermic processes where close temperature control is important: the oxidation of naphthalene into phthalic anhydride, the ammoxidation of propylene into acrylonitrile, the oxychlorination of ethylene into ethylene dichloride—the first step of vinylchloride manufacture, and synthetic gasoline production by the Fischer–Tropsch process.

662

The design and operation of fluidized bed reactors is by no means an easy task, especially when the recirculation of solids is involved. There is much more technological operation knowledge involved than with fixed beds. Before studying the conversion problem along the lines of the preceding chapters we discuss some aspects of this technology, without going into great detail, however. Several books have been written on the subject [1, 2, 3, 4, 10, 28, 29]. To begin, the technological aspects are illustrated by a brief description of a fluidized bed catalytic cracker.

## 13.2 Fluid Catalytic Cracking

Catalytic cracking of gasoil for the production of gasoline is carried out at tempertures of the order of 525°C. The catalyst containing 1 or 2 wt % coke is regenerated with air around 580°C by reducing the coke to 0.4 to 0.8 wt %. An early type of the reactor-regenerator system is shown schematically in Fig. 13.2-1. The oil is fed at the bottom of the reactor through a perforated plate distributor and the gasoline and gases are taken off at the top. The top of the fluidized bed

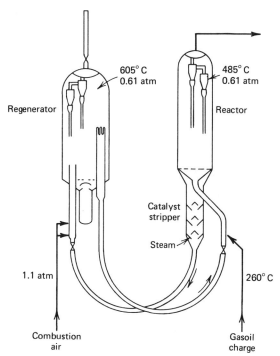

*Figure 13.2-1 Reactor-regeneration system for catalytic cracking of gasoil (after Zenz and Othmer [10]).*

has a more or less clearcut surface but some solids are entrained and a certain freeboard is necessary to minimize this. In order to avoid catalyst loss and elutriation the exit stream flows through a two-stage cyclone. The catalyst is fed back to the bed through pipes, which are called diplegs, and which have a seal at their bottom for preventing leakage of bed-fluid.

The catalyst is allowed to leave the reactor through a bottom standpipe. It is first stripped with steam in order to remove adsorbed hydrocarbons and then moves through the transfer line, under the influence of a static head, to the regenerator, where the static pressure is lower. The lower static pressure in the riser leading to the regenerator is due to the aeration of the gas–solid mixture with air required for burning off the coke. More air for the regenerator is injected in the regenerator itself, through a distributor plate. The regenerator also has a two-stage cyclone in order to reduce catalyst loss. The regenerated catalyst flows into a downcomer or standpipe and back to the reactor. The difference in static pressure required for this is realized by injection of the oil into the riser of the reactor. The oil rapidly evaporates in the pipe and reduces the catalyst bulk density. Slide valves in the lines permit additional adjustment of the flow rates.

The rate of circulation of the solids is dictated by the heat balance and activity level of the catalyst: the heat produced by the regeneration is carried to the reactor by the catalyst and there it evaporates, heats, and cracks the oil. The transfer lines have to be designed in such way that they are not eroded by the catalyst. The catalyst also has to withstand attrition.

*Some typical operating figures for fluid catalytic crackers [from 10]*

---

Medium size capacity: 15,000 barrels/day $= 2390$ m$^3$ gasoil/day
Catalyst: silica-alumina zeolite catalyst or molecular sieves 20–80 $\mu$
Total catalyst inventory: 250 tons
Amount in regenerator: 100 tons
Catalyst bulk densities
   Reactor and regenerator: 320–560 kg/m$^3$
   Stripper: 480–640
   Standpipes: 560–720
   Risers: 80–480
   Diplegs: 240–560
Catalyst circulation rate: 24 tons/min
Catalyst flow rate to cyclones: 7 tons/min
Catalyst loss: 2 tons/day
Superficial velocity in reactor and regenerator: 0.5–1.3 m/s
Velocities in standpipes: 1.7 m/s
        Risers: 7–10 m/s
        Diplegs: 1.7 m/s

---

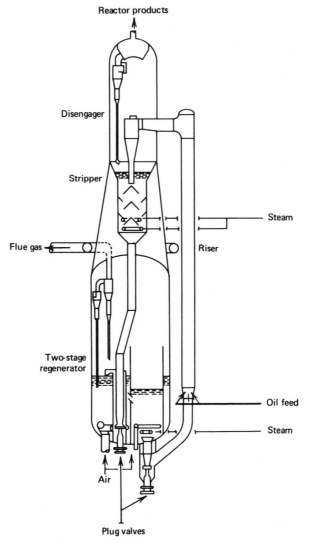

*Figure 13.2-2  Kellogg orthoflow model F convertor with riser cracking and two-stage regeneration ( from Murphy and Soudek [30]).*

**665**

Modern technology is considerably different, particularly since the introduction, in the early 1960s of crystalline zeolite catalysts. These are so active that the cracking mainly or entirely takes place in the riser. To avoid overcracking of the gasoline a very rapid separation of the products and the catalyst is required and this is achieved by means of a cyclone. The former reactor vessel is now mainly reduced to a vessel containing cyclones and a stripping section. Figure 13.2-2 shows such a recent version of a catalytic cracker.

## **13.3** Some Features of the Design of Fluidized Bed Reactors

Consider a packed bed and the pressure drop over the total bed height as a function of the flow rate of a gas flowing through it. The pressure drop will increase as shown in Fig. 13.3-1 along the line 1-2. From the velocity corresponding to the point 3 onwards the pressure drop will decrease slightly due to resettling of the catalyst in the loosest arrangement. Beyond that point, the pressure drop remains practically constant. Upon decreasing the velocity a certain hysteresis is found. The break in the curve, corresponding to the point 4 is the point of minimum fluidization.

The point of minimum fluidization is easily observed. The voidage at minimum fluidization, which is an important value for design, is calculated from:

$$\Delta p_t = L_{mf}(1 - \varepsilon_{mf})(\rho_s - \rho_g) \tag{13.3-1}$$

In the absence of experimental data for a catalyst that is being considered it is possible to predict the minimum fluidization velocity from an equation resulting

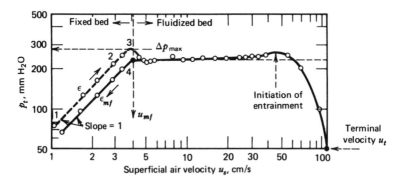

*Figure 13.3-1 Pressure drop versus gas velocity for a bed of uniformly sized sand particles. From Shirai [17] (after Kunii and Levenspiel [3]).*

from equating the pressure drop given by (13.3-1) and by the pressure drop equation for flow through packed beds. The resulting equation contains a shape factor and the void fraction at minimum fluidization, which may not be available. This $\varepsilon_{mf}$ may be correlated with respect to the particle diameter. Finally this leads, according to Leva [4], to

$$u_{mf} = 1.118 \times 10^{-13} \frac{d_p^{1.82} (\rho_s - \rho_g)^{0.94}}{\rho_g^{0.06} \mu^{0.88}} \tag{13.3-2}$$

with $u_{mf}$ in m/s, $d_p$ in $\mu$m, $\rho$ in kg/m$^3$, $\mu$ in N·s/m$^2$. As the fluid velocity is increased above $u_{mf}$ the bed expansion increases but there are no really satisfactory general correlations in this range for predicting the void fraction or bed height.

As the velocity is further increased the bed becomes less dense and finally the particles may be blown out. This maximum permissible velocity is the terminal or free-fall velocity. At this velocity the drag exerted on a particle by the upflowing gas equals the gravity force, so that

$$u_t = \sqrt{\frac{4gd_p(\rho_s - \rho_g)}{3\rho_g C_D}} \tag{13.3-3}$$

For spheres and laminar flow $C_D = 24/\text{Re}$ where $\text{Re} = d_p \rho_g u_t/\mu$. For $\text{Re} > 0.4$ (i.e., outside the laminar flow regime), $C_D$ can only be obtained experimentally. Figure 13.3-2 is a correlation for the terminal velocity as a function of particle diameter.

Pressure drop versus velocity diagrams do not always have the exact appearance of Fig. 13.3-1. Figure 13.3-3 shows such a diagram when slugging occurs. Slugging occurs when the bubble size equals the tube diameter and can be avoided by reducing the height/diameter ratio. Figure 13.3-4 is a diagram revealing channeling. This occurs when the fluid has preferential paths through the reactor and may be avoided by a better distributor and increasing the height/diameter ratio. One rule of thumb is that the pressure drop over the distributor should be at least 0.1 of the pressure drop over the bed.

With fine solids and the velocities used in industrial conditions the fluid velocity often exceeds the terminal velocity, at least for a fraction of the size distribution of the solids. Solids will then be entrained above the bed. Cyclones are installed to retain them. The height at which the cyclone inlet is placed depends on its size and efficiency. In order to determine this height the designer requires information concerning the rate of entrainment and the size distribution of the solids in the entrained fraction as a function of the height above the bed. The entrainment becomes approximately constant from a certain height above the bed onward, called "transport disengaging height." As far as the cyclone size is concerned, nothing is gained by placing the inlet higher than the transport disengaging height. This height has been correlated empirically by Zenz and Weil for fluid catalytic cracking [12]. They have also correlated the rate of entrainment at this height

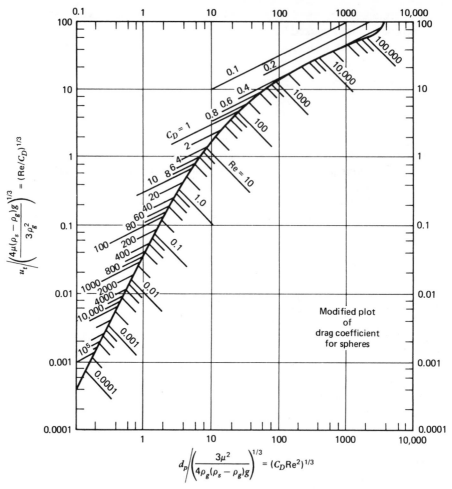

*Figure 13.3-2 Modified plot of drag coefficient for spheres (from Zenz and Othmer* [10].

as a function of the particle size and fluid velocity. Elutriation is the selective removal of fines by entrainment from a bed of particles with a distribution in particle size. Some work has been published on predicting rates of elutriation. The rate of elutriation of fines with particle diameter $d_p$ is considered to be proportional to the fraction of the bed consisting of fines with diameter $d_p$, so that

$$-\frac{1}{\Omega}\frac{dW(d_p)}{dt} = k_1 \frac{W(d_p)}{W} \tag{13.3-4}$$

668 _____ CHEMICAL REACTOR DESIGN

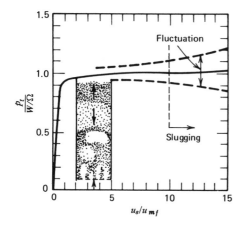

*Figure 13.3-3 Pressure drop diagram for fluctuating and slugging fluidized beds. (from Kunii and Levenspiel [3]).*

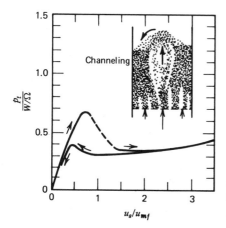

*Figure 13.3-4 Pressure drop diagram for channeling fluidized beds (from Kunii and Levenspiel [3]).*

where $W$ is the total weight of solids in the bed and $W(d_p)$ the weight of the amount of catalyst with diameter $d_p$. The elutriation rate coefficient has been correlated by Osberg and Charlesworth [27] as shown in Fig. 13.3-5 and by Wen and Hashinger [14]. The cyclone design is a problem in itself that is treated in detail in Zenz and Othmer [10] or in standard texts of chemical engineering (e.g., Perry [15]).

In some cases it is necessary to exchange heat with the fluidized bed as, for example, in the catalytic oxidation processes mentioned in the Introduction. Heat may be exchanged with the bed through the wall or through internal heat exchangers. The data available to date have been correlated and are briefly mentioned here. Figure 13.3-6 shows the most complete correlation available for heat

FLUIDIZED BED REACTORS _____ **669**

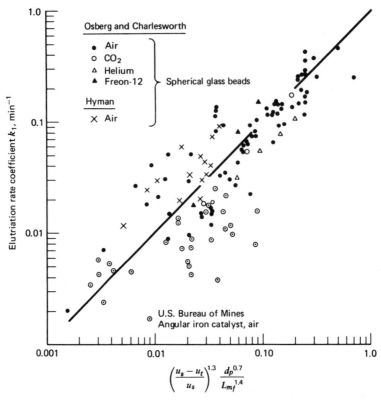

*Figure 13.3-5 Generalized correlation of elutriation rate coefficients for two-component systems (after Leva [4]).*

transfer between the bed and the wall, based on a pseudo-single-phase model (Wender and Cooper [13]). This diagram contains the usual groups involved in such correlations: on the abscissa the Reynolds number, and on the ordinate the Nusselt number, $\alpha d_p/\lambda_g$ multiplied by a group that accounts for the other variables involved. Further work on this problem has been reported by Fritz [19]. Figure 13.3-7 shows the correlation for heat exchange between the bed and internal heat exchangers (whereby $\lambda_g/c_p\rho_g$ is in ft²/hr). In this figure, $C_R$ is a correction factor depending on the location of the heat exchanger with respect to the longitudinal axis of the bed, as shown in Fig. 13.3-8. It follows from the above correlations that the heat transfer coefficient between a fluidized bed and surfaces is large, often in the range of 200 to 600 kcal/m² hr °C (0.232 − 0.697 kw/m²k).

Finally, one has to check whether or not there is a temperature difference between the gas and the catalyst surface. The following correlation has been

CHEMICAL REACTOR DESIGN

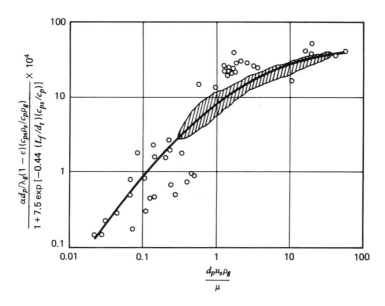

Figure 13.3-6 Wender and Cooper's correlation of fluid bed to external surface heat transfer coefficients [13], (from Zenz and Othmer [10]).

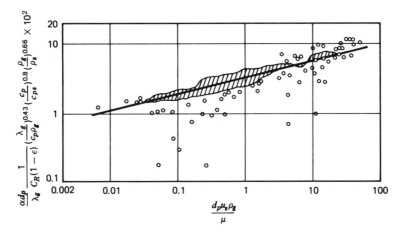

Figure 13.3-7 Wender and Cooper's correlation for fluid bed-to-internal surface heat transfer coefficients [13], (from Zenz and Othmer [10]).

671

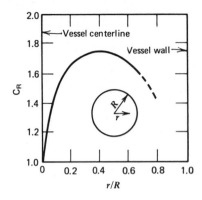

Figure 13.3-8 *Correction factor for nonaxially located internal heat transfer surface (based on data of Vreedenberg* [18], *from Zenz and Othmer* [10]).

proposed for air:

$$\frac{\alpha d_p}{\lambda_g} = 0.017 \left(\frac{d_p G}{\mu}\right)^{1.21} \tag{13.3-5}$$

It can be generalized somewhat more by accounting for the Prandtl number dependence power. The correlation is based on a pseudo-single-phase model. It follows from the correlation that the gas approaches the temperature of the solid in the very first centimeters of the bed. This is confirmed by a recent correlation of Balakrishnan and Pei [20]

$$j_H = 0.043 \left[\frac{d_p g}{u_s^2} \frac{(\rho_s - \rho_g)(1 - \varepsilon)^2}{\rho_g}\right]^{0.25} \tag{13.3-6}$$

where $\varepsilon$ is the void fraction of the fluidized bed (see Balakrishnan and Pei [20]).

As already mentioned in the description of a fluid catalytic cracker the circulation rate of solids is determined by the heat balance between the reactor and the regenerator. Equating heat inputs and outputs over the reactor leads to the following ratio for the circulation rate between the reactor and the regenerator, $\dot{m}_s$(kg/hr) and the gasoil feed rate, $(\dot{m}_r)_i$:

$$\frac{\dot{m}_s}{(\dot{m}_r)_i} = \frac{(-\Delta H)_w - (1 - \Psi)c_{pr} T_r + c_{pi} T_i}{(T_r - T_g)c_{ps}} \tag{13.3-7}$$

where $T_i$ and $c_{pi}$ are the gasoil feed temperature and specific heat, $T_r$ and $c_{pr}$ the temperature and specific heat of the gases leaving the reactor, $c_{ps}$ the specific heat of the solid and $T_g$ the temperature in the regenerator. The solids in the reactor and regenerator are assumed to be completely mixed and to be at the same temperature as the gas. No heat is lost in the transfer lines. $(-\Delta H)_w$ is the heat of cracking on a weight basis (kcal/kg gasoil cracked) and $\Psi$ is the weight fraction of gasoil converted into carbon so that $(\dot{m}_r)_e = (1 - \Psi)(\dot{m}_r)_i$. $T_r$ is dictated by the optimum reaction conditions and $T_i$ by the preheater, so that the circulation rate of solids is fixed by the temperature of the regenerator flue gases, $T_g$. With $T_g$ between 540 and 580°C the ratio $\dot{m}_s/(\dot{m}_r)_i$ is of the order of 25 to 30 [3]. When catalyst deactivation is rapid, however, the circulation rate is dictated by the rate of deactivation. This problem has been analyzed by Petersen [11]. Petersen used a deactivation function for the catalyst in the reactor, which is based on the residence time $t_r$, rather than the coke content: $\phi_r = e^{-\alpha_r t_r}$. The state of the catalyst in the regenerator is expressed by means of the activation function $\phi_g = 1 - e^{-\alpha_g t_g}$, where $t_g$ is the residence time of a catalyst particle in the regenerator. Accounting for the distribution of residence times of the catalyst in both the reactor and the regenerator, and using the methods of Chapter 12, the following expression is derived for the average catalyst activity in the reactor, that is, for the ratio of the actual reaction rate to that in the absence of coking:

$$\bar{A}_r = \frac{\dot{m}_s}{W_r} \frac{1 - \mathcal{L}\{T_g'(t_g)\}}{1 - \mathcal{L}\{T_r'(t_r)\}\mathcal{L}\{T_g'(t_g)\}} \mathcal{L}\left\{\frac{W_r}{\dot{m}_s} T_r'(t_r)\right\} \tag{13.3-8}$$

where $\dot{m}_s$ is the catalyst circulation rate (kg/hr), $W_r$ and $W_g$ the catalyst inventories in the reactor and regenerator respectively (kg). $\mathcal{L}\{T_r'(t_r)\}$ and $\mathcal{L}\{T_g'(t_g)\}$ are the Laplace transforms of the frequency distribution of residence times of the catalyst in the reactor and regenerator respectively. $T_r'(t_r)$ and $T_g'(t_g)$ are the derivatives of the corresponding cumulative distribution functions $T_r(t_r)$ and $T_g(t_g)$.

For complete mixing of the catalyst in the reactor:

$$T_r(t_r) = 1 - e^{-[(\dot{m}_s/W_r)t_r]}$$

and consequently

$$T_r'(t_r) = \frac{\dot{m}_s}{W_r} e^{-[(\dot{m}_s/W_r)t_r]}$$

Analogous formulas are derived for the residence time distribution of the catalyst in the regenerator. Equation 13.3-8 then becomes:

$$\bar{A}_r = \frac{1 - b_g}{1 - b_g b_r} b_r \tag{13.3-9}$$

with

$$b_r = \frac{\dot{m}_s}{W_r} \frac{1}{\alpha_R + \dfrac{\dot{m}_s}{W_r}}$$

and

$$b_g = \frac{\dot{m}_s}{W_g} \frac{1}{\alpha_g + \dfrac{\dot{m}_s}{W_g}}$$

When the activity is chosen, together with the catalyst inventories, the solids circulation rate may be calculated from Eq. 13.3-9. Further aspects of this problem, together with other aspects of modeling, may be found in Tigrel and Pyle [22].

## 13.4 Modeling of Fluidized Bed Reactors

It was soon recognized that the simple plug flow and complete mixing models were inadequate for predicting conversion in fluidized beds. Around 1960 a model with effective axial diffusion superposed on the plug flow was tried, with little more success. Neither of these models can explain experimental conversion, $x$ versus $W/F_0$ curves below those for complete mixing. Such an observation can only be explained by assuming that a fraction of the gas bypasses the catalyst. This is quite logical, since an important fraction of the gas flows through the bed in the form of bubbles. The amount of catalyst in the bubbles is very small and its contribution to the total conversion is usually negligible. Yet, in industrial reactors the fraction of gas flowing through the bed in the form of bubbles is so large that, in order to fit experimental data, one must conclude that this gas does not completely bypass the catalyst. Because of coalescence or bubble growth some interchange of gas was postulated between the bubble phase and the dense phase surrounding it, also called emulsion phase. These considerations led to the two-phase models, still the basis of the recent developments, which have, however, substantially clarified and refined the interchange phenomenon. Figure 13.4-1 schematically represents such a two-phase model. A fraction of the total flow rate

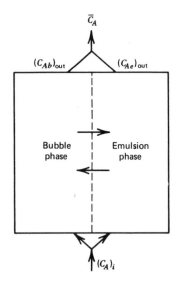

*Figure 13.4-1 Basic two-phase model for fluidized bed.*

through the bed is considered to be in the bubble phase, the rest in the emulsion phase. Between both phases there is a certain interchange or crossflow. At the outlet both streams, with their respective conversions, are hypothetically mixed to give the exit stream, with its mean conversion or concentration. Since there is no reaction in the bubble phase and because of its high velocity the flow through that phase is usually taken to be of the plug flow type. In the emulsion phase various degrees of deviation from plug flow can be postulated. The violent motion of the solid leads to mixing in the gas phase, which is generally described by an effective diffusivity model. Some authors have preferred complete mixing in that phase, based on the observation that there is a definite pattern for the flow of solids: downward near the wall, upward in the central core.

Such a two-phase model therefore contains three parameters: the cross section of the bed taken by either the bubble or the emulsion phase, the interchange coefficient, and the axial effective diffusivity in the emulsion phase. The cross section of the volume fraction of the bed that is occupied by bubbles is easily derived from the now generally accepted observation that essentially all the gas flowing through the bed in excess of that required for minimum fluidization goes into the bubble phase. Based on this postulate the fraction of the bed volume in the bubble phase can easily be derived from the bed heights and void fractions at minimum fluidization and the flow rate considered. The effective diffusivity of the gas in the emulsion phase, $D_e$, was determined by May [5] from the residence time distribution of tagged solids in the bed, assuming the effective diffusivity of the gas to be equal to that of the solid. Knowing $D_e$ the interchange coefficient was determined by fitting of gas residence time distribution measurements. Van Deemter combined the results of gas residence time distribution measurements and steady-state tracer experiments to obtain $k_I$ and $D_e$ [9]. The model expressed mathematically in the following is one of the earliest and most complete i.e. that of May [5], discussed in detail by Van Deemter in order to make the determination of the parameters more convenient [9].

It is clear that for steady-state considerations, only continuity equations have to be considered, since fluidized beds are essentially isothermal—the energy equation is not coupled with the continuity equation.

For a single reaction the equations are:
*Bubble phase*:

$$f_b u_b \frac{dC_{Ab}}{dz} + k_I(C_{Ab} - C_{Ae}) + r_A \rho_b f_b = 0 \qquad (13.4\text{-}1)$$

where $f_b$ is the fraction of the bed volume taken by bubbles and $k_I$ is the interchange coefficient expressed in $[m^3/(m^3 \text{ total bed volume) hr}]$. Notice that the terms in (13.4-1) have dimensions $[kmol/(m^3 \text{ total bed) hr}]$. May and Van Deemter neglect the term for reaction in the bubble phase.

*Emulsion phase*:

$$f_e u_e \frac{dC_{Ae}}{dz} - k_I(C_{Ab} - C_{Ae}) - f_e D_e \frac{d^2 C_{Ae}}{dz^2} + r_A \rho_e (1 - f_b) = 0 \quad (13.4\text{-}2)$$

where $f_e$ is the fraction of the bed volume taken by the emulsion *gas* (not by the emulsion *phase*, which also includes the catalyst) and $u_e$ is the velocity of the emulsion gas, on an interstitial basis—the linear velocity $u_{mf}/\varepsilon_{mf}$.

The concentration $\bar{C}_A$ measured in the gas flow at the exit is given by:

$$u_s \bar{C}_A = f_b u_b C_{Ab} + f_e u_e C_{Ae} \quad (13.4\text{-}3)$$

where $u_s$ is the superficial velocity, based on the total bed cross section. The system Eqs. 13.4-1 to 13.4-2 has to be integrated with the boundary conditions:

$$z = 0 \text{ bubble phase} \qquad C_{Ab} = (C_A)_i$$

$$\text{emulsion} \qquad -D_e \frac{dC_{Ae}}{dz} = u_e(C_{Ai} - C_{Ae})$$

$$z = L \qquad \frac{dC_{Ae}}{dz} = 0$$

Figure 13.4-2 shows some of the calculations of May for a first-order reaction according to this model and compared with idealized models. Curve 1 is for plug flow, curve 2 for complete mixing. The curves 3, 3', 4, and 4' are obtained from the present model for a fixed value of the interchange coefficient, but 3 and 4 with $D_e = 0$ and 3' and 4' with $D_e = \infty$ (i.e., plug flow and complete mixing for the gas in the emulsion phase, respectively). We see that the predicted curves can lie well below that for the ideal complete mixing model. The effect of $D_e$ is rather small when the interchange coefficient has low values (4 and 4'). The higher the values of $k_I$ the closer plug flow is approximated and the more important it is to know $D_e$ accurately (3 and 3').

Van Swaay and Zuiderweg [23] extensively tested the May–van Deemter model as to its ability to scale up the Shell chlorine process (air oxidation of HCl into $Cl_2$). To do so they investigated the decomposition of ozone on an $Fe_2O_3$ catalyst deposited on sand or silica and carried out gas pulse tracer tests. The bed diameter ranged from 10 to 60 cm, the bed heights from 50 to 300 cm. The authors concluded that the van Deemter model adequately describes fluidized bed reactor performance for first-order reactions with a rate coefficient smaller than 2.5 m$^3$/kg cat.hr or 1 s$^{-1}$. It follows from their work that with the silica-based catalyst $u_s/k_I$ linearly increases from 0.25 m to 0.4 m as $u_s$ varies from 6 to 20 cm/s. For fluid beds with a height of up to 1 m, $u_s/k_I$ is proportional to the bed height so that $n_k = L_f k_I/u_s$ is always around 3. For the larger beds the influence of the bed height is less pronounced, but a diameter influence appears: $u_s/(k_I \sqrt{L_f})$ increases

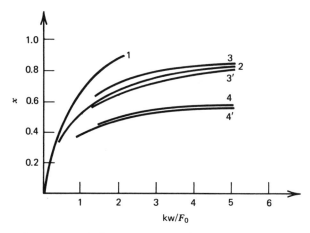

*Figure 13.4-2 Conversion in a fluidized bed reactor. Curve 1: single-phase model with plug flow ($k_I = \infty$, $D_e = 0$); curve 2: single-phase model with complete mixing ($k_I = \infty$, $D_e = \infty$); curve 3: two-phase model $k_I = 3$, $D_e = 0$; curve 3': two-phase model $k_I = 3$, $D_e = \infty$; curve 4: two-phase model $k_I = 1$, $D_e = 0$; curve 4': two-phase model $k_I = 1$, $D_e = \infty$ (after May [5]).*

from 0.37 at $d_t = 0.1$ m to 1 at $d_t = 4$ m. The following correlation was obtained at $u_s/u_{mf} = 15$ for various types of sand:

$$\frac{u_s}{k_I} = 1.5L_f^{0.50}d_t^{0.42} \text{ (m)} \tag{13.4-4}$$

Mireur and Bischoff [6] correlated data on $k_I$ and $D_e$ versus easily accessible parameters like $u_s/u_{mf}$ and $d_t/L_f$; the results are shown in Figs. 13.4-3 and 4. The curve "RTD data" was obtained from residence time distribution experiments. These are performed with a nonadsorbable tracer like helium. The reaction experiments leading to the curve "conversion data" obviously involves adsorbable species. This may explain the difference between the two curves. The correlation is not meant to be definitive since it does not account for the effect of the particle-size distribution pointed out by de Groot [2], by van Swaay and Zuiderweg [23], and by de Vries et al. [24]. The particle-size distribution is known to affect the quality of fluidization. De Vries et al. found that $n_k = L_f k_I/u_s$ varies linearly as a function of the percentage of fines from 4 at 7 percent fines to about 1.8 at 30 percent fines. Also, $n_D = u_s L_f/D_e$ is markedly affected by this variable. Nevertheless

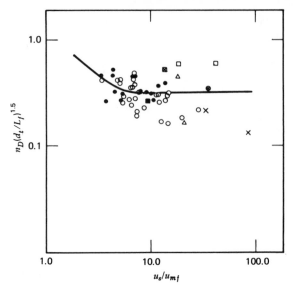

*Figure 13.4-3 Correlation for axial diffusivity in emulsion phase.* $n_D$ = *number of eddy diffusivity units,* $u_s L_f/D_e$. (*From Mireur and Bischoff* [6]).

the values of de Vries et al. for $n_k$ and $n_D$ and those of van Swaay and Zuiderweg for $n_k$ all fall in the central portions of the bands given by Mireur and Bischoff.

It is clear that the design of fluidized beds still requires extensive investigation on different scales in order to set up reliable correlations for extrapolation.

In recent years great efforts were undertaken in order to predict the interchange coefficients from fundamentals. Using hydrodynamic theory it has been possible to calculate the streamlines of gas and particles in the vicinity of a single bubble. The flow patterns of course depend strongly on the relative velocity of the bubble with respect to the emulsion gas. For all practical situations the bubble rises faster than the emulsion gas and Fig. 13.4-5 shows gas streamlines as calculated by Murray [16] from the Davidson theory (on the left side of the figure) [1]. Bubble gas is seen to leave the bubble in the upper part and to enter again in the lower part. Yet, the gas does not leave a sphere that surrounds the bubble. It is in this space between the two spheres, which has been given several names in fluidization jargon, but which we will call the interchange zone that the bubble gas comes into contact with the catalyst and is converted. The interchange zone decreases in importance as the relative bubble velocity, $u_{br}$, increases, so that the "contacting efficiency" of the bed decreases.

In reality the bubble is not spherical, but more like shown on the right-hand side of Fig 13.4-5. Partridge and Rowe found experimentally that the wake oc-

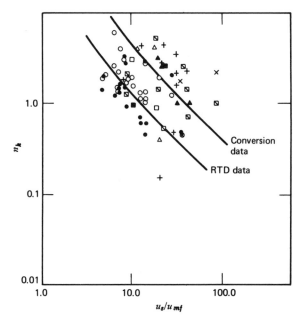

*Figure 13.4-4 Correlation for interchange coefficient from conversion and residence time distribution data. $n_k$ = number of interchange units, $k_I L_f / u_s$. (From Mireur and Bischoff [6]).*

cupies 25 percent of the spherical bubble volume. They also found that the ratio of the sum of the volumes of the bubble and the interchange zone to the volume of the bubble itself is given by $\alpha/(\alpha - 1)$, where $\alpha = u_b \varepsilon_{mf}/u_{mf}$ [7]. From this fundamental picture the interchange coefficient has now been derived in several ways.

Partridge and Rowe [7] have derived an interchange coefficient between the ensemble of bubble plus interchange zone on one hand and the emulsion phase on the other. They used the analogy with mass transfer between a drop and a surrounding fluid for which experimental correlations are available:

$$\text{Sh}_c = 2 + 0.69 \, \text{Sc}^{1/3} \text{Re}_c^{1/2} \tag{13.4-5}$$

where

$$\text{Sh}_c = \frac{k_g d_c}{D_A} \quad \text{and} \quad \text{Re}_c = \frac{u_{br} \rho_g d_c}{\mu}$$

and $d_c$ is the total diameter of bubble and interchange zone. The mass transfer coefficient $k_g$ has the dimensions: $m_f^3/m_c^2$ hr. From it, an interchange coefficient

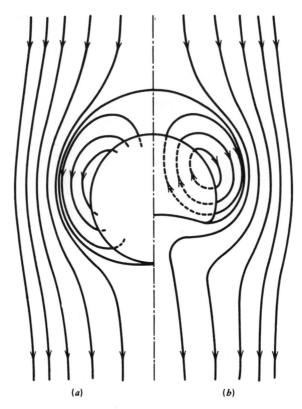

*Figure 13.4-5 Probable form of gas streamlines for a real three-dimensional bubble when $\alpha = u_b/(u_{mf}/\varepsilon_{mf}) = 1.5$. (a) Murray streamlines for spherical bubble. (b) Probable form of streamlines for bubble with particle wake. (From Rowe and Partridge [7]).*

based on the volume of the ensemble may be calculated by means of the relation $(k_{ce})_c = k_g(6\varepsilon_c/d_c)$ and this is related to the interchange coefficient $k_I$ by the relation: $k_I \simeq (k_{ce})f_b$.

In Eq. 13.4-5 and in the following relations for interchange coefficients the bubble diameter or the diameter of the bubble + interchange zone is a critical parameter. There are no generally valid correlations for this diameter to date. An estimate may be obtained from

$$u_{br} = 0.711\sqrt{gd_b} \tag{13.4-6}$$

which is an equation for the relative bubble velocity of a *single* bubble (Davidson and Harrison [1]). The variable $d_b$ is an equivalent bubble diameter and may be calculated when the superficial gas velocity, $f_b$, $u_{mf}$, and $\varepsilon_{mf}$ are known. Partridge and Rowe corrected the bubble volume for the volume taken by the wake by multiplying it with 0.75. The volume of the bubble + interchange zone is then obtained from the empirical relation:

$$V_c = V_b\left(\frac{\alpha + 0.17}{\alpha - 1}\right), \quad \text{where} \quad V_b = 0.75\frac{\pi d_b^{3}}{6} \qquad (13.4\text{-}7)$$

Chiba and Kobayashi [21], Van Swaay and Zuiderweg [23] and de Vries et al. [24] found that Eq. 13.4-6, derived from single-bubble experiments, does not adequately describe the true bubble-rising velocity, probably because of interaction between the bubbles. The first authors therefore derived empirical correlations for the bubble size.

A further refinement is to distinguish between the bubble and the interchange zone. The transfer coefficients are then as follows:

Transfer coefficient from bubble to interchange zone [3]:

$$(k_{bc})_b = 4.5\left(\frac{u_{mf}}{d_b}\right) + 5.85\left(\frac{D_A^{1/2}g^{1/4}}{d_b^{5/4}}\right) \qquad (13.4\text{-}8)$$

from interchange zone to emulsion phase [3]

$$(k_{ce})_b = 6.78\left(\frac{\varepsilon_{mf}D_A u_b}{d_b^{3}}\right)^{1/2} \qquad (13.4\text{-}9)$$

Both coefficients are referred to unit bubble volume ($m_f^3/m_b^3$ hr). The first term in the right-hand side of Eq. 13.4-8 results from the convective transfer illustrated in Fig. 13.4-5, the second term is calculated from the Higbie penetration theory for mass transfer, like Eq. 13.4-9. The coefficients may be combined by the rule of addition of resistances, since both steps are purely in series

$$\frac{1}{(k_{be})_b} = \frac{1}{(k_{bc})_b} + \frac{1}{(k_{ce})_b} \qquad (13.4\text{-}10)$$

Pyle [8] has compared some of the available data and correlations concerning $(k_{be})_b$—see Fig. 13.4-6. The spread is very large. Large discrepancies were also obtained by Chavarie and Grace in their comparison of $(k_{be})_b$ values calculated from 13 different correlations [25]. The coefficient $(k_{be})_b$ is related to $k_I$ by the equation;

$$k_I = (k_{be})_b f_b$$

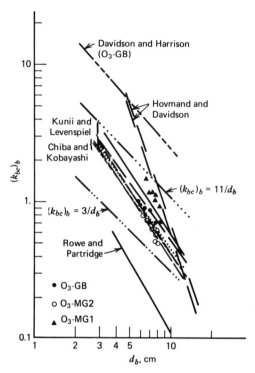

*Figure 13.4-6 Gas exchange from single bubbles. (Data from Chiba and Kobayashi [21], compiled by Pyle [8]).*

In line with the emphasis on fundamentals and bubble behavior Partridge and Rowe wrote the continuity equation (13.4-1) on a unit volume of bubble + interchange zone, rather than on the bubble phase alone. Their model equations are as follows:

Bubble + interchange phase or "cloud":

$$u_c \frac{dC_{Ac}}{dz} + (k_{ce})_c(C_{Ac} - C_{Ae}) + r_A\rho_e \frac{V_{iz}}{V_c} = 0 \qquad (13.4\text{-}11)$$

The terms in this equation have the dimensions $kmol/m_c^3$ hr. The reaction rate terms takes into account that the reaction in the ensemble of bubble and interchange zone really only takes place in the interchange zone, which has a volume $V_{iz}$ and in which one assumes that the catalyst bulk density is the same as that in the emulsion phase, $\rho_e$. According to Partridge and Rowe's experimental correlation $V_{iz}/V_c = 1.17/(0.17 + \alpha)$. Also, $\rho_e = \rho_B/(1 - f_b)$ when the fraction of the

CHEMICAL REACTOR DESIGN

total volume occupied by the bubble + interchange zone simply equals that taken by the bubble, and the amount of catalyst in the bubble phase is neglected.

Emulsion phase:

$$(u_e)_s \frac{dC_{Ae}}{dz} + (k_{ce})_c (C_{Ae} - C_{Ac}) \frac{f_b}{1 - f_b} + r_A \rho_e = 0 \qquad (13.4\text{-}12)$$

In this equation the transport by effective diffusion has been neglected. $(u_e)_s$ is a superficial velocity based on the emulsion phase cross section. The terms are expressed per unit volume of emulsion phase. Since bubbles grow as they move through the bed, Partridge and Rowe let $d_c$, which determines $(k_{ce})_c$, vary with bed height according to an experimental correlation.

Kunii and Levenspiel [3] have concentrated their attention on beds operated well above minimum fluidization, which is the usual industrial practice. In such beds there are definite gross mixing patterns for the solid: downward near the wall, upward in the central core. This has a marked effect on the gas flow in the emulsion phase, which is also forced downward near the wall. Kunii and Levenspiel showed that when $u/u_{mf}$ exceeds 6 to 11 there is practically no net gas flow through the emulsion phase. Since Kunii and Levenspiel distinguish between the bubble itself and the interchange zone Eq. 13.4-1 is written in this case:

$$u_b \frac{dC_{Ab}}{dz} + (k_{bc})_b (C_{Ab} - C_{Ac}) + k C_{Ab} \rho_b = 0 \qquad (13.4\text{-}13)$$

The terms in Eq. 13.4-13 have the dimensions [kmol/m$_b{}^3$ hr]. The very small amount of reaction in the bubble itself is accounted for. First-order reaction is assumed. The continuity equation for the component $A$ in the emulsion phase—as compared to Eq. 13.4-12—now becomes

$$(k_{ce})_b (C_{Ac} - C_{Ae}) = k C_{Ae} \rho_e \frac{(1 - f_b)}{f_b} \qquad (13.4\text{-}14)$$

provided there is no net flow and the transport by effective diffusion is neglected.

Both equations are linked by the equation for the interchange zone:

$$(k_{bc})_b (C_{Ab} - C_{Ac}) = k C_{Ac} \rho_c \frac{V_{iz}}{V_b} + (k_{ce})_b (C_{Ac} - C_{Ae}) \qquad (13.4\text{-}15)$$

In Eqs. 13.4-14 and 13.4-15 the terms are based on unit volume of bubble phase. In Eq. 13.4-10 the interchange zone is calculated according to Partridge and Rowe's correlation; Kunii and Levenspiel include the wake in this zone. There is no evidence, however, that the bubble gas enters the wake. Eliminating $C_{Ac}$ and $C_{Ae}$

from Eqs. 13.4-13, 13.3-14, and 13.4-15 leads to:

$$-u_b \frac{dC_{Ab}}{dz} = k \left[ \rho_b + \cfrac{1}{\cfrac{k}{(k_{bc})_b} + \cfrac{1}{\rho_c \cfrac{V_{iz}}{V_b} + \cfrac{1}{\cfrac{k}{(k_{ce})_b} + \cfrac{1}{\rho_e \cfrac{(1-f_b)}{f_b}}}}} \right] C_{Ab} \qquad (13.4\text{-}16)$$

The right-hand side of Eq. 13.4-16 could also be obtained directly by the proper combination of resistances, since we deal only with first-order processes here. Equation 13.4-16 can also be written in a more concise way:

$$-u_b \frac{dC_{Ab}}{dz} = K_r C_{Ab} \qquad (13.4\text{-}17)$$

Integration between $z = 0$ and $z = L_f$ leads to the expression for the concentration of $A$ in the exit stream:

$$\frac{\overline{C_A}}{C_{Ai}} = \exp\left( -\frac{K_r L_f}{u_b} \right) \qquad (13.4\text{-}18)$$

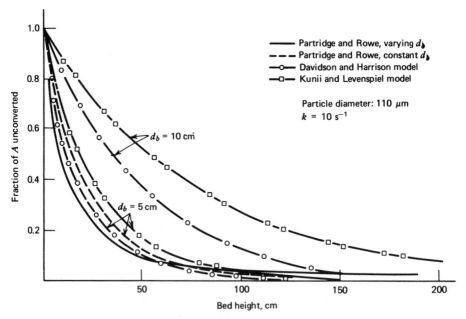

Figure 13.4-7 *Comparison of reactor models (from Pyle [8]).*

Figure 13.4-7 [8] compares conversions calculated according to Davidson and Harrison's model, to Partridge and Rowe's model, and to that of Kunii and Levenspiel. A comparison of the two Partridge and Rowe curves for $d_b = 5$ cm shows the influence of accounting for variations in bubble size with bed height. Far more reliable and general correlations will have to be developed for this variation. Note also the large difference between the prediction of the Kunii and Levenspiel model and that of Davidson and Harrison for the 10-cm bubbles.

Further models for fluidized beds are reviewed and discussed by Yates [26]. It is evident, however, that there is a far greater need for sound hydrodynamic analysis of large-scale fluidized beds than for further models, which are different only in details and which are insufficiently confronted with large-scale results. Pending further progress, model predictions for fluidized bed reactors have to be viewed with caution and more as a help toward a better understanding of the relative importance and effect of the parameters than as a rigorous and sufficient basis for design.

### Example 13.4-1  Modeling of an Acrylonitrile Reactor (after Kunii and Levenspiel [3])

Design a fluidized bed reactor for the production of acrylonitrile by ammoxidation of propylene, with air as the oxidizing agent. The required production of acrylonitrile is 40,400 tons/year (count on approximately 8000 hr or 340 days).

The process achieves a 78 percent conversion of propylene at 400°C and atmospheric pressure. The rate constant of the reaction considered as a first-order process in $k = 1.44$ m$^3$/kg cat. hr at 400°C. The volume fraction of propylene in the feed is 0.24. The catalyst used has a mean particle diameter $\overline{d}_p = 51$ $\mu$ and the following size distribution:

$$d_p(m)\ 5 \cdot 10^{-6}\ 15\ 25\ 35\quad 45\quad 55\quad 65\quad 75\ 85\ 95\ 105\ 115 \cdot 10^{-6}$$
$$p(d_p)(m^{-1})\ 6 \cdot 10^{-2}\ 22\ 46\ 76\ 118\ 170\ 200\ 152\ 99\ 63\quad 36\quad 12 \cdot 10^{-2}$$

The solid density is 2500 kg/m$^3$; the specific heat: 0.2 kcal/kg°C. The void fraction $\varepsilon$ of the packed bed is 0.5. At the minimum fluidization velocity $u_{mf} = 7.2$ m/hr the void fraction $\varepsilon_{mf}$ is 0.6. Gas properties: $\rho_g = 1$ kg/m$^3$; $c_p = 0.25$ kcal/kg°C; $\mu = 0.144$ kg/hr m; $D_e = 0.14$ m$^2$/hr.

*Solution*
Removing the heat of reaction necessitates an internal heat exchanger. This exchanger will also help to limit the bubble diameter. Take vertical tubes of 0.06 m OD on a 0.14-m triangular pitch. This limits the effective diameter of the bubbles to 0.1 m. Note that this is a very crude way of determining the (average) bubble diameter, which is the main variable in the Kunii and Levenspiel model.

Select the superficial velocity of the feed to be: 1800 m/hr. The feed rate of propylene is: $(40.4 \times 10^6)/(8000 \times 53 \times 0.78) = 120$ kmol/hr. and the total gas flow at standard conditions $= 120/0.24 \times 22.4 = 11,200$ m$^3$/hr. The bed diameter is found to be 3.45 m and the number of vertical tubes 552.

With $u_s = 1800$ m/hr, $u_{br}$ and $u_b$ can be calculated:

$$u_{br} = 0.711 (12.70 \times 10 \times 0.1)^{1/2} = 2530 \text{ m/hr}$$
$$u_b = 1800 - 7.2 + 2530 = 4322.8 \text{ m/hr}$$

The mass transfer coefficients are calculated from (13.4-7) and (13.4-8):

$$(k_{bc})_b = 4.5 \times \frac{7.2}{0.1} + 5.85 \frac{0.14^{1/2} \times (12.70 \times 10^7)^{1/4}}{(0.1)^{5/4}} = 4464 \text{ hr}^{-1}$$

$$(k_{ce})_b = 6.78 \left( \frac{0.6 \times 0.14 \times 4322.8}{0.1^3} \right)^{1/2} = 4100 \text{ hr}^{-1}$$

The bulk density of the emulsion phase is found from

$\rho_s = 2500$ kg/m$^3$
$\rho_e = $ bulk density of bed at $mf = \rho_s(1 - \varepsilon_{mf}) = 2500 \times 0.4 = 1000$ kg/m$^3$

and, since

$$f_b = \frac{u - u_{mf}}{u_b} = \frac{1800 - 7.2}{4322.8} = 0.414$$

it follows that

$$\rho_e \frac{1 - f_b}{f_b} = 1000 \times \frac{0.586}{0.414} = 1420 \text{ kg/m}^3$$

According to Rowe and Partridge,

$$\frac{V_{iz}}{V_b} = \frac{1.17}{\alpha - 1}$$

so that, with

$$\alpha = \frac{u_b}{(u_{mf}/\varepsilon_{mf})} = \frac{4322.8}{(7.2/0.6)} = 360$$

and $\rho_c = \rho_e$

$$\rho_c \frac{V_{iz}}{V_b} = 1000 \times \frac{1.17}{359} = 3.26 \text{ kg/m}^3$$

Since only 1.5 percent of the bed solids are in the bubbles, the volume fraction of solids in the bubble is obtained from

$$\gamma_b = \frac{(1 - \varepsilon_{mf})(1 - f_b)}{f_b} \times 0.015 = 0.01 \frac{m^3{}_s}{m^3{}_b}$$

and the catalyst bulk density in the bubble phase is found to be

$$\rho_b = \gamma_b \cdot \rho_s = 0.01 \times 2500 = 25 \text{ kg/m}^3{}_b$$

With

$$k = 1.44 \frac{m^3}{\text{kg cat. hr}}$$

Eq. 13.4-15 finally becomes

$$-4322.8 \frac{dC_{Ab}}{dz} = 1.44 \left( 25 + \cfrac{1}{\cfrac{1.44}{4460} + \cfrac{1}{3.26 + \cfrac{1}{\cfrac{1.44}{4100} + \cfrac{1}{1.420}}}} \right) C_{Ab}$$

After integration:

$$\frac{\bar{C}_A}{C_{Ai}} = 1 - x_A = 0.22 = \exp\left( -\frac{1080 \times L_f}{4322.8} \right) = \exp(-0.25 L_f)$$

from which $L_f = 6.04$ m. The total bed volume may be calculated to be 56.5 m³, the emulsion volume 33 m³, and the amount of catalyst, $W$, roughly 33,000 kg.

Assuming plug flow, one would obtain from $1 - x = e^{-k(W/F')}$, where $F'$ would be the total feed rate in m³/hr at 400°C, an amount of catalyst of 29,031 kg. Assuming complete mixing of the gas phase one would obtain from $F'\Delta x = k(1 - x)W$ an amount of 67,980 kg.

## Problems

13.1 A 0.1-m column is loaded with 5 kg of sand. The size distribution is as follows

| $d_p \times 10^2$ (cm) | Weight fraction $X$ |
|---|---|
| 7.8 and larger | 0.2 |
| 6.5 | 0.25 |
| 5.2 | 0.4 |
| 3.9 | 0.1 |
| 2.6 and smaller | 0.04 |

The density of the sand by displacement methods was measured to be 2.600 kg/m³. The sand is to be fluidized with air at 100°C. Atmospheric pressure is 772 mm Hg. Estimate the minimum fluidization velocity. Calculate the composite particle diameter by means of

$$d_p = \frac{1}{\sum \dfrac{x_i}{d_{pi}}}$$

13.2 Tigrel and Pyle [22] used the following expression for the catalyst deactivation function in a fluidized bed

$$\Phi = \left(2\lambda_R \theta + \frac{1}{\Phi_0{}^2}\right)^{-1/2}$$

where $\theta = t/t_R$, $\bar{t}_R = W_R/F_S$, $\lambda_R = k_R{}^0 \exp -(E'/RT)\bar{t}_R$, and $\Phi_0$ is the value of the deactivation function for the catalyst entering the fluidized bed. When the particles are perfectly mixed the average of $\bar{\Phi}$ in the reactor, $\bar{\Phi}$ is given by:

$$\bar{\Phi} = \exp\left(\frac{1}{2\lambda_R \Phi_0{}^2}\right) \mathrm{erfc}\left(\frac{1}{\Phi_0\sqrt{2\lambda_R}}\right) \sqrt{\frac{\pi}{2\lambda_R}}$$

and the rate of catalytic cracking of gasoil, for example, is represented by

$$r = kc^2\bar{\Phi}$$

where $c$ is the gasoil concentration.

Tigrel and Pyle adopted a two-phase model for the fluidized bed, assuming perfect mixing of gas in the emulsion phase and plug flow in the bubble phase and a transfer of gas between bubble and emulsion phase.

Considering a mean bubble size, derive the following expression for the fraction of feed component $A$ unconverted at the top of the bed

$$y_T = \frac{\exp(-Z)}{1 - \gamma} + \frac{(1 - \beta e^{-Z} - \gamma)^2}{2(1 - \gamma)k^*\bar{\Phi}}\left(-1 + \sqrt{1 + \frac{4k^*\bar{\Phi}}{1 - \beta e^{-Z} - \gamma}}\right)$$

where

$$\beta = 1 - \frac{u_{mf}}{u} \qquad \gamma = \frac{u_s}{u} \qquad k^* = \frac{kH}{u} \qquad Z = \frac{KH}{u_a V}$$

and

$u$ = superficial gas velocity (cm/s)
$u_a$ = absolute bubble-rising velocity (cm/s)
$u_s$ = downward velocity of catalyst particle (cm/s)
$K$ = transversal mass transfer coefficient (cm³/s)
$H$ = total bed height (cm)
$V$ = bubble volume (cm³)

Plot the conversion as a function of the fluid velocity for solids velocities of 1.0, 3.0, and 10 cm/s for a mean bubble diameter of 30 cm, a value of $\phi_0 = 0.9$, and a reactor height of 9 m.

# References

[1] Davidson, J. F. and Harrison, D. *Fluidized Particles*, Cambridge Universities Press (1963).

[2] De Groot, J. H. *Proceedings Int. Symp. on Fluidization*, Netherlands University Press, Eindhoven (1967), p. 348.

[3] Kunii, D. and Levenspiel, O. *Fluidization Engineering*, Wiley, New York (1969).

[4] Leva, M. *Fluidization*, McGraw-Hill, New York (1960).

[5] May, W. G. *Chem. Engng. Progr.*, **55**, 49 (1959).

[6] Mireur, J. P. and Bischoff, K. B. *Am. Inst. Chem. Engrs. J.*, **13**, 839 (1967).

[7] Partridge, B. A. and Rowe, P. N. *Trans. Instn. Chem. Engrs.*, **44**, T335 (1966); **44**, T349 (1966).

[8] Pyle, D. L. *Proceedings 1st Int. Symp. Chem. React. Engng.*, Washington (1970). *Advances in Chem. Series*, **109**, 106, A.C.S. Washington (1972).

[9] Van Deemter, J. J. *Chem. Eng. Sci.*, **13**, 143 (1961).

[10] Zenz, F. A. and Othmer, D. F. *Fluidization and Fluid-particle Systems*, Reinhold Publishing Co., New York (1960).

[11] Petersen, E. E. *A.I.Ch.E. J.*, **6**, 488 (1960).

[12] Zenz, F. A. and Weil, N. A. *A.I.Ch.E. J.*, **4**, 472 (1958).

[13] Wender, L. and Cooper, G. T. *A.I.Ch.E. J.*, **4**, 15 (1958).

[14] Wen, C. Y. and Hashinger, R. F. *A.I.Ch.E. J.*, **6**, 220 (1960).

[15] Perry, R. H. and Chilton C. H. *Chemical Engineers Handbook*, 5th ed., McGraw-Hill, New York (1973).

[16] Murray, J. D. *J. Fluid Mech.*, **21**, 465 (1965); **22**, 57 (1965).

[17] Shirai, L. *Fluidized Beds* (in Japanese *Kagaku Gyntsusha*), Kanazawa (1958), (quoted in [3]).

[18] Vreedenberg, H. A. *Chem. Eng. Sci.*, **11**, 274 (1960).

[19] Fritz, W. *Chem. Ing. Techn.*, **41**, 435 (1969).

[20] Balakrishnan, A. R. and Pei, D. C. T. *Can J. Chem. Eng.*, **53**, 231 (1975).

[21] Chiba, T. and Kobayashi, T. *Proc. Int. Conf. Fluid.*, Toulouse (1973), p. 468.

[22] Tigrel, A. Z. and Pyle, D. L. *Chem. Eng. Sci.*, **26**, 133 (1971).

[23] van Swaay, W. P. M. and Zuiderweg, F. J. *Proc. 5th Eur./2nd Intl. Symp. Chem. React. Eng.*, Amsterdam (1972), Elsevier, Amsterdam (1972), p. B-9-25.

[24] de Vries, R. J., van Swaay, W. P. M., Mantovani, C. and Heijkoop, A. *Proc. 5th Eur./2nd Int. Symp. Chem. React. Eng.*, Amsterdam (1971), Elsevier Amsterdam (1972), p. B-9-59.

[25] Chavarie, C. and Grace, J. R. *Chem. Eng. Sci.*, **31**, 741 (1976).

[26] Yates, J. G. *The Chemical Engineer*, p. 671, November (1975).

[27] Osberg, G. L. and Charlesworth, D. H. *Chem. Eng. Prog.*, **47**, 566 (1951).

[28] Davidson, J. F. and Harrison, D., eds *Fluidization*, Academic Press, New York (1971).

[29] Angelino, H., Couderc, J. P., Gibert, H., and Laguerie, C., eds. *Fluidization and Its Applications*, Proc. Intl. Symp., Sté Chimie Industrielle (1974).

[30] Murphy, J. F. and Soudek, M. in *Proc. Third Intl. Symp. Large Chemical Plants*, p. 81, Uitg. Sprugt, Van Mantgen en De Dees b.v., Leiden (1976).

# 14

## MULTIPHASE
## FLOW
## REACTORS

## 14.1 Types of Multiphase Flow Reactors _____

Reactions between components of a gas and a liquid, the kinetics of which were discussed in Chapter 6, are carried out in a variety of equipment, often having confusing names. The variety stems from a number of conditions that have to be fulfilled simultaneously: efficient contact between gas and liquid—and eventually a solid catalyst, limitation of pressure drop, ease of removal of heat, low cost of construction and operation. Depending on whether the main mass transfer resistance is located in the gas or in the liquid, multiphase reactors or absorbers are operated either with a distributed gas phase and continuous liquid phase or vice versa. Whether co- or countercurrent flow of gas and liquid is used depends on the availability of driving forces for mass and heat transfer and reaction.

Table 14.1-1 classifies various types of equipment for gas-liquid reactions. It is based on geometric aspects, the presence or absence of a solid catalyst, and the flow directions. Before discussing model equations and specific design aspects the various types of reactors of Table 14.1-1 will be briefly characterized in a qualitative way.

### 14.1.a Packed Columns

Packed columns are frequently encountered in industry. Their construction is simple and they can be easily adapted by replacing the packing. They permit rather large variations in flow rates and the pressure drop is relatively low. The packing is often staged to avoid maldistribution of the fluid. Sometimes staging is required to provide intermediate heat exchange, either in external heat exchangers or by means of direct cooling by liquid injection. Packed columns for gas purification, often called absorbers, always operate with countercurrent flow. Typical examples are the absorption of carbon dioxide and hydrogen sulfide by ethanolamines, potassium carbonate or sodium hydroxide in steam reforming or thermal cracker effluents and the absorption of ammonia by sulfuric acid.

*Table 14.1-1  Classification of multiphase flow reactors*

| Locus of reaction | Column | | | Stirred vessel | Miscellaneous |
| | Packed | Plate | Empty | | |
|---|---|---|---|---|---|
| Fluid phase only | Countercurrent flow "Absorber" | Countercurrent flow "Absorber" | Countercurrent flow "Spray tower" Co-current or countercurrent "Bubble column" | "Absorber" or "Reactor" | Venturi Static mixers Falling film, etc. |
| Solid catalyst | Countercurrent Cocurrent downward "Trickle bed reactor" "Packed bubble reactor" Cocurrent upward | | Co- or countercurrent "Bubble reactor" | "Slurry reactor" | |

When the packing is also a catalyst, both countercurrent and cocurrent flow are applied. In the latter case both upflow and downflow operation are encountered. With upflow operation the contacting between gas and liquid is superior, but the pressure drop is higher and there are restrictions on flow rates and packing diameter because of flooding. The downflow cocurrent column, packed with catalyst, may operate in two distinct flow regimes: the "trickle flow" regime when the gas phase is continuous and the liquid phase dispersed or the "bubble flow" regime when the gas phase is dispersed and the liquid phase continuous. For a given gas flow rate both regimes may lead to pulsed flow when the liquid flow rate is increased.

Trickle bed reactors have grown rapidly in importance in recent years because of their application in hydrodesulfurization of naphtha, kerosene, gasoil, and heavier petroleum fractions; hydrocracking of heavy gasoil and atmospheric residues; hydrotreating of lube oils; and hydrogenation processes. In trickle bed operation the flow rates are much lower than those in absorbers. To avoid too low effectiveness factors in the reaction, the catalyst size is much smaller than that of the packing used in absorbers, which also means that the overall void fraction is much smaller.

The fixed bed is preferred to a slurry-type operation when the gas flow rate is relatively low because it leads to a gas and liquid flow pattern that approximates plug flow better. Only for high gas-flow rates would an operation with suspended catalyst be preferred—when the catalyst size permits it—to avoid the pulsed flow regime that might be encountered in fixed bed operation.

## 14.1.b Plate Columns

Plate columns are only used in processes that do not require a solid catalyst and for which relatively long contact times are needed. Since the liquid flow is evenly distributed over the complete height of the column, large diameters can be used. The interfacial area per unit volume of gas liquid mixture is larger than in packed columns, but on the other hand plate columns only have gas–liquid mixtures on the plates themselves. Whether there is more interfacial area per unit total volume of column in the plate column depends on the plate spacing, which is determined by the presence or absence of downcomers, foaming, entrainment, and so on.

A very important industrial example of a plate-column reactor is the so-called absorber in nitric acid production, in which NO, dissolved in dilute acid, is transformed into nitric acid by means of air-oxygen.

## 14.1.c Empty Columns

Empty columns are characterized by the absence of materials or devices for the continuous dispersion of the phases, wnich does not mean that internal heat exchangers are excluded. In fact, the insertion of heat exchangers in such reactors easily permits a continuous and efficient temperature control.

In spray towers the liquid is the dispersed phase and the interfacial area is large. This type of column is used for fast reactions requiring only very short contact times. Although a large volume is needed the investment cost is low. The pressure drop is also low.

Bubble columns, in which the liquid is the continuous phase, are used for slow reactions. Drawbacks with respect to packed columns are the higher pressure drop and the important degree of axial and radial mixing of both the gas and the liquid, which may be detrimental for the selectivity in complex reactions. On the other hand they may be used when the fluids carry solid impurities that would plug packed columns. In fact, many bubble column processes involve a finely divided solid catalyst that is kept in suspension, like the Rheinpreussen Fischer-Tropsch synthesis, described by Kölbel [1], or the former I. G. Farben coal hydrogen process, or vegetable oil hardening processes. Several oxidations are carried out in bubble columns: the production of acetaldehyde from ethylene, of acetic acid from $C_4$ fractions, of vinylchloride from ethylene by oxychlorination, and of cyclohexanone from cyclohexanol.

### 14.1.d Stirred Vessel Reactors

Stirred vessel reactors are preferentially used for reactions involving rather large ratios of liquid to gas, for rather exothermic reactions, because the agitation improves the heat transfer and internal heat exchangers are easily built in. They also permit achieving high interfacial areas. The agitation is favorable also when a finely divided catalyst (e.g., Raney-nickel) has to be kept in suspension. The reactor is then of the "slurry" type. There are examples, however, of nonstirred operation (e.g., when the reaction has to be carried out under very high pressure and shaft leakage may be a problem).

There are many examples of hydrogenations, oxidations, and chlorinations that are carried out in stirred tank reactors, either batch, semibatch, or continuous.

### 14.1.e Miscellaneous Reactors

Table 14.1-1 also mentions some less common types, used for very specific tasks. Venturi reactors are used, for instance, in antipollution devices to wash out small amounts of remaining $SO_2$ by means of caustic. Their advantage resides in their low pressure drop, since the Venturi exhausts the gas into the liquid.

Falling film or wetted wall reactors can be used for very exothermic reactions. Furthermore, the limited and well defined interfacial area permits excellent control of very rapid reactions.

## 14.2 Design Models for Multiphase Flow Reactors _____

In this section some general models, which could be used for the design of any of the reactors or absorbers of the classification of Table 14.1-1, are derived and their solution briefly described. The models are mostly based on plug flow or complete mixing of one or both phases, but effective diffusion and two zone models are also presented. More specific models are discussed in later sections.

Referring to Chapter 6, in which various rate equations were derived for gas–liquid reactions, a distinction is made between reactions taking place in the film only—whereby the bulk concentration of $A$ is zero—and reactions that extend into the bulk of the liquid, whereby this concentration differs from zero.

When the liquid flow rate, $L$, the gas flow rate, $F$, the inlet and outlet partial pressure of the reacting component $A$ of the gas phase, $(p_A)_{in}$ and $(p_A)_{out}$, respectively, and the inlet concentration of the reacting component $B$ of the liquid phase are given, the problem is to find the outlet concentration of $B$, written $(C_B)_{out}$ and the volume, $V$, of the reactor. This is the kind of problem commonly encountered in absorbers. In reactors, on the other hand, conditions are often imposed on the liquid component—in hydrodesulphurization, for example, but also in hydrogenations, oxidations or chlorinations—and $L$, $F$, $(C_B)_{in}$, $(C_B)_{out}$, $(p_A)_{in}$ would be given or imposed through certain constraints like flooding rates, to be discussed in later sections. The unknowns would then be $(p_A)_{out}$ and the volume $V$.

The situation mainly thought of in the next three sections (14.2.a, 14.2.b, and 14.2.c) is that of a gas–liquid reaction, without solid catalyst. Sections 14.2.d and 14.2.e deal with catalytic reactors.

### 14.2.a Gas and Liquid Phase Completely Mixed

This is the simplest situation from the computational point of view, since the concentrations of $A$ and $B$ are uniform and no differential equations are involved. The continuity equation for $A$ may be written, for the total liquid volume:

$$\frac{F}{p_t}[(p_A)_{in} - (p_A)_{out}] = N_A|_{y=0} A_v V(1 - \varepsilon_G) \qquad (14.2.a-1)$$

where $\varepsilon_G$ is the gas holdup and $V$ the total reactor volume (i.e., including liquid and gas). To avoid too cumbersome a notation the index $b$ used in Chapter 6 to refer to bulk conditions is dropped here: $C_B$, $p_A$ ... are bulk concentrations and partial pressures. $N_A$ depends, of course, on the order of reaction. In Chapter 6 various cases were discussed and the corresponding $N_A$ derived—analytically only for (pseudo) first-order and instantaneous second-order reactions, but approximately also for other orders. $N_A$ could even be a tabulated function of the gas and liquid composition. In what follows $N_A$ will be kept general.

From Chapter 6 it is clear that $N_A$ depends on $C_{Ai}$, $C_A$, and $C_B$. The interfacial concentration is no problem, since it may be calculated from $p_A$ when the gas phase transfer coefficient is known or $N_A$ can be referred directly to $p_A$ as in Eq. 6.3.b-5. Two more relations are required, however, to calculate $V$, $(C_A)_{out}$ and $(C_B)_{out}$ or $V$, $(p_A)_{out}$ and $(C_A)_{out}$:

1. An overall material balance:

$$\frac{F}{p_t}[(p_A)_{in} - (p_A)_{out}] = \frac{a}{b}L[(C_B)_{in} - (C_B)_{out}] + L(C_A)_{out} \quad (14.2.a\text{-}2)$$

whereby, with complete mixing $(C_B)_{out} = C_B$ and $(C_A)_{out} = C_A$.

2. A balance on $A$ in the bulk liquid:

$$N_A|_{y=y_L} A_v V(1 - \varepsilon_G) = r_A(1 - A_v y_L)V(1 - \varepsilon_G) + L(C_A)_{out} \quad (14.2.a\text{-}3)$$

where $r_A = r_A(C_A, C_B, T)$.

In the special case that the reaction is completed in the film $C_A = 0$ and one of the above equations drops out, leaving the two Eqs. 14.2.a-1 and 14.2.a-2.

When the reaction is very slow and takes place entirely in the bulk, the mass transfer and reaction are purely in series. As mentioned in Chapter 6, $N_A$ in Eq. 14.2.a-1 and Eq. 14.2.a-3 then equals $k_L(C_{Ai} - C_A)$, but since $C_A$ again differs from zero the complete system Eqs. 14.2.a-1, 14.2.a-2, and 14.2.a-3 has to be solved. An example of application of these design equations is given in a later section on stirred tank gas–liquid reactors.

### 14.2.b  Gas and Liquid Phase in Plug Flow

This is a situation more likely to be approximated in a packed tower. The continuity equation for $A$ in a differential tower volume may then be written:

$$-\frac{F}{p_t} dp_A = N_A|_{y=0}\, a_v'\, dV \quad (14.2.b\text{-}1)$$

where $a_v'$ is the gas–liquid interfacial area per cubic meter reactor volume. Notice that $a_v'$ is usually not the geometrical surface area of the packing and has to be determined from experiments. This is why Eq. 14.2.b-1 has not been written in terms of $A_v$, used in Eq. 14.2.a-1 and which is simply the geometrical gas–liquid interface. Correlations for $a_v'$ are given in a later section.

The relation between $p_A$ and $C_B$ is derived from a balance on $A$ over the top or bottom section of the column, depending on the nature of the problem: absorber or reactor design. In the first, for example, when $(p_A)_{out}$ and $(C_B)_{in}$ are given and for countercurrent operation, a balance on the upper part of the column may be written, provided that the entering liquid does not contain any $A$:

$$\frac{F}{p_t}[p_A - (p_A)_{\text{out}}] = \frac{a}{b}L[(C_B)_{\text{in}} - C_B] + LC_A \qquad (14.2.b\text{-}2)$$

This relation introduces $C_A$, however, so that a third relation is required: a balance on $A$ in a differential volume $dV$ of the bulk:

$$N_A|_{y=y_L}\, a_v' dV = r_A(1 - A_v y_L)(1 - \varepsilon_G)dV + L\, dC_A \qquad (14.2.b\text{-}3)$$

This is Eq. (6.3.a-5).

The solution proceeds as follows in the case of an absorber problem. Choose increments in $\Delta p_A$ for the numerical integration of Eqs. 14.2.b-1 to 14.2.b-3 and calculate the corresponding $\Delta C_A$, $\Delta C_B$ and $\Delta V$ or $\Delta z$. Start from the top, where $(p_A)_{\text{out}}$ and $(C_B)_{\text{in}}$ are known and $(C_A)_{\text{in}} = 0$. Continue until $p_A = (p_A)_{\text{in}}$, when $C_A = (C_A)_{\text{out}}$; $C_B = (C_B)_{\text{out}}$ and $z = Z$, the total height. Compared with the mixed–mixed case the difference is that the system of balance equations has to be solved, not once but in each increment chosen for the integration of the continuity equation for $A$ in the gas phase.

For the reactor problem $(p_A)_{\text{in}}$ at $z = 0$ and $(C_A)_{\text{in}}$ at the top, where $z = Z$ are known, while $(p_A)_{\text{out}}$ at $z = Z$ and $(C_A)_{\text{out}}$ at $z = 0$ are unknown, together with the total height $Z$ or the total volume $V$, of course. An interation is inevitable. Preference has to be given to start the integration from $z = 0$, with a guessed value of $(C_A)_{\text{out}}$. A reasonable guess for $(C_A)_{\text{out}}$ may be based upon an overall material balance on the complete column, knowing that $(C_A)_{\text{out}}$ has to be smaller than $(p_A)_{\text{in}}/H$ and considering that $(p_A)_{\text{out}}$ has to be positive. Then $C_B$ is eliminated from the differential Eqs. 12.2.b-1 and 14.2.b-3 by means of a balance on the lower part of the reactor:

$$\frac{F}{p_t}[p_A - (p_A)_{\text{in}}] = \frac{a}{b}L[(C_B)_{\text{out}} - C_B] + L[C_A - (C_A)_{\text{out}}] \qquad (14.2.b\text{-}4)$$

The differential equations are now integrated until a height $Z_1$ has been reached where $C_A = (C_A)_{\text{in}}$. If $C_B$, calculated from Eq. 14.2.b-4, corresponds to the given $(C_B)_{\text{in}}$ then the guessed value of $(C_A)_{\text{out}}$ is correct and the height at which the integration was stopped is the correct and final value, $Z = Z_1$. If $C_B(Z_1) \neq (C_B)_{\text{in}}$ another value of $(C_A)_{\text{out}}$ has to be tried.

Another way would be to guess $(C_A)_{\text{out}}$ and $Z$ or $V$. The integration would then be stopped when the assumed value of $Z$ is reached. Both $C_A$ and $C_B$ would then have to be compared with the inlet values. If they would not correspond, better values would have to be guessed for $(C_A)_{\text{out}}$ and $Z$. Notice that the latter method leads to a two-dimensional iteration problem. The iteration can be systematized by means of the quasi-linearization approach (see Lee [72]) whereby the two-dimensional problem is reduced to a sequence of two one-dimensional iterations.

Another method of solution of the above nonlinear, two-point boundary-value problem would be based on the invariant imbedding concept.

When the reaction is completed in the film ($C_{Ab} = 0$) Eqs. 14.2.b-1 and 14.2.b-4 are correspondingly simplified and suffice to determine the unknown $(p_A)_{out}$ and $V$ or $Z$. In this case the reactor problem does not require any iteration.

### 14.2.c  Gas Phase in Plug Flow. Liquid Phase Completely Mixed

The continuity equation for $A$ in the gas phase is nothing but Eq. 14.2.b-1 of course:

$$-\frac{F}{p_t} dp_A = N_A|_{y=0} \, a'_v \, dV \qquad (14.2.c\text{-}1)$$

The second relation is the overall balance:

$$\frac{F}{p_t} [(p_A)_{in} - (p_A)_{out}] = \frac{a}{b} L[(C_B)_{in} - (C_B)_{out}] + L(C_A)_{out} \qquad (14.2.c\text{-}2)$$

The balance of $A$ in the bulk now takes the form, when $(C_A)_{in} = 0$:

$$\int_0^V N_A|_{y=y_L} a'_v dV = r_A(1 - A_v y_L)V(1 - \varepsilon_G) + L(C_A)_{out} \qquad (14.2.c\text{-}3)$$

Note that there is no way of applying Eq. 14.2.c-3 to an increment since the liquid is completely mixed. In this case, $(C_B)_{out} = C_B$ and $(C_A)_{out} = C_A$ have to be guessed before the integration of Eq. 14.2.c-1 can be performed. The integration starts from the bottom, for example, although this is not essential. When $p_A = (p_A)_{out}$ the integration is stopped and only at that point can the assumed values of $(C_B)_{out}$ and $(C_A)_{out}$ be checked by means of Eqs. 14.2.c-2 and 14.2.c-3. If the calculated values do not agree with the assumed, the whole cycle has to be repeated. In guessing values for $(C_A)_{out}$ and $(C_B)_{out}$ the overall balance may be of help, together with the constraint $(C_A)_{out} < (p_A)_{out}/H$.

In the reactor problem the same approach can be followed. In that case $(p_A)_{out}$ and $(C_A)_{out} = C_A$ have to be guessed before starting the integration of Eq. 14.2.c-1. Since $(p_A)_{out}$ is a limit of the integral of Eq. 14.2.c-1 and is guessed, it is preferable to integrate from the top to the bottom. The integration is stopped at $p_A = (p_A)_{in}$, a given value, and yields $V$. The corresponding values of $(p_A)_{out}$ and $(C_A)_{out}$ follow from Eqs. 14.2.c-2 and 14.2.c-3 and these values are compared with the assumed ones.

### 14.2.d  An Effective Diffusion Model

Effective diffusion models have also been used to account for intermediate degrees of mixing in the axial direction—see Pavlica and Olson [74] for a useful comprehensive survey. An example of such a model is developed here for the case of a reaction catalyzed by a solid and no reaction in the liquid. Steady-state continuity equation for $A$ in the gas phase:

$$(\varepsilon - \varepsilon_L)D_{eG}\frac{d^2C_{AG}}{dz^2} - (\varepsilon - \varepsilon_L)u_{iG}\frac{dC_{AG}}{dz} - K_L a'_v\left(\frac{p_A}{H} - C_{AL}\right) = 0 \quad (14.2.\text{d-}1)$$

with $1/K_L = 1/k_L + 1/Hk_G$, an overall mass transfer coefficient in terms of the liquid concentration gradient, and where $\varepsilon_L$ is the liquid holdup. The first term arises from the effective diffusion, the second from the plug flow, and the third from the transfer of $A$ from the gas to the liquid phase.

Steady-state continuity equation for $A$ in the liquid phase:

$$\varepsilon_L D_{eL}\frac{d^2C_{AL}}{dz^2} - \varepsilon_L u_{iL}\frac{dC_{AL}}{dz} + K_L a'_v\left(\frac{p_A}{H} - C_{AL}\right) - k_l a''_v(C_{AL} - C_{As}{}^s) = 0$$

$$(14.2.\text{d-}2)$$

The fourth term represents the transfer of $A$ from the liquid to the catalyst surface, $a''_v$ the liquid–solid interfacial area per unit reactor volume ($m_i^2/m_r^3$).

Transfer from liquid to catalyst surface and reaction:

$$k_l a''_v(C_{AL} - C_{As}{}^s) = r_A \rho_B \quad \text{with} \quad r_A(\text{kmol/kg cat. hr}) \quad (14.2.\text{d-}3)$$

When internal concentration gradients have to be accounted for, the right-hand side of Eq. 14.2.d-3 would have to be multiplied by $\eta$, the effectiveness factor, computed as described in Chapters 3 and 11. Accounting for temperature gradients in the axial direction would require an additional differential heat balance, analogous in structure to Eq. 14.2.d-2.

The axial effective diffusivity of $A$ in the gas and liquid phases, $D_{eG}$ and $D_{eL}$ have been determined for various modes of operation and will be reported in the sections related to specific design aspects. Although the Peclet numbers are lower than for a single-phase fixed bed reactor—which means that the effective diffusivities are higher—it would seem again that this effect is negligible compared with that of plug flow. Deviations from plug flow are mainly caused by insufficient contacting between the gas and liquid resulting from preferential paths in the packing or by stagnant zones. The effect of these phenomena cannot adequately be described by effective diffusion, and a more appropriate description could be given by a two-zone model.

### 14.2.e  A Two-Zone Model

The underlying idea for such a model is that only a fraction of the liquid flows in a more or less ordered way through the packing, while at each height there is a stagnant zone in which the liquid is well mixed and that exchanges mass with the flowing fraction.

The continuity equation for $A$ in the gas phase is either the same as in Eq. 14.2.d-1 or is simplified by neglecting the effective diffusion term. The continuity equations for $A$ in the liquid phases may be written, neglecting effective diffusion: For the flowing fraction:

$$-\varepsilon_L{}^f u'_{iL} \frac{dC_{AL}{}^f}{dz} - k_T(C_{AL}{}^f - C_{AL}{}^d) + K_L a'_v\left(\frac{p_A}{H} - C_{AL}{}^f\right) - k_l a''_v(C_{AL}{}^f - C_{As}{}^s) = 0$$

(14.2.e-1)

where $u'_{iL}$ is the interstitial velocity of the flowing fraction of the liquid, $k_l$ the mass transfer coefficient between the *flowing* fraction of the liquid and the solid, and $k_T$ the mass transfer coefficient between the zones containing flowing and stagnant liquid.

For the well-mixed liquid in the corresponding slice of stagnant liquid:

$$k_T(C_{AL}{}^d - C_{AL}{}^f) = k'_l(C_{AL}{}^d - C_{As}{}^s)$$

(14.2.e-2)

Reaction on the catalyst:

$$k_l a''_v(C_{AL}{}^f - C_{As}{}^s) + k'_l(C_{AL}{}^d - C_{As}{}^s) = \eta r_A \rho_B$$

(14.2.e-3)

The transfer coefficient between the two zones, $k_T$, and that between the stagnant fluid and the solid, $k'_l$, contain interfacial areas that are not presently well established.

### 14.2.f An Alternate Approach

In the preceding sections the design of gas–liquid reactors has been based on the fundamental equations following the line of thought set in Chapter 7 and developed throughout this book. An alternate approach, far more empirical, has been advocated by Danckwerts and Gilham [2] and has been discussed more recently by Charpentier and Laurent [3] and Merchuk [4]. In this approach the design is based on a direct comparison of the industrial equipment with a laboratory "model," which does not necessarily bear any resemblance with the industrial equipment but has similar contact times. A packed column, for example, would be "experimentally simulated" by a stirred vessel, calibrated against the large-scale packed column. The characteristics of plate columns are such that the most appropriate laboratory model for simulation would be a rotating drum or a laminar jet.

The approach may be fast in generating partial and approximate answers for existing equipment when the absorbent composition has to be modified, for example, yet will probably not withstand progress in the knowledge of the hydro-dynamics of two-phase flow, which will increase the accuracy of and the confidence in the design along the fundamental lines discussed in the preceding chapters and sections.

## 14.3 Specific Design Aspects

### 14.3.a Packed Absorbers

The design of a packed column absorber starts with the choice of a particle diameter. To avoid bypassing liquid along the wall the ratio of column to particle diameter should exceed a value of 15 to 20. As a rule of thumb it can be said that for gas flow rates of 15 m³/min the particle diameter should exceed 2.5 cm. When the gas flow rate exceeds 50 m³/min the particle should have an equivalent diameter of 5 cm at least. Other technological aspects should not be overlooked. If the height of the column is too large with respect to the diameter, a redistribution of the liquid at one or more intermediate positions should be provided. Furthermore, enough space has to be maintained above the packing to enable the separation of entrained liquid droplets.

Internals, liquid, and gas distribution are practical aspects of column design discussed by Zenz [5]. Since in absorbers the liquid and the gas generally flow in countercurrent directions, there is a close interaction between the column diameter and the liquid and gas flow rates. If, for a given column diameter and liquid flow rate, the gas flow rate is too high, the liquid will be blown to the top of the column, which is said to be flooded. Zenz [5] derived the following relation for the maximum allowable gas and liquid flow rates above which flooding occurs:

$$\left( \frac{0.76F'/\Omega}{\sqrt{\rho_G/\rho_L}} \sqrt{\frac{a_v}{\varepsilon^3}} \mu_L^{0.2} \right)^{1/3} + \left( \frac{L}{580} \sqrt{\frac{a_v}{\varepsilon^3}} \mu_L^{0.2} \right)^{1/2} = 18.91 \qquad (14.3.a\text{-}1)$$

In this equation $F'$ and $L$ are the gas and liquid flow rates in m³/hr, $\varepsilon$ is the void fraction of the packing, and $\mu_L$ is the viscosity of the liquid in kg/m hr. Leva [19] and Lobo, et al. [7] have published the generalized dimensionless correlation for flooding rates shown in Fig. 14.3.a-1, in which $\mu_W$ represents the viscosity of water.

A generalized pressure drop correlation for packed columns with countercurrent flow has been derived by Sherwood et al. [6] and adapted by Lobo et al. [7]. It is shown in Fig. 14.3.a-2. The flooding line in this diagram corresponds with that of Fig. 14.3.a-1. $\mathcal{P}$ is a so-called packing factor, the value of which is given in tabular material provided by the producers of packing material to replace $a_v/\varepsilon^3$, as used by Sherwood. The interfacial area, $a'_v$, differs from the packing surface area, $a_v$, because the packing is not always completely wetted, so that a fraction of the surface is not active in the mass transfer, or because of the presence of stagnant pockets that are less effective than flowing streams. Shulman et al. [8] have established correlations between $a'_v$ and $a_v$ for Raschig rings and Berl saddles for various values of the liquid and gas flow rates. They also determined and correlated the hold up (total and operating) in terms of the operating variables [9, 10].

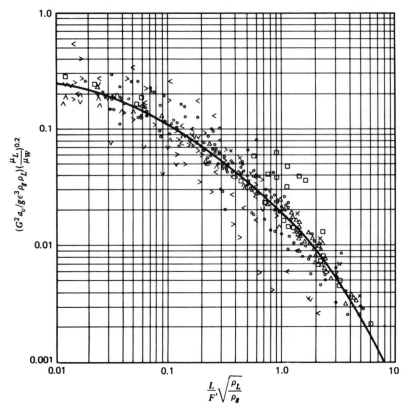

*Figure 14.3.a-1 Generalized correlation for flooding rates in packed columns (after Lobo, et al. [7]).*

In a review of the subject, Laurent and Charpentier [11] recommend the following equation by Onda et al. [12]:

$$\frac{a_v'}{a_v} = 1 - \exp\left[ -1.45 \left( \frac{\sigma_{L,c}}{\sigma_L} \right)^{0.75} \frac{\rho_L^{0.3} L^{0.4} g^{0.05}}{\Omega^{0.4} a_v^{0.35} \mu_L^{0.1} \sigma_L^{0.2}} \right] \qquad (14.3.a\text{-}2)$$

where $\sigma_{L,c}$ represents the critical surface tension above which the packing cannot be wetted. This correlation underestimates the interfacial area obtained with Pall rings by about 50 percent.

Many equations have been proposed for the mass transfer coefficients. For the liquid side coefficient, Laurent and Charpentier [11] recommend Mohunta's equation [13].

$$k_L a_v' = 0.0025 \left( \frac{L^3 a_v^3 \mu_L}{\Omega^3 g^2 \rho_L} \right)^{1/4} \left( \frac{\mu_L}{\rho_L D_{AL}} \right)^{-(1/2)} \left( \frac{a_v \mu_L}{g \rho_L} \right)^{-(2/3)} \left( \frac{\mu_L}{g^2 \rho_L} \right)^{-(1/9)} \qquad (14.3.a\text{-}3)$$

702 _____ CHEMICAL REACTOR DESIGN

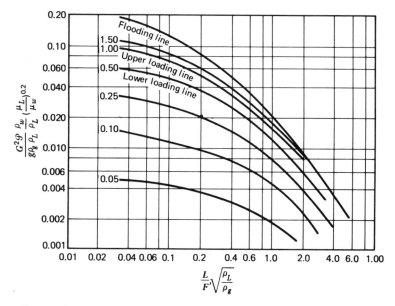

*Figure 14.3.a-2 Generalized pressure drop correlation for packed columns with countercurrent flow. Parameter curves: pressure drop, in inches water per foot (from Eckert [76]).*

The range of validity of Onda's equation (14.3.a-2) and of Mohunta's equation is as follows: liquid mass flow velocities, $\rho_L L/\Omega$: 360–151,200 kg/m² hr; $\mu_L$: 2.62–5.33 kg/m·hr; $Sc_L = \mu_L/\rho_L D_{AL}$: 142–1030; $d_p$: 0.006–0.05 m; column diameter: 0.06–0.5 m; ratio of column to packing diameter: 5–40. Shulman et al. [9] correlated the data of several authors with the following equation:

$$\frac{k_L d_p}{D_{AL}} = 25.1 \left(\frac{d_p \rho_L L}{\mu_L \Omega}\right)^{0.45} \left(\frac{\mu_L}{\rho_L D_{AL}}\right)^{0.5} \tag{14.3.a-4}$$

Ramm [14] mentions the following equation:

$$\frac{k_L d_p}{D_{AL}} = 0.02 \left(\frac{d_p \rho_L L}{\mu_L \Omega}\right)^{0.67} \left(\frac{\mu_L}{\rho_L D_{AL}}\right)^{0.33} \left(\frac{g \rho_L^2 d_p^3}{\mu_L^2}\right)^{0.33} \left(\frac{Z}{d_p}\right)^{-0.33} \tag{14.3.a-5}$$

For the gas-side mass transfer coefficient, Laurent and Charpentier [11] derived the correlation

$$\frac{k_G p_t \Omega}{F} = c(a_v d_p)^{-1.7} \left(\frac{d_p G}{\mu_G}\right)^{-0.3} \left(\frac{\mu_G}{\rho_g D_{AG}}\right)^{-0.5} \tag{14.3.a-6}$$

MULTIPHASE FLOW REACTORS ⎯⎯⎯⎯⎯⎯⎯⎯⎯⎯ **703**

where the constant $c$ equals 2.3 when the equivalent packing diameter is smaller than 15 mm and 5.2 when it exceeds that value. $k_G$ is expressed in kmol/m$^2$ atm hr. Ramm [14] mentions

$$\frac{4k_G \varepsilon}{a_v D_{AG}} \frac{RTp_t}{p_{fA}} = 0.027 \left(\frac{4G}{a_v \mu_G}\right)^{0.8} \left(\frac{\mu_G}{\rho_G D_{AG}}\right)^{0.33} \tag{14.3.a-7}$$

with $k_G$ in kmol/m$^2$ atm hr. The technical information sheets of the producers of packing material contain extremely important information, generally based on large-scale experimentation.

### Example 14.3.a-1  Design of a Packed Column for Carbon Dioxide Absorption (after Danckwerts and Sharma [15])

Carbon dioxide is to be absorbed at 20 atm abs (20.3 bars) and 30°C from a gas stream of 900 kmol/hr with the following molar proportions: $N_2/H_2/CO_2$: 1/3/1.33 by means of a 2.5-molar monoethanolamine (MEA) solution

$$CO_2 + 2R_2NH \longrightarrow R_2NCOO^- + R_2NH_2^+ .$$

The $CO_2$ content is to be reduced to $2.10^{-3}$ percent. The content of dissolved $CO_2$ in the MEA will be 0.15 and 0.40 mol $CO_2$/mol MEA at the top and bottom of the column, respectively. A column packed with $1\frac{1}{2}$ in. Rashig rings will be used.

The model is that of Sec. 14.2.b, which assumes the gas and liquid to be in plug flow. Furthermore, since the reaction is very rapid, the bulk concentration of $A$ is taken to be zero. The amount of $CO_2$ to be absorbed is

$$900 \times \frac{1.33}{5.33} = 225 \text{ kmol/hr}$$

so that the liquid flow rate has to be

$$L = \frac{225}{(0.40 - 0.15)2.5} = 360 \text{ m}^3/\text{hr}$$

The choice of the column diameter is based on the gas flow rate that would give flooding at the bottom. From Fig. 14.3.a-1 it follows that for $(L/F')\sqrt{(\rho_L/\rho_g)} = 2.66$ (with $\rho_g$ calculated from the ideal gas law) the gas flow rate leading to flooding is obtained from

$$\frac{G_f^{\,2}(a_v/\varepsilon^3)(\mu_L/\mu_W)^{0.2}}{g\rho_g\rho_L} = 0.007$$

Since $(a_v/\varepsilon^3) = 400 \text{ m}_i^2/\text{m}_r^3$ and $\mu_L/\mu_W = 1.5$, it follows that $G_f = 5400 \text{ kg/m}^2$ hr. A column diameter of 2.75 m is chosen, giving a gas mass flow rate of about half this value, 2628 kg/m$^2$ hr, and a liquid mass flow rate of 60,840 kg/m$^2$ hr.

From experimental results of Danckwerts and co-workers, $k_L$ is taken to be 0.792 m/hr and $a'_v$, for the given $L$, 140 $m_i^2/m_r^3$. Henry's coefficient depends on the ionic strength of the solution. Van Krevelen and Hoftijzer [16] related the coefficient to that in pure water and to the ionic strength. The correction for the ionic strength is weak in this case: from a value of 35.7 $m^3$ atm/kmol in pure water, $H$ for the MEA solution is increased to 40. $D_B$, the diffusivity of MEA in a 2.5-$M$ solution at 30°C was estimated by Danckwerts and Sharma to be 2.77 $\times$ $10^{-6}$ $m^2$/hr and that of $CO_2$, $D_A$ to be 5.04 $\times$ $10^{-6}$ $m^2$/hr.

The rate coefficient of the reaction at 30°C is 36.7 $\times$ $10^6$ $m^3$/kmol hr.

It is safe to check at this point whether the back pressure of $CO_2$ above the liquid is negligible. This requires data on the stoichiometric equilibrium constant (kmol/$m^3$) of the reaction:

$$K_c = \frac{[R_2NH]^2[CO_2]}{[R_2NCOO^-][R_2NH_2^+]}$$

Provided the concentration of free $CO_2$ is less than that of free amine, the concentration of free MEA is $C_{B0}(1 - 2\alpha)$ and that of $R_2NCOO^-$ and of $R_2NH_2^+$ is $C_{B0}\alpha$, where $\alpha$ is the number of moles of $CO_2$ absorbed per mole of total amine and $C_{B0}$ is the initial concentration of MEA. The equilibrium concentration of $CO_2$ is then given by

$$C_{CO_2} = K_c \frac{\alpha^2}{(1 - 2\alpha)^2}$$

Danckwerts and Sharma selected a value of 9.0 $\times$ $10^{-6}$ kmol/$m^3$ for $K_c$. At the top of the column, where $\alpha = 0.15$, the equilibrium concentration of $CO_2$ is 4.1 $\times$ $10^{-7}$ kmol/$m^3$. The corresponding back pressure is found, using Henry's law, to be 1.6 $\times$ $10^{-5}$ atm while the partial pressure of $CO_2$ at the top of the column is $2.10^{-5} \times 20 = 40.10^{-5}$ atm; therefore, the back pressure is negligible. The same conclusion would be arrived at for the conditions at the bottom of the column.

The $k_G a'_v$ value was taken from a correlation mentioned in Norman [17] for air in ammonia at 20°C and 1 atm total pressure.

$$k_G a'_v = c \times G^{0.72}\left(\frac{\rho_L L}{\Omega}\right)^{0.38}$$

When $k_G a'_v$ is expressed in kmol/$m^3$ hr atm and $G$ and $\rho_L L/\Omega$ in kg/$m^2$ hr, the constant $c$ has the value 0.03975. With S.I. units (bar and s, instead of atm and hr) $c = 0.09015$. This correlation can be used provided it is corrected for the total pressure and adapted to the system $CO_2$–MEA by accounting for the Schmidt numbers. These contain the viscosities of mixtures $N_2$, $H_2$, $CO_2$, and the diffusivity of $CO_2$ in such mixtures, and the corresponding values for the air-ammonia mixtures. Using the correlations developed in Reid and Sherwood [18] the

Schmidt number for the gas mixture is calculated to be 0.80 and that for the air-ammonia mixture 0.65. Now $k_G a'_v$ is calculated as follows, for an average $G$ of 1758 kg/m² hr and an average $\rho_L L / \Omega$ of 60543 kg/m² hr:

$$k_G a'_v = 0.03975(1758)^{0.72}(60543)^{0.38} \left(\frac{0.65}{0.80}\right)^{2/3} \left(\frac{1}{20}\right)$$

$$= 24.65 \frac{\text{kmol}}{\text{m}^3 \text{ hr atm}} = 0.00685 \text{ kmol/m}^3 \text{ s bar}$$

The rate of the global phenomenon can be written as

$$N_A = k_G p_t(y_{AG} - y_{AGi}) = F_A k_L C_{Ai}$$

Setting up a continuity equation for $B$ in a differential volume element of the column, assuming plug flow for both phases, leads to

$$L d C_B = 2 N_A a'_v \Omega \, dz$$

which becomes, after integration,

$$\int_{C_B = 0.5 \, M}^{C_B = 1.75 \, M} \frac{d C_B}{N_A} = \frac{2 a'_v \Omega Z}{L}$$

with $N_A = F_A k_L C_{Ai}$.

The integral has to be calculated numerically. This requires a relation between $N_A$ and $C_B$. The latter is obtained as follows. To find $N_A$ at any point in the column, set $p_{Ai} = p_t y_{AGi} = p_A$ as a first approximation. Calculate $C_{Ai}$ from $C_{Ai} = p_{Ai}/H$ and the free amine concentration $C_B$ from a material balance. Then the groups $(a/b)(C_B/C_{Ai})(D_B/D_A)$ (with $a = 1$ and $b = 2$) and $\gamma = \sqrt{kD_{AL}/k_L}$ are computed. Entering these values into the enhancement factor diagram yields $F_A = N_A/k_L C_{Ai}$ and therefore $N_A$. Knowing $N_A$ the assumed value of $p_{Ai}$ may be checked from $N_A/k_G p_t = y_{AG} - y_{AGi}$. If there is no agreement, iteration is necessary until two successive values of $p_{Ai}$ coincide. For example, the detailed calculations are given for a position in the column where $p_A = 4.0$ atm (i.e., where $y_{AG} = 0.20$). The flow rate of inerts $= 900 \times (1 + 3)/(1 + 3 + 1.33) = 675$ kmol/hr. The flow rate of $CO_2$ at the chosen position is $0.25 \times 675 = 169.2$ kmol/hr. At the top the $CO_2$ flow rate is $(675 \times 2.10^{-5})/(1 - 2.10^{-5}) = 13.5 \times 10^{-3}$ kmol/hr. The amount of $CO_2$ absorbed above the selected position is 169.2 kmol/hr. A balance on the top yields the concentration of free amine at the chosen position, $C_B$.

$$C_B = \frac{360 \times 2.5(1 - 0.30) - 2 \times 169}{360} = 0.81 \frac{\text{kmol}}{\text{m}^3}$$

To calculate the value of $k_G a'_v$ at that position, the mass flow rate of gas has to be calculated:

$$\frac{675 \times \dfrac{28 + 6}{4} + 169.2 \times 44}{\dfrac{3.14}{4} \times (2.75)^2} = 2232 \frac{\text{kg}}{\text{m}^2 \text{ hr}}$$

The value of $k_G a'_v$ becomes

$$k_G a'_v = \left(\frac{2232}{1758}\right)^{0.72} \times 24.65 = 29.27 \frac{\text{kmol}}{\text{m}^3 \text{ hr atm}} = 0.00813 \text{ kmol/m}^3 \text{ s bar}$$

Since $p_{Ai}$ is taken to equal $p_A$ the concentration at the interface is found to be

$$C_{Ai} = \frac{4}{40} = 0.1 \text{ kmol/m}^3$$

Now $\gamma$ can be calculated:

$$\gamma = \frac{\sqrt{k' C_B D_A}}{k_L} = \frac{\sqrt{36.7 \times 10^6 \times 0.81 \times 5.04 \times 10^{-6}}}{0.792} = 15.4$$

Such a high value of $\gamma$ justifies setting the concentration of $CO_2$ in the bulk liquid equal to zero.

The parameter curves in the Van Krevelen and Hoftijzer diagram were originally, according to the film theory, $(a/b)(D_B/D_A)(C_B/C_{Ai})$. To account for the results of the penetration theory, indicating that the mass transfer coefficients are proportional to the square roots of the diffusivities, Danckwerts and co-workers have used $(a/b)(C_B/C_{Ai})\sqrt{D_B/D_A}$ as the parameter group in the Van Krevelen and Hoftijzer diagram. The value of the latter group amounts to

$$\frac{C_B}{2C_{Ai}} \cdot \sqrt{\frac{D_B}{D_A}} = \frac{0.81}{2 \times 0.1} \sqrt{\frac{2.77 \times 10^{-6}}{5.04 \times 10^{-6}}} = 3.0$$

From Fig. 6.3.b-1 the enhancement factor is found to be 3.5. Using the original parameter group, the value of which would be 2.2, $F_A$ would amount to 3.0.

Finally, $N_A$ can be calculated from

$$N_A a'_v = \frac{p_{Ab}}{\dfrac{1}{k_G a'_v} + \dfrac{H}{F_A k_L a'_v}} = \frac{4.0}{\dfrac{1}{29.27} + \dfrac{40}{3.5 \times 0.729 \times 140}} = 29.2 \frac{\text{kmol}}{\text{m}_r^3 \text{ hr}}$$

With this value $p_{Ai}$ is found to be 3.0 atm instead of 4.0 as assumed. A second cycle of calculations is necessary, leading to $N_A a'_v = 33.22 \text{ kmol/m}_r^3$ hr and a third to $N_A a'_v = 33.98 \text{ kmol/m}_r^3$ hr. Further iterations are unnecessary, so that the calculations can proceed to the next increment in $p_A$. The height of the column is found to be 4.60 m. Note that, as the top of the column is approached, the $CO_2$ content becomes so low that $C_B$ varies little and pseudo-first-order behavior is achieved. In that part of the column the calculations could be based on Eq. 6.3.b-3, with $C_A = 0$, of course.

***

### Example 14.3.a-2 Design Aspects of a Packed Column for the Absorption of Ammonia in Sulfuric Acid (after Ramm [14])

This example features the calculational aspects encountered with instantaneous second-order reactions. The inlet partial pressure of ammonia in the gas entering the absorber at essentially atmospheric pressure is 0.05 atm (0.051 bar) the exit partial pressure is to be 0.01 atm (0.0101 bar). The total gas flow rate is 45 kmol/hr. The liquid phase enters the column at the top and flows countercurrently with the gas at a rate of 9 m³/hr. The inlet concentration of sulfuric acid is 0.6 kmol/m³. Consider the operation to be isothermal at 25°C. Determine the exit concentration of the sulfuric acid and the required interfacial area. The irreversible reaction

$$2\,NH_3 + H_2SO_4 \longrightarrow (NH_4)_2SO_4$$
$$\phantom{2\,N}A \phantom{H_3 +} B$$

corresponds to the situation discussed in Sec. 6.3.c, whereby the reaction occurs in a plane in the liquid film or coincides with the interface. Depending on the case, the equation for $N_A$ is given by Eqs. 6.3.c-3 or 6.3.c-8 and it first has to be checked which one should be used.

Let $k_G = 0.35$ kmol/m² hr atm $= 9.6 \times 10^{-5}$ kmol/m² s bar, $k_L = 0.005$ m/hr and the Henry coefficient 0.0133 kmol/m³ atm $= 0.0131$ kmol/m³ bar. Whether Eqs. 6.3.c-3 or 6.3.c-8 are applicable depends on $C_B' = bD_A N_A/aD_B k_L$ or, since $N_A = k_G p_A$, when the reaction plane coincides with the interface $C_B' = (bD_A k_G/aD_B k_L)p_A$. Suppose, as a first approximation, $D_A = D_B$, then

$$C_B' = \frac{0.35}{2 \times 0.005}\,p_A = 35p_A$$

At the bottom of the column, where $p_A = 0.05$ atm

$$C_B' = 35 \times 0.05 = 1.75 \text{ kmol/m}^3$$

and at the top, where $p_A = 0.01$ atm

$$C_B' = 35 \times 0.01 = 0.35 \text{ kmol/m}^3$$

which is lower than $C_B = 0.6$ kmol/m$^3$. The $C_B$ at the bottom can be calculated from an overall balance:

$$\frac{F[(p_A)_{in} - (p_A)_{out}]}{2p_t} = L[(C_B)_{in} - (C_B)_{out}] = 45 \times \frac{0.05 - 0.01}{2} = 9[0.6 - (C_B)_{out}]$$

so that $C_B = 0.5$ kmol/m$^3$, which is lower than $C'_B = 1.75$ kmol/m$^3$.

Consequently, the column has to be calculated in two parts: the top part in which $C_B > C'_B$ so that Eq. 6.3.c-8 has to be applied, and the bottom part in which $C_B < C'_B$ and in which Eq. 6.3.c-3 is valid. The location of the dividing line may also be obtained by means of a material balance, on the bottom part, for example:

$$F\frac{(p_A)_{in} - p_A}{p_t} = \frac{a}{b}L(C'_B - C_B) = 45(0.05 - p_A) = 2 \times 9 \times (35p_A - 0.5)$$

from which $p_A = 0.0166$ atm $= 0.0168$ bar and the corresponding $C'_B = 35p_A = 0.58$ kmol/m$^3$.

*Calculation of the Bottom Part*

In this part

$$C_B < C'_B \qquad \text{and} \qquad N_A = K_G\left(p_A + \frac{a}{b}\frac{D_A}{D_B}\frac{C_B}{H}\right)$$

$$\frac{1}{K_G} = \frac{1}{k_G} + \frac{H}{k_L} = \frac{1}{0.35} + \frac{0.0133}{0.005} = \frac{1}{0.18}$$

so that

$$K_G = 0.180\frac{\text{kmol}}{\text{m}^2\text{ hr atm}} = 4.94 \times 10^{-5} \text{ kmol/m}^2\text{ s bar}$$

The continuity equation for $A$ in a differential element is written as

$$-F\frac{dp_A}{p_t} = N_A\, dO$$

where $O$ is the total interfacial area.

After substitution of $N_A$ and integration:

$$-\int_{0.05}^{0.0166} \frac{dp_A}{p_A + \dfrac{a}{b}\dfrac{D_A}{D_B}\dfrac{C_B}{H}} = \frac{p_t K_G}{F}O$$

To perform the integration $C_B$ has to be related to $p_A$ by means of a material balance on the bottom part. This leads to

$$-\int_{0.05}^{0.0166} \frac{dp_A}{\left(1 - \frac{\varphi F}{2 L}\right) p_A + \frac{\varphi F}{2 L}(p_A)_{in} + (C_B)_{out}} = p_t \frac{K_G}{F} O$$

where $\varphi = (a/b)(D_A/HD_B)$. This equation is easily integrated and yields $O = 180$ m$^2$.

*Calculation of the Top Part*

In this part $C_B > C'_B$ and $N_A = k_G p_A$ so that

$$-\int_{0.0166}^{0.01} \frac{dp_A}{p_A} = p_t \frac{k_G}{F} O$$

from which $O = 65$ m$^2$. The total required interfacial area is 245 m$^2$. The total volume of the column follows from $V = O/a'_v$ where $a'_v$ would be calculated from the correlation by Onda et al. Eq. 14.3.a-2.

---

### 14.3.b Two-Phase Fixed Bed Catalytic Reactors with Cocurrent Downflow. "Trickle" Bed Reactors and Packed Downflow Bubble Reactors

This section deals with problems that bear considerable relation to those dealt with in Chapter 11 on fixed bed catalytic reactors with a single fluid phase, the main difference being in the hydrodynamics, because of the existence of two fluid phases. In addition, the mass and heat transfer phenomena are more complex, since resistances in the gas phase, the liquid phase, and the solid catalyst, where the reaction takes place, have to be considered. Figure 14.3.b-1 illustrates concentration and partial pressure profiles around a catalyst particle and defines the notation. $A$ is the reacting component of the gas phase, $B$ that of the liquid phase.

The advantage of downflow operation with respect to upflow lies in the fact that there is no limitation on the flow rates imposed by flooding limits. The flow rates are only limited by the available pressure head at the inlet. Furthermore, the liquid is much more evenly and thinly distributed then with upward flow. Depending on the respective flow rates of gas and liquid different flow regimes may be obtained, however. Figure 14.3.b-2 is a schematic representation of the different possibilities, derived from diagrams presented by Hofmann [42], by Sato et al. [20], and by Charpentier and Favier [73]. The trickle flow regime

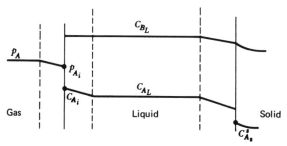

*Figure 14.3.b-1 Concentration and partial pressure profiles in fluid phase and catalyst.*

corresponds to rather low flow rates of gas and liquid; the gas phase is continuous and the liquid phase dispersed. Increasing the gas flow rate leads to pulsed flow. If, for a given liquid flow rate, the gas flow rate is too much increased, spray flow will be obtained, however. For higher liquid throughputs the liquid phase may be continuous and the gas phase dispersed: this is called bubble flow. Increasing the gas flow rate will lead first to dispersed bubble flow and then to pulsed flow. The lines separating the areas corresponding to different flow regimes should not be taken too rigorously, since the composition of the liquid phase is also important.

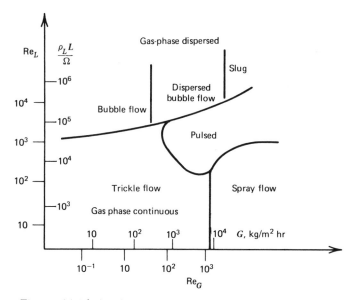

*Figure 14.3.b-2 Flow regimes in downflow packed beds.*

Charpentier and Favier, [73] have determined a similar diagram for foaming fluids (kerosine, gasoil). The transition between gas phase continuous and dispersed agrees with that for nonfoaming liquids. At sufficiently high liquid flow rates an increase of the gas flow rates leads to foaming, foaming and pulsing, and pulsing flow with transitions more or less in agreement with those between bubble, dispersed bubble, and pulsed flow in Fig. 14.3.b-2. In hydrodesulfurization, liquid flow rates range from 1 to 10 kg/m$^2$ s for heavy fractions to 10 to 25 for naphtha, while hydrogen flow rates range from 470 Nm$^3$/ton for heavy gasoil to 840 Nm/ton for heavy residue. In hydrocracking this ratio may be 1700.

There is relatively little information on the pressure drop in two-phase co-current downflow packed beds. Larkins et al. [21] have proposed the following correlation for the two-phase frictional pressure drop:

$$\log \frac{\delta_2}{\delta_L + \delta_G} = \frac{0.416}{\left(\log \sqrt{\frac{\delta_L}{\delta_G}}\right)^2 + 0.666} \qquad (14.3.b-1)$$

where $\delta_2$ is the two-phase frictional pressure drop. The measured pressure drop, $\Delta p_{t2}/\Delta z$, in kilogram force per square meter cross section and unit length, is related to $\delta_2$ by

$$\frac{\Delta p_{t2}}{\Delta z} = \delta_2 - g\varepsilon_L \rho_L$$

where $g\varepsilon_L \rho_L$ represents the liquid head per unit height. $\delta_L$ and $\delta_G$ are the frictional pressure drops per unit length for liquid flow and gas flow only. It has been argued that the term $\varepsilon_L \rho_L$ should not be accounted for, since the liquid is supported by the packing, but in the experiments of Larkins et al. the difference between $\Delta p_{t2}/\Delta z$ and $\delta_2$ was negligible anyway. Larkins et al. derived their equation from the concept of the "equivalent fluid," which considers the gas and the liquid to be quasi homogeneous. This is strictly correct only when the gas and liquid flow rates $u_{sG}$ and $u_{sL}$ are identical.

Sweeney [22] distinguished between the two phases. He assumed that the liquid flows uniformly over the packing surface, that both the liquid and the gas phase are continuous, and that the pressure drops through both phases are identical. He came to the following relation:

$$\left(\frac{\delta_L}{\delta_2}\right)^{1/3} + \left(\frac{\delta_G}{\delta_2}\right)^{1/3} = 1 \qquad (14.3.b-2)$$

The Larkins and Sweeney equations were developed for downward bubble flow. In trickle bed operation, the liquid and gas flow rate are not as high as in packed absorbers, so that there is much less interaction. Single-phase flow pressure drop equations could be used as a first approximation, with the void fraction reduced to account for the liquid holdup. Charpentier et al. [23] recommend, on the basis of an energy balance instead of a force balance, to replace $\delta_2$ in Eqs. 14.3.b-1 and 14.3.b-2 by $(u_{sm}/\rho_L g)(\Delta p_{t2}/\Delta z) + u_{sL}\varepsilon_L + u_{sG}(1 - \varepsilon)(\rho_g/\rho_L)$ whereby $u_{sm}$ is the mean of the liquid and gas velocities [i.e., $u_{sL}\varepsilon + u_{sG}(1 - \varepsilon)$]. Midoux et al. [71] recently proposed the following correlation for $\delta_2/\delta_L$:

$$\left(\frac{\delta_2}{\delta_L}\right)^{0.5} = 1 + \left(\frac{\delta_L}{\delta_G}\right)^{-0.5} + 1.14\left(\frac{\delta_L}{\delta_G}\right)^{-0.27} \tag{14.3.b-3}$$

for any type of flow with nonfoaming liquids and for trickle flow with foaming liquids only.

The wetting of the packing depends on the nature of the packing surface and on the surface tension of the liquid. For the wetting to be complete, an efficient liquid distribution at the top is required, while the column to particle diameter should exceed 20 to 25 to avoid liquid bypassing along the wall. Henry and Gilbert [24] recommend that $d_p \rho_L L/\Omega\mu_L > 10$.

The liquid holdup (volume of liquid/volume of empty reactor) consists of the liquid held in the pores of the catalyst and of that outside the catalyst particles, which is called external holdup. The latter is frequently divided into the free draining and the residual or static holdup. The static holdup has been related by Charpentier et al. [25] to the Eötvös number $\delta_L g d_p^2/\sigma_L$. It varies between 0.02 and 0.05. Several correlations have been given for the dynamic or free draining holdup. They are:

$$\varepsilon_L' = c\left(\frac{\rho_L d_p L}{\mu_L \Omega}\right)^{\alpha}\left(\frac{d_p^3 g \rho_L^2}{\mu_L^2}\right)^{\beta}(a_v d_p)^{\gamma} \tag{14.3.b-4}$$

According to Otake and Okada [26] for nonporous spherical packing $c = 1.295$, $\alpha = 0.676$, $\beta = -0.44$, and $\gamma = 1.0$, whereas Satterfield et al. [27] arrived at $c = 1.0$, $\alpha = 0.333$, $\beta = -0.33$, and $\gamma = 0$. Note that the dynamic holdup is independent of the gas flow rate, but varies with the liquid flow rate. Goto and Smith [28] observed agreement between their experimental data and Otake and Okada's correlation for large-particle diameters only and between their data and Satterfield et al.'s correlation for small-particle diameters only. The following external or total holdup equation is proposed by Midoux [71]:

$$\varepsilon_L = \frac{0.66\left(\dfrac{\delta_L}{\delta_G}\right)^{0.40}}{1 + 0.66\left(\dfrac{\delta_L}{\delta_G}\right)^{0.40}} \tag{14.3.b-5}$$

for any type of flow with nonfoaming liquids and for trickle flow only with foaming liquids. The internal holdup, finally, depends on the porosity of the catalyst.

There are no effective interfacial area correlations in the literature for the specific cases discussed here. The correlation that comes closest to that required for trickle bed operation is that of Puranik and Vogelpohl [29], which is for a continuous gas phase and a dispersed liquid phase, but in a countercurrent packed column, well below the loading point. They derived the following correlation (for $\rho_L L/\Omega = 1.5$ kg/m$^2$ s):

$$\frac{a'_v}{a_v} = 1.05 \, Re_L^{0.04} We^{0.135} \left(\frac{\sigma_{L,c}}{\sigma_L}\right)^{0.18} \qquad (14.3.b\text{-}6)$$

where the Weber number $= We = \rho_L L^2 d_p/\Omega^2 \sigma_L$. It follows that, in the investigated range, $a'_v$ is proportional to $L^{0.31}$. At higher flow rates the equation of Onda et al. mentioned in the section on absorbers might be used—Eq. 14.3.a-2. Charpentier [41] recommends from a literature survey of trickle bed results for spheres and pellets:

$$\frac{a'_v}{a_v} = c \left(\frac{\Delta p_{t2}}{\Delta z} \frac{\varepsilon}{a_v}\right)^{1.2} \qquad (14.3.b\text{-}7)$$

with $c = 0.81$ or $5.23 \times 10^4$ when $\Delta p_{t2}$ is expressed in kgf/m$^2$ or in bars.

The axial effective diffusivity for the liquid phase has been correlated by Elenkov and Kolev [30] as follows:

$$\frac{u_{sL}}{D_{eL} a_v} = 0.068 \left(\frac{4\rho_L L}{\Omega a_v \mu_L}\right)^{0.78} \left(\frac{g\rho_L^2}{a_v^3 \mu_L^2}\right)^{-0.33} \qquad (14.3.b\text{-}8)$$

Mears [31] has given the minimum $Z/d_p$ ratio required to hold the deviation from the reactor length as calculated on the basis of plug flow below 5 percent:

$$Z/d_p > \frac{20n}{Pe_{aL}} \ln \frac{(C_A)_{in}}{(C_A)_{out}}$$

where $n$ is the order of the reaction and $Pe_{aL} = d_p u_{iL}/D_{eL}$. Hochman and Effron [32] correlated the axial mixing in the gas phase in terms of the Peclet and Reynolds numbers for the gas and liquid phase as follows:

$$Pe_{aG} = 1.8 \, Re_G^{-0.7} 10^{-0.005 \, Re_L} \qquad (14.3.b\text{-}9)$$

Deviations from plug flow in the gas phase are not ordinarily of concern in trickle bed operation.

As previously mentioned, the mass transfer from the gas to the active sites of the catalyst involves several steps. For trickle bed operation, in which the interaction between gas and liquid is limited, the values for $k_G$ and $k_L$ are of the same order of magnitude as those given for countercurrent operation in the previous section. Some specific results for cocurrent operation are available in the literature.

Reiss' correlation for $k_G a_v'$ and $k_L a_v'$ in air-ammonia and air-water systems are as follows [33]:

$$k_G a_v' = 2.0 + c_1 E_G^{0.66} \qquad (14.3.\text{b-}10)$$

where $E_G$ is an energy dissipation term for gas flow $= (\Delta p_{t2}/\Delta z)\, u_{sG}$. The constant $c_1$ equals 0.3 when $E_G$ is expressed in kgf/m² s and 0.0665 when $E_G$ is in W/m³. The mass transfer coefficient $k_G$ is in m³/m² s.

$$k_L a_v' = c_2 E_L^{0.5} \qquad (14.3.\text{b-}11)$$

where $E_L$ is an energy dissipation term for liquid flow $= (\Delta p_{t2}/\Delta z)\, u_{sL}$, in kgf/m² s($c_2 = 0.054$) or in W/m³($c_2 = 0.017$). $k_L$ is in m³/m² s. Reiss' results were probably obtained in the pulse flow and spray flow regime. Charpentier [41] extended the range of validity to low liquid and gas flow rates and proposed the following correlation:

$$k_L a_v' = 0.0011 E_L \frac{D_{AL}}{2.4 \times 10^{-9}} \quad (\text{s}^{-1}) \qquad (14.3.\text{b-}12)$$

where $D_A$ is the diffusivity m²/s in the liquid, assuming that the liquid viscosity does not differ too much from that of water. $E_L$ is in W/m³.

The liquid–solid mass transfer coefficient may be obtained, in first approximation, from the $j_D$ correlations for single-phase flow, mentioned in Chapters 3 and 11, although the gas phase exerts a certain influence, as shown by Mochizuki and Matsui for cocurrent upwards flow [34]. Specific results for the situation considered here have been derived by Van Krevelen and Krekels [35], who correlated their data as follows:

$$\text{Sh} = 1.8 \, \text{Re}^{0.5} \, \text{Sc}^{0.33} \qquad (14.3.\text{b-}13)$$

with

$$\text{Sh} = \frac{k_l}{a_v'' D_{AL}} \quad \text{and} \quad \text{Re} = \frac{\rho_L L}{\Omega a_v'' \mu_L}$$

Also see Satterfield, et al [75].

The resistance to mass transfer inside the catalyst particle is dealt with as outlined in Chapters 3 and 11. In trickle bed hydrodesulfurization, the gas film resistance is practically zero, since the gas phase is mainly hydrogen. The liquid side and liquid–solid side resistances are negligible with respect to that inside the catalyst, since hydrogen is very soluble in the liquid. The effectiveness factor is generally around 0.5 to 0.6. An additional complication arises when a fraction of the liquid feed is vaporized, such as in hydrodesulfurization of light petroleum fractions (naphtha, kerosene) or in hydrocracking. In such a case the pores of the catalyst are filled with both liquid and vapor. The theory of the effectiveness factor for such a situation still has to be worked out.

For lack of sufficient knowledge of the hydrodynamics, and because of the considerable difference between small-scale and large-scale results, the design practice for trickle bed reactors is still in a rather early stage of development. Several approaches have been discussed by Satterfield [36] and are briefly outlined below. With industrial hydrotreating operations in mind, suppose the gaseous reactant is present in great excess with respect to the reacting component of the liquid. Since hydrogen is very soluble in petroleum fractions and since the operation is carried out at high pressures (35 to 100 atm) the liquid may be considered to be saturated with hydrogen. If plug flow is assumed for the liquid, the continuity equation for a reacting component $B$ of the liquid may be written:

$$LC_{B0} \, dx = r_B \, dW$$

If the reaction were of first order, and if there were no diffusional limitations in the liquid film or in the catalyst particle,

$$r_B = kC_B$$

Integration then leads to

$$\ln \frac{C_{B0}}{C_B} = k \frac{W}{L}$$

When experiments are carried out in trickle bed reactors and first-order kinetics are assumed, a rate coefficient is derived, represented by $k_{obs}$, which is found to be different from the true rate coefficient $k$. It is found that when both $L$ and $W$ are doubled the conversion is also increased. This means, in the jargon of the trickle bed literature that the contacting effectiveness is $<1$. There may be several reasons for this. First, it may be argued that $k_{obs}$ may include effects of diffusion in the liquid film and in the catalyst particles:

$$\frac{1}{k_{obs}} = \frac{1}{k\eta} + \frac{1}{k_L a'_v}$$

where $k_L a'_v$ is, of course, dependent on $F'$ or $L$—according to correlations given above, to $L^{0.5} - L^{1.0}$.

Furthermore, if the reaction is not truly first order, $k$ values compared at different conversions will be different. When the species are adsorbed on the catalyst, first-order kinetics would be rather unlikely. Even if each species reacted according to first-order kinetics, the lumping of a spectrum of species with different reactivities into one pseudo-component would lead to an overall order higher than 1.

Bondi [37] related $k_{obs}$ to $k$ and the superficial liquid flow rate:

$$\frac{1}{k_{obs}} = \frac{1}{k} + \frac{A'}{(L\rho_L/\Omega)^b} \quad \text{where} \quad 0.5 < b < 0.7$$

For hydrodesulfurization of a heavy gasoil, $k_{obs}/k$ was found to be 0.12 to 0.2 at $L\rho_L/\Omega = 288$ kg/m² hr and 0.6 at $L\rho_L/\Omega = 1080$ kg/m² hr. The power would seem to drop at higher liquid velocities, however. According to Satterfield, the "contacting effectiveness" would become almost 1 at $L\rho_L/\Omega$ values of 1 to 5 kg/m² s. Henry and Gilbert [24] associated the effect of the liquid flow rate with the free-draining holdup:

$$k_{obs} \sim k\varepsilon'_L$$

$\varepsilon'_L$ varies as $L^{1/3}$ and Henry and Gilbert could indeed correlate their results in this way. There is, however, no theoretical justification for taking the rate to be proportional to the liquid free-draining holdup. Mears [31] therefore proposed to consider the rate to be proportional to the external wetted area of the catalyst. Using the Puranik and Vogelpohl correlation he derived the following relation:

$$\log_{10}\frac{C_{\text{in}}}{C_{\text{out}}} \sim Z^{0.32}(\text{LHSV})^{-0.68}d_p{}^{0.18}\left(\frac{\mu_L}{\rho_L}\right)^{-0.05}\left(\frac{\sigma_{L,c}}{\sigma_L}\right)^{0.21}\eta_G \quad (14.3.\text{b-}14)$$

where LHSV is the liquid hourly space velocity. This relation would be valid for $\rho_L L/\Omega < 54$ kg/m² s. At higher flow rates Onda's correlation Eq. 14.3.a-1 should be used for the wetted area.

Little is known on the effect of the gas flow rate on $k_{obs}$. The effect should be small if the liquid is saturated with the gas. According to Charpentier et al. [36], an increase in $G$ decreases $\varepsilon_L$, but favors the exchange between dynamic holdup and stagnant liquid.

Deviations from $k$ could also be explained in terms of deviations from plug flow, because of axial mixing. This can be accounted for either directly through the residence time distribution or through an axial effective diffusivity model as discussed earlier in Chapter 11 for single-phase fixed beds. For trickle bed operation $Pe_{aL}$ is about 0.2 at a Re based on particle diameter of 10, whereas for single-phase operation $Pe_{aL} = 2$. For bench scale operation the minimum reactor length for absence of significant axial mixing could therefore be an order of magnitude greater than for single-phase fixed beds. But, again, for industrial reactors, axial mixing is completely negligible: Henry and Gilbert estimated the minimum length to be 30 cm when $d_p = 1.6$ mm, for a Reynolds number of 10 and a conversion of 90 percent and 70 cm for 99 percent conversion.

### 14.3.c Two-Phase Fixed Bed Catalytic Reactors with Cocurrent Upflow. "Upflow Packed Bubble Reactors"

Generally, the flow rates in these reactors are such that the gas phase is dispersed and the liquid phase continuous. In the bubble flow regime, the bubbles rise at a slightly higher velocity than the liquid. As the gas flow velocity is increased, pulse flow is obtained. For certain ranges of $u_{sL}$ and $u_{sG}$, the spray flow regime, with continuous gas phase and a liquid mist, may be experienced. The regimes shown in Fig. 14.3.b-2 are also encountered here, the only difference being that pulsing is initiated at slightly lower gas velocities and persists to slightly higher gas velocities for upward flow, at a given liquid rate, than for downflow.

The pressure drop for upward flow is the sum of the liquid head and of the friction of gas and liquid, between each other and with the packing and wall. Heilman [38] derived the following semiempirical correlation for air and water and for various packing material and shapes:

$$
\frac{\Delta p_{t2}}{\Delta z} = g\varepsilon_L\rho_L - (1 - 0.25u_{sL})\left(\frac{u_{sG}}{u_{iG}} \cdot \frac{0.85g\varepsilon_L\rho_L}{\varepsilon - \varepsilon_G^0} - \alpha g\varepsilon_G\rho_g\right)
$$

$$
+ \left(0.292 + \frac{4.16}{\mathrm{Re}_L}\right)\frac{\Omega(1 - \varepsilon)}{\varepsilon^3}\rho_L u_{sL}^2 \qquad (14.3.c\text{-}1)
$$

($u_{sG}$ up to 0.03 m/s; $u_{sL}$ up to 0.04 m/s.) Again, $\Delta p_{t2}/\Delta z$ is in kgf/m²/m. The first term on the right represents the liquid head, the second the bubble friction, the third the liquid friction. $\rho_g$ is the gas density at the top of the column. $u_{sL}$ and $u_{sG}$ are superficial liquid and gas velocities, $u_{iG}$ is the interstitial gas velocity, and $\varepsilon_G^0$ is the gas holdup at zero gas flow rate. $\alpha$ accounts for the roughness of the catalyst and varies between 300 and 700.

The following correlation for the gas holdup, $\varepsilon_G$, was also derived by Heilman:

$$
\varepsilon_G = 0.01\left[0.012d_r + 8\left(\frac{\varepsilon}{d_p}\right) - 2 + 4.45u_{sG}^{0.66} + 11\left(\frac{\mu_L}{\mu_W} - 1\right)^{1.25}\right]\left(\frac{\sigma_W}{\sigma_L}\right)^{1.8}
$$

$$
(14.3.c\text{-}2)
$$

This formula is valid only for $\varepsilon_G < 0.45$ and for constant total pressure. Therefore, high columns would have to be divided into sections before the above formulas could be applied to them. The rising velocity of the bubbles has been correlated as follows by Heilman:

$$
u_G = \left(24.5 - 8\frac{\mu_L}{\mu_W}\right) + (2.2 + 1.32d_p)u_{sG} \text{ (in cm/s)}
$$

No general correlation is available for the interfacial area. From Saada's measurements [39] it would seem that $a_v'$ varies approximately as $u_{sG}^{0.5}$, regardless of packing size and type, column diameter, or liquid superficial velocity.

Since the liquid flow rates are generally rather low it may be necessary to account for axial mixing in the liquid phase. This is done in terms of axial effective diffusion. The axial effective diffusivity for the liquid phase is given by Böxkes [40]:

$$
\frac{D_{eL}}{\nu_L} = 3.63 \times 10^3\,\mathrm{Re}_L^{0.07}\,\mathrm{Re}_G^{0.13}\left(\frac{Z}{d_r}\right)^{0.63}(1 - \varepsilon)^{3.6} \qquad (14.3.c\text{-}3)
$$

For the gas phase Böxkes found

$$\frac{D_{eG}}{v_G} = 6.02 \times 10^2 \, \text{Re}_L^{0.15} \, \text{Re}_G^{0.55} \qquad (14.3.\text{c-4})$$

in which $\text{Re}_L$ and $\text{Re}_G$ are based on interstitial velocities. The mass transfer coefficients for gas–liquid transfer and liquid–solid transfer have not been investigated in detail. The correlations for the gas–liquid transfer coefficient may be different from those derived from countercurrent packed absorbers.

Mochizuki and Matsui [34] correlated the liquid–solid transfer coefficient for $\text{Re}_L > 10$ as follows:

$$\frac{\text{Sh}}{\text{Sh}'} = 1 + 4 \, \frac{\text{Re}_G^{0.55}}{\text{Re}_L^{0.7}} \qquad (14.3.\text{c-5})$$

with $\text{Sh}' = 0.75 \, \text{Re}^{0.5} \, \text{Sc}^{0.33}$. $\text{Sh}'$ is the Sherwood number for single-phase flow. The Reynolds numbers are based on interstitial velocities.

### 14.3.d  Plate Columns

Technological details and operating characteristics of various types of plate columns are extensively discussed in textbooks on mass transfer (e.g., see Treybal [44]). The results of extensive research on bubble tray design is reported in the American Institute Chemical Engineers *Bubble Tray Manual* [45]. Flooding and weeping limits and further aspects of sizing are mentioned in a review by Zenz [5].

The formulas required to calculate the number of trays will only be mentioned here without derivation. For dilute gas mixtures Henry's law often applies so that, in terms of mole fractions, $y = mx$. Also, the total flow of liquid, $L'$ (in kmol/hr), and of gas, $F$ (in kmol/hr) is substantially constant. The operating line and the equilibrium line are then both straight and the relation between ingoing and outgoing flow and the number of trays are obtained analytically, for absorption and in terms of mole fractions:

$$\frac{y_{N+1} - y_1}{y_{N+1} - mx_0} = \frac{A'^{N+1} - A'}{A'^{N+1} - 1}$$

$$N_P = \frac{\log\left[\dfrac{y_{N+1} - mx_0}{y_1 - mx_0}\left(1 - \dfrac{1}{A'}\right) + \dfrac{1}{A'}\right]}{\log A'}$$

where $A' = L'/mF$ is the absorption factor (i.e., the ratio of the slope of the operating line to that of the equilibrium line). The notation is clear from Fig. 14.3.d-1.

When the above conditions are not fulfilled, plate-to-plate calculations are required. Total and solute balances on the lower part of the column up to plate $n$

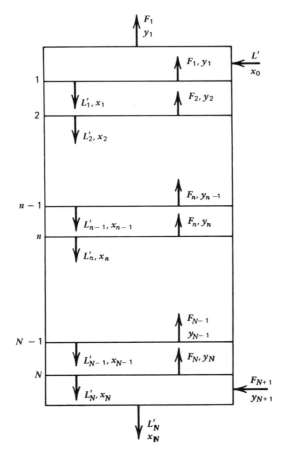

*Figure 14.3.d-1 Tray absorber.*

can be written as follows:

$$L'_n + F_{N+1} = L'_N + F_{n+1}$$
$$L'_n x_n + F_{N+1} y_{N+1} = L'_N x_N + F_{n+1} y_{n+1}$$

$L_n$ and $x_n$ are obtained from these equations. For nonisothermal operation an enthalpy balance is required. For adiabatic operation:

$$L'_n H_{L,n} + F_{N+1} H_{G,N+1} = L'_N H_{L,N} + F_{n+1} H_{G,n+1}$$

This equation yields the temperature of stream $L'_n$. The gas flow $T_n$ is then at the temperature of $L'_n$ and in composition equilibrium with it. Then the equations are applied on that fraction of the column comprised between the bottom and plate $n - 1$ and so on toward the top.

CHEMICAL REACTOR DESIGN

What is obtained in this way is the number of theoretical plates, which means that the liquid is assumed to be completely mixed and that the gas and liquid are in equilibrium. Departure from these conditions are accounted for by means of the Murphree efficiency. When the operating and equilibrium lines are straight, the overall tray efficiency, $\bar{E}$, which is the ratio of the number of ideal to real plates, can be computed analytically from

$$\bar{E} = \frac{\log\left[1 + E_C\left(\frac{1}{A'} - 1\right)\right]}{\log \dfrac{1}{A'}}$$

where $E_C$ is the Murphree tray efficiency corrected for entrainment. The uncorrected Murphree tray efficiency is given by

$$E' = \frac{y_n - y_{n+1}}{y_n{}^* - y_{n+1}}$$

$E'$ is the integral over the tray surface of the point efficiency along a gas streamline, represented by $\dot{E}$:

$$\dot{E} = \frac{\dot{y}_n - \dot{y}_{n+1}}{\dot{y}_n{}^* - \dot{y}_{n+1}}$$

where $\dot{y}_n{}^*$ is the concentration in equilibrium with $\dot{x}_n$. When the gas entering the plate is uniform in composition and the liquid is completely mixed:

$$E' = \dot{E}$$

but when the liquid is in plug flow:

$$E' = A'(e^{\dot{E}/A'} - 1)$$

Intermediate degrees of mixing in the liquid are accounted for by an effective diffusion mechanism.

The Murphree efficiency $E'$ must still be corrected for entrainment to yield the efficiency $E_C$, given by

$$E_C = \frac{E'}{1 + E'\left(\dfrac{\mathscr{E}}{1 - \mathscr{E}}\right)}$$

where $\mathscr{E}$ is the fractional entrainment, moles entrained liquid per mole net liquid flow.

The Murphree efficiency is a convenient concept for correlating experimental results on the mass transfer on trays, but a more fundamental way to do this is to express the results in terms of mass transfer coefficients and effective diffusivities.

MULTIPHASE FLOW REACTORS _____ 721

The A.I.Ch.E. design manual contains correlations for $k_G$, $k_L$, and $D_e$ for bubble caps and perforated trays. Sharma et al. [46] derived the following correlations for bubble cap plates:

$$k_G A_v = 6.2 u_{sG}^{0.75} Z^{-0.67} D_{eG}^{0.5} \qquad (\text{s}^{-1})$$

$$k_L A_v = 7 u_{sG}^{0.75} Z^{-0.67} D_{eL}^{0.5} \qquad (\text{s}^{-1})$$

$$k_G = 11.5 u_{sG}^{0.25} Z^{-0.5} D_{eG}^{0.5} \qquad \left(\frac{\text{cm}^3}{\text{cm}_i^2 \, \text{s}}\right)$$

$$A_{vt} = 0.54 u_{sG}^{0.5} Z^{0.83}$$

with all dimensions in cm and s. $A_v$ is the interfacial area per unit volume of liquid + gas and $A_{vt}$ the interfacial area per unit tray surface. For perforated trays (Sharma and Gupta [47]):

$$k_G A_v = 0.261 \left(\frac{\Omega_h}{\Omega}\right)^{-1.75} \left(\frac{\rho_L L}{\Omega}\right)^{0.6} u_{sG}^{1.2} \left(\frac{\text{kmol}}{\text{s atm m}^3}\right)$$

$$k_L A_v = 4.2 \times 10^{-2} \left(\frac{\Omega_h}{\Omega}\right)^{-2.2} \left(\frac{\rho_L L}{\Omega}\right)^{0.6} u_{sG}^{1.2} \; (\text{s}^{-1})$$

where $\Omega_h$ is the total surface of the holes, $\Omega$ the total surface of the plate, and $u_G$ is in m/s. The interfacial area per unit tray surface is given by

$$A_{vt} = 30 G^{1/2} \rho_g^{-1/4} \qquad \text{with } G \text{ in} \qquad \frac{\text{kg}}{\text{m}^2 \, \text{s}}$$

(See Nonhebel [48].)

## Example 14.3.d-1 Gas Absorption with Reaction in a Plate Column

Figure 1 schematically represents an oxidizer-absorption tower used in nitric acid production. In this tower the nitric oxides generated by high-temperature air oxidation of ammonia on platinum gauze are oxidized and absorbed to yield nitric acid. Koukolik and Marek [49] simulated the behavior of the tower on a digital computer. The following reaction scheme was used:

$$2 NO_2 \rightleftharpoons N_2O_4 \qquad (1)$$

$$2 NO + O_2 \rightleftharpoons 2 NO_2 \qquad (2)$$

$$3 NO_2 + H_2O \rightleftharpoons 2 HNO_3 + NO \qquad (3)$$

Reaction 1 is a third-order reaction taking place in the gas phase. It is slow and irreversible at temperatures below 150°C. Reactions 2 and 3 are fast and reversible. The kinetic parameters and equilibrium constants may be found in the literature.

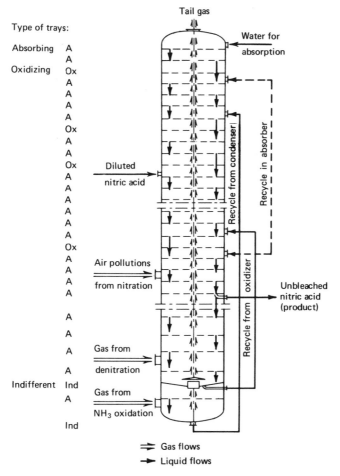

Type of trays:

Absorbing    A
              A
Oxidizing    Ox
              A
              A
              A
              A
              Ox
              A
              A
              Ox
              A
              A
              A
              A
              A
              Ox
              A
              A
              A
              A
              A
              A
              A
Indifferent   Ind
              A
              Ind

Tail gas

Water for absorption

Recycle in absorber

Recycle from condenser

Diluted nitric acid

Air pollutions from nitration

Gas from denitration

Gas from NH₃ oxidation

oxidizer

Recycle from

Unbleached nitric acid (product)

⟹ Gas flows
➝ Liquid flows

*Figure 1 Flows in nitric oxide absorption and oxidation tower (from Koukolik and Marek [49]).*

The following material balance around plate $n$ was set up for the liquid

$$\rho_L L_{n+1} - \rho_L L_n \pm M_n + R_{n1} - R_{n2} + X_{HNO_3} - X_{H_2O} \pm \Delta X_{HNO_3} \pm \Delta X_{H_2O} = 0$$

$\rho_L L_{n+1}$ and $\rho_L L_n$ are the inside liquid flow rates from the plate $n+1$ and to the plate $n-1$. ($L$ in m³/hr), $M_n$ is a side liquid inlet (in kg/hr), $R_{n1}$ and $R_{n2}$ are recycle streams entering and leaving the plate, and $X_{HNO_3}$ is the amount of acid formed on the plate from gaseous components. $X_{H_2O}$ the water consumption by reaction 3, $\Delta X_{HNO_3}$ and $\Delta X_{H_2O}$ are the amounts of $HNO_3$ and $H_2O$ vaporized on the plate (kg/hr).

MULTIPHASE FLOW REACTORS ———————————————— **723**

The enthalpy balance for the liquid on the $n$th plate was as follows:

$$H_{G,n-1} + H_{L,n+1} \pm H_{L,M} + H_{L,R} - H_{G,n} - H_{L,n} + Q_{abs} + Q_{ox}$$
$$+ Q_{ass} + Q_d - Q_c \mp Q_{(HNO_3)} \mp Q_{(H_2O)} = 0.$$

where $H_G$ is the enthalpy of a gas stream and $H_L$ that of a liquid stream. $Q_{abs}$, $Q_{ox}$, $Q_{ass}$, $Q_d$, $Q_c$, $Q_{(HNO_3)}$, and $Q_{(H_2O)}$ are heats of absorption, oxidation, association, dilution, cooling, and latent heats.

Material and enthalpy balances for the gas space are similar. The calculation consists of two sequentially connected loops, the first one including the material balance and the second the enthalpy balance.

Figure 2 shows a comparison between simulated and experimental results for a TVA absorption tower.

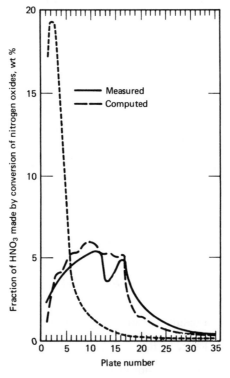

*Figure 2 Fraction of nitric acid produced on each plate. Comparison of experimental and simulated results. (The dashed line is what would be computed from the "TVA model.") (From Koukolik and Marek [49]).*

## 14.3.e Spray Towers

In gas-liquid spray towers the liquid is atomized and enters as a fine spray at the top and the gas is introduced at the bottom. The gas flow rate has to be kept sufficiently low to permit the liquid to fall. It is generally chosen in such way that the liquid drops of mean diameter fall at 20 percent of their free-fall velocity, as calculated from Stokes' law. An efficient dispersion of the liquid requires the openings of the distributor to be small and the pressure high. Thereby a fraction of the drops hits the wall and flows down the wall as a film. Furthermore, a certain degree of coalescence of the drops is inevitable, so that the drop size, velocity, and therefore residence time vary strongly with position. A rigorous hydrodynamic analysis of such a situation is extremely complicated so that only the overall behavior has been studied.

Mehta and Sharma [50, 51] studied spray towers with a maximum diameter of 38 cm. For the interfacial area they found

$$A_v = \alpha\left(\frac{L\rho_L}{\Omega}\right)^\beta u_{sG}{}^{0.28} Z^{-0.38} \tag{14.3.e-1}$$

in which $u_{sG}$ is the superficial gas velocity in cm/s, and $Z$ the effective height in cm. The mass transfer rates are, of course, different from those encountered with single bubbles in rectilinear flow and the following relation was obtained for $k_G$:

$$k_G = \gamma u_{sG}{}^{0.54} \tag{14.3.e-2}$$

In Eqs. 14.3.e-1 and 14.3.e-2 $\alpha$, $\beta$, and $\gamma$ depend on the type of distributor (cone, shower) and the column diameter. With a shower $\alpha = 2.46 \times 10^{-2}$, $\beta = 0.38$ and $\gamma = 1.02 \times 10^{-5}$; with a cone $\alpha$, $\beta$, and $\gamma$ depend on the tower diameter, but average values may be $5.10^{-4}$, $0.65$, and $2.10^{-5}$, respectively.

Liquid–liquid spray towers are less likely to find application as chemical reactors. In this case the dispersed phase flows upward and the continuous downward. Such towers have been thoroughly investigated by Letan and Kehat [52, 53, 54].

## 14.3.f Bubble Reactors

Bubble reactors do not contain any packing and are fed by gas and liquid streams that may be cocurrent or countercurrent. The gas holdup in bubble columns has been measured by Van Dierendonck [55], who obtained the following correlation:

$$\varepsilon = 1.2\left(\frac{\mu_L u_{sG}}{\sigma_L}\right)^{1/4}\left[\frac{u_{sG}}{\left(\frac{\sigma_L g}{\rho_L}\right)^{1/4}}\right]^{1/2} \tag{14.3.f-1}$$

for $\varepsilon \leq 0.45$, $0.03 < u_{sG} < 0.4$ m/sec, $0 \leq u_{sL} < 0.02$ m/sec, $d_r > 0.15$ m, and $0.3 < Z/d_r < 3$. The correlation was also tested in industrial equipment for low-pressure polyethylene production, toluene oxidation, and cyclohexane oxidation, but was less reliable for the prediction of hydrogen holdup. The correlation leads to values of $\varepsilon$ that agree with those of Towell et al. [56], Calderbank and Moo-Young [57], Yoshida and Miura [58], and Reith [59].

According to Van Dierendonck, the bubble diameter is given by

$$\text{Eö}_b = \frac{d_b{}^2 \rho_L g}{\sigma_L} = c \left[ \frac{u_{sG}}{\sqrt[4]{\dfrac{\sigma_L g}{\rho_L}}} \right]^{-(1/2)} M^{-(1/8)} \qquad (14.3.\text{f-}2)$$

where $\text{Eö}_b$ is the Eötvös number, $c = 6.25$ for pure liquids and 2.1 for electrolytes, and $M = \sigma_L{}^3 \rho_L / \mu_L{}^4 g$. The bubble diameters calculated from this formula agree with those predicted by Yoshida and Miura, Marucci and Nicodemo [60], and Calderbank [61]. The interfacial area is then obtained from

$$A_v' = \frac{6\varepsilon}{d_b} = 2 \left[ \frac{u_{sG}}{\sqrt[4]{\dfrac{\sigma_L g}{\rho_L}}} \right] \left( \frac{\rho_L g}{\sigma_L} \right)^{1/2} \qquad (14.3.\text{f-}3)$$

from which it can be seen that $A_v'$ is independent of viscosity, but varies linearly with the superficial gas velocity.

The mass transfer coefficient for the liquid phase is given by Calderbank [62]:

$$k_L = 0.42 \sqrt[3]{\frac{\mu_L g}{\rho_L}} \sqrt{\frac{\rho_L D_{AL}}{\mu_L}} \qquad \text{for } d_b \geq 2 \text{ mm} \qquad (14.3.\text{f-}4)$$

so that $k_L$ is independent of bubble size and velocity and depends only on the physical properties of the system. When $d_b < 2$ mm, Van Dierendonck recommends

$$k_L = k_L(2 \text{ mm}) \times 500 d_b \qquad (\text{m/s}) \qquad (14.3.\text{f-}5)$$

The liquid flow pattern in a bubble column of 15 cm diameter was investigated by Kojima et al. [63] at superficial liquid and gas velocites of the order of 1 cm/s. The liquid flow was found to be complicated and to vary continuously with time. The flow was mainly upward in the central part, but in other parts both upward and downward, although mainly downward near the wall. Radial flow was present across the entire cross section as a result of the pumping effect of the wakes of bubble swarms: when the volume of liquid pumped per unit time by the wakes of the bubbles exceeds that of the liquid fed, downward flow has to occur, inducing radial flow. These observations confirm those of Towell and Ackerman [64] who, in addition, expressed the mixing in terms of axial effective diffusivities for the liquid and gas phase. The authors add, however, that a more realistic model would have to contain a circulation pattern with a superimposed eddy diffusivity. The following dimensional correlations were derived from experiments in 16-in. and 42-in. bubble columns:

$$\text{Pe}_{aL} = \frac{u_{sL} Z}{(1 - \varepsilon) D_{eL}} = \frac{u_{sL} Z}{73.5 (1 - \varepsilon) d_r^{1.5} u_{sG}^{0.5}} \tag{14.3.f-6}$$

$$\text{Pe}_{aG} = \frac{u_{sG} Z}{\varepsilon D_{eG}} = \frac{Z}{19.7 \varepsilon d_r^{2}} \tag{14.3.f-7}$$

Little effect of sparger type was observed, but the influence of the column diameter was very pronounced: the 42-in. column data were an order of magnitude higher than those of the 16-in. column. A draft tube increased the axial mixing at least two to threefold, while a horizontal disc and donut baffles reduced it by a factor of 3.

Typical Peclet numbers for the gas phase, based on the length as characteristic dimension, range from 2 to 15 and for the liquid phase from 0.10 to 0.16. The $u_{sG}$ and $u_{sL}$ are such that $D_{eL}$ and $D_{eG}$ are of the same magnitude. In fact, they are both high, so that the gas phase and the liquid phase are close to complete mixing. The axial mixing characteristics of small laboratory reactors are vastly different from large-scale columns. The conditions in large reactors are close to complete mixing in both phases, while in small diameter reactors they are almost invariably in plug flow with respect to the gas phase and can be either in plug flow or well mixed with respect to the liquid phase.

Scaling up of bubble columns is generally based on the requirement of keeping $k_L A_v'$ constant. Since $k_L A_v'$ is proportional to $u_{sG}$, this implies keeping the superficial gas velocity constant. Some design aspects of bubble reactors will be illustrated in an example following the section on stirred vessel reactors.

### 14.3.g Stirred Vessel Reactors

Agitated absorbers or reactors contain one or more stirrers mounted on a common shaft, depending on the height of the vessel. The influence of a stirrer extends to a height roughly equal to the vessel diameter. The stirrer diameter is generally taken to be one-third of the vessel diameter. Stirred vessels also contain a certain number of vertical baffles, extending into the liquid for about one tenth of the tank diameter. Gas is distributed through some appropriate device into the liquid, underneath the lowest stirrer, but a certain amount of gas can also be aspirated into the liquid from the gas phase above the liquid level by means of a special stirrer. An example of such a gas–liquid reactor is shown schematically in Fig. 14.3.g-1. This reactor, described by Van Dierendonck and Nelemans [65] is used

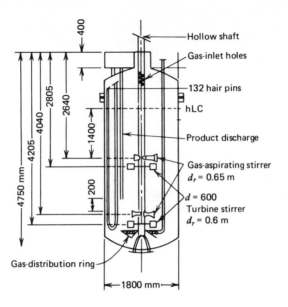

Figure 14.3.g-1 Stirred vessel for hydrogenation (from van Dierendonck and Nelemans [65]).

for the hydrogenation of $\alpha$-nitrocaprolactam into aminolactam, a step of the Dutch State Mines L-lysine process. This is a semibatch reaction carried out at 20 to 30 atm and 85°C with a Raney-nickel catalyst in a vessel with a volume of 9 m³. The gas phase was shown to be completely mixed.

Zwietering [66] determined the maximum gas flow rate $F'_{max}$ (in m³/hr) that can be handled by a stirrer of given geometry rotating at a certain speed. The relation is as follows:

$$F'_{max} = 16 \frac{N^3 d_r^4}{g} \left(\frac{d_s}{d_r}\right)^{3.3} \qquad (14.3.g\text{-}1)$$

According to Van Dierendonck [55, 67], the bubble diameter can be obtained from the following relations:

1. For pure liquids:

$$\text{Eö}_b = \frac{d_b^2(\rho_L - \rho_g)g}{\sigma_L} = 0.41 \qquad (14.3.g\text{-}2)$$

——————————————————— CHEMICAL REACTOR DESIGN

and for stirrer speeds $N > 2.5N_0^*$, where $N_0^*$ is the characteristic speed for bubble aspiration and dispersion from the gas atmosphere above the liquid level. $N_0^*$ is given by

$$\frac{N_0^* d_s^2}{\sqrt[4]{\dfrac{\sigma_L g}{\rho_L}}\, d_r} = 2\left(\frac{H_L - H_s}{d_r}\right)^{1/2} \tag{14.3.g-3}$$

where $H_L - H_s$ is the liquid height above the stirrer in absence of gas flow.

2. For solutions of electrolytes and surface active agents:

$$\text{Eö}_b = \left[1.2 + 260\,\frac{\mu_L(N - N_0)d_s}{\sigma_L}\right]^{-2} \tag{14.3.g-4}$$

for $0 < (N - N_0)d_s < 1.5$ m/s, where $N_0$ is the minimum stirrer speed for efficient dispersion (i.e., the stirrer speed at which the bubble trajectories are markedly influenced).

$N_0$ is given by the following relations:
1. For pure liquids:

$$\frac{N_0 d_s^2}{d_r \sqrt{g d_r}} = 0.07 \tag{14.3.g-5}$$

2. For solutions of electrolytes and surface active agents:

$$\frac{N_0 d_s^2}{\sqrt[4]{\dfrac{\sigma_L g}{\rho_L}}\, dr} = 1 \tag{14.3.g-6}$$

According to Calderbank [62] in the case of pure liquids the bubble diameter is given by:

$$d_b = 4.15\left[\frac{\sigma_L^{0.6}}{\left(\dfrac{P}{\varepsilon_L V}\right)^{0.4} \rho_L^{0.2}}\right]\varepsilon^{1/2} + 0.09 \text{ (in cm)} \tag{14.3.g-7}$$

The gas holdup in pure liquids is obtained from (Calderbank [62].)

$$\varepsilon_G = \left(\frac{\varepsilon u_{sG}}{u_b}\right)^{1/2} + 0.0216\left[\frac{(P/\varepsilon_L V)^{0.4}\rho_L^{0.2}}{\sigma_L^{0.6}}\right]\left(\frac{u_{sG}}{u_b}\right)^{1/2} \tag{14.3.g-8}$$

in which $P$ is the power input, $P/\varepsilon_L V$ the power input per unit liquid volume and $u_b$ the free rising velocity of the bubble, calculated from the Davies and Taylor equation (Davies and Taylor [68]):

$$u_b = 0.711(gd_b)^{1/2}$$

Van Dierendonck proposed for the holdup:
1. With pure liquids:

$$\varepsilon_G = 0.31 \left( \frac{u_{sG}}{\sqrt[4]{\dfrac{\sigma_L g}{\rho_L}}} \right)^{2/3} + 0.45 \frac{(N - N_0{}^*)d_s{}^2}{d_r\sqrt{gd_r}} \tag{14.3.g-9}$$

provided $\varepsilon < 0.25$, $0 < u_{sG} < 0.05$ m/s, $Z/dr = 1$, and $H_L - H_s = 0.5H_L$.
2. With solutions:

$$\varepsilon_G = 0.075 \frac{(N - N_0{}^*)d_s{}^2}{\sqrt[4]{\dfrac{\sigma_L g}{\rho_L}}\, dr} \left( \frac{d_r}{H_L} \right) \tag{14.3.g-10}$$

provided $\varepsilon \leq 0.3$, $0.003 < u_{sG} < 0.03$ m/s, $0.7 < H_L/d_r < 1.4$, and $H_L - H_s = 0.5H_L$. The interfacial area per unit dispersion volume is then easily derived from the relation $A_v' = 6(\varepsilon/d_b)$. Reith [69] compared interfacial areas determined by different methods and found quite a discrepancy between the results.

With pure liquids the mass transfer coefficient for the component $A$ in the liquid is practically independent of the stirrer speed, since the bubble size is practically unaffected by the stirrer speed. Van Dierendonck proposed:

$$k_L = 0.42 \sqrt[3]{\frac{\mu_L g}{\rho_L}}\, Sc^{-(1/2)} \quad (m_L{}^3/m_i{}^2 \text{ s}) \tag{14.3.g-11}$$

The bubble sizes does vary with the stirrer speed for solutions of electrolytes and surface active agents. In that case, $k_L$ is given by

$$\frac{1}{k_L} = \frac{1}{(k_L)_0} \left[ 1.2 + 260 \frac{\mu_L}{\sigma_L} (N - N_0)d_s \right] \tag{14.3.g-12}$$

with

$$(k_L)_0 = 1.13 \sqrt{\frac{D_{A_L} u_b}{d_{bo}}}$$

$$d_{bo} = 0.8 \sqrt{\frac{\sigma_L}{\rho_L g}}$$

$(k_L)_0$ is the mass transfer coefficient of $A$ on the liquid side when $N = N_0$ (i.e., at zero stirring or at very low stirrer speed) and $d_{bo}$ is the corresponding bubble diameter. $u_b$, the bubble rising velocity, is calculated from

$$u_b = \frac{1}{18} \frac{\rho_L g d_b^2}{\mu_L} \quad \text{(Stokes' law) when } \mathrm{Re}_b < 1$$

$$= \frac{1}{4} d_b \sqrt[3]{\frac{(\rho_L - \rho_g)^2 g^2}{\rho_L \mu_L}} \quad \text{when } 30 < \mathrm{Re}_b < 10^3$$

$$= 1.76 \sqrt{\frac{(\rho_L - \rho_g) g d_b}{\rho_L}} \quad \text{when } \mathrm{Re}_b > 10^3 \qquad (14.3.\text{g-13})$$

Since the expression between the square brackets in the right-hand side of the equation for $1/k_L$ is nothing but $(\text{Eö}_b)^{-(1/2)}$, it follows that $k_L/d_b$ is a constant, at least when $0 < (N - N_0)d_s < 1.5$ m/s. Hughmark obtained the following correlation for $(k_L)_0$ for single bubbles, however [70]

$$\mathrm{Sh} = 2.0 + 0.061 \left( \mathrm{Re}_b^{0.48} \mathrm{Sc}^{0.34} \frac{d_b g^{1/3}}{D_{A_L}^{2/3}} \right)^{1.61} \qquad (14.3.\text{g-14})$$

where

$$\mathrm{Sh} = \frac{k_L d_b}{D_{A_L}} \qquad \mathrm{Re}_b = \frac{d_b G}{\mu_G} \qquad \mathrm{Sc} = \frac{\mu_L}{\rho_L D_{A_L}}$$

The turbulence in the bubbles is usually very high so that the mass transfer resistance in the gas phase may be neglected. For the usual stirrer speeds the mixing in the gas and the liquid phase is practically complete. The mixing time in the liquid phase is independent of the gas holdup and given by (Van Dierendonck [55].)

$$N t_m \left( \frac{d_s}{d_r} \right)^2 = 1.7 \qquad (14.3.\text{g-15})$$

for $u_{sG} < 0.05$ m/s, $d_r < 1$ m, and $0.2 < d_s/d_r < 0.5$. The mixing time has to be smaller than the effective absorption time $(k_L A_v)^{-1}$ if the liquid phase is to be considered as completely mixed.

The power requirements are given by:

$$P = c \rho_L N^3 d_s^5$$

with $c$ a characteristic constant depending on the stirrer construction.

According to Van Dierendonck [55] scaling up has to be based on a constant value of the dimensionless effective stirring speed, $(N - N_0)d_s^2/d_r$, which ensures a constant value for the gas holdup, $\varepsilon$. A second condition is to avoid overloading the stirrer (see Eq. 14.3.g-1). Therefore, $u_{sG}$ should generally not exceed 0.03 m/s. According to Reith, however, [59] a safe design rule is to scale up on the basis of constant specific power input, to ensure constant specific interfacial area. Van Dierendonck discusses those two rules and refers to the design of a stirred vessel for the hydrogenation of α-nitrocaprolactam. The effective stirring speed criterion led to an industrial reactor with a volume of 6 m³, a stirrer diameter of 0.5 m and 3.5 revolutions per second, whose performance was successful. The constant specific power input criterion would have led to a volume of 20 m³.

From the correlations given above it is also evident that the liquid composition has an important effect on the interfacial area. All other conditions being equal, the area may be a factor 10 larger in electrolytes than in pure liquids. Evidently, the design and scaling up of stirred gas-liquid reactors still relies on model experiments involving the liquids actually used in the reaction.

## Example 14.3.g-1 Design of a Liquid-Phase o-Xylene Oxidation Reactor

### A. Stirred Tank Reactor

The liquid-phase oxidation of o-xylene into o-methylbenzoic acid by means of air

$$\text{B} \qquad\qquad \text{A}$$

is to be carried out in a continuous stirred tank reactor at 13.6 atm abs (13.8 bars) and 160°C. A yearly production of o-methylbenzoic acid of 30,000 tons is required. For reasons of selectivity the o-xylene conversion per pass has to be limited to 16 percent. An excess of 25 percent with respect to the stoichiometric requirements is chosen for the oxygen feed rate. The rate equation is of pseudo first order with respect to oxygen:

$$r_B = 2.4 \times 10^3 \, C_A$$

where $r_B$, in kmol/m³ hr, is the rate of reaction of o-xylene and $C_A$, in kmol/m³, is the oxygen concentration. From the stoichiometry it follows that the rate based on oxygen consumption is given by $r_A = 1.5 r_B$.

Calculate the reactor dimensions and the stirrer speed.

Basic data:

For o-xylene: $M = 106.16$ kg/kmol, $\rho_L = 750$ kg/m³, $\sigma_L = 16.5$ dyne/cm or $16.5 \times 10^{-3}$ kg/s², $\mu_L = 0.23$ cps or 0.828 kg/m hr, $D = 2.45 \times 10^{-6}$ m²/hr.

For oxygen: $D(O_2/\text{xylene}) = 5.2 \times 10^{-6}$ m²/hr. Henry's constant, $H = 125.3$ m³ atm/kmol = 126.6 m³ bar/kmol.

*General procedure*:

The calculation of the reactor volume requires the oxygen mass transfer coefficient and the interfacial area to be known. This in turn necessitates knowledge of the stirrer speed. It was shown in Sec. 14.3.g, however, how the choice of the stirrer speed depends on the reactor dimensions and geometry, so that the design has an iterative character: a reactor diameter is chosen first; then the stirrer speed is derived from this and finally the reactor volume required to achieve the desired conversion is calculated; the resulting reactor diameter is compared with the initially chosen value.

*Design model*

The model worked out in Sec. 14.2.a, with the gas and liquid phase completely mixed, is adopted here.

*Initial choice of reactor diameter*

The initial choice of the reactor diameter can be oriented as follows, by way of example. Suppose the concentration of $A$ in the liquid is in equilibrium with that in the gas phase, that is, $(C_A)_{out} = (p_A)_{out}/H$. Consider the overall material balance for gas and liquid completely mixed, Eq. 14.2.a-2:

$$\frac{F}{p_t}[(p_A)_{in} - (p_A)_{out}] = \frac{a}{b}L[(C_B)_{in} - (C_B)_{out}] + L(C_A)_{out} \tag{a}$$

whereas a balance on $B$ can be written

$$L[(C_B)_{in} - (C_B)_{out}] = r_B \cdot V(1 - \varepsilon) \tag{b}$$

First determine the flow rates to be substituted into these equations.

*Liquid phase*

Yearly production: $30.10^3$ tons (for 8000 hr)
Hourly production: 3750 kg/hr = 27.5 kmol/hr
$o$-xylene feed rate (fresh + recycle): $27.5/0.16 = 172$ kmol/hr, from which $L = (172 \times 106.16)/750 = 24.4$ m$^3$/hr.

*Gas phase*

Oxygen consumption: 1.5 mol/mol xylene converted, so the total consumption is: $27.5 \times 1.5 = 41.25$ kmol/hr
Oxygen feed rate (25 percent excess): $41.25 \times 1.25 = 51.5$ kmol/hr
Air feed rate at reaction conditions: $F' = (51.5/0.21) \times 22.4 \times (433/273) \times (1/13.6) = 640$ (m$^3$/hr); also $F = 245$ kmol/hr.

At this point, however, $\varepsilon$ is not yet known. One way out is to drop the term $(1 - \varepsilon)$ from Eq. (b) for this first estimation. With $L = 24.4$ m$^3$/hr, $L[(C_B)_{in} - (C_B)_{out}] = 27.5$ kmol/hr, $F' = 640$ m$^3$/hr, $p_t = 13.6$ atm, $(p_A)_{in} = 13.6 \times 0.21 = 2.86$ atm = 2.9 bar, Eq. (a) yields $(p_A)_{out} = 0.5476$ atm = 0.5547 bar.

Assuming equilibrium between the gas and liquid phase $(C_A)_{out} = 4.37 \times 10^{-3}$ kmol/m$^3$. Equation (b) then yields $V = 2.62$ m$^3$. When the height equals the diameter $d_r = 1.49$ m and the cross section $\Omega = 1.75$ m$^2$. Choose the stirrer diameter according to $d_s = d_r/3$, so that $d_s = 0.5$ m and the position of the stirrer $H_L - H_s = \frac{1}{2}$ (liquid height without gas holdup).

*Calculation of stirrer speed*

The characteristic speed for bubble aspiration, $N_0{}^*$, is obtained from Eq. 14.3.g-3:

$$N_0{}^* = 2\sqrt[4]{\frac{\sigma_L g}{\rho_L}\frac{d_r}{d_s{}^2}}\left(\frac{H_L - H_s}{H_L}\right)^{1/2}$$

$$= 2\sqrt[4]{\frac{16.5 \times 10^{-3} \times 9.81}{750}} \times \frac{1.49}{(0.5)^2} \times \left(\frac{1}{2}\right)^{1/2} = 1.01 \text{ rev/s}$$

The minimum stirrer speed that can effectively disperse the gas flow rate is given by Zwietering's relation Eq. 14.3.g-1:

$$N_{min} = \left[\frac{F'}{16}\frac{g\left(\frac{d_r}{d_s}\right)^{3.3}}{d_s{}^4}\right]^{1/3}$$

$$= \left[\frac{640 \times 9.81 \times 3^{3.3}}{16 \times 3600 \times (0.5)^4}\right]^{1/3} = 4.028 \text{ rev/s}$$

Van Dierendonck advises taking the stirrer speed higher than either $N_0{}^*$ or $N_{min}$, so that a value of 4.1 rev/s will be chosen in this case.

*Bubble diameter*

According to Van Dierendonck, for stirrer speeds $>2.5N_0{}^*$ the Eötvös number equals 0.41, so that, from Eq. 14.3.g-2:

$$d_b = \sqrt{\frac{\text{Eö}_b\sigma_L}{g(\rho_L - \rho_g)}} = \sqrt{\frac{0.41 \times 16.5 \times 10^{-3}}{9.81 \times 737.8}} = 0.965 \times 10^{-3} \text{ m}$$

*Gas holdup*

For pure liquids, Eq. 14.3.g-9 can be used:

$$\varepsilon = 0.31\left(\frac{u_{sG}}{\sqrt[4]{\frac{\sigma_L g}{\rho_L}}}\right)^{2/3} + 0.45\frac{(N - N_0{}^*)d_s{}^2}{d_r\sqrt{gd_r}}$$

$$u_{sG} = \frac{640}{3600 \times 2.05} = 0.1016 \text{ m}_G{}^3/\text{m}_r{}^2 \text{ s}$$

so that

$$\varepsilon = 0.31 \left(\frac{0.1016}{0.121}\right)^{2/3} + 0.45 \frac{(4.1 - 1.01)(0.805)^2}{1.49\sqrt{9.81 \times 1.49}} = 0.336$$

*Mass transfer coefficient*

From Eq. 14.3.g-11 it follows that

$$k_L = 0.42 \sqrt[3]{\frac{\mu_L g}{\rho_L}} \sqrt{\frac{D_{A_L} \rho_L}{\mu_L}} = 0.42 \sqrt[3]{\frac{0.23 \times 10^{-3} \times 9.81}{750}} \sqrt{\frac{5.2 \times 10^{-6} \times 750}{3600 \times 0.23 \times 10^{-3}}}$$

$$k_L = 4.14 \times 10^{-4} \, m_L^3/m_i^2 s = 1.485 \, m_L^3/m_i^2 \, hr$$

*Interfacial area*

$$A_v' = \frac{6\varepsilon}{d_b} = A_v(1 - \varepsilon) = \frac{6 \times 0.336}{0.965 \times 10^{-3}} = 2089 \, m_i^2/m_{L+G}^3$$

*Calculation of reactor diameter*

The three equations to be used when the gas and liquid are completely mixed are Eqs. 14.2.a-1, 14.2.a-2, and 14.2.a-3:

$$\frac{F}{p_t}[(p_A)_{in} - (p_A)_{out}] = N_{A|_{y=0}} A_v V(1 - \varepsilon) \tag{c}$$

$$\frac{F}{p_t}[(p_A)_{in} - (p_A)_{out}] = \frac{a}{b} L[(C_B)_{in} - (C_B)_{out}] + L(C_A)_{out} \tag{d}$$

$$N_{A|_{y=y_L}} A_v V(1 - \varepsilon) = r_A(1 - A_v y_L)V(1 - \varepsilon) + L(C_A)_{out} \tag{e}$$

For a pseudo-first-order irreversible reaction:

$$N_{A|_{y=0}} = \frac{k_L \gamma}{\sinh \gamma}(C_{Ai} \cosh \gamma - C_A)$$

which is Eq. 6.3.b-3, while $N_{A|_{y=y_L}}$ is easily derived from Eq. 6.3.b-2:

$$N_{A|_{y=y_L}} = \frac{k_L \gamma}{\sinh \gamma}(C_{Ai} - C_{Ai} \cosh \gamma)$$

Equations (c), (d), and (e), allow $V$, $(C_A)_{out}$ and $(p_A)_{out}$ to be determined. From Eq. (d) it follows that:

$$(p_A)_{out} = 0.57 - 1.35(C_A)_{out} \tag{f}$$

It is assumed that there is no partial pressure drop in the gas phase, so that in Eqs. (c) and (e) $C_{Ai} = (p_A)_{out}/H$. Furthermore, $\gamma = \sqrt{kD_{A_L}/k_L} = 9.2 \times 10^{-2}$ so that $\sinh \gamma = \gamma$ and $\cosh \gamma = 1$. Now substituting Eq. f into Eqs. (c) and (e) leads to:

$$\frac{245}{13.6}[2.29 + 1.35(C_A)_{out}] = 1.485\left[\frac{0.57 - 1.35(C_A)_{out}}{125} - (C_A)_{out}\right]2089V$$

and

$$1.485\left[\frac{0.57 - 1.35(C_A)_{out}}{125} - (C_A)_{out}\right]2089V = 1.5 \times 2.4 \times 10^3(C_A)_{out} \times 0.664V$$

$$+ 24.4(C_A)_{out}$$

from which

$$(C_A)_{out} = \frac{0.7833V - 2.29}{1.35 + 173.9V} \tag{g}$$

and

$$(C_A)_{out} = \frac{14.145V}{5526V + 24.4} \tag{h}$$

From Eqs. (g) and (h), a quadratic in $V$ is obtained, leading to $V = 6.776$ m$^3$. $(C_A)_{out}$ is found to be $2.56 \times 10^{-3}$ kmol/m$^3$ and $(p_A)_{out} = 0.566$ atm $= 0.573$ bar. With the diameter $=$ height of the dispersion, a value of 2.05 m is found for the diameter. These values are the initial values for a second iteration. The final result, corresponding to a reasonable convergence, is $V = 7.1$ m$^3$, $d_r = 2.08$ m, $N = 2.7$ rev/s, $(C_A)_{out} = 2.24 \times 10^{-3}$ kmol/m$^3$ and $(p_A)_{out} = 0.567$ atm $= 0.574$ bar.

## B. Bubble Reactor

Suppose the $o$-xylene oxidation were carried out in a nonstirred vessel—a bubble reactor.

### Design model

The model worked out in Sec. 14.2.c, with the gas phase in plug flow and the liquid phase completely mixed will be applied here.

### Initial choice of cross section, liquid height, and oxygen concentration

Let $d_r = 2$ m so that $\Omega = 3.14$ m$^2$ and $(C_A)_{out} = 2.04 \times 10^{-3}$ kmol/m$^3$. From the approximate relation:

$$L[(C_B)_{in} - (C_B)_{out}] = r_B V(1 - \varepsilon) \tag{i}$$

$$V(1 - \varepsilon) = \frac{27.5}{2.4 \times 10^3 \times 2.04 \times 10^{-3}} = 5.616 \text{ m}^3,$$

which is the volume of the liquid phase only. The superficial gas velocity would then be:

$$u_{sG} = \frac{640}{3600 \times 3.14} = 0.0566 \text{ m/s}$$

*Bubble diameter*

From Eq. 14.3.f-2

$$d_b = \sqrt{\frac{\sigma_L}{\rho_L g} 6.25 \left(\frac{u_{sG}}{\sqrt[4]{\frac{\sigma_L g}{\rho_L}}}\right)^{-(1/2)} \left(\frac{\rho_L \sigma_L^3}{g \mu_L^4}\right)^{-(1/8)}}$$

it follows that

$$d_b = 0.516 \times 10^{-3} u_{sG}^{-(1/4)} = 1.05 \times 10^{-3} \text{ m}$$

*Gas holdup*

From Eq. 14.3.f-1

$$\varepsilon = 1.2 \left(\frac{\mu_L}{\sigma_L}\right)^{1/4} \left(\frac{\sigma_L g}{\rho_L}\right)^{-(1/8)} u_{sG}^{3/4}$$

$$\varepsilon = 1.18 u_{sG}^{3/4} = 0.137$$

*Mass transfer coefficient*

From Eqs. 14.3.f-4 and 14.3.f-5

$$k_L = 500 d_b \times 0.42 \sqrt[3]{\frac{\mu_L g}{\rho_L}} \left(\frac{D_{AL} \rho_L}{\mu_L}\right)^{1/2}$$

$$= 0.207 d_b \text{ m}_L^3 / \text{m}_i^2 \text{ s}$$

$$= 2.17 \times 10^{-4} \text{ m}_L^3 / \text{m}_i^2 \text{ s}$$

$$= 0.781 \text{ m}_L^3 / \text{m}_i^2 \text{ hr}$$

*Interfacial area*

From Eq. 14.3.f-3

$$A_v' = A_v(1 - \varepsilon) = \frac{6\varepsilon}{d_b} = \frac{6 \times 0.137}{1.05 \times 10^{-3}} = 782 \text{ m}_i^2 / \text{m}_r^3$$

*Reactor dimensions*

Equations 14.2.c-1, 14.2.c-2, and 14.2.c-3 are written

$$-\frac{F}{p_t} dp_A = N_{A|_{y=0}} A_v(1 - \varepsilon)\Omega \, dz$$

$$\frac{F}{p_t}[(p_A)_{in} - (p_A)_{out}] = \frac{a}{b} L[(C_B)_{in} - (C_B)_{out}] + L(C_A)_{out}$$

$$\int_0^Z N_{A|_{y=y_L}} A_v\Omega(1 - \varepsilon)dz = r_A(1 - A_v y_L)V(1 - \varepsilon) + L(C_A)_{out}$$

Equation 14.2.c-1 becomes, with $N_A$ from Eq. 6.3.b-2 and in the absence of a $\Delta p$ in the gas phase:

$$\int_{(p_A)_{in}}^{(p_A)_{out}} \frac{dp_A}{p_A - \frac{H(C_A)_{out}}{\cosh \gamma}} = \frac{k_L \gamma \cosh \gamma p_t A_v(1 - \varepsilon)\Omega Z}{HF \sinh \gamma}$$

$(p_A)_{out}$ is not yet known, but may be calculated from Eq. 14.2.c-2:

$$(p_A)_{out} = (p_A)_{in} - \frac{p_t}{F}\left\{\frac{a}{b} L[(C_B)_{in} - (C_B)_{out}] + L(C_A)_{out}\right\}$$

so that

$$(p_A)_{out} = 2.86 - \frac{13.6}{245}(41.25 + 24.4 \times 2.04 \times 10^{-3}) = 0.57 \text{ atm}$$
$$= 0.577 \text{ bar}$$

Since

$$\gamma = \frac{\sqrt{kD_{A_L}}}{k_L} = \frac{\sqrt{1.5 \times 2.4 \times 10^3 \times 5.2 \times 10^{-6}}}{0.781} = 0.175,$$

$Z$ can now be calculated:

$$Z = \frac{125 \times 245 \times 0.175}{0.781 \times 0.175 \times 1.010 \times 13.6 \times 782 \times 3.14} \ln \frac{2.86 - \frac{2.04 \times 10^{-3} \times 125}{1.01}}{0.57 - \frac{2.04 \times 10^{-3} \times 125}{1.01}}$$

$$Z = 2.43 \text{ m} \tag{10}$$

This is a first estimate for the height of the dispersion, gas + liquid. The total volume is 7.63 m³. A better estimate for $(C_A)_{out}$ is obtained from Eq. (i) in which $\varepsilon$ can now be better approximated:

$$(C_A)_{out} = \frac{27.5}{2.4 \times 10^3 \times 3.14 \times 0.863 \times 2.43} = 1.74 \times 10^{-3} \frac{\text{kmol}}{\text{m}^3}$$

A more accurate value for $Z$ is now calculated from Eq. 10:

$$Z = 1.16 \ln \frac{2.86 - 0.215}{0.57 - 0.215} = 2.32 \text{ m}$$

A further estimate for $(C_A)_{out}$ is $1.82 \times 10^{-3}$ kmol/m³, from which $Z = 2.35$ m. These values may be considered as final.

The total volume of gas and liquid is 7.38 m³—only slightly more than in the stirred case, while $(C_A)_{out}$ is even lower. This is easily explained on the basis of the model, which assumes plug flow for the gas in the nonstirred case, against complete mixing in the stirred case. Some design aspects of a continuous stirred tank for the liquid phase oxidation of toluene into phenol have been discussed by Van Dierendonck et al. [43].

# Problems

14.1 Calculate the pressure drop for a trickle bed reactor in which air and water flow down at room temperature through a packed bed of glass beads (0.2 cm diameter). The superficial gas and liquid velocities are 5 and 0.2 cm/s, respectively. The void fraction of the bed is 0.4. Use the correlations presented in Sec. 14.3.b and compare the results.

14.2 Calculate the liquid holdup for the case considered in the previous problem.

14.3 Check by means of Fig. 14.3.b-2 in which regime the reactor described in Problem 1 operates.

# References

[1] Kölbel, H. *Dechema Monographien*, **68**, 35 (1971).

[2] Danckwerts, P. V. and Gilham, A. J. *Trans. Instn. Chem. Engrs.*, **44**, T42 (1966).

[3] Charpentier, J. C. and Laurent, A. *A.I.Ch.E. J.*, **20**, 1029 (1974).

[4] Merchuk, J. C. *A.I.Ch.E. J.*, **21**, 815 (1975).

[5] Zenz, F. A. *Chem. Eng.*, **13**, Nov. 13 120 (1972).

[6] Sherwood, T. K., Shipley, C. H., and Holloway, F. A. *Ind. Eng. Chem.*, **30**, 765 (1938).

[7] Lobo, W. E., Friend, L., Hashmall, F., and Zenz, F. A. *Trans. A.I.Ch.E.*, **41**, 693 (1945).

[8] Shulman, H. L., Ulbrich, C. F., Proulx, A. Z., and Zimmerman, J. O. *A.I.Ch.E. J.*, **1**, 253 (1955).

[9] Shulman, H. L., Ulbrich, C. F., and Wells, N. *A.I.Ch.E. J.*, **1**, 247 (1955).

[10] Shulman, H. L., Ulbrich, C. F., Wells, N., and Proulx, A. Z. *A.I.Ch.E. J.*, **1**, 259 (1955).

[11] Laurent, A. and Charpentier, J. C. *Chem. Eng.*, **8**, 85 (1974).

[12] Onda, K., Sada, E., and Okumoto, Y. *J. Chem. Eng. Japan*, **1**, 63 (1968).

[13] Mohunta, D. M., Vaidyanathan, A. S., and Laddha, G. S. *Indian Chem. Eng.*, **11**, 39 (1969).

[14] Ramm, W. M. *Absorptionsprozesse in der chemischen Technik*, VEB Verlag, Berlin (1953).

[15] Danckwerts, P. V. and Sharma, M. M. *The Chemical Engineer*, No. 202, CE 244 (1966).

[16] Van Krevelen, D. W. and Hoftijzer, P. J. *Chim & Ind. Proc. 21 Congr. Ind. Chem.*, Brussels (1948).

[17] Norman, W. S. *Distillation, Absorption and Cooling Towers*, Longmans Green, London (1961).

[18] Reid, R. C. and Sherwood, T. K. *The Properties of Gases and Liquids*, McGraw-Hill, New York (1958).

[19] Leva, M. *Tower Packings and Packed Tower Design*, U.S. Stoneware, Akron, Ohio (1953).

[20] Sato, Y., Hirose, T., Takahashi, F., Toda, M., and Hashigushi, Y. *J. Chem. Eng. Japan*, **6**, 315 (1973).

[21] Larkins, R. P., White, R. R., and Jeffrey, D. W. *A.I.Ch.E. J.*, **7**, 231 (1961).

[22] Sweeney, D. *A.I.Ch.E. J.*, **13**, 663 (1967).

[23] Charpentier, J. C., Prost, C., and Le Goff, P. *Chem. Eng. Sci.*, **24**, 1777 (1969).

[24] Henry, H. C. and Gilbert, J. B. *Ind. Eng. Chem. Proc. Des. Devpt.*, **12**, 328 (1973).

[25] Charpentier, J. C., Prost, C., Van Swaay, W. and Le Goff, P. *Chim. & Ind.-Gén. Chim.*, **99**, 803 (1968).

[26] Otake, T. and Okada, K. *Chem. Eng. Japan*, **17**, 176 (1953).

[27] Satterfield, C. N., Pelossof, A. A., and Sherwood, T. K. *A.I.Ch.E. J.*, **15**, 226 (1969).

[28] Goto, S. and Smith, J. M. *A.I.Ch.E. J.*, **21**, 706 (1975).

[29] Puranik, S. S. and Vogelpohl, A. *Chem. Eng. Sci.*, **29**, 501 (1974).

[30] Elenkov, D. and Kolev, M. *Chem. Ing. Techn.*, **44**, 845 (1972).

[31] Mears, D. E. *Adv. Chem. Ser.*, **133**, 218 (1974).

[32] Hochman, J. M. and Effron, E. *Ind. Eng. Chem. Fund.*, **8**, 63 (1969).

[33] Reiss, L. P. *Ind. Eng. Chem. Proc. Des. Devpt.*, **6**, 486 (1967).

[34] Mochizuki, S. and Matsui, S. *Chem. Eng. Sci.*, **29**, 1328 (1974).

[35] Van Krevelen, D. W. and Krekels, J. T. C. *Rec. Trav. Chim. Pays-Bas*, **67**, 512 (1948).

[36] Satterfield, C. N. *A.I.Ch.E. J.*, **21**, 209 (1975).

[37] Bondi, A. *Chem. Tech.*, **1**, 185 (March 1971).

[38] Heilman, W. "Zur Hydrodynamik zweiphasig Durchströmten Schüttschichten." Ph.D. Thesis, Erlangen University (1969).

[39] Saada, H. *Chim & Ind.-Gén. Chim.*, **105**, 1415 (1972).

[40] Böxkes, W. "Systemanalyse bei chemischen Reaktoren durchgeführt am Beispiel einer Zweiphasenaufstromkolonne," Ph.D. Thesis, Erlangen University, Germany (1969).

[41] Charpentier, J. C. *Chem. Eng. J.* (1977).

[42] Hofmann, H. *Chem. Ing. Techn.*, **47**, 823 (1975).

[43] Van Dierendonck, L., de Jong P., Van den Hoff, J., and Voncken, H. *Adv. Chem. Ser.*, **133**, 432 (1974).

[44] Treybal, R. E. *Mass Transfer Operations*, McGraw-Hill, New York (1955).

[45] *Bubble Tray Manual*, A.I.Ch.E., New York (1958).

[46] Sharma, M. M., Mashelkar, R. A., and Mehta, V. D. *Brit. Chem. Eng.*, **1**, 70 (1969).

[47] Sharma, M. M. and Gupta, R. K. *Trans. Instn. Chem. Engrs.*, **45**, T169 (1967).

[48] Nonhebel, G. *Gas Purification Process for Air Pollution Control*, Butterworth, London (1972).

[49] Koukolik, M. and Marek, J. *Proc. 4th Eur. Symp. Chem. React. Eng.*, Brussels 1968, Pergamon Press (1971).

[50] Metha, K. C. and Sharma, M. M. *Brit. Chem. Eng.*, **15**, 1440 (1970).

[51] Metha, K. C. and Sharma, M. M. *Brit. Chem. Eng.*, **15**, 1556 (1970).

[52] Letan, R. and Kehat, E. *A.I.Ch.E. J.*, **13**, 443 (1967).

[53] Letan, R. and Kehat, E. *A.I.Ch.E. J.*, **15**, 3 (1969).

[54] Letan, R. and Kehat, E. *Brit. Chem. Eng.*, **14**, 803 (1969).

[55] Van Dierendonck, L. L. "Vergrotingsregels voor gasbelwassers," Ph.D. Thesis, Twente University, Netherlands (1970).

[56] Towell, G. D., Strand, C. P., and Ackerman, G. H. *A.I.Ch.E. J.*, Inst. Chem. Engrs. Symp., Symp. Ser. 10, Trans. Instn. Chem. Engrs., London (1965).

[57] Calderbank, P. H. and Moo-Young, M. B. *Chem. Eng. Sci.*, **16**, 39 (1961).

[58] Yoshida, F. and Muira Y. *Ind. Eng. Chem. Proc. Des. Devpt.*, **2**, 263 (1963).

[59] Reith, T. "Physical Aspects of Bubble Dispersions in Liquids," Ph.D. Thesis, Delft University, Netherlands (1968).

[60] Marucci, G. and Nicodemo, L. *Chem. Eng. Sci.*, **22**, 1257 (1967).

[61] Calderbank, P. H., in Uhl, V. W., and Gray, J. B. *Mixing, Theory and Practice*, Vol. II, Academic Press, New York (1967).

[62] Calderbank, P. H. *The Chemical Engineer*, No. 212, CE209 (1967).

[63] Kojuna, E., Akehata, T., and Shirai, T. *Adv. in Chem. Ser.*, **133**, 231 (1974).

[64] Towell, G. D. and Ackerman, G. H. *Proc. 2nd Intl. 5th Eur. Symp. Chem. React. Eng.*, p.B.3-1 Elsevier, Amsterdam (1972).

[65] Van Dierendonck, L. L., and Nelemans, J. *Proc. 2nd Intl. 5th Eur. Symp. Chem. React. Eng.*, p.B.6-45 Elsevier Amsterdam (1972).

[66] Zwietering, T. *De Ingenieur*, **6**, 60 (1963).

[67] Van Dierendonck, L. L., Fortuin, J. M. H., and Vandenbos, D. *Proc. 4th Eur. Symp. Chem. React. Eng.*, Brussels 1968. Pergamon Press (1971).

[68] Davies, R. M. and Taylor, G. I. *Proc. R. Soc.*, **A200**, 375 (1950).

[69] Reith, T. *Brit. Chem. Eng.*, **15**, 1559 (1970).

[70] Hughmark, G. A. *A.I.Ch.E. J.*, **17**, 1295 (1971).

[71] Midoux, N., Favier, M., and Charpentier, J. C. *J. Chem. Eng. Japan*, **9** (5), 350 (1976).

[72] Lee, E. S. *Quasilinearization and Invariant Imbedding*, Academic Press, New York (1968).

[73] Charpentier, J. C. and Favier, M. *A.I.Ch.E. J.*, **21**, 1213 (1975).

[74] Pavlica, R. T. and Olson, J. H. *Ind. Eng. Chem.*, **62**, No. 12, 45 (1970).

[75] Satterfield, C. N., van Eek, M. W., and Bliss, G. S. *A.I.Ch.E.J.*, **24**, 709 (1978).

# Acknowledgments

## CHAPTER 1

Figure 1.3-1: Originally appeared in *Chem. Eng. Prog.*, **47**, 333 (1951) and is reprinted here with the permission of the copyright owners, the American Institute of Chemical Engineers.

Figure 1 in Example 1.4-1: With permission, *Advances in Catalysis*, **13**, Academic Press, New York (1962).

Figures 1, 2 in Example 1.4-2: Reprinted with permission from *Ind. Eng. Chem. Proc. Des. Devt.*, **10**, 530 (1971). Copyright by the American Chemical Society.

Figures 3 in Example 1.4-2: ibid., p. 538.

Table in Example 1.4-5: From S. W. Benson, *Foundations of Chemical Kinetics*. Copyright 1960. Used with permission of McGraw-Hill Book Co.

Figures 1, 2, 3 in Example 1.6.2-2: With permission, *Ind. Eng. Chem. Proc. Des. Devt.*, **10**, 250 (1971).

Figure 1 in Example 1.7-1: With permission, *Chem. Eng. Sci.*, **18**, 177 (1963).

## CHAPTER 2

Table 2.1-1: With permission, *The Chemical Engineer* (*I. Ch. E.*), No. 6, CE114 (1966), The Institution of Chemical Engineers, London.

Table 2.1-2: From A. G. Oblad, T. H. Milliken, and G. A. Mills, *The Chemistry of Petroleum Hydrocarbons* Reinhold, New York (1955). Reprinted by permission of Van Nostrand Reinhold Company.

Figure 2.1-2: Reprinted by permission from *Ind. Eng. Chem.*, **41**, 2573 (1949). Copyright by the American Chemical Society.

Figure 2.1-3: With permission, *Advances in Catalysis*, **13**, 137 (1962).

Table 2.1-3: With permission, *Ind. Eng. Chem.*, **59**, No. 9, 45 (1967).

Table 2.2-1: With permission, *Chem. Eng. Prog.*, **46**, 146 (1950).

Figures 1, 2 in Example 2.2-2: Originally appeared in *A. I. Ch. E. J.*, **17**, 856 (1971) and is reprinted here with the permission of the copyright owners, the American Institute of Chemical Engineers.

Figure 2.3.d.2-1: With permission, *Advances in Chemical Engineering*, **8**, 97 (1970), Academic Press, New York.

Figure 2.3.d.2-2: With permission, *A. I. Ch. E. J.*, **21**, 1041 (1975).

# CHAPTER 3

Figure 3.4-1: Reprinted with permission from *Ind. Eng. Chem. Fund.*, **7**. 535 (1968). Copyright by the American Chemical Society.

Figure 3.5.a-1: Reprinted with permission from *Chem. Techn.*, **504** (1973). Copyright by the American Chemical Society.

Figures 1, 2, 3, 4 in Example 3.5a-1: With permission, *J. Catal.*, **31**, 13 (1973), Academic Press, New York.

Figure 3.5.c-1: From J. M. Smith, *Chemical Engineering Kinetics*. Copyright 1970. Used by permission of McGraw-Hill Book Co.

Figure 1, 2 in Example 3.5.e-1: With permission, *A. I. Ch. E. J.*, **20**, 728 (1974).

Figure 3.6.a-1: With permission, O. Levenspiel, *Chemical Reaction Engineering*, Wiley, New York (1962).

Figure 3.6.a-2: With permission, R. Aris, *Elementary Chemical Reaction Analysis*, Prentice-Hall, Englewood Cliffs, N.J. (1969).

Figure 3.6.a-3: With permission, R. Aris, *The Mathematical Theory of Diffusion and Reaction in Permeable Catalysts*, Vols. I and II, Oxford University Press, London (1975).

Figure 1 in Example 3.6.b-3: With permission, *A. I. Ch. E. J.*, **16**, 817 (1970).

Figure 3.7.a-1: With permission, *Chem. Eng. Sci.*, **17**, 265 (1962).

Table 3.7.a-1: With permission, *J. Catal.*, **15**, 17 (1969).

Figure 3.7.a-2, 3: With permission, *A. I. Ch. E. J.*, **18**, 347 (1972).

Figure 3.7.a-4: With permission, R. Aris, *Introduction to the Analysis of Chemical Reactors*, Prentice-Hall, Englewood Cliffs, N.J., (1965) and E. E. Petersen, *Chemical Reaction Analysis*, Prentice Hall, Englewood Cliffs, N.J. (1965).

Figure 3.8-1: With permission, *Chem. Eng. Sci.*, **27**, 1409 (1972).

Figure 1, Table 1 in Example 3.8-1: With permission, *Ind. Eng. Chem. Proc. Des. Devt.*, **11**, 454 (1972).

Figures 3.8-2, 3: With permission, *J. Catal.*, **1**, 526, 538 (1962).

Figures 1, 2, 3, 4 in Example 3.9.c-1: With permission, J. C. Spry and W. H. Sawyer, Paper Presented at 68th Annual A. I. Ch. E. Meeting, Los Angeles, (November 1975).

# CHAPTER 4

Figures 4.1-1, 2, 3: With permission, *Ind. Eng. Chem.*, **60**, No. 9, 34 (1968).

Figures 4.2-1, 2: ibid.

Figure 4.2-3: With permission, *A. I. Ch. E. J.*, **14**, 311 (1968).

Figures 1–6 in Example 4.3-1: With permission, *J. Catal.*, **2**, 297 (1963).

Table 1 in Example 4.3-1: ibid.

Figure 4.4-1: With permission, *Chem. Eng. Sci.*, **27**, 763 (1972).

Figures 4.5-1, 2, 3: With permission, *Chem. Eng. Sci.*, **25**, 1091 (1971).

Figures 4.6-1, 2: With permission, *A. I. Ch. E. J.*, **18**, 1231 (1972).

## CHAPTER 5

Figure 5.2.a-1: With permission, *J. Chem. Soc.*, **603**, 1004 (1937), The Chemical Society, London.

Figure 5.2.a-2: Reprinted with permission from *J. Am. Chem. Soc.*, **72**, 1554 (1950). Copyright by the American Chemical Society.

Figure 5.2.c-1: With permission, *J. Catal.*, **5**, 529 (1966).

Figures 5.2.d-1, 2: With permission, *Chem. Eng. Sci.*, **22**, 559 (1967).

Figure 5.3.a-1: With permission, *Ind. Eng. Chem. Proc. Des. Devt.*, **1**, 102 (1962).

Figure 5.3.c-1: With permission, *A. I. Ch. E. J.*, **16**, 397 (1970).

## CHAPTER 6

Figure 6.3.c-2: From P. V. Danckwerts, *Gas-Liquid Reactions.* Copyright 1970. Used by permission of McGraw-Hill Book Co.

Table 6.3.d-1: With permission, *Brit. Chem. Eng.*, **15**, 522 (1970).

Table 6.3.d-2: With permission, *Proc. 20th Anniv. Dept. Chem. Eng.*, University of Houston (1973).

Figure 6.3.f-2: With permission, *Chem. Eng. Sci.*, **21**, 631 (1966).

Figure 6.3.f-3: With permission, *J. Chem. Eng. Japan*, **1**, 132 (1968).

Figure 6.4.a-1: With permission, *Ind. Eng. Chem.*, **45**, 1247 (1953).

Table 6.4.b-1: With permission, W. J. Beek, "Stofoverdracht met en zonder Chemische Reactie," notes, University of Delft (1968).

Figure 6.5-1: From P. V. Danckwerts, *Gas-Liquid Reactions.* Copyright 1970. Used by permission of McGraw-Hill Book Co.

Figure 6.5-2: With permission, W. J. Beek, "Stofoverdracht met en zonder Chemische Reactie," notes, University of Delft (1968).

Figure 6.5-3: From P. V. Danckwerts, *Gas-Liquid Reactions.* Copyright 1970. Used by permission of McGraw-Hill Book Co.

## CHAPTER 8

Figure 1 in Example 8.2-1: With permission, A. R. Cooper and G. V. Jeffreys, *Chemical Kinetics and Reactor Design*, Prentice-Hall, Englewood Cliffs, N.J. (1971).

Figure 8.3.a-1: With permission, R. Aris, *Elementary Chemical Reactor Analysis*, Prentice-Hall, Englewood Cliffs, N.J. (1969).

Figures 1, 2 in Example 8.3.b-1: With permission, *Chemical Engineering*, **77**, No. 3, 121 (1970), McGraw-Hill.

Figures 1–4 in Example 8.3.b-2: With permission, *Ind. Eng. Chem. Proc. Des. Devt.*, **6**, 447 (1967).

## CHAPTER 9

Figures 1 to 6, Example 9.2-1: With permission, *Chem. Eng. Sci.*, **13**, 173, 180 (1961).

## CHAPTER 10

Figure 10.2.b-1: With permission, *Dechema Monographier*, **21**, 203 (1952), DECHEMA, Frankfurt/Main.

Figure 10.3.a-3: With permission, *Chem. Eng. Sci.*, **19**, 994 (1964).

Figure 10.3.b-1a, b: With permission, O. Levenspiel, *Chemical Reaction Engineering*, Wiley, New York (1962).

Table 10.3.b-1: With permission, *Proc. 4th Intl. Symp. Chem. Reac. Engng.*, DECHEMA (1976).

Figure 10.3.b-2: ibid.

Figures 10.3.b-3, 4: With permission, *Chem. Eng. Sci.*, **25**, 1549, 1559 (1970).

Figures 1, 2 in Example 10.4.a-1: With permission, *A. I. Ch. E. J.*, **16**, 410 (1970).

Figure 1 in Example 10.4.b-1: With permission, *Proc. Roy. Soc.*, **A309**, 1 (1969).

Figure 2 in Example 10.4.b-1: ibid.

## CHAPTER 11

Table 11.1-1: With permission, *Chemie Ing. Tech.*, **46**, 374 (1974), Verlag Chemie GmbH, Weinheim.

Figure 11.1-1: ibid.

Figure 11.3-1: With permission, *Chem. Eng. Prog.*, **55**, No. 6, 76 (1959).

Figures 11.3-2, 3: With permission, *Chemie Ing. Tech.*, **46**, 374 (1974).

Figure 11.3-4: With permission, *Chem. Eng. Prog.*, **64**, 59 (1968).

Figure 11.3-5: With permission, C. A. Vancini, *Synthesis of Ammonia*, Macmillan, London (1971).

Figure 11.3-6: With permission, H. Suter, *Phthalsäureanhydrid*, Steinkopf Verlag, Darmstadt (1972).

Figure 11.3-7: With permission, *Chemie Ing. Tech.*, **46**, 374 (1974).

Figure 11.3-8: "Advanced Ammonia Technology," *Hydrocarbon Processing*, **51**(4), 127 (1972). Copyrighted Gulf Publishing Co., Houston. Used with permission.

Figure 11.5.a-1: With permission, *Chemical Engineering*, **56**, 115 (1949).

Figures 11.5.a-2, 3, 4: With permission, *Chem. Eng. Prog.*, **46**, 415 (1950).

Figures 11.5.b-1, 2: With permission, *Chem. Eng. Sci.*, **25**, 1503 (1970).

Figure 11.5.c-2: ibid.

Figures 11.5.e-6, 7: With permission, *Ind. Eng. Chem.*, **59**, No. 1, 72 (1976).

Figures 11.5.e-9, 10: With permission, *Chem. Eng. Sci.*, **20**, 281 (1965).

Figures 11.5.f-1–5: With permission, *Chem. Eng. Sci.*, **16**, 189 (1961).

Figure 11.5.f-9: With permission, *Chem. Eng. Sci.*, **17**, 105 (1962).

Figure 11.6-1: With permission, *Ind. Eng. Chem.*, **59**, No. 2, 18 (1967).

Figures 11.7.c-1, 2, 3, 4, 5, 7: ibid.

Figure 11.7.c-6: With permission, *Periodica Polytechnica*, **15**, 219 (1971). Copyright by Technical University of Budapest, Hungary.

Figures 11.8.a-1, 2: With permission, *Chemical Reaction Engineering*, Advances in Chemistry Series No. 109, Amer. Chem. Soc. (1972).

Figure 11.9.a-1: ibid.

Figure 11.9.a-2: ibid.

Figures 1, 2 in Example 11.9.a-1: With permission, *Ind. Eng. Chem. Proc. Des. Devt.*, **11**, 184 (1972).

## CHAPTER 12

Figure 12.4-1: With permission, H. Kramers and K. R. Westerterp, *Elements of Chemical Reaction Design and Operation*, Academic Press, N.Y. (1963).

Figure 12.5.a-1: With permission, O. Levenspiel, *Chemical Reaction Engineering*, 2nd ed., Wiley, New York (1972).

Figure 2 in Example 12.5.b-2: With permission, *Chem. Eng. Sci.*, **28**, 1021 (1973).

Figure 3 in Example 12.6-1: With permission, *A. I. Ch. E. J.*, **15**, 843 (1969).

Figure 4 in Example 12.6-1: With permission, *12th Symp. (Intl.) Combustion*, p. 90 (1969), The Combustion Institute, Pittsburgh.

Figure 1 in Example 12.6-2: With permission, *A. I. Ch. E. J.*, **15**, 339 (1969).

## CHAPTER 13

Figure 13.2-1: From F. A. Zenz and D. F. Othmer, *Fluidization and Fluid-Particle Systems*. Copyright 1960 by Litton Educational Publishing, Inc. Reprinted by permission of Van Nostrand Reinhold Co.

Figure 13.2-2: With permission, *Proc. Third Intl. Symp. Large Chemical Plants*, p. 81, Uitg. Sprugt, Van Mantgen en De Dees, b.v., Leiden (1976).

Figure 13.3-1: With permission, D. Kunii and O. Levenspiel, *Fluidization Engineering*. Wiley, New York (1969).

Figure 13.3-2: From F. A. Zenz and D. F. Othmer, *Fluidization and Fluid-Particle Systems.* Copyright 1960 by Litton Educational Publishing, Inc. Reprinted by permission of Van Nostrand Reinhold Co.

Figures 13.3-3, 4: With permission, D. Kunii and O. Levenspiel, *Fluidization Engineering*, Wiley, New York (1969).

Figure 13.3-5: From *Fluidization* by M. Leva. Copyright 1960. Used by permission of McGraw-Hill Book Co.

Figures 13.3-6, 7, 8: From F. A. Zenz and D. F. Othmer, *Fluidization and Fluid-Particle Systems.* Copyright 1960 by Litton Educational Publishing, Inc. Reprinted by permission of Van Nostrand Reinhold Co.

Figures 13.4-3, 4: With permission, *A. I. Ch. E. J.*, **13**, 839 (1967).

Figure 13.4-5: With permission, *Trans. Instn. Chem. Engrs.*, **44**, T335 (1966).

Figures 13.4-6, 7: With permission, *Proc. 1st Intl. Symp. Chem. Reac. Eng.*, Advances in Chemistry Series No. 109, Amer. Chem. Soc. (1972).

## CHAPTER 14

Figure 14.3.a-1: Originally appeared in *Trans. A. I. Ch. E.* **41**, 693 (1945) and is reprinted here (in adapted form) with the permission of the copyright owners, the American Institute of Chemical Engineers.

Figure 14.3.a-2: With permission, *Chem. Eng. Prog.*, **54**, No. 9, 57 (1961).

Figures 1, 2 in Example 14.3.d-1: With permission, *Proc. 4th Europ. Symp. Chem. Reac. Eng.*, Brussels, 1968, Pergamon Press (1971).

Figure 14.3.g-1: With permission, *Proc. 2nd Intl. 5th Europ. Symp. Chem. Reac. Eng.*, p. B.6-45, Elsevier, Amsterdam (1972).

# Author Index

Butt, J. B., 170, 174, 204, 208, 210, 212, 224, 229, 271, 294, 525
Buzzelli, D. T., 436

Caddell, J. R., 8
Cadle, P. J., 169, 174, 189, 191
Cairns, E. J., 527
Calderbank, P. H., 544, 726, 729
Calvelo, A., 187
Campbell, D. R., 297
Cappelli, A., 493, 500, 562, 564, 566, 567, 568
Carberry, J. J., 185, 205, 210, 213, 221, 222, 251, 277, 435, 436, 545, 578
Carslaw, H. S., 332, 618
Chang, M., 453
Chapman, F. S., 370
Charlesworth, D. H., 669
Charpentier, J. C., 700, 702, 703, 710, 712, 713, 714, 715, 717
Chauhan, S. P., 616
Chavarie, C., 681
Chen, M. S. K., 602, 646
Chen, N. Y., 164, 167
Chiba, T., 681
Chilton, T. H., 370
Cholette, A., 633
Chou, A., 525
Chu, L. C., 602
Clark, A., 85, 88
Clayton, J. O., 566
Clements, W. C., 643
Cohen, E. S., 417
Collina, A., 486, 493, 500, 562
Collinge, C. N., 644
Colthart, J. D., 655
Colton, C. K., 177
Conolly, J. R., 634
Converse, A. O., 510
Cooper, A. R., 370
Cooper, G. T., 670
Corbett, W. E., 229
Corbetta, D., 493, 500
Costa, E. C., 264
Coughlin, R. W., 85
Coulaloglou, C. A., 634
Crank, J., 618
Cresswell, D., 212, 577
Crider, J. E., 544
Cunningham, R. E., 169, 187
Cunningham, R. S., 161, 169

Curl, R. L., 646

Dalla Lana, I. G., 149
Dallenbach, H., 370
Danckwerts, P. V., 318, 322, 327, 330, 332, 334, 335, 337, 338, 339, 340, 341, 593, 594, 616, 617, 624, 700, 704, 705
Davidson, B., 93
Davidson, J. F., 681, 685
Davies, R. M., 729, 730
Dayan, J., 637
Deans, H. A., 546
De Boer, J. H., 85
Deckwer, W. D., 632
Dedrick, R. L., 175
Defay, R., 4
De Groot, J. H., 677
Delborghi, M., 243
Delgado-Diaz, S., 229
Delmon, B., 242
Deming, L. S., 87
Deming, W. E., 87
Denbigh, K. G., 4, 11, 382, 383, 436, 437
Denn, M. M., 454
Dente, M., 486, 562, 566
De Pauw, R. P., 118, 286, 292, 296, 521
Devillon, J. C., 616
De Vries, R. J., 677, 681
De Wasch, A. P., 476, 534, 538, 578, 579, 629
De Wilt, H. G. J., 184
Dickey, C. A., 478
Di Napoli, N. M., 169
Dogu, G., 170
Dombrowsky, H. S., 478
Doraiswamy, L. K., 155, 415
Dorokhov, I. N., 632
Douglas, J. M., 408, 453
Drew, T. B., 370
Drott, D. W., 207, 559, 560
Dudukovic, M. P., 243
Dukler, A. E., 335
Dullien, F. A. L., 147, 169
Dumez, F. J., 121, 184, 289, 294, 296, 297, 520, 521, 571, 577
Dunn, J. C., 243
Durbin, L. D., 623, 624
Dwyer, F. G., 82
Dyson, D. C., 512

Eagleton, L. C., 82, 408
Eakman, J. M., 50

Eberly, P. E., 520
Eckert, C. A., 63, 64
Effron, E., 632, 714
Eigenberger, G., 549
Elenkov, D., 714
Emig, G., 118, 400, 641
Emmett, P. H., 77
Emptage, M. R., 61
Ergun, S., 477, 481, 482
Ertl, H., 147
Eschenbrenner, G. P., 472, 473
Evangelista, J. J., 646, 649, 651, 658
Evans, H. C., 272
Evans, J. E., 417
Evans, J. W., 259, 262
Eymery, J. P., 510
Eyring, H., 61

Fair, J. R., 413
Fan, L. T., 593, 602, 607, 619, 621, 625, 626, 627, 633, 640, 646
Favier, M., 710, 712
Feng, C. F., 168, 170, 172, 174
Fenn, J. B., 175
Finlayson, B. A., 539, 625, 626
Finneran, J. A., 472, 473
Fleming, H., 371
Flood, E. A., 85
Flory, P., 40
Fogler, H. S., 654
Foss, A. S., 544
Fournier, C. D., 380, 381
Franckaerts, J., 110, 114
Frank-Kamenetski, D., 143
Franklin, J. L., 35
Friedrich, F., 118, 400
Fritz, W., 670
Froment, G. F., 47, 57, 93, 104, 110, 114, 116, 118, 120, 121, 131, 184, 286, 288, 289, 290, 291, 292, 293, 294, 299, 400, 401, 402, 403, 411, 414, 417, 476, 486, 490, 516, 518, 520, 521, 522, 523, 534, 538, 539, 545, 552, 562, 571, 573, 577, 578, 579, 629
Frost, A. A., 46
Frouws, M. J., 184

Gaitonde, N. Y., 453
Gamson, B. W., 145, 147
Gates, B. C., 77
Gavalas, G. R., 35, 49

Gay, 50
Geankoplis, C. J., 161, 169
Gear, C. W., 413
Georgakis, C., 208
Germain, J. E., 81
Gerrens, H., 437, 441
Ghai, R. K., 147
Giaucque, W. F., 566
Gibilaro, L. G., 633
Gibson, J. W., 285
Gidley, J. L., 653, 654
Gilbert, J. B., 713, 717
Gilham, A. J., 700
Gill, W. N., 621, 625
Gilles, E. D., 451
Gillespie, B. M., 435, 436
Gilliland, E. R., 175, 187
Gioia, F., 323
Glaser, M. B., 147
Glasser, D., 644
Glasstone, S., 61
Goettler, L. A., 323
Goldfinger, P., 34
Goldstein, A. M., 637, 639
Good, G. M., 285
Goodwin, R. D., 252
Goto, S., 652, 713
Gorring, R. L., 164, 277
Grace, J. R., 681
Graessley, W. W., 442, 615
Gray, J. B., 634
Greensfelder, B. S., 80
Gregg, S. J., 85, 161
Groves, F. R., 380, 381
Grummit, O., 371
Günkel, A. E., 634
Gunn, D. J., 169, 194, 631, 643, 644
Gupta, R. K., 719
Gwyn, J. E., 639, 655

Haensel, V., 83
Haller, G. L., 175
Hammes, G. G., 64
Handley, D., 147, 477, 576
Harriott, P., 101, 335
Harrison, D., 681, 685
Hartman, J., 184
Hashimoto, K., 325, 326
Hashinger, R. F., 669
Haskins, D. E., 434
Hatcher, W., 552

Haynes, H. W., 169, 174
Hayward, D. O., 85
Heertjes, P. M., 167
Hegedus, L., 187, 296
Heggs, P. J., 147, 477, 576
Heilman, W., 718
Henry, H. C., 713, 717
Henry, J. P., 169
Henson, T. L., 120
Herzfeld, K. F., 31, 36, 44
Hiby, J. W., 625
Hickman, J. B., 468
Hicks, J. S., 202
Hicks, R. E., 477, 481, 482, 557, 560
Higbie, R., 327, 330, 332
Hikita, H., 322, 327
Hill, W. J., 120
Himmelblau, D. M., 49, 50, 116, 353, 451,
  593, 594, 598, 602, 607, 619, 621, 626,
  633, 645, 646
Hinze, J. O., 353
Hlavacek, V., 203, 209, 486, 530, 531, 544
Hochman, J. M., 632, 636, 637, 714
Hoffmann, U., 116
Hofmann, H., 116, 118, 400, 423, 451, 530,
  531, 615, 633, 641, 644, 710
Hoftijzer, P. J., 313, 317, 321, 322, 327, 705
Holland, F. A., 370
Horn, F., 380
Hosten, L. M., 104, 116, 120, 121, 131, 286,
  522, 641
Hottel, H. C., 417
Hougen, J. O., 640
Hougen, O. A., 12, 90, 98, 99, 109, 110, 146,
  191, 297, 394, 401
Hsu, H. W., 149
Huang, C. J., 322, 335
Hudgins, R. R., 195
Hughes, R., 266
Hughmark, G. A., 731
Hull, D. E., 598
Hunt, B., 620
Hurt, D. M., 8
Hyun, J. C., 442

Innes, W. B., 78
Iordache, C., 538
Ishida, M., 246, 247
Ishino, T., 145

Jackson, R., 525

Jacob, S. M., 27, 655
Jaeger, J. C., 332, 618
Jebens, R. H., 370
Jeffreys, G. V., 370
Johanson, L. N., 288
Johnson, A. I., 413
Johnson, M. F. L., 172
Jones, C. R., 49, 50
Jungers, J. C., 94
Juusola, J. A., 131

Kafarov, V. V., 632
Kallenback, R., 170
Kasaoka, S., 229
Kattan, A., 652, 653
Katz, D. L., 415, 437
Katz, S., 378, 383, 384, 387, 644, 646, 649
  651, 658
Katzer, J. R., 77
Keane, T. R., 437
Kehat, E., 725
Kehoe, J. P. G., 204, 208, 210, 212, 237
Kennedy, A. M., 322, 340
Kenson, R. E., 101
Kermode, R. I., 424, 426
Khang, S. J., 634
Kim, K. K., 243
King, C. J., 169
Kishinevski, M. K., 321
Kistiakowski, G., 65
Kittrell, J. R., 47, 110, 128
Kjaer, J., 512, 545
Klinkenberg, A., 624
Klugherz, P. D., 101
Knudsen, C. W., 182
Knutzen, J. G., 415
Kobayashi, T., 681
Kölbel, H., 694
Kojima, E., 726
Kolev, M., 714
Komiyama, H., 175
Kondrat'ev, V. N., 28
Koros, R. M., 199, 633
Kostrov, V. V., 170, 172
Koukolik, M., 722
Krambeck, F. J., 164
Kramers, H., 436, 612, 615, 622
Krekels, J. T. C., 715
Kropholler, H. W., 633
Kubicek, M., 203
Kubota, H., 149

Küchler, L., 36
Kunii, D., 243, 527, 534, 535, 538, 541, 578, 683, 685
Kunugita, E., 294
Kuo, C. H., 322, 335
Kuo, J. C. W., 27

Laidler, K. J., 11, 31, 32, 34, 37, 38, 45, 61
Lamb, D. E., 27
Lambrecht, G., 400
Langmuir, I., 624
Languasco, J. M., 187
Lapidus, L., 47, 116, 546, 624, 628, 640, 641
Larkins, R. P., 712
Larson, M. A., 645
Laurent, A., 700, 702, 703
Lee, E. S., 697
Lee, J. C. M., 184, 208, 209, 213, 560
Lee, K. Y., 497, 501
Lelli, U., 624
Lerou, J., 545
Letan, R., 725
Letort, M., 34
Leva, M., 476, 478, 482, 667, 701
Levenspiel, O., 47, 175, 251, 289, 296, 364
423, 428, 429, 527, 593, 617, 619, 627
633, 634, 637, 651, 683, 685
Lewis, W. K., 306
Lichtenstein, T., 413
Lightfoot, E. N., 144, 147
Liljenroth, F. G., 445
Lim, H. C., 243
Lindeman, F. A., 30
Littman, H., 629
Liu, S. L., 207, 538, 547
Livbjerg, H., 229
Lobo, W. E., 417, 701
Logeais, B. A., 510
Lucas, H. L., 129
Lucki, S. J., 164
Luss, D., 27, 184, 207, 208, 209, 212, 213,
229, 243, 265, 428, 490, 529, 559, 560

Mc Cord, J. R., 636, 637
Mc Cracken, E. A., 607
Mc Cune, L. K., 145
Mc Greavy, C., 212, 560, 561, 577
Mc Guire, M., 546
Maeda, S., 476
Magelli, F., 624
Mah, R. S. H., 639

Mann, U., 644
Manogue, W. H., 169, 174
Marchello, J. M., 335
Marcinkowsky, A. E., 101
Marek, J., 722
Marek, M., 203, 209, 544
Marguardt, D. W., 115
Marrero, T. R., 147
Marucci, G., 726
Masamune, S., 293, 294
Mason, E. A., 147, 168
Matsubara, M., 653
Matsuyama, H., 609
Matsui, S., 715, 719
Maxted, E. B., 272
May, W. G., 675, 676
Mayer, F. X., 609
Mears, D. E., 195, 212, 549, 626, 714, 717
Mehta, B. N., 200
Mehta, K. C., 725
Menon, P. G., 521
Mercer, M. C., 213
Merchuk, J. C., 700
Methot, J. C., 617
Metzner, A. B., 205, 612
Mezaki, R., 110, 116
Michelson, M. L., 641, 643, 644
Mickley, H. S., 527
Midoux, N., 713
Mihail, R., 538
Mikhail, R. Sh., 162
Miller, G. A., 629
Miller, R. S., 625
Milliken, T. H., 80
Millman, M. C., 378, 383, 384, 387
Mills, G. A., 80, 273
Min, K. W., 437
Mingle, J. O., 221
Mireur, J. P., 644, 677, 678
Miura, Y., 726
Mixon, F. O., 643
Miyauchi, T., 609
Mochizuki, S., 715, 719
Mohunta, D. M., 702
Moo-Young, M. B., 726
Morgan, C. Z., 217
Morita, Y., 294
Moser, J. H., 639
Moss, R. L., 78
Mower, E. B., 164
Muno, W. E., 229

Roberts, H. L., 510
Roberts, S. M., 495
Rodigin, N. M., 19
Rodigina, E. N., 19
Roemer, M. H., 623, 624
Rony, P. R., 229
Rosenbrock, H. H., 116
Ross, J., 61
Rossini, F., 58
Rothfeld, L. B., 169
Rowe, P. N., 679, 682, 683, 685, 686
Rudd, D. F., 655
Rudershausen, C. G., 288
Rudnitsky, L. A., 88
Russell, T. W. F., 436
Rutgers, A. J., 39, 63
Rys, P., 174

Saada, H., 718
Sada, E., 279, 280, 281
Sakata, Y., 229
Sama, C., 624
Sampath, B. S., 266
Sato, Y., 633, 710
Satterfield, C. N., 167, 169, 170, 171, 174,
    177, 178, 184, 189, 191, 201, 557, 632,
    713, 715, 716
Sawyer, W. H., 225
Schechter, R. S., 116, 384, 653, 654
Schertz, W. W., 527
Schlünder, E. U., 535, 536
Schmitz, R. A., 446, 453, 454
Schneider, P., 175, 190, 631
Schoenemann, K., 423
Schügerl, K., 644
Schuette, W. L., 53, 55
Schuit, G. C. A., 77, 476
Schultz, G. V., 40
Schwartz, A. B., 170
Schwartz, C. S., 527, 545
Scott, D. S., 169
Seinfeld, J. H., 116, 624, 628, 640, 641
Sen Gupta, A., 147
Shah, B. H., 645
Shah, M. J., 93, 413, 507
Sharma, M. M., 318, 337, 341, 704, 705, 722
Sherry, H. S., 164
Sherwood, T. K., 152, 317, 415, 701, 705
Shimizu, F., 243
Shinnar, R., 437, 602, 624, 625, 644, 646,
    649, 651, 658

Shipman, L. M., 468
Shirai, L., 666
Shulman, H. L., 701, 703
Shutt, H. C., 413
Silberberg, I. H., 654
Simon, J. M., 512
Sinclair, C. G., 633
Sinex, W. E., 654
Sinfelt, J. H., 83
Sing, K. S. W., 85, 161
Singer, E., 578
Sladek, K. J., 175
Slattery, J., 353
Sliger, G., 259
Smith, J. M., 151, 170, 175, 201, 221, 243,
    264, 293, 294, 370, 527, 535, 541, 545,
    578, 631, 713
Smith, R. B., 465
Smith, T. G., 213
Snow, R. H., 413
Sohn, H. Y., 256, 259, 265
Sørensen, B., 229
Spaulding, D. E., 594
Spielman, L. A., 651
Spikins, D. J., 633
Spry, J. C., 225
Sreeramamurthy, R., 521
Steacie, E. W. R., 38
Steisel, N., 224
Stevens, W. F., 424, 426
Stewart, W. E., 144, 146, 147, 148, 168, 170,
    172, 182, 191, 195
Storey, C., 116
Stout, L. E., 617, 634
Strek, F., 370
Strider, W., 163
Suga, K., 294
Sundaram, K. M., 57, 59, 414
Suter, H., 470
Sweeney, D., 712
Swift, S. T., 654
Szekely, J., 256, 259, 262, 384
Szepe, S., 288, 289

Taecker, F. G., 145
Takeuchi, M., 296
Tang, Y. P., 50
Taniyama, I., 633
Tavlarides, L. L., 634
Taylor, G. I., 620, 729, 730
Teller, E. T., 87

Temkin, M. I., 511
Theile, H., 36
Theofanous, T. G., 243
Thiele, E. W., 179
Thodos, G., 145, 147, 477
Thomas, C. L., 78
Thomas, J. M., 77
Thomas, W. J., 77, 194
Thomson, S. J., 77
Thornton, J. M., 560, 561
Tigrel, A. Z., 674
Tinkler, J. D., 205
Toor, H. L., 146, 147, 335, 652
Towell, G. D., 726
Trapnell, D. O., 85
Treybal, R. E., 634, 719
Truong, K. T., 617
Turian, R. M., 626
Turner, J. C. R., 382, 436

Uppal, A., 454

Valstar, J., 545
Van Brakel, J., 167
Van Cauwenberghe, A. R., 624
Vancini, C. A., 469
Van Damme, P. S., 57, 400, 402
Van Deemter, J. J., 639, 675, 676
Van der Laan, E. Th., 630, 641
Van de Vusse, J. G., 324, 325, 327, 433, 434, 435, 436, 633
Van Dierendonck, L. L., 725, 726, 727, 728, 730, 731, 732, 734, 739
Van Heerden, C., 445, 447, 449, 503
Van Krevelen, D. W., 313, 317, 321, 322, 327, 705, 715
Van Swaay, W. P. M., 632, 676, 677, 678, 681
Van Welsenaere, R. J., 486, 490, 562
Van Zoonen, D. D., 521
Vassilatos, G., 652
Vejtasa, S. A., 446
Vellenga, K., 184
Venuto, P. B., 81
Vercammen, H., 417
Verschoor, H., 476
Villadsen, J., 195, 229, 294
Villermaux, J., 615, 616, 632
Viville, L., 552
Voetter, H., 436
Voge, G. M., 80, 81, 217
Vogelpohl, A., 714

Voltz, S. E., 24, 26
Von Rosenberg, D. U., 598
Voorhies, A., 286, 518, 609

Wagner, G. A., 472, 473
Wakao, N., 170, 527, 629, 631
Walen, N., 472, 473
Walker, P. L., 196
Wang, S. C., 264
Watson, C. C., 246, 288
Watson, K. M., 12, 90, 109, 191, 288, 297, 300, 394, 401, 413
Wauquier, J. P., 94
Webb, G., 77
Weber, M. E., 634
Weekman, V. W., 24, 26, 109, 291, 296, 297, 524, 609
Wehner, J. F., 624
Wei, J., 19, 20, 23, 27, 47, 48, 49, 82, 208, 218, 229, 317, 433, 609
Weil, N. A., 667
Weinstein, H., 616
Weisz, P. B., 82, 83, 84, 163, 167, 170, 171, 174, 187, 191, 192, 194, 202, 252, 557, 560
Wen, C. Y., 240, 243, 244, 245, 246, 247, 264, 279, 280, 281, 593, 602, 607, 619, 621, 625, 626, 627, 633, 640, 646, 669
Wender, L., 670
Wentz, C. A., 477
Westerterp, K. R., 436, 447, 612, 615
Wheeler, A., 214, 273
Whitaker, D. R., 643
White, D., 545, 578
White, D. E., 251
White, E. T., 642
Whitman, W. G., 305, 306
Wicke, E., 170, 547, 625
Widom, B., 23
Wilde, D. G., 116
Wilhelm, R. H., 145, 578, 624
Wilke, C. R., 145, 147
Williams, J. A., 643
Williams, R. J. J., 169
Winnacker, K., 466, 467
Winter, R. L., 634
Wojciechowski, B. W., 45, 289, 297
Wolfe, 78

Yagi, S., 527, 534, 535, 538, 541, 578
Yamanaka, Y., 149
Yang, K. H., 98, 99, 110

# Subject Index

Absorption of gases in liquids, 305
  experimental equipment for, 336
  surface renewal theory, 327
  two-film theory, 308
Absorption in industrial equipment, 691
  design examples of, 704, 708, 732
  design models for, 695
  specific design aspects for, 701
Acetone, thermal cracking of, 402
Acrylonitrile reactor, design example, 685
Activation energy, 42
  for complex reactions, 44
  determination of, 43, 400
  diffusional falsification of, 43, 143, 186
Active site, 83, 90, 95, 98, 102, 104
Activity, of catalysts, 289
Adiabatic reactor, with axial diffusion, 528
  batch, 367
  effect of micromixing on conversion in, 651
  multibed, 493
  stirred tank, 446
  tubular with plug flow, 408
Adsorption, classification of metals for, 86
  equilibrium constant, 87
  heat of, 87
  isotherms, 86
Adsorption groups, in catalytic reactions, 98
Advancement, degree of, 5
Age distribution functions, 593
Ammonia, absorption in sulfuric acid, 708
  synthesis, 466, 510
Arrhenius law, 42
Autocatalytic reaction, 13, 15
Autothermic operation of catalytic reactors, 501
Axial dispersion, criterion for importance of, 626. *See also* Axial mixing
Axial mixing, in fixed bed reactors, 525
  multiple steady states due to, 528
  Peclet number for, 527

Bartlett's chi-square test, 120
Batch data, use in kinetic analysis, 364

Batch reactor, design procedures, isothermal, 370
  non-isothermal, 367
  optimal operating conditions, 373
Bellman's optimum principle, 496
Bifunctional catalysts, 82
  diffusion in, 192
Biot number, 198
Bosanquet formula, 169
Boundary conditions, for steady flow reactor with axial dispersion, 624
Bubble flow, 711
Bubble phase (in fluidized beds), 674
  characteristics of, 678
Bubble reactor, 692
  design aspects, 725
  design example, 736
Butane, thermal cracking of, 37
Butene dehydrogenation, coking in, 294, 297
  effect of particle size in, 217
  model discrimination for, 121
  reactor simulation for, 571

Carbon-carbon dioxide reaction, diffusional resistance in, 196
Carbon deposition, *see* Coke deposition in chemical reactors
Carbon dioxide absorption column, design example, 704
Catalysts, activity of, 289
  classification of, 79
  concentration gradients in pellets, 178
  deactivation of, by coking, 284, 515
  by poisoning, 271
  interfacial gradients to, 143
  internal structure of, 159
  reoxidation of, 551
  temperature gradients in, 200
Catalytic cracking, effect of pore diffusion on, 164
Catalytic cracking of gasoil, coking in, 291
  industrial performance of, 663, 672
  kinetic model for, 24

Catalytic cracking of n-hexadecane, 80
Catalytic reaction, comparison with chain
  reactions, 78
  kinetic equations for, 89
  mechanisms for, 77
Catalytic reactors, fixed bed, 465
  fluidized bed, 663
  for kinetic analysis, 106
Chain reactions, 31
Chain transfer, 39
Chemisorption of gases on metals, 86
Chi-square test, 120
Codimerization of olefins, 53
Coke combustion, 252, 264
Coke deposition in chemical reactors,
    examples of, 521, 571
  modeling of, 516
Coking, kinetics of, in butene hydrogenation,
    294, 297
  inside catalyst particle, 292
  in catalytic cracking of gas oil, 291
  in n-pentane isomerization, 292
Complete segregation, 613
Complex reaction, activation energy of, 44
  in gas-liquid systems, 323
  kinetic equations for, 16
  see also Consecutive reactions; Parallel
    reactions
Complex reaction networks, 19
  HJB method for, 49
  in perfectly mixed reactors, 433
  Wei-Prater treatment of, 19, 47
Confidence region, 128
Configurational diffusion, 177
  reaction accompanied by, 224
Consecutive reactions, 18
  in batch reactors, 382, 384
  with catalyst deactivation, 282, 523, 526
  gas-liquid, 323
  in perfectly mixed reactors, 430
  with pore diffusion, 215
  in stirred tanks, 430
Continuity equations, general form of, 350
  simplified forms of, 353
Continuous flow stirred tank reactor, 420
Conversion, 5
  and segregation, 616
Cracking, see Catalytic cracking; Thermal
    cracking

Damköhler number, 530

for poisoning, 276
Danckwerts' age distribution function, 308,
    327
Danckwerts' boundary condition, 624
Deactivation of catalysts, by coking, 284
  influence on autothermic operation, 506
  influence on selectivity, 291
  non-steady state reactor behavior due to, 515,
    571
  by poisoning, 271
Deactivation constant, 288
Deactivation function, for coking, 288
  for poisoning, 273
Decomposition of ester, batch reactor design
    for, 370
Degree of advancement, 5
Degree of polymerization, 40
Dehydrogenation, of 1-butene, see Butene
    dehydrogenation
  of ethanol, see Ethanol dehydrogenation
Design, see Reactor design
Differential method of kinetic analysis, with
    batch data, 8, 46, 366
  for catalytic reactions, 109
  with tubular reactor data, 398
Differential reactor, 108
Diffusional falsification of kinetic parameters,
    187, 192
Diffusion resistance, in gas-liquid reactions, 305
  in gas-solid reactions, external, 146, 197
  internal, see Pore diffusion
  in methanol synthesis, 189
    with bifunctional catalysts, 192
    in carbon to carbon-dioxide reaction, 196
    with complex reactions, 214
Diffusivity, see Effective diffusivity
Discrimination, among rival models, 119
Dispersion coefficient, 618
  determination of, 618, 621, 640
Divergence criterion, in sequential planning,
    119
Drag coefficient, in fluidized beds, 667
Driving force groups in, catalytic reactions, 98
Dual function catalysts, see Bifunctional
    catalysts
Dynamic programming, application to multibed
    adiabatic reactor design, 496

Effective binary diffusivity, 148
Effective diffusion model, in fixed bed reactors,
    532

in multiphase flow reactors, 698
Effective diffusivity, in catalyst particle, 168
  in packed beds, in axial direction, 527
    in radial direction, 533
  in turbulent flow, 352
Effectiveness factor, 179, 557
  global, 197, 209
Effective surface diffusivity, 175
Effective thermal conductivity, inside catalyst
    pellets, 201
  in packed beds, in axial direction, 528
    in radial direction, 534
Energy of activation, *see* Activation energy
Energy equation, general form of, 357
  simplified forms of, 358
Enhancement factor for gas absorption, 312
Eötvös number, 713
  based on bubble diameter, 726
Equivalent penetration, 262
Equivalent reactor volume, 401
Ethane, thermal cracking of, activation energy, 44
    kinetics, 35
    simulation, 410
Ethanol dehydrogenation, calculations for, 151
  model discrimination for, 125
Ethylene oxidation, catalytic, 101
Expansion, of reaction mixture, 363
Experimental reactors, 106, 336
Extent of reaction, 5
External film, heat transfer, 146
  mass transfer, 143
  pressure and temperature drop calculation,
    150

Fanning pressure drop equation, 396
Fauser-Montecatini reactor, 562
Fick's law for binary system, 163
  extension to multicomponent systems, 147,
    351
Fictitious component, 21, 47
Film pressure factor, 144
Film theory, in gas-liquid reactions, 308
Fixed bed reactors, 465
  modeling of, 474
Flooding rate, 701
Flow pattern, non ideal, axial dispersion
    model, 620
    combined models, 627
    flow models, 617
    tanks in series model, 621
Fluid catalytic cracking, 663

Fluidized bed, age distribution functions in,
    598
  free-fall velocity, 667
  heat transfer in, 670
  interchange coefficient, 675
  mean value of rate constant in, example, 609
  minimum fluidization velocity, 667
  terminal velocity, 667
  transport disengaging height, 667
Fluidized bed reactor, 662
  two-phase model for, 675
Fluid stagnancy, detection of regions of, 606
Fractional conversion, 5
Fractional coverage, of catalysts, 87, 88
Fractional entrainment, 721
Free radical, mechanisms, 31
  in polymerization reactions, 38
  in steady-state approximation, 33
  types, 34
Frequency factor, 42
Freundlich isotherm, 89
Friction factor, 478

Gas-liquid reactions, 305
  determination of kinetics, 336
  kinetics of, 322
  models for, 308
Gasoil, catalytic cracking of, 24, 291, 663
Gas-solid reactions, catalytic, 76
  non-catalytic, 239
    general model, 242
    grain model, 256
    pore model, 259
    shrinking core model, 249
    two-stage model, 245
Graham's law, 169
Grain model, in gas-solid reactions, 256

Hatta number, 310
Heat of adsorption, 87
Heat transfer, in batch reactors, 369
  in film surrounding a particle, 151
  in fixed beds, 532
  in fluidized beds, 670
  in plug flow fixed bed reactors, 476
  inside solid particles, 151, 200, 264
Henry's law, 306
Heterogeneous catalysts, 79
Himmelblau, Jones and Bischoff method
    (HJB) for reaction networks,
    49

Hot spot, in fixed bed reactors, 483, 515, 541, 575
Hougen-Watson rate equations, 90
Hydrocarbon oxidation, 482, 539, 550
Hydrogenation reactions, 94, 715

Initial rate method, 110
Initiation step, 31
Integral method of kinetic analysis, with batch data, 8, 366
  for catalytic reactions, 115
  with tubular reactor data, 398
Integral reactor, 108
Interchange coefficient, correlations for, 677
  in fluidized beds, 675
Interfacial area, correlations for, 714, 722, 725
Interfacial resistance, determination of, 150
  for heat transfer, 151, 546, 671
  for mass transfer, 141, 150, 306, 546
Internal age distribution, 597
  of perfectly mixed vessel, 595
Internal heat exchange, reactor with, 509
Internal recycle, reactor with, 108
Internal void fraction, of catalyst pellets, 161
Intraparticle gradients, diffusional, 178
  thermal, 200
Isomerization, mechanism, 83
  of n-heptane, 83
Isomerization of n-pentane, influence of coking on, 286, 292
  kinetics of, 104
  parameter estimation for, 116, 129

Kinetic analysis, differential method of, 8, 109, 366, 398
  integral method of, 8, 115, 366, 398
  of non-isothermal data, 399
Kinetic groups, for catalytic reactions, 98
Kinetic parameters, determination of, diffusional falsification of, 143, 186
  in gas-liquid reactions, 336
  in heterogeneous catalysis, 106
  from tubular reactor data, 397
Knudsen diffusivity, in catalytic reactions, 163

Laminar flow, axial dispersion model for, 620
Laminar jet, 338
LHHW rate equations, 90
Lewis number (modified), 202
Limestone reaction, 259
Lindemann theory, 30

Macropores, 161, 170
  diffusion and reaction in, 221
Mass transfer, transient, in packed column, 631
Mass transfer coefficient, correlations for, 145, 702, 715, 722, 725, 726, 730
  definitions, 141, 307
Maximum principle, Bellman's, 496
Methanol synthesis, diffusion limitations in, 189
  reactor simulation, 562
Micropores, 161, 170
  diffusion and reaction in, 221
Mixing, dispersion coefficient, 618
  effects on large scale, 634
  macro, 613
  micro, 613
    coalescence-redispersion model for, 646
    effect on conversion of adiabatic combustion, 651
    model for intermediate levels of, 616
    parameter for, 649
    population balance model for, 646
Model discrimination, 106
Modulus, see Thiele modulus
Molar expansion of reaction mixture, 363
Momentum equation, see Pressure drop equation
Multibed adiabatic reactor, 493
Multicomponent diffusion, 146
Multiple steady states, in catalyst pellets, 202, 207, 213, 252
  in stirred tank reactors, 443
Multiple steady states in packed beds, with autothermic operation, 505
  with axial mixing, 528
  with interfacial gradients, 547
Multitubular reactor, 468, 470
Murphree tray efficiency, 721

Networks, with pore diffusion, 218
  of reactions, 19, 47
Nickel catalyst reoxidations, 551
Nitric oxide absorption, 722

Observable modulus, 192, 195
Optimal batch operating time, 374
Optimal design, of experiments, 118
  of multibed adiabatic reactors, 495
Optimal temperature, in batch reactors, 377
Optimum principle, in dynamic programming, 496

762 _____

Overall order for free radical mechanisms, 34
Oxidation, of ethylene, 101
　of hydrocarbons, 482, 539, 550
　of nitric oxide, 722
　of sulfur dioxide, 466, 493
　of o-xylene, 732

Packed column, transient mass transfer in, 631
Packed column absorber, 701
Parallel crosslinked pore model, 172
Parallel reactions, in batch reactors, 382, 384
　in gas-liquid systems, 323
　perfectly mixed reactors, 430
　with pore diffusion, 214, 216
Parameter estimation, 106, 640
Parametric sensitivity, 483, 543
Partial order, 7, 46
Partial pressure, 7
　drop at catalyst surface, 150
Particle transfer function, 629
Peclet number, for axial transport, 527
　for radial heat transport, 540
　for radial mass transport, 533
Penetration theory, 307, 327
Perfectly mixed reactors, 420
　complex reactions, 433
　consecutive reactions, 430
　economic optimization for series of reactors, 426
　parallel reactions, 430
　polymerization reactions, 437
　stability of, 443
　steady state operation, 422
　transient operation, 449
Plate column absorber, 719
Plug flow, 354, 392
Plug flow reactor, adiabatic, 408
Poisoning of catalyst, shell progressive, 275
　uniform, 273
Polymerization, degree of, 40
Polymerization reactions, kinetics of, 38
　in stirred tank reactors, 437
Population balance models, 644
　for micromixing, 646
　for pore size distribution changes, 653
Pore diffusion, 163
　with adsorption, 174
　and complex reactions, 214
　configurational, 177, 224
　with reaction, 178, 221
　with surface diffusion, 175

Pore model, in non catalytic gas-solid reactions, 259
　parallel crosslinked, 172, 323
　random, 170, 221
Pore size distribution, of catalysts, 161
　changes by surface reaction, 653
Prandtl number, 146
Prater number, 201
Pressure drop, in empty tubes, 396, 415
　in fluidized beds, 666
　in packed beds, 478
　in packed columns, 701, 712, 718
Pressure drop equation, in tubular reactors, 413
Profit optimization, in batch reactor, 374
　in multibed adiabatic reactor, 495
Propagation step, 32
Propane, thermal cracking of, 57, 397
Pseudo steady state approximation, gas-solid reactions, 243
Pulsed flow, 14-30

Radial gradients, in fixed beds, 474
Radicals, *see* Free radical
Random pore model, 170
Rate coefficient, definitions, 7
　determination of, 46
　variation with temperature, 42
Rate coefficient matrix, 20
Rate determining step, 27
　in catalytic reactions, 92
　determination from initial rate data, 110
Rate of reaction, 1, 15
Reaction, extent of, 5
　networks, 19
　order of, 6
　path, 23
Reactor design, for acrylonitrile synthesis, 685
　batch, 367, 370, 373
　for ethane cracking, 410
　fixed bed, 474
　fluidized bed, 675, 682, 685
　for hydrocarbon oxidation, 482, 539, 550
　multibed adiabatic, 493
　multiphase flow, 695, 701
　for propane cracking, 397
　stirred tank, 637
　tubular, with plug flow, 408
Recycle type model, application of, 637
　for large scale mixing effects, 634
Reoxidation of catalysts, 551
Residence time, in batch reactor, 362

in tubular reactor, 393
Residence time distribution, application of,
    for combination of non-interacting
      regions, 605
    for detection of regions of fluid stagnancy,
      606
    for determination of dispersion coefficient,
      618, 621
    for reactor design, 607, 611, 612
    definition, 594
    experimental determination of, 596
    of perfectly mixed vessel, 595
    in non-steady state, 601
    of n perfectly mixed tanks, 603
Reversible reaction, 10
Rice-Herzfeld mechanisms, 31
Runaway, criteria for, 485
    in fixed bed reactors, 483, 541

Schmidt number, 145
Segregation, effect on conversion, 615
Selectivity, 17
    effects of pore diffusion on, 214
    in gas-liquid reactions, 323
    influence of catalyst deactivation on, 279,
      291, 523
Shell progressive poisoning, of catalysts, 275
Sherwood number, 715
    modified, 311
    for poisoning, 276
Shrinking core model, 249
Simulation of reactors, for acetone cracking,
    404
    with axial dispersion, 624
    with catalyst deactivation, 515
    for ethane cracking, 410
    for hydrocarbon oxidation, 482, 539, 550
    with internal heat exchange, 509
    for methanol synthesis, 562
    for nitric oxide absorption, 722
    at non-steady state operation, 551
    for propane cracking, 57, 397
    with transient behavior, 549
    with two distinct regions, 627
Spinning basket reactor, 108
Spray flow, 711
Spray tower, 725
Stability in perfectly mixed reactors, 443, 449
Stefan-Maxwell equations, 148
Stirred tank, models for, 633, 634
Stoichiometric coefficients, 2, 15

Sulfur dioxide oxidation, 466, 493
Surface coverage, 87, 88
Surface diffusion, 175
Surface reaction, with reactive solid, 141
    on solid catalysts, 90
Surface renewal theory, 327

Tanks in series model, for non-ideal flow
    pattern, 621, 623
Temkin adsorption isotherm, 89
Temperature, optimal in batch reactors,
    377
Temperature drop, over film, 151
Temperature gradients, external to solid
    particles, 151, 208, 546
    inside solid particles, 200, 264, 535
Termination step, 32
Thermal conductivity, see Effective thermal
    conductivity
Thermal cracking, of acetone, 402
    of butane, 37
    of ethane, 35, 44, 410
    of propane, 57, 397
Thiele modulus, for first order reaction,
    180
    general, 184
Tortuosity, factor, 167
    tensor, 172
Transient behavior of reactor, due to catalyst
    deactivation, 515
    simulation of, 549
Trickle bed reactor, 710
    model for flow pattern in, 632
Trickle flow, 711
Tubular reactor with plug flow, design of,
    408
    isothermal, 394
    non-isothermal, 399, 410
Two-film theory, 308
Two stage model, in gas-solid reactions,
    245
Two zone model, for multiphase reactors,
    699

Uniform poisoning, of catalysts, 273
Unimolecular rate theory, 30
Utilization factor, global, 312
    liquid side, 311

Van Knevelen-Hoftijzer diagram, 313
Van't Hoff equation, 42

Velocity profile, in tubular reactors, 360, 545

Voorhies relation, 286, 518

Wei-Prater treatment of complex reaction networks, 19

Weisz-Prater criterion, 191
  extended, 194

use of, 196, 252

Wetted wall column, 338

Wilke equation, 148

O-Xylene oxidation, in gas phase, 539
  in liquid phase, 732

Zeolites, diffusion in, 164